Metallomics and the Cell

Metal Ions in Life Sciences

Volume 12

Guest Editor:

Lucia Banci

Series Editors:

Astrid Sigel, Helmut Sigel, and Roland K.O. Sigel

For further volumes:
http://www.springer.com/series/8385 and http://www.mils-series.com

Astrid Sigel • Helmut Sigel • Roland K.O. Sigel
Series Editors

Lucia Banci
Editor

Metallomics and the Cell

Springer

Editor
Lucia Banci
Centro Risonanze Magnetiche
University of Florence
Via Luigi Sacconi 6
I-50019 Sesto Fiorentino (Florence), Italy
banci@cerm.unifi.it

Department of Chemistry
University of Florence
Via della Lastruccia 5
I-50019 Sesto Fiorentino (Florence), Italy

Series Editors
Astrid Sigel
Department of Chemistry
Inorganic Chemistry
University of Basel
Spitalstrasse 51
CH-4056 Basel
Switzerland
astrid.sigel@unibas.ch

Helmut Sigel
Department of Chemistry
Inorganic Chemistry
University of Basel
Spitalstrasse 51
CH-4056 Basel
Switzerland
helmut.sigel@unibas.ch

Roland K.O. Sigel
Institute of Inorganic Chemistry
University of Zürich
Winterthurerstrasse 190
CH-8057 Zürich
Switzerland
roland.sigel@aci.uzh.ch

ISSN 1559-0836
ISSN 1868-0402 (electronic)
ISBN 978-94-007-5560-4
ISBN 978-94-007-5561-1 (eBook)
DOI 10.1007/978-94-007-5561-1
Springer Dordrecht Heidelberg New York London

Library of Congress Control Number: 2013933450

Printed on acid-free paper

Springer is part of Springer Science+Business Media (www.springer.com)

Dedication to Ivano Bertini

Ivano Bertini passed away untimely on July 7, 2012, at the age of 71. With him the Bioinorganic Community lost one of its giants. Ivano left his traces all the way from Inorganic Chemistry through Magnetic Resonance and Biochemistry to the Medical Sciences with more than **650** papers. He and his group in Florence, known for their pioneering NMR work, solved over **150** protein structures. Thanks to his enthusiasm, driving personality, and the way he viewed international relationships, he was initiator and cofounder of the *International Conferences on Bioinorganic Chemistry* (ICBICs), the *European Biological Inorganic Chemistry Conferences* (EuroBICs), the *Eurasia Conferences on Chemical Sciences* (EuAsC2S), the *European Conferences on Chemistry for Life Sciences* (ECCLS), the *Chianti Workshops*, and more. Ivano was also one of the founders of the *Society for Biological Inorganic Chemistry* (SBIC) and the first Editor of the *Journal of Biological Inorganic Chemistry*

(JBIC). To name just a few of his initiatives and impacts, which last but not least included also the foundation of CERM, the Magnetic Resonance Center of the University of Florence.

This volume *Metallomics and the Cell* is devoted to Ivano, who coauthored Chapter 1 of this book, and who was fascinated by the "omics" sciences and the possibilities they offer for understanding life. It is thus no surprise to read in the mentioned chapter ..." It is the duty of the bioinorganic chemistry community to convince funding agencies that the research on metallomes and metallomics deserves attention" and support. He was a true protagonist of new ideas!

There is no way to list here all the achievements of Ivano Bertini and therefore, we simply close by saying: THANK YOU, Ivano, for what you did for us, the chemical sciences and the scientific community in general – you will never be forgotten!

Sesto Fiorentino (Florence), Italy — Lucia Banci
Basel, Switzerland — Astrid Sigel
Basel, Switzerland — Helmut Sigel
Zürich, Switzerland — Roland K.O. Sigel

Historical Development and Perspectives of the Series

Metal Ions in Life Sciences*

It is an old wisdom that metals are indispensable for life. Indeed, several of them, like sodium, potassium, and calcium, are easily discovered in living matter. However, the role of metals and their impact on life remained largely hidden until inorganic chemistry and coordination chemistry experienced a pronounced revival in the 1950s. The experimental and theoretical tools created in this period and their application to biochemical problems led to the development of the field or discipline now known as *Bioinorganic Chemistry*, *Inorganic Biochemistry*, or more recently also often addressed as *Biological Inorganic Chemistry*.

By 1970 *Bioinorganic Chemistry* was established and further promoted by the book series *Metal Ions in Biological Systems* founded in 1973 (edited by H.S., who was soon joined by A.S.) and published by Marcel Dekker, Inc., New York, for more than 30 years. After this company ceased to be a family endeavor and its acquisition by another company, we decided, after having edited 44 volumes of the *MIBS* series (the last two together with R.K.O.S.) to launch a new and broader minded series to cover today's needs in the *Life Sciences*. Therefore, the Sigels new series is entitled

Metal Ions in Life Sciences.

After publication of the first four volumes (2006–2008) with John Wiley & Sons, Ltd., Chichester, UK, and the next five volumes (2009–2011) with the Royal Society of Chemistry, Cambridge, UK, we are happy to join forces now in this still new endeavor with Springer Science & Business Media B.V., Dordrecht, The Netherlands; a most experienced Publisher in the *Sciences*.

---

*Reproduced with some alterations by permission of John Wiley & Sons, Ltd., Chichester, UK (copyright 2006) from pages v and vi of Volume 1 of the series *Metal Ions in Life Sciences* (MILS-1).

The development of *Biological Inorganic Chemistry* during the past 40 years was and still is driven by several factors; among these are (i) the attempts to reveal the interplay between metal ions and peptides, nucleotides, hormones or vitamins, etc., (ii) the efforts regarding the understanding of accumulation, transport, metabolism and toxicity of metal ions, (iii) the development and application of metal-based drugs, (iv) biomimetic syntheses with the aim to understand biological processes as well as to create efficient catalysts, (v) the determination of high-resolution structures of proteins, nucleic acids, and other biomolecules, (vi) the utilization of powerful spectros-copic tools allowing studies of structures and dynamics, and (vii), more recently, the widespread use of macromolecular engineering to create new biologically relevant structures at will. All this and more is and will be reflected in the volumes of the series *Metal Ions in Life Sciences*.

The importance of metal ions to the vital functions of living organisms, hence, to their health and well-being, is nowadays well accepted. However, in spite of all the progress made, we are still only at the brink of understanding these processes. Therefore, the series *Metal Ions in Life Sciences* will endeavor to link coordination chemistry and biochemistry in their widest sense. Despite the evident expectation that a great deal of future outstanding discoveries will be made in the interdisciplinary areas of science, there are still "language" barriers between the historically separate spheres of chemistry, biology, medicine, and physics. Thus, it is one of the aims of this series to catalyze mutual "understanding".

It is our hope that *Metal Ions in Life Sciences* proves a stimulus for new activities in the fascinating "field" of *Biological Inorganic Chemistry*. If so, it will well serve its purpose and be a rewarding result for the efforts spent by the authors.

Astrid Sigel and Helmut Sigel
Department of Chemistry, Inorganic Chemistry,
University of Basel, CH-4056 Basel, Switzerland

Roland K.O. Sigel
Institute of Inorganic Chemistry,
Unversity of Zürich, CH-8057 Zürich, Switzerland

October 2005,
October 2008,
and August 2011

Preface to Volume 12

Metallomics and the Cell

In this volume of the *Metal Ions in Life Sciences* series a broad and complete coverage of the "omics" approach to the field of metal ions in biological systems is presented and discussed in a cellular context, i.e., in the frame of a cell or of living systems, underlining the complexity of the interactions between living systems and inorganic elements.

In Chapter 1 general concepts are elaborated and the definition of the metallome is provided, as the ensemble of all the biomolecules in a system which bind a given metal ion or an inorganic element in general, or are affected by that element. The metallome is therefore a dynamic property which depends on the conditions of the system. Metallomics is the systematic study of a metallome and of the functional connections and interactions of the metal ion with the genome, proteome, metabolome, and other biomolecules in an organism. The role of bioinformatics and information technology are also discussed, and this knowledge framed in a cellular context. The need of this perspective is required in order to describe the functional processes. The ultimate goal of reaching a systems biology vision from the knowledge of the interlinks and connections among the various pathways and processes involving metal ions in the cell is discussed.

Chapter 2 presents an overview of the most suitable methods for the analysis of metal distributions in a cell, and how this distribution is affected by the cellular status and conditions.

In the following, from Chapter 3 through 15, the metallome relative to various metal ions is described, limited to the essential metal ions for at least one domain of life. So the chapters encompass sodium and potassium and range to magnesium, calcium, and then to manganese, iron, cobalt, nickel, copper, zinc, as well as molybdenum and tungsten. When needed and/or enough knowledge is available, the metallome description has been subdivided based on the type of organism, i.e., bacteria and eukaryotic ones, as well on the type of cofactors, e.g., heme for iron. An overview of the most relevant biomolecules (mostly proteins) functionally interacting with each of the selected metal ions is provided for the analyzed metals, together with data on their metal binding mode, structural and functional properties, and cellular localization.

Finally, Chapter 16 presents recent work on comparative genome analysis, through bioinformatics tools, with respect to metal-dependent proteins and metal utilization in general. This type of analysis provides insights on the evolution of the metal utilization, and on the general features of metal handling by organisms in the various domains of life. The distribution and relevance of each individual metal ion along evolution is also evaluated, analyzed, and presented.

Lucia Banci
CERM and Department of Chemistry
University of Florence
I-50019 Sesto Fiorentino (Florence)
Italy

Contents

Contributors to Volume 12

Numbers in parentheses indicate the pages on which the authors' contributions begin.

Wafaa S.M. Aly The School of Biological Sciences, The University of Reading, Whiteknights, Reading, RG6 6AJ, UK (203)

Simon Andrews The School of Biological Sciences, The University of Reading, Whiteknights, Reading, RG6 6AJ, UK, s.c.andrews@reading.ac.uk (203)

Michael Aschner Department of Pediatrics, 2215-B Garland Avenue, 11415 MRB IV, Vanderbilt University Medical Center, Nashville, TN 37232-0414, USA, michael.aschner@vanderbilt.edu (169)

Lucia Banci Centro Risonanze Magnetiche, University of Florence, Via Luigi Sacconi 6, I-50019 Sesto Fiorentino (Florence), Italy, banci@cerm.unifi.it (1)

Ruma Banerjee Department of Biological Chemistry, University of Michigan, 3220B MSRBIII, 1150 W. Medical Center Dr., Ann Arbor, MI 48109-0600, USA, rbanerje@umich.edu (333)

David R. Benson Department of Chemistry, University of Kansas, Multidisciplinary Research Building, 2030 Becker Dr., Lawrence, KS 66047, USA, drb@ku.edu (279)

Ivano Bertini Centro Risonanze Magnetiche, University of Florence, Via Luigi Sacconi 6, I-50019 Sesto Fiorentino (Florence), Italy (1)

Marisa Brini Department of Comparative Biomedicine and Food Science, University of Padova, Viale G. Colombo 3, I-35131 Padova, Italy, marisa.brini@unipd.it (119)

Tito Calì Department of Comparative Biomedicine and Food Sciences, University of Padova, Viale G. Colombo 3, I-35131 Padova, Italy, tito.cali@unipd.it (119)

Ernesto Carafoli Venetian Institute of Molecular Medicine (VIMM), Via G. Orus, I-35129 Padova, Italy, ernesto.carafoli@unipd.it (119)

Michael Jakob Voldsgaard Clausen Centre for Structural Biology, Department of Molecular Biology and Genetics, University of Aarhus, Science Park, Gustav Wieds Vej 10c, DK-8000 Aarhus C, Denmark, mjc@mb.au.dk (41)

Paul A. Cobine Department of Biological Sciences, 101 Rouse Life Sciences Building, Auburn University, Auburn, AL 36849, USA, pac0006@auburn.edu (451)

Pierre Cornelis Department of Molecular and Cellular Interactions, Laboratory of Microbial Interactions, Flanders Interuniversity Institute of Biotechnology VIB6, Vrije Universiteit Brussel, Brussels, Belgium (203)

Valentin Cracan Department of Biological Chemistry, University of Michigan 3220B MSRBIII, 1150 W. Medical Center Dr., Ann Arbor, MI 48109-0600, USA, vfcracan@umich.edu (333)

Adrienne C. Dlouhy Department of Chemistry and Biochemistry, University of South Carolina, Columbia, SC 29208, USA (241)

Sylvia Franke McDevitt Department of Biology, Skidmore College, 815 North Broadway, Saratoga Springs, NY 12866, USA, sfranke@skidmore.edu (417)

Vadim N. Gladyshev Division of Genetics, Department of Medicine, Brigham and Women's Hospital and Harvard Medical School, New Research Building, 77 Av. Pasteur, Boston, MA 02115, USA, vgladyshev@rics.bwh.harvard.edu (529)

Helen Goodluck The School of Biological Sciences, The University of Reading, Whiteknights, Reading, RG6 6AJ, UK (203)

Hayaa F. Hashemi Department of Biological Sciences, 101 Rouse Life Sciences Building, Auburn University, Auburn, AL 36849, USA (451)

Wolfgang Maret King's College London, School of Medicine, Diabetes and Nutritional Sciences Division, Metal Metabolism Group, Franklin Wilkins Bldg, 150 Stamford St, London SE1 9NH, UK, wolfgang.maret@kcl.ac.uk (479)

Ralf R. Mendel Institute of Plant Biology, Braunschweig University of Technology, Humboldtstr.1, D-38106 Braunschweig, Germany, r.mendel@tu-bs.de (503)

Hanna Mourad-Agha The School of Biological Sciences, The University of Reading, Whiteknights, Reading, RG6 6AJ, UK (203)

Ian Norton The School of Biological Sciences, The University of Reading, Whiteknights, Reading, RG6 6AJ, UK (203)

Denis Ottolini Department of Biomedical Sciences, University of Padova, Viale G. Colombo 3, I-35131 Padova, Italy, denis.ottolini@unipd.it (119)

Caryn E. Outten Department of Chemistry and Biochemistry, University of South Carolina, Columbia, SC 29208, USA, outten@mailbox.sc.edu (241)

James E. Penner-Hahn Department of Chemistry and Biophysics, University of Michigan, 930 N. University Avenue, Ann Arbor, MI 48109-1055, USA, jeph@umich.edu (15)

Silvia Ponzoni Departamento de Ciências Fisiológicas, CCB, Universidade Estadual de Londrina, Caixa Postal 6001, 85051-970 Londrina, PR, Brazil (169)

Hanne Poulsen Centre for Structural Biology, Department of Molecular Biology and Genetics, University of Aarhus, Science Park, Gustav Wieds Vej 10c, DK-8000 Aarhus C, Denmark, hp@mb.au.dk (41)

Christopher Rensing Department of Plant and Environmental Sciences, University of Copenhagen, Thorvaldsensvej 40, DK-1870 Frederiksberg C, Denmark (417)

Mario Rivera Department of Chemistry, University of Kansas, Multidisciplinary Research Building, 2030 Becker Dr., Lawrence, KS 66047, USA, mrivera@ku.edu (279)

Andrea M.P. Romani Department of Physiology and Biophysics, School of Medicine, Case Western Reserve University, 10900 Euclid Avenue, Cleveland, OH 44106-4970, USA, andrea.romani@case.edu, amr5@po.cwru.edu (69)

Jerome Roth Department of Pharmacology and Toxicology, 11 Cary Hall, University at Buffalo School of Medicine, Buffalo, NY 14214, USA (169)

Arvindkumar S. Salunkhe The School of Biological Sciences, The University of Reading, Whiteknights, Reading, RG6 6AJ, UK (203)

Andrew M. Sydor Department of Chemistry, University of Toronto, 80 St. George St., Toronto, ON M5S 3H6, Canada (375)

Katherine E. Vest Department of Biological Sciences, Auburn University, 101 Rouse Life Sciences Building, Auburn, AL 36849, USA (451)

Deborah B. Zamble Department of Chemistry, University of Toronto, 80 St. George St., Toronto, ON M5S 3H6, Canada, dzamble@chem.utoronto.ca (375)

Yan Zhang Key Laboratory of Nutrition and Metabolism, Institute of Nutritional Sciences, Shanghai Institutes for Biological Sciences, Chinese Academy of Sciences, University of Chinese Academy of Sciences, 294 Tai Yuan Road, Shanghai 200031, China, yanzhang01@sibs.ac.cn (529)

Titles of Volumes 1-44 in the *Metal Ions in Biological Systems* Series

edited by the SIGELs
and published by Dekker/Taylor & Francis (1973–2005)

Contents of Volumes in the *Metal Ions in Life Sciences* Series

edited by the SIGELs

Volumes 1–4
published by John Wiley & Sons, Ltd., Chichester, UK (2006–2008)
<http://www.Wiley.com/go/mils>

Volumes 5–9
by the Royal Society of Chemistry, Cambridge, UK (2009–2011)
<http://www.rsc.org/shop/metalionsinlifesciences>

and from Volume 10 on
by Springer Science & Business Media BV, Dordrecht, The Netherlands (since 2012)
<http://www.mils-series.com>

Volume 4 Biomineralization. From Nature to Application

Volume 5 Metallothioneins and Related Chelators

Volume 7 Organometallics in Environment and Toxicology

Volume 10 Interplay between Metal Ions and Nucleic Acids

Comments and suggestions with regard to contents, topics, and the like for future volumes of the series are welcome.

Chapter 1
Metallomics and the Cell: Some Definitions and General Comments

Lucia Banci and Ivano Bertini

Contents

Abstract In this chapter we aim at underlining the complexity of the interactions between living systems and inorganic elements. Attempts are made to move this field towards an "omics" approach through bioinfomatics and information technology. The metallome can be defined as the ensemble of all the biomolecules in a system which bind a given metal ion or an inorganic element at broad, or are affected by that element. A number of subsets of a metallome can then be defined based on the nature of the biomolecules interacting with the metal ions and the inorganic elements in general. The most relevant and most studied subset is the metalloproteome. This field now needs to be framed in a cellular context. The interlinks and connections among the various pathways and processes involving metal ions in the cell have to be described in an integrated way so to reach a systems biology vision.

Keywords metallome • metalloproteome • metals in cellular processes • systems biology

Please cite as: *Met. Ions Life Sci.* 12 (2013) 1–13

L. Banci (✉) • I. Bertini
CERM and Department of Chemistry, University of Florence,
I-50019 Sesto Fiorentino (Florence), Italy
e-mail: banci@cerm.unifi.it

L. Banci (ed.), *Metallomics and the Cell*, Metal Ions in Life Sciences 12,
DOI 10.1007/978-94-007-5561-1_1, © Springer Science+Business Media Dordrecht 2013

1 Introduction

In the year 1995 Craig Venter and coworkers sequenced the first genome, that of *Haemophilus influenzae* [1]. A new era started. In the year 2000 the human genome was sequenced [2,3]. The US President and the British Prime Minister met to discuss the consequences in terms of intellectual property and bioethical problems. Thereafter, projects of structural genomics started in various countries worldwide with the goal of determining the 3D structure of the products of genomes. Already in 2005, funding agencies implemented postgenomic programs and finally, the old term Systems Biology was reused in the new frame of modeling an entire system with the structural, thermodynamic, and kinetic information regarding all the players of a system, as identified from the knowledge of the genome [4].

As a consequence of these developments, the study of expression of the genes became a must, and thus proteomics and transcriptomics florished. Then the realization that some bases of DNA could be modified as a result of exposure to environmental factors prompted the epigenomics development. Finally, the complete coverage of the metabolites of an organism initiated the field of metabolomics. This new approach, collectively referred to as "omics" approach, completely changed the methods and ways to study biological molecules and their interactions and functional processes.

The word "omics" does not have a unique origin and meaning but now it is accepted to mean the coverage and study of a complete ensemble of biological entities, e.g., of genes, proteins, metabolites, etc. A complete ensemble is full of information, which goes beyond the sum of the single entities. Indeed, knowledge at the "omics" level provides, for an organism, the knowledge of all metabolic pathways and allows comparisons and correlations among them and among various organisms. The amount of data obtained from experimental analyses is increasing exponentially, and requires the organization and curation of databases and of tools for data browsing, analysis, and comparison. Bioinformatics is more and more an essential tool for any research in the field of Life Science with an "omics" approach.

Of course, metal ions are part of this world, but have rarely been conjugated with the omics world. Bioinorganic chemistry has tended to characterize the structure and function of metalloproteins, to design and study metal-based drugs, and to study single aspects, such as synthesizing models of metal binding sites in biomolecules. And this has consequences on our contribution to the development of science and on our position with respect to the frontiers of science.

In the frame of the genomic revolution, first, and of the other "omics" sciences later, we need to introduce metal ions (or bioinorganic chemistry) to the "omics" world. And for this the new "omics" field, metallomics and metallome, have been proposed and established. A shared definition of metallome is difficult to reach by itself, for the special complexity of the matter. Indeed, the definition of "metallome" is still an open question and extensive debate is on going [5–10]. Metals in the metallome could, at a first glance, be taken as the inorganic elements which interact with living systems. According to this definition, most metals enter this pool. Other inorganic elements could also be taken into account, such as selenium, phosphorus, iodine, etc. Of all these inorganic elements, some are necessary for life, i.e., they are essential elements, some are toxic, some could be used for metal-based drugs, and some occur only in traces. The metallome therefore deals with the molecular mechanisms and the metabolic pathways of metal-dependent life processes and the entirety of metals and inorganic species at broad in cells and in living organisms, and metallomics is related to the study of these aspects.

Metallomics addresses various aspects and components, such as metal-induced proteomic and metabolomic profiles and their metal-dependent alterations. Therefore, the concept of the metallome is expansive, and is generally subdivided in various classes, depending on the nature of the biomolecules interacting with the metal ion or the inorganic element. The most common subset is that of the metalloproteome, i.e., the complete set of proteins or the subset of the proteome that interacts or is affected by a metal ion. Also the extracellular metal ions need to be taken into account, as well as their interactions with extracellular biomolecules and cellular receptors on the cell surface. Metal-based drugs interact with a subset of all biomolecules at the cellular and extracellular levels. These interactions obviously affect the metabolic pathways which need to be understood in order to have a complete picture of the metallome in a living organism.

The aim of this volume is to present a critical and detailed analysis of the cellular location, role and function of metal ions and the proteins and biomolecules they are bound to, in an "omics" perspective, i.e., a comprehensive and complete (as much as possible) coverage of the metal-related biomolecules, in a cellular context, by taking into account the compartment(s) of the metal-binding biomolecules and of the process(es) in which they are involved. These two pieces of information are the basis on which, through their integration, a description of processes at the system level can be elaborated. Indeed, in order to describe and understand the various cellular processes and metabolic pathways involving metal ions, we need to know not only all the players involved in a process, but also how the processes interconnect, and therefore, are affected, by one another.

The following chapters in this volume of the *Metal Ions in Life Sciences* series enlighten connections between different metal ions, their metabolism, and how the various metal ion homeostasis affect each other, towards a systems biology description of the metal handling by living organisms.

2 The "Omics" Perspective

The goal is now to frame the chemistry and biology of metal ions in an omics perspective. In this context, the omics perspective is defined as a view of the ensembles of metal ions and other inorganic elements in living organisms, i.e., the metallome. The term "metallome" was initially proposed by R. J. P. Williams, who defined it as "an element distribution" in a cell [11]. The first question is therefore, already addressed: How do we define a metallome? The concept of metallome goes beyond its links to the genome and the proteome, as indeed all the molecules in a system which bind a given metal ion or an inorganic element at broad, or are affected by that element, should be taken into account. Therefore, the metallome is dynamic, as it is affected by changes in living organisms, such as changes in the metabolome and/or in the proteome. From this definition it is clear that we can define a metallome for each metal ion or inorganic element, even if they do not exist in isolation but are tightly interlinked and affected by one another (see later).

Most of the studies reported up to now, and also most of what is reported and discussed in the following chapters, is devoted to metalloproteomes, i.e., the ensemble of proteins which bind a specific metal ion. A following question is then, whether and how it is possible to define the metalloproteome for each metal for each living organism? There have been attempts on some bacteria to define, at the proteomic level, the metalloproteome of some, most relevant metal ions [7,12–14]. However, the problem is that any experimental approach is limited in one aspect or another because they are affected by experimental and/or accidental variability due to the environment, sample conditions, sensitivity of the approach, detectability levels of the experiments, and so on [7,9,10]. Outcomes of experimental analyses can be affected by both false positive and false negative results. Among the experimental problems, one is related to the metalation status of a given protein *in vitro* or in heterologous cells. This experimental condition could lead to the acquisition of the wrong metal as, outside the correct cellular environment, selection for metal binding by a biomolecule is simply driven by the highest affinity. Also challenging is the maintainance of non-denaturating conditions throughout the entire duration of the experiment.

On the other hand, if we learn from the other omics, a fundamental role is played by bioinformatics. Browsing databases such as those of genomes, proteomic data, interactomics, protein structures, and so on, is indeed an invaluable source of information. Of course, the data archived there come from experiments which must be properly controlled. This is one of the reasons for investing efforts in defining standard operating procedures and ontologies, i.e., a standard and unified definition of nomenclature. The best source of data for the analysis of metal-binding biomolecules is the Protein Data Bank (PDB). Most of the attempts to define the ensembles of the metal-binding proteins indeed started from the PDB, possibly after a reevaluation and analysis of the data. The PDB, particularly after the structural genomics projets, is enough representative of all metalloproteins, and starts to structurally cover a significant part of the proteomes and of the fold types. Thus, the PDB can constitute an effective starting point for the analysis of metalloproteomes. From the identification of the metal coordination environment in the PDB, metal binding patterns, which can include also residues in the sorrounding of the metal-binding residues, can be identified in the gene sequence [15]. By combining this knowledge with metal-binding domain classification, as defined in the Pfam library, which identifies protein domains classified as potential metal-binding domains [16], complete genomes can be screened for potential metal-binding protein sequences. From this type of analysis we have learned that the use of the various metal ions differs among the various species and the various kingdoms of life. It appears that the share of the proteome that binds zinc, i.e., the number of zinc-binding proteins with respect to the total number of proteins, steadily increases from bacteria to archaea to eukaryotes, and, among them, with the complexity of the organism, humans being the largest of these organisms [17,18]. The opposite is true for other metal ions, such as iron [19,20] and manganese [21].

From this type of analysis we can also learn how many metalloproteins are enzymes, and how they work. Indeed, exploiting a bioinformatic approach, a database has been developed for metalloenzymes, Metal-MACiE [22], which reports also on the mechanism of action of the enzymes. So, we learn that about half of the known enzymatic reactions involve metal ions, and that 13 different metal ions, even with quite different levels of occurrence, are involved in catalysis. The metal ion present in the largest set of enzymes is magnesium, although it is often loosely interacting with phosphate-containing molecules such as ATP and with nucleic acids, followed by zinc and iron. In terms of types of catalytic reactions, magnesium, calcium, zinc, and manganese are mostly used for substrate activation and electrostatic stabilization, iron and copper are present as redox centers. Cobalt and molydenum perform their function in association with organic cofactors.

Contributions to the description of the involvement of metal ions in biological processes can also come from the analysis of large datasets derived from experimental proteomics or DNA or RNA levels as a function of the addiction or restriction of a given metal ion [12,23,24]. Obviously, through this type of analysis all proteins involved in pathways and networks requiring one or more metal-binding proteins are altered, and therefore the knowledge on the metalloproteome, or of its interaction with nucleic acids, is neither direct nor simple.

From all these types of studies, we can make comparative genetics or genomics analyses and try to compare the whole metalloproteome from one living organism with another [25]. The omic approach also allows researchers to develop general criteria for correlations between sequences and the metal-binding ability of the protein scaffold and to propose evolution patterns regarding binding and functions of metal ions [21,25,26]. Indeed, metal specificity for proteins and other biomolecules has been influenced during evolution by metal availability in the environment and in the habitat and by living conditions, such as anaerobic environments *versus* aerobic atmosphere [21,27]. Some functions and proteins evolved to adapt to the new environment and are now performed by different protein scaffolds and different metal ions with respect to ancient times [28,29].

However, while through these approaches, a broad, somehow complete, picture is available for the metalloproteomes, the situation is worse for the rest of the metallome, i.e., all the other molecules which interact with metal ions in living systems. In this context, we have to address processes and pathways, such as trafficking of metal ions into, out, and within the cell. Are there small ligands which accompany metal ions through membranes? Is there a pool of metal ions in the matrix of mitochondria? Metal ions are either necessary or toxic and even used as drugs. What are the selection rules both for essential metal ions and for trace metal ions? How is metal regulation performed by metal sensors? Too many data are still missing regarding expression and regulation of proteins involved in metal trafficking, and too many questions are still without answer. It appears that the omics perspective for metal ions is quite broad and fragmented and requires many more reflection: This is why several chapters here start with "metallome" and eventually end with "metalloproteome".

3 The "Cellular" Perspective

The cellular perspective in molecular studies is a frontier field which is not limited to metal ions. The cell is a complex entity that a "classical" bioinorganic chemist may not fully appreciate. In a cell there are many compartments and organelles, which are dynamic entities. So cellular organization is not something static. Within this frame, metal ions need to enter the cell and to move from one compartment or organelle to another, in most cases after binding to proteins, essentially all expressed in the nucleus. Mechanisms are now discovered for these trafficking processes which are organelle-specific [30]. For mitochondria, for example, metal carrier proteins are often unfolded in order to pass a membrane, then they fold to be trapped in a given compartment [31–34].

Metal selectivity is an essential aspect for life. Thermodynamically, following the Irving-Williams series for divalent metal ions [35], copper and zinc form the most stable complexes with respect to the other divalent metal ions. So, in an ideal, isolated solution, all the biomolecules, if presented with equal amounts of metal ions, would preferentially bind copper or zinc. However, in real life, metal uptake occurs in cells, where kinetic factors can concur with thermodynamics for the selection of a particular reaction and where different compartments provide different levels of metal availability, as well as the presence of other necessary molecules, thus favoring the binding of the right metal to the right biomolecule.

Furthermore, metal transfer from a protein to its partner occurs through an associative mechanism (Figure 1), so molecular recognition between the two interacting biomolecules is fundamental to the correct route of metal handling and to the selectivity of metal uptake, preventing the binding of the wrong metal ion to the wrong biomolecule [36,37]. It has been also shown that the cellular compartment where protein folding and metal uptake occur, which defines the availability of metal ions, can overcome the opposite, "incorrect" thermodynamic properties [30]. To achieve these thightly controlled metal levels in the various cellular compartments, living organisms have developed complex systems of metal importers, metal exporters, and metal sensors which maintain the correct metal homeostasis [38].

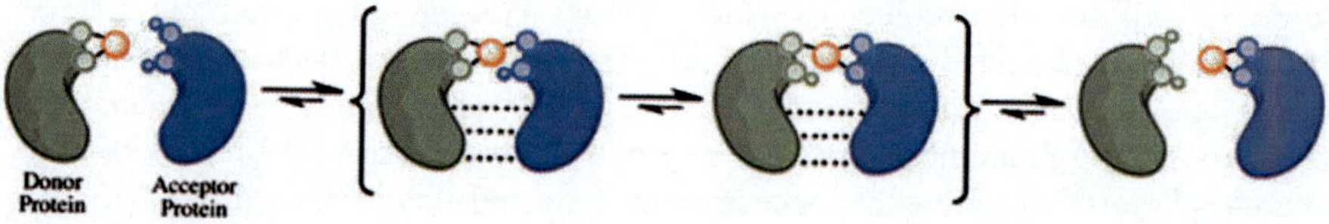

Figure 1 Copper transfer from a donor to an acceptor protein occurs through an associative mechanism: The coordination by ligands of the acceptor protein weakens the interaction with the acceptor, thus leading to the metal transfer. Protein-protein recognition occurs only in the presence of the metal ion and is modulated by a fine tuning of the interactions between the two proteins facing surfaces.

Experimental tools have been developed and are under development to characterize, at atomic resolution, the metal distribution in cells and the processes involving metal ions [9,10]. For example, fluorescent probes and correlative microscopy have provided valuable data on the localization of metalloproteins during their maturation from unfolded polypetide to the complete metalloproteins [39–41]. Synchrotrones are developing beam lines optimized to detect specific metal ions in cellular organelles [42–45]. In-cell NMR is also exploited to monitor metal uptake processes and their effects on the protein properties directly in living cells [46]. This is a very difficult challenge, which requires overexpression of the proteins of interest, possibly in homologous cells, to overcome sensitivity problems, and major advancements in NMR methodology.

4 Towards Systems Biology

Systems biology aims at describing the functional processes of biological systems based on the knowledge of their molecular constituents, their patterns of interactions, and the thermodynamics and kinetics parameters of the processes. Systems biology, therefore, requires the integration of various types of knowledge at the "omics" level, i.e., genome-, proteome-, and transcriptome-wide data, which provide a potentially complete list of the molecular components of a given organism, with data on the mode, pattern and properties of their interactions, and with information on their cellular localization. Processes can be localized to a single cellular compartment or can occur within more than one. So, in order to have a complete description of the system, all these cellular aspects need to be taken into account. When we have all the players, i.e., all the biomolecules, we can locate them in their cellular context.

A systems-wide approach is clearly needed for the comprehension of the mechanisms exploited by cells to overcome the simple coordinating properties and affinities of biomolecules for metal ions, so to place the correct metal to the correct biomolecule at the correct time in the cell life cycle. Too few studies are available with this approach and within this perspective to elaborate a comprehensive description of metal-dependent processes. The need of this type of approach is evident from the few studies where metallochaperones or other metal-binding proteins have been systematically knocked out. The results showed that several proteins and processes are affected, thus indicating the interlinks among the various processes, which therefore cannot meaningfully be studied in isolation [24,47,48].

For the metal transport processes, several studies have addressed the issue of protein-protein recognition. One protein carries the metal ion and delivers it to a partner through molecular recognition, as the metal transfer occurs through an associative mechanism. Of course, the transfer is thermodynamically allowed but may not be favored. The affinity constants of copper for a large number of copper proteins have been determined in conditions of pH, ionic strength, presence of small ligands, and redox conditions so to match as closely as possible the physiological ones, and in a consistent, coherent way that they can be meaningfully compared [49]. From these values, we learn that the processes of copper transfer are driven by gradients in affinity between the metal transporters and the metal-receiving proteins. But it is the kinetics of the process which determines its occurrence. From a molecular characterization of these metal transfer processes, it appears that the protein-protein interaction is metal-mediated, as already anticipated. This means that the two proteins interact only when the metal is coordinated and that they interact with each other through the metal ion, which shares ligands from the two proteins. Metal transfer occurs with an associative process where the metal-receiving protein invades the coordination sphere of the metal-transporting protein, weakens the latter bonds, thus inducing the transfer of the metal ion (Figure 1).

A large share of copper transport proteins coordinates the metal ion through cysteine ligands which have to be in the reduced state in order to bind the metal. A number of proteins, depending on the cellular compartment where the metal transfer process takes place, are responsible for maintaining the metal-transporting proteins in the reduced, functional state. This indicates that these copper transport processes are interconnected with those modulating the cellular redox properties and therefore they need to be taken into account simultaneously to the copper transfer process. Also, some of the copper-transporting proteins have a different folding state depending on the cellular compartment. Reaching the correct fold requires other players and other pathways. Among the various systems, we have shown that in the inter membrane space (IMS) of mitochondria, copper is delivered to cytochrome *c* oxidase (CcO), the terminal oxidase of the respiratory chain, through a series of transfers from Cox17 to Sco1/Sco2, to subunit II of CcO [50–57]. Cox17, which contains three cysteine pairs, two structural ones and one for copper binding, is unfolded in the cytoplasm but acquires a CHCH fold in the IMS, through interaction and oxidation of the two structural cysteine pairs by the oxidoreductase Mia40 [58]. Only folded Cox17 is able to bind copper and to transfer it to its partner proteins. After Cox17 oxidation, Mia40 is reduced and therefore needs to transfer the extra electrons to acquire again its functional state. This occurs through reduction of its partner ALR, which in turn transfers electrons to cytochrome *c* [59–61]. The latter also exchanges electrons with CcO, thus indicating how several processes are interlinked with each other and cannot be analyzed isolated and independently from one another (Figure 2).

Furthermore, import and transport of a given metal ion, and the regulation of these processes, can be interlinked and dependent on the homeostasis processes of another metal ion. This is clearly evident for iron and copper. Iron import in the cell is strictly copper-dependent, as iron is imported in the 3+ oxidation state and, to reach it, it is oxidized by a multicopper oxidase (ceruloplasmin in humans and Fet3 in yeast). Consistently with these tight links, the copper chaperones to the multicopper oxidase are regulated in the nucleus by iron-binding proteins [62,63]. This is an example which indicates that not only biomolecules interacting with a metal ion cannot be studied in isolation, thus requiring a description at the metallome level, but also that the metallome of one given element is interlinked with that of another one.

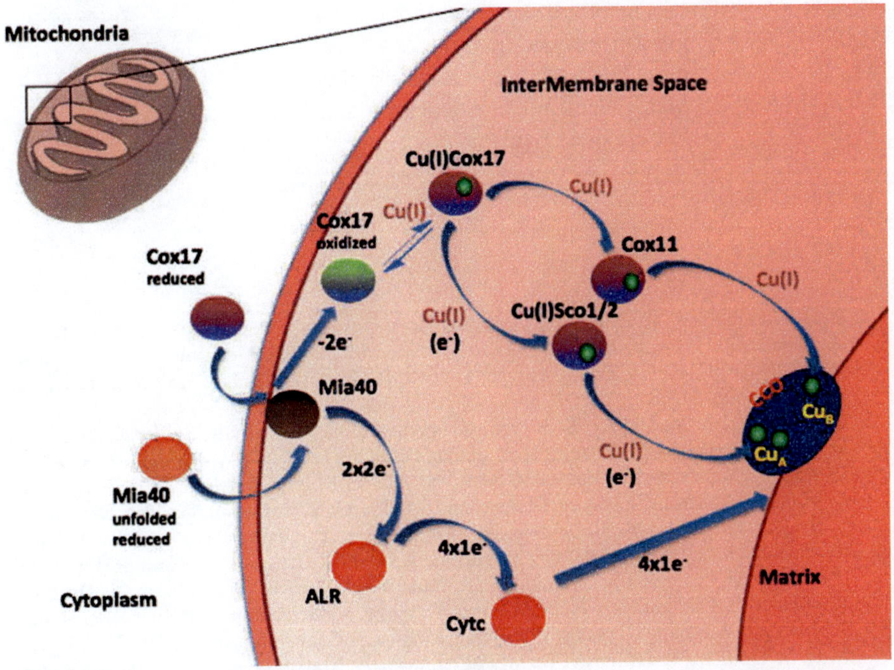

Figure 2 A schematic representation of the connection between the copper incorporation process in the CuA site of CcO (from Cox17 to Sco1/2 to CuA) and that responsible for Cox17 import in the IMS and its oxidation and folding. Cox17 is reduced in the cytoplasm. Once entered in the IMS, it interacts with Mia40 oxidizing it, thus inducing Cox17 folding and its trapping in the IMS. Cox17 then binds copper(I) and transfers it to CcO through Sco1/2 and Cox11. On another route, Mia40, which is now reduced, is oxidized back to its functional state through a series of redox reactions with ALR, cytochrome *c*, to CcO.

5 Concluding Remarks

The discussed interactions and interconnections among the metallomes of various metal ions make our field of research even more exciting and intriguing. It is the duty of the bioinorganic chemistry community to convince funding agencies that the research on metallomes and metallomics deserves attention as any frontier field of research.

Abbreviations

ALR	augmenter of liver regeneration
ATP	adenosine 5'-triphosphate
CcO	cytochrome *c* oxidase

CHCH	coil-helix-coil-helix fold
IMS	inter-membrane space
Mia40	mitochondrial import and assembly
Metal-MACiE	Metal Mechanism, Annotation and Classification in Enzymes
PDB	Protein Data Bank

References

1. R. D. Fleischmann, M. D. Adams, O. White, R. A. Clayton, E. F. Kirkness, A. R. Kerlavage, C. J. Bult, J. F. Tomb, B. A. Dougherty, J. M. Merrick, K. McKenney, G. G. Sutton, W. FitzHugh, C. Fields, J. D. Gocayne, J. Scott, R. Shirley, L.-I. Liu, J. C. Venter, *Science* **1995**, *269*, 496–512.
2. E. S. Lander, L. M. Linton, B. Birren, C. Nusbaum, M. C. Zody, J. Baldwin, K. Devon, K. Dewar, M. Doyle, W. FitzHugh, R. Funke, D. Gage, K. Harris, A. Heaford, J. Howland, L. Kann, J. Lehoczky, R. Levine, P. McEwan, K. McKernan, J. Meldrim, J. P. Mesirov, C. Miranda, W. Morris, J. Naylor, C. Raymond, M. Rosetti, R. Santos, A. Sheridan, C. Sougnez, N. Stange-Thomann, N. Stojanovic, A. Subramanian, D. Wyman, J. Rogers, J. Sulston, R. Ainscough, S. Beck, D. Bentley, J. Burton, C. Clee, N. Carter, A. Coulson, R. Deadman, P. Deloukas, A. Dunham, I. Dunham, R. Durbin, L. French, D. Grafham, S. Gregory, T. Hubbard, S. Humphray, A. Hunt, M. Jones, C. Lloyd, A. McMurray, L. Matthews, S. Mercer, S. Milne, J. C. Mullikin, A. Mungall, R. Plumb, M. Ross, R. Shownkeen, S. Sims, R. H. Waterston, R. K. Wilson, L. W. Hillier, J. D. McPherson, M. A. Marra, E. R. Mardis, L. A. Fulton, A. T. Chinwalla, K. H. Pepin, W. R. Gish, S. L. Chissoe, M. C. Wendl, K. D. Delehaunty, T. L. Miner, A. Delehaunty, J. B. Kramer, L. L. Cook, R. S. Fulton, D. L. Johnson, P. J. Minx, S. W. Clifton, T. Hawkins, E. Branscomb, P. Predki, P. Richardson, S. Wenning, T. Slezak, N. Doggett, J. F. Cheng, A. Olsen, S. Lucas, C. Elkin, E. Uberbacher, M. Frazier, R. A. Gibbs, D. M. Muzny, S. E. Scherer, J. B. Bouck, E. J. Sodergren, K. C. Worley, C. M. Rives, J. H. Gorrell, M. L. Metzker, S. L. Naylor, R. S. Kucherlapati, D. L. Nelson, G. M. Weinstock, Y. Sakaki, A. Fujiyama, M. Hattori, T. Yada, A. Toyoda, T. Itoh, C. Kawagoe, H. Watanabe, Y. Totoki, T. Taylor, J. Weissenbach, R. Heilig, W. Saurin, F. Artiguenave, P. Brottier, T. Bruls, E. Pelletier, C. Robert, P. Wincker, A. Rosenthal, M. Platzer, G. Nyakatura, S. Taudien, A. Rump, H. M. Yang, J. Yu, J. Wang, G. Y. Huang, J. Gu, L. Hood, L. Rowen, A. Madan, S. Z. Qin, R. W. Davis, N. A. Federspiel, A. P. Abola, M. J. Proctor, R. M. Myers, J. Schmutz, M. Dickson, J. Grimwood, D. R. Cox, M. V. Olson, R. Kaul, C. Raymond, N. Shimizu, K. Kawasaki, S. Minoshima, G. A. Evans, M. Athanasiou, R. Schultz, B. A. Roe, F. Chen, H. Q. Pan, J. Ramser, H. Lehrach, R. Reinhardt, W. R. McCombie, M. de la Bastide, N. Dedhia, H. Blocker, K. Hornischer, G. Nordsiek, R. Agarwala, L. Aravind, J. A. Bailey, A. Bateman, S. Batzoglou, E. Birney, P. Bork, D. G. Brown, C. B. Burge, L. Cerutti, H. C. Chen, D. Church, M. Clamp, R. R. Copley, T. Doerks, S. R. Eddy, E. E. Eichler, T. S. Furey, J. Galagan, J. G. R. Gilbert, C. Harmon, Y. Hayashizaki, D. Haussler, H. Hermjakob, K. Hokamp, W. H. Jang, L. S. Johnson, T. A. Jones, S. Kasif, A. Kaspryzk, S. Kennedy, W. J. Kent, P. Kitts, E. V. Koonin, I. Korf, D. Kulp, D. Lancet, T. M. Lowe, A. McLysaght, T. Mikkelsen, J. V. Moran, N. Mulder, V. J. Pollara, C. P. Ponting, G. Schuler, J. R. Schultz, G. Slater, A. F. A. Smit, E. Stupka, J. Szustakowki, D. Thierry-Mieg, J. Thierry-Mieg, L. Wagner, J. Wallis, R. Wheeler, A. Williams, Y. I. Wolf, K. H. Wolfe, S. P. Yang, R. F. Yeh, F. Collins, M. S. Guyer, J. Peterson, A. Felsenfeld, K. A. Wetterstrand, A. Patrinos, M. J. Morgan, *Nature* **2001**, *409*, 860–921.
3. J. C. Venter, M. D. Adams, E. W. Myers, P. W. Li, R. J. Mural, G. G. Sutton, H. O. Smith, M. Yandell, C. A. Evans, R. A. Holt, J. D. Gocayne, P. Amanatides, R. M. Ballew, D. H. Huson, J. R. Wortman, Q. Zhang, C. D. Kodira, X. H. Zheng, L. Chen, M. Skupski, G. Subramanian, P. D. Thomas, J. Zhang, G. L. Gabor Miklos, C. Nelson, S. Broder, A. G. Clark, J. Nadeau,

V. A. McKusick, N. Zinder, A. J. Levine, R. J. Roberts, M. Simon, C. Slayman, M. Hunkapiller, R. Bolanos, A. Delcher, I. Dew, D. Fasulo, M. Flanigan, L. Florea, A. Halpern, S. Hannenhalli, S. Kravitz, S. Levy, C. Mobarry, K. Reinert, K. Remington, J. Abu-Threideh, E. Beasley, K. Biddick, V. Bonazzi, R. Brandon, M. Cargill, I. Chandramouliswaran, R. Charlab, K. Chaturvedi, Z. Deng, V. Di Francesco, P. Dunn, K. Eilbeck, C. Evangelista, A. E. Gabrielian, W. Gan, W. Ge, F. Gong, Z. Gu, P. Guan, T. J. Heiman, M. E. Higgins, R. R. Ji, Z. Ke, K. A. Ketchum, Z. Lai, Y. Lei, Z. Li, J.Li, Y. Liang, X. Lin, F. Lu, G. V. Merkulov, N. Milshina, H. M. Moore, A. K. Naik, V. A. Narayan, B. Neelam, D. Nusskern, D. B. Rusch, S. Salzberg, W. Shao, B. Shue, J. Sun, Z. Wang, A. Wang, X. Wang, J. Wang, M.Wei, R. Wides, C. Xiao, C. Yan, A. Yao, J. Ye, M. Zhan, W. Zhang, H. Zhang, Q. Zhao, L. Zheng, F. Zhong, W. Zhong, S. Zhu, S. Zhao, D. Gilbert, S. Baumhueter, G. Spier, C. Carter, A. Cravchik, T. Woodage, F. Ali, H. An, A. Awe, D. Baldwin, H. Baden, M. Barnstead, I. Barrow, K. Beeson, D. Busam, A. Carver, A., Center, M. L. Cheng, L. Curry, S. Danaher, L. Davenport, R. Desilets, S. Dietz, K. Dodson, L. Doup, S. Ferriera, N. Garg, A. Gluecksmann, B. Hart, J. Haynes, C. Haynes, C. Heiner, S. Hladun, D. Hostin, J. Houck, T. Howland, C. Ibegwam, J. Johnson, F. Kalush, L. Kline, S. Koduru, A. Love, F. Mann, D. May, S. McCawley, T. McIntosh, I. McMullen, M. Moy, L. Moy, B. Murphy, K. Nelson, C. Pfannkoch, E. Pratts, V. Puri, H. Qureshi, M. Reardon, R. Rodriguez, Y. H. Rogers, D. Romblad, B. Ruhfel, R. Scott, C. Sitter, M. Smallwood, E. Stewart, R. Strong, E. Suh, R. Thomas, N. N. Tint, S. Tse, C.Vech, G. Wang, J. Wetter, S. Williams, M. Williams, S. Windsor, E. Winn-Deen, K. Wolfe, J. Zaveri, K. Zaveri, J. F. Abril, R. Guigo, M. J. Campbell, K. V. Sjolander, B. Karlak, A. Kejariwal, H. Mi, B. Lazareva, T. Hatton, A. Narechania, K. Diemer, A. Muruganujan, N. Guo, S. Sato, V. Bafna, S. Istrail, R. Lippert, R. Schwartz, B. Walenz, S. Yooseph, D. Allen, A. Basu, J. Baxendale, L. Blick, M. Caminha, J. Carnes-Stine, P. Caulk, Y. H. Chiang, M. Coyne, C. Dahlke, A. Mays, M. Dombroski, M. Donnelly, D. Ely, S. Esparham, C. Fosler, H. Gire, S. Glanowski, K. Glasser, A. Glodek, M. Gorokhov, K. Graham, B. Gropman, M. Harris, J. Heil, S. Henderson, J. Hoover, D. Jennings, C. Jordan, J. Jordan, J. Kasha, L. Kagan, C. Kraft, A. Levitsky, M. Lewis, X. Liu, J. Lopez, D. Ma, W. Majoros, J. McDaniel, S. Murphy, M. Newman, T. Nguyen, N. Nguyen, M. Nodell, S. Pan, J. Peck, M. Peterson, W. Rowe, R. Sanders, J. Scott, M. Simpson, T. Smith, A. Sprague, T. Stockwell, R. Turner, E. Venter, M. Wang, M. Wen, D. Wu, M. Wu, A. Xia, A. Zandieh, X. Zhu, *Science* **2001**, *291*, 1305–1351.

4. T. Ideker, T. Galitski, L. Hood, *Annu. Rev. Genomics Hum.Genet.* **2001**, *2*, 343–372.
5. H. Haraguchi, *J. Anal. At. Spectrom.* **2004**, *19*, 5–14.
6. J. Lopez-Barea, J. L. Gomez-Ariza, *Proteomics* **2006**, *6, Suppl 1*, S51–S62.
7. W. Shi, M. R. Chance, *Cell Mol. Life Sci.* **2008**, *65*, 3040–3048.
8. D. W. Koppenaal, G. M. Hieftje, *J. Anal. At. Spectrom.* **2007**, *22*, 855–855
9. S. Mounicou, J. Szpunar, R. Lobinski, *Chem. Soc. Rev.s* **2009**, *38*, 1119–1138.
10. H. Sun, Z.-F. Chai, *Annu. Rep. Prog. Chem.: Inorg. Chem.* **2010**, *106*, 20–38.
11. R. J. P. Williams, *Coord. Chem. Rev.* **2001**, *583*, 216–217.
12. A. M. Sevcenco, M. W. H. Pinkse, H. T. Wolterbeek, P. D. E. M. Verhaert, W. R. Hagen, P. L. Hagedoorn, *Metallomics* **2011**, *3*, 1324–1330.
13. W. Shi, M. R. Chance, *Curr. Opin. Chem. Biol.* **2011**, *15*, 144–148.
14. A. Cvetkovic, A. L. Menon, M. P. Thorgersen, J. W. Scott, F. L. Poole, F. E. Jenney, Jr., W. A. Lancaster, J. L. Praissman, S. Shanmukh, B. J. Vaccaro, S. A. Trauger, E. Kalisiak, J. V. Apon, G. Siuzdak, S. M. Yannone, J. A. Tainer, M. W. Adams, *Nature* **2010**, *466*, 779–782.
15. C. Andreini, I. Bertini, A. Rosato, *Bioinformatics* **2004**, *20*, 1373–1380.
16. R. D. Finn, J. Mistry, B. Schuster-Bockler, S. Griffiths-Jones, V. Hollich, T. Lassmann, S. Moxon, M. Marshall, A. Khanna, R. Durbin, S. R. Eddy, E. L. Sonnhammer, A. Bateman, *Nucleic Acids Res.* **2006**, *34*, D247–D251.
17. C. Andreini, L. Banci, I. Bertini, A. Rosato, *J. Proteome Res.* **2006**, *5*, 196–201.
18. C. Andreini, L. Banci, I. Bertini, A. Rosato, *J. Proteome Res.* **2006**, *5*, 3173–3178.
19. I. Bertini, G. Cavallaro, *J. Biol. Inorg. Chem.* **2008**, *13*, 3–14.
20. C. Andreini, L. Banci, I. Bertini, S. Elmi, A. Rosato, *Proteins-Structure Function and Bioinformatics* **2007**, *67*, 317–324.

21. C. L. Dupont, S. Yang, B. Palenik, P. E. Bourne, *Proc. Natl. Acad. Sci. USA* **2006**, *103*, 17822–17827.
22. C. Andreini, I. Bertini, G. Cavallaro, G. L. Holliday, J. M. Thornton, *J. Biol. Inorg. Chem.* **2008**, *13*, 1205–1218.
23. J. Szpunar, *Anal. Bioanal. Chem.* **2004**, *378*, 54–56.
24. L. Banci, I. Bertini, S. Ciofi-Baffoni, A. D'Alessandro, D. Jaiswal, V. Marzano, S. Neri, M. Ronci, A. Urbani, *J. Proteomics* **2011**, *74*, 2522–2535.
25. I. Bertini, L. Decaria, R. J. P. Williams, *Metallomics* **2010**, *2*, 706–709.
26. A. Y. Mulkidjanian, A. Y. Bychkov, D. V. Dibrova, M. Y. Galperin, E. V. Koonin, *Proc. Natl. Acad. Sci. USA* **2012**, *109*, E821–E830.
27. K. J. Waldron, N. J. Robinson, *Nat. Rev. Microbiol.* **2009**, *7*, 25–35.
28. M. H. Saier, Jr., R. Tam, A. Reizer, J. Reizer, *Mol. Microbiol.* **1994**, *11*, 841–847.
29. A. Anton, C. Grosse, J. Reissman, T. Probyl, D. H. Nies, *J. Bacteriol.* **1999**, *181*, 6876–6881.
30. S. Tottey, K. J. Waldron, S. J. Firbank, B. Reale, C. Bessant, K. Sato, T. R. Cheek, J. Gray, M. J. Banfield, C. Dennison, N. J. Robinson, *Nature* **2008**, *455*, 1138–1142.
31. L. Banci, I. Bertini, F. Cantini, S. Ciofi-Baffoni, *Cell. Mol. Life Sci.* **2010**, *67*, 2563–2589.
32. A. Atkinson, D. R. Winge, *Chem. Rev.* **2009**, *109*, 4708–4721.
33. P. A. Cobine, F. Pierrel, D. R. Winge, *Biochim. Biophys. Acta* **2006**, *1763*, 759–772.
34. W. Neupert, J. M. Herrmann, *Annu. Rev. Biochem.* **2007**, *76*, 723–749.
35. H. Irving, R. J. P. Williams, *Nature* **1948**, *162*, 746–747.
36. L. Banci, I. Bertini, K. S. McGreevy, A. Rosato, *Nat. Prod. Rep.* **2010**, *27*, 695–710.
37. L. Banci, I. Bertini, F. Cantini, I. C. Felli, L. Gonnelli, N. Hadjiliadis, R. Pierattelli, A. Rosato, P. Voulgaris, *Nat. Chem. Biol.* **2006**, *2*, 367–368.
38. K. J. Waldron, J. C. Rutherford, D. Ford, N. J. Robinson, *Nature* **2009**, *460*, 823–830.
39. T. Terai, T. Nagano, *Curr. Opin. Chem. Biol.* **2008**, *12*, 515–521.
40. D. W. Domaille, E. L. Que, S. J. Chang, *Nat. Chem. Biol.* **2008**, *4*, 168–175.
41. D. O'Malley, in *Biophysical Tools for Biologists*, Vol. 2: *In vivo Techniques*, Eds J. J. Correia, H. W. Detrich, Elsevier Academic Press, Amsterdam, Boston, 2008, pp 95.
42. R. Ortega, P. Cloetens, G. Deves, A. Carmona, S. Bohic, *Plos ONE* **2007**, *9*, e925.
43. E. Kosior, S. Bohic, H. Suhonen, R. Ortega, G. Deves, A. Carmona, F. Marchi, J. F. Guillet, P. Cloetens, *J. Struct. Biol.* **2012**, *177*, 239–247.
44. H. M. Kim, B. R. Cho, *Acc. Chem. Res.* **2009**, *42*, 863–872.
45. L. Yang, R. McRae, M. M. Henary, R. Patel, B. Lai, S. Vogt, C. J. Fahrni, *Proc. Natl. Acad. Sci. USA* **2005**, *102*, 11179–11184.
46. L. Banci, L. Barbieri, I. Bertini, F. Cantini, E. Luchinat, *Plos ONE* **2011**, *6*, e23561.
47. R. Lill, U. Muhlenhoff, *Annu. Rev. Biochem.* **2008**, *77*, 669–700.
48. S. C. Leary, *Antioxid. Redox. Signal.* **2010**, *13*, 1403–1416.
49. L. Banci, I. Bertini, S. Ciofi-Baffoni, T. Kozyreva, K. Zovo, P. Palumaa, *Nature* **2010**, *465*, 645–648.
50. L. Banci, I. Bertini, V. Calderone, S. Ciofi-Baffoni, S. Mangani, M. Martinelli, P. Palumaa, S. Wang, *Proc. Natl. Acad. Sci. USA* **2006**, *103*, 8595–8600.
51. L. Banci, I. Bertini, S. Ciofi-Baffoni, I. Leontari, M. Martinelli, P. Palumaa, R. Sillard, S. Wang, *Proc. Natl. Acad. Sci. USA* **2007**, *104*, 15–20.
52. L. Banci, I. Bertini, S. Ciofi-Baffoni, T. Hadjiloi, M. Martinelli, P. Palumaa, *Proc. Natl. Acad. Sci. USA* **2008**, *105*, 6803–6808.
53. L. Banci, I. Bertini, S. Ciofi-Baffoni, R. Boelens, A. M. Bonvin, A. D. J. van Dijk, *J. Proteome Res.* **2007**, *6*, 1530–1539.
54. Y. C. Horng, S. C. Leary, P. A. Cobine, F. B. Young, G. N. George, E. A. Shoubridge, D. R. Winge, *J. Biol. Chem.* **2005**, *280*, 34113–34122.
55. A. B. Maxfield, D. N. Heaton, D. R. Winge, *J. Biol. Chem.* **2004**, *279*, 5072–5080.
56. T. Nittis, G. N. George, D. R. Winge, *J. Biol. Chem.* **2001**, *276*, 42520–42526.
57. K. Rigby, P. A. Cobine, O. Khalimonchuk, D. R. Winge, *J. Biol. Chem.* **2008**, *283*, 15015–15022.

58. L. Banci, I. Bertini, C. Cefaro, L. Cenacchi, S. Ciofi-Baffoni, I. C. Felli, A. Gallo, L. Gonnelli, E. Luchinat, D. P. Sideris, K. Tokatlidis, *Proc. Natl. Acad. Sci. USA* **2010**, *107*, 20190–20195.
59. L. Banci, I. Bertini, V. Calderone, C. Cefaro, S. Ciofi-Baffoni, A. Gallo, E. Kallergi, E. Lionaki, C. Pozidis, K. Tokatlidis, *Proc. Natl. Acad. Sci. USA* **2011**, *108*, 4811–4816.
60. S. Allen, V. Balabanidou, D. P. Sideris, T. Lisowsky, K. Tokatlidis, *J. Mol. Biol.* **2005**, *353*, 937–944.
61. K. Bihlmaier, N. Mesecke, N. Terziyska, M. Bien, K. Hell, J. M. Herrmann, *J. Cell Biol.* **2007**, *179*, 389–395.
62. S. J. Lin, R. Pufahl, A. Dancis, T. V. O'Halloran, V. C. Culotta, *J. Biol. Chem.* **1997**, *272*, 9215–9220.
63. Y. Yamaguchi-Iwai, R. Stearman, A. Dancis, R. D. Klausner, *EMBO J.* **1996**, *15*, 3377–3384.

Chapter 2
Technologies for Detecting Metals in Single Cells

James E. Penner-Hahn

Contents

J.E. Penner-Hahn (✉)
Departments of Chemistry and Biophysics, University of Michigan,
930 N. University Avenue, Ann Arbor, MI 48109-1055, USA
e-mail: jeph@umich.edu

L. Banci (ed.), *Metallomics and the Cell*, Metal Ions in Life Sciences 12,
DOI 10.1007/978-94-007-5561-1_2, © Springer Science+Business Media Dordrecht 2013

Abstract In order to fully understand the metallomics of an organism, it is essential to know how much metal is present in each cell and, ideally, to know both the spatial and chemical distributions of each metal (i.e., where within the cell is a metal found, and in what chemical form). No single technique provides all of this information. This chapter reviews the various methods that can be used and the strengths and weaknesses of each.

Keywords fluorophore • laser ablation • mass spectrometry • nanoprobe • PIXE • secondary ion • SIMS • X-ray fluorescence

Please cite as: *Met. Ions Life Sci.* 12 (2013) 15–40

1 Introduction and Scope

The goal of this chapter is to present an overview of the methods that are in common use for detecting, quantitating, and characterizing metal ions in cells. The ideal technique would be able to determine the three-dimensional distribution of all of the metal ions in a cell with molecular resolution and would be able to distinguish between different chemical environments – at a minimum this would involve distinguishing between free and protein-bound metals, but would ideally include the ability to distinguish between the different chemical and/or biological environments of each metal. In the ideal world, this method would also be non-destructive, allowing measurements to be made on a single cell over a period of time, following the evolution of the cells' metallome over a period of time – for example, in response to a toxin or a nutrient, or as a function of the cell cycle. No single technique can provide all of these capabilities, although each of these ideals is approached by at least one method.

In the short length of this chapter, it is not possible to do justice to even one of the techniques used for cellular metal detection. Rather, I will attempt to present an overview of the physical principles behind the three main physical methods that are used to image and, sometimes, to quantitate metal ions in cells: mass spectrometric imaging, microscopy coupled to metal-specific fluorophores, and intrinsic X-ray fluorescence. Importantly, this chapter will also attempt to highlight both the advantages and some of the limitations of the different analytical tools.

Detailed discussion of the various applications of each of these methods is left for subsequent chapters. In addition, there are a variety of other methods that can be used for metal analysis that will not be discussed. Classically, metals have long been identified by colorimetric methods [1]. These rely on the fact that many inorganic complexes are distinguished by their intense visible absorption spectrum, and indeed many histochemical methods rely on such changes (see, for example, [2] for a discussion of various metal-specific histochemical stains). Although useful for studying metal distributions in tissue, these methods do not generally have sufficient sensitivity for single-cell studies. Similarly, radioactive tracers have long been used to follow the uptake and distribution of various substances, including trace metals, in biological samples (for a list of relevant isotopes, see [2]). One attraction of radiotracer studies is that they can, at least in principle, be done *in vivo*. Although autoradiography has recently been extended to single-cell studies (e.g., [3]), it has not, to date, been used in studies of cellular-level metal distributions.

1.1 General Issues for Metal Detection

It is impossible to understand in detail the metallome of a cell if you do not have some way to measure what metals are present in the cell. This can most readily be accomplished either by converting the metals to ions which can then be counted using some form of mass spectrometry or by causing the metal to absorb or emit some sort of light, with emission being much more sensitive. The most useful light emissions are either in the visible region or in the X-ray region. Since visible light emission is seldom metal-specific, at least under cellular conditions, visible light methods generally rely on the presence of a metal-specific fluorophore that emits light (ideally, light of a specific wavelength) only when the fluorophore is bound to a particular metal. In contrast, X-ray emission methods generally have intrinsic metal specificity – the X-ray that is emitted is characteristic of the metal ion that is present. Emissive methods differ in the way that the metal ion is excited, but all need some method to create an excited state. It is possible to learn about metals using other wavelengths. For example, perturbations of radiowave absorption, more commonly known as magnetic resonance imaging (MRI), can be used to localize paramagnetic metals in a sample. However, such techniques do not, at least at present, have sufficient sensitivity to permit cellular imaging.

The key concerns in any metal detection/imaging scheme are specificity and sensitivity – that is, how do you know that only the metal(s) of interest are being detected and how much of the metal (and in what chemical form) must be present in order to be detected. These are discussed below. Other concerns include the achievable spatial resolution (for imaging studies), the ability to compare more than one metal simultaneously, the ability to study living samples, and the ability to correlate metal concentrations with other biological properties.

1.2 Importance of Single Cell and Spatially Resolved Measurement

Until relatively recently, most of the available data for cellular metallomes has been limited to bulk studies of millions (or more) cells. Such experiments determine how much metal is present *on average* in an ensemble of cells and show how metal composition changes as a function of chemical and/or biological treatment. However, it has long been known that biological populations are heterogeneous and thus that it is much better to obtain information about individual cells, rather than settling for average information [4–6]. Thus, for example, if the average Zn concentration is found to double following a particular treatment, is this because all of the cells have doubled their Zn, or is it because 10% of the cells have increased their Zn 11-fold while the remaining cells are unaffected? That is, as with virtually all measurements, the average value alone provides only part of the story; details of the population distribution are often as important as the mean value. One of the exciting recent developments in metallome studies has been the range of new methods that can provide cell-specific metal quantitation, and thus can begin to probe details of the population distribution. For this reason, the present chapter focused on methods that can be applied to single cells.

In many cases, it is possible to obtain details not only about the cell-specific metal composition, but also about the subcellular distribution of metals. It is unlikely that metal ions would be uniformly distributed within a cell and, indeed, where evidence is available, it seems likely that metal ions are highly compartmentalized. As with any advance in microscopy, the ability to "see" more has led to better understanding (and also, often, to new questions). In the case of metallome studies, the ability to study the distribution of metals across different cellular compartments and to follow the evolution of intracellular concentration gradients has provided exciting new insights into the roles played by metals in biology.

1.3 Sample Preparation

Although the focus of this chapter is on the physical principles of metal determination rather than the details of particular experiments, it is important to note a few details on sample preparation. The ideal method would require no sample preparation and would thus permit study of living systems. Some methods do so. Others, particularly those measurements that are made under vacuum conditions, require some sort of sample fixation prior to measurement. Although chemical fixation is occasionally used, cryofixation, in which a sample is cooled very rapidly (perhaps 10^3 degrees/second) is often preferred. To achieve such rapid cooling, samples (often living samples) are plunged into a cryogenic liquid such as liquid propane held at liquid nitrogen temperature. Freezing in this way generally seems to preserve internal structure [7], including the location of mobile ions [8], as detected by electron microscopy. In some cases, samples are vacuum-dehydrated after freezing. This has the advantage of allowing subsequent measurements to be made at room temperature, but has at least the potential of disrupting metal distributions. While some techniques (see below) provide three-dimensional resolution, in others, three-dimensional metal distributions are determined by preparing thin slices of the frozen sample using a cryomicrotome.

2 Mass Spectrometry

Conceptually, mass spectrometry (MS) is relatively straightforward – one needs some way to isolate a small portion of sample, some way to ionize the sample, and some way to separate and detect the resulting ions [9]. Since modern mass spectrometers generally have excellent mass resolution, MS is not only able to provide unique elemental identification, but even to resolve different isotopes of each element (a feature that can be useful in, for example, tracer studies of metal uptake). Given this resolution, there is seldom any question regarding the specificity of MS measurements. Mass spectrometry has exceptional detection limits, being able to detect even single ions, and generally has a wide dynamic range, which can be important for studies of very heterogeneous samples [10]. These mean that MS generally has good sensitivity, although this depends on the efficiency with which a sample can be converted into ions. In practice, observed detection limits range from μg/g to pg/g.

For cellular studies, intense ion or laser beams are the most useful methods for removing a small amount of sample (see Sections 2.1 and 2.2, respectively), although other techniques such as electrospray ionization have also been used. Spatial resolution can be as good as 50 nm for ion beam imaging, while laser beam imaging generally gives much lower resolution of a few μm to hundreds of μm [11–13]. Although most studies have been done in probe mode, where the spatial resolution results from the use of a small probe beam (see Sections 2.1 and 2.2), it is also possible to do MS in microscope mode using ion optics to image the emitted ions [14].

Much of the recent work on MS imaging has focused on "metabolomics", where the emphasis is on identifying intact cellular metabolites. Such studies generally strive for gentle ionization in order to ensure that the metabolite remains intact. For metallome studies the focus instead is on energetic ionization methods that remove metal ions completely from any surrounding protein environment.

2.1 *Secondary Ion Mass Spectrometry*

In secondary ion mass spectrometry (SIMS), an ion beam strikes the sample. If the energy of this beam is sufficiently high, it will be able both to eject a portion of the sample and to ionize and fragment the ejected sample, releasing atoms, molecules, and molecular fragments. The ions formed in this way are secondary ions, as distinct from the primary ions that make up the initial beam. Ionization is a relatively rare process, with most of the ejected material being neutral and thus undetectable by MS [15].

Since ions are not able to penetrate more than a few monolayers of a sample, SIMS is inherently surface-sensitive and, for the same reason, requires that measurements be performed in vacuum. Since the material ejection process, known as "sputtering", removes a layer of material from the sample, it is possible to do depth profiling by making successive scans over an area. Although SIMS is a key tool for surface analysis and is widely used in studies of inorganic materials, it is just now being adopted for biological samples. This is in part because of the requirement that SIMS studies be performed under vacuum, thus preventing *in vivo* studies, but is perhaps more strongly driven by the fact that until recently there were no commercial spectrometers for nm-scale SIMS imaging.

One of the limitations of SIMS is that it can suffer from strong "matrix" effects, meaning that secondary ion yield can vary by orders of magnitude depending on the chemical environment, or "matrix", that contains the metal [13]. Since the ejected ions need to make their way into the mass spectrometer, SIMS yields can also be sensitive to the sample "topography" [16]. Finally, since biological samples are generally insulating, it is possible for the ion beam to cause a build-up of charge on the sample, interfering with secondary ion ejection and detection. All of these can, in principle, be addressed experimentally: calibration curves can be prepared by spiking a sample of the same material with known amounts of metal of interest [17]; a microtome can be used to prepare completely flat samples; and samples can be treated to avoid charging effects, for example by applying a thin layer of metal to the surface [18]. Nevertheless, these effects can make it difficult to determine absolute metal concentrations using SIMS and it is possible that chemical variations within a cell (e.g., between metal that is tightly bound to a protein and metal that is "free" in the cytosol) could make certain metal populations effectively invisible to SIMS.

The primary ion beam can be focused electrostatically to a spot size well under 1 μm [18], and in some cases as small as 50 nm [9,12,13,18], making SIMS ideal for imaging subcellular structures. However, since higher spatial resolution comes at the expense of a smaller volume of ejected ions and thus lower sensitivity, measurements are not always made at the highest possible resolution. For imaging experiments, the incident ion beam can be rastered across the sample so that there is no need to change the sample position (this is not the case for most other methods – the resulting need to physically position samples with high precision can substantially complicate an experiment).

The mass spectrometry field distinguishes between "static" and "dynamic" SIMS measurements. The former refers to conditions in which the primary ion dose is kept lower than ~10^{13} impacts/cm^2 [18]. At this dose, less than ~1% of the surface is impacted and there is, in principle, no change to the surface, hence the name "static". At higher doses, the entire surface begins to erode, giving greater fragmentation and a higher yield of ions but with significant sample damage.

The fact that an entire layer of sample is removed in dynamic SIMS opens the possibility of depth profiling without the need to cut successive sections using a microtome. In this approach, successive dynamic SIMS scans across the same area allow one to sample successively lower areas of the sample. This has worked better for inorganic materials than for softer biological samples because of the damage that the ion beam can cause to underlying layers. However, recent work with polyatomic ion beams such as C_{60}^+ have shown better ion ejection and less sample damage for biological samples [16].

Static SIMS would not seem compatible with this sort of depth profiling since, by definition, in static SIMS there is no significant change in the sample surface. One variant is to use two ion beams – one for the SIMS measurements (typically a pulsed beam for TOF-SIMS – see below) and a second high-intensity continuous beam of, for example, C_{60}^+ to remove a layer of sample between successive image measurements [12]. Alternatively, one can use a focused beam of ions directed tangentially to mill off a defined layer of a cell [19] or to do high-resolution sectioning of a sample [20] between static SIMS measurements.

The ions produced in SIMS can be resolved in one of two ways. The simplest is to use a magnetic field to bend the beam of ejected ions. The radius of curvature depends on the charge/mass ratio (*m/q*) of an ion and thus by adjusting the position of the detector one can tune which *m/q* is detected. In this so-called "magnetic sector" mass spectrometer, the user predetermines which elements, or more precisely which *m/q* values, will be detected by selecting the position (or positions) of the detector(s). This is the method that is used with dynamic SIMS, and the development of a commercial high-spatial resolution magnetic sector SIMS instrument (the NanoSIMS from CAMECA) has done much to popularize this method [9].

In static SIMS the incident ion beam is generally pulsed, permitting use of a time-of-flight (TOF) mass spectrometer. In TOF, each pulse of secondary ions is accelerated by an electric field and the time required for the ions to cover a fixed distance is measured. Since the kinetic energy, and thus the velocity, of each ion depends on its *m/q*, time of arrival can be used to determine mass. One of the advantages of TOF measurements is that they permit simultaneous resolution of all of the ions in a sample. Static TOF-SIMS thus has the advantage of permitting simultaneous mapping of not only all of the metal ions in a sample but also other (e.g., organic) molecular ions that may be helpful in understanding cellular structure. In contrast, dynamic SIMS is limited to a relatively small number of pre-selected elements, but gives much better sensitivity and thus allows better spatial resolution [12].

2.2 *Laser Ablation Mass Spectrometry*

In laser-based methods, an intense laser rather than an ion beam is used to remove sample material. One common variant involves adding an organic "matrix" to the sample, with the matrix chosen so that it will strongly absorb the laser light (typically in the UV), causing a layer of both matrix and sample to be ejected and ionized. This matrix-assisted laser desorption/ionization, or MALDI, has been widely used in metabolomic studies because it is a "soft" ionization method and is thus able to create large molecular ions without significant fragmentation. This is essential if one needs information about protein distribution within a cell, but is not especially useful for metallome studies, where the goal is to obtain the free metal ion. An alternative that works well for wet samples is to use an intense IR laser for the irradiation, although the laser power that is required is much higher than with MALDI.

Infrared radiation excites the O-H vibrations in the water molecules, with the resulting rapid heating causing a small explosion that ejects particles from the surface. The ejected material then needs to be ionized since, in contrast with UV or ion beam excitation, the IR laser typically lacks sufficient energy to ionize molecules. Much of the work on laser ablation MS imaging has used electrospray ionization since this is, again, a relatively soft ionization method appropriate for studies of intact metabolites. For metallome studies a harder ionization is needed, for example an inductively coupled plasma (ICP). In laser ablation ICP-MS, the ablated molecules are pulled into a high-temperature ionized plasma that is able to completely ionize the sample, thus permitting the detection of metals independent of their initial protein environment.

The sampling depth of the ablating laser is typically a few to tens of micrometers depending on the intensity of the laser and the absorption coefficient of the material [10]. As with SIMS, it is possible to make successive scans in order to analyze successively deeper layers of the sample. The spatial resolution is generally limited by the size of the laser spot, although it may be worse, depending on the size of the ablation pit. State of the art spatial resolution is 1–2 μm [21], although reported resolution is often 30–50 μm or worse [12,15]. At this resolution, it is not possible to determine sub-cellular distributions. However, laser ablation can be used to analyze single cells that are, for example, chosen for analysis based on videomicroscopy of the sample. This allows one to pick out a single cell from a complex mixture such as a natural population of bacteria or a tissue sample [5,22–27]. Most of this work, to date, has focused on identification of metabolites rather than ions.

One promising improvement in spatial resolution is to use near-field optical methods [13]. In this approach, an object such as a silver needle is placed immediately adjacent to the sample and used to enhance the local laser ablation field. This has been reported to give ablation craters as small as 200 nm to 2 μm [13].

Quantitation of ICP-MS requires suitable reference materials, since the probability of ICP ionization is different for each element and depends somewhat on the chemical environment. One way to accomplish this is to homogenize a sample of biologically equivalent tissue and add well-defined amounts of the metals of interest [17]. This gives reported limits of detection ranging from 1 ng/g to 1 μg/g.

2.3 Related Methods

There is a wide variety of other ways in which mass spectrometry can potentially be used to interrogate biological samples. For example, a flow-cytometer can be coupled to an ICP-MS for high-throughput multiplex detection [28] of the elemental composition of millions of cells. This lacks the imaging capability of the methods discussed above, but has the advantage of allowing study of a large population of cells. To date, the focus of flow-cytometry has been on non-biological lanthanide tags that have been used to label antibodies, but this could be applied to study, for example, metal uptake using isotopically labeled samples.

2.4 Examples

One example that illustrates the sort of information that can be obtained from SIMS imaging is a study by Smart et al. looking at metal distributions in the leaf tissue of *Alyssum lesbiacum*, a plant known to accumulate high concentrations of Ni [29].

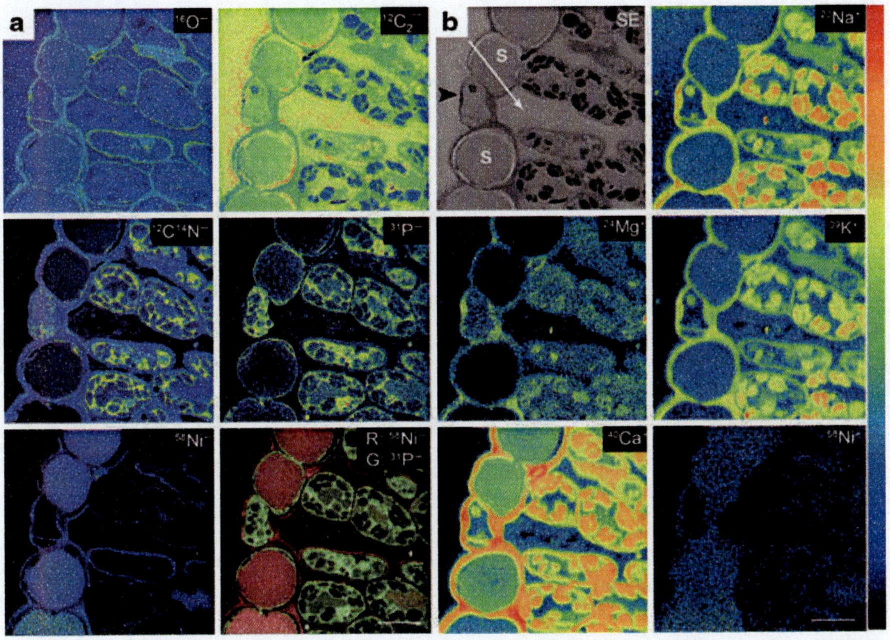

Figure 1 NanoSIMS images of a cross-section of a nickel-rich *Alyssum lesbiacum* leaf (**a**) Image obtained using a Cs^+ primary ion beam (**b**) Image obtained using an O^- primary ion beam. A complementary image of secondary electron (SE) is shown in the upper left of (b). Scale bar: 10 μm. Reproduced from [68] with permission from John Wiley & Sons; copyright (2010).

These data (Figure 1) show images measured both with a Cs^+ primary beam to generate anionic secondary ions and an O^- primary beam to generate positive ions mode (the former has higher resolution). These data show that nickel tends to be sequestered within epidermal cell vacuoles and to be excluded from the protoplast of stomatal cells. Nickel and phosphorus show rather striking inverse correlations, as shown by the red-green overlay in (a). There is an increasing flood of applications of SIMS to various aspects of cellular composition, and many more studies like this are to be expected in the future [9]. Although absolute quantitation may be challenging, work such as this provide unprecedented detail about the local distribution of metals in cells.

Laser ablation MS has also been used in metallomics, particularly a series of studies by Becker and coworkers imaging metals in the brain [11,13,17]. An intriguing application of laser ablation to explore metal trafficking was a study looking at otoliths from salmon and using the imaging capability of laser ablation to track the changes in Sr:Ca ratios through the life of the fish in order to tell when an individual fish had lived in fresh *versus* sea water, taking advantage of the fact that Sr concentration depends on salinity. As with SIMS, there is every reason to expect an explosion of similar studies in the future.

3 Visible Light

Since its invention, microscopy has been providing important new insights into biological structure, with better and better spatial resolution as microscopy methods have improved. A challenge for many biological samples is finding a contrast mechanism that is sensitive to the structure of interest. One of the most powerful solutions to the contrast problem has been the use of fluorescence probes to label the site(s) of interest, which can then be studied using fluorescence microscopy. Although there have been exciting recent developments in improving the resolution of microscopy beyond the so-called diffraction limit (i.e., to resolutions better than the wavelength of the probe light, typically ~500 nm) [30], in general fluorescence microscopy can be described as a mature technology. The conventional resolution limit of 500 nm – 1 μm is sufficient for a wide range of cellular studies, as evidenced by the enormous literature on using fluorescence microscopy to investigate cellular function. With regard to metallomic studies, current interest lies in the development of new metal-responsive fluorescent probes [31–35].

In comparison with the other methods discussed in this chapter, fluorescence microscopy has two key advantages for studies of biological samples. This first is that it is the only method that can truly be used to probe living samples, and thus to follow the evolution of metal concentrations and localization over time. There are some caveats – depending on the wavelength and power of the illumination, even visible light microscopy can cause damage, as can the added fluorophore. Nevertheless, the ability to study living samples is a unique advantage in comparison with mass spectrometry and X-ray methods. Secondly, the use of confocal microscopy (again, a mature technique) permits relatively straightforward extension of fluorescence microscopy into three-dimensional measurements [36]. Confocal microscopy takes advantage of the fact that light (unlike ions) is readily transmitted, at least for some depth, through biological samples. It is thus possible to probe more than just the surface of the sample. With appropriate optics, visible light (although not X-rays) is easily focused, allowing collection of fluorescence signals from just one portion of the sample.

3.1 Design Considerations for Metal-Specific Fluorophores

Although a few inorganic complexes are strongly luminescent (for example, zinc porphyrins), this is not generally the case. Consequently, in order to create a system in which the visible fluorescence is metal specific, it is generally necessary to add a metal-specific fluorophore. In the most general sense, this is a metal chelator with fluorescent properties that are modulated when the metal binds. The metal may change either the electronic structure, for example, turning on or turning off a relaxation mechanism, or it may change the molecular structure, for example, changing the distance between a donor-acceptor pair (see, for example, [33,34,37,38] for more details on the different photophysical mechanisms relevant to the design of metal-specific fluorophores).

Binding of a paramagnetic metal may enhance intersystem crossing and quench the emission from a fluorophore (a "turn-off" sensor), while in a different design, metal binding may lower the energy of a lone-pair that would otherwise quench emission via photoinduced electron transfer, leading to an increase in emission (a "turn-on" sensor). Ideally, metal binding would change not only the intensity but also the wavelength of the emission. This can happen if metal binding causes a conformational change that brings a donor and an acceptor fluorophore close together such that Förster (or fluorescence) resonance energy transfer (FRET) occurs only when the metal is bound. In this case, the metal-bound sensor will fluoresce at the lower-energy acceptor wavelength while the metal-free sensor will fluoresce at the higher-energy donor wavelength.

Although the term "fluorophore" is used throughout this chapter to refer to any emissive molecule, it is not necessarily the case that the molecule must fluoresce in order to be a useful sensor. Thus, Lippard et al. have recently demonstrated that chelators that emit by phosphorescence rather than fluorescence can give improved rejection of background signals [39]. The advantage of a phosphorescent sensor is that the emissive lifetime is much longer, allowing photoluminescence lifetime measurements to better distinguish different emitters.

The ideal fluorophore would have a large molar absorption coefficient so that it could be readily excited, a high quantum efficiency for fluorescence so that it would give an as high as possible signal-to-noise ratio, and excellent specificity, so that only the metal of interest is able to modulate its emission properties. Unfortunately, these properties alone may not be sufficient for quantitative studies. If the fluorophore emits only in the presence of the metal, then it can be challenging to interpret observed changes in fluorescence intensity: has the intensity increased because more metal is bound to the fluorophore or because more fluorophore is present in the cell (or in a particular compartment within the cell)? For this reason, "ratioable" fluorophores are preferred, when possible. These are fluorophores that emit light at one wavelength when metal-free and at a second, ideally non-overlapping, wavelength when the metal is bound. This might be, for example, a FRET-based sensor where donor emission is seen in the absence of metal, while metal binding causes a conformational change that favors FRET, resulting in acceptor emission. Much of the recent work on metal-specific fluorophores has focused on development of ratioable sensors.

3.1.1 Metal-Binding Equilibria

It is relatively straightforward to describe metal binding to a metal chelator *in vitro*. For a 1:1 metal-chelator complex (ML, equation 1), the concentration of metal is, at least in principle, simply related to the metal-ligand dissociation constant (K_d) and the ratio of metal-bound to metal-free chelator (equation 2).

$$\mathrm{ML} \rightleftharpoons \mathrm{M} + \mathrm{L} \tag{1}$$

$$[\mathrm{M}] = K_d \frac{[\mathrm{ML}]}{[\mathrm{L}]} \tag{2}$$

Equation 2 illustrates the value of a ratioable fluorophore. The concentration of chelator complex, [ML], is determined directly from the intensity of the emission. However, it is impossible to use this to determine the metal concentration unless one knows either the free chelator concentration, [L] or the total chelator concentration, [ML] + [L] (since the latter can be used to determine [L] by difference). With a ratioable fluorophore one has a way to measure [L] directly. Without this, the best that one can do is to make some plausible assumption about the total amount of chelator that is present. This is fine for many qualitative assessments, but is not adequate for quantitative determinations.

If [M] << K_d, most of the fluorophore will be in the metal-free form, while if [M] >> K_d, most of the metal will be bound to the fluorophore and the signal will be saturated. This means that the range of metal concentrations that a fluorophore can sense is determined by K_d. For this reason, there is a great deal of effort devoted to constructing families of fluorophores with ranges of K_d values; if K_d is too small, the fluorophore will be saturated under most conditions, while if K_d is too large, the fluorophore will bind metal too weakly to be useful.

In the discussion above, it was implicitly assumed that there is only one relevant equilibrium, between ML and M + L, and that this equilibrium is fast. Unfortunately, neither assumption is necessarily true. In addition to binding to the fluorophore, the metal is likely to also bind to proteins, nucleic acids, and various small molecules present in the cell, and at least some metals may be physically sequestered within a cell. If the fluorophore cannot get to the cellular compartment that contains the metal, it will not be able to report on metal concentration.

Many of the fluorophores that have been developed bind the metal of interest quite tightly (i.e., have a small K_d). In contrast, many cellular binding sites, particularly those related to endogenous small molecules, have relatively weak binding. As long as an endogenous site has a dissociation constant larger than the K_d for the fluorophore, it should release metal to the fluorophore. Other sites, particularly those associated with protein-specific metal binding, may have K_d values smaller than that for the fluorophore. This metal will be invisible to the fluorophore. For this reason, the fluorophore-detected metal concentration is often referred to as the "free", or "loosely bound" pool of metal. It is important to remember that there may be other pools in which a metal is tightly bound (or slowly exchanging). This metal cannot generally be detected using metal-specific fluorophores.

Given that sensors are often designed to bind metals tightly, one needs to be certain that the sensor does not alter the cell, particularly for studies that are attempting to follow the evolution of metal concentrations in a living cell. It is possible that an added fluorophore could decrease the availability of essential metals. Thus, it has been shown in at least one case that changes in fluorophore concentration from 2 to 20 μM could change the calculated "free" zinc concentration in the cell by a factor of two [40]. This sort of change could, at least in principle, affect cell physiology.

3.1.2 Other Considerations

Even an ideal fluorophore will be useless if it cannot be delivered to the appropriate location in a cell. If the fluorophore is permeable (typically, this means neutral) it may be possible to simply add it to the cellular medium. If the metal-fluorophore complex is charged, this should limit the ability of the fluorophore to diffuse out of the cell. The process can be made more rigorous by using a fluorophore that is made membrane permeable by esterification of carboxylate groups; when the esters are cleaved by intracellular esterases, the fluorophore is trapped in the cell [41]. It is possible to inject fluorophores directly by microinjection, although this is difficult to implement in a high-throughput model. An alternative is to genetically encode a protein-based sensor such that the cell is programmed to synthesize its own sensor (see Section 3.1.4).

Although fluorescence is often a highly probable process, it is never the only way in which the excited state of the fluorophore can return to the ground state. It is always possible for the excited state to react rather than undergoing fluorescence. Depending on the details of the reaction, this usually results in either temporary or permanent photobleaching. This means that after some period of time, most fluorophores stop responding and it can therefore be difficult to collect high resolution images for some samples. It is important to guard against this and other sorts of photodamage when making measurements.

3.1.3 Exogenous Fluorophores

There is a large literature on the design of metal-specific fluorophores. The current state of the art for design of different metal-specific probes has been reviewed by many authors [31,32,34,35,37] and is further illustrated by other chapters in this volume. The best sensors are those that combine careful control of the photophysical properties that determine the sensor's emission with the classic coordination chemistry that determines the specificity of metal binding. The most progress has been made on sensors for diamagnetic metals (Ca^{2+}, Zn^{2+}, Cu^{+}), since these lend themselves to turn-on sensors. Paramagnetic metals (Fe^{3+}, Mn^{2+}) are generally limited to turn-off sensors, which are more susceptible to background signals. Weakly bound metals (Na^{+}, K^{+}) remain a challenge. Many sensors show a pH dependence which may interfere with measurements, particularly in certain cellular compartments. All sensors have at least some susceptibility to interference from other metals, and this can be severe given the relatively high concentrations of Ca^{2+} and Mg^{2+} compared to the concentration of trace elements. This and other concerns have led some groups to explore protein-based sensors.

3.1.4 Endogenous Fluorophores

In some cases, it is possible to engineer a cell so that it produces a fluorophoric sensor endogenously, avoiding the need to add an exogenous fluorophore. This approach avoids some of the difficulties associated with exogenous sensors. There is no need to worry about the permeability of a sensor, although this may be replaced by concerns over the stability of the protein and the level of expression of the mature protein. The metal-sensing component of such sensors is often a naturally occurring metal-binding protein. This has the significant advantage of often providing much better metal specificity than can be obtained using small molecule chelators.

One such example is carbonic anhydrase, a natural metal-binding protein that has both high affinity and high specificity for Zn^{2+}. When Zn^{2+} is bound to carbonic anhydrase, the protein will bind a variety of ligands, including in particular aryl sulfonamides. If carbonic anhydrase is modified to add a fluorophore, either by covalent attachment [42], or by fusing it with a naturally fluorescent protein [43], FRET between the sulfonamide and the fluorophore can be used as a ratioable zinc sensor, since sulfonamide binding only occurs if Zn^{2+} is bound.

Although the carbonic anhydrase example involves endogenous production of the sensor protein, it still requires addition of the sulfonamide. An alternative approach is to incorporate both the donor and the acceptor function as part of the protein, for example as cyan fluorescent protein and yellow fluorescent protein [44], thus avoiding the need to add a fluorescent molecule. A variety of metal sensors can be combined with these fluorescent protein domains to produce an overall sensor [37,44–46]. One attraction of genetically encoded endogenous sensors is that they can be combined with fairly traditional cell-biology methods for targeting particular subcellular regions, thus giving greater control over intracellular localization [47].

3.2 *Two-Photon Excitation*

Two-photon excitation refers to the non-linear phenomenon in which two low-energy photons can effectively be absorbed simultaneously, exciting a transition equal to twice their energy [48]. For excited states that emit in the visible, this means using an excitation source in the near-IR. Since IR light penetrates better than visible light in most biological samples, this allows *in vivo* studies at significant depth (up to hundreds of microns, see Figure 2). A second advantage is that two-photon photo-absorption depends quadratically on photon flux, and thus two-photon absorption occurs in a very small focal volume, allowing precise optical sectioning of a sample [36]. Finally, the low energy (IR) of two-photon probes generally cause less photodamage and is less likely to stimulate tissue autofluorescence. The development of two-photon probes for metal ions has recently been reviewed [49] and has shown the utility of these for studying metal distributions within brain tissue, providing cell-specific information by virtual sectioning.

3.3 *Examples*

There are literally thousands of examples of metal-specific fluorophores being used to probe the distribution and evolution of metal concentrations in cells, a few of which are noted in this section, and many more of which are described elsewhere in this volume, and in numerous other reviews [2,31–34,37,50]. It is not an exaggeration to say that metal-specific fluorophores have revolutionized our understanding of the distribution of "free" metal in a cell. However, in contrast with mass

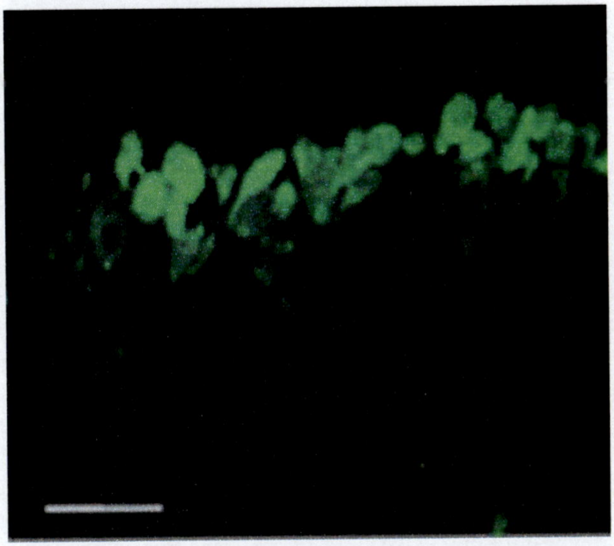

Figure 2 Two-photon images of a mouse hippocampal slice stained with a magnesium-specific fluorophore showing CA1pyramidal neurons at a depth of <150 μm. Scale bar is 30 μm. Reproduced with permission from [49]; copyright (2009) American Chemical Society.

spectrometry (Section 2) and X-ray methods (Section 4), fluorophoric methods will always be limited by the need to move the metal from its initial site to the fluorophore in order to detect the metal.

4 Intrinsic X-Ray Fluorescence

High energy excitation can be used to eject an electron from a core shell (typically 1*s*, although occasionally 2*s* or 2*p*) of an atom. The resulting highly excited state rapidly relaxes, typically via X-ray fluorescence (XRF). For 1*s* excitation, the dominant relaxation pathway is $2p \rightarrow 1s$, which is accompanied by emission of a so-called Kα X-ray. To an excellent approximation, the Kα X-ray energy and quantum yield are purely atomic phenomena, and are unaffected by the chemical environment of the excited atom. This atomic-like character makes X-ray emission an ideal tool for determining the total amount of an element in a sample, since it is not necessary to prepare the sample in any way, and no exogenous fluorophore is needed. As a bonus, the technique is largely insensitive to matrix effects.

Although all elements show X-ray fluorescence, practical applications are largely limited to phosphorus and heavier elements since for lighter elements the emitted X-rays fall in the soft X-ray region (< 2 keV), which is difficult experimentally due to the lack of windows that are transparent to low energy X-rays. A major challenge to the wider application of XRF in biology is its relatively low sensitivity – in comparison with visible light, X-rays interact relatively weakly with matter and thus have relatively low molar absorption coefficients. The typical concentration detection limits for X-ray fluorescence are in the micromolar range. On the other hand, it is possible to use very tightly focused excitation sources for XRF, giving the technique mass detection limits as low as a few hundred atoms, thus making XRF compatible with MS detection limits (as noted above, detection limits have a somewhat different meaning for fluorophore studies). By using a focused excitation source, it is relatively straightforward to measure spatially resolved metal distributions. All elements are detectable in an XRF experiment, providing the excitation source is sufficiently energetic to excite the characteristic X-ray emission line. This intrinsic multichannel capability can be particularly important in studies of the interaction between different metal ions.

X-ray excited states can be generated using either particle beam excitation (typically electrons, although protons can also be used) or X-ray excitation. The energy of the emitted Kα X-ray increases monotonically with increasing atomic number. For metallome studies, the relevant energies start at a low of ~1 keV for sodium and magnesium or 2–3 keV for phosphorus through chlorine. The former requires a vacuum environment, while the latter can be studied at atmospheric pressure. The energies for the heavier metals run from 3.3 keV for potassium to 17.5 keV for Mo. Higher atomic number elements are generally studied using "L" rather than K emission lines (2*s* or 2*p* core hole initial states rather than 1*s*) which also fall in the experimentally more accessible ~2 keV to ~20 keV window. For X-rays in this energy range, biological samples are relatively transparent, with the absorption path length (the thickness required to absorb 1/e of the incident beam) ranging from ~20 μm at 2 keV to>1 cm at 20 keV. This relative transparency means that XRF measurements lack the intrinsic surface sensitivity of mass spectrometry and, as noted above, because X-rays are not easily focused there is no X-ray equivalent of a high-resolution confocal microscope. As a consequence, XRF imaging typically gives only a two-dimensional projection of the three-dimensional distribution of metal ions unless a surface sensitive excitation is used (see Section 4.1) or complex tomographic imaging is undertaken (see Section 4.3).

XRF measurements are typically made on dried cells, as described in Section 1.3. This is required for particle beam excitation and is often preferred for X-ray excitation. Like mass spectrometry, XRF imaging is destructive. While the elemental distribution in the sample will generally survive irradiation (particularly when the sample is dried), the intense beam of ionizing radiation is not compatible with measurements on living samples.

4.1 *Particle Excitation*

Particle beams, particularly electrons, can be focused to an extremely small spot size, and thus provide the best possible resolution for XRF studies. As noted in Section 2, ion beams are strongly scattered on passing through matter, including gas, and therefore particle excitation requires the use of vacuum, which in turn generally requires the use of dried samples. An electron microscope is a convenient source of focused electron beams, and most electron microscopes are equipped with energy dispersive X-ray detectors. A wide range of names have been used for electron-excited XRF, with electron probe microanalysis (EPMA) being one of the more common. It is also sometimes referred to as energy dispersive spectroscopy (EDS) although this is not a particularly informative name, since electron, proton, and X-ray excited XRF all use energy dispersive spectrometers. A third name that is sometimes used is PIXE which, in some accounts, stands for particle induced X-ray emission. Unfortunately, the PIXE label is also used to represent proton induced X-ray emission, and that is the sense in which it is used here (see below).

When using an electron beam, the spatial resolution can be as good a 1 nm and thus better than any of the other techniques discussed in this chapter. However, the resolution is often degraded by scattering of the electrons, particularly if the sample is thick [2]. Because electrons do not penetrate far into the sample, only a small number of atoms can be excited. In addition, there can be significant background for particle-excited EPMA due to the electron excited Brehmstrahlung background [51–53]. As a consequence, the detection limits for EPMA are typically 100–1000 μg/g. This is worse than can be achieved with proton or X-ray excitation; however, electron excitation is often better than X-ray excitation for lower atomic number elements.

Proton microprobe beams cannot be focused as tightly as electron beams, with the result that resolution in PIXE is typically ~1 μm although it may be as good as a few hundred nm in the best case [52]. Proton beams are more effective than electron beams at exciting core-electron excited states, and thus PIXE is generally more sensitive than electron microprobe with a sensitivity for transition metal (and heavier) elements of 1–10 μg/g.

In addition to measuring the particle-excited XRF, it is also possible to measure the transmitted and scattered particles. With electron excitation, and with a thin enough sample, one can measure energy dependence of transmitted absorption, a spectroscopy known as electron energy loss spectroscopy (EELS) [54], which can give insight into the chemical environment of the metal. With proton excitation one can measure both the back-scattered proton beam and the transmitted proton beam. Proton back-scattering gives direct information about the organic composition of the sample, while proton absorption gives the density of the sample. The two together can be used to determine the absolute thickness of a sample [52]. This is important because, without this, XRF measurements (see below) can only be used to determine so-called areal concentration – that is, the "concentration" in g/cm^2. If the absolute thickness is known, the areal concentration can be converted to ppm.

4.2 X-ray Excitation

Although X-ray fluorescence has been measured for over 100 years, the practical application of XRF as a tool for studying biological samples required the development of very intense synchrotron X-ray sources, which has increased by many orders of magnitude the available X-ray flux density, and thus the sensitivity of the XRF measurement. A second advantage of synchrotron excitation, in comparison with a conventional X-ray tube, is that all of the X-ray power can be concentrated in a single narrow band of energies. This dramatically reduces the X-ray scattering background. Synchrotron excited XRF is sometimes referred to by the acronym SXRF for synchrotron-excited X-ray fluorescence. This is a largely unnecessary acronym since it is irrelevant whether the incident X-ray beam arises from a synchrotron or another X-ray source, although there is some value in distinguishing X-ray excited XRF from proton or electron-excited XRF. A third abbreviation, used in some of the literature, is X-ray fluorescence microscopy (XFM), reflecting the micrometer to nanometer resolution that is possible with modern synchrotron sources and distinguishing XRF imaging from more conventional XRF quantitative analysis.

With synchrotron excitation, detection limits can be as good as 10^{-18} g in a spot size of 90 nm [11,55]. The physics of X-ray excitation and emission are well known and the only matrix effects arise from background X-ray absorption which, for most biological samples, is negligible. Consequently, given a standard reference material with well-defined composition, XRF can provide absolute metal composition. None of the other methods discussed in this chapter can provide absolute composition as easily. However, as noted above, XRF "concentrations" are limited to areal concentrations, or the mass of metal per cm^2 of sample that is illuminated. Areal concentrations are fine for comparing relative concentrations of different metals in an area (or, more accurately, within a projected volume), but do not directly allow determination of the chemically more useful molar concentration. If one assumes a thickness for the sample, it is possible to estimate the concentration. As noted above, absolute concentrations can be obtained by combining XRF with proton scattering. Alternatively, it has recently been shown that X-ray phase contrast microscopy can be used to measure the thickness of a sample [56]. This relies on the fact that at high X-ray energy, the real part of the refractive index of a biological sample is dominated by the C, H, N, and O in the sample, and is approximately proportional to the density of the object. Thus, the phase change is proportional to sample thickness.

4.3 Examples

The relatively limited availability of intense proton beam sources and intense synchrotron nanoprobe sources means that XRF has seen far fewer applications than fluorescence microscopy in metallome studies. There have, nevertheless, been hundreds of biological XRF and XFM studies. These have been reviewed recently, for example in [2,11,53,57–59] and will not be rehearsed here. One representative study is the use of XRF to investigate the localization of Fe in rat PC12 cells (Figure 3). Iron has been implicated in the selective loss of dopaminergic neurons in Parkinson's disease, and Fe has been shown to accumulate in the substantia nigra of Parkinson's patients. By combining XFM with conventional epifluorescence imaging

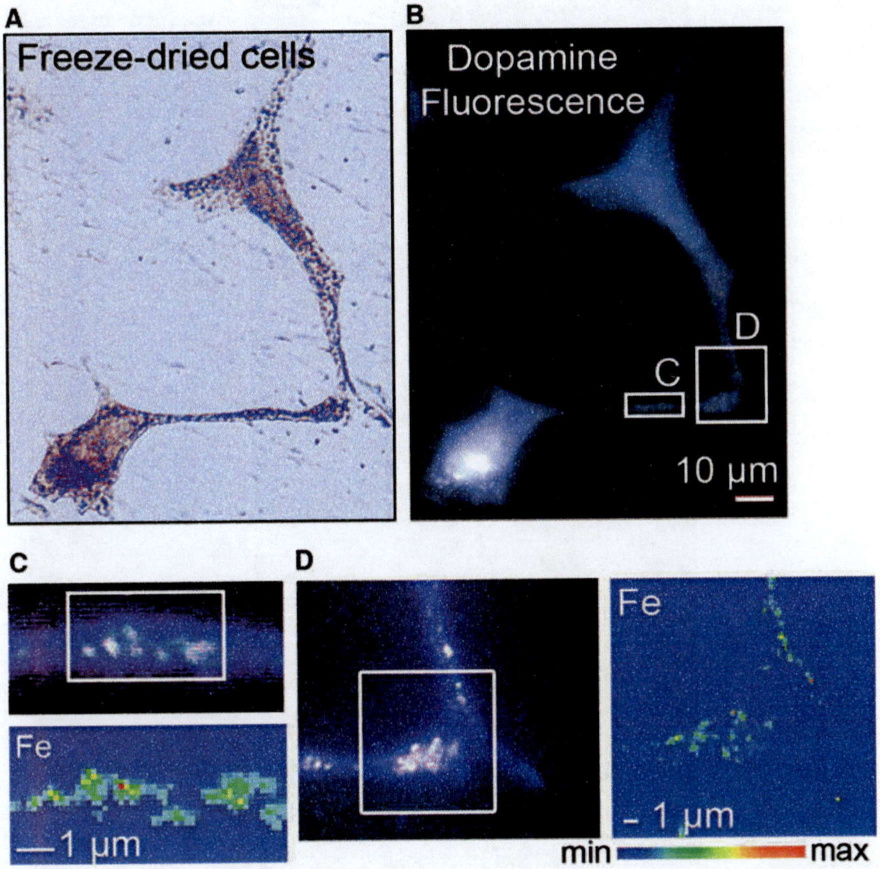

Figure 3 Dopamine and iron localization in a neurovesicles. Visible light and epifluoresence images of freeze-dried cells are shown in (**A**) and (**B**), respectively. High resolution comparisons of dopamine and Fe distributions are shown for regions (**C**) and (**D**) (white boxes in (**B**) show the regions used for high resolution scans; white boxes in (**C**) and (**D**) mark the regions scanned by Fe XFM). Reproduced with permission from [55] doi:info:doi/10.1371/journal.pone.0000925.g005.

to localize dopamine, the authors were able to show that Fe and dopamine are colocalized, presumably in dopamine neurovesicles [55]. This report used an X-ray beam with 90 nm resolution, which was sufficient for detailed subcellular localization. Although not shown here, this study compared distributions of Fe, Zn, and K and found that only the Fe showed the unusual distribution pattern seen in Figure 3. This multielement capability, together with the ability to correlate element distributions with conventional visible-light fluorescence microscopy, is one of the advantages of XFM. It is important to note, however, that from the data in Figure 3 it is impossible to determine the relative three-dimensional arrangement of the different metals (see below, however).

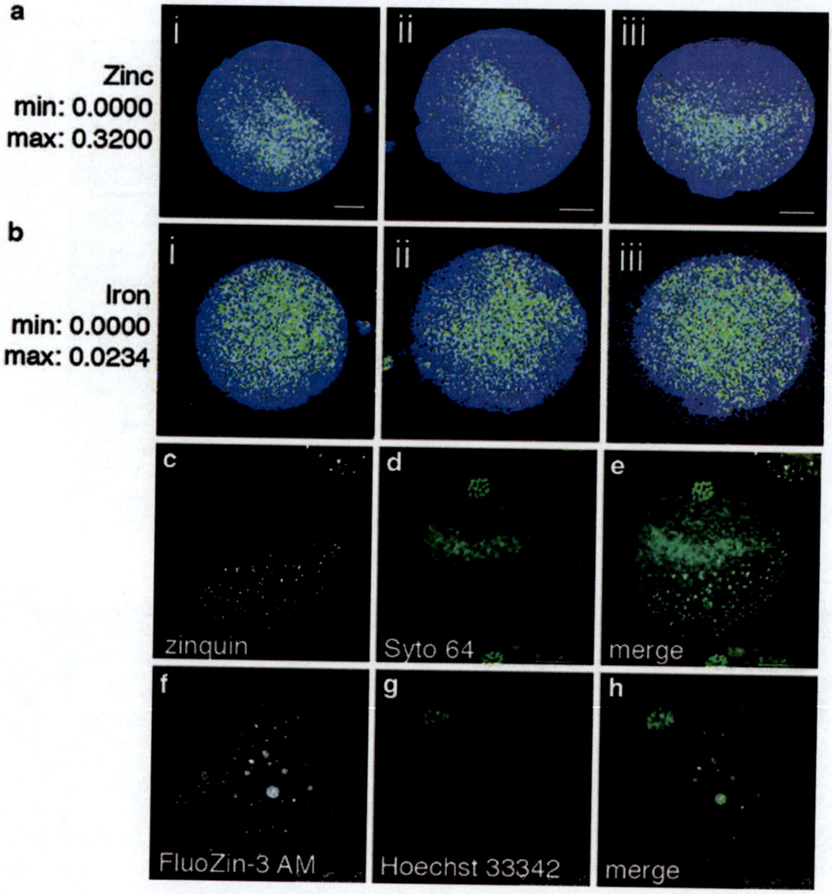

Figure 4 Chemical composition for three different mouse eggs. Total Zn and Fe (top two rows) were measured by XRF. Labile Zn was detected by confocal microscopy (images c and f) using two different fluorophores and shown by overlay either with different DNA markers (Syto 64 and Hoechst 33342) at the vegetal pole away from the meiotic spindle. Reproduced with permission from [60]; copyright (2011) American Chemical Society.

The difference between sensitivity to total metal (MS and XRF) and "free" metal (visible-light fluorescence microscopy) is important. However, it is not clear which of these measures is most important biologically. If the goal is to understand the available metal stores, total metal is more useful, and XRF, with its ability to probe the complete cell depth in one experiment, is likely to be more useful than MS, unless one is able to complete a full three-dimensional depth profiling of a sample. On the other hand, if the goal is to understand processes that are controlled by metal binding to metal-responsive receptors, "free" metal is more relevant. The ideal may be to combine the two measurements. One example of this is shown in Figure 4, where total Zn and Fe (by XRF) is compared with "free" Zn in mouse oocytes.

This study revealed that substantial amounts of Zn are released from mouse oocytes shortly after fertilization [60], raising the question of how the Zn was stored initially. The XFM images revealed that although Fe is uniformly distributed, Zn is cortically polarized. Of particular interest for this chapter is the value of combining measurements of total Zn (XFM) with measurements of free Zn (fluorophores, panels c and f of Figure 4). These showed that although the free zinc occupies the half of the oocyte that has elevated total Zn, the detailed distribution of "free" Zn is much more punctate than that for total Zn.

As noted above, one of the limitations of SXRF is that it provides only a two-dimensional projection of the three-dimensional distribution of metals, as compared to MS and microscopy, which can both give the full three-dimensional distribution. One way around this limitation is to measure XRF images for a series of different orientations of a sample, and use these via computed tomography to reconstruct the three dimensional distribution. An early example of this capability is shown in Figure 5, where XRF computed tomography was used to determine the localization of Fe in a single seed [61] with a resolution of ~5 μm, allowing resolution of subcellular structure, albeit for a fairly large cell. This work demonstrated that the iron localization depends on a vacuolar iron uptake transporter known as Vit1. Although Vit1 mutants showed no change in *total* iron, XRF imaging revealed a dramatic change in metal localization. More recent work on sperm at a so-called third-generation synchrotron demonstrates even better resolution [62] allowing, for example, localization of Co within a single human keratinocyte [63].

4.3.1 Metal Speciation

As noted in Section 1, the ideal metal analysis would include an ability to determine metal speciation. That is – is the metal bound to a protein or "free" in solution? If it is bound to a protein, which protein and with what ligands? What is the oxidation state of the metal? In general, none of the methods discussed to date are able to provide any direct information regarding the chemical environment of a metal ion. To some extent, it may be possible to infer chemical environment from some MS methods, particularly if softer ionization is used to limit protein fragmentation, although even this would provide only indirect information.

The combination of metal-specific fluorphores with XFM can distinguish the labile pool of metal from the tightly-bound or otherwise inaccessible pool of metal, but does not provide direct information on speciation. The traditional bioinorganic spectroscopies (UV-visible, electron paramagnetic resonance, nuclear magnetic resonance, etc.) are often used to determine metal speciation, but generally lack sufficient sensitivity to be used on single cells. One exception is XRF which can, at least in principle, provide direct insight into the chemical speciation of a metal. This is based on the fact that the X-ray absorption spectrum (XAS) for an element depends on energy of the incident beam.

A particularly straightforward example is to use X-ray absorption near-edge structure (XANES) imaging to image different oxidation states of an element

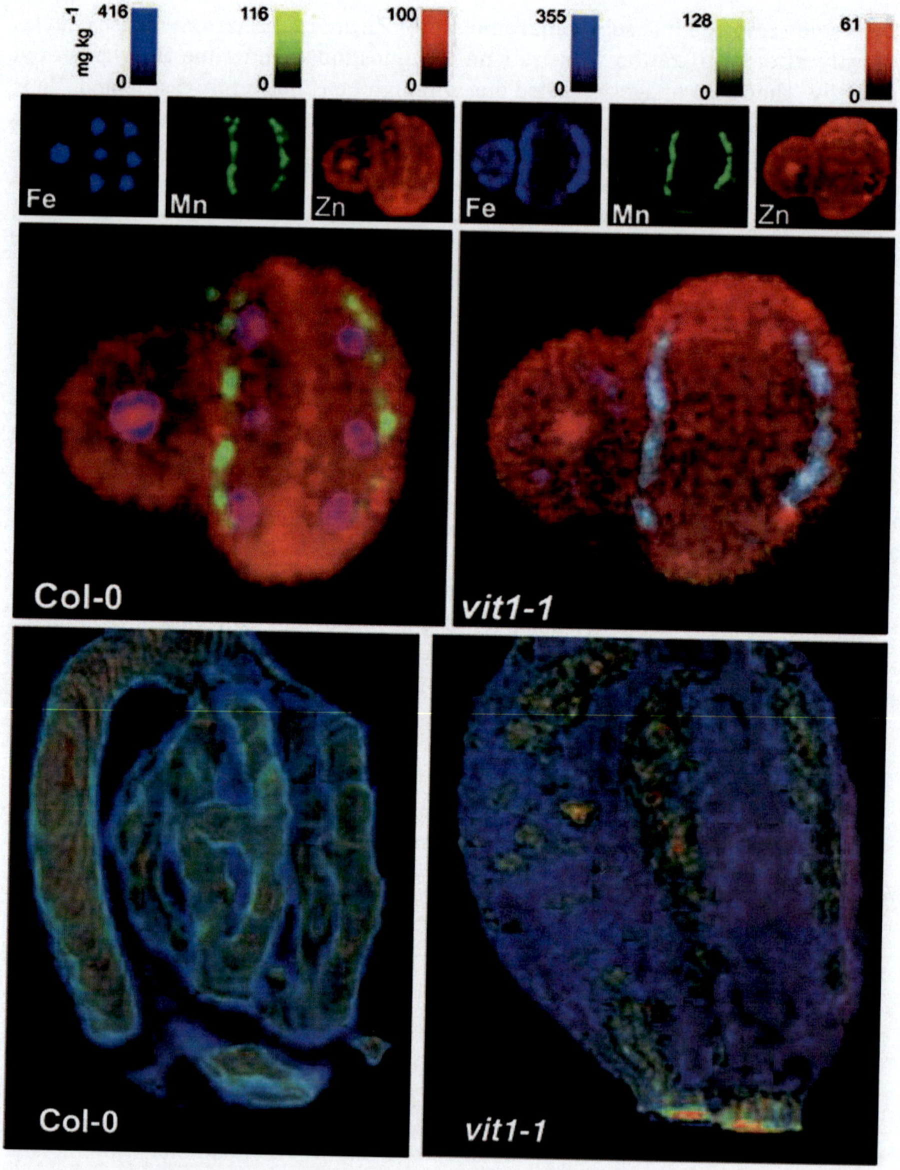

Figure 5 Tomographic reconstructions of the distribution of Fe, Mn, and Zn in single arabadopsis seeds. Images on the left are for the wild-type seed; images on the right is for a mutant lacking Vit1, a putative vacuolar iron transporter. **Top**: slices through the reconstructed distributions; **middle**: overlay of all three elements; **bottom**: three-dimensional rendering of iron distributions. Reproduced with permission from [61]; copyright (2006) The American Association for the Advancement of Science.

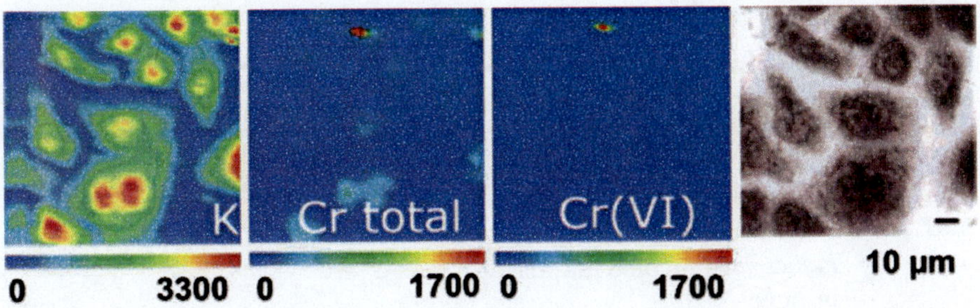

Figure 6 Chemical mapping of cells following exposure to particulate $PbCrO_4$. Potassium and total Cr were obtained by exciting at 6020 eV, Cr(VI) was obtained by exciting at 5993 eV. Optical micrograph shown for comparison. Reproduced with permission from [66]; copyright (2005) American Chemical Society.

[64,65]. This relies on the fact that the XANES spectrum of carcinogenic Cr^{6+} has an intense pre-edge peak at 5993.5 eV while Cr^{3+} has negligible absorption at this energy. Both oxidation states are uniformly excited by a higher energy X-ray beam (e.g., 6020 eV). Consequently, if XRF is measured with excitation at 5993.5 eV and at 6020 eV, it is possible to distinguish the fluorescence due to Cr^{6+} and thus, by difference, the signal due to Cr^{3+}. As shown in Figure 6, the XANES-XRF images allowed the authors to show that after cells were exposed to the insoluble $PbCrO_4$ the Cr^{6+} was found in particles on the cell surface while Cr^{3+} was uniformly distributed in the cytosol [66].

It is possible to use the X-ray nanoprobe beam coupled with traditional XAS measurements to measure the full XANES spectrum, and maybe even the full XAS spectrum of a sample in order to determine the local structural details. However, such measurements remain extremely challenging, not only as a consequence of low signal/noise, but also because of concerns about radiation damage [67].

5 Concluding Remarks and Future Directions

Perhaps the key conclusion is that there is no one method that can answer all of the questions that one would like to ask about cellular metallomes. Where possible, the combination of two or more methods is probably ideal. Among the factors to consider are what element(s) need(s) to be measured, whether one needs to know "free" metal or total metal, what resolution is needed, whether three-dimensional data is needed, and whether measurements need to be made on living samples. Rapid experimental progress is being made on all three of the major techniques, and there is every reason to believe that the next decade will see increasingly detailed characterization of the distribution, quantitation, and speciation of metals in cells.

Abbreviations and Definitions

EDS	electron dispersive spectroscopy
EELS	electron energy loss spectroscopy
EPMA	electron probe microanalysis
ESI	electrospray ionization
FRET	Förster resonance energy transfer
ICP	inductively coupled plasma
IR	infrared
MS	mass spectrometry
MALDI	matrix-assisted laser desorption/ionization
MRI	magnetic resonance imaging
PIXE	proton induced X-ray emission (sometimes particle instead of proton)
SE	secondary electron
SIMS	secondary ion mass spectrometry
SXRF	synchrotron induced X-ray fluorescence
TOF	time-of-flight
XANES	X-ray absorption near edge structure
XAS	X-ray absorption spectroscopy
XFM	X-ray fluorescence microscopy
XRF	X-ray fluorescence

Acknowledgments Some of the measurements described in this chapter were supported by the National Institutes of Health, Grants GM-38047 and GM-70545. The XRF imaging work that was described was all measured at one or more national synchrotron laboratories, much of it at laboratories supported by the US Department of Energy Office of Science.

References

1. A. I. Vogel, *A Text-Book of Macro and Semimicro Qualitative Inorganic Analysis*, Longmans, Green, London, New York, 1954, pp. xv, 663 p.
2. R. McRae, P. Bagchi, S. Sumalekshmy, C. J. Fahrni, *Chem. Rev.* **2009**, *109*, 4780–4827.
3. J. L. Nielsen, P. H. Nielsen, in *Environ. Microbiol.*, Vol. 397 of *Methods in Enzymology*, Ed J. R. Leadbetter, Elsevier Academic Press, Inc., San Diego, CA, 2005, pp. 237–256.
4. B. F. Brehm-Stecher, E. A. Johnson, *Microbiol. Mol. Biol. Rev.* **2004**, *68*, 538–559.
5. A. Amantonico, P. L. Urban, R. Zenobi, *Anal. Bioanal. Chem.* **2010**, *398*, 2493–2504.
6. A. Oikawa, K. Saito, *Plant J.* **2012**, *70*, 30–38.
7. H. Plattner, L. Bachmann, *International Review of Cytology-a Survey of Cell Biology* **1982**, *79*, 237–304.
8. K. Zierold, *J. Microscopy* **1991**, *161*, 357–366.
9. S. G. Boxer, M. L. Kraft, P. K. Weber, *Annu. Rev. Biophys.* **2009**, *38*, 53–74.
10. P. Nemes, A. Vertes, *TRAC-Trends. Anal. Chem.* **2012**, *34*, 22–34.
11. Z. Y. Qin, J. A. Caruso, B. Lai, A. Matusch, J. S. Becker, *Metallomics* **2011**, *3*, 28–37.
12. J. S. Hammond, in *Imaging Mass Spectrometry: Protocols for Mass Microscopy*, Ed M. Setou, Springer, 2010, pp. 235–257.

13. B. Wu, J. S. Becker, *Intl. J. Mass. Spect.* **2011**, *307*, 112–122.
14. S. L. Luxembourg, T. H. Mize, L. A. McDonnell, R. M. A. Heeren, *Anal. Chem.* **2004**, *76*, 5339–5344.
15. H. Ye, T. Greer, L. J. Li, *Bioanalysis* **2011**, *3*, 313–332.
16. J. S. Fletcher, J. C. Vickerman, N. Winograd, *Curr. Opin. Chem. Biol.* **2011**, *15*, 733–740.
17. J. S. Becker, A. Matusch, B. Wu, C. Palm, A. J. Becker, D. Salber, *Intl. J. Mass. Spect.* **2011**, *307*, 3–15.
18. T. A. Zimmerman, E. B. Monroe, K. R. Tucker, S. S. Rubakhin, J. V. Sweedler, in *Biophysical Tools for Biologists, Vol 2: In Vivo Techniques*, Eds J. J. Correia. H. W. Detrich, *Methods in Cell Biology*, Elsevier Academic Press, Inc., San Diego, CA, 2008, Vol. 89, pp. 361–390.
19. C. Szakal, K. Narayan, J. Fu, J. Lefman, S. Subramaniam, *Anal. Chem.* **2011**, *83*, 1207–1213.
20. P. K. Weber, G. A. Graham, N. E. Teslich, W. M. Chan, S. Ghosal, T. J. Leighton, K. E. Wheeler, *J. Microscopy* **2010**, *238*, 189–199.
21. A. Svatos, *Anal. Chem.* **2011**, *83*, 5037–5044.
22. P. L. Urban, T. Schmid, A. Amantonico, R. Zenobi, *Anal. Chem.* **2011**, *83*, 1843–1849.
23. A. Amantonico, P. L. Urban, S. R. Fagerer, R. M. Balabin, R. Zenobi, *Anal. Chem.* **2010**, *82*, 7394–7400.
24. H. Mizuno, N. Tsuyama, T. Harada, T. Masujima, *J. Mass Spectrom.* **2008**, *43*, 1692–1700.
25. B. Shrestha, J. M. Patt, A. Vertes, *Anal. Chem.* **2011**, *83*, 2947–2955.
26. N. Musat, H. Halm, B. Winterholler, P. Hoppe, S. Peduzzi, F. Hillion, F. Horreard, R. Amann, B. B. Jorgensen, M. M. M. Kuypers, *Proc. Natl. Acad. Sci USA* **2008**, *105*, 17861–17866.
27. T. Li, T.-D. Wu, L. Mazeas, L. Toffin, J.-L. Guerquin-Kern, G. Leblon, T. Bouchez, *Environ. Microbiol.* **2008**, *10*, 580–588.
28. S. D. Tanner, D. R. Bandura, O. Ornatsky, V. I. Baranov, M. Nitz, M. A. Winnik, *Pure Appl. Chem.* **2008**, *80*, 2627–2641.
29. K. E. Smart, M. R. Kilburn, C. J. Salter, J. A. C. Smith, C. R. M. Grovenor, *Intl. J. Mass. Spect.* **2007**, *260*, 107–114.
30. S. W. Hell, *Science* **2007**, *316*, 1153–1158.
31. T. Terai, T. Nagano, *Curr. Opin. Chem. Biol.* **2008**, *12*, 515–521.
32. D. W. Domaille, E. L. Que, C. J. Chang, *Nature Chem. Biol.* **2008**, *4*, 168–175.
33. P. J. Jiang, Z. J. Guo, *Coord. Chem. Rev.* **2004**, *248*, 205–229.
34. M. Formica, V. Fusi, L. Giorgi, M. Micheloni, *Coord. Chem. Rev.* **2012**, *256*, 170–192.
35. N. Boens, V. Leen, W. Dehaen, *Chem. Soc. Rev.* **2012**, *41*, 1130–1172.
36. D. O'Malley, in *Biophysical Tools for Biologists, Vol 2: In Vivo Techniques*, Eds. J. J. Correia, H. W. Detrich, *Methods in Cell Biology*, Elsevier Academic Press, Inc., San Diego, CA, Vol. 89, 2008, pp. 95.
37. K. M. Dean, Y. Qin, A. E. Palmer, *Biochim. Biophys. Acta* **2012**, doi:10.1016/j.bbamcr.2012.1004.1001.
38. J. R. Lakowicz, *Anal. Biochem.* **2001**, *298*, 1–24.
39. Y. You, S. Lee, T. Kim, K. Ohkubo, W.-S. Chae, S. Fukuzumi, G.-J. Jhon, W. Nam, S. J. Lippard, *J. Am. Chem. Soc.* **2011**, *133*, 18328–18342.
40. A. Krezel, W. Maret, *J. Biol. Inorg. Chem.* **2006**, *11*, 1049–1062.
41. R. Y. Tsien, *Nature* **1981**, *290*, 527–528.
42. C. A. Fierke, R. B. Thompson, *Biometals* **2001**, *14*, 205–222.
43. D. Wang, T. K. Hurst, R. B. Thompson, C. A. Fierke, *J. Biomed. Optics* **2011**, *16*, http://dx.doi.org/10.1117/1.3613926.
44. E. M. W. M. van Dongen, L. M. Dekkers, K. Spijker, E. W. Meijer, L. W. J. Klomp, M. Merkx, *J. Am. Chem. Soc.* **2006**, *128*, 10754–10762.
45. P. J. Dittmer, J. G. Miranda, J. A. Gorski, A. E. Palmer, *J. Biol. Chem.* **2009**, *284*, 16289–16297.
46. W. Qiao, M. Mooney, A. J. Bird, D. R. Winge, D. J. Eide, *Proc. Natl. Acad. Sci USA* **2006**, *103*, 8674–8679.

47. J. L. Vinkenborg, T. J. Nicolson, E. A. Bellomo, M. S. Koay, G. A. Rutter, M. Merkx, *Nature Methods* **2009**, *6*, 737–U710.
48. W. Denk, J. H. Strickler, W. W. Webb, *Science* **1990**, *248*, 73–76.
49. H. M. Kim, B. R. Cho, *Acc. Chem. Res.* **2009**, *42*, 863–872.
50. N. Johnsson, K. Johnsson, *ACS Chem. Biol.* **2007**, *2*, 31–38.
51. M. Cholewa, C. Dillon, P. Lay, D. Phillips, T. Talarico, B. Lai, D. X. Balaic, Z. Barnea, Z. Cai, G. B. Deacon, P. Ilinski, D. Legnini, S. Rainone, G. Shea-McCarthy, A. P. J. Stampfi, L. K. Webster, W. Yun, *Nucl. Instrum. Methods. B.* **2002**, *198*, 108–109.
52. R. Ortega, G. Deves, A. Carmona, *J. Royal Soc. Interface* **2009**, *6*, S649–S658.
53. M. Ralle, S. Lutsenko, *Biometals* **2009**, *22*, 197–205.
54. P. Duhutrel, C. Bordat, T. D. Wu, M. Zagorec, J. L. Guerquin-Kern, M. C. Champomier-Verges, *Appl. Environ. Microbiol.* **2010**, *76*, 560–565.
55. R. Ortega, P. Cloetens, G. Deves, A. Carmona, S. Bohic, *PLOS One* **2007**, *2*.
56. E. Kosior, S. Bohic, H. Suhonen, R. Ortega, G. Deves, A. Carmona, F. Marchi, J. F. Guillet, P. Cloetens, *J. Struct. Biol.* **2012**, *177*, 239–247.
57. E. H. Bigio, T. Paunesku, M. Mishra, S. Vogt, B. Lai, J. Maser, G. E. Woloschak, "X-ray fluorescence spectroscopy in situ quantification of metals/elements in Alzheimer disease", *Proceedings of the 81st Annual Meeting of the American-Association-of-Neuropathologists*, Arlington, VA, 2005, pp. 48.
58. T. Paunesku, S. Vogt, J. Maser, B. Lai, G. Woloschak, *J. Cell. Biochem.* **2006**, *99*, 1489–1502.
59. S. Vogt, B. Lai, L. Finney, B. Palmer, L. E. Wu, H. Harris, T. Paunesku, M. de Jonge, D. Legnini, J. Maser, D. Glesne, P. Lay, G. Woloschak, *Microscop. Microanal.* **2007**, *13*, 40–41.
60. A. M. Kim, M. L. Bernhardt, B. Y. Kong, R. W. Ahn, S. Vogt, T. K. Woodruff, T. V. O'Halloran, *ACS Chem. Biol.* **2011**, *6*, 716–723.
61. S. A. Kim, T. Punshon, A. Lanzirotti, L. Li, J. M. Alonso, J. R. Ecker, J. Kaplan, M. L. Guerinot, *Science* **2006**, *314*, 1295–1298.
62. S. Kehr, M. Malinouski, L. Finney, S. Vogt, V. M. Labunskyy, M. V. Kasaikina, B. A. Carlson, Y. Zhou, D. L. Hatfield, V. N. Gladyshev, *J. Mol. Biol.* **2009**, *389*, 808–818.
63. R. Ortega, C. Bresson, A. Fraysse, C. Sandre, G. Devès, C. Gombert, M. Tabarant, P. Bleuet, H. Seznec, A. Simionovici, P. Moretto, C. Moulin, *Toxicol. Lett.* **2009**, *188*, 26–32.
64. I. J. Pickering, E. Y. Sneeden, R. C. Prince, E. Block, H. H. Harris, G. Hirsch, G. N. George, *Biochemistry* **2009**, *48*, 6846–6853.
65. I. J. Pickering, R. C. Prince, D. E. Salt, G. N. George, *Proc. Natl. Acad. Sci USA* **2000**, *97*, 10717–10722.
66. R. Ortega, B. Fayard, M. Salomé, G. Devès, J. Susini, *Chem. Res. Toxicol.* **2005**, *18*, 1512–1519.
67. T. Bacquart, G. Devès, A. Carmona, R. Tucoulou, S. Bohic, R. Ortega, *Anal. Chem.* **2007**, *79*, 7353–7359.
68. K. E. Smart, J. A. C. Smith, M. R. Kilburn, B. G. H. Martin, C. Hawes, C. R. M. Grovenor, *Plant J.* **2010**, *63*, 870–879.

Chapter 3
Sodium/Potassium Homeostasis in the Cell

Michael Jakob Voldsgaard Clausen and Hanne Poulsen

Contents

M.J.V. Clausen • H. Poulsen (✉)
Centre for Structural Biology, Department of Molecular Biology and Genetics, University of Aarhus, Science Park, Gustav Wieds Vej 10c, Aarhus C, Denmark
e-mail: mjc@mb.au.dk; hp@mb.au.dk

L. Banci (ed.), *Metallomics and the Cell*, Metal Ions in Life Sciences 12,
DOI 10.1007/978-94-007-5561-1_3, © Springer Science+Business Media Dordrecht 2013

Abstract All animals are characterized by steep gradients of Na^+ and K^+ across the plasma membrane, and in spite of their highly similar chemical properties, the ions can be distinguished by numerous channels and transporters. The gradients are generated by the Na^+,K^+-ATPase, or sodium pump, which pumps out Na^+ and takes up K^+ at the expense of the chemical energy from ATP. Because the membrane is more permeable to K^+ than to Na^+, the uneven ion distribution causes a transmembrane voltage difference, and this membrane potential forms the basis for the action potential and for much of the neuronal signaling in general. The potential energy stored in the concentration gradients is also used to drive a large number of the secondary transporters responsible for transmembrane carriage of solutes ranging from sugars, amino acids, and neurotransmitters to inorganic ions such as chloride, inorganic phosphate, and bicarbonate. Furthermore, Na^+ and K^+ themselves are important enzymatic cofactors that typically lower the energy barrier of substrate binding.

In this chapter, we describe the roles of Na^+ and K^+ in the animal cell with emphasis on the creation and usage of the steep gradients across the membrane. More than 50 years of Na^+,K^+-ATPase research has revealed many details of the molecular machinery and offered insights into how the pump is regulated by post-translational modifications and specific drugs.

Keywords action potential • ion gradients • membrane potential • Na^+ and K^+ homeostasis • ouabain • palytoxin • renal tubular system • secondary transporters • sodium pump • voltage-gated channels

Please cite as: *Met. Ions Life Sci.* 12 (2013) 41–67

1 Introduction

Studies of the ionic gradients in animals have been fundamental for the development of modern day molecular biology. Already in the 18th century, Luigi Galvani recognized that electrical currents can activate the muscles in a frog leg [1], and in 1902 Ernest Overton found that muscles lose their ability to contract after incubation in an isotonic solution of sucrose, but regain excitability upon addition of sodium. At this time it was already recognized that muscles leak some of their potassium content when stimulated, and after discovering the requirement of extracellular sodium, Ernest Overton very cleverly noted that some counter mechanism that restores these ion gradients must exist [2]. During the Second World War, blood stored cold for transfusions was found to gradually lose potassium, a process that could be reverted by the addition of glucose to restore cellular ATP [3,4]. In 1953, the constant regeneration of the sodium and potassium gradients were found to be sensitive to the cardiotonic glycoside strophanthin [5], and in 1957 the ratio of the energy dependent transport was found to be two K^+ for three Na^+ [6]. In 1957, Jens Christian Skou finally demonstrated that the Na^+/K^+ exchange is due to a Mg^{2+}-dependent ATPase, now recognized as the Na^+,K^+-ATPase or simply the sodium pump [7], a discovery awarded the Nobel prize in 1997.

The sodium pump is responsible for the steep ionic gradients across the plasma membrane with high concentrations of intracellular potassium and extracellular sodium (~140 mM) and low concentrations of intracellular sodium and extracellular potassium (~10 mM). A fascinating hypothesis is that the high intracellular potassium concentration is a remnant of the very first proto-cells that evolved on Earth [8]. These cells had neither ion-tight membranes nor membrane pumps, so their intracellular environment would resemble that of the surroundings. The early, most basic cellular machineries thus evolved under ionic conditions similar to those in the primordial pond, but as life spread to different environments, cells evolved sophisticated pumps and transporters that allowed the maintenance of an intracellular solution similar to that found where life originated [8]. This theory also explains why it is a common trait across the kingdoms of life to have an intracellular concentration of potassium much higher than that of sodium.

Life thrives on Earth under exceedingly varying ionic conditions; even in the Dead Sea with NaCl concentrations approaching saturation, archaea, bacteria, and fungi have been described. Organisms that tolerate extreme salt concentration (>2.5 M NaCl) are known as halophilic microorganisms, and they rely on halophilic proteins to cope with the unusual ionic concentrations [9]. To maintain the osmotic balance, some of the microbes synthesize small molecules, e.g., glycerol and amino acids [10]. Others build up high cytosolic levels of K^+ [11] and have evolved highly acidic halophilic proteins with an increased number of small hydrophobic residues to favor a negative surface [12].

2 Sodium and Potassium as Enzymatic Cofactors

2.1 Ionic Properties

Sodium and potassium both belong to the group 1 (previously named IA) metals and have similar chemical properties with a relatively small ionic radius and a single valence electron in the s-orbital that is easily lost. In solution, Na^+ interacts with three or four water molecules, while the slightly larger K^+ optimally coordinates four or five water molecules, but less strongly [13]. An illustrative example of how this difference is utilized in biological systems is the K^+-channel, which is highly selective for K^+ over Na^+ even though Na^+ is the smaller ion (Figure 1). The selectivity is achieved by the pore having carbonyl oxygen atoms perfectly substituting for the waters that coordinate free K^+ [14] (Figure 1).

When bound within proteins, Na^+ and K^+ are mostly coordinated by oxygen atoms from amino acid side chains or the backbone, but cases of cation-π interactions with Tyr, Phe or Trp have also been reported (Figure 2a) [15].

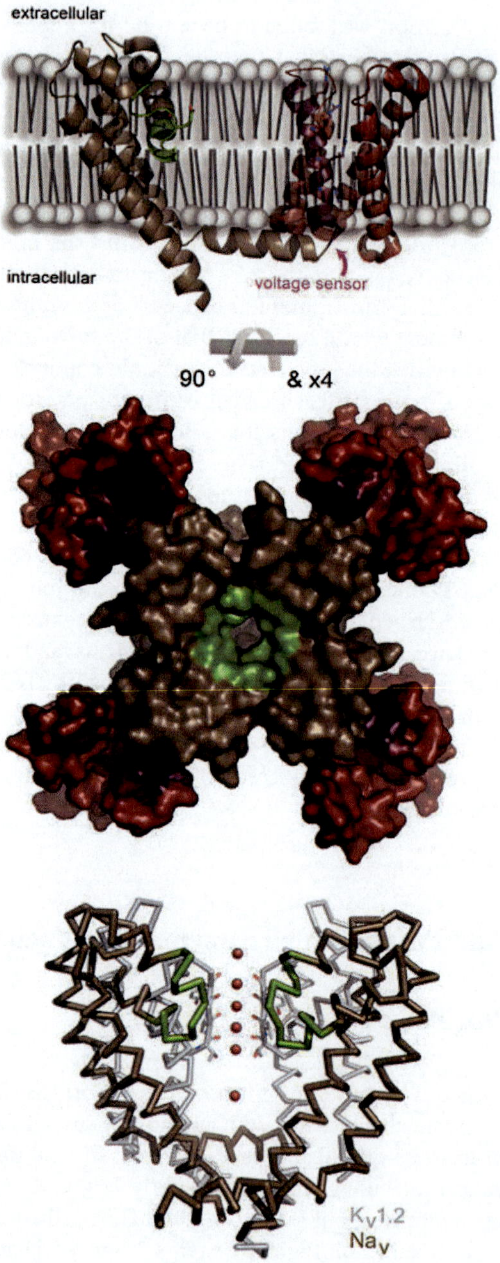

Figure 1 The structure of Na^+ and K^+ selective channels. At the **top**, the topology within the membrane of a single subunit of a tetrameric bacterial Na_v is shown. In mammals, Na_v channels are expressed as a single polypeptide with four repeats of structurally similar domains. The selectivity filter is green with the central glutamate shown in stick, and the voltage sensor is purple with arginines shown in sticks. In the **middle**, an extracellular view of the sodium channel in surface

The cellular Na^+ and K^+ have their lead roles in transmembrane transport and signalling rather than as enzyme regulators. Nonetheless, there are examples of them serving as important facilitators of substrate binding by lowering energy barriers, and there is a strong correlation between the localization of an enzyme and its preferred monovalent ion; intracellular enzymes generally use K^+, extracellular enzymes use Na^+ [16].

2.2 *Sodium- and Potassium-Dependent Enzymes*

The chemical properties of Na^+ and K^+ are utilized by several enzymes. In diol and glycerol dehydratases, K^+ is coordinated by five enzyme oxygens and directly interacts with two oxygens from the substrate (Figure 2b). In accordance with the tight substrate-ion interaction, these enzymes are functionally dependent on K^+, and in the apo state, the ion remains bound with two water molecules substituting for substrate [17]. Glycerol dehydrogenase is gaining increased interest from the industry because of its crucial role in the production of 1,3-propanediol from fermentation of glycerol. 1,3-propanediol has a potential use in the fabrication of, e.g., polyesters, cosmetics, and lubricants, and glycerol is readily available, since it is a by-product of biodiesel extraction [18].

The phosphoryl transfer enzymes, e.g., pyruvate kinase, also depend on K^+ for optimal substrate binding. Together with K^+, the enzyme binds a divalent cation, usually Mg^{2+}, and the ions are located at the substrate binding site where they create an optimal docking pocket for phosphoenolpyruvate (Figure 2c). Pyruvate kinases catalyse the final reaction of glycolysis where phosphoenolpyruvate and ADP are converted into pyruvate and ATP. The phosphate part of the substrate couples electrostatically to R49 and K240, but is also stabilized by the positively charged cations. When the phosphate group from the substrate is transferred to ADP, the enzyme uses R49, K240, Mn^{2+}, and K^+ as activators [16]. Pyruvate kinases can use Na^+ as a substitute for K^+, but only with an 8% activity, even though the structures of K^+ and Na^+ bound enzyme are indistinguishable, emphasizing the importance of their different electrostatic contributions [19].

Dialkylglycine decarboxylase is an example of an enzyme that requires K^+, but not in the direct coordination of the substrate. The enzyme catalyzes two coupled

Figure 1 (continued) representation. Ions pass through the central pore, and the four voltage-sensing domains are on the rim of the channel ring. At the **bottom**, the selectivity filter of the sodium channel is overlayed with that of the potassium channel $K_v1.2$ (grey). For both channels, only two of the four subunits are shown. In $K_v1.2$, the potassium-coordinating oxygens are indicated with sticks, and the potassium ions are indicated with red spheres. Within the selectivity filter, there are four potassium binding sites of which 1 and 3 or 2 and 4 will be occupied in a conducting channel. The Na_v selectivity filter is shorter with a central glutamate compensating sodium dehydration. The figure was made with pymol (www.pymol.org) using PDB IDs 3RVY (*Arcobacter butzleri* Na_v) and 2A79 (*Rattus norvegicus* $K_v1.2$).

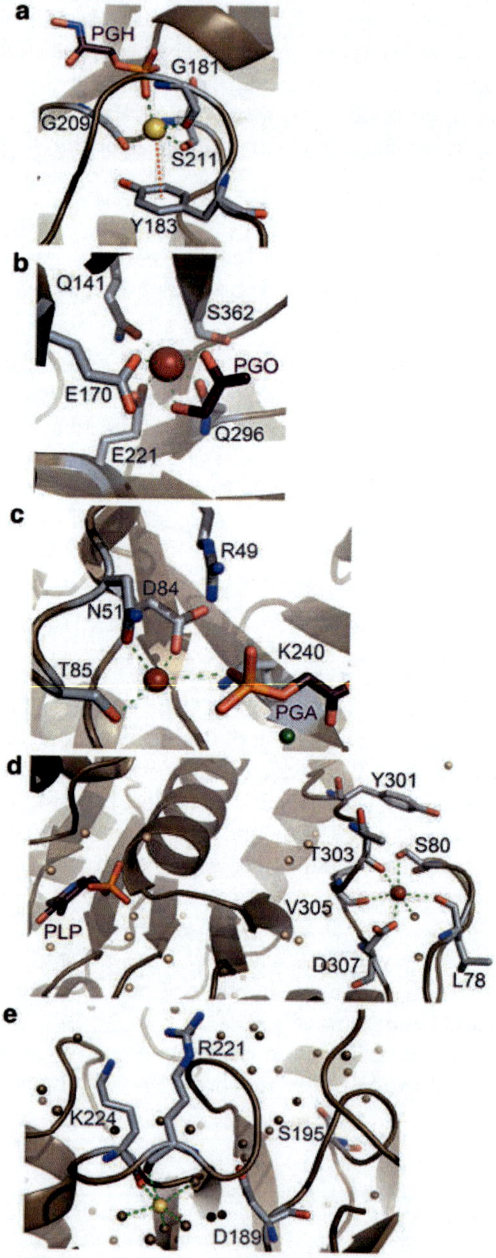

Figure 2 Coordination of Na^+ and K^+ in enzymes. (**a**) Tagatose-1,6-bisphosphate aldolase coordinates a Na^+ (yellow sphere) with four oxygen ligands (dashed green lines) and a π interaction (dashed red line) with a tyrosine. PGH: Phosphoglycolohydroxamic acid. PDB code 1GVF. (**b**) Diol dehydratase uses five enzyme oxygens to coordinate K^+ which is then used to capture the substrate propanediol (PGO) via coordination to the two hydroxyl oxygen atoms. PDB code 1DIO.

half-reactions: the decarboxylation of 2,2-dialkylglycine to CO_2 and dialkyl ketone, and the formation of L-alanine from pyruvate. The enzyme is a tetrameric homodimer of dimers with two alkali metal ion binding sites, one at the surface where Na^+ binds, and one near the active site where various alkali metals can bind, but the activity is only optimal with K^+ [20]. When K^+ is bound at the site close to the active site, it is coordinated by five oxygens from amino acids and a single water molecule (Figure 2d), while a Na^+ bound at the same site lacks the interactions with T303 and S80. The smaller ionic radius of Na^+ therefore leads to a steric clash between S80 and Y301, which reduces enzyme activity.

As many other serine proteases [16], the clotting protease thrombin requires Na^+ for activation. Na^+ binds in the vicinity of the substrate-binding pocket where it is coordinated by backbone oxygens of K224 and R221 together with four water molecules (Figure 2e). From this location the positive charge of the ion orients D189 for correct engagement with the substrate, and the catalytic S195 is affected through a network of buried water molecules [21].

3 The Membrane Potential

The plasma membrane is practically impermeable to charged solutes such as K^+, Na^+, and Cl^-, but it contains a multitude of protein channels and transporters that allow specific ions to pass under certain conditions. In general, the membrane is most permeable to K^+ because of the "two-pore domain" potassium channels that allow a K^+ leak through the plasma membrane [14,22,23]. This outward flow of positively charged ions polarizes the membrane, making the inside negative with respect to the outside, until a resting potential is reached, where the chemical driving force of the ion gradient is exactly counterbalanced by the force of the electric field. The equilibrium between the chemical energy associated with movement of ions in a concentration gradient and the electrical energy associated with movement of a charged particle in an electric field is described by the Nernst equation:

$$E = \frac{RT}{zF} \ln \frac{[X_1]}{[X_2]}$$

where E is the membrane potential required to counterbalance the driving force of the chemical concentration gradient between $[X_1]$ and $[X_2]$. R, T. z, and F are the gas

Figure 2 (continued) (**c**) Pyruvate kinase coordinates K^+ (red sphere) with three oxygen ligands from protein residues and one from the substrate PGA, 2-phosphoglycolic acid (dashed green lines). There is also a Mn^{2+} ion (green sphere) in the active site. PDB code 1A3W. (**d**) Dialkylglycine decarboxylase coordinates K^+ with five oxygens from amino acids and a single water molecule (dashed green lines). The ion is not directly coordinated by the substrate pyridoxal-5'-phosphate (PLP), but the potassium ion is required for the structure of the active site. PDB code 1DKA. (**e**) Thrombin coordinates Na^+ (yellow sphere) with two oxygen ligands from protein residues and four water molecules (dashed green lines). PDB code 1SG8.

constant, the temperature, the valence of the charged particle, and the Faraday number, respectively. If *R* and F are constant and T is room temperature, the Nernst equation simplifies to:

$$E = \frac{-58}{z} \log \frac{[X_1]}{[X_2]}$$

with E in mV.

The Nernst equation shows that the equilibrium potential of an ion depends on the concentration ratio of its chemical gradient and on its valence. The equilibrium potentials for K^+ and Na^+ differ between cell types, but they are typically around –80 to –100 mV and +40 to +60 mV, respectively. In addition to K^+, the membrane of most cells is also slightly permeable to Na^+, and in some instances also to Cl^-. The relationship between the membrane potential and the chemical gradients of these ions, when their respective permeabilities are considered, is described in a modified Nernst equation known as the Goldman equation:

$$E = -58 \log \frac{P_K[K_{in}] + P_{Na}[Na_{in}] + P_{Cl}[Cl_{out}]}{P_K[K_{out}] + P_{Na}[Na_{out}] + P_{Cl}[Cl_{in}]}$$

where P_{ion} is the membrane permeability to that ion. The Goldman equation predicts that the resting membrane potential of a cell depends on the ionic gradients and on the ionic permeabilities, and since the plasma membrane is more permeable to K^+, the resting potential is close to the equilibrium potential of K^+, resulting in a membrane potential of –50 to –90 mV in most cells.

3.1 *The Action Potential*

One of the most illustrative examples of how the cells can utilize the steep gradients of sodium and potassium is the action potential. In neuronal networks, fast intercellular communication occurs via electrical signals made possible by the action of the Na^+,K^+-ATPase, and the majority of ATP hydrolyzed in the brain is used to fuel the sodium pump.

Neurons are not directly connected to each other, but communicate over structures know as synapses. A synapse has two terminals separated by the synaptic cleft; the transmitting cell signals via the pre-synaptic terminal, and the target cell receives the signal at the post-synaptic terminal. The typical neuron receives the majority of electrical inputs via synapses localized at subcellular domains called spines. Neuronal spines are confined compartments connected to the dendritic shaft via a narrow spine neck. Due to diffusion limitations through the neck, spines are relatively isolated chemically, allowing ion gradients to exist between the spine head and the dendritic shaft [24]. A synapse can be either inhibitory (hyperpolarizing the membrane potential) or excitatory (depolarizing the membrane potential).

Inhibitory synapses increase Cl^- permeability, causing an influx of negative charge, while excitatory synapses typically increase Na^+ permeability to allow an inward flow of Na^+, and the Na^+ concentration within a spine may rise dramatically during periods of intense activity to levels around 100 mM [25].

The approximately –70 mV resting potential of a neuron is constantly influenced by excitatory and inhibitory signals from other neurons. If the membrane potential becomes depolarized to a certain threshold, an action potential is initiated (Figure 3). The threshold potential is defined by voltage-sensitive channels that are closed at the resting potential, but open when the membrane becomes sufficiently depolarized. Voltage-gated Na^+ channels (Na_v) open first, so the initial phase is dominated by increased Na^+ permeability and thereby further depolarization towards E_{Na} (cf. Figure 3), initiating a positive feedback loop where Na^+ permeability increases progressively. The time spent near E_{Na} is limited by two other voltage-dependent events, namely inactivation of Na_v and opening of voltage-gated K^+ channels (K_v) that let K^+ flow out of the cell. The re-polarization phase of an action potential may drive the membrane voltage below the resting potential because the increased K^+ permeability does not return to the resting state immediately (Figure 3).

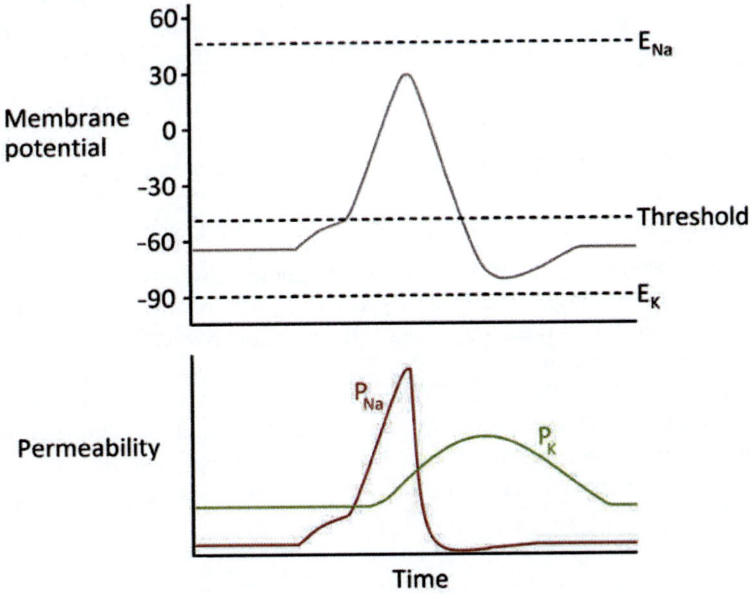

Figure 3 An action potential is initiated when the threshold voltage is reached. This causes a rapid increase in the Na^+ permeability because Na_vs open with fast kinetics and in a positive feedback loop. Before the voltage reaches the equilibrium potential for Na^+ the Na_vs are inactivated in a voltage-dependent manner immediately reducing the Na^+ permeability. K_vs open and close with slower kinetics than Na_vs, and therefore allow the voltage to depolarize before the increased K^+ permeability hyperpolarizes the membrane to near the equilibrium potential of K^+.

The phenomenon where the membrane hyperpolarizes with respect to the resting potential is known as after-hyperpolarization and is functionally implicated in the regulation of action potential firing patterns. Recently, two independent studies have linked the after-hyperpolarization that follows a period of high frequency stimulation to the activity of the Na^+,K^+-ATPase. The calyces of Held are large pre-synaptic terminals that fire with high precision at 1 kHz. In rats, large quantities of the sodium pump α3 isoform in the terminals cause hyperpolarization after a period of high activity, which may function to increase firing reliability [26]. A similar discovery was made in *Drosophila melanogaster* larval motor neurons, where Na^+,K^+-ATPase activity mediated long lasting hyperpolarization following a train of action potentials, and this was speculated to be involved in a novel short term cellular memory mechanism [27].

The crucial feature of the action potential is the ability of Na_v and K_v to sense changes in the membrane potential and respond accordingly. The molecular architecture of the channels is comparable, but while K_v is a homotetramer of subunits with six transmembrane helices (TMs), Na_v is a single polypeptide with four similar 6TM domains (Figure 1). The fourth TM helix of a subunit or domain contains several positively charged residues that respond to changes in the membrane potential by moving within the membrane, making TM4 the voltage-sensing domain. Selectivity between K^+ and Na^+ is obtained by a filter towards the extracellular side, where dehydrated ions are optimally coordinated. Since this compensation is structurally tight, only ions of an exact size fit perfectly and are efficiently allowed to pass.

4 Na^+ and K^+ Gradients and Secondary Transporters

The potential energy stored in the Na^+ gradient is also used to drive a multitude of secondary transporters that translocate solute molecules across the membrane against their concentration gradients. By coupling transport to the Na^+ gradient, solute flux can occur at rates 10^5 times faster than passive diffusion and against concentration gradients up to 10^6 [28,29]. The Na^+ gradient is therefore essential for household cellular tasks like uptake of nutrients as well as for specialized uptake, e.g., in the nervous system. Several atomic structures of members from the different transporter families have now been solved, so we are starting to glimpse the mechanism for how the transporters can be energized by sodium co-transport as described in the following.

4.1 Uptake of Amino Acids, Sugars, and Transmitters

Communication between neurons occurs when an action potential from a transmitting cell reaches a presynaptic terminal and triggers release of neurotransmitters that bind receptors in the post synaptic terminal of the receiving cell. Termination of the signal is facilitated by fast re-uptake of released neurotransmitters by neuronal and glial cells. The majority of transport proteins involved in neurotransmitter re-uptake utilizes energy from the electrochemical Na^+ gradient and belong to the family of neurotransmitter sodium symporters (NSSs), which includes GABA, glycine, noradrenaline, serotonin, and dopamine transporters. Malfunctions of these transport systems are linked to disorders like depression, schizophrenia, epilepsy, and Parkinson's disease [30]. Norepinephrine and serotonin transporters are targets for several antidepressants, including the rather non-selective tricyclic agents and more modern selective serotonin re-uptake inhibitors (SSRIs), as well as psychostimulants like cocaine and amphetamine that also target dopamine transporters [30].

The dominant excitatory neurotransmitter of the central nervous system is the amino acid glutamate, and its re-uptake is carried out by a family of transporters, excitatory amino acid transporters (EAATs), that are structurally distinct from the NSSs, but also couple to the ionic gradients with co-import of three Na^+ and a H^+ and export of a K^+ [31].

Uptake of ions and nutrients is carried out by polarized epithelial cells with an apical membrane that faces the lumen and a basolateral membrane oriented toward the tissue fluid. For instance, in kidney, colon, and small intestine, Na^+,K^+-ATPases are found exclusively in the basolateral membrane, where they vigorously keep the intracellular Na^+ concentration low, while the counter imported K^+ rapidly diffuses out of the cells through inwardly rectifying K^+ channels, either basolaterally (e.g., Kir4.1/5.1) [32] or apically (e.g., ROMK/Kir1.1) [33].

Nutrient uptake in the small intestine and re-uptake in the kidney are largely carried out by transporters that are structurally and mechanistically similar to the NSSs. Following digestion, peptides and amino acids are absorbed by enterocytes lining the small intestine [34], and in the kidney, the proximal tubule is responsible for the re-absorption of approximately 95–99% of amino acids. Rather than having specific transporters for each amino acid, more general and partly overlapping systems handle uptake of neutral, cationic, and anionic groups. Such redundancy provides backup in case of mutational inactivation of a transport system [35].

In the small intestine and the proximal tubules of the kidneys, sodium-dependent glucose transporters (SGLTs) situated in the apical membrane drive simultaneous uptake of Na^+ and glucose from the lumen at the expense of energy stored in the electrochemical Na^+ gradient [36], and in the basolateral membrane, facilitative transporters (GLUTs) use the diffusion gradient of glucose to transport sugars into the blood.

X-ray crystal structures of Na^+-coupled transporters have revealed common structural architecture, even between transporters with no identifiable sequence conservation. For example, the structures of sugar, amino acid, and the NSSs are evolutionary conserved from bacteria to higher eukaryotes with a characteristic core of two sets of 5TM helices that are interwoven in a 10TM domain with the subdomains related by a pseudo-two-fold axis of symmetry along the plane of the membrane. Structural information about NSSs comes to a large part from studies of a bacterial leucine transporter (LeuT) for which structures of several conformations in the catalytic cycle have been solved, and this has provided unprecedented insight into the molecular details of how the NSS family members work [37,38].

In LeuT, the two subdomains with an inverted topology in the plane of the membrane are comprised by TM1–TM5 and TM6–TM10. The transporter operates in an alternating access fashion, where the open to the outside conformation first binds two Na^+ and then the substrate leucine. Upon substrate binding, the access tunnel to the extracellular solution closes, and the transporter transiently adopts an occluded conformation, where Na^+ and leucine are shielded from both internal and external solutions, before an internal gate opens to allow the two Na^+ and leucine to diffuse into the cell. The transitions from outward over occluded to inward open states and back again require large scale conformational changes, and the energy to drive this process comes solely from the electrochemical Na^+ gradient. In some NSSs, the charge transferred during the translocation cycle is partly compensated by co-import of Cl^-, and in Cl^--independent transporters, H^+ antiport presumably has a similar role [39]. The inverted topology has suggested a 'rocking-bundle' theory, where symmetrical movement of the two subdomains during the cycle provides alternating access to either the cytoplasm or the extracellular space [40].

4.2 *Co-transporters in Cell Volume and Ion Balances*

4.2.1 Cell Volume

The volume of a mammalian cell is relentlessly confronted by changes in extracellular tonicity, transmembrane transfer of osmotically active substances and construction or withdrawal of intracellular osmoles [41]. Since alterations of cell volume can interfere with the structural integrity and the intracellular environment of cells, complex mechanisms continuously operate to maintain the required influx or efflux of water.

Most cells are water-permeable because of specialized water channels known as aquaporins that selectively allow water molecules to cross in single file while charged species, including protons, are impermeable. Since the water flow passively follows the net movement of ions, cell volume regulation is mediated primarily by ion transport systems that affect the overall osmolarity of the intracellular solution.

The gradients of Na^+ and K^+ drive import and export of Cl^- via cation-chloride co-transporters (CCCs), including K^+-coupled Cl^- exporters (KCCs), Na^+-coupled Cl^- importers (NCCs), and Na^+-coupled K^+ and Cl^- importers (NKCCs) [42]. Two isoforms of NKCC have been described, NKCC1, which is found in almost all mammalian cells, and NKCC2, which is restricted to the apical membrane of epithelial cells of the thick ascending loop of the kidney [43]. Recent data suggest that while NKCC2 is a dedicated ion transporter, NKCC1 may additionally co-import water along with ions, even against the osmotic gradient [44].

Cell swelling activates a process known as regulatory volume decrease, where the activation of KCCs [45] relieves pressure by exporting K^+ and Cl^-. Conversely, shrinkage results in activation of regulatory volume increase by Na^+/H^+ exchangers, NKCCs, and Na^+ channels. A specific mechanism in cell volume is regulation of transporters by phosphorylation to activate NKCC1 and inhibit KCC [46]. The consensus within the field is that Cl^-/volume sensitive kinases regulate the phosphorylation state of KCC and NKCC1 and hence control the fluxes of water and Cl^- [47].

4.2.2 Inhibitory or Excitatory Action of γ-Aminobutyric Acid

In neurons, KCC2 and NKCC1 play important roles in the tuning of the chloride gradient. During development, the expression of NKCC1 decreases, and that of KCC2 increases, effectively resulting in a reversion of the chloride gradient from being highest intracellularly to being lowest. The effects of neurotransmitters like GABA and glycine that open chloride channels are therefore reversed during development from excitatory during pre-natal development to inhibitory in adulthood.

4.2.3 Re-uptake of Inorganic Solutes

CCCs play a pivotal role in the reabsorption of salts from renal corpuscle filtrate in the renal tubules. Nephrons, the functional units of kidneys, include a tubular system that starts with the proximal convoluted tubule which is followed by the loop of Henle that contains a descending and an ascending limp that continues to the distal convoluted tubule and finally the collecting duct. Figure 4 gives an overview of some specialized epithelial cell types along the renal tubules and highlights some of the transport systems that are involved in the reabsorption of inorganic solutes and that are associated with the gradients of Na^+ and K^+.

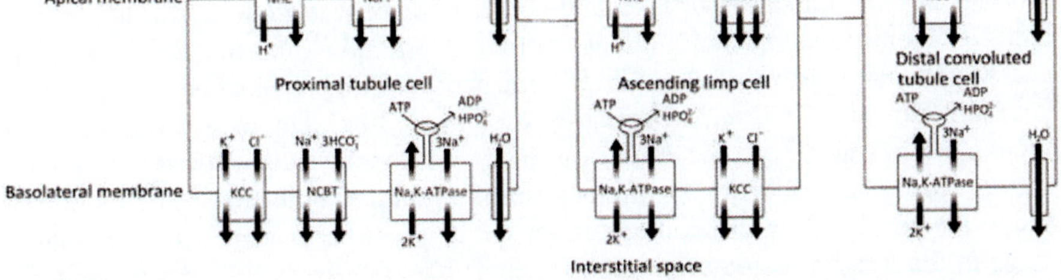

Figure 4 The renal tubular system contains several specialized epithelial cell types that expresses different transporters and utilize the cell polarity with an apical and a basolateral membrane to reabsorb a huge amount of solutes from the tubular filtrate. The Na,K-ATPase in the basolateral membrane creates chemical and electrical gradients of Na^+ and K^+ that drive numerous secondary transporters, some of which are exemplified in this figure.

Phosphate is the most abundant anion in the human body and is implicated in numerous biological functions as bone mineralization, signaling, nucleotide and energy metabolism, and preservation of membrane integrity [48]. In the kidney, re-uptake of inorganic phosphate (P_i) depends on Na^+-coupled symporters (NaPi) that are located in the apical membrane of proximal tubule epithelial cells [49].

Sodium-coupled bicarbonate transporters (NCBTs) are important regulators of intra- and extracellular pH, and exist as both electrogenic (NBCe1 and NCBe2) and electroneutral (NBCn1, NBCn2, and NDCBE) forms. In the proximal tubule of the kidney nephrons, NCBe1 is localized in the basolateral membrane where it couples with apically localized Na^+-coupled H^+ exporters (NHEs) and aquaporins to drive the reabsorption of approximately 80% of the filtered HCO_3^- which is of crucial importance for maintenance of, e.g., blood pH. The remaining 20% is reabsorbed in more distal parts of the renal tubules so virtually no HCO_3^- escapes with the urine. HCO_3^- that in the filtrate is titrated to CO_2 and H_2O by H^+ that is secreted by NHEs. Both CO_2 and H_2O then enter the proximal tubule epithelium at the apical membrane via aquaporins and are converted into HCO_3^- by the enzyme carbonic anhydrase before it is exported by the basolateral NCBe1 together with Na^+ in a 3:1 ratio [50]. Since NCBe1 is co-exporting Na^+ against the electrochemical gradient, the driving force of this transporter is partly the membrane potential [51].

The proximal convoluted tubule is connected to the distal convoluted tubule via the loop of Henle which encompasses a descending and an ascending loop that dives into the medulla of the kidney where the interstitial fluid has a high concentration of solutes. Since water permeability in the descending limp is high, massive amount of water is drawn out of the tubular fluid and into the interstitium, while solutes remain due to a low population of transporters in the descending limp epithelium. The fluid that reaches the ascending limp is consequently highly concentrated and, because the epithelia that line this part of the tubular system does not express aquaporins, only salts are re-absorbed via, e.g., NKCCs and NHEs. In the distal convoluted tubule NCCs are responsible for the reabsorption of approximately 5% of the renal corpuscle filtrate [52].

Electroneutral cation-chloride co-transporters are targets for some of the most prescribed drugs in the world, and are used to treat hypertension and heart failure because they inhibit salt re-uptake, and hence increase water output through urination. NCC represents the major target site for thiazide-type diuretics that are used to treat hypertension, extracellular fluid volume overload, and renal stones. Furosemide is an inhibitor of NKCC and is primarily used in the treatment of hypertension and edema, but has also been the center of some controversy due to its usage on race horses. Fast or intense exercise often causes pulmonary hemorrhage in race horses due to the combination of capillary and alveolar pressure increments. The hemorrhage in the lungs is often seen after a race as bleeding from the nostrils, and the condition results in poor athletic performance. Pre-race administration of furosemide is known to increase performance because it decreases the incidence and severity of exercise-induced pulmonary hemorrhage [53].

5 Homeostasis of the Na^+ and K^+ Gradients by the Na^+,K^+-ATPase

5.1 *The Subunits of the Na^+,K^+-ATPase*

The Na^+,K^+-ATPase is an oligomeric protein that, as a minimum, contains an α-subunit with ten transmembrane spanning helixes and a 1TM type II glycosylated β-subunit [54]. All principal ion pump activity is managed by the α-subunit, which contains the translocation channel with ion-binding sites, the ATP hydrolysis core, and the target site for the specific Na^+,K^+-ATPase inhibitors cardiotonic glycosides. The β-subunit associates with the α-subunit in the endoplasmic reticulum (ER) and is required for stability and sorting to the plasma membrane [55–57]. In addition to the essential α- and β-subunits, the Na^+,K^+-ATPase may also associate with a 1TM regulatory subunit of the FXYD family of membrane proteins (Figure 5) [58].

Animal cells depend strongly on the large gradients of Na^+ and K^+ that Na^+,K^+-ATPases create and maintain, and the great diversity of cell types within the animal kingdom call for an assortment of various sodium pump subtypes. To fulfill the different requirements, multiple isoforms of the Na^+,K^+-ATPase subunits have evolved, in mammals there are four different types of α-, three of β-, and seven of FXYD-subunits. Since all isoform combinations yield functional pumps, the potential number of different Na^+,K^+-ATPases in mammals is 84.

The subunit isoforms of the Na^+,K^+-ATPase display very distinct expression profiles depending on cell type, tissue and developmental state. The α1 isoform is ubiquitously expressed, α2 is found predominantly in muscle, adipose, heart and lung tissues, and in glial cells [59], the α3 isoform is mostly found in neurons [60], but has also been detected in other tissues, including the developing rat cardiac ventricle [61] and erythroid cells [62]. The most restricted Na^+,K^+-ATPase isoform is α4, which is only expressed in sperm cells [63,64]. Both β- and FXYD- subunits are viewed as regulators of the principal α-subunit, and their complex expression profiles further illustrate the fine-tuning of pump function required for different cell types [65].

Diverse subunit compositions yield Na^+,K^+-ATPases optimized to operate under different cellular conditions. The α3 isoform for example is associated with low affinity for intracellular Na^+ and high affinity for ATP [65]. In essence, this gives a pump that remains relatively dormant under basal conditions where the level of intracellular Na^+ is approximately 10 mM, but in neurons, α3 accumulates in dendritic spines, and the concentration of Na^+ increases dramatically during high neuronal activity and will not be limiting. The high affinity for intracellular ATP further ensures that α3 remains active even when energy is low. Another adaptive technique is seen for the α2-subunit isoform, which is the isoform most sensitive to the membrane potential, being inhibited by hyperpolarization.

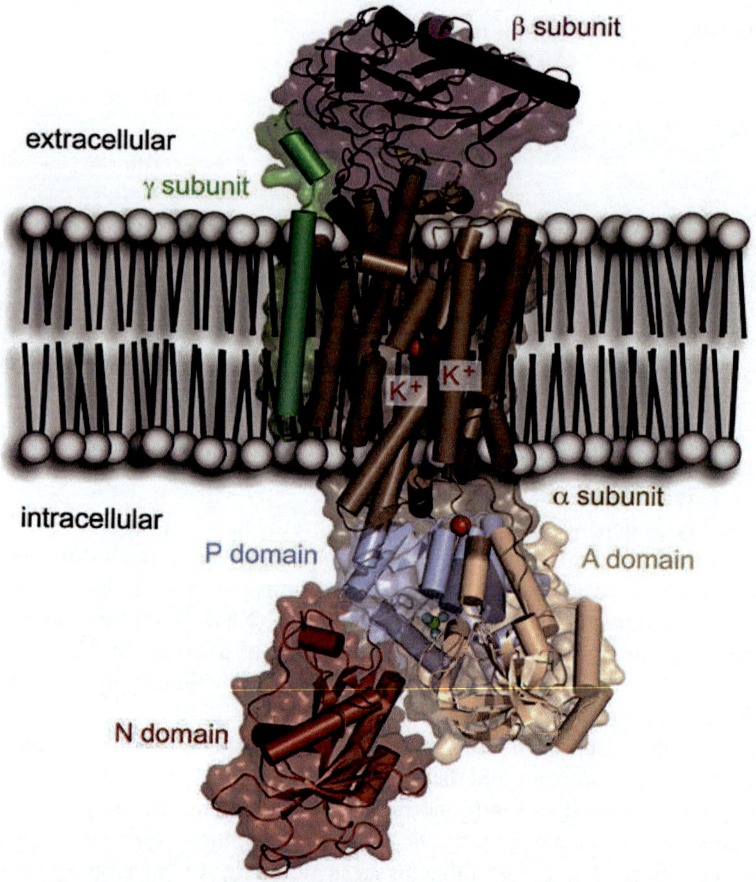

Figure 5 Structural overview of the sodium pump in the membrane. In the α-subunit, the transmembrane part is brown, the phosphorylation (P) domain is light blue, the actuator (A) domain is light brown, and the nucleotide (N) domain is red. The β-subunit is in purple and the γ-subunit is in green. The two occluded K^+ and the cytoplasmic regulatory K^+ are indicated with red spheres. The phosphorylation is mimicked by magnesium fluoride shown as green and light blue spheres. PDB code 2ZXE.

5.2 *The Structure of the Na^+,K^+-ATPase*

The Na^+,K^+-ATPase belongs to a family of membrane proteins known as the P-type ATPases, where the defining term "P-type" comes from the notion that all members are phosphorylated and dephosphorylated at a conserved aspartate during the reaction cycle (Figure 6) [66]. All P-type ATPases have three intracellular domains that together constitute the engine of the pump, transforming energy in the form of ATP into mechanical energy that opens and closes the enzyme toward the cytoplasm and the extracellular solution in an alternating fashion. The three intracellular domains are named the actuator domain (A-domain), the nucleotide binding domain (N-domain) and the phosphorylation domain (P-domain) [67]. The transmembrane part of the Na^+,K^+-ATPase α-subunit contains three ion-binding sites within the lipid bilayer. Two of these sites are shared between Na^+ and K^+, while the third site binds only Na^+. The C-terminal tail of the Na^+,K^+-ATPase is implanted within the transmembrane part of the pump in proximity to the exclusive Na^+-binding site. On the extracellularly exposed part of the α-subunit, a cavity serves as an ion entrance/exit vestibule and as a binding site for the specific Na^+,K^+-ATPase inhibitors cardiotonic glycosides (cf. Figures 7 and 8 in Section 5.5.1).

The β-subunit is glycosylated and provides the majority of the extracellular architecture. It plays an important role in cell-cell adhesion because two opposing β-subunits can interact [68].

5.3 The Mechanism of the Na^+,K^+-ATPase

Since the discovery of the Na^+,K^+-ATPase, much has been learned about the cycle of conformational shifts required to translocate both Na^+ and K^+ against their respective chemical gradients, and for Na^+ also against the electrical potential of the cell. The reaction scheme is named "The Post-Albers Model" after two of the pioneers in its formulation, and it is based on a cycle where the pump alternates between two major states; an E1 state open to the inside of the cell with high affinity for Na^+, and an E2 state open to the outside with a high K^+ affinity (Figure 6) [69,70]. From the combined biochemical, structural, and electrophysiological data, much of the mechanics driving the pump have been elucidated. High resolution crystal structures of both the Na^+,K^+-ATPase [71,72] and the closely related sarcoplasmic reticulum Ca^{2+}-ATPase SERCA [73–79], the latter in several conformational states, have

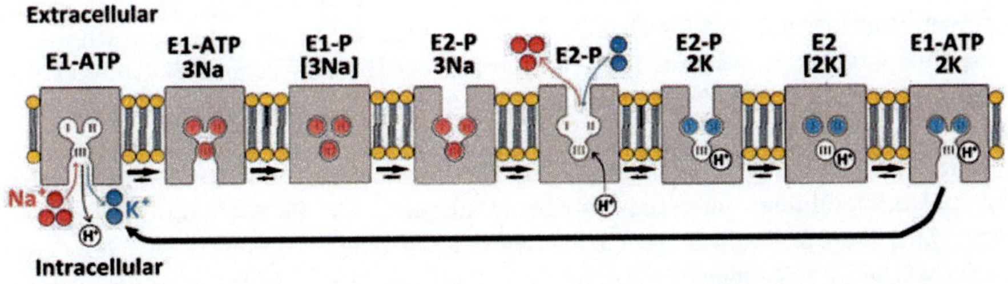

Figure 6 The catalytic cycle of the sodium pump. The Na^+,K^+-ATPase exists in two principal conformations; E1 with high Na^+ affinity and opened inwards, and E2 with high K^+ affinity and opened outward. Each complete cycle results in the export of three Na^+, import of two K^+ and the hydrolysis of a single molecule of ATP. Recent speculations suggest that a cytoplasmic proton enters the pump and stabilizes it in the K^+ bound state, and then returns to the cytoplasm when K^+ is released [104].

given astonishing insight into how the different domains of the pumps move as the cycle progresses.

The N-domain promotes the E2 to E1 conformational shift by binding of ATP. For the sodium pump, this means intracellular release of two K^+ and binding of three Na^+. The N-domain then phosphorylates the P-domain, whereby the three bound Na^+ are occluded within the pump. Release of Na^+ is presumably initiated by protonation of a membrane-embedded aspartate by an intracellular proton, and the opening of the structure to the external side is forced by a 90° rotation of the A-domain. In relation to the P-domain, the A-domain rotation situates it in a favorable position for acting as a phosphatase, and dephosphorylation induces occlusion of the two newly bound K^+ (Figure 6).

The activity of a membrane-embedded molecular engine such as the Na^+,K^+-ATPase that catalyzes an electrogenic reaction will be influenced by the transmembrane potential. Energetically, the transport of positive charge against the electrical potential becomes increasingly expensive with hyperpolarization, and since the 3:2 stoichiometry is maintained at all investigated potentials, the pump's turnover rate will cease if the electrochemical energetic cost of the ion transport approaches the energy of ATP hydrolysis [80].

5.4 *Regulation of the Na^+,K^+-ATPase*

The activity of the Na^+,K^+-ATPase can be regulated by cellular interaction partners and by posttranslational modifications that affect the pump's kinetic properties or its cellular distribution.

5.4.1 Posttranslational Modifications

The best-studied posttranslational modifications are phosphorylations. In particular, phosphorylation of FXYD1 in heart and skeletal muscle increases pump activity, probably because unphosphorylated FXYD1 lowers sodium and potassium affinities, and phosphorylation releases the inhibitory effect [81]. PKA and PKC target the C-terminal part of FXYD1, and as indicated by its alternative name, phospholemman, it is phosphorylated to a large extent [82].

The α-subunit is also phosphorylated. The most conserved site is a serine in the cytoplasm/membrane interface, which is targeted by PKA. With the use of phosphospecific antibodies, the site has been shown to be phosphorylated in response to activation of a number of G-protein coupled receptors, and the phosphorylation generally leads to inhibition of the pump without affecting the level of pumps at the plasma membrane, suggesting direct inactivation [83]. The α isoforms can all be phosphorylated by PKA; for example, mouse studies showed that α2 phosphorylation levels respond to morphine administration [84]. Yet a detailed knowledge of the cellular pathways and roles of the phosphorylation remain uncertain.

Recently, the sodium pump has also been shown to be subjected to glutathionylation. In cardiac cells experiencing oxidative stress, a cysteine in the transmembrane part of β1 is modified by glutathione, which inhibits the pump [85], but FXYD proteins can reverse the inhibition [86]. During, e.g., heart failure, it may be an important pathophysiological consideration that oxidative stress inhibits the sodium pump, but since it is reversible, pump glutathionylation could in general serve a signaling role.

5.4.2 Cellular Interactions

A rapidly increasing number of protein interaction partners are being described for the sodium pump. In many cases, the interaction is believed to control pump activity by regulating, e.g., its subcellular localization [87,88], but intriguingly, there is also emerging evidence that the pump interacts with channels that functionally couple directly to the ion clearance, so far mainly sodium-selective channels.

The atypical Na_x channels are expressed in glia cells in parts of the brain that control salt intake. The Na_x channels open in response to high extracellular sodium levels, which leads to increased lactate production in the cell, and lactate functions as a signal to GABAergic neurons. The channel was shown to interact with the P-domains of α1 and α2, but not α3. The interaction was required for elevated sodium levels causing upregulated glucose intake and thereby lactate signalling. Tight coupling of Na_x with the Na^+,K^+-ATPase secures efficient pump activation, since the substrate is directly loaded to the pump, but a direct physical interaction may serve additional functions [89].

The α1 sodium pump subunit was also shown to co-immunoprecipitate with AMPA receptor subunits GluR1 and 4 in lysates from rat cerebellum [90] and with GluR1 and 2 in rat cortical lysates [91]. The majority of excitatory neurotransmission in the CNS is attributable to the AMPA subgroup of the glutamate receptors. The channels are heterotetramers of the GluR1-4 subunits that allow influx of sodium upon activation and thus depend strongly on sodium pump activity. In cortical neurons, inactivation of the sodium pump by ouabain caused up to 80% of the AMPAR subunits to be internalized and subsequently degraded. Like the Na_x channel, AMPAR function depends on the sodium gradient created by the Na,K-ATPase, and the interaction study further indicated that the sodium pump may play an active part at the synapse by determining the number of AMPARs and thereby the strength of the synapse [91].

In neuromuscular junctions, the Na^+,K^+-ATPase interacts with the nicotinic acetylcholine receptor (nAchR) at the postsynapse. The nAchRs are ligand-gated cation channels, so agonist binding leads to an influx of sodium, but also to potassium release. In mammals, the sodium pump α2 colocalizes and copurifies with nAchR, and the desensitized channel enhances pump activity [92]. In *Caenorhabditis elegans*, the Na^+,K^+-ATPase (EAT-6) is required for proper neuromuscular localization of the nAchR, again suggesting a critical role for the pump in synaptic organization [93].

It thus seems a recurring theme that the sodium pump associates with some of the channels that depend on it in a regulatory network. We are only now beginning to glimpse the higher-order complexes, but they will undoubtedly be in the focus for a fuller understanding of how the cell maintains and utilizes sodium and potassium gradients.

5.5 *Na⁺,K⁺-ATPase Toxins*

5.5.1 Cardiotonic Glycosides

Inhibitors of the Na^+,K^+-ATPase are produced by both plants and animals, and for centuries it has been recognized that low doses of the toxins can be used to treat cardiac insufficiency, hence their name, cardiotonic glycosides (Figures 7 and 8). Na^+,K^+-ATPase inhibition leads to higher intracellular sodium levels and, via the sodium-calcium exchangers, to higher calcium levels, which increases the force of myocardial contractions. However, the therapeutic index is quite narrow, and higher doses are lethal. Apart from accidental intake, human poisoning has mainly been fictional (e.g., in Agatha Cristie novels and a James Bond movie), but African hunters have used extracts with cardiotonic glycosides as arrow poison [94,95].

At least 14 different plant species and nine *Bufo* toads produce cardiotonic steroids. Furthermore, the larvae of the Monarch butterfly live solely on a plant that

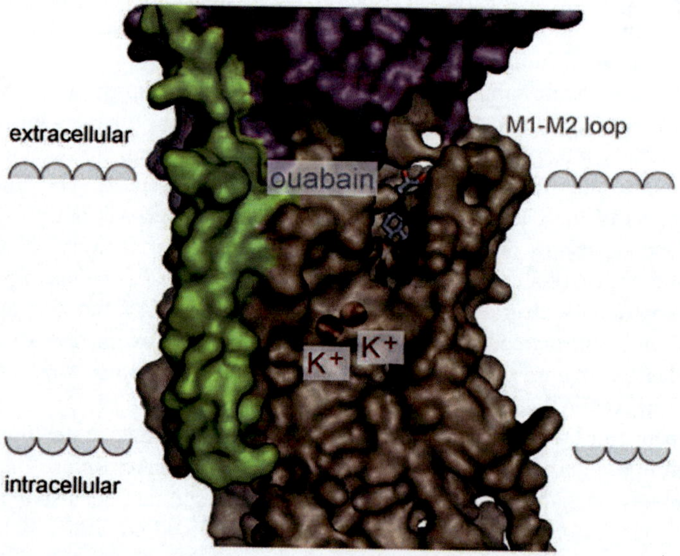

Figure 7 Membrane view of a low-affinity Na^+,K^+-ATPase-ouabain complex. Indicated are ouabain, the occluded potassium ions, and the loop between TM1 and TM2, the M1-M2 loop. Surface representations of the subunits α (brown), β (purple), and γ (green). The figure was made with pymol using PDB ID 3A3Y (a shark Na,K-ATPase). In the high-affinity complex, ouabain most likely binds similarly, but helix movements are expected to close the binding site further.

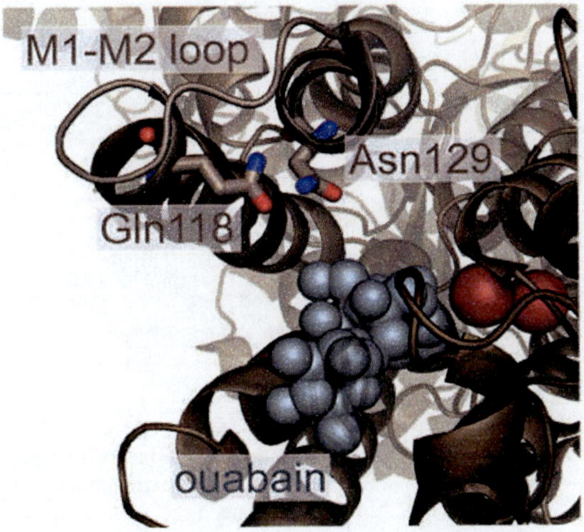

Figure 8 Extracellular view of the structure shown in Figure 7 with similar color coding and with indication of the loop between TM1 and TM2, the M1-M2 loop. The two residues known to be important for ouabain binding are indicated.

produces cardiotonic steroids, and both the larvae and the butterfly retain sufficient amounts of poison for it to serve as a defense against predators [96]. Similarly, two species of the leaf beetles *Chrysochus* eat plants that produce cardiotonic steroids, making the beetles toxic to predators [97]. Plants energize their membranes with a proton gradient and are therefore not affected by sodium pump inhibitors, but the animals that produce or retain the poison need to protect their own pumps. They avoid inhibition by having mutations in the extracellular part of their α subunits where the cardiotonic steroids otherwise bind (Figure 9).

The cardiotonic glycosides inhibit the Na^+,K^+-ATPase by locking it in an E2P state, thus putting a break in the catalytic cycle. Surprisingly, the cardiotonic steroid ouabain is also being recognized to function as a hormone at very low concentrations (nM), causing intracellular signalling events to change gene expression and cell survival. There is no measurable effect on the Na^+,K^+-ATPase activity at these concentrations, but it is suggested that locking of a sub-portion of the pumps in the E2P state causes release of a Src kinase interacting with the pump, thus allowing it to start the signaling cascade via phosphorylations [98].

5.5.2 Palytoxin

While cardiotonic glycosides block the Na^+,K^+-ATPase, palytoxin transforms the pump into a cation channel with an open probability of around 90%. Compared to the normal pump, the palytoxin bound pump therefore allows sodium and potassium to flow in the opposite directions at speeds many thousandfold higher than pumping, so the ionic gradients are destroyed. Loss of the ionic gradients is

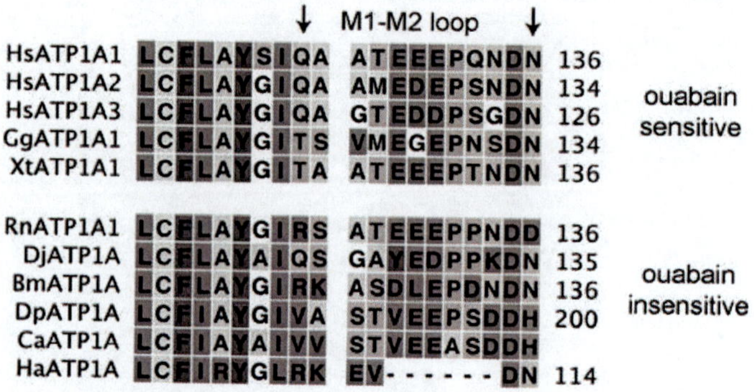

Figure 9 Alignment of residues involved in ouabain binding. Arrows indicate residues known to affect ouabain affinity at the end of TM1 (corresponding to Gln118 in Figure 8) and at the beginning of TM2 (corresponding to Asn129 in Figure 8). The extracellular loop between TM1 and TM2, the M1-M2 loop, is indicated. Abbreviations used are *Homo sapiens* (Hs), *Gallus gallus* (Gg), *Xenopus tropicalis* (Xt), *Rattus norvegicus* (Rn), *Dugesia japonica* (Dj), *Bufo marinus* (Bm), *Danaus plexippus* (Dp), *Chrysochus auratus* (Ca), and *Heterosigma akashiwo* (Ha). In rodents, α1 is insensitive to ouabain. Cardiotonic glycosides are used for protection by *Bufo* toads. The monarch butterfly (Dp) and a leaf beetle (Ca) take up cardiotonic glycosides produced by the plants they feed on. Sequences of Na^+,K^+-ATPases from non-animal species, e.g., a marine algae (Ha), suggest that they are not sensitive to ouabain.

extremely lethal, causing rapid heart failure upon intravenous exposure, and palytoxin is the second deadliest non-peptide molecule known with an LD_{50} in mice of less than 50 ng/kg [99].

Palytoxin was first isolated in 1971 from the Hawaiian coral *Palythoa toxica*. The molecule is huge with 64 stereogenic centers, and the mechanism of action remains for the most part mysterious, but somehow palytoxin recognizes the extracellular part of the pump and causes opening of both intracellular and extracellular ion pathways at the same time [99].

6 Pathophysiology of Na^+,K^+-ATPase Disturbance

Maintaining the sodium and potassium gradients is essential to all animal cells, so deleterious mutations in the gene for the ubiquitous α1 subunit are not compatible with life. For α2 and α3, haploinsufficiency is tolerated, but can cause the rare human diseases familial hemeplegic migraine 2 (FHM2 [100]) and rapid-onset dystonia parkinsonism (RDP [101]), respectively. FHM2 patients suffer from attacks of migraine with aura and hemiparesis, and the attacks can be accompanied by other symptoms such as ataxia, epilepsy/seizure, and loss of consciousness. In some FHM2 families, there are also reports of, e.g., weakened cognitive functions and psychiatric disorders [102].

Similar symptoms are described for patients with mutations in genes encoding the voltage-gated calcium channel $Ca_V2.1$ (causing FHM1) or the voltage-gated sodium channel $Na_V1.1$ α1 subunit (causing FHM3), which suggests that the phenotype reflects an at least partly common mechanism where the ionic balances in the brain are misregulated.

RDP patients very abruptly (within hours to weeks) develop dystonia, where the muscles remain involuntarily contracted, causing twisting of the body, but there are also parkinsonian symptoms, e.g., slow movements and postural instability. The severity of the symptoms follows a rostro-caudal gradient with the face affected the most and the legs the least, and typical medications for parkinsonism have no effect. The onset is triggered by physical or emotional stress, often involving alcohol.

36 persons with RDP from ten families have been identified to have mutations in the ATP1A3 gene, while maybe an order of magnitude more have been identified with FHM2. For both syndromes, healthy family members carrying a disease-causing mutation have been identified, so there is incomplete penetration of the phenotype, and whether a mutation causes disease will depend on other genetic and environmental variations.

7 Concluding Remarks and Future Directions

The sodium and potassium gradients are absolutely required for animal life, since they form the basis for numerous essential cellular processes, and the sodium pump is required to regulate osmolarity, which is vital to the animal cells that, in contrast to, e.g., plant cells, are not protected by a cell wall. Although a Na^+,K^+-ATPase was probably not the primoridal P-type ATPase [103], it has been essential for metazoan evolution. The original Na^+,K^+-ATPase may have depended on a single subunit, and in, e.g., algae and archaea, genes predicted to encode the α- and not the β-subunit have been identified, but the β-subunit's role in cell-cell interactions combined with the ionic gradients created by the α-subunit have likely been prerequisites for the earliest assemblies of polarized cells that evolved into animals.

The studies of sodium and potassium gradients began centuries ago, and the sodium pump has been known for more than fifty years. Nonetheless, there are still many unanswered questions regarding the mechanism, ion pathways and regulation, and the coming years will undoubtedly bring new perspectives on the sodium pump's role in cellular networks.

Abbreviations

A-domain	actuator domain
ADP	adenosine 5'-diphosphate
AMPA	2-amino-3-(5-methyl-3-oxo-1,2-oxazol-4-yl)propanoic acid

ATP	adenosine 5'-triphosphate
CCCs	cation-chloride co-transporters
CNS	central nervous system
EAATs	excitatory amino acid transporters
ER	endoplasmic reticulum
FHM2	familial hemeplegic migraine 2
GABA	γ-aminobutyric acid
GLUTs	glucose transporters
KCCs	K^+-coupled Cl^- exporters
K_v	voltage-gated K^+ channels
LD_{50}	lethal dose, 50%
LeuT	leucine transporter
nAchR	nicotinic acetylcholine receptor
NaPi	Na^+-coupled P_i symporter
Na_v	voltage-gated Na^+ channels
Na_x	subfamily of voltage-gated sodium channels (formerly Nav2.1 in humans)
NCBTs	sodium-coupled bicarbonate transporters
NCCs	Na^+-coupled Cl^- importers
N-domain	nucleotide-binding domain
NHEs	Na^+-coupled H^+ exporters
NKCCs	Na^+-coupled K^+ and Cl^- importers
NSSs	neurotransmitter sodium symporters
P-domain	phosphorylation domain
P_i	inorganic phosphate
PKA	protein kinase A
PKC	protein kinase C
RDP	rapid-onset dystonia parkinsonism
SGLTs	sodium-dependent glucose transporters
SSRIs	selective serotonin re-uptake inhibitors
TMs	transmembrane helices

Acknowledgments We are grateful to Poul Nissen for advice and support. MJC and HP were funded by the Danish National Research Center PUMPKIN and HP by The Lundbeck Foundation, The Carlsberg Foundation, and L'Oréal/UNESCO.

References

1. C. McCaig, A. Rajnicek, B. Song, M. Zhao, *Physiol. Rev.* **2005**, *85*, 943–1021.
2. E. Overton, *Pflügers Arch.* **1902**, *92*, 346–386.
3. T. Danowski, *J. Biol. Chem.* **1941**, *139*, 693–705.
4. J. Harris, *J. Biol. Chem.* **1941**, *141*, 579–595.
5. H. Schatzmann, *Helv. Physiol. Pharmacol. Acta* **1953**, *11*, 346–400.
6. R. Post, P. Jolly, *Biochim. Biophys. Acta* **1957**, *25*, 118–146.
7. J. Skou, *Biochim. Biophys. Acta* **1957**, *23*, 394–795.

8. A. Mulkidjanian, A. Bychkov, D. Dibrova, M. Galperin, E. Koonin, *Proc. Nat. Acad. Sci. USA* **2012**, *109*, 30.
9. D. Madern, C. Ebel, G. Zaccai, *Extremophiles: Life under Extreme Conditions* **2000**, *4*, 91–99.
10. P. Yancey, *J. Exper. Biol.* **2005**, *208*, 2819–2849.
11. R. Vreeland, *Crit. Rev. Microbiol.* **1987**, *14*, 311–367.
12. S. Kennedy, W. Ng, S. Salzberg, L. Hood, S. DasSarma, *Genome Res.* **2001**, *11*, 1641–1691.
13. K. Collins, *Biophys. J.* **1997**, *72*, 65–141.
14. D. Doyle, J. Morais Cabral, R. Pfuetzner, A. Kuo, J. Gulbis, S. Cohen, B. Chait, R. MacKinnon, *Science* **1998**, *280*, 69–146.
15. D. Hall, C. Bond, G. Leonard, C. Watt, A. Berry, W. Hunter, *J. Biol. Chem.* **2002**, *277*, 22018–22042.
16. M. Page, E. Di Cera, *Physiol. Rev.* **2006**, *86*, 1049–1141.
17. N. Shibata, J. Masuda, T. Tobimatsu, T. Toraya, K. Suto, Y. Morimoto, N. Yasuoka, *Structure* **1999**, *7*, 997–2005.
18. E. Wilkens, A. Ringel, D. Hortig, T. Willke, K.-D. Vorlop, *Appl. Microbiol. Biotechnol.* **2012**, *93*, 1057–1120.
19. T. Larsen, M. Benning, I. Rayment, G. Reed, *Biochemistry* **1998**, *37*, 6247–6302.
20. M. Toney, E. Hohenester, J. Keller, J. Jansonius, *J. Mol. Biol.* **1995**, *245*, 151–230.
21. A. Pineda, C. Carrell, L. Bush, S. Prasad, S. Caccia, Z.-W. Chen, F. Mathews, E. Di Cera, *J. Biol. Chem.* **2004**, *279*, 31842–31895.
22. S. Brohawn, J. del Mármol, R. MacKinnon, *Science* **2012**, *335*, 436–477.
23. A. Miller, S. Long, *Science* **2012**, *335*, 432–438.
24. K. Svoboda, D. Tank, W. Denk, *Science* **1996**, *272*, 716–725.
25. C. Rose, A. Konnerth, *J. Neurosci.* **2001**, *21*, 4207–4221.
26. J. Kim, I. Sizov, M. Dobretsov, H. von Gersdorff, *Nature Neuroscience* **2007**, *10*, 196–401.
27. S. Pulver, L. Griffith, *Nature Neuroscience* **2010**, *13*, 53–62.
28. A. Chakrabarti, D. Deamer, *Biochim. Biophys. Acta* **1992**, *1111*, 171–178.
29. M. Roux, S. Supplisson, *Neuron* **2000**, *25*, 373–456.
30. M. Hahn, R. Blakely, *Pharmacogenomics J.* **2002**, *2*, 217–252.
31. N. Zerangue, M. Kavanaugh, *Nature* **1996**, *383*, 634–641.
32. S. Lachheb, F. Cluzeaud, M. Bens, M. Genete, H. Hibino, S. Lourdel, Y. Kurachi, A. Vandewalle, J. Teulon, M. Paulais, *Am. J. Physiology. Renal Physiology* **2008**, *294*, 407.
33. P. Welling, K. Ho, *Am. J. Physiology. Renal Physiology* **2009**, *297*, 63.
34. S. Adibi, S. Gray, E. Menden, *Am. J. Clin. Nutrition* **1967**, *20*, 24–57.
35. S. Bröer, *Physiol. Rev.* **2008**, *88*, 249–335.
36. E. Wright, D. Loo, B. Hirayama, *Physiol. Rev.* **2011**, *91*, 733–827.
37. H. Krishnamurthy, E. Gouaux, *Nature* **2012**, *481*, 469–543.
38. A. Yamashita, S. Singh, T. Kawate, Y. Jin, E. Gouaux, *Nature* **2005**, *437*, 215–238.
39. Y. Zhao, M. Quick, L. Shi, E. Mehler, H. Weinstein, J. Javitch, *Nature Chem. Biol.* **2010**, *6*, 109–125.
40. L. Forrest, R. Krämer, C. Ziegler, *Biochim. Biophys. Acta* **2011**, *1807*, 167–255.
41. F. Lang, G. Busch, H. Völkl, *Cell. Physiol. Biochem.: Int. J. Exper. Cell. Physiol., Biochem., Pharmacol.* **1998**, *8*, 1–46.
42. C. Lytle, *J. Biol. Chem.* **1997**, *272*, 15069–15146.
43. J. Russell, *Physiol. Rev.* **2000**, *80*, 211–287.
44. T. Zeuthen, N. Macaulay, *J. Physiol.* **2012**, *590*, 1139–1193.
45. P. Dunham, G. Stewart, J. Ellory, *Proc. Nat. Acad. Sci. USA* **1980**, *77*, 1711–1716.
46. G. Gamba, *Physiol. Rev.* **2005**, *85*, 423–516.
47. K. Kahle, J. Rinehart, A. Ring, I. Gimenez, G. Gamba, S. Hebert, R. Lifton, *Physiology* **2006**, *21*, 326–361.
48. A. Alizadeh Naderi, R. Reilly, *Nature Rev. Nephrology* **2010**, *6*, 657–722.

49. I. Forster, N. Hernando, J. Biber, H. Murer, *Kidney Int.* **2006**, *70*, 1548–1607.
50. W. Boron, *J. Am. Soc. Nephrol.: JASN* **2006**, *17*, 2368–2450.
51. I. Choi, H. Soo Yang, W. Boron, *J. Physiol.* **2007**, *578*, 131–173.
52. S. Hebert, D. Mount, G. Gamba, *Pflügers Arch.: Eur. J. Physiol.* **2004**, *447*, 580–673.
53. K. Hinchcliff, P. Morley, A. Guthrie, *J. Am. Vet. Med. Assoc.* **2009**, *235*, 76–158.
54. J. Kyte, *J. Biol. Chem.* **1971**, *246*, 4157–4222.
55. E. Cayanis, H. Bayley, I. Edelman, *J. Biol. Chem.* **1990**, *265*, 10829–10864.
56. K. Geering, *FEBS Lett.* **1991**, *285*, 189–282.
57. K. Geering, J. Kraehenbuhl, B. Rossier, *J. Cell Biol.* **1987**, *105*, 2613–2622.
58. G. Crambert, K. Geering, *Science's STKE: Signal Transduction Knowledge Environment* **2003**, *2003*.
59. K. McGrail, J. Phillips, K. Sweadner, *J. Neurosci.* **1991**, *11*, 381–472.
60. P. Bøttger, Z. Tracz, A. Heuck, P. Nissen, M. Romero-Ramos, K. Lykke-Hartmann, *J. Compar. Neurol.* **2011**, *519*, 376-780.
61. P. Lucchesi, K. Sweadner, *J. Biol. Chem.* **1991**, *266*, 9327–9358.
62. J. F. Hoffman, *Proc. Nat. Acad. Sci. USA* **2002**, *99*.
63. J. Hlivko, S. Chakraborty, T. Hlivko, A. Sengupta, P. James, *Mol. Reprod. Devel.* **2006**, *73*, 101–116.
64. A. Woo, P. James, J. Lingrel, *J. Membr. Biol.* **1999**, *169*, 39–83.
65. G. Blanco, *Seminars in Nephrology* **2005**, *25*, 292–595.
66. P. L. Pedersen, E. Carafoli, *Trends Biochem. Sci.*, **1987** , *12*, 146–296.
67. M. Palmgren, P. Nissen, *Annu. Rev. Biophys.* **2011**, *40*, 243–309.
68. O. Vagin, L. Dada, E. Tokhtaeva, G. Sachs, *Am. J. physiol. Cell Physiol.* **2012**.
69. R. Albers, *Annu. Rev. Biochem.* **1967**, *36*, 727–783.
70. R. Post, S. Kume, T. Tobin, B. Orcutt, A. Sen, *J. Gen. Physiol.* **1969**, *54*, 306–332.
71. J. Morth, B. Pedersen, M. Toustrup-Jensen, T. Sørensen, J. Petersen, J. Andersen, B. Vilsen, P. Nissen, *Nature* **2007**, *450*, 1043–1052.
72. T. Shinoda, H. Ogawa, F. Cornelius, C. Toyoshima, *Nature* **2009**, *459*, 446–496.
73. C. Olesen, M. Picard, A.-M. L. Winther, C. Gyrup, J. Morth, C. Oxvig, J. Møller, P. Nissen, *Nature* **2007**, *450*, 1036–1078.
74. C. Olesen, T. Sørensen, R. Nielsen, J. Møller, P. Nissen, *Science* **2004**, *306*, 2251–2256.
75. T. Sørensen, J. Clausen, A.-M. L. Jensen, B. Vilsen, J. Møller, J. Andersen, P. Nissen, *J. Biol. Chem.* **2004**, *279*, 46355–46363.
76. C. Toyoshima, T. Mizutani, *Nature* **2004**, *430*, 529–564.
77. C. Toyoshima, M. Nakasako, H. Nomura, H. Ogawa, *Nature* **2000**, *405*, 647–702.
78. C. Toyoshima, H. Nomura, *Nature* **2002**, *418*, 605–616.
79. C. Toyoshima, H. Nomura, T. Tsuda, *Nature* **2004**, *432*, 361–369.
80. R. Rakowski, D. Gadsby, P. De Weer, *J. Gen. Physiol.* **1989**, *93*, 903–944.
81. S. Despa, J. Bossuyt, F. Han, K. Ginsburg, L.-G. Jia, H. Kutchai, A. Tucker, D. Bers, *Circulat. Res.* **2005**, *97*, 252–261.
82. C. Palmer, B. Scott, L. Jones, *J. Biol. Chem.* **1991**, *266*, 11126–11156.
83. H. Poulsen, P. Morth, J. Egebjerg, P. Nissen, *FEBS Lett.* **2010**, *584*, 2589–2684.
84. Z.-Q. Wu, J. Chen, Z.-Q. Chi, J.-G. Liu, *Mol. Pharmacol.* **2007**, *71*, 519–549.
85. H. Rasmussen, E. Hamilton, C.-C. Liu, G. Figtree, *Trends Cardiovasc. Med.* **2010**, *20*, 85–175.
86. S. Bibert, C.-C. Liu, G. Figtree, A. Garcia, E. Hamilton, F. Marassi, K. Sweadner, F. Cornelius, K. Geering, H. Rasmussen, *J. Biol. Chem.* **2011**, *286*, 18562–18634.
87. D. Alves, G. Farr, P. Seo-Mayer, M. Caplan, *Molecul. Biol. Cell* **2010**, *21*, 4400–4408.
88. H. Blom, D. Rönnlund, L. Scott, Z. Spicarova, J. Widengren, A. Bondar, A. Aperia, H. Brismar, *BMC Neuroscience* **2011**, *12*, 16.
89. H. Shimizu, E. Watanabe, T. Hiyama, A. Nagakura, A. Fujikawa, H. Okado, Y. Yanagawa, K. Obata, M. Noda, *Neuron* **2007**, *54*, 59–131.
90. S. Santos, B. Manadas, C. Duarte, A. Carvalho, *J. Proteome Res.* **2010**, *9*, 1670–1752.

91. D. Zhang, Q. Hou, M. Wang, A. Lin, L. Jarzylo, A. Navis, A. Raissi, F. Liu, H.-Y. Man, *J. Neuroscience* **2009**, *29*, 4498–5009.
92. J. Heiny, V. Kravtsova, F. Mandel, T. Radzyukevich, B. Benziane, A. Prokofiev, S. Pedersen, A. Chibalin, I. Krivoi, *J. Biol. Chem.* **2010**, *285*, 28614–28640.
93. M. Doi, K. Iwasaki, *Mol. Cell. Neurosci.* **2008**, *38*, 548–606.
94. B. Cassels, *J. Ethnopharmacol.* **1985**, *14*, 273–354.
95. D. Watt, J. Simard, P. Mancuso, *Comp. Biochem. Physiol. A, Comp. Physiol.* **1982**, *71*, 375–457.
96. S. Zhan, C. Merlin, J. Boore, S. Reppert, *Cell* **2011**, *147*, 1171–1256.
97. E. Labeyrie, S. Dobler, *Mol. Biol. Evolut.* **2004**, *21*, 218–239.
98. Z. Li, Z. Xie, *Pflügers Arch.: Eur. J. Physiol.* **2009**, *457*, 635–679.
99. D. Hilgemann, *Proc. Nat. Acad. Sci. USA* **2003**, *100*, 386–394.
100. M. De Fusco, R. Marconi, L. Silvestri, L. Atorino, L. Rampoldi, L. Morgante, A. Ballabio, P. Aridon, G. Casari, *Nature Genetics* **2003**, *33*, 192–198.
101. P. de Carvalho Aguiar, K. Sweadner, J. Penniston, J. Zaremba, L. Liu, M. Caton, G. Linazasoro, M. Borg, M. Tijssen, S. Bressman, W. Dobyns, A. Brashear, L. Ozelius, *Neuron* **2004**, *43*, 169–244.
102. P. Bøttger, C. Doğanlı, K. Lykke-Hartmann, *Neurosci. Biobehav. Rev.* **2012**, *36*, 855–926.
103. K. Axelsen, M. Palmgren, *J. Mol. Evolut.* **1998**, *46*, 84–185.
104. H. Poulsen, H. Khandelia, J. Morth, M. Bublitz, O. Mouritsen, J. Egebjerg, P. Nissen, *Nature* **2010**, *467*, 99–201.

Chapter 4
Magnesium Homeostasis in Mammalian Cells

Andrea M.P. Romani

Contents

A.M.P. Romani (✉)
Department of Physiology and Biophysics, School of Medicine,
Case Western Reserve University, 10900 Euclid Avenue,
Cleveland, OH 44106-4970, USA
e-mail: amr5@po.cwruy.edu

L. Banci (ed.), *Metallomics and the Cell*, Metal Ions in Life Sciences 12,
DOI 10.1007/978-94-007-5561-1_4, © Springer Science+Business Media Dordrecht 2013

Abstract Magnesium, the second most abundant cation within the cell, plays an important role in numerous biological functions. Experimental evidence indicates that mammalian cells tightly regulate cellular magnesium ion content through specific mechanisms controlling Mg^{2+} entry and efflux across the cell membrane and the membrane of various cellular organelles as well as intracellular Mg^{2+} buffering under resting conditions and following hormonal and metabolic stimuli. This chapter will provide an assessment of the various mechanisms controlling cellular Mg^{2+} homeostasis and transport, and the implications changes in cellular Mg^{2+} content play under physiological and pathological conditions.

Keywords cell membrane • cytoplasm • endoplasmic reticulum • Mg^{2+} homeostasis • Mg^{2+} transport • mitochondria

Please cite as: *Met. Ions Life Sci.* 12 (2013) 69–118

1 Introduction

Magnesium is the second most abundant cellular cation after potassium. Total magnesium ion concentrations ranging between 16 to 20 mM have been consistenly measured within mammalian cells by various techniques [1]. These Mg^{2+} levels regulate numerous cellular functions and enzymes, including ion channels, metabolic cycles, and signaling pathways. Despite significant progress, our understanding of how cells regulate Mg^{2+} homeostasis still remains incomplete for conceptual and methodological reasons. The slow turn-over rate for Mg^{2+} movement across the plasma membrane or other biological membranes in the absence of metabolic and hormonal stimuli, the absolute abundance of total and free Mg^{2+} within the cell, and the limited changes in free [Mg^{2+}] have fostered the assumption that cellular Mg^{2+} concentration does not undergo significant changes over time as its level is more than adequate for acting as a cofactor for various cellular enzymes. As a consequence, there has been limited or not interest in developing techniques and methodologies able to rapidly and accurately measure changes in cellular Mg^{2+} content.

In the last twenty-five years, however, an increasing number of experimental and clinical observations have challenged this assumption. More than one thousand entries in the literature indicate the occurrence of major fluxes of Mg^{2+} in either direction across the plasma membrane of mammalian cells following metabolic or hormonal stimuli, and point at Mg^{2+} as a key regulatory cation for a variety of cellular functions. In turn, the movement of Mg^{2+} across the cell membrane has resulted in appreciable changes in total and free Mg^{2+} level within the cell and specific cellular organelles. Genetic and electrophysiological approaches have identified several Mg^{2+} transport mechanisms that operate in the plasma membrane or in the membrane of cellular organelles such as mitochondria and Golgi. The increased interest in Mg^{2+} as a regulator of biological functions has advanced the development of new methodologies able to detect and measure changes in cellular Mg^{2+} content under specific physiologic and pathologic conditions in both animal models and human patients.

2 Cellular Mg^{2+} Distribution

Determinations of cellular Mg^{2+} content by electron probe X-rays microanalysis (EPXMA), ^{31}P NMR, selective Mg^{2+}-electrode, ^{13}C NMR citrate/isocitrate ratio or fluorescent indicators (Table I in ref. [1], and [3]) consistently indicate that total Mg^{2+} concentration ranges between 16 to 20 mM within mammalian cells [1,2], equivalent Mg^{2+} concentrations being localized within nucleus, mitochondria, and endo-(sarco)-plasmic reticulum. The presence of such a high Mg^{2+} concentration within the organelles has been explained with the tendency of Mg^{2+} to bind to phospholipids, proteins, nucleic acids, chromatin and nucleotides within the organelles. Consequently, between 0.8 to 1.2 mM, or 15% to 22% of the cellular Mg^{2+} content, is free within the cytoplasm and the lumen of organelles such as cardiac and liver mitochondria [4,5]. These values are not too different from the level of Mg^{2+} present as free in the extracellular space [1–3].

Similar free Mg^{2+} concentrations are postulated to be present within the nucleus and the lumen of the endo-(sarco)-plasmic reticulum although no direct determinations have been successfully carried out due to technical limitations. In the case of the nucleus, this assumption is based upon the porous structure of the nuclear envelope, which does not support the presence of an ionic gradient between cytoplasm and intranuclear environment. In the case of the endoplasmic (sarcoplasmic) reticulum, the luminal free $[Mg^{2+}]$ cannot be reliably measured because of the competing effect of elevated millimolar Ca^{2+} concentration present within the organelle [6], and the high affinity of this cation (~50 μM) as compared to the affinity of Mg^{2+} (~1.5 mM) for Mag-Fura and Mag-Indo fluorescent dyes [7]. Cytoplasmic Mg^{2+} represents the last large and well detectable cellular Mg^{2+} pool. The majority of this Mg^{2+} pool (~4–5 mM) forms a complex with ATP, phosphonucleotides, and phosphometabolites [8]. Because of its abundance (~5 mM) and Mg^{2+}-binding affinity (K_d ~78 μM), ATP represents the largest metabolic pool able to bind Mg^{2+} within the cytoplasm or the mitochondrial matrix [9], and maintain cytosolic *free* [Mg^{2+}] between 0.5–1 mM within these compartments [1]. Hence, it would appear that there is a very limited chemical gradient for Mg^{2+} between intracellular and extracellular environment, and between cytoplasm and the lumen of cellular organelles.

In cells devoid of cellular compartmentation (e.g., erythrocytes), Mg^{2+} buffering depends exclusively on ATP, phosphonucleotides and phosphometabolites, proteins, and metabolic pools. Three kinetically distinct Mg^{2+} binding pools have been observed in erythrocytes [10]: one low-capacity, high-affinity pool represented by cell proteins, including hemoglobin, and two pools that correspond reasonably well to ATP and 2,3-diphosphoglycerate (2,3-DPG) content, respectively [11]. Clearly, small changes in binding distribution can occur based upon the oxygenated or not oxygenated state of hemoglobin within the erythrocytes [12].

As for Mg^{2+} binding by other cellular proteins little information is available. Aside from hemoglobin [12], consensus sequence for Mg^{2+} binding has been observed in calmodulin [13], troponin C [14], parvalbumin [15], and S100 protein [16]. It is presently undefined whether other cytosolic or intra-organelle proteins bind Mg^{2+} and contribute to the total Mg^{2+} concentrations measured within mitochondria or specific regions of endoplasmic or sarcoplasmic reticulum. Bogucka and Wojtczak [17] have suggested the presence of two proteins in the inter-membraneous space of the mitochondrion binding Mg^{2+} with high affinity/low capacity and high capacity/low affinity, respectively. However, neither of these proteins has been identified. The presence of Mg^{2+}-binding sites has been reported for other cellular proteins, but we lack information about the actual role of these proteins in binding Mg^{2+} under physiological or pathological conditions. We also lack evidence for a consistent Mg^{2+}-binding sequence in these proteins. The basic assumption has been that asparagine and glutamate residues are the most likely amino acids able to form coordination bonds with Mg^{2+}. Lastly, the physiological relevance of Mg^{2+} binding by proteins has been challenged by the observation that parvalbumin *null* mice do not present hypomagnesemia or detectable changes in tissue Mg^{2+} homeostasis [18].

Finally, Mg^{2+} concentration in plasma and extracellular fluid is approximately 1.2–1.4 mM, one third of which is bound to proteins (e.g., albumin) or other biochemical moieties [19]. The comparison between this concentration and those reported within the cell supports the notion that chemical free [Mg^{2+}] across any mammalian biological membrane is at, or near *zero trans* condition. Because the electrochemical equilibrium potential for cellular free [Mg^{2+}] is ~50 mM under resting conditions [20], it is evident that mechanisms must operate in the cell membrane to maintain cytosolic *free* Mg^{2+} and total cellular Mg^{2+} content within the measured levels.

3 Mg^{2+} Transport Mechanisms

Many mammalian cells cultured in the presence of low or virtual *zero* $[Mg^{2+}]_0$ do not present a significant decrease in total or free Mg^{2+} content despite the large Mg^{2+} gradient imposed across the cell membrane [2,21]. Turnover rates ranging from

1 hour in adipocytes to several days in lymphocytes have been observed and attributed to structural and functional cellular differences [21]. Discrepancies can also be observed in the same cell types depending on experimental conditions or modality of isolation (e.g., *in situ versus* in culture, or freshly isolated). For example, ^{28}Mg equilibrium in cardiac ventricular myocytes can vary from 3 hours in the whole animal to 72–80 hours in dispersed cells incubated at 37 °C, to even longer periods of time for cells maintained at 20 °C [22–24], consistent with the notion that hormonal or humoral factors can influence Mg^{2+} transport, and the transport mechanisms have a specific Q10. Similar differences have been observed in freshly isolated [25] *versus* cultured [26] lymphocytes.

Clearly, the occurrence of slow Mg^{2+} turn-over in various cells has generated the erroneous impression that cellular Mg^{2+} content does not change, or changes at such a slow pace that it has no physiological relevance. This impression has been challenged and completely reversed by a large body of experimental evidence documenting the occurrence of large fluxes of Mg^{2+} across the plasma membrane of mammalian cells within minutes from the application of a metabolic or hormonal stimulus [21,27–29]. For example, lymphocytes [25,30], erythrocytes [31], cardiac myocytes [32,33], and liver cells [34–36] are several of the mammalian cells able to extrude a significant percentage (between 10% to 20%) of their total cellular Mg^{2+} content in less than 10 min from the application of an adrenergic stimulus. The flux amplitude and its rate support the presence and operation of powerful Mg^{2+} transport mechanisms in the cell membrane (see [29] for a list of cells). Surprisingly, these fluxes while large, have resulted in relatively small changes in free $[Mg^{2+}]_i$ in the majority of cells examined, suggesting that the operation of these Mg^{2+} transport mechanisms is tightly coupled with the ability of the cell to rapidly buffering the amount of Mg^{2+} extruded or accumulated [37,38].

These Mg^{2+} transport mechanisms can be cursorily divided into channels, which allow Mg^{2+} accumulation, and exchange mechanisms, which allow Mg^{2+} extrusion. The majority of the Mg^{2+} entry mechanisms are located at the cell membrane level. Two of the entry mechanisms favor Mg^{2+} accumulation into mitochondria and Golgi cisternae, respectively. Most of these entry mechanisms show modest selectivity for Mg^{2+} over other divalent cations. No information is available about the mechanisms favoring Mg^{2+} entry in the endoplasmic or sarcoplasmic reticulum, or Mg^{2+} extrusion across the cell membrane or the membrane of cellular organelles.

3.1 Channels

Mg^{2+} entry through channels was first observed in prokaryotes [39,40] and protozoan [41]. More recently, Mg^{2+} entry through channels or channel-like features has been observed in eukaryotic cells. The characterization of these Mg^{2+} channels, however, is far from being complete, and information about their regulation is still fragmentary, limiting our understanding of their relative contribution in regulating Mg^{2+} entry in specific cells or physio-pathological conditions.

3.1.1 TRPM Channels

TRPM7 [42] and TRPM6 [43] were the first Mg^{2+} channels identified in mammalian cells. While TRPM7 is ubiquitous and more in control of Mg^{2+} homeostasis in individual cells, TRMP6 is localized in the colon and the distal convolute tubule of the nephron, a distribution that emphasizes the role of this channel in controlling whole body Mg^{2+} homeostasis through intestinal absorption and renal resorption.

These two channels share several similarities in terms of structure and operation. Yet, they differ in various aspects ranging from location to hormonal modulation.

3.1.1.1 TRPM7:

Fleig's group was the first to report the involvement of TRPM7 in Mg^{2+} accumulation and cell growth [42]. Originally identified as LTRPC7, or long TRP channel 7, owing to the presence of a long amino acid extension outside the channel sequence [44], this channel was already known as CHAK1 (channel kinase 1) [45] because of the presence of an α-kinase domain at its C-terminus [45], and its functional homology to eEF2-kinase [46]. Shortly after the original report [42], Runnels et al. [47] confirmed that TRPM7 combines a channel structure with an α-kinase domain at the C-terminus.

Located at the locus 15q21 of the human chromosome 15, TRPM7 is formed by 1865 amino acids arranged in 10 trans-membrane domains with both the C- and N-termini internalized. Ubiquitously expressed, the channel would carry preferentially Mg^{2+} and Ca^{2+} [42], but also divalent cations such as Ni^{2+} and Zn^{2+} [48,49]. The functional structure of the protein is supposed to be a tetramer, but it is unclear whether the channel is a homotetramer or a hetero-tetramer with varying stoichiometry combination of TRPM7 and TRPM6 monomers. Voets and colleagues [50], for example, reported the functional expression of TRPM6 channels in HEK-293 cells with electrophysiological properties similar to those of TRPM7. In contrast, Chubanov et al. [51] reported no electrical conductance through TRPM6 when this channel is expressed by itself in HEK-293 cells or in *X. laevis* oocytes, and suggested that TRPM7 needs to be co-expressed with TRPM6 for the latter to be incorporated into channel complexes in the cell membrane. Although the association of TRPM6 and TRPM7 channel proteins in a functional structure has been confirmed by Schmitz et al. [52], the functional characterization of TRPM6/TRPM7 chimeras has remained controversial [53] until Yue's group demonstrated that (i) TRPM6 and TRPM7 do form a heterotetramer, and (ii) pure TRPM6, pure TRPM7, and TRPM6/TRPM7 chimeras constitute three distinct ion channel entities with different divalent cation permeability, pH sensitivity, and unique single channel conductance [54,55]. This group also reported that the activities of TRPM6, TRPM7, and TRPM6/TRPM7 can be differentiated by using 2-2-aminoethoxydiphenyl-borate (2-APB), which markedly increases Mg^{2+} and Ca^{2+} entry through TRPM6 [54]. These results support the notion that TRPM6 can form functional homotetrameric channels but also heterotetrameric channels with TRPM7 [56]. As the tissue distribution of these channels is still incomplete, it is possible that homotetrameric TRPM6, homotetrameric TRPM7, and heterotetrameric TRPM6/TRPM7 channels play different roles in diverse tissues under physiological or pathological conditions.

Recently, our understanding of TRPM7 regulation has registered some progress. Protons markedly enhance TRPM7 inward current by competing with Ca^{2+} and Mg^{2+} for binding sites within the channel pore, releasing the blockade of inward monovalent currents by divalent cations [48,49]. At physiological pH, Ca^{2+} or Mg^{2+} bind to TRPM7 and inhibit monovalent cation currents. At higher H^+ concentrations, instead, the protons decrease the affinity of TRPM7 for Ca^{2+} and Mg^{2+}, allowing monovalent cations to permeate the channel [57]. Another level of regulation is provided by PIP2. Initially reported by Clapham's group [58], this regulation has been confirmed by other groups [59–61], which have established that addition of exogenous PIP2 decreases TRPM7 run-down whereas activation of phospholipase C (PLC) by phenylephrine accelerates it [60]. TRPM7 activity is also modulated by ATP [61] and non-hydrolyzable GTP analogs [60], most likely through the formation of $Mg(ATP)^{2-}$ [61] and accelerated PLC-mediated channel run-down, respectively [60]. Bradykinin or angiotensin-II, which also activates PLC, exhibit a similar PIP2-mediated modulation of TRPM7 [59,62]. Activation of TRPM7, however, only occurs in the presence of a physiological cellular $[Mg^{2+}]_i$, any reduction in this concentration resulting in inactivation of TRPM7 activity via PIP2 depletion [59]. All together, these results suggest that PLC-activation accelerates TRPM7 'run-down' via PIP2 depletion. Alternatively, PIP2 depletion would play a feed-back regulation on TRPM7 activation by PLC [59].

A functional TRPM7 is required for a sustained phosphoinositide-3-kinase (PI3K)-mediated signaling in lymphocytes to the point that TRPM7-deficient cells rapidly down-regulate their rate of growth as a result of signaling deactivation downstream PI3-Kinase even in the presence of physiological $[Mg^{2+}]_0$ [63]. Furthermore, TRPM7 regulates the transition of lymphocytes from quiescent to proliferative metabolic state [64]. In contrast, TRPM7-deficient cells upregulate p27, exit cell cycle, and enter quiescence [64]. Because TRPM7 is widely expressed in immuno-competent cells, these results suggest that TRPM7 is essential to regulate rapid cell proliferation and possibly malignancy development.

TRPM7 has originally been identified based on the α-kinase activity present at its C-terminus [45], which phosphorylates serine and threonine residues within an α-helix [65]. Initially, this kinase domain was considered to be essential to modulate TRPM7 activity and gating [58]. Subsequent studies, however, indicated that TRPM7 channels lacking the kinase domain could still be activated by internal Mg^{2+} depletion [66]. Presently, it is undefined which signaling components induce channel opening and/or modulate the α-kinase domain. It is however, clear that kinase autophosphorylation plays a significant role in target recognition by this domain [67]. Massive autophosphorylation of the TRPM7 kinase domain in a region rich in serine and threonine residues located immediately upstream the kinase catalytic domain increases the rate of substrate phosphorylation [68]. Deletion of this region does not affect the intrinsic catalytic activity of the kinase but prevents substrate phosphorylation, supporting the role of this region in substrate recognition [68].

This Ser/Thr region is poorly conserved in TRPM6 in terms of amino acid sequence. However, the kinase domain of TRPM6 appears to require a similar autophosphorylation of its Ser/Thr residues for proper recognition and efficient phosphorylation of the substrates [68].

One obvious consequence of generating TRPM7 lacking the kinase domain is the inability of TRPM7 to properly phosphorylate and activate downstream cellular targets. In support of this notion, homozygous TRPM7-deficient mice carrying the deletion of the kinase domain (TRPM7$^{\Delta Kinase}$) presented early embryonic lethality [69,70]. The heterozygous mice, instead, were viable but presented defective intestinal Mg^{2+} absorption and hypomagnesemia. Cells derived from these heterozygous mice presented reduced TRPM7 currents with an increased sensitivity to inhibition by Mg^{2+} [70]. Embryonic stem cells lacking the TRPM7 kinase domain showed arrest in proliferation and could be rescued by Mg^{2+} supplementation, validating the report by Scharenberg's group [64]. The relevance of the kinase domain in mediating TRPM7 signaling is confirmed by a recent publication by Perraud et al. [70]. These authors reported a role of TRPM7 in regulating the rate of protein synthesis based upon Mg^{2+} availability, and observed that phosphorylation of Thr^{56} residue on eEF2, which inhibits the protein activity, increases under hypomagnesemia. These authors also indicated that Mg^{2+} regulation requires an active and viable kinase domain in the TRPM7 protein [70]. The regulation of eEF2 by TRPM7 is indirect, occurring through eEF2 cognate kinase (eEF2-k), which becomes phosphorylated by TRPM7 kinase on Ser^{77} [70].

At the substrate level, myosin IIA heavy chain [69,71], calpain [72], and annexin I [73,74] have been identified as targets phosphorylated by the TRPM7 kinase domain. Thus, it appears that TRPM7 plays an important role in regulating cell adhesion, contractility or inflammation in different cells, in addition to its role in Mg^{2+} homeostasis. For example, TRPM7 regulates neuronal function and survival under hypoxia or ischemia-reperfusion conditions. Because it can transport either Ca^{2+} or Mg^{2+}, TRPM7 exhibits an ambivalent role based upon the permeating cation. Following activation by reactive oxygen/nitrogen species and prolonged oxygen and glucose deprivation, TRPM7 favors Ca^{2+} fluxes that result in a toxic event for neurons [75]. In contrast, Mg^{2+} permeation enhances anti-apoptotic and cell survival mechanisms, preventing anoxic death of the neurons [67,76]. Following 1 hour occlusion of middle cerebral artery, TRPM7 expression in ipsilateral hippocampus is enhanced, with deleterious consequences for the neurons [77]. Pretreatment of neurons with nerve growth factor activated the TrkA pathway and counteracted both the increase in TRPM7 expression and its harmful consequences [77]. Cell death can be prevented by blocking TRPM7 current via 5-lipoxygenase inhibitors [78]. This block occurs without changes in protein expression and cell membrane concentration [78]. TRPM7 also promotes the specific secretion of acetylcholine at the synaptic level by favoring the fusion of cholinergic vesicles with the pre-synaptic membrane of parasympathetic fibers without any effect on dense core vesicle secretion [79].

The effect of TRPM7 on cell proliferation and differentiation as observed in endothelial cells [76] also extends to osteoblasts [80,81]. Expression of TRPM7 increases during osteoblast differentiation, suggesting a role of cellular Mg^{2+} on cell differentiation. Culturing osteoblasts in low extracellular Mg^{2+} or Ca^{2+} significantly reduces their differentiation [82]. Matrix mineralization is also reduced under these conditions while expression of collagen type I, predominant in the extracellular matrix, increases [80,81]. Osteoblastic differentiation and extracellular matrix mineralization are affected to a comparable extent by TRPM7 silencing during the differentiation stage, further connecting cellular Mg^{2+} homeostasis with TRPM7 expression and activity. Expression of the osteoblastic transcription factor Runx2 was also reduced in cells maintained in the presence of low $[Mg^{2+}]_0$, or by TRPM7 silencing [80]. All together, these results indicate that cellular Mg^{2+} and Ca^{2+} homeostasis via TRPM7 are important for osteoblastic differentiation. It remains to be determined to which extent Mg deficiency in the general population, which is more common than anticipated, associates with altered osteoblastic differentiation and inadequate bone formation and osteoporosis development.

Data by the Clapham's group, however, have casted some doubts about the effective role of TRPM7 in regulating Mg^{2+} entry and homeostasis [82]. This group has observed altered embryonic development and tissue specific deletion of the channel in T cell lineage in TRPM7 *null* mouse, with disrupted thymopoiesis and progressive depletion of thymic medullary cells in the absence of significant changes in acute Mg^{2+} accumulation or total Mg^{2+} content in the T cells. Absence of TRPM7, however, significantly dysregulated the synthesis of several growth factors altering thymic epithelial cells differentiation [82]. Hence, it appears that TRPM7 is the first TRP channel with an essential, non-redundant role in embryogenesis and thymopoiesis. It is still unclear, however, how TRPM7 absence alters T cells differentiation.

3.1.1.2 TRPM6:

The TRPM6 channel is uniquely localized in the colon and the renal distal convolute tubule, two epithelia highly impermeable to salt re-absorption. This specific localization supports the specific role of this channel in controlling whole-body Mg^{2+} homeostasis by regulating intestinal Mg^{2+} absorption and renal Mg^{2+} reabsorption.

The *TRPM6* gene was originally identified by genetic analysis as the site of various mutations responsible for Hypomagnesemia with Secondary Hypocalcemia (HSH, OMIM 602014). A rare autosomal recessive disease, HSH is characterized by Mg^{2+} and Ca^{2+} wasting, a serum Mg^{2+} level around 0.5–0.6 mmol/L, or half the physiological level despite massive intravenous and oral Mg^{2+} administration [43]. Because the primary defect is at the level of the TRPM6 channels expressed in the intestine [43], any excess in Mg^{2+} supplementation is rapidly filtered at the glomerular level and results in increased passive renal absorption via paracellin-1 (see Section 3.1.2). Transcellular absorption via apical TRPM6 channels in the renal epithelium, however, remains depressed and unable to restore physiological serum Mg^{2+} level [43].

Experimental evidence suggests that TRPM6 also forms a functional tetramer at the plasma membrane level [54]. Several point mutations in the TRPM6 amino acid sequence have been identified [83], which result in the expression of a truncated and non-functional channel [83]. The missense mutation $S^{141}L$, for example, which occurs at the N-terminus of the channel, prevents the correct assembly of TRPM6 as a homotetramer, or a heterotetramer with TRPM7 [83]. The missense mutation $P^{1017}R$, instead, occurs in the pore region of the channel, but affects negatively and more significantly TRPM7 when it is co-expressed with TRPM6 [83]. TRPM6 *null* mice have also been developed [84]. Aside for low Mg^{2+} level in plasma (~0.67 *versus* 0.75 mM), heterozygous *Trpm6*$^{+/-}$ mice present normal electrolyte levels, whereas the majority of the homozygous *Trpm6*$^{-/-}$ mice die by embryonic day 12.5 [84,85]. Few animals survive to term, and the majority of them present exencephaly, spina bifida occulta, and other significant neural tube defects. Administration of a high Mg diet to dams allows for offspring survival to weaning [86], althought this aspect has not been confirmed in a more recent study [87].

Similarly to TRPM7, TRPM6 presents an α-kinase domain at the C-terminus with functional homology to eEF2-kinase, whereby the protein was originally termed CHAK2 (channel kinase 2) [46]. This kinase domain phosphorylates serine and threonine residues located within an α-helix [45,46,65]. Owing to their dual function as a channel and a kinase, TRPM6 and TRPM7 are referred to as *chanzymes*. As for TRPM7, removal of the kinase domain does not abolish TRPM6 activity but modulates the extent to which the channel is regulated by intracellular free Mg^{2+} or $Mg(ATP)^{2-}$ complex [51–54,86], and affects the ability of the chanzyme to phosphorylate downstream targets. At variance of what was reported for TRPM7, no specific substrate phosphorylated by TRPM6 kinase has been identified, with the exception of TRPM7 itself. While the TRPM6 kinase domain can phosphorylate residues on the TRPM7 channel within a heterotetramer structure, TRPM7 kinase cannot phosphorylate residues on TRPM6 [52]. Hence, it remains to be clarified whether the TRPM6 kinase domain phosphorylates substrates similar to, or different from those associated with the TRPM7 kinase domain.

At variance with TRPM7, TRPM6 appears to be sensitive to changes in estrogen level and dietary Mg^{2+} intake. Estrogens (17β-estradiol) selectively up-regulate TRPM6 mRNA in both colon and kidney, leaving unaffected TRPM7 mRNA in other tissues [87,88]. In the absence of estrogen, the repressor of estrogen receptor activity (REA) binds to the 6th, 7th, and 8th β-sheets of the TRPM6 kinase domain in a phosphorylation-dependent manner, and inhibits its activity [90]. Estrogen administration rapidly dissociates REA binding, resulting in increased TRPM6 activity [90]. Dietary Mg^{2+} restriction also up-regulates TRPM6 mRNA in both colon and kidney, but it has no effect on TRPM7 mRNA [88,89]. In contrast, exposure to Mg^{2+}-enriched diet up-regulates TRPM6 mRNA only in the colon, consistent with increased intestinal absorption [88]. These changes in TRPM6 expression and Mg^{2+} levels do not occur in mice exhibiting normal or high erythrocyte and plasma Mg^{2+} levels [89]. Therefore, it is becoming progressively apparent that genetic factors control TRPM6 expression and activity, and that dietary Mg^{2+} restriction promotes a compensatory increase in Mg^{2+} absorption and reabsorption by enhancing TRPM6 expression in intestine and kidney, respectively [88,89].

Cellular ATP decreases TRPM6 current [51–54], as observed also for TRPM7. The site of inhibition resides in the conserved ATP-binding motif GxG(A)xxG within the α-kinase domain [89]. Full deletion of the kinase domain and point mutations within the ATP-binding motif ($G^{1955}D$) completely abolish the inhibitory effect of ATP. The effect of ATP, however, is independent of α-kinase autophosphorylation activity [86].

TRPM6 activity is modulated by cellular signaling molecules. Over-expression of RACK1 (receptor for activated protein kinase C) directly binds to the α-kinase domain of TRPM6, and possibly TRPM7 due to the >84% homology between the two kinase domains [90]. RACK1 binding site is located in the region corresponding to the 6th, 7th, and 8th β sheets, the same sheets involved in REA regulation [87]. Following RACK1 binding, TRPM6 (and possibly TRPM7) activity is inhibited. Channel activity is not suppressed when RACK1 is co-expressed with the α-kinase deleted TRPM6 mutant. The inhibitory effect of RACK1 depends on the autophosphorylation of threonine 1851 (T^{1851}), which is localized at the end of the 4th α-helix adjacent to the RACK1 binding site. Mutating T^{1851} to alanine ($T^{1851}A$) or aspartate ($T^{1851}D$) decreases TRPM6 autophosphorylation but does not affect RACK1 binding. The inhibitory effect of RACK1 on channel activity is abolished by $T^{1851}A$ mutation, while it is unaffected by $T^{1851}D$ mutation [90]. The latter mutation renders the kinase autophosphorylation directly proportional to the Mg^{2+} concentration, with a steady increase in the 0.1 to 1 mM range. The $T^{1851}A$ mutant, instead, is less sensitive to intracellular Mg^{2+} concentrations as compared to the wild-type (IC_{50} ~0.7 *versus* 0.5 mM, respectively). Activation of protein kinase C (PKC) by phorbol-myristate acetate (PMA) prevents the inhibitory effect of RACK1 on channel activity [90] whereas the PKC inhibitor chelerythrine restores the inhibition [90]. All together, these results suggest a competing effect of PKC for RACK1.

The epidermal growth factor (EGF) also acts as an autocrine/paracrine magnesiotropic hormone [91]. Following the receptor engagement in the basolateral domain of the distal convolute tubule, EGF activates TRPM6 at the cellular apical domain and promotes cellular Mg^{2+} accumulation. Point mutation in the pro-EGF sequence ($P^{1070}L$) retains EGF to the apical membrane of the cell and inhibits Mg^{2+} accumulation, resulting in Mg^{2+} wasting (isolated recessive renal hypomagnesemia or IRH syndrome, OMIM 611718). The axis EGF/TRPM6/Mg^{2+} reabsorption becomes altered in cancer patients undergoing treatment with antibodies anti-EGFR [92,93] as the block of the EGF receptor prevents TRPM6 activation and results in renal Mg^{2+} wasting [92,93]. EGF promotes TRPM6 activity and/or expression via ERK1/2 phosphorylation [94] and adaptin protein-1 (AP-1) signaling [95]. The process is prevented by antagonists for integrin $\alpha_v\beta3$ and MEK1/MEK2 activity, or siRNA for TRPM6 [94]. It is undefined whether this signaling axis releases RACK1-mediated inhibition of TRPM6 activity through PKC activation [90].

As TRPM6 is located in the apical domain of the intestinal and renal epithelium, it is unresolved how apically accumulated Mg^{2+} is transported across the cytoplasm to be delivered to the basolateral domain and be extruded into the blood stream. The general consensus is that baso-lateral Mg^{2+} extrusion occurs via a Na^+/Mg^{2+} exchanger (see Section 3.2.1). Uncertain is also as to whether Mg^{2+} is transported through the cytoplasm bound to proteins or diffuses freely. Parvalbumin and calbindin-D_{28k}, two proteins abundantly present within the cells of the distal convolute tubule of the nephron, could mediate trans-cellular Mg^{2+} transport, accelerating the delivery rate of the cation to the basolateral domain. However, no detectable defects in Mg^{2+} excretion or homeostasis are observed in parvalbumin *null* mice [17], raising some doubts on whether parvalbumin does play a role in transporting Mg^{2+} under physiological conditions, or other proteins can compensate for its absence in the *null* model.

3.1.2 Claudins

Genetic analysis of patients affected by Familial Hypomagnesaemia with Hypercalciuria and Nephrocalcinosis (FHHNC, OMIM 248250) identified paracellin-1 as the first mammalian protein able to transport Mg^{2+} [96]. FHHNC is characterized by massive renal Mg^{2+} and Ca^{2+} wasting that leads rapidly and irreversibly to renal failure [96] as symptoms and renal deterioration are not ameliorated by Mg^{2+} supplementation [96]. The gene responsible for the disease was termed *Paracellin-1* (*PCLN-1*) [96], which encodes for paracellin-1 (PCLN-1), a protein now renamed claudin-16. This protein is in fact a member of the claudin family [97], a group of tight junction proteins with 4 trans-membrane spans coordinated by 2 extracellular loops, and with both C- and N-termini on the cytoplasm side. More than 20 mutations affecting trafficking or permeability of claudin-16 have been currently identified [98].

Claudin-16 mediates paracellular Ca^{2+} and Mg^{2+} fluxes throughout the nephron. Yet, the modality by which these fluxes are generated is still controversial. Data obtained in LLC-PK1 (a renal cell line of porcine origin) indicate that claudin-16 mediates paracellular Na^+ permeation which, in turn, generates a positive potential within the lumen of the nephron that acts as driving force for Mg^{2+} and Ca^{2+} reabsorption [99]. Data in MDCK cells, instead, point to an increase in Mg^{2+} and a decrease in Na^+ permeability [100]. It is unclear whether these discrepancies reflect a different *modus operandi* in cell lines of differing origin, or depend on the experimental conditions utilized in the two studies. Either study, however, support the evidence that *PCLN-1* expression is modulated by $[Mg^{2+}]_0$ [101].

To properly function, claudin-16 has to be delivered to the tight junction where it interacts with the scaffolding protein ZO-1 [102]. Claudin-16.ZO-1 association and dissociation are regulated via PKA-mediated phosphorylation of Ser^{217} within claudin-16 sequence [102]. Activation of the Calcium Sensing Receptor (CaSR) dephosphorylates this residue [103], whereby claudin-16 dissociates from ZO-1 and accumulates within the lysosomal compartment [100]. Mutations of Ser^{217} accelerate claudin-16 turn-over and modulate its function. Mutation of Thr^{233} ($T^{233}R$) also impairs the claudin-16/ZO-1 interaction, and favors claudin-16 accumulation into lysosomes [100,102].

Recent evidence indicates the involvement of claudin-19 isoform in mediating Mg^{2+} and Ca^{2+} reabsorption [104]. Claudin-19 forms a head-to-head complex with claudin-16 at the level of the tight junction, increasing cation selectivity. While claudin-16 function as a channel does not appear to depend on its association with claudin-19, claudin-19 is necessary to recruit claudin-16 and form a co-polymer at the tight junction level of the nephron and to switch channel selectivity from anion to cation [104]. The heteromeric association between claudin-16 and claudin-19 is affected by point mutations in claudin-16 ($L^{145}P$, $L^{151}F$, $G^{191}R$, $A^{209}T$, and $F^{232}C$) and claudin-19 ($L^{90}P$ and $G^{123}R$). Each of these mutations abolishes the physiological synergism between the two proteins, and results in FHHNC development.

3.1.3 MagT1

Human epithelial cells up-regulate MagT1 encoding gene following exposure to low $[Mg^{2+}]_0$ [105]. This protein has an estimated molecular weight of 38 KDa and 5 trans-membrane domains in its immature form. Following the cleavage of the first trans-membrane segment located near the C-terminus, the mature protein contains only 4 trans-membrane spans. At variance of SLC41 (Section 3.3.1) and Mrs2 (discussed in the next Section), MagT1 does not present any significant degree of homology to prokaryotic Mg^{2+} transporters, but it exhibits some similarities with the oligosaccharide transferase complex OST3/OST6 that regulates protein glycosylation in the endoplasmic reticulum of yeast [106]. The murine orthologue of MagT1 is highly expressed in liver, heart, kidney, and colon, with detectable levels in lung, brain, and spleen [105]. For the most part, MagT1 levels in these tissues are consistent with the mRNA levels, the only exception being the liver in which a low protein level is detected [105]. MagT1 appears to be highly specific for Mg^{2+} ($K_m = 0.23$ mM), and the Mg^{2+}-elicited currents are inhibited by Ni^{2+}, Zn^{2+}, and Mn^{2+} but not Ca^{2+}, although the inhibiting concentrations of any of these cations are >0.2 mM, thus exceeding the physiological concentration present in extra-cellular fluids. Nitrendipine at a concentration of ~10 μM, but not nifedipine, inhibits MagT1-mediated Mg^{2+} current [105]. Limited information is available about N33, a second member of the MagT family. Although able to transport Mg^{2+}, N33 exhibits a much lower specificity for Mg^{2+} than MagT1. In addition, N33 can also transport Fe^{2+}, Mn^{2+}, and Cu^{2+} [105].

MagT1 appears to possess channel-like characteristics and high selectivity for Mg^{2+}, suggesting that this transporter is essential to regulate Mg^{2+} homeostasis in mammalian cells. This hypothesis is supported by the observation that knocking out of MagT1 and its human homolog TUSC3 in HEK-293 cells markedly reduces cellular Mg^{2+} content [106]. Either MagT1 or TUSC3 can complement the yeast Mg^{2+} transporter ALR1 [106]. Exposure of HEK-293 cells to low $[Mg^{2+}]_0$ for 1–2 days increases the mRNA level of MagT1 but not TUSC3, whereas incubation in high $[Mg^{2+}]_0$ does not affect the expression of either protein [106].

3.1.4 Mrs2

Mrs2 was identified during a screening aimed at isolating nuclear genes suppressing RNA splicing defects in yeast mitochondrial introns [107]. Mrs2-deficient yeasts present: (i) a splicing phenotype, (ii) a significant reduction in cytochromes content, and (iii) a deficit in mitochondria respiration to the point that the yeasts become unable to grow on non-fermentable substrates [108].

Structurally, Mrs2 shows short regions of homology to the bacterial transporter CorA [109], and a similar membrane topology with 2 trans-membrane domains. Mutant yeasts lacking Mrs2 present a decrease in total mitochondrial and matrix free Mg^{2+} content [110], and can be rescued by mitochondrial targeted CorA. In contrast, Mrs2 over-expression results in a marked increase in matrix free Mg^{2+} [110]. Hence, suggestion is there for an essential role of Mrs2 in regulating mitochondrial Mg^{2+} homeostasis acting as a channel modulated by mitochondrial membrane potential ($\Delta\psi$). Inhibitors of F0-F1-ATPase or ANT also modulate Mrs2 activity, decreasing the amplitude of Mg^{2+} influx. Highly conserved motifs in the coiled-coil middle region of the channel are essential to gate it and form a functional unit. Knocking out Mrs in HEK-293 cells impairs mitochondrial complex I expression, reduces the level of mitochondrial Mg^{2+}, affects cell morphology and promotes apoptosis [111], to the point that cell viability is completely lost within 2 weeks [111]. It is unclear whether the decrease in mitochondrial Mg^{2+} depends on the absence of Mrs2, or is related to complex I absence, which affects mitochondrial $\Delta\psi$ and consequently Mg^{2+} retention within the organelle [112].

A single Mrs2 orthologue is expressed in mammalian cells, in which it mediates mitochondrial Mg^{2+} entry as the yeast homologue [113]. Under conditions in which Mrs2p is absent or not functional, an alternative but much slower mitochondrial Mg^{2+} entry mechanism becomes operative in restoring or maintaining Mg^{2+} homeostasis, ensuring the survival of Mrs2-deficient yeast. The identity, abundance, and regulation of this alternative transporter in mitochondria are presently unknown. All together, these data suggest that Mrs2 is essential but not indispensable to regulate mitochondrial Mg^{2+} level which, in turn, plays an essential role in modulating mitochondrial dehydrogenases and oxygen consumption [114,115].

3.1.5 MMgTs

This gene family comprehends two proteins termed MMgT1 and MMgT2 (for membrane Mg^{2+} transporter 1 and 2) [116]. The chromosomal locations of these proteins are XA5 for MMgT1 and 11B2 for MMgT2 in the mouse, and Xq36, and 10q23, respectively, in the rat. The human MMgT1 orthologue is located on Xq26.3 [117]. MMgT1 and MMgT2 are located in the Golgi complex and post-Golgi vesicles, in which they contribute to regulate Mg^{2+}-dependent enzymes involved in protein assembly and glycosylation [117]. The localization in the Golgi system, however, does not exclude that these proteins may be transported to the cell membrane or to other destinations downstream the Golgi network, where they can play a role in modulating Mg^{2+} homeostasis. These proteins are formed by 131 (MMgT1) and 123 (MMgT2) amino acids, assembled into two trans-membrane domains in a wide variety of tissues.

The small size of these proteins suggests that they form homo- and/or hetero-oligomeric channels to favor Mg^{2+} permeation. MMgT-mediated Mg^{2+} uptake exhibits a $K_m \simeq 1.5$ mM for MMgT1, and $\simeq 0.6$ mM for MMgT2, and these values do not vary with voltage. Both MMgT1 and MMgT2 are not specific for Mg^{2+} as they can transport other cations with some slight differences in cation permeation. In addition to Mg^{2+}, MMgT1 transports Sr^{2+}, Fe^{2+}, Co^{2+}, and Cu^{2+}, while MMgT2 transports Sr^{2+}, Co^{2+}, Cu^{2+}, Ba^{2+}, and Mn^{2+} [116].

Limited information is currently available about the specifics of MMgT1 expression and operation, which can be summarized as follows: (i) Mg^{2+}-generated currents in MMgT1 are inhibited by Mn^{2+} (~0.2 mM) but not Gd^{3+} or Ni^{2+} [116], and (ii) MMgT1 mRNA increases ~2.5 fold in the kidney cortex of mice on low-Mg^{2+} diet and ~3.5 fold in MDCT epithelial cells culture in low Mg^{2+} medium whereas MMgT2 mRNA increases ~1.5-fold in kidney cortex and ~3-fold in MDCT cells under similar experimental conditions [116]. These changes in expression are not specific for these transport mechanism as similar increases have been observed for other Mg^{2+} entry mechanisms.

The audience interested in a more detailed description of the intrinsic characteristics of the various Mg^{2+} entry mechanisms described in this section is referred to recent reviews by Touyz et al. [62], Schmitz et al. [117], Bindels et al. [118], and Quamme [119].

3.2 *Exchangers*

While Mg^{2+} entry is mediated by channels or channel-like mechanisms, Mg^{2+} extrusion is mediated by two exchange mechanisms. The specific electrochemical requirements favoring Mg^{2+} extrusion indicate that these mechanisms operate as a Na^+-dependent and Na^+-independent Mg^{2+} exchanger, respectively. Most of the information about the operation, abundance and tissue specificity of these mechanisms is largely circumstantially or indirectly based upon experimental conditions or pharmacological inhibition.

3.2.1 Na^+-Dependent Exchanger (Na^+/Mg^{2+} Exchanger)

Günther, Vormann, and Forster provided the first evidence for the presence and operation of a Mg^{2+} transport mechanism in erythrocytes [120], and indicated that this transport mechanism elicits Mg^{2+} extrusion in a Na^+-dependent, amiloride-inhibited manner [121]. The operation of such a mechanism has been subsequently confirmed by other groups in a large variety of mammalian cells and tissues [12,122–126] (see also [29] for a list). Several laboratories [25,33,127,128] including ours

[32,34,129–134] have provided compelling evidence that this Na^+-dependent, amiloride-inhibited Mg^{2+} extrusion mechanisms is specifically activated by cAMP, irrespective of the modality by which the cAMP level is increased. Stimulation of β-adrenergic, glucagon, or PGE2 receptors, or administration of forskolin or cell-permeant cAMP analogs all results in Mg^{2+} extrusion via cAMP-mediated phosphorylation of the Na^+-dependent mechanism [127–134]. Inhibition of adenylyl cyclase by Rp-cAMP or PKA by PKI, instead, blocks Mg^{2+} extrusion irrespective of the receptor or the modality utilized to increase cAMP level [25].

To mediate Mg^{2+} extrusion this exchanger requires a physiological concentration of Na^+ in the extracellular milieu [129,131], suggesting that the mechanism occurs through a Na^+/Mg^{2+} exchanger. A recent report [135] infers that this Na^+/Mg^{2+} exchanger is the SLC41 transporter identified by Wabakken et al. [136] and discussed in detail in one of the next sections. However, no detailed information about membrane abundance, proximity to and interaction with other cellular transporters, and stoichiometry of this transporter is currently available. Günther, Vormann, and Forster suggested that this exchanger operates on electroneutral basis ($2Na^+_{in}:1\ Mg^{2+}_{out}$) at least in chicken or turkey erythrocytes [120,121]. Data obtained in mammalian erythrocytes including human red blood cells have confuted this hypothesis, supporting an electrogenic operation ($1Na^+_{in}:1\ Mg^{2+}_{out}$) [123–125]. The reason behind this discrepancy is not apparent, although it is possible that experimental models (i.e., cell isolation *versus* cultured cells), incubation medium composition, or modality of cellular Mg^{2+} loading can all contribute. Recently, we have reported that Mg^{2+} extrusion via the Na^+/Mg^{2+} exchanger is coupled to the outward movement of Cl^- ions [137]. In the absence of Cl^-, the exchanger switches from electrogenic ($1Na^+_{in}:1\ Mg^{2+}_{out}$) to electroneutral ($2Na^+_{in}:1\ Mg^{2+}_{out}$) [137]. Interestingly, only inhibitors of the Na^+/Mg^{2+} exchanger block Cl^- extrusion while specific inhibitors of Cl^- transporters (e.g., flufenamic acid, DNDS, or DIDS) are ineffective. A role of cellular Cl^- in stimulating the Na^+/Mg^{2+} exchanger in erythrocytes has been reported by Ebel and Günther [138]. Moreover, Rasgado-Flores and collaborators have observed Cl^- transport following reverse activation of the Na^+/Mg^{2+} exchanger in dialyzed squid axon [139]. Hence, the extrusion of Cl^- can be interpreted as an attempt to equilibrate charge movement across the hepatocyte or the axon membrane. However, it is unclear whether Cl^- extrusion occurs through the Na^+/Mg^{2+} exchanger directly, or through Cl^- channels present in the hepatocyte membrane [140] and activated by the exchanger.

Irrespective of the stoichiometry of exchange and the experimental model utilized, the results obtained by the various groups consistently indicate a K_m for Na^+ between 15 to 20 mM [141–143]. In terms of pharmacological inhibition, amiloride, imipramine, and quinidine represent the three most commonly utilized inhibitors of the Na^+-dependent Mg^{2+} extrusion [120,127,144]. Because of their limited specificity, however, it is unclear as to whether they inhibit the Na^+/Mg^{2+} exchanger directly, or indirectly by blocking other transport mechanisms including Na^+ and K^+ channels, and altering the cell membrane potential and driving force for Mg^{2+} transport across the plasma membrane.

Using a hybridoma screening procedure, the group of Schweigel, Vormann, and Martens has generated inhibiting monoclonal antibodies against the Na^+/Mg^{2+} exchanger present in porcine red blood cells. Western Blot analysis using these antibodies has evidenced a ~70 KDa protein band [145]. This is the first time that information about the molecular size of the elusive Na^+/Mg^{2+} exchanger has been obtained. Hence, these antibodies could represent an ideal tool to identify and recognize this transporter in mammalian tissue.

3.2.2 Na^+-Independent Exchanger

In the absence of extracellular Na^+ to support the operation of the Na^+/Mg^{2+} exchanger, or in the presence of amiloride, imipramine or quinidine, which block the exchanger, Mg^{2+} extrusion occurs via an alternative, Na^+-independent mechanism. The specificity of this mechanism, however, is not defined. Cations such as Ca^{2+} or Mn^{2+}, and anions such as HCO_3^-, Cl^-, or choline [146,147], have all been observed to promote Mg^{2+} extrusion through this mechanism. Hence, it is unclear whether we are in the presence of a transporter that can operate as an antiporter for cations or a sinporter for cations and anions based upon the experimental conditions. Ebel and collaborators [147] have suggested that this Na^+-independent Mg^{2+} extrusion mechanism is the choline transporter based upon its inhibitability by cinchona alkaloids [147]. It is also unclear whether the Na^+-independent pathway is activated by hormonal stimulation. Stimulation of liver cells by epinephrine, a mixed adrenergic agonist, elicits an extrusion of Mg^{2+} that is equivalent to the sum of the amounts of Mg^{2+} mobilized by the separate stimulation of α_1- and β-adrenergic receptors [131,132,148]. The stimulation of α_1-adrenergic receptors by phenylephrine requires the presence of physiological concentrations of both Na^+ and Ca^{2+} in the extracellular medium to elicit Mg^{2+} extrusion from liver cells [131,132]. Since phenylephrine stimulation would activate Ca^{2+}-CaM signaling and capacitative Ca^{2+} entry to induce Mg^{2+} extrusion [132], it might be possible that the Ca^{2+}-CaM signaling pathway represents an alternative modality of activation of the Na^+/Mg^{2+} exchanger. Alternatively, it is possible that it activates a different Mg^{2+} extrusion mechanism identifiable with the Na^+-independent mechanism.

It is also controversial whether ATP is required for the operation of the Na^+-independent and Na^+-dependent mechanisms. Reports by Günther and collaborators [149,150] indicate a dependence of the Na^+/Mg^{2+} exchanger on the presence of a physiological concentration of cellular ATP to the point that cellular Mg^{2+} efflux is reduced under conditions that markedly decrease cellular ATP level [149,150]. In the case of red blood cells, Mg^{2+} homeostasis and transport are affected by changes in both ATP and 2,3-diphosphoglycerate levels [11]. The absence of a regulatory effect of ATP on Mg^{2+} extrusion has been confirmed in purified liver plasma membrane vesicles [143].

3.2.3 Mg^{2+}/H^+ Exchange

This exchange mechanism, originally identified in *A. thaliana* and termed AtMHX, is ubiquitous in plants [151]. It presents 11 putative trans-membrane domains, it is exclusively localized in the plant vacuolar membrane, and it electrogenically exchanges Mg^{2+} or Zn^{2+} for protons. Following ectopic over-expression of the transporter, tobacco plants become able to grow in the presence of elevated concentrations of Mg^{2+} (or Zn^{2+}) [151]. Presently, no corresponding gene and encoded protein has been identified in mammalian cells, although some experimental evidence suggests direct or indirect exchange of Mg^{2+} for H^+ under well defined conditions [152]. An enhanced extrusion of cellular Mg^{2+} has been reported to occur in cells incubated in the presence of an acidic extracellular environment, which imposes an inwardly oriented H^+ gradient, as long as extracellular Na^+ is present [152,153]. Amiloride derivates, which inhibit the Na^+/H^+ exchanger with high affinity, are ineffective at blocking Mg^{2+} extrusion under these experimental conditions [154], thus excluding the involvement of the Na^+/H^+ exchanger in mediating Mg^{2+} extrusion either directly or indirectly through a coupling of this exchanger with the Na^+/Mg^{2+} antiporter.

3.3 *Carriers*

Several novel Mg^{2+} transport mechanisms of murine or human origin have been identified as a result of exposure to low Mg^{2+} in the diet (diet restriction) or in the culture medium (medium restriction). The *modus operandi* of these transport mechanisms is plagued by limited information. For practical reasons, these transport mechanisms are non-descriptively classified here as carriers.

3.3.1 SLC41 (Solute Carrier Family 41)

This family of Mg^{2+} transport mechanisms includes three members (A1, A2, and A3), all distantly related to prokaryotic MgtE channel [155]. Because no study has addressed function and structure of the SLC41A3 isoform, all the available information provided here refers to the SLC41A1 and A2 isoforms.

SLC41A1 was the first member of this family to be identified [136]. The hydrophobic profile of this protein (~56 kDa Mr) predicts the presence of 10 transmembrane domains, two of which presenting a discrete level of homology with MgtE [136]. Northern blot analysis indicates a broad distribution of the *SLC41A1* gene, but its abundance varies markedly among tissues, the highest expression being in heart and testis and the lowest being in hematopoietic tissues and cells [136]. The expression of this gene is modest under basal conditions, but becomes markedly up-regulated in the renal cortex of mice fed low Mg^{2+} diet for several days [156]. Functional expression of mouse SLC41A1 in *X. laevis* oocyte indicates that this protein can transport Mg^{2+} but also Fe^{2+}, Cu^{2+}, Zn^{2+}, and Cd^{2+}. In contrast, Ca^{2+} is not transported nor does it inhibit Mg^{2+} transport [156]. The initial observation suggested the presence of a Mg^{2+}-generated current, which would be consistent with SLC41A1 operating as a channel [156], or alternatively as an electrogenic antiporter. A recent report [157], however, strongly suggests that SLC41A1 operates as a carrier in promoting Mg^{2+} efflux. This hypothesis is supported by another recent report [135] indicating this transporter as the putative Na^+/Mg^{2+} exchanger previously described.

Incubation of HEK-293 cells in Mg^{2+}-free media resulted in a significant reduction of total Mg^{2+} content and free cellular Mg^{2+} concentration ($[Mg^{2+}]_i$,), the amplitude of Mg^{2+} loss depending on the number of SLC41A1 molecules expressed in the membrane and the induction time. Lastly, the changes in $[Mg^{2+}]_i$ were sensitive to the experimental temperature but insensitive to the Mg^{2+} channel blocker CoHexamine [157]. Kolisek and collaborators also suggested that SLC41A1 forms high-molecular-weight complexes within the cell membrane with molecular masses ranging between 720 and 1236 kDa [157]. Addition of SDS resulted in the progressive degradation of the complexes in a step-wise manner until a protein band of ~56 kDa is obtained, which corresponds to the molecular mass of the SLC41A1 monomer [157]. It is presently undefined whether the SLC41A1 monomer aggregates to form large multimeric complexes or interacts with auxiliary proteins. The reason for the absence of Mg^{2+}-generated currents in this study as compared to the original observation by Goytain and Quamme [156] is also not clear. One possibility could be that the murine [156] and human orthologs [157] operate differently although they are expected to operate in a similar manner based upon their high degree (>90%) of homology. Yet, the possibility that point mutations can dramatically alter SLC41A1 ion specificity and modality of function cannot be completely dismissed. Another point of difference is that while Goytain and Quamme did not report a dependency of SLC41A1 operation on Na^+ or other cations or anions following expression in *X. laevis* oocytes [156], Kolisek and collaborators reported a marked Cl^- conductance following expression in HEK-293 cells, which was abolished by DIDS [157]. Whether this reflects the operation of additional transport mechanisms or the presence of structural differences in the cell membrane of HEK-293 cells [157] as compared to *X. laevis* oocyte [156] are possibilities that need further investigation.

A second isoform labeled SLC41A2 has been identified in both humans and mice. SLC41A2 transports Mg^{2+} as well as other divalent cations albeit with a different selectivity and inhibition profile than SLC41A1 [158]. In addition to Mg^{2+}, SLC41A2 can carry Ba^{2+}, Ni^{2+}, Co^{2+}, Fe^{2+}, and Mn^{2+} but not Ca^{2+}, Cu^{2+} or Zn^{2+}. At variance of SLC41A1, Mg^{2+} transport via SLC41A2 is inhibited by Ca^{2+} [158]. Both SLC41A1 and SLC41A2 generate Mg^{2+} currents in *X. laevis* oocyte, and the ionic uptake is voltage-dependent with an apparent affinity of 0.75 mM and 0.31 mM, respectively [156,158]. SLC41A2 is also widely expressed in mammalian tissues, but its expression is not affected by low Mg^{2+} diet [158]. At the structural level, SLC41A2 shares >70% homology with SLC41A1 and it is supposed to present 10 trans-membrane domains as well. A recent study by Scharenberg's group, however, suggests a structural arrangement in 2 spans of five trans-membrane motifs each linked together by a supplementary spanning motif [159]. Hydrophobicity analysis indicates that the C- and N- termini are located on different sites of the cell membrane [159], a configuration that will be consistent with a total of 11 trans-membrane segments.

3.3.2 ACDP2

The human *ACDP* gene family was identified by Wang and collaborators [160] as a possible candidate of the urofacial syndrome. Mapped to the 10q23-10q24 chromosome, this gene family comprises 4 isoforms differentially located in human tissues. ACDP1 is essentially restricted to the brain. ACDP2 is more widely expressed, but still retains the highest expression in the brain while being absent in skeletal muscles. ACDP3 and ACDP4 are both ubiquitous, but have the highest expression in the heart [161]. The murine distribution of ACDP isoforms is very similar to that observed for the human orthologues [162]. Termed ancient conserved domain protein because all isoforms share one domain phylogenetically conserved from bacteria to man [160], these proteins are >50% homologous to the CorC transporter, which together with CorB and CorD plays a role in Mg^{2+} efflux in prokaryotes [163]. Over-expression of ACDP2 in *X. laevis* oocytes indicates that this protein can transport a variety of divalent cations including Mg^{2+}, Co^{2+}, Mn^{2+}, Sr^{2+}, Ba^{2+}, Cu^{2+}, and Fe^{2+}, whereas Zn^{2+} inhibits its activity [164]. Mg^{2+} transport via ACDP2 is voltage-dependent, occurs with a K_m of ~0.5 mM, and does not require the presence of extracellular Na^+ or Cl^- [164]. Similarly to SLC41A1, *ACDP2* gene becomes over-expressed following exposure to Mg^{2+}-deficient diet [164].

3.3.3 NIPA

Located in the SPG6 locus of chromosome 15q11-q13, the *NIPA1* gene is so called for 'non-imprinted in Prader-Willi/Angelman syndrome, a disease characterized by a complex developmental disorder that affects numerous organs and systems [165]. Located among a set of approximately 30 genes linked to the disease [165], *NIPA1* has also been implicated in autosomal dominant hereditary spastic paraplegia (HSP, OMIM 182600). The human and mouse genomes contain four members of the NIPA family, termed NIPA1 through NIPA4, with an overall similarity of ~40%. Homology between human and mice proteins is ~98%. NIPA1 [166] and NIPA2 [167] can both operate as Mg^{2+} transporters. Presenting a sequence of 323 (NIPA1) and 359 amino acids (NIPA2) arranged to form 9 and 8 trans-membrane spans, respectively, these two proteins transport Mg^{2+} in a saturable fashion, with different K_m and specificity. NIPA1 has a K_m ~0.66 mM for Mg^{2+} [166], but can also transport Sr^{2+}, Fe^{2+} or Co^{2+}, albeit to a lesser extent [166]. NIPA2, instead, is highly specific for Mg^{2+} with K_m ~0.31 mM [167]. Neither NIPA3 nor NIPA4 transports Mg^{2+}; NIPA3 transports Sr^{2+}, Ba^{2+}, Fe^{2+}, and Cu^{2+} while NIPA4 transport Sr^{2+} and Ba^{2+}.

The insurgence of autosomal dominant HSP is based on specific point mutations in NIPA1 (i.e., $G^{100}R$ or $T^{45}R$) [168]. Both glycine and threonine residues are conserved among ortholog NIPA1 channels in different species. There are no similar consensus sites in NIPA2, NIPA3, and NIPA4 paralogs, implying that the folding of these proteins might be different. In HSP patients, NIPA2 appears to be normal but it cannot functionally replace NIPA1 to ameliorate HSP symptoms, nor can NIPA3 or NIPA4 substitute for the defective NIPA1. This is surprising for NIPA2 as the encoding gene is part of the 30 genes cluster associated with the Prader-Willi/Angelman syndrome together with *NIPA1*. Presently, there is no indication as to whether the Prader-Willi/Angelman syndrome presents alteration in Mg^{2+} homeostasis.

3.3.4 Huntingtin

Huntingtin-interacting protein 14 (HIP14) and its related protein HIP14-like (HIP14L) are significantly up-regulated (~3-fold) by low extracellular Mg^{2+} [169]. Formed by ~532 amino acids arranged in 6 trans-membrane spans, HIP14 presents 69% homology to HIP14L and a strong sequence similarity to the ankyrin repeat protein Akr1p [170]. HIP14 also possesses a cytoplasmic DHHC cysteine-rich domain. Defined by an Asp-His-His-Cys sequence motif this domain confers palmitoyl-acyltransferase activity to the protein, and gives it the ability to palmitoylate membrane components whereby modulating their structure. Mg^{2+} accumulation via HIP14 and HIP14L appears to be electrogenic, voltage-dependent, and saturable, with K_m ~0.87 and ~0.74 mM, respectively [169]. Inhibition of palmitoylation activity by 2-Br-palmitate, or deletion of the DHHC domain decreases HIP14-mediated Mg^{2+} accumulation by ~50%, suggesting that palmitoylation is not required for basal Mg^{2+} transport.

The widespread tissue distribution and intracellular localization of HIP14 (nuclear and perinuclear regions, Golgi complex, mitochondria, microtubules, endosomes, clathrin-coated and non-coated vesicles, and plasma membrane [171]) implicates this protein in numerous cellular processes including transcriptional regulation, mitochondrial bioenergetics, structural scaffolding, vesicle trafficking, endocytosis, and dendrite formation [171]. Golgi and post-Golgi vesicles, however, appear to be the primary location of HIP14 [169,171]. Hence, it can be hypothesized that Mg^{2+} accumulation via this protein is associated with the role HIP14 plays in the physiological functioning of the cellular compartments in which the protein is located. The neuropathological manifestation of Huntington disease is characterized by progressive neurodegenerative disorders, cognitive deficits, and choreic movements. All these manifestations are linked to the abnormal expansion of glutamine residues from less than 34 to more than 37 at the 18th amino acid position [170]. Presently, the mechanism responsible for the insurgence of these defects is unknown [170]. Similarly unknown is whether the poly-glutamine expansion alters Mg^{2+} transport, and whether perturbation of Mg^{2+} homeostasis plays any role in the insurgence of the neuronal defects typical of Huntington disease.

3.4 *Mg^{2+} Transport in Purified Plasma Membrane Vesicles*

Several laboratories including ours have used plasma membrane vesicles to better characterize how different Mg^{2+} extrusion mechanisms operate in particular cell types. The plasma membrane model presents several advantages including: (i) a well defined ionic extra- and intra-vesicular milieu composition to determine the modality of operation of the various Mg^{2+} transporters under rigorous experimental conditions, and (ii) the ability to investigate the operation of different Mg^{2+} extrusion mechanisms in the absence of Mg^{2+} buffering by ATP, proteins or other cytosolic components, and partitioning within intracellular organelles. By purifying total liver plasma membrane or cardiac sarcolemmal vesicles as well as specific subpopulations enriched in basolateral or apical domains, our laboratory has been able to provide a better understanding of the selective location and specificity of operation of the Na^+-dependent and Na^+-independent Mg^{2+} extrusion mechanisms in liver cells and cardiac myocytes.

In the hepatocyte, the Na^+-dependent extrusion mechanism is specifically located in the basolateral domain [143], is selectively activated by Na^+ [143,172], and is inhibited only by imipramine [172], and not by amiloride and amiloride derivates [172]. Moreover, the operation of the exchanger is completely abolished by pretreatment of basolateral vesicles with alkaline phosphatase, but it can be restored by loading the vesicles with ATP and PKA-catalytic subunit [133,134], leaning further support to the notion that the Na^+/Mg^{2+} exchanger is activated upon phosphorylation by cAMP. As this exchanger continues to operate in the presence of *zero trans* Mg^{2+} across the plasma membrane (i.e., 20 mM Mg^{2+} inside and outside the vesicles) [143], indication is there that Mg^{2+} extrusion strictly depends on the Na^+ transmembrane gradient, with a $K_m \leq 20$ mM [143], in good agreement with kinetic data obtained in isolated hepatocytes [130] and other cell types [141]. Experiments based on TPP^+ distribution have confirmed the electrogenicity of this exchange mechanism in plasma membrane vesicles, supporting a $1Na^+_{in}:1Mg^{2+}_{out}$ exchange ratio under the majority of experimental conditions tested [133,143,172]. Upon removal of intravesicular Cl^-, the stoichiometry ratio of the exchanger switches from electrogenic to electroneutral (i.e. $2Na^+_{in}:1Mg^{2+}_{out}$) [137]. Interestingly, in the presence of intravesicular Cl^-, a Cl^- extrusion is observed in concomitance with the extrusion of Mg^{2+} and the accumulation of external Na^+ into the vesicles [137]. This Cl^- extrusion is not inhibited by inhibitors of anion transport (e.g., DNDS, DIDS, or niflumic acid), of the Na/K/Cl cotransporter NKCC1 (e.g., bumetanide or furosemide) [137], excluding that Cl^- movement occurs via one of these mechanisms. The only agent able to block Cl^- extrusion is imipramine [137], which specifically blocks the operation of the Na^+/Mg^{2+} exchanger in the basolateral domain of the hepatocyte [172]. Hence, it would appear that Cl^- is extruded via the Na^+/Mg^{2+} exchanger or, alternatively, via Cl^- channels for partial charge compensation [137]. The possibility that Cl^- is extruded via the Na^+/Mg^{2+} exchanger has been suggested by Rasgado-Flores et al. [139] in dialyzed squid axons, and it would be in good agreement with the observation by Ebel and Günther that intracellular Cl^- has a stimulatory role on the activity of the Na^+/Mg^{2+} antiport in red blood cells [138].

Experiments carried out in liver plasma membrane vesicles enriched in apical domain indicate the presence of two apparently distinct and unidirectional Mg^{2+} transport mechanisms, which extrude intravesicular Mg^{2+} for extravesicular Na^+ and Ca^{2+}, respectively [172]. The apical Na^+-dependent Mg^{2+} transporter presents several similarities to the basolateral transporter: (i) its K_m for Na^+ is comparable at ~15–20 mM; (ii) it selectively uses Na^+ over other monovalent cations; (iii) it electrogenically exchanges $1Na^+_{in}:1Mg^{2+}_{out}$ [172]. This apical exchanger can be pharmacologically distinguished from the basolateral one due to its specific inhibition by amiloride [172], although it retains a significant level of inhibition by imipramine. In contrast, imipramine only blocks the basolateral antiport [172]. The apical exchanger can also be distinguished from the basolateral antiport based on its inability to operate in reverse mode [172] and the non-requirement for cAMP-mediated phosphorylation to become active [133]. Based on preliminary observation, it would appear that this apical antiport does not transport Cl^- as part of its operation (A. M. P. Romani, personal observation).

The apical domain of the hepatocyte also possesses a Ca^{2+}-dependent Mg^{2+} extrusion mechanism [133]. Specifically located in this domain, this exchanger is activated by micromolar Ca^{2+} concentrations ($K_m \geq 50$ μM), and is insensitive to alkaline phosphatase pre-treatment [133,134]. The Mg^{2+} extrusion through this antiport occurs on electroneutral basis (i.e., $1Ca^{2+}_{in}:1\ Mg^{2+}_{out}$) [133]. This exchanger is not Ca^{2+}-specific, as Mg^{2+} extrusion can occur following the extravesicular addition of micromolar concentrations of various divalent cations ($Ca^{2+} \gg Co^{2+} = Mn^{2+} > Sr^{2+} \gg Ba^{2+} > Cu^{2+} \gg Cd^{2+}$) [172]. Similarly to the apical Na^+/Mg^{2+} antiport, the Ca^{2+}-dependent mechanism is inhibited by amiloride or imipramine [172]. This observation raises the question as to whether we are in the presence of two distinct apical mechanisms, modulated by Na^+ and cations, respectively. Several lines of evidence, however, do not fully support this possibility. First, the co-addition of Na^+ and Ca^{2+} to a purified subpopulation of apical plasma membrane vesicles does not appear to enlarge Mg^{2+} extrusion to a significant extent (A. M. P. Romani, personal observation). Second, amiloride inhibits both Na^+- and Ca^{2+}-dependent Mg^{2+} extrusion processes to a comparable extent at a similar concentration [172]. Third, alkaline phosphatase treatment does not affect the Mg^{2+} extrusion elicited by either exchanger in apical liver plasma membrane vesicles [133]. Fourth, neither of these exchangers can operate in reverse at variance of the basolateral Na^+/Mg^{2+} antiport. Taken together, these observations suggest the operation of a non-selective exchange mechanism able to utilize monovalent or divalent cations to promote Mg^{2+} extrusion. At the present time, the physiological implication for the operation of such an exchanger in the apical domain of the hepatocyte is not clear.

The operation of functionally similar Na^+- and Ca^{2+}-dependent Mg^{2+} extrusion mechanisms has also been observed in cardiac sarcolemma vesicles [134]. As in the case of liver plasma membrane vesicles, cardiac sarcolemma vesicles do not require intravesicular ATP to support the operation of Mg^{2+} transporters [134], and pretreatment of the vesicles with alkaline phosphatase specifically inhibits the reversible Na^+-dependent Mg^{2+} extrusion mechanism but not the Ca^{2+}-dependent Mg^{2+} extrusion pathway [134]. For technical reasons, it is presently undefined whether cardiac myocytes also possess two distinct Na^+-dependent Mg^{2+}-extrusion mechanisms in the sarcolemma, and whether the Ca^{2+}/Mg^{2+} exchanger in sarcolemmal vesicles can also utilize Na^+ to promote Mg^{2+} extrusion.

The operation of specific Mg^{2+} transport mechanisms has been observed in plasma membrane vesicles from brush border cells of rabbit ileum [173] and from rat duodenum and jejunum [174]. In these models, however, Mg^{2+} accumulation rather than extrusion has been observed. By using membrane vesicles from rabbit ileum and cell permeant and non-permeant Mag-Fura, Juttner and Ebel have observed the operation of a saturable Mg^{2+} uptake mechanism when the intracellular Na^+ concentration is higher than the extracellular concentration [173]. The process becomes inoperative when the Na^+ gradient is reversed (i.e., $[Na^+]_i < [Na^+]_o$), the vesicles are in *zero trans* condition for Na^+, or external Na^+ is removed. At variance with the Na^+-Mg^{2+} antiporter operating in liver plasma membrane vesicles, the pathway in ileum vesicles is not reversible and appears to be electroneutral. Yet, it possesses a K_m for Na^+ of 16 mM, a value similar to the K_m calculated in liver plasma membranes [143], in smooth muscle cells from guinea pig tenia caecum [141], and in chicken erythrocytes [175]. Another similarity with the transporter operating in basolateral liver plasma membranes is the lack of inhibition by amiloride analogs [173]. In good agreement with other reports [138], the transporter characterized by Juttner and Ebel is modulated by intravesicular anions, especially Cl^- and SCN^-, and is markedly stimulated by antagonists of anion transport (e.g., H_2-DIDS) [173].

The main difference between plasma membrane vesicles from duodenum and jejunum [174,175] is that a single Mg^{2+} uptake mechanism operates in the duodenum with a K_m ~0.8 mM, whereas two transporters operate in the jejunum with K_m values of 0.15 mM and 2.4 mM, respectively. In both these experimental models, Mg^{2+} but not Ca^{2+} accumulation is reduced in the presence of alkaline phosphatase inhibitors [176], suggesting that Ca^{2+} and Mg^{2+} are transported via distinct pathways. This hypothesis is further supported by the observation that Mg^{2+} accumulation is inhibited by amiloride but not by Ca^{2+} channel antagonists. Consistent with the report by Juttner and Ebel [173], Mg^{2+} accumulation is stimulated by an intravesicular electronegative potential or an alkaline pH_o [174]. The effect of external pH, however, is lost when $[Mg^{2+}]_0 > 1$ mM [174]. Under the latter condition (i.e. $[Mg^{2+}]_0 > 1$ mM), Mg^{2+} accumulation is enhanced by the presence of Na^+ or K^+ but it is inhibited by the presence of divalent cations ($Co^{2+} > Mn^{2+} > Ca^{2+} > Ni^{2+} > Ba^{2+} > Sr^{2+}$) in the extravesicular space [174]. The molecular aspects of these processes, however, have not been elucidated and remain highly speculative.

4 Regulation of Mg^{2+} Transport and Homeostasis

The majority of mammalian cells retains their basal Mg^{2+} content virtually unchanged under resting conditions even when a major trans-membrane gradient is artificially imposed [1–3]. At the same time, compelling evidence supports the notion that different hormones induce the movement of large amounts of Mg^{2+} in either direction across eukaryotic cell membranes. As a result of these movements, changes in serum, total and – to a lesser extent – free Mg^{2+} content have been observed. These changes have resulted in detectable variations in Mg^{2+} level within organelles, especially mitochondria, with significant repercussions on cellular bioenergetics. Hence, a picture is slowly emerging, which relates changes in total Mg^{2+} content to energetic substrate utilization (e.g., glucose), cell cycle progression [64] or meaningful changes in Mg^{2+} content within discrete portions of the cell or cellular organelles with consequent modulation of the activity of specific enzymes located therein.

4.1 Mg^{2+} Extrusion

Hormones like catecholamine or glucagon induce Mg^{2+} extrusion from various cell types or perfused tissues. The majority of these hormones have in common their ability to increase cellular cAMP level. While the Mg^{2+} extrusion elicited by these hormones depletes to a varying extent the Mg^{2+} pools present within the cytoplasm and the cellular compartments, the physical outward transport of Mg^{2+} across the cell membrane primarily occurs via the Na^+/Mg^{2+} exchanger previously described. Magnesium extrusion has also been observed following metabolic treatments that decrease cellular ATP content, the main Mg^{2+} buffering component. Interestingly, several of the hormones that induce Mg^{2+} extrusion from liver cells also elicit glucose output from the hepatocyte. Conversely, hormones that promote glycogen synthesis stimulate Mg^{2+} accumulation rather than extrusion (discussed in Section 4.2). Hence, it would appear that at least in liver cells Mg^{2+} extrusion is functionally associated with glucose transport and utilization.

4.1.1 Cyclic AMP-Dependent Mg^{2+} Extrusion

Elliot and Rizack were the first to report Mg^{2+} transport across the plasma membrane of adipocytes stimulated by adrenocorticotrophic hormone [177], but the authors did not elucidate the modality of transport or the mechanism involved. The first extensive characterization of hormone-induced Mg^{2+} transport was provided by Maguire and colleagues in S49 lymphoma cells and primary lymphocytes stimulated by β-adrenergic receptor agonist or PGE1 [178–181]. Maguire and Erdos [182] also provided the first observation that protein kinase C (PKC) activation enhances Mg^{2+} influx in S49 cells at variance of β-adrenergic receptor stimulation, which inhibits the process. Observation carried out in S49 cells lacking protein kinase A (PKA) or adenylyl cyclase (AC), however, indicated that the inhibitory effect of β-adrenergic agonists was not mediated by cAMP [183,184]. At variance of what was reported for primary lymphocytes [25], Mg^{2+} transport in S49 cells appears to be independent of extracellular Na^+ concentration or membrane potential (Maguire, unpublished observation). Further, Mg^{2+} turnover in S49 requires more than 40 hours as compared to the much faster Ca^{2+} turnover, which is completed in less than 3 hours [184].

These initial observations were followed by a long series of reports all supporting the notion that β-adrenergic agonists and other hormones control Mg^{2+} homeostasis in mammalian cells. In the majority of eukaryotic cells, hormones or agents that increase cellular cAMP level elicit a significant extrusion of Mg^{2+} into the extracellular space or the circulation [32–34]. This effect has been observed in cardiac ventricular myocytes [32,33,129,185], liver cells [34,35,130–132], red blood cells [31], lymphocytes [30], and Ehrlich ascites cells [186] among other cells (see [27] for a more comprehensive list), as well as in anesthetized animals [187,188]. In all these cell types, Mg^{2+} extrusion is a fast process that reaches the maximum within 8 min from the application of the stimulus irrespective of the hormone (catecholamine, isoproterenol, glucagon, PGE1, or arachidonic acid) [30–35,185,186] or agent (i.e., forskolin or cell permeant cyclic AMP analogs) [30–34,130–132] utilized to increase cellular cAMP level. The key role of cAMP in modulating Mg^{2+} extrusion is further corroborated by the observation that pre-treatment of cells with hormones or agents that decrease cAMP production (e.g., carbachol [30–34,130–132], insulin [189]) or prevent PKA activation (e.g., Rp-cAMP [25]) completely prevents cellular Mg^{2+} mobilization. In an open perfusion system, the amount of Mg^{2+} extruded from the organ (i.e. heart or liver) returns towards baseline level within 8 min from the application of the agonist irrespective of its dose or persistence in the perfusate [32,34]. This temporally limited extrusion suggests that Mg^{2+} is rapidly mobilized from a well defined cellular pool(s) that is (are) rapidly depleted. This notion is supported by the observation that sub-maximal doses of agonist sequentially infused within a few minutes from each other elicit Mg^{2+} extrusions of progressively decreasing amplitudes [30].

Under all these conditions, limited changes in cytosolic free $[Mg^{2+}]_i$ have been observed [37,190], suggesting that Mg^{2+} is rapidly released from binding and buffering sites within the cytoplasm or cellular organelle(s) and extruded across the cell membrane. Irrespective of the hormone utilized, cAMP-mediated Mg^{2+} extrusion occurs via the putative Na^+/Mg^{2+} exchanger described previously. In fact, either the removal of extra-cellular Na^+ [142] or the cell pre-treatment with non-selective Na^+ transport inhibitors like amiloride or imipramine [33,142], abolishes the Mg^{2+} extrusion almost completely. Under either inhibitory condition the reduced Mg^{2+} extrusion across the cell membrane originates a more sustained rise in cytosolic free $[Mg^{2+}]_i$ [37,190], suggestive of the concept that blocking the Na^+-dependent transport mechanism prevents Mg^{2+} from being extruded across the cell membrane but not from being released from binding/buffering sites such as ATP or proteins, and/or from cellular organelles (i.e., mitochondria and endoplasmic reticulum) into the cytoplasm. Two corollaries of this observation are that: (i) cAMP operates on at least two different levels (i.e., cellular organelle(s) and plasma membrane) to mobilize Mg^{2+} from the cell, and (ii) only Mg^{2+} transport across the cell membrane is Na^+-dependent whereas the mobilization from cellular organelle(s) is largely Na^+-independent. Alternatively, it has to be postulated that cytosolic Na^+ concentration, which ranges between 15 to 20 mM in most cell types) is more than sufficient to favor Mg^{2+} transport across the membrane of cellular organelles.

4.1.2 Cyclic AMP-Independent Mg^{2+} Extrusion

In 1989, Jakob and collaborators reported that phenylephrine administration also promotes Mg^{2+} extrusion from liver cells via α_1-adrenergic stimulation [36]. In addition to confirming this observation, our laboratory has provided the first evidence that co-stimulation of α_1- and β-adrenergic receptor are additive and complementary processes to induce Mg^{2+} extrusion from liver cells [131,148]. This event is of particular relevance especially when the two classes of adrenergic receptors are stimulated by mix-adrenergic agonists such as epinephrine or norepinephrine [131,148]. Pre-infusion of insulin only abolishes β-adrenergic receptor mediated Mg^{2+} extrusion from liver cells, leaving unaffected the mobilization of Mg^{2+} mediated via α_1-adrenergic receptors [148]. The inhibitory effect of insulin persists even in cells treated with cell-permeant cAMP analogs [148]. A similar inhibitory effect of insulin on β-adrenergic receptor mediated, cAMP-modulated, Mg^{2+} extrusion has been observed in cardiac myocytes [189]. These results have been attributed to an inhibitory effect of insulin on β-adrenergic receptor activation [191], and a stimulatory effect of the hormone on the cytosolic phosphodiesterase that degrades cAMP [192]. Experimental evidence also suggests a direct modulating effect of insulin on the Na^+/Mg^{2+} exchanger, at least in erythrocytes [193].

Fagan and Romani [131,132] further investigated the modality of Mg^{2+} extrusion following α_1-adrenergic receptor stimulation in liver cells. Their results indicate that phenylephrine-induced Mg^{2+} extrusion strictly depends on the activation of capacitative Ca^{2+} entry [132]. Inhibition of IP3-induced Ca^{2+} release from the endoplasmic reticulum, chelating of cytosolic Ca^{2+}, or inhibition of Ca^{2+} entry at the plasma membrane level all result in the complete inhibition of Mg^{2+} extrusion from the hepatocyte [132]. The scant information available about possible binding of Mg^{2+} by cellular proteins prevented the authors from ascertaining whether Mg^{2+} extruded from the hepatocyte was mobilized from the ER, or displaced from cytosolic binding sites following the massive entry of Ca^{2+} across the cell membrane ([132] and refs. therein). Interestingly, extracellular Na^+ and Ca^{2+} are both required for the phenylephrine-induced Mg^{2+} extrusion to occur [132]. In the absence of extracellular Ca^{2+}, in fact, the amplitude of Mg^{2+} extrusion is decreased by ~15% to 20% whereas extracellular Na^+ is responsible for the remaining 80%–85% of the extrusion. It is presently unclear whether Mg^{2+} extrusion occurs via the Ca^{2+}-activated, Na^+-dependent mechanism observed in the apical domain of the hepatocyte, or whether Na^+ is required to maintain membrane potential and facilitate Ca^{2+} entry across the hepatocyte cell membrane. In the absence of receptor activation, thapsigargin administration can mimic phenylephrine stimulation and elicit Mg^{2+} extrusion from the hepatocyte, even in the absence of extracellular Ca^{2+} [132], although to a lesser extent. Hence, it would appear that an optimal level of cytosolic Ca^{2+} has to be attained in order for Mg^{2+} extrusion to occur via displacement from cellular binding sites or via a Ca^{2+}-calmodulin-activated mechanism [132]. Interestingly, the group of Schweyen has provided evidence that in yeast Mg^{2+} deprivation accelerates Ca^{2+} accumulation. In turn, this translates into a more rapid activation of Ca^{2+}-mediated signaling [194].

4.1.3 Mg^{2+} Homeostasis and Glucose

The presence of redundant Mg^{2+} extrusion mechanisms or modalities of activation of a common Mg^{2+} extrusion pathway raises the question: What is the physiological significance of Mg^{2+} mobilization in mammalian cells?

The general answer is that Mg^{2+} extrusion can have a different significance in different cells due to the physiological differentiation and function of the various cell types. In the case of cardiac myocytes, an increase in extracellular Mg^{2+} level has been associated with a modulating effect on the open probability of the L-type Ca^{2+}-channels [195] and a temporary decrease in sino-atrial node action potential [185]. In the case of liver cells, instead, Mg^{2+} transport appears to be associated with a regulatory role on glucose transport and utilization. Catecholamine [131,148] or glucagon [131], and adrenergic agonists like isoproterenol or phenylephrine [131,148], which elicit Mg^{2+} extrusion from liver cells, all activate glycogenolysis and promote release of hepatic glucose into the blood-stream within a similar time frame [131]. Interestingly, the presence of amiloride or imipramine inhibits both Mg^{2+} extrusion and hepatic glucose output [131]. The converse is also true. Inhibition of glucose transporter activity by phlorethin results in a qualitatively similar inhibition of Mg^{2+} extrusion from liver cells [131]. The presence of a close functional 'link' between glucose and Mg^{2+} homeostasis is corroborated by the observation that overnight starvation completely depletes hepatic glycogen and glucose, and concomitantly decreases to a significant extent (minus 15%) total hepatic Mg^{2+} content as a consequence of pro-glycemic hormones (i.e., catecholamine and glucagon) activation [196]. Noteworthy, this decrease in hepatic Mg^{2+} content is equivalent in amplitude to that elicited via *in vitro* stimulation of perfused livers by the same hormones [196], or that observed to occur in livers of type-I diabetic animals [197], which are markedly depleted in cellular glycogen. This functional link between glucose and Mg^{2+} homeostasis can also be observed under conditions in which glucose accumulation and glycogen synthesis are stimulated by insulin administration to cardiac ventricular myocytes [189] or pancreatic β cells [198]. In both experimental models, the amount of Mg^{2+} accumulated within the cells is directly proportional to the amplitude of glucose accumulation. Conversely, decreasing extracellular Mg^{2+} concentration directly reduces the amount of glucose accumulated within the cells [131,189]. A role of Mg^{2+} in regulating glucose homeostasis is underlined by the observation that several glycolytic enzymes, including hexokinase, phosphofructokinase, phosphoglycerate mutase, phosphoglycerate kinase, enolase and pyruvate kinase, show activation at low, and inhibition at high Mg^{2+} concentrations [199,200].

4.1.4 Mg^{2+} Homeostasis and ATP

Mg^{2+} extrusion also occurs following exposure to various agents or conditions that markedly decrease cellular ATP content and production including cyanide [153,201], mitochondrial uncouplers [38,112], fructose [202], ethanol [203], or hypoxia [204]. All these agents, in fact, decrease ATP content by preventing the mitochondrial electron chain from generating ATP (cyanide or uncouplers), by acting as an ATP trap (fructose), or by altering the redox state of pyridine nucleotide within the cytoplasm and the mitochondrion (ethanol). Because ATP represents the major Mg^{2+} buffering component within the cell [8,9], a decrease in its content or its degradation into ADP or AMP results in an increased dissociation of Mg^{2+} from the binding and an increase in cytosolic free $[Mg^{2+}]_i$. Ultimately, such an increase in cytosolic Mg^{2+} level originates in a detectable Mg^{2+} extrusion from the cell [153,201–204].

Such an extrusion can be observed in erythrocytes, which possess limited cellular buffering capacity for Mg^{2+} and no compartmentation [205], as well as in cells that possess additional Mg^{2+} buffering due to the presence of proteins or cellular organelles in addition to ATP and phosphonucleotides [153,201–204]. In several cases, such as fructose addition [202], the changes in cytosolic $[Mg^{2+}]_i$ can elicit glycogenolysis via activation of glycogen phosphorylase and glucose utilization to restore cellular ATP levels [202]. The majority of these experimental conditions promote a modest increase in cytosolic free $[Mg^{2+}]_i$, which is considerably lower than the increase expected to occur based upon the corresponding decrease in ATP level. This observation strongly supports the notion that the majority of Mg^{2+} released from ATP and other binding sites is extruded from the cell. Furthermore, because the ATP level decreases following changes in pyrimidine nucleotide ratio or mitochondria poisoning, it would appear that not phosphorylation but the rise in cytosolic Mg^{2+}, even if modest, is sufficient to activate Mg^{2+} extrusion and limit the rise in cytosolic free Mg^{2+} concentration to approximately 100–200 μM at the most [201]. Hence, it can be presumed that such an increase is sufficient to activate enzymes and metabolic reactions controlled by Mg^{2+}.

On the other hand, cellular ATP regulates Mg^{2+} extrusion in ways other than acting as a buffering component. Evidence for an additional role of ATP has been provided by experiments in giant squid axon [206], mammalian hepatocytes [207] or erythrocytes [11]. In squid axon, the Na^+-dependent Mg^{2+} extrusion requires a physiological level of ATP to operate, and as the level of ATP decreases so does the amplitude of Mg^{2+} extrusion [207]. In erythrocytes and hepatocytes, instead, ATP appears to regulate the Na^+-independent Mg^{2+} extrusion process [11,207]. The modality by which ATP regulates the Mg^{2+} extrusion process is unclear, but it appears to be unrelated to the operation of an ATPase mechanism. This notion is supported by the observation that a decrease in cellular ATP level as it occurs for example under diabetic or alcoholic conditions paradoxically results in an increased extrusion of Mg^{2+} via the Na^+-dependent mechanism in a manner directly proportional to the decrease in ATP level [197,203]. Because ATP predominantly acts as a ligand for Mg^{2+} both in the cytoplasm and the mitochondrial matrix [8,9], any decrease in ATP level will result in an increase in free Mg^{2+} and the consequent extrusion of Mg^{2+} from the cell.

4.2 Mg^{2+} Accumulation

The identification of several Mg^{2+} entry mechanisms strongly support the hypothesis that cellular Mg^{2+} is dynamically maintained through the operation of entry and exit mechanisms that are differentially regulated by hormones and metabolic conditions. A striking difference is there, however, between the Mg^{2+} exit and the Mg^{2+} entry mechanisms. In the case of Mg^{2+} extrusion mechanisms we have a good understanding of the signaling activating their operation but we lack any information about the structure of the mechanisms themselves. In the case of Mg^{2+} entry mechanisms, instead, we do have structural information about several of these mechanisms but for the most part we lack detailed information about their individual activation by hormones or second messenger, and their possible cooperation under specific conditions.

4.2.1 Role of Protein Kinase C

Experimental evidence indicates that mammalian cells can accumulate large amounts of Mg^{2+} as a result of hormonal stimulation. Administration of hormones like carbachol, vasopressin, angiotensin-II, or insulin to various cell types results in the inhibition of cAMP-mediated Mg^{2+} extrusion and/or the reversal of Mg^{2+} extrusion into Mg^{2+} accumulation [32,189]. The list of cells that respond to hormonal stimulation by accumulating Mg^{2+} is quite long (see [29] for a list), and includes cardiac myocytes [32,189], smooth muscle cells [208], hepatocytes [34,209], platelets [210], lymphocytes [211], fibroblasts [212], and pancreatic β cells [198]. In addition to inhibiting cAMP production, several of the hormones indicated above activate protein kinase C (PKC) as part of their cellular signaling. Evidence supporting a role of PKC in mediating Mg^{2+} accumulation has been provided by several laboratories. Maguire and collaborators have reported that administration of phorbol-myristate acetate (PMA), which directly activates PKC, elicits a marked accumulation of Mg^{2+} in S49 lymphoma cells [180]. A similar effect of PMA has been reported in thymocytes [213], cardiac myocytes [209] and hepatocytes [209].

Furthermore, our group has reported that down-regulation of PKC by exposure to a large dose of PMA for 3 hours completely abolishes the ability of cardiac and liver cells to accumulate Mg^{2+} while leaving unaffected the responsiveness of these cells to adrenergic agonists [209]. A similar inhibition of Mg^{2+} accumulation has been observed following treatment of cells with the PKC inhibitors calphostin [208] or staurosporine [214]. Alteration in PKC distribution and activity associated with a defective accumulation of Mg^{2+} have been observed in arterial smooth muscle cells [215] and hepatocytes [216] isolated from animals exposed to alcohol, or in liver cells of diabetic animals [217].

Protein kinase C activation is only part of the integral response of hormones like angiotensin-II or vasopressin. The interaction of these hormones with their receptor, in fact, activates phospholipase C which, in turn hydrolyses PIP2 to generate diacyl-glycerol (DAG) and IP3. These two molecules would then activate PKC and IP3 receptor in the ER, respectively. Activation of IP3 receptor results in a marked but transient increase in cytosolic Ca^{2+} followed by a more sustained Ca^{2+} entry across the plasma membrane through store-operated channels (SOC). Thus, Ca^{2+} signaling is an integral component of the cellular response elicited by these hormones. Yet, the contribution of Ca^{2+} increase and signaling in mediating Mg^{2+} accumulation is poorly defined. Liver cells loaded with Bapta-AM, which chelates cytosolic Ca^{2+}, are unable to extrude and accumulate Mg^{2+} following stimulation by phenylephrine and PMA, respectively [130]. Administration of thapsigargin, which inhibits the SRCA pumps and increases cytosolic Ca^{2+} by favoring its release from the endoplasmic reticulum, also prevents Mg^{2+} accumulation [130] and actually induces a Mg^{2+} extrusion from the liver cell if applied for more than 3-5 min [130,131]. Because of the different time-scale and amplitude of the changes in cellular Ca^{2+} and Mg^{2+} content [130], it is difficult to properly correlate these variations. Cytosolic free Ca^{2+} transiently increases several orders of magnitude above its resting level. In contrast, cytosolic free Mg^{2+}, which is already in the millimolar or sub-millimolar range, increases by 10% to 15% [37] at the most, although in absolute terms this amount far exceeds the overall change in cytosolic Ca^{2+} mass.

An unresolved point of inconsistency in the role of Ca^{2+} and PKC signaling in regulating Mg^{2+} accumulation is provided by the reports that the administration of phenylephrine, which activates PKC signaling in addition to inositol 1,4,5-trisphosphate and Ca^{2+} signaling, does not elicit Mg^{2+} accumulation but induces a Mg^{2+} extrusion from liver cells [132]. These results raise the question as to what modulates the different cellular response to the administration of phenylephrine or vasopressin. One possibility could be that different PKC isoforms are activated under one condition but not the other. For example, hepatocytes possess 3 classical and at least 2 novel PKC isoforms [217]. Thus, it is possible that one isoform (or class of isoforms) is involved in mediating Mg^{2+} accumulation while another isoform (or class of isoforms) is involved in modulating Mg^{2+} extrusion. Consistent with this hypothesis, recent data from our laboratory suggests that PKCε is essential for Mg^{2+} accumulation to occur [216]. Under conditions in which the expression of this isoform is inhibited by antisense, or its translocation to the cell membrane is prevented, for example, by ethanol administration, no Mg^{2+} accumulation is observed in liver cells [216]. Interestingly, this PKC isoform has the highest affinity for Mg^{2+} among all PKC isoenzymes, with a K_m ~1 mM [218], close to the physiological free $[Mg^{2+}]_i$ measured in the cytoplasm of the hepatocyte [37,219] and other mammalian cells as well [208]. Although the mechanism ultimately responsible for the accumulation of Mg^{2+} within the hepatocyte has not been identified, it is worth considering the recent observation by Bindels and collaborators that in the absence of PKC activation or following RACK1 over-expression, RACK1 can bind to TRPM6, and possibly TRPM7, at the level of the kinase domain and inhibit the channel activity [118].

4.2.2 Role of MAPKs

Several lines of evidence indicate that additional signaling pathways (e.g., MAPKs) are involved in determining differing cellular responses under seemingly similar stimulatory conditions. Reports from Altura and collaborators in arterial muscle cells [220], Touyz's laboratory in vascular smooth muscle cells [221], and our group in liver cells [222] indicate that pharmacological inhibition of ERK1/2 and p38 MAPKs abolishes PKC-mediated Mg^{2+} accumulation [222]. In addition, inhibition of MAPKs signaling hampers Mg^{2+} accumulation and affects cyclin activity in vascular smooth muscle cells [221], preventing the cells from progressing in the cell cycle [221]. This effect may occur via changes in nuclear functions directly regulated by Mg^{2+}, as proposed by Rubin [223], and/or changes in nuclear signaling by ERK2, which depends on Mg^{2+} level to properly dimerize, translocate, and activate specific nuclear targets [224]. The role of ERK1/2 in regulating Mg^{2+} homeostasis is further emphasized by the evidence that increased ERK1/2 phosphorylation and TRPM6 expression have been observed following EGF administration to renal epithelial cells [94,95]. The role of MAPKs in Mg^{2+} homeostasis, however, is far from clear as ERK1/2 appears to be involved in mediating also Mg^{2+} extrusion [220,225].

4.2.3 Role of the Epidermal Growth Factor

EGF is also implicated in regulating Mg^{2+} accumulation, at least in kidney cells. The administration of EGF controls TRPM6 channel expression and operation in the apical domain of renal epithelial cells to promote Mg^{2+} accumulation [94,95,226]. Point mutations in the EGF sequence limit TRPM6 functioning and Mg^{2+} accumulation within the cells [227]. The modulation of TRPM6 expression appears to occur via ERK1/2 signaling coupled to activating protein-1 (AP-1) [95]. Indirect evidence that EGF regulates Mg^{2+} homeostasis is provided by the observation that antibodies against EGF used in several forms of colon cancer [92,93,228] induce Mg^{2+} wasting and hypomagnesemia.

5 Serum Mg^{2+} Level and Mg^{2+}-Sensing Mechanism

Humans and many mammals present a circulating Mg^{2+} level of ~1.2–1.4 mEq/L (~0.8 mM) [19,229]. Clinical and experimental evidence indicates that serum Mg^{2+} level decreases in humans and animals in several chronic diseases [197]. Yet, there is a remarkable lack of information as to whether serum Mg^{2+} undergoes circadian fluctuations following hormonal or non-hormonal stimuli (e.g., fasting or exercise). The infusion of catecholamine [229–231] or isoproterenol [177,187,232] results in a marked dose- and time-dependent increase in circulating Mg^{2+} content. This increase is maximal within 20 min from the agent administration [187], remaining unchanged for up to 2 hours following the removal of the agonist [187]. Considering this time frame of changes, the pre-infusion level of serum Mg^{2+}, the glomerular filtration rate (1.62 mL/min), and the fractional excretion (17%) [233], it is evident that the increase in serum Mg^{2+} level is independent of the hemodynamic changes elicited by the β-adrenergic agonist [187] and renal excretion [233]. Consistent with the whole body distribution of β_2 versus β_1 adrenergic receptors [234,235], the increase in serum Mg^{2+} occurs can be mimicked by specific β_2-adreno-ceptor agonist and inhibited by specific β_2-blocker [187]. The amplitude of the increase in circulating Mg^{2+} level suggests that the adrenergic agonist mobilizes Mg^{2+} from various tissues [187], including bone [188]. The latter hypothesis is supported by the observation that the infusion of carbonic anhydrase inhibitor prevents the increase in serum Mg^{2+} level elicited by isoproterenol administration in anesthetized rats [188]. It is interesting to note that the hormones that increase plasma Mg^{2+} by mobilizing the cation from different organs or tissues are also responsible for increasing Mg^{2+} reabsorption in the Henle's loop, thus preventing a net Mg^{2+} loss.

Presently, no specific Mg^{2+}-sensing mechanism in the circulation has been identified. However, the Ca^{2+}-sensing receptor [236] can detect changes in circulating Mg^{2+} level in a range of concentrations higher than those of Ca^{2+} [237] and consistent with the increase in serum Mg^{2+} levels reported in the literature [187,188]. The observation that in cells of the distal convoluted tubule (MDCT) of the mouse the Ca^{2+}-sensing receptor can be activated by extracellular Ca^{2+} and Mg^{2+} with comparable sensitivity [238] suggests interesting hypotheses in terms of whole body physiology. The activation of this sensor mechanism would inhibit glucagon- or vasopressin-mediated Mg^{2+} accumulation into the cells [239] and favor its urinary elimination, possibly explaining the clinical and experimental evidence that hypomagnesemia and hypocalcemia inhibit hormone-stimulated cAMP-mediated reabsorption of Mg^{2+} and Ca^{2+} along the different segments of the nephron [240]. In addition, the Ca^{2+} sensing receptor would represent a distal regulatory mechanism to restore magnesemia to a physiological level following the increase observed in anesthetized animals infused with adrenergic agonists [187,188]. It is still an open question as to whether this sensing mechanism or associated modulating components are altered under diabetic conditions in which a significant loss of tissue Mg^{2+} content and increased magnesuria are observed.

At variance of hypercalcemia, which is associated with muscle weakness and arrhythmia, an increase in serum Mg^{2+} level appears to be well tolerated under *in vivo* conditions. Rats infused with boluses of Mg^{2+} that increase serum Mg^{2+} level by 50% do not exhibit significant systemic hemodynamic changes but show a marked increase in coronary artery flow [241]. Baboons infused with pharmacological doses of Mg^{2+} sufficient to prevent epinephrine-induced cardiac arrhythmias show a significant attenuation of epinephrine-induced increase in mean arterial pressure and systemic vascular resistance [242]. It would appear, therefore, that an increase in extracellular Mg^{2+} concentration regulates catecholamine release from peripheral and adrenal sources [243] and consequently cardiac contractility [184]. Taken together, these observations suggest that an increase in serum Mg^{2+} level following adrenergic stimulation can: (i) act as a feed-back mechanism to modulate catecholamine release and activity, and (ii) contribute to improved blood flow and O_2 delivery to the heart and possibly other tissues at a time when an increase in energy production is expected.

The presence of a Mg^{2+} *sensor* at the cell level is also debated. The presence and operation of such a *sensor* mechanism is supported by several lines of evidence. First, prolonged exposure to 0 mM $[Mg^{2+}]_0$ decreases cytosolic *free* Mg^{2+} concentration by approximately 50% in cardiac ventricular myocytes [244], MDKC [245], or MDCT cells [246]. This reduced cytosolic Mg^{2+} level is maintained as long as the cells are incubated in the presence of 0 mM $[Mg^{2+}]_0$, but returns to normal level as soon as $[Mg^{2+}]_0$ is increased in a time-frame that is directly proportional to the extracellular Mg^{2+} concentration utilized [244–246]. The presence of L-type Ca^{2+}-channel inhibitors (e.g., verapamil or nifedipine) or La^{3+} in the extracellular milieu prevents the restoration of Mg^{2+} level [244]. The concomitant absence of significant changes in cytosolic $[Ca^{2+}]_i$ excludes that Ca^{2+} may act as a regulatory mechanism, and suggests a direct effect of these inhibitory agents on the Mg^{2+} entry mechanism [244]. As TRPM7 operation is affected by gadolinium [54], these results anticipate the presence and operation of the Mg^{2+}-specific channels TRPM7 [42] and TRPM6 [43].

A second line of evidence for the presence of a Mg^{2+} sensor in eukaryotic cells is provided by the occurrence of Mg^{2+} extrusion in all the conditions in which cellular ATP decreases as a result of chemical hypoxia [201], or exposure to fructose [202], ethanol [203] or cyanide [153]. Under all these conditions, Mg^{2+} extrusion only occurs when extracellular Na^+ is available to be exchanged for cellular Mg^{2+}. In the absence of external Na^+, almost no Mg^{2+} extrusion occurs [153,203,247], and a significant increase in cytosolic Mg^{2+} can be detected [153,247]. Hence, a scenario can be envisage whereby release of Mg^{2+} from cellular organelles or from binding moieties such as ATP [153,203] results in an increase in cytosolic Mg^{2+} content that is detected by the sensor which, in turn, activates the Mg^{2+} extrusion mechanism. The nature of this sensor is still undefined. Because almost all the metabolic conditions mentioned above are characterized by changes in the ratio between reduced and oxidized pyrimidine nucleotide levels (e.g., ethanol [203]), it is an appealing albeit unproved hypothesis that the concentrations of these nucleotides (or their ratio) act as a Mg^{2+} sensor in eukaryotic cells.

Similarly to Mg^{2+} extrusion, cellular Mg^{2+} accumulation also requires proper ion distribution, especially phosphate [246] and potassium [248], across the cell membrane. The role of potassium is of particular relevance as it suggests that Mg^{2+} is accumulated for charge compensation as the result of changes in membrane potential [249–251]. Especially in polarized epithelia (e.g., nephron and intestine) Mg^{2+} entry mechanisms such as TRPM7 and TRPM6 are located on the apical side, counterbalancing the operation of the Na^+/Mg^{2+} exchanger and the Na^+/K^+-ATPase on the basolateral domain of the cell (see [118] for a review). Whether the effect on K^+ occurs through changes in membrane potential, or indirectly via a reduced operation of the Na^+/K^+-ATPase coupled to the operation in reverse of the Na^+/Mg^{2+} exchanger [252] is topic for future investigation. In the particular case of K^+, it has also to be noted that pathological conditions characterized by a marked decrease in tissue

Mg^{2+} content (e.g., diabetes, [197]) are also characterized by an inability of the tissue to properly transport potassium [253,254]. This effect is the direct result of insulin absence or ineffectiveness coupled to a reduced activity rate of the Na^{+}/K^{+}-ATPase. It remains to be determined as to whether changes in pyrimidine nucleotide levels (or ratio), this time in an opposite direction, promote Mg^{2+} accumulation.

6 Physiological Role of Intracellular Mg^{2+}

One of the conclusions generated by the data presented in the previous sections is that Mg^{2+} acts as an indispensable regulatory cation for enzymes, phosphometabolites, and channels [1,255]. Several glycolytic enzymes, including hexokinase, phosphofructokinase, phosphoglycerate mutase, phosphoglycerate kinase, enolase and pyruvate kinase, show activation at low, and inhibition at high Mg^{2+} concentrations [199,200]. Adenylyl cyclase represents the best example of an enzyme directly regulated by Mg^{2+}. As suggested by Maguire's data (reviewed in [256]), Mg^{2+} exerts this effect by acting at two different sites: one site is on the guanine nucleotide coupling protein, where it regulates agonist affinity as well as the interaction with the catalytic subunit. The second site is on the catalytic subunit and regulates the activity of this subunit.

The regulation of adenylyl cyclase and other cellular enzymes (such as those involved in glucose homeostasis [199,200]) occurs at Mg^{2+} concentrations between 0.5 to 1 mM, which are well within the fluctuations in free $[Mg^{2+}]_i$ measured in the cytoplasm of various cells including hepatocyte [219]. With the exception of the glycolytic enzymes, however, studies attempting to evidence *in vitro* or *in situ* a regulatory role of Mg^{2+} for cytosolic enzymes have been disappointing, mostly because of the underlying assumption that Mg^{2+} would operate as Ca^{2+} in modulating enzyme activity. While Ca^{2+} presents a major concentration gradient between cytoplasm and extracellular space and between cytoplasm and endoplasmic (and sarcoplasmic) reticulum lumen, free Mg^{2+} concentrations in the cytoplasm and the extracellular fluid are very similar, both being in the millimolar or sub-millimolar range. Consequently, an increase or a decrease in cytosolic Mg^{2+} level of an amplitude equivalent to those observed for Ca^{2+} will remain largely undetected by fluorescent or ^{31}P NMR techniques. Heretofore, a role of Mg^{2+} as transient regulator of cytosolic enzymes appears to be unlikely. It has to be noted that even under conditions in which hormonal and non-hormonal stimuli elicit major fluxes of Mg^{2+} across the cell plasma membrane in either direction, massive translocations of Mg^{2+} that increase or decrease total cellular Mg^{2+} content by 1–2 mM (equivalent to 5%–10% of the total cell Mg^{2+} content) result in limited or no changes in cytosolic $[Mg^{2+}]_i$ [189,201]. This disconnection can be explained by assuming that the source or destination of the transported Mg^{2+} is a cellular organelle, or a major binding site, or that Mg^{2+} is rapidly buffered by phosphonucleotides, phospholipids, or G proteins. Therefore, regulation of cellular functions by Mg^{2+} should not be necessarily expected to occur in the cytosol, like for Ca^{2+}, but within organelles and plasma where Mg^{2+} concentration can rapidly increase or decrease by more than 20% [187,188].

The following pages will highlight what is known about the regulatory effect of extracellular or intracellular Mg^{2+} on cation channels activity at the plasma membrane level, as well as on mitochondria respiration and integrity following changes in Mg^{2+} concentration within the organelle.

6.1 Ca^{2+} and K^+ Channels

White and Hartzell were the first to report a regulatory effect of intracellular free Mg^{2+} on calcium channels [257]. These authors observed that increasing intracellular free $[Mg^{2+}]_i$ from 0.3 to 3.0 mM by internal perfusion in cardiac ventricular myocytes resulted in a small decrease of basal L-type Ca^{2+} channel current (I_{Ca}) while it decreased by more than 50% the cAMP-mediated enhancement in I_{Ca} amplitude [257]. This effect was due to a direct action of Mg^{2+} on the phosphorylated channel or on the dephosphorylation rate of the channel rather than to changes in cAMP concentration or cAMP-dependent phosphorylation [257]. Similar results were reported in guinea pig cardiac myocytes by Agus and Morad who observed a Mg^{2+}-induced block on Ca^{2+} current by direct effect on the inactivation state of the channel [258]. The block persisted in the presence of cAMP, and was not reversed by elevation of extracellular Ca^{2+} concentration or addition of catecholamine [258]. Similar effects of Mg^{2+} on Ca^{2+} channels have been observed in vascular smooth muscle cells and endothelial cells from human placenta [259], in which $MgCl_2$ (but also $MgSO_4$) acts at an extracellular site of the voltage-gated Ca^{2+} channels, and on T-type Ca^{2+}-channels [260]. Recent evidence by Catterall and his group proposes a modulating effect of Mg^{2+} on the EF-hand motif located in the C-terminus of $Ca_v1.2$ channels [261].

Additional Ca^{2+} channels modulated by extracellular Mg^{2+} are the store-operated Ca^{2+} channels (SOC) and the store-operated calcium release-activated Ca^{2+} (CRAC) channels. In the case of SOCs, Mg^{2+} prevents or reverses the vasoconstriction elicited by phenylephrine administration but not that induced by K^+ depolarization [262]. This observation would suggest that Mg^{2+} contributes to regulate both the myogenic tone and the α_1-adrenoceptor-induced, Ca^{2+}-mediated vasoconstriction occurring through SOCs. This effect on the vasculature could be lost to a significant extent under hypertensive conditions, in which a decrease in plasma Mg^{2+} and a vasoconstriction hypertone have been observed.

As for CRACs, the effect of Mg^{2+} is more at the intracellular level [263]. CRAC channels are highly selective for Ca^{2+} under physiological conditions whereas removal of extracellular divalent cations makes them freely permeable to monovalent cations, in particular Na^+. Experimental evidence indicates that intracellular Mg^{2+} can modulate the activity and selectivity of these channels therefore affecting monovalent cation permeability. A report by Prakriya and Lewis [263], however, argues that the channels modulated by intracellular Mg^{2+} are not CRAC channels, but a different class of channels that open when Mg^{2+} is washed out of the cytosol. These channels have been termed Mg^{2+}-inhibited cation (MIC) channels, and could be distinguished by CRAC channels based upon modality of inhibition, regulation, ion permeation and selectivity [262]. These results, however, do not exclude the possibility of an inhibitory effect of intracellular Mg^{2+} on CRAC channels.

Potassium channels are also targets for Mg^{2+}. Matsuda [264] has reported that cytosolic Mg^{2+} blocks the outward currents of inwardly rectifying K^+ channels without affecting the inward currents. However, the Mg^{2+} block is achieved at a half-saturating concentration of 1.7 μM, a concentration far from the physiological Mg^{2+} level in the cytoplasm. Hence, it is difficult to envision the occurrence of a similar regulatory effect under normal conditions without invoking Mg^{2+} micro compartmentalization. More realistic would be the occurrence of a regulatory role of intracellular Mg^{2+} on voltage-regulated potassium channels (K_v channels) in vascular smooth muscle cells [265]. In this case, in fact, an increase in intracellular Mg^{2+} – in a range of concentrations consistent with its physiological variations – slows down the kinetic of activation of the K_v channel, causing also inward rectification at positive membrane potentials and a shift in voltage-dependent inactivation [265]. Intracellular Mg^{2+} also modulates large-conductance (BK-type) Ca^{2+}-dependent K^+ channels either by blocking the pore of BK channels in a voltage-dependent manner, or by activating the channels independently of changes in Ca^{2+} and voltage through preferential binding to the channel open conformation at a site different from Ca^{2+} sites. Interestingly, Mg^{2+} may also bind to Ca^{2+} sites and competitively inhibit Ca^{2+}-dependent activation [266].

The inhibitory effect of Mg^{2+} is not restricted to channels in the cell membrane. Experimental evidence by Bednarczyk et al. [267] indicates that Mg^{2+} within the mitochondrial matrix can modulate gating and conductance of mitochondrial K_{ATP} channels, which play a key role in promoting mitochondrial recovery and cell survival under ischemia/reperfusion conditions.

6.2 *Mitochondrial Dehydrogenases*

Mitochondria represent one of the major cellular Mg^{2+} pools. The concentration of Mg^{2+} within the organelle ranges between 14 to 16 mM [268], and circumstantial evidence from this [269] and other laboratories [247,270,271] suggests that Mg^{2+} can be mobilized from mitochondria under various conditions including hormonal stimulation through a not fully elucidated mechanism. Regulation of mitochondrial Mg^{2+} homeostasis has been analyzed in detail in several recent reviews [16,20,268], and we direct the interested reader to them for further information. In this section, we will focus on the role of intra- and extra-mitochondrial Mg^{2+} in modulating the activity of specific proteins within the organelle.

It is commonly accepted that changes in matrix Ca^{2+} can affect the activity rate of mitochondrial dehydrogenases and consequently the respiration rate [272,273]. Experimental evidence supports a similar role for Mg^{2+} as the activity of several mitochondrial dehydrogenases has been observed to increase within minutes from the application of hormonal or metabolic stimuli in the absence of a detectable increase in mitochondrial Ca^{2+} [274,275]. In particular, the results indicate that a decrease in mitochondrial Mg^{2+} increases several fold the activity of succinate and glutamate dehydrogenases while leaving unaffected the activity of α-ketoglutarate dehydrogenase and pyruvate dehydrogenase [114,115]. This evidence would support the concept that changes in matrix Mg^{2+} content (in combination with, or in alternative to changes in mitochondrial Ca^{2+}) can control mitochondrial respiration, at least under well defined conditions. In this respect, mitochondrial Mg^{2+} content appears to change quite significantly during transition from state 3 to state 4 [276], affecting the amplitude of mitochondria respiration. In addition, data from our laboratory [269], from Zhang and Melvin [271], and Kubota et al. [247] all suggest that catecholamine stimulation can mobilize mitochondrial Mg^{2+} via a direct effect of cAMP on mitochondria. Hence, catecholamine administration will enhance mitochondrial respiration via cAMP-mediated modulation of mitochondrial Mg^{2+}, which, in turn, will directly stimulate succinate and glutamate dehydrogenases while sensitizing other dehydrogenases to changes in mitochondrial Ca^{2+} concentrations.

Additional mitochondrial function modulated by changes in Mg^{2+} within the organelle are anion channels present in the mitochondrial membrane [277] as well as the opening of the permeability transition pore [278]. The mitochondrial inner membrane anion channel (IMAC) transports various anions, and is involved in regulating the organelle volume in conjunction with the K^+/H^+ antiporter. Although its fine regulation is not fully elucidated as yet, experimental evidence suggests that matrix Mg^{2+} and protons maintain the channel in its closed state [277]. Kinetic studies by Beavis and Powers support a main role of Mg^{2+} in maintaining the channel in a conformation that would allow fine modulation by small changes in pH and proton distribution under physiological conditions [277]. The end results will be the maintenance of an optimal proton gradient and $\Delta\psi$ across the mitochondrial membrane, essential to retain proper organelle function and intra-mitochondrial Mg^{2+} content [112].

Perturbance of mitochondrial $\Delta\psi$, Ca^{2+} content or ATP level all result in the opening of the permeability transition pore (PTP) in the inner mitochondrial membrane [278] and the rapid re-equilibration of intra-mitochondrial ions and solutes down their concentration gradient. While it is well established that an increase in mitochondrial Ca^{2+} content facilitates PTP opening, an increase in mitochondrial Mg^{2+} antagonizes it. This effect can be appreciated well in yeasts, which do not possess a canonical PTP [279]. Creatine kinase also regulates PTP opening by tightly associating to the mitochondrial membrane and remaining in an active state [280]. Both the binding and activity state of the protein are Mg^{2+}-dependent, and removal of Mg^{2+} from the extra-mitochondrial environment results in a decline in creatine kinase activity and PTP opening [280].

Hence, it appears that Mg^{2+} regulates volume, ion composition, and ATP production within the mitochondrion, modulating the metabolic interaction between the organelle and the hosting cell.

6.3 *Reticular Glucose 6-Phosphatase*

The endoplasmic reticulum (ER) represents another major Mg^{2+} pool within the cell, with a total concentration estimated to be between 14 to 18 mM [1]. Yet, no information is available about the modality by which Mg^{2+} ions enter and exit the organelle and how it is buffered within the ER lumen. Limited information is also available about any major role of luminal Mg^{2+} on reticular functions other than protein synthesis [223].

Work by Volpe and collaborators [281,282], Gusev and Niggli [283], and Laver and Honen [284] suggests that cytosolic and perhaps luminal Mg^{2+} concentrations have a major effect in limiting Ca^{2+} uptake into the ER/SR and its release from the organelle via IP3 [282] and ryanodine receptor (RyR) [284]. While a direct effect of Mg^{2+} on RyR opening has been observed [283,284], it is unclear whether a similar effect takes place on the IP_3 receptor.

Recently, our laboratory has reported that cytosolic Mg^{2+} can have a regulatory role on the activity of reticular glucose 6-phosphatase (G6Pase) in liver cells [285]. This effect is biphasic, with an optimal stimulatory effect at ~0.5 mM $[Mg^{2+}]_i$ and an inhibitory effect at higher Mg^{2+} concentrations [285]. The Mg^{2+} effect appears to be at the level of the glucose 6-phosphate (G6Pi) transport component of the G6Pase enzymatic complex in that it is abolished by EDTA (as Mg^{2+} chelating agent) or taurocholic acid, which permeabilizes the ER membrane allowing for the direct delivery of G6Pi to the catalytic site of the G6Pase within the ER lumen bypassing the transport mechanism [285]. This effect of Mg^{2+} on G6Pase hydrolysis rate also occurs in purified microsomes [286] isolated from livers of animals exposed for 2 weeks to a Mg^{2+}-deficient diet [286]. Also in microsomes, the G6Pi hydrolysis rate is dynamically decreased by addition of Mg^{2+} at a concentration similar to that reported to be present in the hepatocyte cytoplasm, or increased by EDTA addition [286]. It is presently undetermined whether Mg^{2+} exerts a similar modulating effect on other reticular enzymatic activities.

6.4 *Cell pH and Volume*

Cells exposed to cyanide [153], fructose [202], hypoxia, [201,204], ethanol [203], or choline chloride [130] undergo a marked cellular acidification, decrease in cellular ATP content, and a large Mg^{2+} extrusion. This extrusion is the consequence of a decrease in buffering capacity (ATP loss) and binding affinity within the cytoplasm. Recently, Yamaguchi and Ishikawa [287] reported that a cytosolic $[Mg^{2+}]_i$ of ~1 mM (a physiological Mg^{2+} concentration measured in the cytosol of various cells [219,255]), inhibits by ~50% the current generated by the electrogenic Na^+-HCO_3^- cotransporter NBCe1-B. Increasing the *free* Mg^{2+} concentration to 3 mM completely abolishes NBCe1-B current. This regulatory effect is exerted by Mg^{2+} and not

$Mg(ATP)^{2-}$, and occurs at the N-terminus of the transporter [287]. It is still unresolved whether Mg^{2+} binds the N-terminus of the transporter directly or exerts its effects via an intermediate, Mg^{2+}-modulated regulatory protein [287].

On the other hand, increasing cellular Mg^{2+} content has a stimulatory role on the expression of aquaporin 3 in CaCo-3 cells [288]. This isoform of aquaporin is highly expressed in the gastrointestinal tract, in which it absorbs water, glycerol, and urea. The effect of Mg^{2+} on aquaporin mRNA expression appears to involve cAMP/PKA/CREB signaling, as well as MEK1/2 and MSK1 [288], suggesting the occurrence of both short- and long-term regulation on the protein activity and expression. As aquaporin 3 is highly expressed in brain, erythrocytes, kidney, and skin, in addition to the gastrointestinal tract, the occurrence of a modulating effect of Mg^{2+} on aquaporin 3 expression in these tissues may be highly relevant for various physiological and pathological conditions including brain swelling following traumatic injury. It remains to be determined whether Mg^{2+} exerts a similar regulatory role on other aquaporin isoforms.

Taken together, these two sets of information emphasize a role of Mg^{2+} in regulating directly pH, volume, and cation concentration, especially Na^+ within the cell, and indirectly fatty acid metabolism via aquaporin 3-mediated glycerol accumulation.

6.5 Cell Cycle

Cell cycle [221,289,290], cell proliferation [291], and cell differentiation [292–294] have all been associated with the maintenance of an optimal Mg^{2+} level. Under conditions in which cellular Mg^{2+} accessibility is restricted or reduced, cell proliferation and cell cycle progression are markedly impaired as is cell differentiation [292–294]. The mechanisms by which a decrease in cellular Mg^{2+} content affects these cellular processes revolve around defective MAPKs [221] and p27 [290] signaling, increased oxidative stress level [292], and decreased $Mg(ATP)^{2-}$ levels [223,294]. Because the cellular $Mg(ATP)^{2-}$ level is at a level optimal for protein synthesis [223], any alteration in this metabolic parameter will have major repercussion on the proper functioning of the cell. In addition, extracellular Mg^{2+} levels regulate integrin signaling, *de facto* modulating the interaction among cells and between cells and extracellular matrix [295]. All together, these observations support the notion that an optimal Mg^{2+} level is essential to guarantee cell cycle progression and retention of proper cell morphology and function, and prevent the undesired progression towards cell death or neoplastic destiny [296].

7 Conclusions

In the last few years, our understanding of the mechanisms regulating cellular and whole body Mg^{2+} homeostasis has advanced significantly. Although in terms of overall understanding the field still lags behind the knowledge available for other ions such as Ca^{2+}, H^+, K^+ or Na^+, the identification of Mg^{2+} channels and transport mechanisms in the membrane of cells and cellular organelles, and a better comprehension of the various signaling pathways and conditions regulating Mg^{2+} transport are providing new tools to address essential questions about the relevance of Mg^{2+} for various cell functions under physiological and pathological conditions.

Abbreviations

$[Mg^{2+}]_0$	extracellular Mg^{2+} concentration
$[Mg^{2+}]_i$	cytosolic free Mg^{2+} concentration
2,3-DPG	2,3-diphosphoglycerate
2-APB	2-2-aminoethyldiphenyl-borate
AC	adenylyl cyclase
ACDP2	ancient conversed domain protein 2
ADP	adenosine 5'-diphosphate
AMP	adenosine monophosphate
AP-1	adaptin protein 1
ATP	adenosine 5'-triphosphate
Bapta-AM	1,2-bis-(o-Aminophenoxy)-ethane-N,N,N',N'-tetraacetic acid, tetraacetoxymethyl ester
BK channel	big potassium channel
CaM	calmodulin
cAMP	cyclic-adenosine monophosphate
CaSR	calcium-sensing receptor
CHAK	channel kinase
CRAC channel	calcium release-activated Ca^{2+} channel
CREB	cAMP response element binding
DAG	diacyl glycerol
DHHC	Asp-His-His-Cys motif
DIDS	4,4'-diisothiocyano-2,2'-stilbenedisulfonic acid
DNDS	4,4'-dinitrostilbene-2,2'-disulfonic acid
EDTA	ethylenediamine-N,N,N',N'-tetraacetate
eEF2-k	eEF2 cognate kinase
EGF	epidermal growth factor
ER	endoplasmic reticulum
ERK	extracellular responsive kinase
EXPMA	electron probe X-ray microanalysis
FHHNC	familial hypomagenesemia with hypercalciuria and nephrocalcinosis
G6Pase	glucose 6-phosphatase
G6Pi	glucose 6-phosphate
GTP	guanosine 5'-triphosphate
HIP	huntingtin-interacting protein
HSH	hypomagnesemia with secondary hypocalemia

HSP	hereditary spastic paraplegia
IC_{50}	inhibitory dose achieving 50% inhibition
IMAC	inner membrane anion channel
IP3	1,4,5, inositol-trisphophate
IRH	isolated recessive renal hypomagnesemia
Mag-Fura	5-oxazolecarboxylic acid, 2-[5-[2-[(acetyloxy)methoxy]-2-oxoethoxy]-6-[bis[2-[(acetyloxy)methoxy]-2-oxoethyl]amino]-2-benzofuranyl]-(acetyloxy)methyl ester
MagT1	magnesium transporter 1
MAPK	mitogen-activated protein kinase 1
MDCT cells	mouse distal convoluted tubule cells
MDKC cells	Madin-Darby canine kidney cells
MEK	MAP (erk1/2) kinase kinase
MIC channel	Mg^{2+}-inhibited cation channel
MMgT	membrane Mg^{2+} transporter
MSK	musculoskeletal
NIPA	non-imprinted in Prader-Willi/Angelman syndrome
OMIM	Online Mendelian Inheritance in Man
OST	oligosaccharide transferase
PCLN-1	paracellin-1 = claudin 16
PG1	polygalacturonase 1
PI3K	phosphoinositide-3-kinase
PIP2	phosphatidyl-inositol bisphohate
PKA	protein kinase A
PKC	protein kinase C
PLC	phospholipase C
PMA	phorbol-myristate acetate
PTP	permeability transition pore
Q10	Q10 temperature coefficient
RACK	receptor for activated protein kinase C
REA	repressor of estrogen receptor activity
RyR	ryanodine receptor
SCN	sodium cyanide
SDS	sodium dodecyl sulfate
siRNA	short inhibitory RNA molecule
SLC41	solute carrier family 41
SOC	store-operated channel
SR	sarcoplasmic reticulum
SERCA	sarcoplasmic-endoplasmic reticulum calcium pump
TRPM6	transient receptor potential melastatin subfamily, isoform 6
TRPM7	transient receptor potential melastatin subfamily, isoform 7

Acknowledgments This work was supported by grant AA-11593.

References

1. A. Romani, A. Scarpa, *Arch. Biochem. Biophys.* **1992**, *298*, 1–12.
2. F. I. Wolf, A. Torsello, S. Fasanella, A. Cittadini, *Mol. Asp. Med.* **2003**, *24*, 11–26.
3. F. I. Wolf, A. Cittadini, *Mol. Asp. Med.* **2003**, *24*, 3–9.
4. D. W. Jung, L. Apel, G. P. Brierley, *Biochemistry* **1990**, *29*, 4121–4128.
5. G. A. Rutter, N. J. Osbaldeston, J. G. McCormack, R. M. Denton, *Biochem. J.* **1990**, *271*, 627–634.
6. A. V. Somlyo, G. McClellan, H. Gonzalez-Serratos, A. P. Somlyo, *J. Biol. Chem.* **1985**, *260*, 6801–6807.
7. A. M. Hofer, T. E. Machen, *Proc. Natl. Acad. Sci. USA* **1993**, *90*, 2598–2602.
8. A. Scarpa, F. J. Brinley, *Fed. Proc.* **1981**, *40*, 2646–252.
9. D. Lüthi, D. Günzel, J.A. McGuigan, *Exp. Physiol.* **1999**, *84*, 231–252.
10. P. W. Flatman, V. L. Lew, *J. Physiol.* **1981**, *315*, 421–446.
11. T. Günther, J. Vormann, J. A. McGuigan, *Biochem. Mol. Biol. Int.* **1995**, *37*, 871–875.
12. J. E. Raftos, V. L. Lew, P. W. Flatman, *Eur. J. Biochem.* **1999**, *263*, 635–645.
13. S. Oki, M. Ikura, M. Zhang, *Biochemistry* **1997**, *36*, 4309–4316.
14. S. Wang, S. E. George, J. P. Davis, J. D. Johnson, *Biochemistry* **1998**, *37*, 14539–14544.
15. D. Allouche, J. Parello, Y. H. Sanejouand, *J. Mol. Biol.* **1999**, *285*, 857–873.
16. Y. Ogoma, H. Kobayashi, T. Fujii, Y. Kondo, A. Hachimori, T. Shimizu, M. Hatano, *Int. J. Biol. Macromol.* **1992**, *14*, 279–286.
17. K. Bogucka, L. Wojtczak, *Biochem. Biophys. Res. Commun.* **1971**, *44*, 1330–1337.
18. H. Belge, P. Gailly, B. Schwaller, J. Loffing, H. Debaix, E. Riveira-Munoz, R. Beauwens, J. P. Devogelaer, J. G. Hoenderop, R. J. Bindels, O. Devuyst, *Proc. Natl. Acad. Sci. USA* **2007**, *104*, 14849–14854.
19. Geigy Scientific Tables. Ed C. Lentner, Ciba-Geigy, Basel, Switzerland, 1984.
20. P. W. Flatman, *J. Membr. Biol.* **1984**, *80*, 1–14.
21. A. Romani, *Arch. Biochem. Biophys.* **2007**, *458*, 90–102.
22. P. I. Polimeni, E. Page, in *Recent Advances in Study in Cardiac Cells and Metabolism*, Ed N. S. Dhalle, Volume IV, University Park Press, Baltimore, 1974, pp. 217–232.
23. T. A. Rogers, F. L. Haven, P. E. Mahan, *J. Natl. Cancer. Inst.* **1960**, *25*, 887–888.
24. T. A. Rogers, *J. Cell. Comp. Physiol.* **1961**, *57*, 119–121.
25. F. I. Wolf, A. Di Francesco, V. Covacci, A. Cittadini, *Arch. Biochem. Biophys.* **1997**, *344*, 397–403.
26. M. E. Maguire, J. J. Erdos, *J. Biol. Chem.* **1978**, *253*, 6633–6636.
27. R. D. Grubbs, M. E. Maguire, *Magnesium* **1987**, *6*, 113–127.
28. A. M. Romani, M. E. Maguire, *Biometals* **2002**, *15*, 271–283.
29. A. Romani, A. Scarpa, *Front. Biosci.* **2000**, *5*, D720–D734.
30. T. Günther, J. Vormann, *Magnes. Trace Elem.* **1990**, *9*, 279–282.
31. T. Matsuura, Y. Kanayama, T. Inoue, T. Takeda, I. Morishima, *Biochim. Biophys. Acta* **1993**, *1220*, 31–36.
32. A. Romani, A. Scarpa, *Nature* **1990**, *346*, 841–844.
33. J. Vormann, T. Günther, *Magnesium* **1987**, *6*, 220–224.
34. A. Romani, A. Scarpa, *FEBS Lett.* **1990**, *269*, 37–40.
35. T. Günther, J. Vormann, V. Hollriegl, *Magnes. Bull.* **1991**, *13*, 122–124.
36. A. Jakob, J. Becker, G. Schottli, G. Fritzsch, *FEBS Lett.* **1989**, *246*, 127–130.
37. M. Fatholahi, K. Lanoue, A. Romani, A. Scarpa, *Arch. Biochem. Biophys.* **2000**, *374*, 395–401.
38. T. Kubota, Y. Shindo, K. Tokuno, H. Komatsu, H. Ogawa, S. Kudo, Y. Kitamura, K. Suzuki, K. Oka, *Biochim. Biophys. Acta* **2005**, *1744*, 19–28.
39. D. G. Kehres, C. H. Lawyer, M. E. Maguire, *Microb. Comp. Genomics* **1998**, *3*, 151–159.

40. M. B. Moncrief, M. E. Maguire, *J. Biol. Inorg. Chem.* **1999**, *4*, 523–527.
41. R. R. Preston, *Science* **1990**, *250*, 285–288.
42. M. J. Nadler, M. C. Hermosura, K. Inabe, A.-L. Perraud, Q. Zhu, A. J. Stokes, T. Kurosaki, J. P. Kinet, R. Penner, A. M. Scharenberg, A. Fleig, *Nature* **2001**, *411*, 590–595.
43. K. P. Schlingmann, S. Weber, M. Peters, N. L. Niemann, H. Vitzthum, K. Klingel, M. Kratz. E. Haddad, E. Ristoff, D. Dinour, M. Syrrou, S. Nielsen, M. Sassen, S. Waldegger, H. W. Seyberth, M. Konrad, *Nat. Genet.* **2002**, *31*, 166–170.
44. H. Yamaguchi, M. Matsushita, A. Nairn, J. Kuriyan, *Mol. Cell.* **2001**, *7*, 1047–1057.
45. L. V. Ryazanova, K. S. Pavur, A. N. Petrov, M. V. Dorovkov, A. G. Ryazanov, *Mol. Biol.* (Moskow) **2001**, *35*, 321–332.
46. A. G. Ryazanov, *FEBS Lett.* **2002**, *514*, 26–29.
47. L. W. Runnels, L. Yue, D. E. Clapham, *Science* **2001**, *291*, 1043–1047.
48. B. F. Bessac, A. Fleig, *J. Physiol.* **2007**, *582*, 1073–1086.
49. M. K. Monteilh-Zoller, M. C. Hermosura, M. J. Nadler, A. M. Scharenberg, R. Penner, A. Fleig, *J. Gen. Physiol.* **2003**, *121*, 49–60.
50. T. Voels, B. Nilliues, S. Hoefs, A. W. van der Kemp, G. Droogmans, R. J. Bindels, J. G. Hoenderop, *J. Biol. Chem.* **2004**, *279*, 19–25.
51. V. Chubanov, S. Waldegger, Y. Mederos, M. Schnitzler, H. Vitzthum, M.C. Sassen, H.W. Seyberth, M. Konrad, T. Gudermann, *Proc. Natl. Acad. Sci. USA* **2004**, *101*, 2894–2899.
52. C. Schmitz, M. V. Dorovkov, X. Zhao, B. J. Davenport, A. G. Ryazanov, A.-L. Perraud, *J. Biol. Chem.* **2005**, *280*, 37763–37771.
53. V. Chubanov, T. Gudermann, K. P. Schlingmann, *Pflugers. Arch.* **2005**, *451*, 228–234.
54. M. Li, J. Jiang, L. Yue, *J. Gen. Physiol.* **2006**, *127*, 525–537.
55. M. Li, J. Du, J. Jiang, W. Ratzan, L. T. Su, L.W. Runnels, L. Yue, *J. Biol. Chem.* **2007**, *282*, 25817–25830.
56. A. Gwanyanya, B. Amuzescu, S. I. Zakharov, R. Macianskiene, K .R. Sipido, V. M. Bolotina, J. Vereecke, K. Mubagwa, *J. Physiol.* **2004**, *559*, 761–776.
57. J. Jiang, M. Li, L. Yue, *J. Gen. Physiol.* **2005**, *126*, 137–150.
58. L. W. Runnels, L. Yue, D. E. Clapham, *Nat. Cell. Biol.* **2002**, *4*, 329–336.
59. M. Langeslag, K. Clark, W. H. Moolenaar, F. N. van Leeuwen, K. Jalink, *J. Biol. Chem.* **2007**, *282*, 232–239.
60. R. Macianskiene, A. Gwanyanya, J. Vereecke, K. Mubagwa, *Cell. Physiol. Biochem.* **2008**, *22*, 109–118.
61. A. Gwanyanya, K. R. Sipido, J. Vereecke, K. Mubagwa, *Am. J. Physiol.* **2006**, *291*, C627–C635.
62. R. M. Touyz, Y. He, A. C. I. Montezano, G. Yao, V. Chubanov, T. Gudermann, G. E. Callera, *Am. J. Physiol.* **2006**, *290*, R73–R78.
63. J. Sahni, A. M. Scharenberg, *Cell. Metab.* **2008**, *8*, 84–93.
64. J. Sahni, R. Tamura, I. R. Sweet, A. M. Scharenberg, *Cell Cycle* **2010**, y, 3565–3574.
65. J. Middelbeek, K. Clark, H. V. Venselaar, M. A. Huynen, F. N. van Leeuwen, *Cell. Mol. Life Sci.* **2010**, *67*, 875–890.
66. C. Schmitz, A.-L. Perraud, C. O. Johnson, K. Inabe, M. K. Smith, R. Penner, T. Kurosaki, A. Fleig, A. M. Scharenberg, *Cell* **2003**, *114*, 191–200.
67. K. Clark, M. Langeslag, B. van Leeuwen, L. Ran, A. G. Ryazanov, C. G. Figdor, W. H. Moolenaar, K. Jalink, F. N. van Leeuwen, *EMBO J.* **2006**, *25*, 290–301.
68. L. V. Ryazanova, L. J. Rondon, S. Zierler, Z. Hu, J. Galli, T. P. Yamaguchi, A. Mazur, A. Fleig, A. G. Ryazanov, *Nature Commun.* **2010**, *1*, 109.
69. T. E. Woudenberg-Vrenken, A. Sukinta, A. W. van der Kemp, R. J. Bindels, *Nephron Physiol.* **2011**, *117*, 11–19.
70. A.-L. Perraud, X. Zhao, A. G. Ryazanov, C. Schmitz, *Cell. Signal.* **2011**, *23*, 587–593.
71. K. Clark, J. Middelbeek, E. Lasonder, N. G. Dulyaninova, N. A. Morrice, A. G. Ryazanov, A. R. Bresnick, C. G. Figdor, F. N. van Leeuwen, *J. Mol. Biol.* **2008**, *378*, 790–803.
72. L.-T. Su, M.A. Agapito, M. Li, W. Simpson, A. Huttenlocher, R. Habas, L. Yue, L. W. Runnels, *J. Biol. Chem.* **2006**, *281*, 11260–11270.

73. M. V. Dorovkov, A. G. Ryazanov, *J. Biol. Chem.* **2004**, *279*, 50643–50646.
74. T. M. Paravicini, A. Yogi, A. Mazur, R. M. Touyz, *Hypertension* **2009**, *53*, 423–429.
75. M. Aarts, K. Iihara, W. L. Wei, Z. G. Xiong, M. Arundine, W. Cerwinski, J. F. MacDonald, M. Tymianski, *Cell* **2003**, *115*, 863–877.
76. W. L. Wei, H. S. Sun, M. E. Olah, X. Sun, E. Czerwinska, W. Czerwinski, Y. Mori, B. A. Orser, Z. G. Xiong, M. F. Jackson, M. Tymianski, J. F. MacDonald, *Proc. Natl. Acad. Sci. USA* **2007**, *104*, 16323–16328.
77. H. Jiang, S. L. Tian, Y. Zeng, L. L. Li, J. Shi, *Brain Res. Bull.* **2008**, *76*, 124–130.
78. H.-S. Chen, J. Xie, Z. Zhang, L.-T. Su, L. Yue, L. W. Runnels, *PLoS One* **2010**, *5*, e11161.
79. S. Brauchi, G. Krapivinsky, L. Krapivinsky, D. E. Clapham, *Proc. Natl. Acad. Sci. USA* **2008**, *105*, 8304–8308.
80. E. Abed, R. Moreau, *Cell, Prolif.* **2007**, *40*, 849–865.
81. E. Abed, C. Martineau, R. Moreau, *Calcif. Tissue Int.* **2011**, *88*, 246–253.
82. J. Jin, B. N. Desai, B. Navarro, A. Donovan, N. C. Andrews, D. E. Clapham, *Science* **2008**, *322*, 756–760.
83. R. Y. Walder, D. Landau, P. Meyer, H. Shalev, M. Tsolia, Z. Borochowitz, M. B. Boettger, G. E. Beck, R. K. Englehardt, R. Carmi, V. C. Sheffield, *Nat. Genet.* **2002**, *31*, 171–174.
84. R. Y. Walder, B. Yang, J.B. Stokes, P. A. Kirby, X. Cao, P. Shi, C. C. Searby, R. F. Husted, V. C. Sheffield, *Human Mol. Gen.* **2009**, *18*, 4367–4375.
85. T. E. Woudenberg-Vrenken, A. Sukinta, A. W. van der Kemp, R. J. Bindels, J. G. Hoenderop, *Nephron. Physiol.* **2011**, *117*, 11–19.
86. S. Thébault, G. Cao, H. Venselaar, Q. Xi, R. J. Bindels, J. G. Hoenderop, *J. Biol. Chem.* **2008**, *283*, 19999–20007.
87. G. Cao, J. van der Wijst, A. van der Kemp, F. van Zeeland, R. J. Bindels, J. G. Hoenderop, *J. Biol. Chem.* **2009**, *284*, 14788–14795.
88. W. M. Groenestege, J. G. Hoenderop, L. van den Heuvel, N. Knoers, R. J. Bindels, *J. Am. Soc. Nephrol.* **2006**, *17*, 1035–1043.
89. L. J. Rondón, W. M. Groenestege, Y. Rayssiguier, A. Mazur, *Am. J. Physiol.* **2008**, *294*, R2001–R2007.
90. G. Cao, S. Thébault, J. van der Wijst, A. van der Kemp, E. Lasonder, R. J. Bindels, J. G. Hoenderop, *Curr. Biol.* **2008**, *18*, 168–176.
91. W. M. Groenestege, S. Thébault, J. van der Wijst, D. van den Berg, R. Janssen, S. Tejpar, L. P. van den Heuvel, E. van Cutsem, J. G. Hoenderop, N. V. Knoers, R. J. Bindels, *J. Clin. Invest.* **2007**, *117*, 2260–2267.
92. D. Cunningham, Y. Humblet, S. Siena, D. Khayat, H. Bleiberg, A. Santoro, D. Bets, M. Mueser, A. Harstrick, C. Verslype, I. Chau, E. van Cutsem, *N. Engl. J. Med.* **2004**, *351*, 337–345.
93. H. Dimke, J. van der Wjist, T. R. Alexander, I. M. Mejier, G. M. Mulder, H. van Goor, S. Tejpar, J. G. Hoenderop, R. J. Bindels, *J. Am. Soc. Nephrol.* **2010**, *21*, 1309–1316.
94. A. Ikari, C. Okude, H. Sawada, Y. Yamazaki, J. Sugatani, M. Miwa, *Biochem. Biophys. Res. Commun.* **2008**, *369*, 1129–1133.
95. A. Ikari, A. Sanada, C. Okude, H. Sawada, Y. Yamazaki, J. Sugatani, M. Miwa, *J. Cell. Physiol.* **2010**, *222*, 481–487.
96. D. B. Simon, Y. Lu, K. A. Choate, H. Velazquez, E. Al-Sabban, M. Praga, G. Casari, A. Bettinelli, G. Colussi, J. Rodrigues-Soriano, D. McCredie, D. Milford, S. Sanjad, R. P. Lifton, *Science* **1999**, *285*, 103–1106.
97. M. Lal-Nag, P. J. Morin, *Genome Biol*, **2009**, *10*, 235.
98. P. J. Kausalya, S. Amasheh, D. Gunzel, H. Wurps, D. Muller, M. Fromm, W. Hunziker, *J. Clin. Invest.* **2006**, *116*, 878–891.
99. J. Hou, D. L. Paul, D. A. Goodenough, *J. Cell. Sci.* **2005**, *118*, 5109–118.
100. A. Ikari, S. Matsumoto, H. Harada, K. Takagi, H. Hayashi, Y. Suzuki, M. Degawa, M. Miwa, *J. Cell. Sci.* **2006**, *119*, 1781–1789.
101. E. Efrati, J. Arsentiev-Rozenfeld, I. Zelikovic, *Am. J. Physiol.* **2005**, *288*, F272–F283.

102. D. Muller, P. J. Kausalya, F. Claverie-Martin, I. C. Meij, P. Eggert, V. Garcia-Nieto, W. Hunziker, *Am. J. Hum. Genet.* **2003**, *73*, 1293–1301.
103. M. A. Khan, A. D. Conigrave, *Br. J. Pharmacol.* **2010**, *159*, 1039–1050.
104. J. Hou, A. Renigunta, A. S. Gomes, M. Hou, D. L. Paul, S. Waldegger, D. A. Goodenough, *Proc. Natl. Acad. Sci. USA* **2009**, *106*, 15350–15355.
105. A. Goytain, G. A. Quamme, *BMC Genomics* **2005**, *6*, 48.
106. T. Shibatani, L. L. David, A. L. McCormack, K. Frueh, W. R. Skach, *Biochemistry* **2005**, *44*, 5982–5992.
107. H. Zhou, D. E. Clapham, *Proc. Natl. Acad. Sci. USA* **2009**, *106*, 15750–15755.
108. G. Wiesenberger, M. Waldherr, R. J. Schweyen, *J. Biol. Chem.* **1992**, *267*, 6963–6969.
109. D. M. Bui, J. Gregan, E. Jarosch, A. Ragnini, R. J. Schweyen, *J. Biol. Chem.* **1999**, *274*, 20438–20443.
110. M. Kolisek, G. Zsurka, J. Samaj, J. Weghuber, R. J. Schweyen, M. Schweigel, *EMBO J.* **2003**, *22*, 1235–1244.
111. M. Piskacek, L. Zotova, G. Zsurka, R. J. Schweyen, *J. Cell. Mol. Med.* **2009**, *13*, 693–700.
112. K. E. Akerman, *J. Bioenerg. Biomembr.* **1981**, *13*, 133–139.
113. G. Zsurka, J. Gregan, R. J. Schweyen, *Genomics* **2001**, *72*, 158–168.
114. A. Panov, A. Scarpa, *Biochemistry* **1996**, *35*, 427–432.
115. A. Panov, A. Scarpa, *Biochemistry* **1996**, *35*, 12849–12856.
116. A. Goytain, G.A. Quamme, *Am. J. Physiol.* **2008**, *294*, C495–502.
117. C. Schmitz, F. Deason, A.-L. Perraud, *Magnes. Res.* **2007**, *20*, 6–18.
118. R. T. Alexander, J. G. Hoenderop, R. J. Bindels, *J. Am. Soc. Nephrol.* **2008**, *19*, 1451–1458.
119. G. A. Quamme, *Am. J. Physiol.* **2010**, *298*, 407–429.
120. T. Günther, J. Vormann, R. Forster, *Biochem. Biophys. Res. Commun.* **1984**, *119*, 124–131.
121. T. Günther, J. Vormann, *Biochem. Biophys.Res. Commun.* **1985**, *130*, 540–545.
122. J.-C. Feray, R. Garay, *Biochim. Biophys. Acta* **1986**, *856*, 76–84.
123. P. W. Flatman, L. M. Smith, *J. Physiol.* **1990**, *431*, 11–25.
124. W. Xu, J. S. Willis, *J. Membr. Biol.* **1994**, *141*, 277–287.
125. H. Ludi, H. J. Schatzmann, *J. Physiol.* **1987**, *390*, 367–382.
126. J. Vormann, K. Magdorf, T. Günther, U. Wahn, *Eur. J. Clin. Chem. Clin. Biochem.* **1994**, *32*, 833–836.
127. T. Günther, J. Vormann, *FEBS Lett.* **1992**, *297*, 132–134.
128. F. I. Wolf, A. Di Francesco, V. Covacci, D. Corda, A. Cittadini, *Arch. Biochem. Biophys.* **1996**, *331*, 194–200.
129. A. Romani, C. Marfella, A. Scarpa, *Circ. Res.* **1993**, *72*, 1139–1148.
130. A. Romani, C. Marfella, A. Scarpa, *J. Biol. Chem.* **1993**, *268*, 15489–15495.
131. T. E. Fagan, A. Romani, *Am. J. Physiol.* **2000**, *279*, G943–G950.
132. T. E. Fagan, A. Romani, *Am. J. Physiol.* **2001**, *280*, G1145–G1156.
133. C. Cefaratti, C. Ruse, *Mol. Cell. Biochem.* **2007**, *297*, 209–214.
134. C. Cefaratti, A. Romani, *Mol. Cell. Biochem.* **2007**, *303*, 63–72.
135. M. Kolisek, A. Nestler, J. Vormann, M. Schweigel-Rontgen, *Am. J. Physiol.* **2012**, *302*, C318-326
136. T. Wabakken, E. Rian, M. Kveine, H. C. Aasheim, *Biochem. Biophys. Res. Commun.* **2003**, *306*, 718–724.
137. C. Cefaratti, A. Romani, *Mol. Cell. Biochem.* **2011**, *351*, 133–142.
138. H. Ebel, T. Günther, *FEBS Lett.* **2003**, *543*, 103–107.
139. H. Rasgado-Flores, H. Gonzalez-Serratos, J. DeSantiago, *Am. J. Physiol.* **1994**, *266*, C1112–C1117.
140. E. C. Aromataris, M. L. Roberts, G. J. Barritt, G. Y. Rychkov, *J. Physiol.* **2006**, *573*, 611–625.
141. M. Tashiro, M. Konishi, *Biophys. J.* **1997**, *73*, 3371–3384.
142. T. Günther, *Magnes. Bull.* **1996**, *18*, 2–6.
143. C. Cefaratti, A. Romani, A. Scarpa, *Am. J. Physiol.* **1998**, *275*, C995–C1008.
144. J.-C. Feray, R. Garay, *Naunyn-Schmied. Arch. Pharmacol.* **1988**, *338*, 332–337.

145. M. Schweigel, J. Vormann, H. Martens, *Am. J. Physiol.* **2000**, *278*, G400–G408.
146. T. Günther, *Miner. Electrolyte Metab.* **1993**, *19*, 259–265.
147. H. Ebel, M. Hollstein, T. Günther, *Biochim. Biophys. Acta* **2002**, *1559*, 135–144.
148. D. Keenan, A. Romani, A. Scarpa, *FEBS Lett.* **1996**, *395*, 241–244.
149. T. Günther, J. Vormann, V. Höllriegl, *Biochim. Biophys. Acta* **1990**, *1023*, 455–461.
150. H. Ebel, M. Hollstein, T. Günther, *Biochim. Biophys. Acta* **2004**, *1667*, 132–140.
151. O. Shaul, D. W. Hilgemann, J. de-Almeida-Engler, M. Van Montagu, D. Inz, G. Galili, *EMBO J.* **1999**, *18*, 3973–3980.
152. T. Günther, J. Vormann, *FEBS Lett.* **1990**, *265*, 55–58.
153. P. Dalal, A. Romani, *Metabolism* **2010**, *59*, 1663–1671.
154. D. Günzel, W. R. Schlue, *J. Physiol.* **1996**, *491*, 595–608.
155. R. L. Smith, L. J. Thompson, M. E. Maguire, *J. Bacteriol.* **1995**, *177*, 1233–1238.
156. A. Goytain, G. A. Quamme, Physiol. *Genomics* **2005**, *21*, 337–342.
157. M. Kolisek, P. Launay, A. Beck, G. Sponder, N. Serafini, M. Brenkus, E. M. Froschauer, H. Martens, A. Fleig, M. Schweigel, *J. Biol. Chem.* **2008**, *283*, 16235–16247.
158. A. Goytain, G. A. Quamme, Biochem. *Biophys. Res. Commun.* **2005**, *330*, 701–705.
159. J. Sahni, B. Nelson, A. M. Scharenberg, *Biochem. J.* **2007**, *401*, 505–513.
160. C. Y. Wang, A. Davoodi-Semiromi, J. D. Shi, P. Yang, Y. Q. Huang, J. A. Agundez, J. M. Moran, B. Ochoa, B. Hawkins-Lee, J.X. She, *Am. J. Med. Genet. A* **2003**, *119*, 9–14.
161. C. Y. Wang, J. D. Shi, P. Yang, P. G. Kumar, Q. Z. Li, Q. G. Run, Y. C. Su, H. S. Scott, K. J. Kao, J. X. She, *Gene* **2003**, *306*, 37–44.
162. C. Y. Wang, P. Yang, J. D. Chi, S. Purohit, D. Guo, H. An, J. G. Gu, J. Ling, Z. Dong, J. X. She, *BMC Genomics* **2004**, *5*, 7.
163. D. G. Kehres, M. E. Maguire, *Biometals* **2002**, *15*, 261–270.
164. A. Goytain, G. A. Quamme, *Physiol. Genomics*. **2005**, *22*, 382–389.
165. M. G. Butler, *Am. J. Med. Genet.* **1990**, *35*, 319–332.
166. A. Goytain, R. M. Hines, A. El-Husseini, G. A. Quamme, *J. Biol. Chem.* **2007**, *282*, 8060–8068.
167. A. Goytain, R. M. Hines, G. A. Quamme, *Am. J. Physiol.* **2008**, *295*, C944–C953.
168. S. Rainier, J. H. Chai, D. Tokarz, R. D. Nicholls, J. K. Fink, *Am. J. Hum. Genet.* **2003**, *73*, 967–971.
169. A. Goytain, R. M. Hines, G. A. Quamme, *J. Biol. Chem.* **2008**, *283*, 33365–33374.
170. S.-H. Li, X.-J. Li, *Trends Genet.* **2004**, *20*, 146–152.
171. A. Yanai, K. Huang, R. Kang, R. R. Singaraja, P. Arstikaitis, L. Gan, P. C. Orban, A. Mullard, C. M. Cowan, L. A. Raymond, R. C. Drisdel, W. N. Green, B. Ravikumar, D. C. Rubinsztein, A. El-Husseini, M. R. Hayden, *Nat. Neurosci.* **2006**, *9*, 824–831.
172. C. Cefaratti, A. Romani, A. Scarpa, *J. Biol. Chem.* **2000**, *275*, 3772–3780.
173. R. Juttner, H. Ebel, *Biochim. Biophys. Acta* **1998**, *1370*, 51–63.
174. M. Baillien, M. Cogneau, *Magnesium* **1995**, y, 331–339.
175. H. J. Schatzmann, *Biochim. Biophys. Acta* **1993**, *1148*, 15–18.
176. M. Baillien, H. Wang, M. Cogneau, *Magnesium* **1995**, *8*, 315–329.
177. D. A. Elliot, M. A. Rizack, *J. Biol. Chem.* **1974**, *249*, 3985–3990.
178. S. J. Bird, M. E. Maguire, *J. Biol. Chem.* **1978**, *253*, 8826–8834.
179. J. J. Erdos, M. E. Maguire, *Mol. Pharmacol.* **1980**, *18*, 379–383.
180. J. J. Erdos, M. E. Maguire, *J. Physiol.* **1983**, *337*, 351–371.
181. R. D. Grubbs, C. A. Wetherill, K. Kutschke, M. E. Maguire, *Am. J. Physiol.* **1984**, *248*, C51–C57.
182. M. E. Maguire, J. J. Erdos, *J. Biol. Chem.* **1978**, *253*, 6633–6636.
183. M. E. Maguire, J. J. Erdos, *J. Biol. Chem.* **1980**, *255*, 1030–1035.
184. R. D. Grubbs, S. D. Collins, M. E. Maguire, *J. Biol. Chem.* **1985**, *259*, 12184–12192.
185. F. C. Howarth, J. Waring, B. I. Hustler, J. Singh, *Magnes. Res.* **1994**, *7*, 187–197.
186. F. I. Wolf, A. Di Francesco, A. Cittadini, *Arch. Biochem. Biophys.* **1994**, *308*, 335–341.
187. D. Keenan, A. Romani, A. Scarpa, *Circ. Res.* **1995**, *77*, 973–983.
188. T. Günther, J. Vormann, *Magnes. Bull.* **1992**, *14*, 122–125.

189. A. M. Romani, V. D. Matthews, A. Scarpa, *Circ. Res.* **2000**, *86*, 326–333.
190. T. Amano, T. Matsubara, J. Watanabe, S. Nakayama, N. Hotta, *Brit. J. Pharmacol.* **2000**, *130*, 731–738.
191. V. Karoor, K. Baltensperger, H. Paul, M. C. Czech, C. C. Malbon, *J. Biol. Chem.* **1995**, *270*, 25305–25308.
192. J. A. Smoake, G.-M. M. Moy, B. Fang, S. S. Solomon, *Arch. Biochem. Biophys.* **1995**, *323*, 223–232.
193. A. Ferreira, A. Rivera, J. R. Romero, *J. Cell. Physiol.* **2004**, *199*, 434–440.
194. G. Wiesenberger, K. Steinleitner, R. Malli, W. F. Graier, J. Vormann, R. J. Schweyen, J. A. Stadler, *Eukaryot. Cell* **2007**, *6*, 592–599.
195. M. Wang, J. R. Berlin, *Am. J. Physiol.* **2006**, *291*, C83–C92.
196. L. M. Torres, J. Youngner, A. Romani, *Am. J. Physiol.* **2005**, *288*, G195–G206.
197. T. E. Fagan, C. Cefaratti, A. Romani, *Am. J. Physiol.* **2004**, *286*, E184–E193.
198. J. C. Henquin, T. Tamagawa, M. Nenquin, M. Cogneau, *Nature* **1983**, *301*, 73–74.
199. D. Garfinkel, L. Garfinkel, *Magnesium* **1988**, *7*, 249–261.
200. M. Otto, R. Heinrich, B. Kuhn, G. Jacobasch, *Eur. J. Biochem.* **1974**, *49*, 169–178.
201. A. W. Harman, A. L. Nieminen, J. J. Lemasters, B. Herman, *Biochem. Biophys. Res. Commun.* **1990**, *170*, 477–483.
202. V. Gaussin, P. Gailly, J.-M. Gillis, L. Hue, *Biochem. J.* **1997**, *326*, 823–827.
203. P.A. Tessman, A. Romani, *Am. J. Physiol.* **1998**, *275*, G1106–G1116.
204. A. Gasbarrini, A. B. Borle, H. Farghali, C. Bender, A. Francavilla, D. van Thiel, *J. Biol. Chem.* **1992**, *267*, 6654–6663.
205. J. Hwa, A. Romani, C. Marfella, A. Scarpa, *Biophys. J.* **1993**, *63*, A307.
206. R. Di Polo, L. Beauge, *Biochim. Biophys. Acta* **1988**, *946*, 424–428.
207. T. Günther, V. Höllriegl, *Biochim. Biophys. Acta* **1993**, *1149*, 49–54.
208. R. M. Touyz, E. L. Schiffrin, *J. Biol. Chem.* **1996**, *271*, 24353–24358.
209. A. Romani, C. Marfella, A. Scarpa, *FEBS Lett.* **1992**, *296*, 135–140.
210. D. L. Hwang, C. F. Yen, J. L. Nadler, *J. Clin. Endocrinol. Metab.* **1993**, *76*, 549–553.
211. R. D. Grubbs, M. E. Maguire, *J. Biol. Chem.* **1986**, *261*, 12550–12554.
212. S. Ishijima, M. Tatibana, *J. Biochem.* **1994**, *115*, 730–737.
213. P. Csermely, P. Fodor, J. Somogyi, *Carcinogenesis* **1987**, *8*, 1663–1666.
214. T. Günther, J. Vormann, *Biochim. Biophys. Acta* **1995**, *1234*, 105–110.
215. Z. W. Yang, J. Wang, T. Zheng, B. T. Altura, B. M. Altura, *Stroke* **2001**, *32*, 249–257.
216. L. M. Torres, B. Konopnika, L. N. Berti-Mattera, C. Liedtke, A. Romani, *Alcohol. Clin. Exp. Res.* **2010**, *34*, 1659–1669.
217. E. Y. Tang, P. J. Parker, J. Beattie, M. D. Houslay, *FEBS Lett.* **1993**, *326*, 117–123.
218. Y. Konno, S. Ohno, Y. Akita, H. Kawasaki, K. Suzuki, *J. Biochem.* **1989**, *106*, 673–678.
219. B. E. Corkey, J. Duszynski, T. L. Rich, B. Matchinsky, J. R. Williamson, *J. Biol. Chem.* **1986**, *261*, 2567–2574.
220. Z. Yang, J. Wang, B. T. Altura, B. M. Altura, *Pflugers Arch.* **2000**, *439*, 240–247.
221. R. M. Touyz, G. Yao, *J. Cell. Physiol.* **2003**, *197*, 326–335.
222. L. M. Torres, C. Cefaratti, B. Perry, A. Romani, *Mol. Cell. Biochem.* **2006**, *288*, 191–199.
223. H. Rubin, *Adv. Cancer Res.* **2005**, *93*, 1–58.
224. W. F. Waas, K. N. Dalby, *Biochemistry* **2003**, *42*, 2960–2970.
225. S. J. Kim, H. S. Kang, M. S. Kang, X. Yu, S. Y. Park, I. S. Kim, N. S. Kim, S. Z. Kim, Y. G. Kwak, J. S. Kim, *Biochem. Biophys. Res. Commun.***2005**, *333*, 1132–1138.
226. S. Thebault, R. T. Alexander, W. M. Tiel Groenestege, J. G. Hoenderop, R. J. Bindels, *J. Am. Soc. Nephrol.* **2009**, *20*, 78–85.
227. J. van der Wjist, J. G. Hoenderop, R. J. Bindels, *Magnes. Res.* **2009**, *22*, 127–132.
228. G. H. Mudge, in *The Pharmacological Basis of Therapeutics*, Eds A. Goodman Gilman, L. S. Godman, T. W. Rall, F. Murad, MacMillian, New York, 1989.
229. H. Joborn, G. Akerstrom, S. Ljunghall, *Clin. Endocrinol.* **1985**, *23*, 219–226.
230. C. Bailly, M. Imbert-Teboul, N. Roinel, C. Amiel, *Am. J. Physiol.* **1990**, *258*, F1224–F1231.

231. Y. Rayssiguier, *Horm. Metab. Res.* **1977**, *9*, 309–314.
232. G. Guideri, *Arch. Intern. Pharmacodyn. Therap.* **1992**, *14*, 122–125.
233. I. M. Shafik, G. A. Quamme, *Am. J. Phyiol.* **1989**, *257*, F974–F977.
234. P. B. Molinoff, *Drugs* **1984**, *28*, 1–14.
235. P. J.Barnes, *Am. J. Respir. Crit. Care Med.* **1995**, *152*, 838–860.
236. E. M. Brown, G. Gamba, D. Riccardi, M. Lombardi, R. Butters, O. Kifor, A. Sun, M. A. Hediger, J. Lytton, S. C. Herbert, *Nature* **1993**, *366*, 575–580.
237. E. F. Nemeth, A. Scarpa, J. Biol. Chem. **1987**, *262*, 5188–5196.
238. B. W. Bapty, L. J. Dai, G. Ritchie, F. Jirik, L. Canaff, G. N. Hendy, G. A. Quamme, *Kidney Int.* **1998**, *53*, 583–592.
239. B. W. Bapty, L. J. Dai, G. Ritchie, L. Canaff, G. N. Hendy, G. A. Quamme, *Am. J. Physiol.* **1998**, *275*, F353–F360.
240. G. A. Quamme, J. H.Dirks, *Am. J. Physiol.* **1980**, *238*, F187–F198.
241. D. J. Dipette, K. Simpson, J. Guntupalli, *Magnesium* **1987**, *6*, 136–149.
242. J. B. Stanbury, *J. Pharmacol. Exp. Ther.* **1984**, *93*, 52–62.
243. M. F. M. James, R. C. Cork, G. M. Harlen, J. F. White, *Magnesium* **1988**, *7*, 37–43.
244. G. A. Quamme, S. W. Rabkin, *Biochem. Biophys. Res. Commun.* **1990**, *167*, 1406–1412.
245. G. A. Quamme, L.-S.Dai, *Am. J. Physiol.* **1990**, *259*, C521–C525.
246. L.-J. Dai, P. A. Friedman, G. A.Quamme, *Kidney Inter.* **1997**, *51*, 1710–1718.
247. T. Kubota, Y. Shindo, K. Tokuno, H. Komatsu, H. Ogawa, S. Kudo, Y. Kitamura, K. Suzuki, K. Oka, *Biochim. Biophys. Acta* **2005**, *1744*, 19–28.
248. L.-J. Dai, P. A. Friedman, G. A. Quamme, *Kidney Inter.* **1991**, *51*, 1008–1017.
249. M. Tashiro, P. Tursun, T. Miyazaki, M. Watanabe, M. Konishi, *Jpn. J. Physiol.* **2002**, *52*, 541–551.
250. M. Schweigel, I. Lang, H. Martens, *Am. J. Physiol.* **1999**, *277*, G976–G982.
251. M. Schweigel, H. Martens, *Am. J. Physiol.* **2003**, *285*, G45–G53.
252. T. Günther, J.Vormann, *Biochim. Biophys. Acta* **1995**, *1234*, 105–110.
253. C. E. Mondon, C. B. Dolkas, J. M. Olefsky, G. M. Reaven, *Diabetes* **1974**, *24*, 225–229.
254. R. Taylor, L. Agius, *Biochem. J.* **1988**, *250*, 625–640.
255. R. D. Grubbs, M. E. Maguire, *Magnesium* **1987**, *6*, 113–127.
256. M. E. Maguire, *Trends Pharmacol. Sci.* **1984**, *5*, 73–77.
257. R. E. White, H. C. Hartzell, *Science* **1988**, *239*, 778–780.
258. Z. S. Agus, M. Morad, *Annu. Rev. Physiol.* **1991**, *53*, 299–307.
259. M. Bara, A. Guiet-Bara, *Magnes. Res.* **2001**, *14*, 11–18.
260. J. Serrano, S. R. Dashti, E. Perez-Reyes, S. W. Jones , *Biophys. J.* **2000**, *79*, 3052–3062.
261. S. Brunet, T. Scheuer, R. Klevit, W. A. Catterall, *J. Gen. Physiol.* **2006**, *126*, 311–323.
262. J. Zhang, G. Wier, M. P. Blaustein, *Am. J. Physiol.* **2002**, *283*, H2692–H2705.
263. M. Prakriya, R. S. Lewis, *J. Gen. Physiol.* **2002**, *119*, 487–507.
264. H. Matsuda, *Annu. Rev. Physiol.* **1991**, *53*, 289–298.
265. P. Tammaro, A. L. Smith, B. L. Crowley, S. V. Smirnov, *Cardiovasc. Res.* **2005**, *65*, 387–396.
266. J. Shi, G. Krishnamoorthy, Y. Wang, L. Hu, N. Chaturvedi, D. Harilal, J. Qin, J. Cui, *Nature* **2002**, *418*, 876–880.
267. P. Bednarczyk, K. Dolowy, A. Szewczyk, *FEBS Lett.* **2005**, 579, 1625–1630.
268. T. Günther, *Magnesium* **1986**, *5*, 53–59.
269. A. Romani, E. Dowell, A. Scarpa, *J. Biol. Chem.* **1991**, *266*, 24376–24384.
270. G. H. Zhang, J. E. Melvin, *J. Biol. Chem.* **1992**, *267*, 20721–20727.
271. G. H. Zhang, J. E. Melvin, *J. Biol. Chem.* **1996**, *271*, 29067–29072.
272. J. G. McCormack, A. Halestrap, R. M. Denton, *Physiol. Rev.* **1990**, *70*, 391–425.
273. R. G. Hansford, *J. Bioenerg. Biomembr.***1994**, *26*, 495–508.
274. C. S. Moravec, M. Bond, *Am. J. Physiol.* **1991**, *260*, H989–H997.
275. C. S. Moravec, M. Bond, *J. Biol. Chem.* **1992**, *267*, 5310–5316.
276. G. P. Brierley, M. Davis, D. W. Jung, *Arch. Biochem. Biophys.* **1987**, *253*, 322–332.

277. A. D. Beavis, M. Powers, *J. Biol. Chem.* **2004**, *279*, 4045–4050.
278. P. Bernardi, *Physiol. Rev.* **1999**, *79*, 1127–1155.
279. P. C. Bradshaw, D. R. Pfeiffer, *BMC Biochem.* **2006**, *7*, 4.
280. M. Dolder, B. Walzel, O. Speer, U. Schlattner, T. Wallimann, *J. Biol. Chem.* **2003**, *278*, 17760–17766.
281. P. Volpe, B. H. Alderson-Lang, G .A. Nickols, *Am. J. Physiol.* **1990**, *258*, C1077–C1085.
282. P. Volpe, L. Vezú, *Magnes. Res.* **1993**, *6*, 267–274.
283. K. Gusev, E. Niggli, *J. Gen. Physiol.* **2008**, *132*, 721–730.
284. D. R. Laver, B. N. Honen, *J. Gen. Physiol.* **2008**, *132*, 429–446.
285. L. Doleh, A. Romani, *Arch. Biochem. Biophys.* **2007**, *467*, 283–290.
286. A. Barfell, A. Crumbly, A. Romani, *Arch. Biochem. Biophys.* **2011**, *509*, 157–163.
287. S. Yamaguchi, T. Ishikawa, *Biochem. Biophys. Res. Commun.* **2008**, *376*, 100–104.
288. M. Okahira, M. Kubota, K. Iguchi, S. Usui, K. Hirano, *Eur. J. Pharmacol.* **2008**, *588*, 26–32.
289. M. E. Maguire, *Ann. N.Y. Acad. Sci.* **1988**, *551*, 215–217.
290. A. Sgambato, F. I. Wolf, B. Faraglia, A. Cittadini, *J. Cell. Physiol.* **1999**, *180*, 245–254.
291. F. I. Wolf, V. Trapani, M. Simonacci, A. Boninsegna, A. Mazur, J. A. Maier, *Nutr. Cancer* **2009**, *61*, 131–136.
292. V. Covacci, N. Bruzzese, A. Sgambato, A. Di Francesco, M. A. Russo, F. I. Wolf, A. Cittadini, *J. Cell. Biochem.* **1998**, *70*, 313–322.
293. F. I. Wolf, V. Covacci, N. Bruzzese, A. Di Francesco, A. Sachets, D. Cord, A. Cittadini, *J. Cell. Biochem.* **1998**, *71*, 441–448.
294. A. Di Francesco, R. W. Desnoyer, V. Covacci, F. I. Wolf, A. Romani, A. Cittadini, M. Bond, *Arch. Biochem. Biophys.* **1998**, *360*, 149–157.
295. A. Trache, J. P. Trzeciakowski, G. A. Meininger, *J. Mol. Recognit.* **2010**, *23*, 316–321.
296. F. I. Wolf, A. R.Cittadini, J. A. Maier, *Cancer Treat. Rev.* **2009**, *35*, 378–382.

Chapter 5
Intracellular Calcium Homeostasis and Signaling

Marisa Brini, Tito Calì, Denis Ottolini, and Ernesto Carafoli

Contents

M. Brini (✉) • T. Calì
Department of Comparative Biomedicine and Food Sciences, University of Padova, Viale G. Colombo 3, I-35131 Padova, Italy
e-mail: marisa.brini@unipd.it; tito.cali@unipd.it

D. Ottolini
Department of Biomedical Sciences, University of Padova, Viale G. Colombo 3, I-35131 Padova, Italy
e-mail: denis.ottolini@unipd.it

E. Carafoli (✉)
Venetian Institute of Molecular Medicine (VIMM), Via G. Orus 2, I-35129 Padova, Italy
e-mail: ernesto.carafoli@unipd.it

L. Banci (ed.), *Metallomics and the Cell*, Metal Ions in Life Sciences 12,
DOI 10.1007/978-94-007-5561-1_5,

Abstract Ca^{2+} is a universal carrier of biological information: it controls cell life from its origin at fertilization to its end in the process of programmed cell death. Ca^{2+} is a conventional diffusible second messenger released inside cells by the interaction of first messengers with plasma membrane receptors. However, it can also penetrate directly into cells to deliver information without the intermediation of first or second messengers. Even more distinctively, Ca^{2+} can act as a first messenger, by interacting with a plasma membrane receptor to set in motion intracellular signaling pathways that involve Ca^{2+} itself. Perhaps the most distinctive property of the Ca^{2+} signal is its ambivalence: while essential to the correct functioning of cells, Ca^{2+} becomes an agent that mediates cell distress, or even (toxic) cell death, if its concentration and movements inside cells are not carefully tuned. Ca^{2+} is controlled by reversible complexation to specific proteins, which could be pure Ca^{2+} buffers, or which, in addition to buffering Ca^{2+}, also decode its signal to pass it on to targets. The most important actors in the buffering of cell Ca^{2+} are proteins that transport it across the plasma membrane and the membrane of the organelles: some have high Ca^{2+} affinity and low transport capacity (e.g., Ca^{2+} pumps), others have opposite properties (e.g., the Ca^{2+} uptake system of mitochondria). Between the initial event of fertilization, and the terminal event of programmed cell death, the Ca^{2+} signal regulates the most important activities of the cell, from the expression of genes, to heart and muscle contraction and other motility processes, to diverse metabolic pathways involved in the generation of cell fuels.

Keywords apoptosis • calcium • calcium buffering proteins • calcium sensor proteins • calmodulin • fertilization • gene expression • ion pumps • mitochondria • protein dephosphorylation • protein phosphorylation

Please cite as: *Met. Ions Life Sci.* 12 (2013) 119–168

1 Introduction

In the course of evolution, Ca^{2+} has been selected as a universal carrier of signals. The selection occurred at the time of the transition from unicellular to multicellular life, when the division of labor among cells of the organisms brought with it the necessity of exchanging signals. As a rule, unicellular organisms do not require Ca^{2+} (although some bacterial functions, e.g., chemotaxis, do require Ca^{2+} and its manipulation) and do not need to exchange signals, their interplay being restricted to the competition for nutrients. The selection of Ca^{2+} as carrier of information has been dictated by coordination chemistry, which makes Ca^{2+} ideally suited to be accommodated within the sites of irregular geometry offered by complex cellular molecules (proteins) [1,2]. A molecule selected to transmit signals within the cell must be tightly regulated. In the case of Ca^{2+}, given its chemical properties, this is optimally achieved by binding it reversibly, and with the appropriate affinity and specificity, to cellular proteins. In the complexing proteins, oxygen is the preferred ligand atom for Ca^{2+}: the introduction of nitrogen in the primary coordination sphere usually decreases the selectivity for Ca^{2+}. In most cases, the coordination number for Ca^{2+} is 8 (in some cases it may be 6 or 7): by comparison, the coordination number for the other abundant cellular divalent cation, Mg^{2+}, is only 6. The coordination stereochemistry of 6 is that of a regular octahedron, implying that the Mg-O bond distances in the primary coordination sphere vary only little (between 0.200 and 0.216 nm), whereas the Ca-O bond distances vary over a much more extended range (between 0.229 and 0.265 nm). It follows that Ca^{2+} can accept binding cavities of irregular shape, in which the ligand oxygen atoms can be at considerably variable distances from it.

The facility with which Ca^{2+} becomes bound permits the lowering of its cell concentration to levels that are too low to trigger its precipitation as an insoluble phosphate salt. This is the extra dividend of the choice of Ca^{2+} as a cellular signaling agent: if it were not possible to maintain its background concentration very low inside cells, phosphate could not be used as the energy currency. In addition to Ca^{2+}, a number of other metals are essential to cell life, such as iron, zinc, copper, manganese. All of them are active-site metals that participate directly in the mechanism of enzyme catalysis. Ca^{2+}, instead, is not an active-site metal, it is an allosteric metal *par excellence*, which binds to (enzyme) proteins at sites different from the active site, modulating their activities, namely, activating (in most cases) or inhibiting them. Modulation of enzyme processes is of utmost importance to cells, thus, the control of cellular Ca^{2+} is of critical importance, as the array of Ca^{2+}-regulated functions covers the entire spectrum of processes that are essential to cell life. The vital importance of the precise control of Ca^{2+} is reflected in the multitude of systems developed by evolution to fulfill the task. Basically, these systems either transport Ca^{2+} across membrane boundaries, or complex it reversibly in the cytosol or in the lumen of the organelles.

The transport of Ca^{2+} across membranes is the ultimate way to buffer it; it is performed by channels, ATPases, exchangers, in which Ca^{2+} is exchanged for another ion (usually Na^{+}), and by an electrophoretic uniporter in the inner mitochondrial membrane. The control of Ca^{2+} by non-membrane proteins is performed within the organelles by low-affinity, large-capacity proteins, that, however, may also fulfill other cellular functions [3]. In the cytosol, Ca^{2+}-binding proteins modulate the Ca^{2+} signal spatially and temporally. Some are pure Ca^{2+} buffers, e.g., parvalbumin, calbindins, and calreticulin, others are classified as Ca^{2+} sensors, since in addition to buffering Ca^{2+}, they also process its signal. The most important and versatile Ca^{2+} sensor protein is calmodulin (CaM), which is expressed ubiquitously in cells, while other Ca^{2+} sensors are tissue specific, e.g., the neuronal Ca^{2+} sensor proteins. The distinction between Ca^{2+} buffering and Ca^{2+} sensor proteins, while justified in principle, is not absolute, as some cytosolic Ca^{2+} buffers, e.g, calbindin D28K, may also have signal processing function [4], and on the other hand even the prototypical Ca^{2+} sensor protein, calmodulin, could under some circumstances act essentially as a Ca^{2+} buffer [5].

The array of processes that are controlled by Ca^{2+} begins with the origin of cell life at fertilization, and ends with the process of programmed death that terminates life once cells have reached the end of their vital cycle. Between these two events, Ca^{2+} controls processes that may be general to all cells, e.g., gene transcription, differentiation, the generation of fuels in a number of metabolic pathways (essentially, by enzyme phosphorylation and dephosphorylation), motility in the cytoplasmic structures, and cell motility and migration in general. Other processes may be cell-specific, e.g., secretion of solutes (of neurotransmitters in neurons), contraction/relaxation of skeletal muscles and heart. Figure 1 offers a comprehensive panorama of the cell processes that are under the control of Ca^{2+}. Some of them demand rapid and transient exposure to large changes of Ca^{2+} in the environment that may even be accomplished by the generation of repetitive substantial increases in the form of oscillations. Others demand instead a more sustained change of Ca^{2+} in their vicinity. In all cases, however, it is of utmost importance that the long-term basal concentration of Ca^{2+} in the bulk cytosol, after the transient elevation demanded by the activation of the target functions, is returned to the low/intermediate nM range. Cells will

Calcium-modulated functions in eukaryotic cells

Generation of fuels	**Miscellaneous functions**
Glycogenolysis (phosphorylase b kinase) Lipases and phospholipases α-glycerophosphate dehydrogenase Pyruvate dehydrogenase phosphate phosphatase NAD-dependent isocitric dehydrogenase α-ketoglutarate dehydrogenase NADH-dehydrogenase (plant mitochondria) β-hydroxybutyrate dehydrogenase	Light emission Cell cycle Some proteolytic enzymes Excitation-transcription coupling Some protein kinases Calcineurin Production of messengers (e.g. NO) Vision
Membrane-linked functions	**Contractile and motile systems**
Excitation-contraction coupling Excitation/secretion coupling (e.g. neurotransmitters) Some action potentials Tight junctions Cell contact	Muscle myofibrils Cilia and flagella Microtubules and microfilaments Cytoplasmic streaming Pseudopod formation
Hormonal regulation	
Formation/degradation of Cyclic AMP and GMP Release of several hormones from storage vesicles	

Figure 1 A comprehensive scheme of the cell processes that are under the control of Ca^{2+}.

not tolerate protracted abnormal increases of Ca^{2+} in the cytosol, where most targets of its signaling function are located. Should this happen, as is frequently the case in disease conditions, the correct functioning of Ca^{2+}-controlled processes becomes compromised and Ca^{2+} regulation comes to an end. Ca^{2+} is thus an ambivalent messenger: while essential to the correct functioning of cell life when tightly controlled, it becomes a conveyor of doom when control fails.

2 Distinctive Properties of the Ca^{2+} Signal

The cellular transmission and processing of signals typically involves the interaction of first messengers, i.e., compounds that interact with receptors on the plasma membrane of cells, e.g., hormones, followed by their processing in a form that activates internal signaling events that are mediated by diffusible molecules, termed second messengers, that are the result of the interaction of first messengers with their own plasma membrane receptors. This is the general rule for the exchange of information among cells, however, cells can communicate with each other in other ways as well, e.g., by direct contacts, in the form of gap junctions, or by means of surface proteins that recognize partner proteins on the surface of adjacent cells. However, first messengers may also bypass the plasma membrane and penetrate directly into cells to interact with receptors in various cell compartments without the intermediation of second messengers. Interesting as they may be, these alternative possibilities are the exception, the typical way to exchange information from cell to cell remaining that based on the first messenger/second messenger pattern of operation. Within this general background, Ca^{2+} appears to be a typical diffusible second messenger generated within cells in response to the interplay of the plasma membrane with external first messengers. However, in looking at the signaling function of Ca^{2+} more closely, peculiarities emerge that cannot be reconciled with an exclusive canonical second messenger role.

The canonical processing of the information of first messengers at the plasma membrane through the interaction with G-proteins and the activation of downstream enzymes does not directly "generate" Ca^{2+}. It generates instead another second messenger, e.g., inositol 1,4,5 trisphosphate ($InsP_3$), which then liberates Ca^{2+} from the endoplasmic reticulum (ER) store. One could thus define Ca^{2+} as a "third" messenger. But at the same time Ca^{2+} could also be defined as a bona fide "first" messenger, as it could penetrate directly into cells through a variety of channels, to modulate intracellular systems without the help of other second messengers. In a strict sense, however, the definition of Ca^{2+} as a first messenger based solely on its direct penetration into the cytoplasm could be questioned, as the opening of the plasma membrane Ca^{2+} channels demands the intervention of external ligands or of physical events like membrane potential changes, that would be formally equivalent to first messengers. But the first messenger role of Ca^{2+} is impeccably demonstrated on the plasma membrane by the existence of a growing number of cell types of a classical G-protein-linked seven-transmembrane domain receptor that recognizes Ca^{2+} as its first

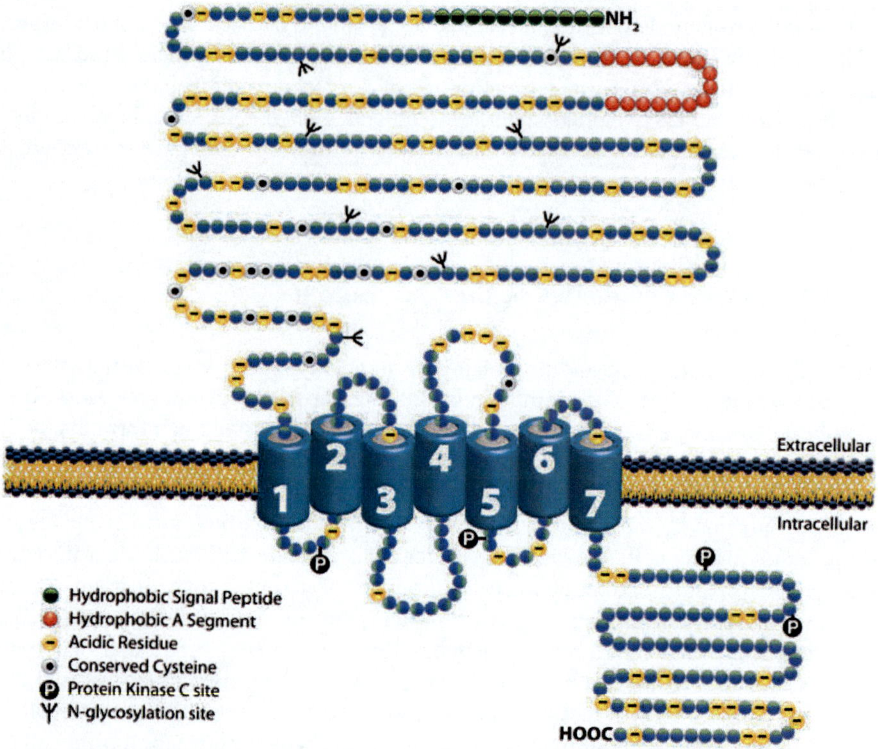

Figure 2 Schematic diagram of the 7-transmembrane domain plasma membrane Ca^{2+}-sensing receptor. Symbols are given in the key at the lower left. The diagram highlights the abundance of negative charged residues in the N- and C-terminal portions of the protein. Adapted from [6].

messenger [6], to set in motion the conventional chain of phospholipase C (PLC) mediated events that results in the elevation of cytosolic Ca^{2+}. The Ca^{2+}-sensing receptor (CaR) is organized in the plasma membrane in three domains (Figure 2): a large (600 residues) extracellular domain that contains a number of acidic regions similar to those of the low affinity Ca^{2+}-binding proteins, and which are likely to form binding sites for Ca^{2+}, a mid-domain with the canonical 7 transmembrane helices of G-protein-linked receptors, and a 200 residue intracellular C-domain. The receptor, commonly called the "Ca^{2+} sensor" was first recognized in the parathyroid cells that secrete the calciotropic hormones that regulate the organismic Ca^{2+} homeostasis, and then discovered also in cells not directly involved in the regulation of organismic Ca^{2+} homeostasis, e.g., the brain. It modifies the release of hormones in response to changes in extracellular Ca^{2+} [7], i.e., it depresses the release of parathormone by parathyroid cells, and activates the release of calcitonin by the C-cells of the thyroid [8]. Clearly, then, Ca^{2+} is not only an intracellular signaling agent, it is also an extracellular carrier of information that transmits signals to cells involved in the production of calciotropic hormones, but possibly to other cell types as well.

Another distinctive property of the Ca^{2+} signal, that sets it apart from other carriers of biological information is autoregulation, i.e., Ca^{2+} itself controls the activity of the actors that transmit its information. Autoregulation occurs at both the transcriptional and post-transcriptional levels. Important early findings on transcriptional regulation are those showing that the long-term survival of cerebellar granule neurons in culture demands a modest increase of Ca^{2+} in the cytosol. To change the set point of cytosolic Ca^{2+} to the new modestly increased level, a complete reprogramming occurs in the transcription of its transporters in the plasma membrane and in the membranes of the organelles (see below) [9–12]. The extensive transcriptional re-programming of the transporters to cope with an altogether minor cytosolic Ca^{2+} increase may at a first glance seem excessive. However, it underlines in a striking way the importance of controlling Ca^{2+}, especially in neurons, with utmost precision, i.e., it demands the concerted work of several systems.

Another important development related to the transcriptional autoregulation of the Ca^{2+} signal is the control of a plasma membrane Ca^{2+} transporter (isoform 3 of the Na^+/Ca^{2+}-exchanger, NCX3 [13]) which is crucial for the regulation of the homeostasis of Ca^{2+} in neurons by the downstream regulatory element antagonistic modulator (DREAM) [14]. DREAM (see below) is a Ca^{2+}-binding protein of the EF hand family that binds to a downstream regulatory element (DRE) site in the promoters of a number of genes, silencing them in the absence of Ca^{2+}. Upon binding Ca^{2+} to the EF hand motifs DREAM leaves the DNA, relieving the genes from inhibition. DREAM is a particularly interesting case of autoregulation of the Ca^{2+} signal: it is itself Ca^{2+}-regulated and it controls the transcription of an important Ca^{2+} transporter. More recent work has actually found that another system that controls cellular Ca^{2+} homeostasis, a plasma membrane voltage-gated channel, is a target of the transcriptional regulation by DREAM [15].

The examples of the post-trancriptional autoregulation of the Ca^{2+} signal are also numerous. A classical case is the plasma membrane Ca^{2+} pump, which is regulated by calmodulin [16]. A more recent autoregulation case is that of the neuronal plasma membrane Na^+/Ca^{2+} exchanger, which is cleaved and inactivated by calpain [17]. Calpain itself is Ca^{2+}-dependent, and becomes activated in response to the penetration of Ca^{2+} induced in the neurons by glutamate to cleave NCX3. It is also worth mentioning that the plasma membrane Ca^{2+} pump has been shown to modulate the activity of the Ca^{2+}-dependent protein phosphatase calcineurin [18], and that Ca^{2+} gates the Ca^{2+} release channels of ER (see below).

3 The Ambivalent Nature of the Ca^{2+} Signal

As briefly mentioned above, depending on a number of factors Ca^{2+} can also transmit negative signals, i.e., signals that activate processes that are detrimental to cells, and that can even lead to cell death. This ambivalence is perhaps the most striking distinctive property of Ca^{2+} as a carrier of information. Its message must be delivered to cells, and processed by them, in an exquisitely controlled way. Its level in the cytoplasm may be allowed to rise to levels above, even much above, the low-middle nM range that characterizes the resting state, but only if this occurs in a carefully controlled spatio-temporal way. This is the essential point: deviations from the physiological Ca^{2+} concentration at rest, even large deviations, may not only be tolerated, they may even be necessary to satisfy the physiological demands of cell processes, but they must be planned, and shaped by space and time coordinates that, one would be tempted to say, cells have learnt to apply intelligently.

The issue is essentially one of time: for instance, as mentioned, cells could use rapid repetitive Ca^{2+} transient, i.e., oscillatory signals, as a device to deliver the message to functions that require Ca^{2+} concentrations much in excess to those of the normal cytosol at rest. The problem of ambivalence sets in when the increase of Ca^{2+} occurs in a way that is not planned, but induced by the interplay of toxicants with cells. Again, the issue is one of time: abnormal increases of Ca^{2+} can be coped when their duration is short. The mitochondrial uptake system (see below) can accommodate them, as mitochondria would accumulate the extra Ca^{2+} together with phosphate, to precipitate insoluble hydroxyapatite within their matrix. Mitochondria are thus safety devices that can buy precious time for the cell, enabling it to survive cytosolic Ca^{2+} storms. But they can only do it for a short time, as they use the same energy to take up Ca^{2+}, which they use to synthesize essential ATP. If mitochondria are forced to use energy to accumulate Ca^{2+} for a protracted time a situation of ATP deprivation would ensue, that would even deprive of energy the ATP-dependent Ca^{2+} pumps that would expel Ca^{2+} from the cytosol. A negative vicious circle would thus be initiated that would lead to a situation of Ca^{2+} overload, and would eventually result in cell death. This is so because all Ca^{2+}-controlled stimulated functions would become activated under this condition, including potentially detrimental functions, like proteases, phospholipases, and nucleases. Their uncontrolled activity would damage the cell irreversibly, eventually ending with its death.

In a sense, then, having chosen Ca^{2+} as a determinant for function, cells are forced to live in a state of permanent controlled risk, in which the possibility of a Ca^{2+} catastrophe, i.e., of the necrotic cell death resulting from the unwanted global and massive cytosolic Ca^{2+} overload, is around the corner. But the Ca^{2+}-mediated cell death can also result from the controlled decision of cells to commit suicide. This is the process of programmed cell death (apoptosis), which is one of the meaningful ways in which cells process the Ca^{2+} signal to control essential processes such as tissue renewal and organ modeling. It has been calculated that a human body of about 70 kg loses (renews) each day a number of cells corresponding to about 1.2 kg. Apoptosis is thus essential to the life of an organism and will be discussed in some more detail later on.

Cell Ca^{2+}, however, may also be deranged in more subtle ways that do not lead to cell death. A number of cell distress conditions exists that may disturb the operation of individual actors in the Ca^{2+} controlling and signaling operation. Most of these conditions are genetic, and affect proteins (enzymes) that process the Ca^{2+} signal and/or transport Ca^{2+} across membrane barriers, thus regulating its homeostasis. These individual defects permit cell life to continue, albeit with various degrees of discomfort that can even be reflected in prominent general disease phenotypes. The area of Ca^{2+} signaling and disease has now become a popular area of research; a recent book [19] covers it comprehensively.

4 Regulation of the Ca^{2+} Signal by Ca^{2+} Buffering and Ca^{2+} Sensor Proteins

As mentioned, the distinction between Ca^{2+} buffers and Ca^{2+} sensors now appears to be less absolute than originally accepted. Distinctive properties of Ca^{2+} sensor proteins, i.e., the presence of large Ca^{2+}-induced conformational changes, their interaction with specific targets, and the ability to modulate their function as a result of the interaction, are known now to be shared by proteins hitherto classified as pure Ca^{2+} buffers, e.g, CB-D28k [4,20] and calreticulin [21–24]. Ca^{2+} buffering proteins are conventionally defined as fast or slow, depending on the rate with which they bind Ca^{2+}. Parvalbumin (PV) is conventionally considered the prototypical slow Ca^{2+} buffering protein (K_d of 4–9 nM), whereas CB-D9k (K_d of 200–500 nM) and calreticulin (K_d of 2 mM) are routinely classified as fast Ca^{2+} buffering proteins. The on rate for Ca^{2+} binding (K_{on}) is 2–3 orders of magnitude faster in CB-D9k than in PV. The specific physiological attitudes and Ca^{2+} signaling demands of cells determine the expression of slow or fast Ca^{2+} buffering proteins. Thus, the expression of CB-D9k appears to be restricted to non-excitable cells involved in Ca^{2+} re-adsorption, e.g., those of various kidney sectors [25], that of PV, in addition to kidneys, to some subsets of neurons [26] and to fast twitch muscles [27]. According to a generally accepted assumption, once Ca^{2+} gains access to the cytosol it is rapidly buffered by (fast) Ca^{2+} buffering proteins; the amount that escapes buffering will then activate the targets of the signaling function, for instance, the calmodulin-modulated processes. The list of Ca^{2+}-binding proteins has now grown very impressively. For instance, the superfamily of EF hand proteins, which are the most important Ca^{2+} sensor proteins, now numbers more than 600 members [28]. Table 1 groups the most important Ca^{2+} buffering and Ca^{2+} sensor proteins.

Proteins that are considered as pure Ca^{2+} sensors may also fulfill an important Ca^{2+} buffering role, particularly because they are routinely present in cells in high concentrations, on the order of 10 μM or more [4]. Recent work [5] has actually shown that calmodulin, the most important and ubiquitous Ca^{2+} sensor protein, buffers Ca^{2+} faster than any other Ca^{2+} buffering protein. This has led to the proposal that calmodulin would rapidly bind incoming Ca^{2+}, and then pass it on to other, slower, Ca^{2+} buffers. The proposal is at sharp variance with the common conception according to which calmodulin would instead sense the lower Ca^{2+} left free by the other Ca^{2+} buffers. Thus, according to the proposal, slow Ca^{2+} buffering proteins like

Table 1 A selection of cellular Ca^{2+}-buffering and Ca^{2+}-sensing proteins in eukaryotic cells. Some Ca^{2+}-buffering proteins have been suggested to have specific roles in Ca^{2+} signaling (indicated in the Table). As expected, however, all Ca^{2+}-buffering proteins influence (albeit indirectly) Ca^{2+} signalling. They interact with Ca^{2+} with widely different affinities: their K_ds vary from the mM to the sub μM range. All K_ds reported in the table are taken from literature sources. They sometimes vary considerably.

Protein[a]	K_d	Tissue distribution/Localization	Function
Calcium-buffering proteins			
Calbindin D9K	200–500 nM 60–300 nM	Intestinal epithelial cells/ cytosol	Mediates the transport of calcium across the enterocytes from the apical side
Calbindin D28K	180–240 nM 410–510 nM	Neuroendocrine cells; cerebellum/cytosol	
Calmegin	–	Testis/ER	Chaperone protein, spermatogenesis and infertility
Calreticulin	2 mM	Ubiquitous/ER (nucleus)	ER Ca^{2+} storage protein, chaperone protein, protein quality control (transcription regulation)
Calsequestrin	400–600 μM	Cardiac and skeletal muscle/SR	SR Ca^{2+} buffering
Calretinins	28 μM	Nervous tissue/cytosol	Intracellular Ca^{2+} buffering
Crystallins	~ 4 μM	Lens and cornea of the vertebrate eye	Increase of the refractive index of the lens, keeping transparency
Oncomodulin	–	Fetal placenta, central nervous system, macrophages, neutrophils/cytoplasm	Axon regeneration, optic nerve regeneration, macrophage-derived growth factor
Parvalbumin	4–9 nM	Fast-contracting muscles, brain, endocrine tissues	Cell-cycle regulation, second messenger production, muscle contraction, organization of microtubules and vision
Reticulocalbin	–	Ubiquitous/ER membrane	Regulation of Ca^{2+}-dependent activities in the lumen of the ER
Sorcin	~ 0.7 μM	Striated and smooth muscle cells; cardiomyocytes/associated with the RyR	Regulation of intracellular Ca^{2+} release; regulation of excitation-contraction coupling
Calcium-sensing proteins			
Alpha-spectrin	–	Brain, erythrocytes/cytoskeleton	Molecular scaffold protein
Alpha-fodrin	–	Non-erythrocyte cells/cytoskeleton	Actin crosslinking; DNA repair; cell cycle regulation

Alpha-actinin	–	Z-discs in myofibrils and cytoskeleton/cytosolic in non-muscle cells	Actin filament crosslinking and bundling protein
Annexins	μM–mM range	Ubiquitous/intracellular and extracellular/ membranes; nucleus	Trafficking and organization of vesicles, exocytosis, endocytosis fibrinolysis, coagulation, inflammation and apoptosis
Calcineurin	sub μM	Ubiquitous/cytosol and nucleus	Serine/threonine protein phosphatase, signal transduction, gene expression
Calmodulin (CaM)	<10 μM 10–100 μM	Ubiquitous/cytosolic (nuclear)	Signal transduction
CaM-dependent kinases	<10 μM 10–100 μM[b]	Ubiquitous/cytosol and nucleus	Serine/threonine protein kinases; signal transduction, mediator of learning and memory
Calnexin	2 mM[c]	Ubiquitous/ER membrane	Chaperone assisting protein folding and quality control
Calpains	μM–mM range	Ubiquitous/cytosolic (also mitochondrial)	Cysteine proteases, remodeling of cytoskeletal/membrane attachments, signal transduction, apoptosis
Calsenilin/DREAM/ KCh3	14 μM	Ubiquitous, abundant in brain/component of K^+ channels, cytosol, nucleus	Regulator of neuronal excitability; transcriptional repressor, interaction with presenilin
Calsensin	–	Invertebrate, leech nervous system/axons, growth cone	Formation or maintenance of specific axonal tracts
Centrin	1.2 μM 160 μM	Centrosome	Duplication of centrioles, other functions not clear
Fimbrin	0.15-1.5 μM	Tissue specific isoforms/cytoskeleton	Actin-crosslinking protein, formation of filopodia
Frequenin/NCS-1	0.3 μM	Ubiquitous/cytosolic	Maturation of neuromuscular synapses; synaptic transmission, role in learning and memory
Gelsolin	10–200 μM	Muscle, numerous other tissues/cytosolic	Actin binding (capping/severing), modulation of metabolism of acidic phospholipids
Hippocalcin	0.6 μM	Mammalian brain, especially hippocampus	Regulation of long term depression and spatial learning interactions
Myosine Light Chain	0.1–1 μM	Tissue specific isoforms/cytoskeleton	Muscle and general cell motility

(continued)

Table 1 (continued)

Protein[a]	K_d	Tissue distribution/Localization	Function
Neurocalcin	2 μM	Mammalian brain; central nervous system, retina adrenals	Signal transduction
Osteonectin/SPARC	80–300 nM; 5 mM	Bone/secreted, extracellular matrix glycoprotein	Initiating mineralization, promoting mineral crystal formation
Recoverin	2.1 μM	Photoreceptors cells/cytosol and disc membrane depending Ca^{2+} binding	Inhibition of rhodopsine kinase, retina sensory adaptation to the light
S-100 proteins	20–500 μM	Tissue specific expression/intracellular but also extracellular	Signal transduction; cell differentiation; transcription; regulation of cell motility; cell cycle progression
STIM1	200–600 μM	Ubiquitous/ER and plasma membrane	ER Ca^{2+} level sensor, activation of the "store-operated" ORAI1 Ca^{2+} channels
Synaptotagmin	~74 μM	Brain, endocrine system/ Membrane-bound	Synaptic vesicle docking and fusion
Troponin C	0.06–3 μM	Tissue specific subtypes/cytoskeleton	Muscle contraction
Vilips	0.3–1.7 μM		
Visinin	–	Retinal cone cell-specific	Phototransduction
Visinin-like proteins	–	Brain, cerebellum/cytosol and membrane depending Ca^{2+} binding	Signal transduction

[a] Membrane transporting proteins are not included. [b] The K_d refers to Ca^{2+} binding to the CaM that activates the kinase. [c] The K_ds reported here for calnexin corresponds to that of calreticulin.

PV would regulate the amount of Ca^{2+} bound to calmodulin, and in this way contribute directly to the regulation of the Ca^{2+} signal.

However, even if the buffering of cell Ca^{2+} by Ca^{2+} sensor proteins may be quantitatively significant, it is not their most important role; as stated above, their primary role is the processing of the Ca^{2+} signal. The transformation process has been studied in great molecular detail only in the most important sensor protein, calmodulin, but its general principles are likely to be valid for at least the hundreds of sensor proteins of the calmodulin family, the EF-hand proteins [29].

Calmodulin is an elongated protein, in which two terminal lobes, each containing two Ca^{2+}-binding helix-loop-helix motifs, are separated by a long 25-residue α helix. When Ca^{2+} becomes bound, the protein undergoes a first conformational change that exposes hydrophobic patches, mostly methionine pockets, on the surface of the two lobes. The protein at this stage still maintains its length of 62 Å, but collapses instead around the binding domains of target proteins that have come in contact with it (Figure 3). At this point, the extended dumbbell-shaped calmodulin molecule has the conformation of a hairpin, and the Ca^{2+} information it originally carried is transferred to the target protein. Calmodulin processes and transmits the Ca^{2+} information to dozens of targets, i.e., it is not a Ca^{2+} sensor committed to a single target partner. It becomes temporarily associated with them in its Ca^{2+}-bound form as a separate subunit, and in a small number of cases the association may occur and persist even in the absence of Ca^{2+} (see below).

Other sensor proteins of the EF hand group, e.g., recoverin or troponin-C, are instead committed to the modulation of the activity of a single target or of a limited number of them. Ca^{2+} sensor proteins not belonging to the EF hand family, e.g., annexins, gelsolin, proteins containing C2 domains, are generally also committed, i.e., they transform Ca^{2+} information for the benefit of a single target. The rule, however, is not absolute: important Ca^{2+} sensor proteins exist that process the Ca^{2+} signal directly and then act on numerous interacting targets, e.g., by phosphorylating them. A prominent example of Ca^{2+} sensors of this type is protein kinase C (PKC).

Coming briefly back to calmodulin and the EF hand proteins, it is interesting that some target proteins of the signaling function of Ca^{2+} possess their own "calmodulin" covalently integrated in the sequence (this is the case, for instance, for calpain). In these cases, the "calmodulin-like" processing of the Ca^{2+} signal occurs directly within the target protein itself. The pattern of transmission and processing of Ca^{2+} information is even more complex in other EF hand proteins, such as calcineurin. This phosphatase, in addition to having its own "calmodulin" as a separate subunit, also has a conventional calmodulin binding domain that senses real exogenous calmodulin with the larger subunit and it is, thus, under dual regulation by the Ca^{2+} signal.

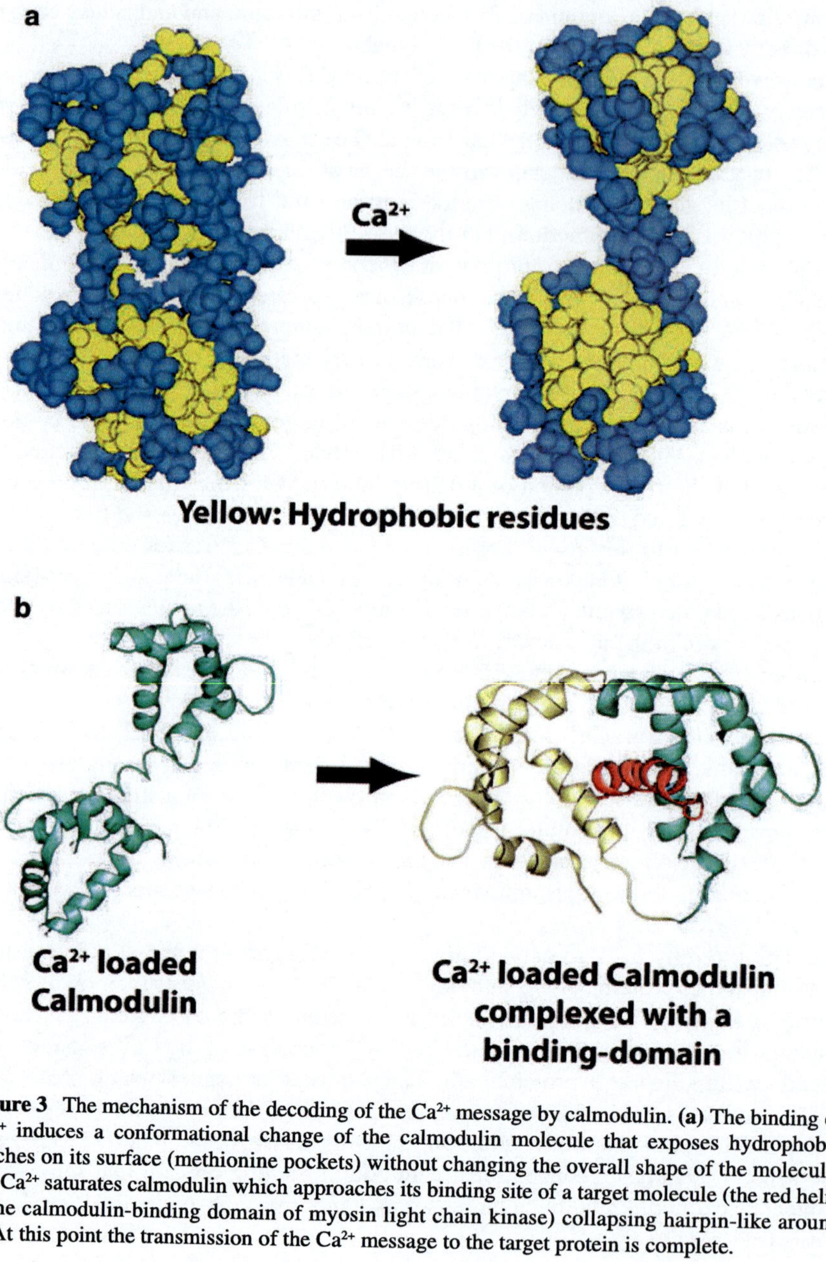

Figure 3 The mechanism of the decoding of the Ca^{2+} message by calmodulin. **(a)** The binding of Ca^{2+} induces a conformational change of the calmodulin molecule that exposes hydrophobic patches on its surface (methionine pockets) without changing the overall shape of the molecule. **(b)** Ca^{2+} saturates calmodulin which approaches its binding site of a target molecule (the red helix is the calmodulin-binding domain of myosin light chain kinase) collapsing hairpin-like around it. At this point the transmission of the Ca^{2+} message to the target protein is complete.

5 Regulation of the Ca^{2+} Signal by Membrane Transport Systems

5.1 Ca^{2+} Channels

Several plasma membrane Ca^{2+} channels have been identified and in recent years the focus has moved towards the identification of their distinctive function. The channels have been historically divided in three major groups: the voltage-gated channels (VOCs), the receptor-operated channels (ROCs), and the store-operated Ca^{2+} entry channels (SOCEs) (Figure 4).

5.1.1 The Voltage-Gated Channels

The VOCs are key transducers of membrane potential changes into intracellular Ca^{2+} transients. They are the best characterized and are divided in subfamilies that have distinct roles in biological processes. Their Ca^{2+} selectivity is high, thus making Ca^{2+} the preferred permeating species even in the presence of other abundant cations, i.e., Na^+ and K^+, in the extracellular ambient. They are complexes of 5 distinct subunits ($\alpha 1$, $\alpha 2$, β, γ, δ) encoded by different genes (Figure 5). $\alpha 1$ is the largest subunit and forms the pore: it is organized in four repeated modules of six transmembrane domains (S1-S6), the fourth of which contains the voltage sensor, in analogy with the S4 domain of Na^+ and K^+ voltage-gated channels [30]. A membrane-associated loop between the 4 S5 and S6 domains forms the channel properly. The β subunit has no transmembrane segments, whereas the γ subunit is a glycoprotein with four transmembrane segments. The $\alpha 2$ subunit is an extracellular extrinsic glycoprotein, bound to the δ subunit through a disulfide linkage that is possibly linked to the membrane through a phosphatidylinositol anchor.

Based on physiological and pharmacological properties of the type of current carried, the VOCs can be divided in six classes, termed L, N, P, Q, R and T, distinguished on the basis of the $\alpha 1$ type subunit. In turn, they can be divided in three structurally and functionally related subfamilies: Cav1, Cav2, and Cav3. The Cav1 subfamily initiates muscle contraction, secretion, regulation of gene expression, integration of synaptic signals, and mediates the L-type current. Cav2 subunits conduct N-type, P/Q and R currents, and are mainly responsible for the initiation of synaptic transmission at fast synapses. The Cav3 subfamily is important for the repetitive firing of action potentials in cardiac myocytes and thalamic neurons and is responsible for the T-type current [31].

The diversity of channels structure and function is further enhanced by the presence of multiple β subunits, that are encoded by four different genes [32,33]. Cav1 channels are more abundant in the cell bodies and proximal dendrites of neurons than Cav2 and Cav3 channels, which are instead predominant in nerve terminals and dendrites, respectively. Their preferential locations, coupled with the selective

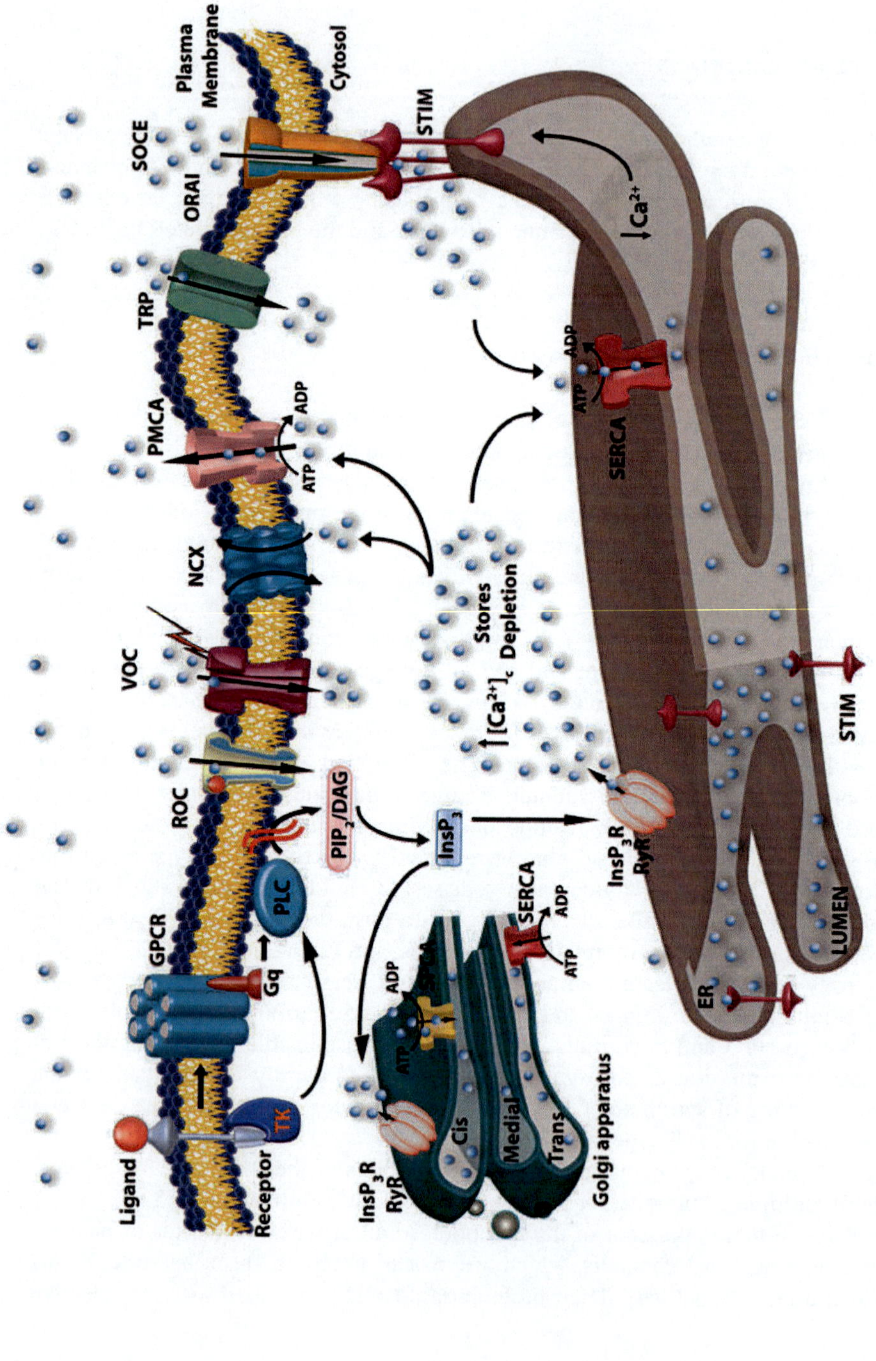

Figure 4 Schematic representation of the different Ca^{2+} transport proteins of the plasma membrane and the main intracellular stores. For details see the text. TK, tyrosine kinase-coupled receptor; GPCR, G-protein-coupled receptor; PLC, phospholipase C, PIP_2, phosphatidyl inositol 4,5-biphosphate, DAG, diacylglycerol.

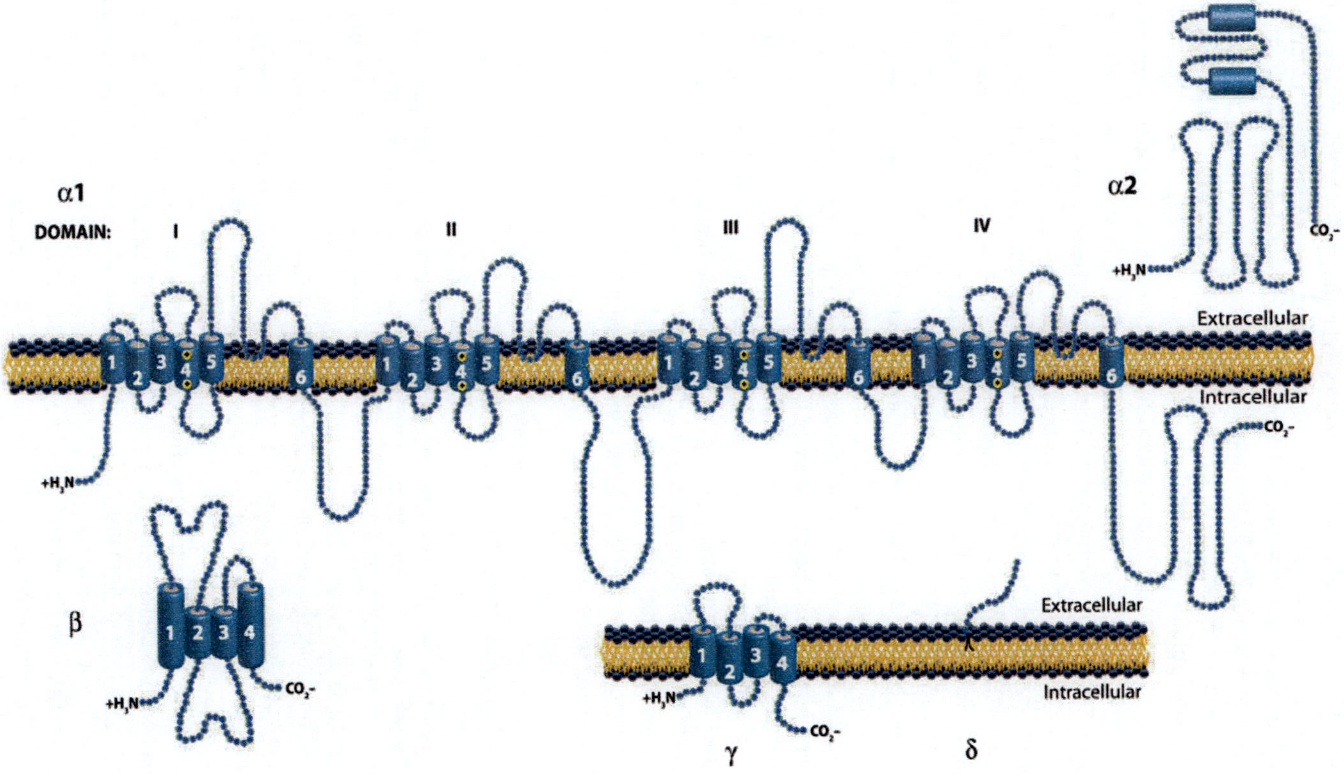

Figure 5 The membrane topology of the 5 subunits of the voltage-dependent plasma membrane Ca^{2+} channels. Details are found in the text. The positive residues in transmembrane helix 4 in each of the 6-helix transmembrane domain 4 of the α1 subunit are the voltage sensor. Adapted from [31].

local regulation of Ca^{2+} by specific buffers, confers specificity to the processes regulated by Ca^{2+} entry, e.g., it confers to the L-type channels a privileged role in the regulation of gene transcription. Calmodulin binding to the proximal C-terminal domain of Cav1.2 channel is required for the regulation of transcription in neurons [34,35], and calcineurin binding to the distal C-terminus acts as a potential transcriptional regulator as well. The distal C-terminus itself has also been proposed as transcriptional regulator [36].

Cav1 channel activity is also involved in the secretion of hormones from endocrine cells, and is specifically required for some type of synaptic transmission, i.e., in photoreceptors, when continuous neurotransmitter release is necessary.

Cav2 channels are the predominant pathways for Ca^{2+} entry initiating synaptic transmission by the release of classical neurotransmitters like glutamate, acetylcholine and GABA. Ca^{2+} entry through presynaptic P/Q and N-type channels initiates exocytosis by triggering the fusion of secretory vesicles with the plasma membrane through the action of the SNARE protein complex of syntaxin, SNAP-25 and VAMP/synaptobrevin [37]. Presynaptic Cav2.1 and Cav2.2 channels interact with SNARE, the interaction being regulated by Ca^{2+} and protein phosphorylation. The interaction has a dual role. It favors the coupling of Ca^{2+} channels with secretory vesicles, and it regulates channels activity.

Interestingly, N-type and P/Q-type Ca^{2+} currents are also regulated by G-proteins, possibly through the activity of Gβγ subunits that shift the voltage dependence of Ca^{2+} channel activation to more positive values of membrane potential through a mechanism of protein-protein interactions, thus slowing channel activation [38].

Cav3 channels conduct a T-type current, which is activated in the same range of negative membrane potential of the Na^{+} channels, and it thus well suited to sustain the rhythmic firing of action potential.

5.1.2 The Receptor-Operated Channels

The second class of Ca^{2+} channels is activated by the interaction with ligands. Most prominent among them is L-glutamate, which is the most important excitatory transmitter in mammalian brain. Glutamate activates two classes of receptors, the ionotropic receptors (iGluRs) and the metabotropic receptors (mGluRs). The iGluRs are ligand-gated non-selective cation channels and are divided in three groups on the basis of the activity of specific agonist: AMPA (2-amino-3-hydroxy-5-lethyl-4-isoxazolepropionic acid), NMDA (N-methyl-D-aspartate), and kainate (KA) [39]. iGluRs are macromolecular complexes composed of four or five subunits, and are predicted to have a bilobar structure, four membrane-spanning helices (however, M2 is not a bonafide transmembrane domain but rather a hairpin loop), with a large extracellular N-terminal domain, and an intracellular C-terminal domain. They depend on ATP for full activity: phosphorylation of their C-terminal domain by PKA, but also PKC and CaMKII, increases currents in all types of glutamate receptors [40–43]. An interesting post-transcriptional mechanism regulates the Ca^{2+} permeability of iGluRs. RNA editing occurs in the GluR2 subunit of AMPA receptors, as well as in subunit GluR5 and GluR6 of KA receptors, leading to a Gln to Arg substitution in the M2 hairpin loop. The replacement of Gln with a positively charged amino acid is evidently essential to confer Ca^{2+} impermeability to the channels. The efficiency of editing is higher in the GluR2 subunit than in the GluR5 and GluR6 subunits, and, in the case of GluR6, the editing of other two residues in the M1 transmembrane domain also controls the Ca^{2+} permeability.

KA and AMPA receptors are the primary receptors for rapid excitatory transmission in the central nervous system and, following glutamate activation, they are primarily permeable to Na^{+} and K^{+}. However, AMPA and KA receptors may also be permeable to Ca^{2+}. NMDA receptors respond to glutamate more slowly than AMPA and KA receptors, possibly because Mg^{2+} inhibits them in a voltage-dependent manner: membrane depolarization following AMPA and KA receptor activation relieves the Mg^{2+} inhibition of NMDA receptors.

The mGluRs, instead, are coupled to G-proteins. Accordingly, they are organized with the canonical seven transmembrane domains. They are encoded by 8 genes (mGluR1-8) and exist as homodimers that generate Ca^{2+} signals through the activation of distinct downstream signaling cascades that activate PLC and activate, or inhibit, adenylyl cyclase. They are expressed in neuronal and glial cells within the brain, spinal cord, and peripheral neurons and are involved in the pathophysiology of a number of diseases.

mGluR1 is the most abundantly expressed metabotropic receptor in the mammalian central nervous system, with highest expression in the Purkinje cells of the cerebellum. mGluR1 produces two types of neuronal depolarization, a rapid transient depolarization related to the release of Ca^{2+} from intracellular stores, and a prolonged and larger depolarization resulting from the activation of transient receptor potential (TRP) channels (see below).

5.1.3 The Store-Operated Ca^{2+} Entry channels

The third class of Ca^{2+} channels is that of the store-operated Ca^{2+} entry channels (SOCEs), which are activated by the release of Ca^{2+} from the ER. They were initially described in non-excitable cells, but they have now been documented in neurons and skeletal muscle cells. The idea that ER Ca^{2+} depletion could represent a signal for Ca^{2+} entry can be traced back to early work that had proposed that the biphasic nature of agonist-activated Ca^{2+}-mobilization was due to an initial emptying of the intracellular Ca^{2+} pool, e.g., by $InsP_3$, followed by the rapid entry of Ca^{2+} into the cytosol in the continued presence of $InsP_3$. The rapid entry of Ca^{2+} from outside the cell continued until the Ca^{2+} content of the store pool reached a level that inactivated it [44].

Later on, it was reported that the depletion of intracellular stores induced by thapsigargin (TG), an inhibitor of the SERCA pump (see below) was *per se* able to induce Ca^{2+} entry in different cell types. The identification of a small store-operated Ca^{2+}current (the Ca^{2+}-release activated current, CRAC) [45] activated in mast and T cells independently of the occupancy of surface receptors or of changes in cytosolic

Ca^{2+} enhanced the interest in the topic, but the pathway involved in the process remained obscure for a long time. Different mechanisms were proposed ranging from the existence of free diffusible messengers to a conformational coupling of CRAC channels and $InsP_3$ receptors. A number of candidate genes were proposed as the putative messenger, among them those of the TRP channels, but the molecules involved in the pathway were identified only recently: a Ca^{2+}-binding transmembrane protein of the EF-hand family (STIM proteins) serves as sensor of Ca^{2+} within the ER. The protein communicates with the plasma membrane store-operated channel that is composed of ORAI subunits. 2 STIM isoforms (1 and 2) were identified and the ORAI protein family was found to be composed of three isoforms [46], in which ORAI1 was demonstrated to be the pore-forming subunit of the channels [47]. Under resting condition, with the STIM proteins fully occupied by Ca^{2+}, STIM1 and ORAI1 would be diffusely localized at the ER and plasma membrane (PM) sites, respectively. When store depletion occurs, STIM1 undergoes a conformational change that redistributes it to specific districts forming "puncta" structures, that correspond to ER-PM junction, i.e., to specialized regions of the ER positioned within 10–20 nm of the PM. At the same time, ORAI1 accumulates at the corresponding PM sites, thus coupling with STIM1 and allowing the opening of the CRAC channels and the generation of localized Ca^{2+} hot spots [48]. In a second phase, store refilling causes the return of STIM1 and ORAI1 proteins to the original states, thus dissolving the "puncta". This model, based on distinctive rearrangements of STIM1 and ORAI1 in the cell, would require seconds for the activation of the channels. This timing has functional consequences, e.g., in T cells the "puncta" assembling and disassembling may generate Ca^{2+} oscillations [49] that would in turn drive gene expression through NFAT and other transcription factors [50,51] (see below).

5.1.4 Transient Receptor Potential Channels

Another class of channels, which had been originally related to the SOCE channels, and which can also generate changes in intracellular Ca^{2+} concentration by mediating its entry across the PM are the TRP channels.

They constitute a large and functionally versatile family of cation-conducting channels, and are generally considered cell sensors. They are expressed in a large number of tissues and cell types (excitable and non-excitable) and, when activated, cause a cell depolarization that in turn may trigger the activation of different voltage-dependent ion channels. In mammals, 28 TRP channels have so far been found, classified according to their homology in 6 different subtypes: TRPC (canonical), TRPV (vanilloid), TRPM (melastatin), TRPA (ankyrin), TRPML (mucolipin), and TRPP (polycistin). The literature on TRP channels has now grown very impressively and only a short overview can be given here. More detailed information can be found in a number of comprehensive reviews, e.g., [52,53].

A common theme that links the TRP channels is their activation or modulation by phosphatidylinositol phosphates, such as phosphatidylinositol 4,5-bisphosphate (PIP_2) [54]. However, they are also modulated by Ca^{2+}, which is responsible for generating both a positive and a negative feedback. They are organized with six transmembrane domains, and most probably assemble into tetramers to form non-selective cationic channels. The TRP channels can contribute to change the intracellular Ca^{2+} concentration either directly by acting as a Ca^{2+} entry pathway (even if their selectivity for Ca^{2+} differs in the different subtypes), or indirectly by changing the membrane polarization. The TRPC-type channels have been claimed to have a special relationship with the SOCEs (and their constituents STIM1 and ORAI1); the issue has been a matter of vigorous discussion in the field of Ca^{2+} signaling. It is now generally accepted that ORAI1 may interact with TRPCs and act as regulatory subunit that confers STIM1-mediated store depletion sensitivity to them [55–57]. Thus, in a sense, the TRPC channels might act as SOCEs, even if they are a distinct type of channel with their own properties: high Ca^{2+} selectivity, very small single channel conductance, and different Ca^{2+} modulation.

A final comment is necessary on TRP channels: even if it is clear that they have different functional effects depending on their strategic localization on the plasma membrane, most of them are also localized in the membrane of the intracellular organelles. Thus, TRPV and TRPP channels have been found on ER and Golgi membranes, and TRPMLs have been proposed to mediate a NAADP-activated intracellular Ca^{2+} release from endosomes and lysosomes (see below).

5.1.5 The Intracellular Ca^{2+} Channels

The endo/sarcoplasmic reticulum (ER/SR) and more recently, the Golgi apparatus (GA), are recognized as the main intracellular Ca^{2+} stores. Two types of Ca^{2+} receptors/channels essentially operate Ca^{2+} mobilization from them, the ubiquitous inositol 1,4,5-trisphosphate receptors ($InsP_3R$) and the ryanodine receptor (RyR), which is not present in all cell types. $InsP_3R$ and RyR are channels with large conductance. They are only relatively selective for Ca^{2+}, at variance with the voltage-gated and store-operated plasma membrane Ca^{2+} channels that are more selective. However, considering that Ca^{2+} is probably the only cation with an appreciable electrochemical gradient across the ER/SR membrane the lack of selectivity does not represent a problem.

The $InsP_3Rs$ are encoded by three different genes that have distinct patterns of tissue expression (however, some overlapping occurs, especially during cell differentiation) and contribute to shaping different Ca^{2+}-linked signaling pathways. The channels are constituted by homo- or hetero-tetramers of a large (2700 residues) protein spanning the membrane with a hydrophobic region containing six helices. A partial 3D structure of the channel has recently become available [58]. The N-terminus and the C-terminus of the protein are in the cytosolic region, the N-terminal region representing the portion of the protein, which contains the $InsP_3$ binding domain and the "regulatory" domain. The opening of the $InsP_3R$ is controlled by the binding of the second messenger $InsP_3$ (generated by activation of PLC enzymes), mainly PLC β-stimulated by G-protein-coupled receptors, and PLCγ by tyrosine kinase receptors.

A flexible linker region, connecting the $InsP_3$ binding domain with the first 200 amino acids (residues 1–223 of isoform 1) of the protein, is essential for the modulation of pore opening, possibly by decreasing $InsP_3$ affinity. Several molecules interact with the $InsP_3R$ and modulate its activity, e.g., homer family adaptor proteins, protein phosphatases (i.e., calcineurin), PKA, PKC, and CaMKII, the tacrolimus-binding immunophilin FKBP12, IRBIT, ATP, and Ca^{2+}. Ca^{2+} is probably the most important interactor as it has both stimulatory and inhibitory effects, however, the structural basis of its regulation is still not understood [59]. Cytosolic Ca^{2+} is a co-agonist of the $InsP_3Rs$, as it strongly increases its activity at concentrations up to about 300 nM. By contrast, at higher concentrations it inhibits the receptor. Luminal Ca^{2+} also sensitizes the $InsP_3Rs$, possibly by tuning its sensitivity to cytosolic $InsP_3$. A Ca^{2+}-mediated inhibition of the receptor is assumed to contribute to the termination of local cytosolic Ca^{2+} signals. However, it is not clear whether this effect depends on the binding of Ca^{2+} to the receptor or to an associated protein [60]. Calmodulin had been initially suggested as a candidate protein, but the suggestion has now lost momentum.

The RyRs are also encoded by three distinct genes with different tissue expression pattern. RyR1 is expressed in skeletal muscle, RyR2 in heart, cerebellum (Purkinije neurons), and cerebral cortex, and RyR3 is more ubiquitous, even if with low levels of expression. The RyRs are formed by homo-tetramers that associate to form the largest known channel (>2 MDa). Cryoelectron microscopy studies have contributed to the understanding of the functioning of this gigantic molecule (reviewed in [61]). The C-terminal portion of the protein forms the pore and the large cytoplasmic region contains the sites where most RyR modulators interact. The major gating mechanism is the excitation-contraction (E-C) mediated coupling with the voltage-dependent Ca^{2+} channel dihydropyridine receptor (DHPR) located in the T-tubules. The molecular mechanism of E-C coupling differs between skeletal and cardiac muscle [62]. In skeletal muscles a physical interaction (electromechanical coupling) between the Cav1.1 DHPR and RyR1 is required; in cardiac muscle Ca^{2+} release by the RyR2 is initiated by Ca^{2+} influx via Cav1.2, i.e., by the Ca^{2+}-induced Ca^{2+} release (CICR). In heart, then, the interaction is functional rather than physical as in the case of RyR1. CIRC can also originate from the flickering of ER Ca^{2+} channels and, even if originally described for the gating of RyR2, is now recognized as the major gating mechanism for RyR3. Other agents can gate RyR2 and 3, i.e., cyclic ADP ribose, cADPR, generated by ADP-ribosyl cyclases, in particular by their ectoenzyme form [63]. cADPR appears to act mainly in smooth muscle cells [64], pancreatic acinar cells [65], and in the nervous system [66].

The $InsP_3R$ the channel activity of the RyR is also modulated by a number of molecules, e.g., PKA, FK506 binding proteins (FKBP12 and 12.6), calmodulin, Ca^{2+}/calmodulin-dependent protein kinase II, calsequestrin, triadin, junctin, Mg^{2+}, ATP, and Ca^{2+} itself.

5.2 Ca^{2+} Pumps

Animal cells express three Ca^{2+} ATPases (pumps) in the PM (PMCAs), in the ER/SR (SERCAs), and in the Golgi membranes (SPCAs). They lower the concentration of cytosolic Ca^{2+} by exporting it to the external medium, or to the internal space of the vesicles of the reticulum and of the Golgi system. The three pumps, like additional Ca^{2+} pumps in plant cells and in cells of lower eukaryotes, which will not be discussed here, belong to the superfamily of P-type ATPases [67] which conserve temporarily the energy liberated by the splitting of ATP in the form of an aspartyl phosphate in their reaction center. The superfamily now contains hundreds of members, sub-grouped in at least 8 subfamilies (Figure 6).

The mammalian ATPases belong to sub-groups II A (the SERCA and SPCA ATPases) and IIB (the PMCA ATPases). They display significant sequence differences in regions not directly related to the catalytic mechanism, i.e., in areas related to regulation and interaction with inhibitors and other partners, but share essential properties, e.g., membrane topography and the general reaction mechanism. The reaction scheme of the three pumps (Figure 7) had initially predicted two functional/conformational states: in the E1 state the pumps would have high affinity for Ca^{2+} and would interact with it at one membrane side, and in the E2 state the affinity for Ca^{2+} would become much lower, causing its release to the opposite side of the membrane [68]. The solution of the 3D structure of the SERCA pump at the atomic level 12 years ago [69,70] has confirmed the basic principle of the E1-E2 reaction scheme, but has greatly increased the complexity of the catalytic mechanism, showing that the binding of Ca^{2+} at one side of the membrane induces a series of large conformational transformations that switch the extra-membrane portion of the pump from a compact to a more open structure. The conformational changes, however, also involve the transmembrane domains of the pump, leading to the phosphorylation of the catalytic aspartic acid by ATP and, in a series of documented conformational transitions, to the change of the high affinity phosphorylated E1 pump to a lower Ca^{2+} affinity state that leads to the dissociation of Ca^{2+}, regenerating the Ca^{2+}-free E2 enzyme.

The cartoon of Figure 8 [71], which is reproduced with minor modifications from a review by Toyoshima [72], offers a pictorial view of the atomic path by which Ca^{2+} crosses the membrane of the SR on its way from the cytosol to its lumen. It contains details on the atomic aspects of the transfer that cannot be described and explained in the context of this review. A full discussion of them can be found in [72]. The 3D structure has confirmed the existence, in the SERCA pump, of the two Ca^{2+} binding sites that had been predicted by mutagenesis work [73,74]. The two sites are a peculiarity of the SERCA pump; the PMCA pump has only one, corresponding to site 2 of the SERCA pump, as it lacks an essential acidic residue in the transmembrane domain 5 [75]. This residue is also absent in the SPCA pump, which also has only one Ca^{2+} binding site.

The three mammalian pumps are all inhibited by the general inhibitors of P-type ATPases La^{3+} and orthovanadate $[(VO_3(OH)]_2^-$, although mechanistic differences in the case of La^{3+} exist for the case of the PMCA pumps. The SERCA pump is also specifically inhibited by compounds that are inactive against the other two pumps, e.g., TG. Inhibitors of similar specificity and potency are not available for the other two pumps. Interestingly, however, 2 peptides of the caloxin family (caloxin 2A1 and caloxin 1A1) have been claimed to inhibit the PMCA pump by interacting with its extracellular domains 2 and 1, respectively [76,77].

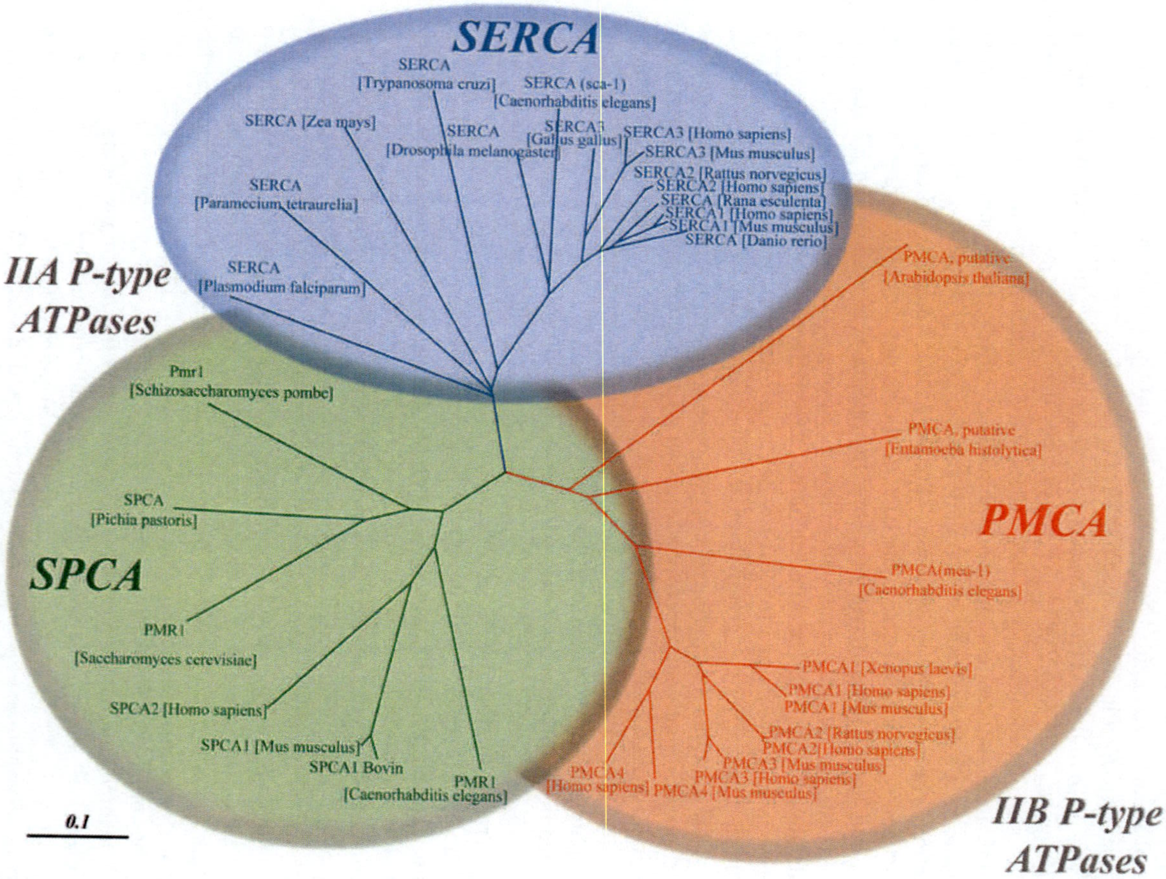

Figure 6 Phylogenetic tree of Ca^{2+} transporting ATPases (pumps). The sequences were aligned with ClustalW software, and the tree was generated using Tree View. The 3 branches represent the 3 Ca^{2+}- ATPases discussed in the text. Adapted from [71].

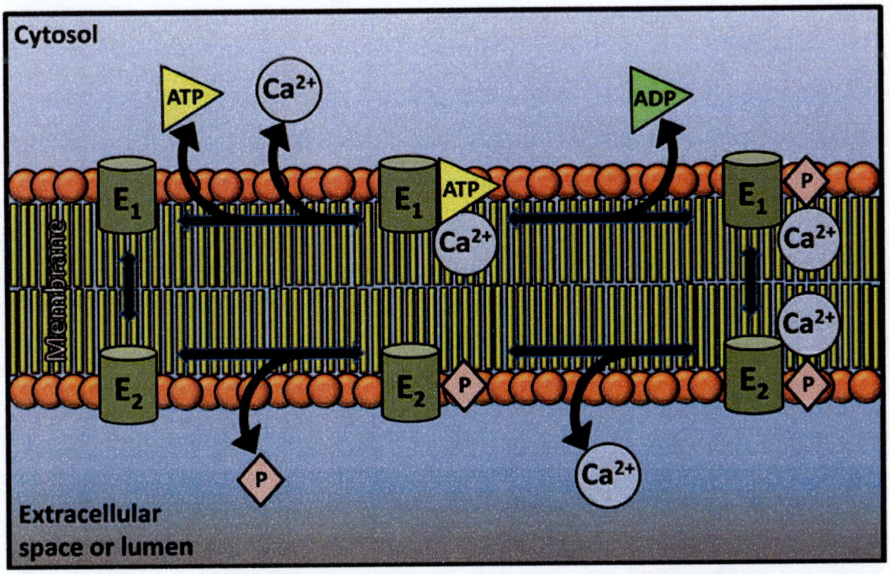

Figure 7 The reaction scheme of Ca^{2+} ATPases. The scheme is simplified, the number of states of the ATPases has been greatly increased by the work on the three dimensional structure of the SERCA pump. See the text for details.

All Ca^{2+} pumps interact with Ca^{2+} with high affinity [71], and are thus the fine tuners of cell Ca^{2+}. Their $K_m(Ca^{2+})$ is well below 1 μM. The affinity for Ca^{2+} is particularly high in the SPCA pumps, whose K_d is about 10 nM in the SPCA1 isoform, and even lower in the SPCA2 isoform [78,79]. The extremely high Ca^{2+} affinity of the SPCA pumps, which have K_ds well below the concentration of Ca^{2+} in the cytosol at rest, ensures that the Golgi vesicles will be always filled with Ca^{2+} even in the absence of agonist-induced cytosolic transients. This is crucial, since Ca^{2+} is required for the activity of enzymes within the Golgi vesicles, most notably the endoproteases that process the pro-hormones. Importantly, the SPCAs also transport Mn^{2+}, which is essential inside the Golgi vesicles for the O- and N-glycosylations of a number of proteins [80,81]. The Ca^{2+} affinity of the PMCA pump requires another comment. The pump is a classical target of calmodulin regulation (see above). In its absence it can still interact with Ca^{2+}, but only with very low affinity, i.e., with a K_d between 10 and 30 μM [82]. The K_d drops to about 0.5 μM in the presence of calmodulin [82] which interacts with a C-terminal domain of the pump [83], with a K_d in the nM range [84]. This domain, however, also plays another role in the regulation of the activity of the pump. In the absence of calmodulin it folds over, binding to two sites in the main body of the enzyme. It keeps the pump auto-inhibited until calmodulin removes it from the binding sites, relieving the autoinhibition [85,86].

Of the 4 basic PMCA isoforms (see below), one, PMCA2, behaves peculiarly in its reaction to calmodulin [87,88]; it expresses very high activity even in its

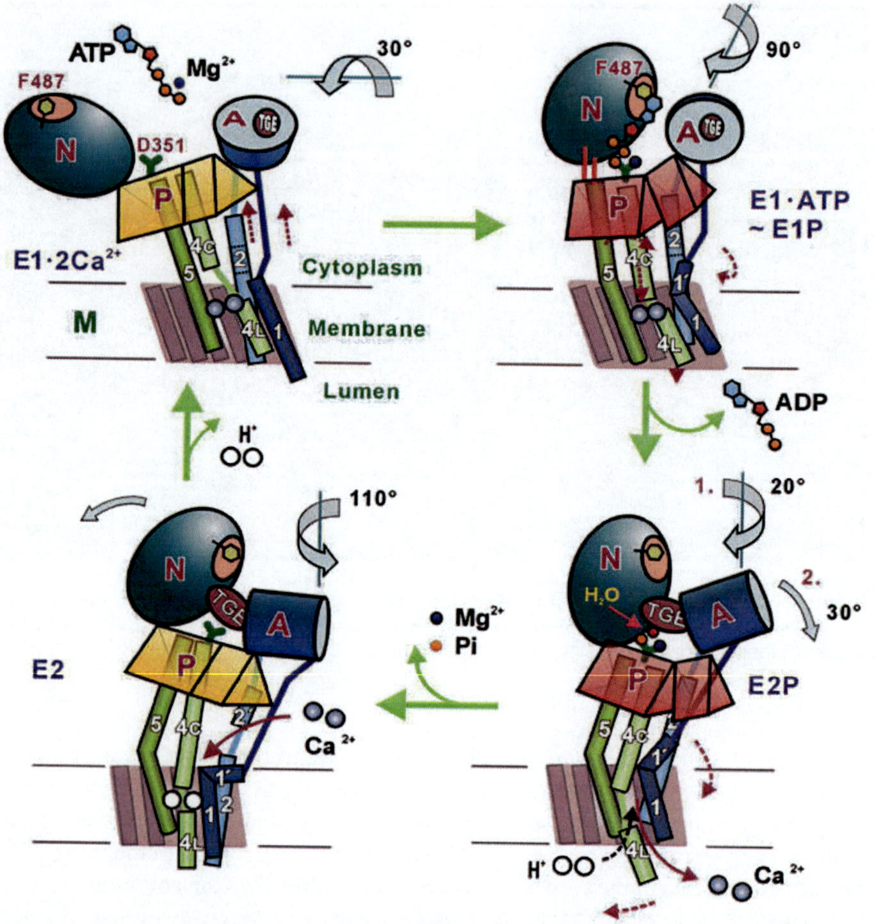

Figure 8 A cartoon illustrating the conformational changes of the main domain of SERCA pump during the reaction cycle. The model is based on the three dimensional structure of the SERCA pump. Adapted from [72].

absence. Since the PMCA2 calmodulin-binding domain does not differ from that of the other 3 basic PMCA isoforms, it is likely that the high activity in the absence of calmodulin reflects the suboptimal ability of the calmodulin-binding domain of PMCA2 to interact with the autoinhibitory site(s) in the main body of the pump. Another important aspect of the Ca^{2+} affinity of the PMCA pump is its stimulation by acidic phospholipids, which decreases the K_d to values even lower than those achieved with optimal calmodulin (about 0.2 μM [89]). The significance of the activation by acidic phospholipids, which bind to the basic calmodulin domain, but also to another domain in the first large cytosolic loop [90,91] is still obscure, but it appears possible that phospholipids could activate the pump *in vivo* in alternative to calmodulin.

The three mammalian Ca^{2+} pumps are the products of multigene families: separate genes express three basic isoforms of the SERCA pump, four of the PMCA pump, two of the SPCA pump. The basic isoforms of each pump type share reaction mechanism and membrane topology, but differ in tissue distribution, regulation properties, and in some details of activity, e.g., the affinity for Ca^{2+}. Their number is greatly increased by the alternative splicing of the primary transcripts of all pumps (SPCA2 is the only exception), the functional differences among the splicing products being in general greater than those among the original basic gene products. A detailed discussion of all splicing variants of the pumps would be out of the scope of this contribution; the short description offered here will thus only underline aspects and variants that are particularly significant.

The transcripts of the SERCA genes are subjected to alternative processing at their 3' end, generating a number of splice variants with specific tissue distribution and activity. That of SERCA1 is spliced to generate the SERCA1a and 1b variants, which are expressed in adult and neonatal fast-twitch skeletal muscles, respectively. The change in expression pattern during development and tissue differentiation indicates that each isoform is adapted to specific functions. The transcript of the SERCA2 gene is alternatively processed to generate the SERCA2a variant, which is expressed selectively in heart, slow-twitch skeletal muscles, and smooth muscle, and to the SERCA2b variant which is expressed ubiquitously and is thus considered as the housekeeping SERCA pump [92]. Interestingly, the extended, 49 amino acid long C-terminal portion of the SERCA2b pump, which contains a highly hydrophobic segment that forms an additional transmembrane domain (11th) [93], confers to the variant higher affinity for Ca^{2+} and lower catalytic turnover rate [92,94]. Both SERCA2a and 2b are sensitive to the membrane protein phospholamban, which regulates their activity by becoming reversibly bound to them in a process that depends on its phosphorylation by PKA (possibly, also by a calmodulin-dependent protein kinase). Unphosphorylated phospholamban binds to the pump maintaining it inhibited, phosphorylated phospholamban leaves the binding site(s), restoring pump activity, for instance during β-adrenergic stimulation [95–97]. SERCA3 is expressed in a limited number of non-muscle cells, and the splicing pattern of the transcript of its gene is complex. All documented variants have lower Ca^{2+} affinity than the other basic isoforms. SERCA3 seems to be specialized for the control of vascular and tracheal smooth muscles, its low Ca^{2+} affinity suggesting that it would only become activated when cytosolic Ca^{2+} reaches abnormally high levels.

The two basic products of the SPCA genes have different tissue distribution. The SPCA1 pump is ubiquitous, and is thus considered the housekeeping isoform. Its expression level varies with the tissue, and is particularly high in human epidermal keratinocytes [98]. The expression of SPCA2 is much more tissue restricted. Its transcript has been found in particularly high amounts in the mucus-secreting goblet cells of human colon [99] indicating its possible role in the regulation of the secretion of mucus. Alternative splicing has only been documented in the primary transcript of SPCA1, resulting in the generation of four transcripts. Very little is known on their possible differential properties, although some kinetic differences have been described.

The four basic PMCA gene products differ in tissue distribution and calmodulin affinity. Pumps 1 and 4 are ubiquitous and have poorer calmodulin affinity than isoforms 2 and 3. They were both considered as housekeeping pumps, but recent work has indicated that PMCA4 could have more specialized roles, e.g., in the testis where it represents more than 90% of the total PMCA protein [100]. The ablation of the PMCA4 gene causes male infertility, due to the inability of the sperms to achieve hyperactivated motility, and thus to reach the eggs to perform the fertilization [101]. PMCA2 and 3 have higher calmodulin affinity and their expression is restricted to a limited number of tissues; PMCA2 is expressed prominently in the nervous system and in the mammary gland, and PMCA3 in the nervous system and skeletal muscles [102]. The transcripts of all PMCA genes undergo alternative processing at two sites. Site A corresponds to the portion of the pump located upstream of the phospholipid binding domain in the first cytosolic loop. Site A insertions lead to a number of variants depending on the number of exons inserted; the most important is variant *w* in which the insertion of 3 exons directs the pump (the isoform tested was PMCA2) to the apical plasma membrane of polarized cells, whereas smaller inserts sort the variants to the basolateral domain [103]. Site C corresponds to the C-terminal calmodulin binding domain of the pump and generates a plethora of variants depending on the type of alternative processing, in which portions of exons can be inserted piecemeal. In most cases the result of the site C insert(s) is a change of the reading frame, and the creation of a premature stop codon that truncates the resulting pump protein. Pumps in which no site C inserts occur are designated as *b* variants, whereas those with various inserts are designated as c, d, e, f, and a. The most important is variant *a*, in which one full exon (PMCA1, 3, and 4), or two full exons (PMCA2) are inserted.

The alternative processing of the transcripts of the PMCA1, 3, and 4 genes occurs in an essentially similar way. That of the gene of PMCA2, however, has peculiar complexities. Together with other properties, for instance the ability of the pump to function at a very high rate in the absence of calmodulin (see above), the complexity of alternative processing singles out PMCA2 from the other 3 basic PMCA isoforms.

5.3 *The Plasma Membrane Na^+/Ca^{2+} Exchanger*

The plasma membrane of most animal cells contains a system with lower Ca^{2+} affinity than the PMCA pump that ejects Ca^{2+} in exchange with Na^+ (NCX). The system belongs to the SLC8 (solute carrier family 8) superfamily of Na^+/Ca^{2+} exchangers. The superfamily also contains SLC24 that transports K^+ as well; a cluster of 23 orthologous genes (COG0530) are named Ca^{2+}/Na^+ antiporters, even if no functional data have so far been produced. The NCX is particularly active in the cells of excitable tissues, and uses the energy of the electrochemical Na^+ gradient to allow Na^+ to flow into the cell across the PM in exchange for the export of Ca^{2+}, with a transport stoichiometry of 3 Na^+ for 1 Ca^{2+}. It has a much larger transport capacity than the PMCA, transporting thousands of Ca^{2+} ions per second [104], a property that, coupled to its low Ca^{2+} affinity, allows it to return the cytosolic Ca^{2+} concentration to its normal resting low levels after large physiological increases, for instance those produced by the neuronal action potentials. Since the operation of the NCX is electrogenic and voltage-sensitive, it can reverse during cell activation and lead to the intake of Ca^{2+} into the cell [105]. The reversal of the canonical direction of the exchanger is also induced by changes in the concentration, i.e., the gradient, of the transported species. This occurs, for instance, in the case of heart cells exposed to the action of digitalis: the inhibition of the Na^+/K^+-pump increases Na^+ in the sarcoplasm, reversing the operation of the exchanger and inducing the well-known positive inotropic effect linked to the influx of Ca^{2+}.

Three genes code for distinct isoforms of the NCX in mammalian cells (NCX 1-3). They have variable tissue distribution and regulatory properties. NCX1 is distributed ubiquitously, whereas the expression of NCX2 and NCX3 is restricted to brain and skeletal muscles [106,107]. All NCXs are predicted to contain nine membrane-spanning domains, separated in two parts of the sequence by a 500-residue intracellular loop [108,109]. Interestingly, the loop can be removed still leaving behind an active exchanger [110], showing that the transmembrane portion of the exchanger is the basic functional transport unit (the 3D structure of a prokaryotic exchanger has recently been published, showing 10 transmembrane domains instead of the 9 of the eukaryotic exchangers [111]). Each exchanger protein contains two conserved homologous α repeats, one on each of the two transmembrane domains, arising from a gene duplication event, that are important for the binding and the translocation of ions [109]. The two motifs are also present in other members of the Na^+/Ca^{2+} exchanger superfamily, e.g., the NCKX, that also exchanges K^+ and was first identified in the retina [112], and in microbial exchangers that lack the large intracellular loop [109,113]. They are not present, however, in NCLX, a philogenetically ancestral branch of the Na^+/Ca^{2+} exchanger superfamily which also exchanges Li^+ [114,115], and which has recently been identified as the long sought mitochondrial Na^+/Ca^{2+} exchanger (see below). The number of exchanger isoforms is increased by alternative processing of the transcripts in a region corresponding to the C-terminal portion of the large cytosolic loop. The processing of the primary transcript of NCX1 gives rise to a complex set of splice variants that differ in the 561–681 stretch of the protein. The transcript of NCX3 also undergoes alternative splicing in a region corresponding to a similar location in the protein as NCX1, whereas no splicing products have so far been described for NCX2.

Operationally, the NCX is activated by the binding of intracellular Ca^{2+} to a Ca^{2+}-binding domain 1 (CBD1) in the main cytosolic loop that triggers a conformational change that transforms the NCX into an activated state. The 3D crystal [116] and NMR [117] structures of CBD1 have recently been solved, and shown to bind 4 Ca^{2+} to an immunoglobulin-like fold. A second Ca^{2+}-binding domain was also identified [117] (CBD2), and a structural model was built of the entire regulatory loop which is given in the cartoon of Figure 9. A detailed discussion of the model is out of the scope of this contribution, but can be found in [117] and [118]. Basically, in the model CBD1 and CBD2 are arranged in antiparallel fashion, the Ca^{2+}-binding region

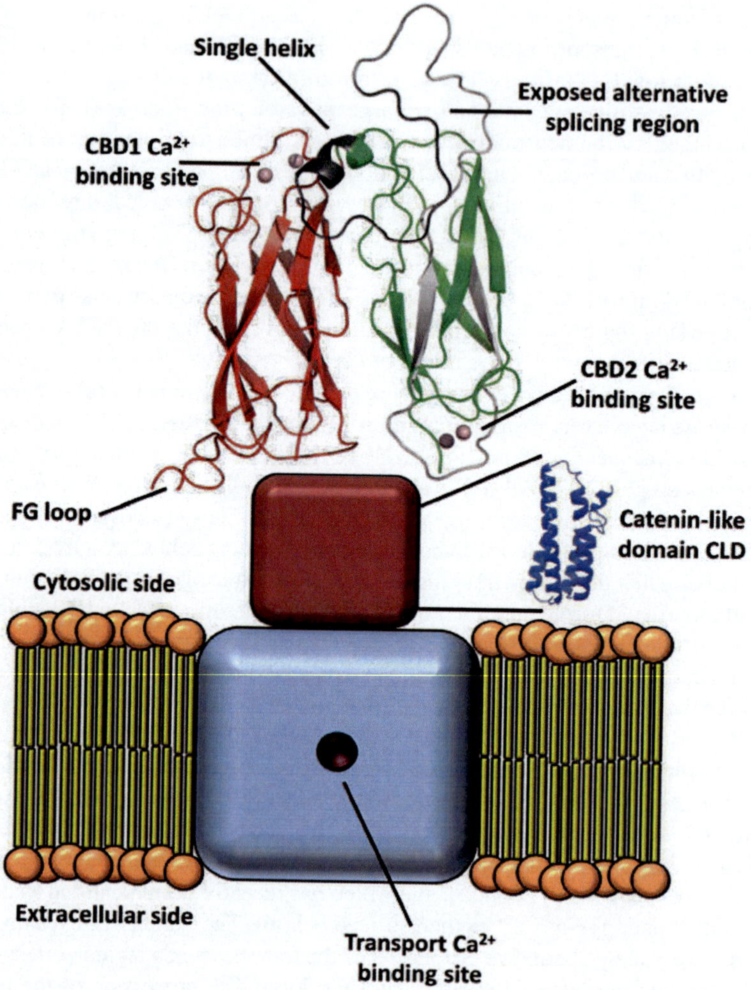

Figure 9 A model for the structure of the plasma membrane Na^+/Ca^{2+} exchanger. The cartoon is a modified version of the model proposed by Hilge et al. [117].

of CBD1 unfolding when Ca^{2+} is removed, possibly moving the exchanger into an inactive state. Hilge et al. [117] have presented a structure for each of the two main splicing variants of the NCX, which contain the mutually exclusive CBD2 encoding exons A and B. At variance with CBD2A, CBD2B has unstructured Ca^{2+}-binding sites under physiological conditions, suggesting lower Ca^{2+} fluxes in non-excitable cells that contain exon B compared with excitable cells that contain exon A.

An interesting recent development relates NCX (or, rather, NCKX) to the color of the skin. The allelic frequency of a single nucleotide polymorphism (SNP) in the coding region of member 5 of SLC24A family (which contains 6 members) varied considerably between populations of Caucasian and African ancestry [119]. The study was extended to zebrafish [120] and other human groups, and has shown that the SLCA24A5 encodes a protein similar to the K^+-dependent members of the Na^+/Ca^{2+} exchanger superfamily.

6 Intracellular Organelles

6.1 Mitochondria

Mitochondria are intracellular organelles endowed with two membrane barriers with decreasing ion permeability properties: the outer mitochondria membrane (OMM) is freely permeable to ions and small molecules, and the inner mitochondrial membrane (IMM), which is folded into the internal invagination called "cristae", contains the multi-enzymatic complexes of the respiratory chain, of the ATP synthase and the Ca^{2+} transport systems.

Mitochondrial Ca^{2+} transport has unique characteristics. The uptake does not need ATP hydrolysis for Ca^{2+} entry, but utilizes a uniporter and the membrane potential ($\Delta\Psi$, negative inside) maintained across the inner membrane by the respiratory chain as the driving force. The uniporter has been proposed to be a gated and highly selective ion channel [121,122]. For Ca^{2+} efflux, mitochondrial exchangers use the concentration gradient of Na^+ (H^+ as well in some mitochondrial types) across the inner membrane to cause the release of Ca^{2+} back into the cytosol [123]. The cycle is then completed thanks to the efflux of Na^+ via the Na^+/H^+ exchanger (NHE) (Figure 10) [124]. Under resting conditions the rates of Ca^{2+} influx and efflux are slow and ensure the maintenance of a low matrix Ca^{2+} concentration. The kinetic equilibrium between influx and efflux thus results in a futile (energy consuming) cycle of Ca^{2+} across the mitochondrial inner membrane [125]. When cytoplasmic Ca^{2+} increases above a given threshold (>10 μM) a rapid Ca^{2+} accumulation by mitochondria is initiated and matrix Ca^{2+} increases dramatically. Finally, excess Ca^{2+} accumulation by mitochondria (mitochondrial Ca^{2+} overload, MCO) may result in the opening of a large non-selective channel in the inner mitochondrial membrane, the mitochondrial permeability transition pore (mPTP) that collapses the membrane potential, induces swelling of the inner membrane and rupture of the outer one, and releases proteins of the intermembrane space (IMS) into the cytoplasm.

After the discovery that isolated respiring mitochondria were capable to sustain Ca^{2+} accumulation [126,127], many aspects of the mitochondrial Ca^{2+} uptake and extrusion mechanisms were clarified. Thus, it was established that Ca^{2+} uptake was an electrogenic process, which was countered by Ca^{2+} efflux so that electrochemical gradient equilibrium did not occur [128].

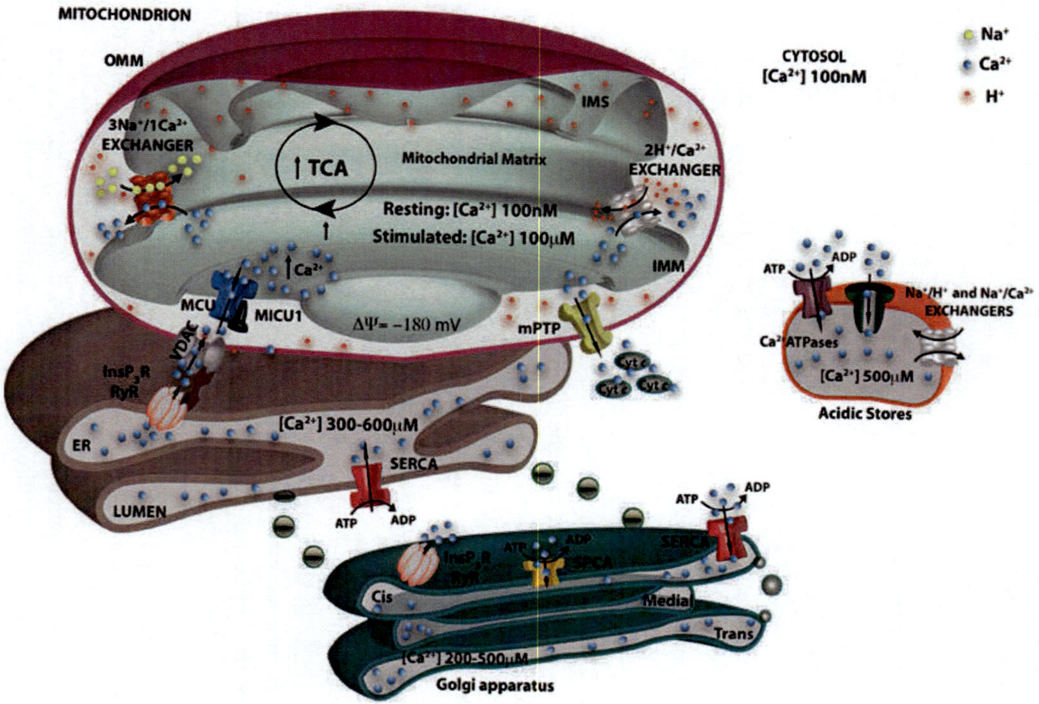

Figure 10 Schematic representation of Ca^{2+}-handling organelles and their molecular toolkit. Mitochondria, endoplasmic reticulum (ER), Golgi apparatus, and acidic stores are shown. For details see the text. ER mitochondria contact sites are shown by the juxtaposition of the MCU and $InsP_3$ receptor. Interestingly, the voltage-dependent cation channels (VDAC) of the OMM clustered at the ER/mitochondrial contact sites play a key role in the rapid transfer of the high Ca^{2+} microdomain from the surface of the mitochondria to the intermembrane space to which the MCU is exposed [228].

However, the finding that the uptake system operated with very low Ca^{2+} affinity appeared difficult to reconcile with its function in the extremely low cytosolic Ca^{2+} concentration. For a while, then, the idea that mitochondria could efficiently control the homeostasis of Ca^{2+} in the cell lost favor, even if research on the topic continued to produce information. Thus, patch clamp experiments on mitoplasts (swollen mitochondria without the OMM) showed that the uniporter (MCU) is a highly selective hardly saturable Ca^{2+} channel with an activation domain and a transport site [121]. Pharmacological studies led to the identification of compounds able to either inhibit or activate the MCU, e.g., ions like lanthanides, Mg^{2+}, ruthenium red (RR) and its derivate Ru360 (reviewed in [129]), and the plasma membrane Na^+/Ca^{2+} exchanger inhibitor KB-R7943 [130]. Physiological concentrations of polyamines, such as spermine and related compounds [131], were instead shown to activate the MCU at Ca^{2+} concentrations that would otherwise be too low to allow the uniporter to operate efficiently: they could thus have a physiological role in intracellular Ca^{2+} handling.

The efflux route of Ca^{2+} from mitochondria was documented as Na^+-dependent pathway by the observation that the addition of Na^+ to isolated mitochondria promoted the efflux of Ca^{2+} [123]. Further work then characterized the pathway as a Na^+/Ca^{2+} antiporter (NCLX) [132]. The transport was later found to be electrogenic with a probable transport stoichiometry of 3 to 1, as in the case of the plasma membrane NCX [133,134]. The NCLX was inhibited competitively by Sr^{2+}, Ba^{2+}, Mg^{2+} or Mn^{2+}, and by many compounds of pharmacological interest including diltiazem, clonazepam, verapamil, tetraphenyl-phosphonium, trifluoperazine amiloride and its derivatives. In particular, the chloro-5-(2-chlorophenyl)-1,5-dihydro-4,1-benzothiazepin-2(3 H)-one (CGP 37157) inhibited it with high specificity and it is now widely used [135]. In the exchange process Na^+ could be replaced by Li^+, an observation that was later used in the work that identified the exchanger protein (see below).

A Na^+-independent Ca^{2+} extrusion mechanism has also been described in liver [136] and some other mitochondrial types. It transports Ca^{2+}, but also Sr^{2+}, or Mn^{2+} from the matrix to the intermembrane space against the Ca^{2+} electrochemical gradient. The rate of efflux via this mechanism decreases with increasing ΔpH (internally alkaline) [137]. The transport is electroneutral and it has been characterized as a 1 Ca^{2+} for 2 H^+ exchanger [138]. Cyanide, low levels of uncouplers, and very high levels of RR inhibit it [139].

In spite of the large mass of information that was becoming available, the low Ca^{2+} affinity problem led to the general assumption that Ca^{2+} sequestration by mitochondria in living cells had no important role in the regulation of Ca^{2+} homeostasis, unless in extreme conditions of Ca^{2+} overload [140]. The very limited Ca^{2+} transport activity (assumed to occur *in vivo*) was essentially only considered important for the activation of 3 matrix deydrogenases that have been found to be controlled by Ca^{2+} [141]. Thus, even if information had become available that mitochondrial Ca^{2+} transport did occur *in vivo* in spite of the insufficient affinity of the system, skepticism prevailed. At the beginning of 1990s the conundrum was solved by specifically targeting a recombinant Ca^{2+} sensor to the mitochondrial matrix. The work clearly demonstrated that in intact cells mitochondria promptly accumulated Ca^{2+} following cell stimulation [142]. The problem of the low affinity of the mitochondrial uptake system was overcome by demonstrating that mitochondria could sense localized microdomains at high Ca^{2+} concentration generated close to the mouth of the ER Ca^{2+} channels [143] in close proximity to mitochondrial Ca^{2+} uptake sites [144], and presumably also by functional coupling with Ca^{2+} entry channels at the plasma membrane [145,146]. The finding renewed interest in mitochondrial Ca^{2+} and on its physiological role and sparked new research aimed at identifying the mitochondrial Ca^{2+} transporters.

Their molecular identity had remained elusive for a long time (for comprehensive reviews see [147,148]). Different actors had been proposed but none of them had survived conclusive tests. Only very recent research has eventually identified molecularly the NCLX and the MCU [149–151]. A uniporter component (named MICU1) that may have a role in Ca^{2+} sensing rather than in Ca^{2+} transport has been identified in silico by developing a MitoCarta database [152] as a 54 kDa protein, associated with the mitochondrial inner membrane. It has one putative transmembrane domain and two canonical EF hands.

Using the same database, two independent groups then identified an integral inner membrane protein that satisfies the criteria for being the pore-forming subunit of the uniporter MCU [149,150]. The 40 kDa protein is ubiquitously expressed in mammals, but missing in yeast (yeast mitochondria do not have a uniporter [153]). It is predicted to have two transmembrane regions connected by an acidic loop, and it forms oligomers in the inner membrane. Its downregulation drastically reduced mitochondrial Ca^{2+} uptake and its overexpression enhanced it in intact cells. Most importantly, the channel activity of purified MCU reconstituted in a planar lipid bilayer revealed properties previously reported for the uniporter, thus definitively demonstrating that MCU represents its pore-forming channel [149].

The Na^+/Ca^{2+} exchanger (NCLX) has also been identified [151] as a mammalian member of the phylogenetically ancestral Ca^{2+}/anion exchanger family that catalyzes Na^+ (or Li^+) dependent Ca^{2+} transport [113]. NCLX was found to be enriched in the mitochondrial cristae. As expected, it transported Li^+ in addition to Ca^{2+} and was sensitive to CGP-35137. Its size was very similar to that of a mitochondrial protein that, when purified and reconstituted, exhibited Na^+/Ca^{2+} exchange activity [154,155].

In addition to the antiporters, the other mechanism of Ca^{2+} transport that may play a role in the mitochondrial Ca^{2+} efflux, especially in conditions in which mitochondrial Ca^{2+} concentration in the matrix reaches threshold levels, is still molecularly unknown; this is the mPTP [156]. In addition to Ca^{2+}, factors such as pH, adenine nucleotides, free radicals, and the mitochondrial membrane potential ($\Delta\Psi m$) modulate its opening. Mitochondrial Ca^{2+} overload and excess increases in reactive oxygen species (ROS) in the matrix would be the "point of no return" that causes permeabilization of the inner membrane, proton electrochemical gradient dissipation, ATP depletion, further ROS production and organelle swelling. These events are collectively termed "mitochondrial permeability transition" (MPT), a process that, in turn, causes the release of cytochrome *c* and culminates in cell death.

6.2 The Acidic Compartments

In addition to the ER/SR and the GA, other acidic organelles, such as the acidic endosomes, lysosomes, and secretory granules are now also considered as possible Ca^{2+} stores in mammalian cells. Their Ca^{2+} transport functions are not yet well characterized and the literature on the "acidic Ca^{2+} stores" has controversial aspects, especially on the possibility of a still unknown, ATP-dependent Ca^{2+} uptake mechanism, at least in lysosomes. The Ca^{2+} uptake through this system appears to rely on the large proton gradient established by the vacuolar proton V-ATPase. A Ca^{2+}/H^{+} exchanger has been proposed, but Ca^{2+}/H^{+} exchangers have so far only been found in protist, yeast and plant vacuoles [157]. The total Ca^{2+} content in the acidic Ca^{2+} stores changes with the organelle type, but has been claimed to be in the mM range. However, direct measurements of free Ca^{2+} in the lumen of the organelles have reported values in the μM range, the discrepancy being possibly due to the presence of Ca^{2+}-binding proteins, such as chromogranins and secretogranins, with a large Ca^{2+}-buffering capacity [158]. A direct measurement of lysosomal luminal Ca^{2+} is difficult, due to the very acidic environment and to the presence of proteolytic enzymes, however, a specific localized probe has revealed a very high Ca^{2+} concentration (about 500 μM [159]).

The acidic organelles have also been reported to be able to release Ca^{2+}, supporting a possible physiological role in modulating specific cell function such as secretion, endosome-lysosome fusion and, possibly, maintenance of osmoregulation. The release of Ca^{2+} has been reported to be promoted by all canonical second messengers described for the ER/SR and the Golgi, i.e., $InsP_3$, ryanodine, caffeine, and cADPR. Importantly, the most efficient Ca^{2+}-releasing agent is the novel Ca^{2+}-linked messenger NAADP (nicotinic acid adeninedinucleotide phosphate) [160], as mentioned, the ectoenzyme ADP-ribosyl cyclase, produces cADPR, but also produces NAADP from NADP and nicotinic acid. NAADP-sensitive Ca^{2+} release has been reported from endosomes [161], lysosomes [162], and secretory granules [163], but the existence of a specific NAADP receptor in the acidic organelles is still not conclusively established. cADPR and NAADP would operate on a non-selective cation channel, the transient receptor potential mucolipin 1 channel (TRPML1) which is present in lysosomes [164]. A new family of channels, called "two pore channels" (TPC) has also been proposed to operate in the membrane of acidic organelles. They are present ubiquitously in mammalian cells and can be divided in three subtypes according to their specific localization: TPC1 and 3 are found in endosomes, TPC1 mainly in lysosomes [165]. An interesting aspect of these channels is that the release of Ca^{2+} by TPC1 generally leads to a spatially restricted Ca^{2+} signal, whereas that operated by TPC2 triggers ER Ca^{2+} release by activating $InsP_3$/ryanodine receptors, enhancing the propagation of a global signal [166].

6.3 Ca^{2+} Regulation in the Nucleus: An Open Problem

The nucleus is the seat of numerous functions that are regulated by Ca^{2+}. Some are specific of the organelle, beginning with the expression of some genes, and which will be discussed in more detail later on. The presence of specific Ca^{2+} regulated functions would *prima facie* demand that Ca^{2+} in the nucleoplasm be regulated independently of the cytosol. The problem is that the nuclear envelope, which is an extension of the ER that separates the nucleoplasm from the cytosol, is interrupted by numerous nuclear pores, complex structures that form openings with a diameter of about 9 nm that allow the traffic of nucleic acids, proteins, and other macromolecules. If no mechanism existed to temporarily occlude the pores, nuclear Ca^{2+} would thus instantaneously equilibrate with Ca^{2+} in the cytosol. In line with this idea, numerous experiments with various Ca^{2+} indicators, including some in which the indicator was selectively targeted to the nucleoplasm [167,168], have indeed shown that the kinetics of cytosolic and nuclear Ca^{2+} increases induced by cell stimulation were temporally indistinguishable, suggesting that the envelope did not represent a barrier to the free diffusion of Ca^{2+}. Others, however, using the same technique found that the Ca^{2+} signals evoked by the stimulation of cells were invariably lower in the nucleus [169]. Persistent gradients of Ca^{2+} between the nucleus and the cytosol were also observed by directly injecting Ca^{2+} dyes into the nucleus of starfish oocytes [170]. Patch clamp experiments on the envelope of isolated nuclei [171], and even on the nuclear envelope *in situ* [172] are also difficult to reconcile with the idea of free diffusion of Ca^{2+} between the cytosol and the nucleoplasm. They showed no flow of current during long recording periods in spite of the presence of hundreds of pores in the patch, but recorded instead the activity of selective K^+ channels with multiple conductance states. The logical conclusion of this electrophysiological work would be that for significant periods of time the pore would remain sealed to ions, including Ca^{2+}. The mechanism of the putative gating of the pores is unknown, but atomic force microscopy work has shown that most pores contain a "plug" that could be part of the gating mechanism [173], and that the conformation of the pores is altered by extranuclear Ca^{2+} and ATP [174]. Thus, the matter of the Ca^{2+} permeability of the nuclear envelope is still an open issue. Perhaps, a conciliatory view could propose that the pores would exist in freely permeable or gated states depending on physiological conditions and demands (see some comprehensive reviews for a full discussion of the issue [104,175,176]).

A recent development on the matter of the role of Ca^{2+} in the nucleus and on its release to it is the demonstration that the envelope folds inside the nucleoplasm forming invaginations (a "nucleoplasmic reticulum") [176–179]. Earlier work had shown that the nuclear envelope contains $InsP_3Rs$ and RyRs [180,181] and a Ca^{2+} pump predictably identical to that of the ER [182]. Early work had also found most enzymes of the phosphoinositide cycle in the nuclear envelope [183–185]. A problem, here, is to understand how plasma membrane agonists that are known to initiate the phosphatidylinositol cycle would become activated in the nuclear envelope. Irrespective of this problem, however, the structural arrangement of the invaginations would facilitate the agonist-induced delivery of Ca^{2+} to selective sub-compartments of the nucleoplasm.

7 Physiology of the Ca^{2+} Signal: A Selection of Cellular Processes Controlled by Ca^{2+}

As repeatedly underlined above, Ca^{2+} controls a very large number of the processes that are essential to cell life. A detailed and comprehensive discussion of the physiology of the signal would evidently be out of the scope of this contribution; here, only a succinct description of the most significant Ca^{2+} regulated functions will thus be presented.

The discussion could be initiated with the process of fertilization, which originates new cell life. Vertebrate eggs remain arrested at the metaphase of the second meiotic division until sperm interacts with them to generate an increase of Ca^{2+} that initiates at the sperm interaction site [186]. This triggers the exit from metaphase II arrest, and initiates the cell divisions which will eventually produce the multicellular organisms. In many invertebrates and non-mammalian vertebrates the Ca^{2+} increase takes the form of a single transient, but in mammals the fertilizing Ca^{2+} signal consists of repetitive transients [187,188]. The mechanism by which the Ca^{2+} increase is generated has been controversial. One proposal suggested the direct flow of Ca^{2+} into the egg during gamete fusion, another the role of a surface receptor activated by a sperm factor that would set in motion an intracellular signaling pathway linked to PLC and $InsP_3$. A third proposal, which has now become generally accepted, suggests instead that the fusion of the sperm with the egg delivers into the latter a sperm-specific new isoform of PLC (PLCζ) which initiates the hydrolysis of PIP_2 to produce $InsP_3$ [189,190]. That $InsP_3$ is involved in the Ca^{2+} release in the fertilized egg is now broadly accepted. However, recent evidence suggests that the $InsP_3$-mediated global increase in Ca^{2+}, at least in echinoderm eggs, could be preceded by a localized increase of Ca^{2+} promoted by the recently discovered Ca^{2+} messenger NAADP, that would be followed by the globalization of the Ca^{2+} wave [191]. It should also be mentioned that recent work has underlined the importance of the dynamic rearrangement of the actin cytoskeleton produced by the increase of Ca^{2+} at fertilization in guiding sperm entry and in modulating the intracellular Ca^{2+} signaling [192].

A second process in which Ca^{2+} regulation is acquiring increasing importance is gene expression. A seminal report by Greenberg and coworkers in 1986 [193] had shown that acetylcholine receptor agonists induced the rapid transcription of the c-fos protooncogene in PC12 pheochromcytoma cells in a process that required Ca^{2+} influx. The work was then extended to neurons, and to numerous other genes involved in neuronal activity [194,195] underlining the special importance of the regulation of gene transcription by Ca^{2+} to neurons (see [196] for a review). Early work showed that the Ca^{2+} regulation of the transcription of (immediate early) genes could be mediated by phosphorylation/dephosphorylation reactions catalyzed by calmodulin-dependent kinases and the also calmodulin-dependent phosphatase calcineurin [197]. The most interesting extension of the work on gene regulation by Ca^{2+} is that on the EF-hand protein DREAM (see above) which acted as a gene silencer on the dynorphin gene [14]. As briefly mentioned above, Ca^{2+}-free DREAM binds to a tandem of DRE sites in the promoter of the gene, repressing its transcription. Binding of Ca^{2+} to the EF motifs of DREAM promotes its detachment from the DRE sites, reactivating transcription. The list of genes controlled by DREAM has now increased substantially, and includes some that code for Ca^{2+} regulating/regulated systems, e.g., one of the Na^{+}/Ca^{2+} exchangers (NCX3) [13], the L-type Ca^{2+} channels [15], and a nucleotidase that plays a role in the protein folding pathway [198].

That Ca^{2+} plays a role in the contraction of muscles has been known for 130 years. It was the finding that Ca^{2+} promoted the contraction of heart cells [199] that officially inaugurated the topic of Ca^{2+} signaling. The story of the role of Ca^{2+} in the regulation of heart, and then skeletal muscle, has progressed from the original days of Ringer through other seminal findings such as, to name only some, the discovery of the Ca^{2+} receptor in the myofibrils (the EF-hand protein troponin C), the findings on the Ca^{2+} fluxes in SR mediated by a Ca^{2+} pump and by ligand-gated channels, the characterization of the regulatory roles of phospholamban and sarcolipin in the uptake of Ca^{2+} in the ER/SR. Some of these aspects of the function of Ca^{2+} in the regulation of muscle contraction have been already discussed in the sections above; appropriate details can be found in a number of recent comprehensive reviews [200].

Protein phosphorylation/dephosphorylation is a universal mechanism by which the activity of enzymes is regulated. The large group of protein kinases and phosphatases includes important members that are activated by Ca^{2+}-calmodulin [201]. While several calmodulin kinases are known, only one protein phosphatase (calcineurin, also known as protein phosphatase 2B [202]) is regulated by Ca^{2+}-calmodulin. Calmodulin (CaM) kinases phosphorylate Ser-Thr residues, however, calcineurin also dephosphorylates phosphorylated Tyr residues. The CaM kinases can have narrow specificity, i.e., they only phosphorylate one substrate. Myosin light chain kinase (MLCK) [203] phosphorylates the light chain of myosin to initiate smooth muscle contraction and potentiate the contraction of skeletal muscles. It exists in two gene products, one only expressed in skeletal muscles and one, termed smooth muscle MLCK, expressed in a number of tissues. Phosphorylase kinase [204] phosphorylates and activates glycogen phosphorylase, thus accelerating glycogen degradation to contributing to blood glucose homeostasis and providing an energy source for muscle contraction. The enzyme consists of 4 catalytic γ subunits which form a holoenzyme complex with α, β, and δ regulatory subunits, each present in 4 copies: the 4 δ subunits are calmodulin molecules which, interestingly, remain stably associated with the holoenzyme even when the concentration of Ca^{2+} in the ambient is very low. The binding of Ca^{2+} to the δ subunits activates the enzyme, which is further activated by the phosphorylation of the α and β subunits by PKA.

The elongation factor 2 kinase (also known as CaMK III [205]) translocates along mRNA during translation and is inhibited by phosphorylation.

CaMK I [206] is a ubiquitous cytosolic enzyme which exists in 3 isoforms, α, β, and γ, which are the product of separate genes, which are processed alternatively to generate additional isoforms (a kinase originally termed CaMK V is actually a spliced variant of CaMK I). CaMK I is initially activated by the binding of calmodulin, and further activated by an upstream kinase, the calmodulin-dependent kinase kinase (CaMKK). Not much is known on the substrates phosphorylated by CaMK I, but *in vitro* experiments have described phosphorylation of synapsin I and of CREB, to activate CREB-dependent gene transcription.

The activity of CaM kinases is not restricted to the phosphorylation of only one substrate. For instance, CaMK II is a ubiquitous enzyme that has been shown to phosphorylate over 50 protein substrates *in vitro* (only relatively few of them, however, have been shown to be phosphorylated within cells under physiological conditions). It regulates diverse important physiological processes, among them neuronal plasticity, gene transcription, learning and memory, and exocytosis [207]. One of the best characterized substrates of CaMK II is the AMPA ionotropic glutamate receptor; its phosphorylation at Ser 831 plays an important role in synaptic transmission. Four genes encode α, β, γ, and δ isoforms of the kinase, and the alternative processing of the transcripts gives rise to nearly 30 variants of the enzyme, many of them with specific tissue distribution. Some tissues contain very high amounts of CaMK II, for instance the brain, where the kinase accounts for 1–2% of the total protein. All CaMK II isoforms contain a catalytic domain, an autoinhibitory domain, a variable segment, and a self-association domain [207]. The autoinhibitory domain binds to the catalytic domain, blocking its activity [208]. Auto-phosphorylation of the autoinhibitory domain (Thr 286) in the presence of Ca^{2+} and calmodulin removes the block, leading to persistent activation of the enzyme [209–212]. The concentration of Ca^{2+} in the vicinity determines the number of subunits that become autophosphorylated on Thr 286, i.e., CaMK II is able to decode the frequency and amplitude of the Ca^{2+} transients. This property may prolong the effects of the signaling after transient Ca^{2+} changes, as could for instance occur in learning and memory. The splice variants of the kinase could have specific intracellular localization. For example, one splice variant of CaMK II δ contains a nuclear localization signal and has been shown to regulate gene transcription.

α CaMK IV [197] is a monomeric enzyme expressed in the nervous tissue, in the testis, and in T-cells, while its β splice variant is expressed in the cerebellum during development. Like CaMK I, it is initially activated by calmodulin binding and further activated by phosphorylation by CaMKK. Combined with N-terminal autophosphorylation, this leads to Ca^{2+}-independent activity of CaMK IV. CaMK IV contains a nuclear localization sequence and is thought to phosphorylate numerous transcription factors.

CaMKK exists in two isoforms (α and β) and increases the activity of CaMK I and IV in the presence of calmodulin. Unlike all other CaM kinases, CaMKK does not contain acidic residues that recognize basic residues close to its preferred phosphorylation site.

Calcineurin, also called protein phosphatase 2B, is the only protein phosphatase whose activity is regulated by Ca^{2+}. It was first identified in extracts of mammalian brain (hence its name) but was later found to be expressed in most tissues of eukaryotes [202]. As mentioned above, it is a heterodimer of a catalytic subunit (CnA) tightly bound to a smaller, calmodulin-like regulatory subunit (CnB). Three basic isoforms of CnA (α, β, γ) and two of CnB (CnB1 and CnB2) exist as the products of separate genes. Splicing variants, however, have only been detected at the transcript level. Structurally, the catalytic domain in the N-terminal two thirds of calcineurin, which contains a binuclear iron-zinc active center, is followed down in the sequence by a CnB-binding, a calmodulin-binding, and an autoregulatory domain. CnB binds 4 Ca^{2+} to canonical EF hand motifs. The activity of CnA requires the binding of CnB, which in turn only occurs if the latter has bound Ca^{2+}. The binding of calmodulin to CnA increases the activity of the phosphatase 50-100 fold.

Importantly, calcineurin is the target of the immunosuppressive drugs cyclosporin A and tacrolimus (FK506) bound to their respective immunophilins [213] and has thus a key role in the transduction pathway from the plasma membrane to the nucleus leading to T-cell activation [214]. This occurs by dephosphorylation of the transcription factor NFAT following Ca^{2+} increase induced by the occupancy of the plasma membrane T-cell receptor. The activation of calcineurin promotes the dephosphorylation and the exposure of a nuclear localization signal in NFAT, promoting its translocation together with calcineurin to the nucleus, where NFAT can then perform its gene regulation tasks. As mentioned, calcineurin was originally discovered in the brain, where it represents about 1% of the total proteins. In the brain, calcineurin dephosphorylates two inhibitors of protein phosphatase-1 (Inhibitor 1 and DARPP32), inhibiting them. This triggers a phosphatase cascade that opposes the effects of cAMP and Ca^{2+}-activated kinases, explaining for instance the antagonistic effects of Ca^{2+} release induced by the occupancy of some receptors, e.g., the NMDA glutamate and dopamine receptors. Calcineurin, however, dephosphorylates a number of other substrates involved in the regulation of important neuronal processes, including the expression and activity of ion channels, the release of neurotransmitters, and the outgrowth of neurites.

The mention of calcineurin's role in the release of neurotransmitters introduces the secretion process, in which Ca^{2+} has a crucial role, which was first described nearly 50 years ago by Katz and Miledi [215,216] for the release of neurotransmitters at the neuromuscular junctions. The role was then extended to release processes in endocrine cells and in other cell types [217]. The basic mechanistic principle of the release process is the storage of the substance to be released in membrane vesicles that will eventually fuse with the plasma membrane in a process mediated by Ca^{2+} penetrating through activated plasma membrane channels, discharging their content, be it a neurotransmitter or a hormone, in the extracellular space; in the case of neurotransmitters, this is the synaptic cleft. The difference between the release by the synaptic terminals, and the release, for instance, by endocrine cells, is essentially one of time; the secretion by endocrine cells is a much slower process, with long latencies [218,219]. Differences in the concentrations of Ca^{2+}-buffering proteins may explain the delayed response in endocrine cells, but the distance between the plasma membrane Ca^{2+} channels and the storage vesicles may have a greater influence on the secretory response.

It seems appropriate to close the description of the cellular processes controlled by Ca^{2+} with a brief description of the role of Ca^{2+} in the controlled termination of cell life. There is growing consensus that the various forms of cell death (necrosis, apoptosis, and autophagy) do not occur through entirely separate pathways, but share molecular effectors and signaling routes. Among them, Ca^{2+} plays a clear role. Apoptosis is the best characterized form of cell death from the standpoint of its relationship to Ca^{2+} signaling.

Apoptosis (programmed cell death) involves the suicide of individual cells to guarantee normal tissue development and homeostasis in both vertebrate and invertebrate species. However, it can also contribute to many forms of pathological cell loss as it can degenerate into necrotic death. Apoptosis probably plays a role in many chronic degenerative processes, for instance in neuron conditions like Alzheimer's and Parkinson's diseases and in heart failure. By contrast, inhibition of apoptosis can be at the basis of the abnormal cell growth in tumors. Apoptotic cells are classified on morphological characteristics that include condensation and margination of chromatin, cytoplasmic vacuolization, cellular shrinkage, increase in cellular density, nuclear fragmentation, and apoptotic body formation [220].

The Ca^{2+} link with the apoptotic pathways is now a large topic, which cannot be covered in detail in this contribution; a number of comprehensive reviews offer a more complete panorama of the topic [221,222]. The idea of the involvement of Ca^{2+} initiated with the *in vitro* demonstration that Ca^{2+} ionophores, i.e., molecules capable of transporting Ca^{2+} across membranes down its electrochemical gradient, are highly toxic to cells, and by the finding that the neurotransmitter glutamate, or related compounds, have the ability to induce neuronal death as a result of excess Ca^{2+} penetration due to receptor overstimulation. Later on, both Ca^{2+} release from the ER and capacitative Ca^{2+} influx through Ca^{2+} release-activated Ca^{2+} channels were shown to be apoptogenic [223–225]. The discovery that important regulators of apoptosis, namely the proteins of the Bcl-2 family, are localized in organelles deeply involved in Ca^{2+} handling (the mitochondria and the ER), and may modulate the ER content/release of Ca^{2+}, definitely established the Ca^{2+} link to apoptosis. The current view is that Ca^{2+} can sensitize cells to apoptotic challenges, acting on the mitochondrial "checkpoint". Mitochondria are the site of several proapoptotic proteins like Smac/DIABLO, Omi/HtrA2, AIF, and EndoG, which are maintained in equilibrium with antiapoptotic proteins like XIAP, cIAP-1, and cIAP-2 to finely regulate the balance between cell death and life. Thus, the role of mitochondria and Ca^{2+} appears to be a key determinant in the molecular events leading to cell death. Ca^{2+} loads in the mitochondrial matrix have been shown to sensitize the mPTP to apoptotic stimuli, inducing its opening, causing mitochondrial changes in morphology and the release of cytochrome *c* [226], followed by caspase activation [221,227].

The overall picture that emerges from a large number of contributions is that the release of Ca^{2+} from the ER and its uptake into mitochondria are pivotal in initiating apoptotic signals, and that a mechanism through which the overexpression of antiapoptotic proteins (or the ablation of proapoptotic proteins) counteracts cell death, is the reduction in the amount of Ca^{2+} available in the ER for the release process and the uptake into mitochondria. The amount of releasable Ca^{2+} – rather than the Ca^{2+} concentration of the ER – appears to be the important parameter for the transduction of the death signal, as it eventually controls the "amplitude" of the signal that reaches mitochondria.

8 Concluding Remarks

Ca^{2+} signaling has become a topic too large to be covered comprehensively in a normal review. This contribution has thus singled out aspects of the Ca^{2+} signal that distinguish it from all the carriers of biological information. Among them are the autoregulatory property, its ability to function both as a second and a first messenger, and, especially, its ambivalence: Ca^{2+} is not only a messenger without which correct cell life would not be possible, it also conveys negative signals, or even death signals, if its concentration and movements within cells are not carefully controlled. However, if correctly controlled and delivered, the Ca^{2+} signal modulates essentially all important aspects of cell life, from its origin at fertilization, to its end in the process of apoptosis.

Abbreviations

AMPA	2-amino-3-hydroxyl-5-ethyl-4-isoxazolepropionic acid
ATP	adenosine 5'-triphosphate
cADPR	cyclic adenosine diphosphate ribose
CaM	calmodulin
CaMK	calmodulin dependent kinase
cAMP	cyclic adenosine monophosphate
CaR	calcium receptor
CBD	Ca^{2+}-binding domain
CICR	Ca^{2+}-induced Ca^{2+} release
CRAC	Ca^{2+}-release activated current
CREB	cAMP response element binding
DHPR	dihydropyridine receptor
DRE	downstream regulatory element
DREAM	downstream regulatory element
ER	endoplasmic reticulum
GA	Golgi apparatus
GABA	γ-amino butyric acid
GluR	glutamate receptor

IMM	inner mitochondrial membrane
IMS	intermembrane space
$InsP_3$	inositol 1,4,5-trisphosphate
$InsP_3R$	inositol 1,4,5-trisphosphate receptor
KA	kainate
MCO	mitochondrial Ca^{2+} overload
MCU	mitochondrial Ca^{2+} uniporter
MICU1	mitochondrial calcium uptake 1
MPT	mitochondrial permeability transition
NAADP	nicotinic acid adenine dinucleotide phosphate
NADP	nicotinamide adenosine diphosphate
NCLX	mitochondrial Na^+/Ca^{2+} exchanger
NCX	Na^+/Ca^{2+} exchanger
NFAT	nuclear factor of activated T cells
NHE	Na^+/H^+ exchanger
NMDA	N-methyl-D-aspartate
OMM	outer mitochondrial membrane
PC12	pheochromcytoma cells
PIP2	phosphatidylinositol 4,5-bisphosphate
PKA	protein kinase A
PKC	protein kinase C
PLC	phospholipase C
PM	plasma membrane
PMCA	plasma membrane Ca^{2+}-ATPase
PTP	permeability transition pore
PV	parvalbumin
ROC	receptor operated Ca^{2+} channels
ROS	reactive oxygen species
RR	ruthenium red
RyR	ryanodine receptor
SERCA	sarco/endoplasmic reticulum Ca^{2+}-ATPase
SLC	solute carrier
SNAP-25	synaptosomal-associated protein 25
SNARE	soluble NSF attachment protein receptor
SNP	single nucleotide polymorphism
SOCE	store operated Ca^{2+} entry channels
SPCA	secretory pathway Ca^{2+} ATPase
SR	sarcoplasmic reticulum
STIM	sensors stromal interaction molecule
TG	thapsigargin
TPC	two pore channel
TRP	transient receptor potential channels
VAMP	vesicle associate membrane protein
VOC	voltage operated Ca^{2+}-channels

Acknowledgments The original work by the authors has been supported over the years by grants from the Italian Ministry of University and Research (FIRB2001 to E.C., PRIN 2003, 2005 and 2008 to M.B), the Telethon Foundation (Project GGP04169 to M.B.), the FP6 program of the European Union (FP6 Network of Excellence NeuroNe, LSH-2003-2.1.3-3 to E.C. and Integrated Project Eurohear to E.C.), the Human Frontier Science Program Organization to E.C., the ERANet-Neuron (nEUROsyn), and CARIPARO Foundation to E.C, the Italian National Research Council (CNR) and the University of Padova (Progetto di Ateneo 2008 CPDA082825) to M.B.

References

1. E. Carafoli, *Proc. Natl. Acad. Sci. USA* **2002**, *99*, 1115–1122.
2. E. Carafoli, *Nat. Rev. Mol. Cell. Biol.* **2003**, *4*, 326–332.
3. D. Prins, M. Michalak, *Cold Spring Harbor Perspect. Biol.* **2011**, *3*.
4. B. Schwaller, *Cold Spring Harbor Perspect. Biol.* **2010**, *2*, a004051.
5. G. C. Faas, S. Raghavachari, J. E. Lisman, I. Mody, *Nat. Neurosci.* **2011**, *14*, 301–304.
6. E. M. Brown, S. M. Quin, P. M. Vassilev, in *Calcium as a Cellular Regulator*, Eds E. Carafoli, C. Klee, Oxford University Press, New York, Oxford, 1999, pp. 295–310.
7. E. F. Nemeth, J. Wallace, A. Scarpa, *J. Biol. Chem.* **1986**, *261*, 2668–2674.
8. D. S. McGehee, M. Aldersberg, K. P. Liu, S. Hsuing, M. J. Heath, H. Tamir, *J. Physiol.* **1997**, *502*, 31–44.
9. D. Guerini, E. Garcia-Martin, A. Gerber, C. Volbracht, M. Leist, C. G. Merino, E. Carafoli, *J. Biol. Chem.* **1999**, *274*, 1667–1676.
10. D. Guerini, X. Wang, L. Li, A. Genazzani, E. Carafoli, *J. Biol. Chem.* **2000**, *275*, 3706–3712.
11. L. Li, D. Guerini, E. Carafoli, *J. Biol. Chem.* **2000**, *275*, 20903–20910.
12. A. A. Genazzani, E. Carafoli, D. Guerini, *Proc. Natl. Acad. Sci. USA* **1999**, *96*, 5797–5801.
13. R. GomezVillafuertes, B. Torres, J. Barrio, M. Savignac, N. Gabellini, F. Rizzato, B. Pintado, A. Gutierrez-Adan, B. Mellstrom, E. Carafoli, J. R. Naranjo, *J. Neurosci.* **2005**, *25*, 10822–10830.
14. A. M. Carrion, W. A. Link, F. Ledo, B. Mellstrom, J. R. Naranjo, *Nature* **1999**, *398*, 80–84.
15. J. J. Ronkainen, S. L. Hanninen, T. Korhonen, J. T. Koivumaki, R. Skoumal, S. Rautio, V. P. Ronkainen, P. Tavi, *J. Physiol.* **2011**, *589*, 2669–2686.
16. E. Carafoli, *FASEB J.* **1994**, *8*, 993–1002.
17. D. Bano, K. W. Young, C. J. Guerin, R. Lefeuvre, N. J. Rothwell, L. Naldini, R. Rizzuto, E. Carafoli, P. Nicotera, *Cell* **2005**, *120*, 275–285.
18. K. Schuh, S. Uldrijan, M. Telkamp, N. Rothlein, L. Neyses, *J. Cell Biol.* **2001**, *155*, 201–205.
19. *Calcium Signalling and Disease*, Ed J. R. Harris, Springer, New York, 2007.
20. B. Schwaller, *Cell Mol. Life Sci.* **2009**, *66*, 275–300.
21. E. E. Saftenku, *Cerebellum* **2012**, *11*, 102–120.
22. M. J. Hubbard, N. J. McHugh, *FEBS Lett.* **1995**, *374*, 333–337.
23. N. J. Hack, M. C. Wride, K. M. Charters, S. B. Kater, T. N. Parks, *J. Neurosci.* **2000**, *20*, RC67.
24. B. Schwaller, I. Durussel, D. Jermann, B. Herrmann, J. A. Cox, *J. Biol. Chem.* **1997**, *272*, 29663–29671.
25. G. S. Lee, K. C. Choi, E. B. Jeung, *Am. J. Physiol. Endocrinol. Metab.* **2006**, *290*, E299–307.
26. M. R. Celio, *Science* **1986**, *231*, 995–997.
27. C. W. Heizmann, M. W. Berchtold, A. M. Rowlerson, *Proc. Natl. Acad. Sci. USA* **1982**, *79*, 7243–7247.
28. S. K. Nakayama, H., Kretsinger, R., in *Calcium Homeostasis*, Eds E. Carafoli, C. Klee, Springer, Vol. 3, 2000, pp. 29–58.
29. H. Kawasaki, S. Nakayama, R. H. Kretsinger, *Biometals* **1998**, *11*, 277–295.
30. W. A. Catterall, *Neuron* **2010**, *67*, 915–928.

31. W. A. Catterall, *Cold Spring Harbor Perspect. Biol.* **2011**, *3*, a003947.
32. F. Hofmann, M. Biel, V. Flockerzi, *Annu. Rev. Neurosci.* **1994**, *17*, 399–418.
33. A. C. Dolphin, *J. Bioenerg. Biomembr.* **2003**, *35*, 599–620.
34. H. Bito, K. Deisseroth, R. W. Tsien, *Cell* **1996**, *87*, 1203–1214.
35. R. E. Dolmetsch, U. Pajvani, K. Fife, J. M. Spotts, M. E. Greenberg, *Science* **2001**, *294*, 333–339.
36. N. Gomez-Ospina, F. Tsuruta, O. Barreto-Chang, L. Hu, R. Dolmetsch, *Cell* **2006**, *127*, 591–606.
37. Z. P. Pang, T. C. Sudhof, *Curr. Opin. Cell Biol.* **2010**, *22*, 496–505.
38. K. D. Wickman, D. E. Clapham, *Curr. Opin. Neurobiol.* **1995**, *5*, 278–285.
39. S. P. H. Alexander, A. Mathie, J. A. Peters, *Br. J. Pharmacol.* **2009**, *158*, S1–254.
40. E. R. Liman, A. G. Knapp, J. E. Dowling, *Brain Res.* **1989**, *481*, 399–402.
41. C. Rosenmund, D. W. Carr, S. E. Bergeson, G. Nilaver, J. D. Scott, G. L. Westbrook, *Nature* **1994**, *368*, 853–856.
42. L. Y. Wang, M. W. Salter, J. F. MacDonald, *Science* **1991**, *253*, 1132–1135.
43. I. M. Raman, G. Tong, C. E. Jahr, *Neuron* **1996**, *16*, 415–421.
44. J. W. Putney, Jr., *Cell Calcium* **1986**, *7*, 1–12.
45. M. Hoth, R. Penner, *Nature* **1992**, *355*, 353–356.
46. Y. Gwack, S. Srikanth, S. Feske, F. Cruz-Guilloty, M. Oh-hora, D. S. Neems, P. G. Hogan, A. Rao, *J. Biol. Chem.* **2007**, *282*, 16232–16243.
47. M. Prakriya, S. Feske, Y. Gwack, S. Srikanth, A. Rao, P. G. Hogan, *Nature* **2006**, *443*, 230–233.
48. R. M. Luik, M. M. Wu, J. Buchanan, R. S. Lewis, *J. Cell Biol.* **2006**, *174*, 815–825.
49. R. E. Dolmetsch, R. S. Lewis, *J. Gen. Physiol.* **1994**, *103*, 365–388.
50. R. E. Dolmetsch, R. S. Lewis, C. C. Goodnow, J. I. Healy, *Nature* **1997**, *386*, 855–858.
51. R. E. Dolmetsch, K. Xu, R. S. Lewis, *Nature* **1998**, *392*, 933–936.
52. C. Montell, *Science Signalling, The Transduction Knowledge Environment* **2005**, *2005*, re3.
53. M. Gees, B. Colsoul, B. Nilius, *Cold Spring Harbor Perspect. Biol.* **2010**, 2, a003962.
54. T. Rohacs, *Pflugers Arch.* **2007**, *453*, 753–762.
55. I. S. Ambudkar, H. L. Ong, X. Liu, B. C. Bandyopadhyay, K. T. Cheng, *Cell Calcium* **2007**, *42*, 213–223.
56. M. Lu, R. Branstrom, E. Berglund, A. Hoog, P. Bjorklund, G. Westin, C. Larsson, L. O. Farnebo, L. Forsberg, *J. Mol. Endocrinol.* **2010**, *44*, 285–294.
57. Y. Liao, C. Erxleben, E. Yildirim, J. Abramowitz, D. L. Armstrong, L. Birnbaumer, *Proc. Natl. Acad. Sci. USA* **2007**, *104*, 4682–4687.
58. J. Chan, H. Yamazaki, N. Ishiyama, M. D. Seo, T. K. Mal, T. Michikawa, K. Mikoshiba, M. Ikura, *J. Biol. Chem.* **2010**, *285*, 36092–36099.
59. C. W. Taylor, S. C. Tovey, *Cold Spring Harbor Perspect. Biol.* **2010**, *2*, a004010.
60. C. W. Taylor, A. J. Laude, *Cell Calcium* **2002**, *32*, 321–334.
61. L. Kimlicka, F. Van Petegem, *Sci. China Life Sci.* **2011**, *54*, 712–724.
62. J. T. Lanner, D. K. Georgiou, A. D. Joshi, S. L. Hamilton, *Cold Spring Harbor Perspect. Biol.* **2010**, 2, a003996.
63. A. De Flora, L. Franco, L. Guida, S. Bruzzone, E. Zocchi, *Cell Biochem. Biophys.* **1998**, *28*, 45–62.
64. N. Bai, H. C. Lee, I. Laher, *Pharmacol. Ther.* **2005**, *105*, 189–207.
65. P. Thorn, O. Gerasimenko, O. H. Petersen, *EMBO J.* **1994**, *13*, 2038–2043.
66. H. Higashida, A. B. Salmina, R. Y. Olovyannikova, M. Hashii, S. Yokoyama, K. Koizumi, D. Jin, H. X. Liu, O. Lopatina, S. Amina, M. S. Islam, J. J. Huang, M. Noda, *Neurochem. Int.* **2007**, *51*, 192–199.
67. P. Pedersen, E. Carafoli, *Trends Biochem. Sci.* **1987**, *12*, 146–150.
68. L. de Meis, A. L. Vianna, *Annu. Rev. Biochem.* **1979**, *48*, 275–292.
69. C. Toyoshima, M. Nakasako, H. Nomura, H. Ogawa, *Nature* **2000**, *405*, 647–655.
70. C. Toyoshima, H. Nomura, *Nature* **2002**, *418*, 605–611.
71. M. Brini, E. Carafoli, *Physiol. Rev.* **2009**, *89*, 1341–1378.
72. C. Toyoshima, *Arch. Biochem. Biophys.* **2008**, *476*, 3–11.

73. D. M. Clarke, T. W. Loo, G. Inesi, D. H. MacLennan, *Nature* **1989**, *339*, 476–478.
74. D. M. Clarke, T. W. Loo, D. H. MacLennan, *J. Biol. Chem.* **1990**, *265*, 6262–6267.
75. D. Guerini, A. Zecca-Mazza, E. Carafoli, *J. Biol. Chem.* **2000**, *275*, 31361–31368.
76. J. Pande, M. M. Szewczyk, I. Kuszczak, S. Grover, E. Escher, A. K. Grover, *J. Cell Mol. Med.* **2008**, *12*, 1049–1060.
77. J. Chaudhary, M. Walia, J. Matharu, E. Escher, A. K. Grover, *Am. J. Physiol. Cell Physiol.* **2001**, *280*, C1027–1030.
78. L. Dode, J. P. Andersen, L. Raeymaekers, L. Missiaen, B. Vilsen, F. Wuytack, *J. Biol. Chem.* **2005**, *280*, 39124–39134.
79. L. Dode, J. P. Andersen, J. Vanoevelen, L. Raeymaekers, L. Missiaen, B. Vilsen, F. Wuytack, *J. Biol. Chem.* **2006**, *281*, 3182–3189.
80. R. J. Kaufman, M. Swaroop, P. Murtha-Riel, *Biochemistry* **1994**, *33*, 9813–9819.
81. A. Varki, *Trends Cell Biol.* **1998**, *8*, 34–40.
82. V. Niggli, E. S. Adunyah, J. T. Penniston, E. Carafoli, *J. Biol. Chem.* **1981**, *256*, 395–401.
83. P. James, M. Maeda, R. Fischer, A. K. Verma, J. Krebs, J. T. Penniston, E. Carafoli, *J. Biol. Chem.* **1988**, *263*, 2905–2910.
84. A. Enyedi, A. K. Verma, R. Heim, H. P. Adamo, A. G. Filoteo, E. E. Strehler, J. T. Penniston, *J. Biol. Chem.* **1994**, *269*, 41–43.
85. R. Falchetto, T. Vorherr, J. Brunner, E. Carafoli, *J. Biol. Chem.* **1991**, *266*, 2930–2936.
86. R. Falchetto, T. Vorherr, E. Carafoli, *Protein Sci.* **1992**, *1*, 1613–1621.
87. N. L. Elwess, A. G. Filoteo, A. Enyedi, J. T. Penniston, *J. Biol. Chem.* **1997**, *272*, 17981–17986.
88. H. Hilfiker, D. Guerini, E. Carafoli, *J. Biol. Chem.* **1994**, *269*, 26178–26183.
89. A. Enyedi, T. Vorherr, P. James, D. J. McCormick, A. G. Filoteo, E. Carafoli, J. T. Penniston, *J. Biol. Chem.* **1989**, *264*, 12313–12321.
90. E. Zvaritch, P. James, T. Vorherr, R. Falchetto, N. Modyanov, E. Carafoli, *Biochemistry* **1990**, *29*, 8070–8076.
91. P. Brodin, R. Falchetto, T. Vorherr, E. Carafoli, *Eur. J. Biochem.* **1992**, *204*, 939–946.
92. J. Lytton, M. Westlin, S. E. Burk, G. E. Shull, D. H. MacLennan, *J. Biol. Chem.* **1992**, *267*, 14483–14489.
93. H. Verboomen, F. Wuytack, L. Van den Bosch, L. Mertens, R. Casteels, *Biochem. J.* **1994**, *303*, 979–984.
94. L. Dode, J. P. Andersen, N. Leslie, J. Dhitavat, B. Vilsen, A. Hovnanian, *J. Biol. Chem.* **2003**, *278*, 47877–47889.
95. A. M. Katz, *Ann. NY Acad. Sci.* **1998**, *853*, 9–19.
96. D. H. MacLennan, E. G. Kranias, *Nat. Rev. Mol. Cell Biol.* **2003**, *4*, 566–577.
97. P. James, M. Inui, M. Tada, M. Chiesi, E. Carafoli, *Nature* **1989**, *342*, 90–92.
98. Z. Hu, J. M. Bonifas, J. Beech, G. Bench, T. Shigihara, H. Ogawa, S. Ikeda, T. Mauro, E. H. Epstein, Jr., *Nat. Genet.* **2000**, *24*, 61–65.
99. J. Vanoevelen, L. Dode, K. Van Baelen, R. J. Fairclough, L. Missiaen, L. Raeymaekers, F. Wuytack, *J. Biol. Chem.* **2005**, *280*, 22800–22808.
100. V. Prasad, G. W. Okunade, M. L. Miller, G. E. Shull, *Biochem. Biophys. Res. Commun.* **2004**, *322*, 1192–1203.
101. K. Schuh, E. J. Cartwright, E. Jankevics, K. Bundschu, J. Liebermann, J. C. Williams, A. L. Armesilla, M. Emerson, D. Oceandy, K. P. Knobeloch, L. Neyses, *J. Biol. Chem.* **2004**, *279*, 28220–28226.
102. E. E. Strehler, D. A. Zacharias, *Physiol. Rev.* **2001**, *81*, 21–50.
103. M. C. Chicka, E. E. Strehler, *J. Biol. Chem.* **2003**, *278*, 18464–18470.
104. E. Carafoli, L. Santella, D. Branca, M. Brini, *Crit. Rev. Biochem. Mol. Biol.* **2001**, *36*, 107–260.
105. M. P. Blaustein, W. J. Lederer, *Physiol. Rev.* **1999**, *79*, 763–854.
106. K. D. Philipson, D. A. Nicoll, S. Matsuoka, L. V. Hryshko, D. O. Levitsky, J. N. Weiss, *Ann. NY Acad. Sci.* **1996**, *779*, 20–28.

107. S. L. Lee, A. S. Yu, J. Lytton, *J. Biol. Chem.* **1994**, *269*, 14849–14852.
108. D. A. Nicoll, M. Ottolia, K. D. Philipson, *Ann. NY Acad. Sci.* **2002**, *976*, 11–18.
109. K. D. Philipson, D. A. Nicoll, *Annu. Rev. Physiol.* **2000**, *62*, 111–133.
110. D. W. Hilgemann, *Nature* **1990**, *344*, 242–245.
111. J. Liao, H. Li, W. Zeng, D. B. Sauer, R. Belmares, Y. Jiang, *Science* **2012**, *335*, 686–690.
112. H. Reilander, A. Achilles, U. Friedel, G. Maul, F. Lottspeich, N. J. Cook, *EMBO J.* **1992**, *11*, 1689–1695.
113. X. Cai, J. Lytton, *Mol. Biol. Evol.* **2004**, *21*, 1692–1703.
114. R. Palty, E. Ohana, M. Hershfinkel, M. Volokita, V. Elgazar, O. Beharier, W. F. Silverman, M. Argaman, I. Sekler, *J. Biol. Chem.* **2004**, *279*, 25234–25240.
115. J. Lytton, *Biochem. J.* **2007**, *406*, 365–382.
116. D. A. Nicoll, M. R. Sawaya, S. Kwon, D. Cascio, K. D. Philipson, J. Abramson, *J. Biol. Chem.* **2006**, *281*, 21577–21581.
117. M. Hilge, J. Aelen, G. W. Vuister, *Mol. Cell* **2006**, *22*, 15–25.
118. D. Noble, A. Herchuelz, *EMBO Rep.* **2007**, *8*, 228–232.
119. The International HapMap Consortium, *A Haplotype Map of the Human Genome*, *Nature* **2005**, *437*, 1299–1320.
120. R. L. Lamason, M. A. Mohideen, J. R. Mest, A. C. Wong, H. L. Norton, M. C. Aros, M. J. Jurynec, X. Mao, V. R. Humphreville, J. E. Humbert, S. Sinha, J. L. Moore, P. Jagadeeswaran, W. Zhao, G. Ning, I. Makalowska, P. M. McKeigue, D. O'Donnell, R. Kittles, E. J. Parra, N. J. Mangini, D. J. Grunwald, M. D. Shriver, V. A. Canfield, K. C. Cheng, *Science* **2005**, *310*, 1782–1786.
121. Y. Kirichok, G. Krapivinsky, D. E. Clapham, *Nature* **2004**, *427*, 360–364.
122. N. E. Saris, A. Allshire, *Methods Enzymol.* **1989**, *174*, 68–85.
123. E. Carafoli, R. Tiozzo, G. Lugli, F. Crovetti, C. Kratzing, *J. Mol. Cell Cardiol.* **1974**, *6*, 361–371.
124. D. G. Nicholls, *Cell Calcium* **2005**, *38*, 311–317.
125. E. Carafoli, *FEBS Lett.* **1979**, *104*, 1–5.
126. F. D. Vasington, J. V. Murphy, *J. Biol. Chem.* **1962**, *237*, 2670–2677.
127. H. F. Deluca, G. W. Engstrom, *Proc. Natl. Acad. Sci. USA* **1961**, *47*, 1744–1750.
128. D. G. Nicholls, M. Crompton, *FEBS Lett.* **1980**, *111*, 261–268.
129. R. Rizzuto, P. Bernardi, T. Pozzan, *J. Physiol.* **2000**, *529 Pt 1*, 37–47.
130. J. Santo-Domingo, L. Vay, E. Hernandez-Sanmiguel, C. D. Lobaton, A. Moreno, M. Montero, J. Alvarez, *Br. J. Pharmacol.* **2007**, *151*, 647–654.
131. I. Rustenbeck, G. Eggers, H. Reiter, W. Munster, S. Lenzen, *Biochem. Pharmacol.* **1998**, *56*, 977–985.
132. M. C. Crompton, E. Carafoli, *Eur. J. Biochem.* **1976**, *69*, 435–462.
133. D. W. Jung, K. Baysal, G. P. Brierley, *J. Biol. Chem.* **1995**, *270*, 672–678.
134. R. K. Dash, D. A. Beard, *J. Physiol.* **2008**, *586*, 3267–3285.
135. D. A. Cox, L. Conforti, N. Sperelakis, M. A. Matlib, *J. Cardiovasc. Pharmacol.* **1993**, *21*, 595–599.
136. G. Fiskum, A. L. Lehninger, *J. Biol. Chem.* **1979**, *254*, 6236–6239.
137. T. E. Gunter, J. H. Chace, J. S. Puskin, K. K. Gunter, *Biochemistry* **1983**, *22*, 6341–6351.
138. T. E. Gunter, L. Buntinas, G. Sparagna, R. Eliseev, K. Gunter, *Cell Calcium* **2000**, *28*, 285–296.
139. D. E. Wingrove, T. E. Gunter, *J. Biol. Chem.* **1986**, *261*, 15166–15171.
140. A. P. Somlyo, M. Bond, A. V. Somlyo, *Nature* **1985**, *314*, 622–625.
141. R. M. Denton, J. G. McCormack, *Nature* **1985**, *315*, 635.
142. R. Rizzuto, A. W. Simpson, M. Brini, T. Pozzan, *Nature* **1992**, *358*, 325–327.
143. R. Rizzuto, M. Brini, M. Murgia, T. Pozzan, *Science* **1993**, *262*, 744–747.
144. R. Rizzuto, P. Pinton, W. Carrington, F. S. Fay, K. E. Fogarty, L. M. Lifshitz, R. A. Tuft, T. Pozzan, *Science* **1998**, *280*, 1763–1766.
145. M. Hoth, C. M. Fanger, R. S. Lewis, *J. Cell Biol.* **1997**, *137*, 633–648.

146. R. Malli, M. Frieden, K. Osibow, W. F. Graier, *J. Biol. Chem.* **2003**, *278*, 10807–10815.
147. G. Hajnoczky, G. Csordas, *Curr. Biol.* **2010**, *20*, R888–891.
148. I. Drago, P. Pizzo, T. Pozzan, *EMBO J.* **2011**, *30*, 4119–4125.
149. D. De Stefani, A. Raffaello, E. Teardo, I. Szabo, R. Rizzuto, *Nature* **2011**, *476*, 336–340.
150. J. M. Baughman, F. Perocchi, H. S. Girgis, M. Plovanich, C. A. Belcher-Timme, Y. Sancak, X. R. Bao, L. Strittmatter, O. Goldberger, R. L. Bogorad, V. Koteliansky, V. K. Mootha, *Nature* **2011**, *476*, 341–345.
151. R. Palty, W. F. Silverman, M. Hershfinkel, T. Caporale, S. L. Sensi, J. Parnis, C. Nolte, D. Fishman, V. Shoshan-Barmatz, S. Herrmann, D. Khananshvili, I. Sekler, *Proc. Natl. Acad. Sci. USA* **2010**, *107*, 436–441.
152. F. Perocchi, V. M. Gohil, H. S. Girgis, X. R. Bao, J. E. McCombs, A. E. Palmer, V. K. Mootha, *Nature* **2010**, *467*, 291–296.
153. E. Carafoli, W. X. Balcavage, A. L. Lehninger, J. R. Mattoon, *Biochim. Biophys. Acta.* **1970**, *205*, 18–26.
154. W. Li, Z. Shariat-Madar, M. Powers, X. Sun, R. D. Lane, K. D. Garlid, *J. Biol. Chem.* **1992**, *267*, 17983–17989.
155. P. Paucek, M. Jaburek, *Biochim. Biophys. Acta* **2004**, *1659*, 83–91.
156. A. Rasola, P. Bernardi, *Cell Calcium* **2011**, *50*, 222–233.
157. S. Patel, R. Docampo, *Trends Cell Biol.* **2010**, *20*, 277–286.
158. S. H. Yoo, S. Y. Chu, K. D. Kim, Y. H. Huh, *Biochemistry* **2007**, *46*, 14663–14671.
159. E. Lloyd-Evans, A. J. Morgan, X. He, D. A. Smith, E. Elliot-Smith, D. J. Sillence, G. C. Churchill, E. H. Schuchman, A. Galione, F. M. Platt, *Nat. Med.* **2008**, *14*, 1247–1255.
160. A. H. Guse, H. C. Lee, *Sci. Signal* **2008**, *1*, re10.
161. A. Menteyne, A. Burdakov, G. Charpentier, O. H. Petersen, J. M. Cancela, *Curr. Biol.* **2006**, *16*, 1931–1937.
162. M. Yamasaki, R. Masgrau, A. J. Morgan, G. C. Churchill, S. Patel, S. J. Ashcroft, A. Galione, *J. Biol. Chem.* **2004**, *279*, 7234–7240.
163. K. J. Mitchell, F. A. Lai, G. A. Rutter, *J. Biol. Chem.* **2003**, *278*, 11057–11064.
164. F. Zhang, P. L. Li, *J. Biol. Chem.* **2007**, *282*, 25259–25269.
165. M. X. Zhu, J. Ma, J. Parrington, A. Galione, A. M. Evans, *FEBS Lett.* **2010**, *584*, 1966–1974.
166. M. X. Zhu, J. Ma, J. Parrington, P. J. Calcraft, A. Galione, A. M. Evans, *Am. J. Physiol. Cell Physiol.* **2010**, *298*, C430–441.
167. M. Brini, M. Murgia, L. Pasti, D. Picard, T. Pozzan, R. Rizzuto, *EMBO J.* **1993**, *12*, 4813–4819.
168. N. L. Allbritton, E. Oancea, M. A. Kuhn, T. Meyer, *Proc. Natl. Acad. Sci. USA* **1994**, *91*, 12458–12462.
169. M. N. Badminton, J. M. Kendall, C. M. Rembold, A. K. Campbell, *Cell Calcium* **1998**, *23*, 79–86.
170. L. Santella, K. Kyozuka, *Biochem. Biophys. Res. Commun.* **1994**, *203*, 674–680.
171. M. Mazzanti, L. J. DeFelice, J. Cohn, H. Malter, *Nature* **1990**, *343*, 764–767.
172. M. Mazzanti, B. Innocenti, M. Rigatelli, *FASEB J.* **1994**, *8*, 231–236.
173. C. Perez-Terzic, L. Stehno-Bittel, D. E. Clapham, *Cell Calcium* **1997**, *21*, 275–282.
174. A. Rakowska, T. Danker, S. W. Schneider, H. Oberleithner, *J. Membr. Biol.* **1998**, *163*, 129–136.
175. L. Santella, E. Carafoli, *FASEB J.* **1997**, *11*, 1091–1109.
176. W. Echevarria, M. F. Leite, M. T. Guerra, W. R. Zipfel, M. H. Nathanson, *Nat. Cell Biol.* **2003**, *5*, 440–446.
177. P. Marius, M. T. Guerra, M. H. Nathanson, B. E. Ehrlich, M. F. Leite, *Cell Calcium* **2006**, *39*, 65–73.
178. C. Cardenas, J. L. Liberona, J. Molgo, C. Colasante, G. A. Mignery, E. Jaimovich, *J. Cell Sci.* **2005**, *118*, 3131–3140.
179. S. Guatimosim, M. J. Amaya, M. T. Guerra, C. J. Aguiar, A. M. Goes, N. L. Gomez-Viquez, M. A. Rodrigues, D. A. Gomes, J. Martins-Cruz, W. J. Lederer, M. F. Leite, *Cell Calcium* **2008**, *44*, 230–242.

180. L. Santella, K. Kyozuka, *Cell Calcium* **1997**, *22*, 11–20.
181. D. J. Hennager, M. J. Welsh, S. DeLisle, *J. Biol. Chem.* **1995**, *270*, 4959–4962.
182. L. Lanini, O. Bachs, E. Carafoli, *J. Biol. Chem.* **1992**, *267*, 11548–11552.
183. N. Divecha, H. Banfic, R. F. Irvine, *Cell* **1993**, *74*, 405–407.
184. A. M. Martelli, L. M. Neri, R. S. Gilmour, P. J. Barker, N. S. Huskisson, F. A. Manzoli, L. Cocco, *Biochem. Biophys. Res. Commun.* **1991**, *177*, 480–487.
185. A. M. Martelli, L. Cocco, R. Bareggi, G. Tabellini, R. Rizzoli, M. D. Ghibellini, P. Narducci, *J. Cell Biochem.* **1999**, *72*, 339–348.
186. S. A. Stricker, *Dev. Biol.* **1999**, *211*, 157–176.
187. D. Kline, J. T. Kline, *Dev. Biol.* **1992**, *149*, 80–89.
188. S. Miyazaki, H. Shirakawa, K. Nakada, Y. Honda, *Dev. Biol.* **1993**, *158*, 62–78.
189. K. Swann, C. M. Saunders, N. T. Rogers, F. A. Lai, *Semin. Cell Dev. Biol.* **2006**, *17*, 264–273.
190. C. M. Saunders, M. G. Larman, J. Parrington, L. J. Cox, J. Royse, L. M. Blayney, K. Swann, F. A. Lai, *Development* **2002**, *129*, 3533–3544.
191. L. Santella, D. Lim, F. Moccia, *Trends Biochem. Sci.* **2004**, *29*, 400–408.
192. L. Santella, J. T. Chun, *Sci. China Life Sci.* **2011**, *54*, 733–743.
193. M. E. Greenberg, E. B. Ziff, L. A. Greene, *Science* **1986**, *234*, 80–83.
194. L. B. Rosen, D. D. Ginty, M. E. Greenberg, *Adv. Second Messenger Phosphoprotein Res.* **1995**, *30*, 225–253.
195. A. Lanahan, P. Worley, *Neurobiol. Learn Mem.* **1998**, *70*, 37–43.
196. C. P. Bengtson, H. Bading, *Adv. Exp. Med. Biol.* **2012**, *970*, 377–405.
197. H. Enslen, P. Sun, D. Brickey, S. H. Soderling, E. Klamo, T. R. Soderling, *J. Biol. Chem.* **1994**, *269*, 15520–15527.
198. T. Cali, L. Fedrizzi, D. Ottolini, R. Gomez-Villafuertes, B. Mellstrom, J. R. Naranjo, E. Carafoli, M. Brini, *J. Biol. Chem.* **2012**.
199. S. Ringer, *J. Physiol.* **1883**, *4*, 29–42 23.
200. S. E. Ebashi, M. Endo, I. Ohtsuki, in *Calcium as a Cellular Regulator*, Eds E. Carafoli, C. Klee, Oxford University Press, New York, Oxford, 1999, pp. 579–595.
201. A. P. Braun, H. Schulman, *Annu. Rev. Physiol.* **1995**, *57*, 417–445.
202. J. Aramburu, A. Rao, C. B. Klee, *Curr. Top. Cell Regul.* **2000**, *36*, 237–295.
203. K. E. Kamm, J. T. Stull, *J. Biol. Chem.* **2001**, *276*, 4527–4530.
204. R. J. Brushia, D. A. Walsh, *Front Biosci.* **1999**, *4*, D618–641.
205. K. S. Pavur, A. N. Petrov, A. G. Ryazanov, *Biochemistry* **2000**, *39*, 12216–12224.
206. J. Goldberg, A. C. Nairn, J. Kuriyan, *Cell* **1996**, *84*, 875–887.
207. A. Hudmon, H. Schulman, *Annu. Rev. Biochem.* **2002**, *71*, 473–510.
208. T. Kanaseki, Y. Ikeuchi, H. Sugiura, T. Yamauchi, *J. Cell Biol.* **1991**, *115*, 1049–1060.
209. L. C. Griffith, C. S. Lu, X. X. Sun, *Mol. Interv.* **2003**, *3*, 386–403.
210. J. Lisman, *Trends Neurosci.* **1994**, *17*, 406–412.
211. R. D. Blitzer, T. Wong, R. Nouranifar, R. Iyengar, E. M. Landau, *Neuron* **1995**, *15*, 1403–1414.
212. K. P. Giese, N. B. Fedorov, R. K. Filipkowski, A. J. Silva, *Science* **1998**, *279*, 870–873.
213. J. Liu, J. D. Farmer, Jr., W. S. Lane, J. Friedman, I. Weissman, S. L. Schreiber, *Cell* **1991**, *66*, 807–815.
214. N. A. Clipstone, G. R. Crabtree, *Nature* **1992**, *357*, 695–697.
215. B. Katz, R. Miledi, *J. Physiol.* **1965**, *181*, 656–670.
216. B. Katz, R. Miledi, *J. Physiol.* **1967**, *189*, 535–544.
217. R. P. Rubin, *Calcium and the Secretory Process*, Plenum Press, New York, 1974.
218. G. J. Augustine, M. P. Charlton, S. J. Smith, *J. Physiol.* **1985**, *367*, 163–181.
219. R. Llinas, I. Z. Steinberg, K. Walton, *Biophys. J.* **1981**, *33*, 323–351.
220. J. F. Kerr, A. H. Wyllie, A. R. Currie, *Br. J. Cancer* **1972**, *26*, 239–257.
221. P. Pinton, C. Giorgi, R. Siviero, E. Zecchini, R. Rizzuto, *Oncogene* **2008**, *27*, 6407–6418.
222. R. Rizzuto, P. Pinton, D. Ferrari, M. Chami, G. Szabadkai, P. J. Magalhaes, F. Di Virgilio, T. Pozzan, *Oncogene* **2003**, *22*, 8619–8627.

223. I. E. Wertz, V. M. Dixit, *J. Biol. Chem.* **2000**, *275*, 11470–11477.
224. P. Pinton, D. Ferrari, E. Rapizzi, F. Di Virgilio, T. Pozzan, R. Rizzuto, *EMBO J.* **2001**, *20*, 2690–2701.
225. S. Jiang, S. C. Chow, P. Nicotera, S. Orrenius, *Exp. Cell Res.* **1994**, *212*, 84–92.
226. P. Pacher, G. Hajnoczky, *EMBO J.* **2001**, *20*, 4107–4121.
227. G. Kroemer, J. C. Reed, *Nat. Med.* **2000**, *6*, 513–519.
228. G. Szabadkai, K. Bianchi, P. Varnai, D. De Stefani, M. R. Wieckowski, D. Cavagna, A. I. Nagy, T. Balla, R. Rizzuto, *J. Cell Biol.* **2006**, *175*, 901–911.

Chapter 6
Manganese Homeostasis and Transport

Jerome Roth, Silvia Ponzoni, and Michael Aschner

Contents

J. Roth
Department of Pharmacology and Toxicology, 11 Cary Hall,
University at Buffalo School of Medicine, Buffalo, NY 14214, USA

S. Ponzoni
Departamento de Ciências Fisiológicas, CCB, Universidade Estadual de Londrina,
Caixa Postal 6001, 85051-970, Londrina, PR, Brazil

M. Aschner (✉)
Department of Pediatrics, 2215-B Garland Avenue, 11415 MRB IV,
Vanderbilt University Medical Center, Nashville, TN 37232-0414, USA
e-mail: michael.aschner@vanderbilt.edu

L. Banci (ed.), *Metallomics and the Cell*, Metal Ions in Life Sciences 12,
DOI 10.1007/978-94-007-5561-1_6, © Springer Science+Business Media Dordrecht 2013

Abstract The review addresses issues pertinent to Mn accumulation and its mechanisms of transport, its neurotoxicity and mechanisms of neurodegeneration. The role of mitochondria and glia in this process is emphasized. We also discuss gene x environment interactions, focusing on the interplay between genes linked to Parkinson's disease (PD) and sensitivity to Mn.

Keywords divalent metal transporter 1 (DMT1) • manganese • mitochondria • parkin • transport

Please cite as: *Met. Ions Life Sci.* 12 (2013) 169–201

1 Introduction

Manganese is an essential trace metal in all forms of life. In mammalians it is required for normal amino acid, lipid, protein, and carbohydrate metabolism. Mn-dependent enzymes include oxidoreductases, transferases, hydrolases, lyases, isomerases, and ligases, to name a few. Mn metalloenzymes include arginase, glutamine synthetase, phosphoenolpyruvate decarboxylase, and Mn superoxide dismutase (MnSOD). Mn plays a critical role in multiple bodily functions including immunity, regulation of blood sugars and cellular energy, blood clotting, reproduction, digestion, and bone growth. Mn-containing polypeptides include arginase, the diphtheria toxin, and MnSOD. The latter is found in eukaryotic as well as bacterial organisms and is likely one of the most ancient enzymes, given that almost all organisms that live in an oxygen environment utilize the enzyme to dismutate superoxide. Several bacteria are an exception; nevertheless, they also use Mn via a non-enzymatic mechanism, involving Mn^{2+} ions complexed with polyphosphate.

Mn is also essential in plants as it is involved in photosynthesis. Oxygen-evolving complexes (OEC) contained within the thylakoid membranes of chloroplasts are responsible for the terminal photooxidation of water during the light reactions of photosynthesis. The chloroplasts contain a metalloenzyme core with 4 Mn atoms.

Surprisingly, no formal Recommended Dietary Allowance (RDA) exists for Mn, but the U.S. National Research Council (NRC) has established an estimated safe and adequate dietary intake (ESADDI) of 2–5 mg/day for adults [1]. Adequate intakes for newborns (<6 months of age) are 3 μg/day and 600 μg/day for infants at 12 months of age to (NAS, 2001) [2]. Children between 1–3 and 4–8 years of age have adequate daily Mn intakes of 1.2 and 1.5 mg/day, respectively.

The most important source of Mn is diet, with most daily intakes approximating 5 mg Mn/kg. Grain, rice, and nuts are highly enriched with Mn along with tea. Water concentrations of Mn typically range from 1 to 100 μg/L. Only a small fraction (1–5%) of ingested Mn is absorbed in normal conditions, the Mn arriving at the liver in the portal circulation is protein-bound. Within the plasma approximately 80% of Mn in the 2+ oxidation state is bound to globulin and albumin, and a small fraction (<1%) of trivalent (3+) Mn is bound to the iron-carrying protein, transferrin (Tf) [3,4]. Mn deficiency in humans is rare.

Generally, irrespective of intake route (oral, inhalation, dermal), animals maintain stable tissue concentrations given tight homeostatic control of Mn absorption and excretion [5–10]. Yet exposure to high Mn levels may lead to increased body-burden of the metal, resulting in adverse neurological, reproductive, and respiratory effects. The predominant site of Mn-induced damage is the brain, with symptoms commonly manifesting in motor dysfunction and psychological disturbances.

2 Manganese Transport

Since Mn is an essential cofactor for a diverse assortment of enzymes, its cellular concentrations are stringently managed by a variety of processes controlling cellular uptake, retention, and excretion. In general, overall systemic homeostatic levels of Mn are maintained, *in vivo*, via its rate of transport across enterocytes lining the intestinal wall and by its efficient removal within the liver [11]. Under typical nutritional consumption, the transport processes within the enterocytes lining the intestinal wall and the subsequent down-stream trafficking systems are efficiently balanced and work in harmony to preserve requisite supplies of Mn for the different cells and organelles within the body. This normally proficient system of checks and balances that control Mn levels *in vivo*, however, appears to fail under conditions of chronic exposure to high atmospheric levels (in various oxidation states; Mn is absorbed though predominantly in the 2+ and 3+ oxidation states) of the metal and thus, vital adjustments in the complex homeostatic processes are no longer adequate to preserve the required status quo. Failure of these systems to adjust to excess exposures infers that the rate limiting step governing systemic levels and ultimately Mn toxicity encompasses the operative biochemical processes necessary for its uptake and elimination within the various compartments of the body.

Like other required divalent metals, there are redundant transport systems for cellular uptake of Mn. The functioning system is dependent on both the ionic species present in biological fluids and by the specific transport proteins present within any given cell. Many of these redundant systems are also capable of transporting other metals suggesting that its role in managing the uptake of Mn may not be their primary responsibility. Figure 1 describes the different transport systems responsible for transport of Mn into cells. These include uptake by (i) the voltage-regulated and the ionotropic glutamate receptor Ca^{2+} channels [12,13], (ii) the transient receptor potential cation channel, subfamily M, member 7 (TRPM7) [14,15],

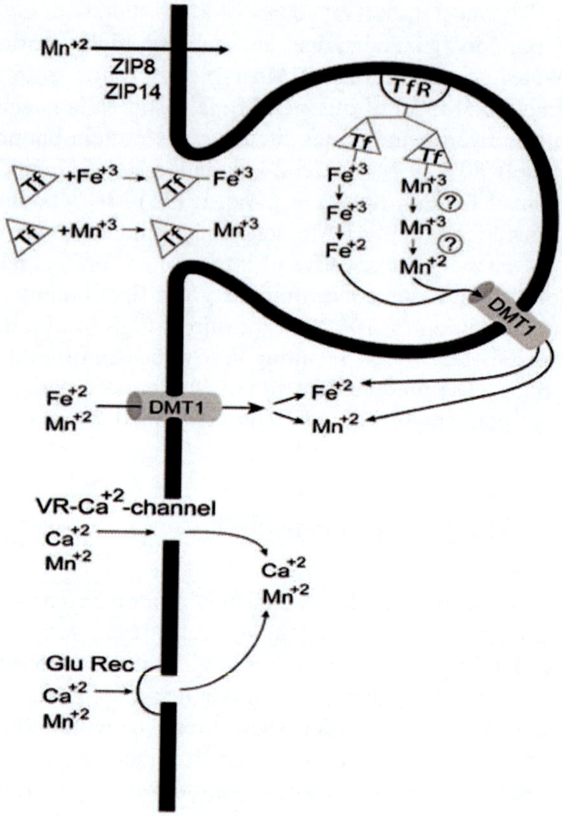

Figure 1 Transport mechanisms responsible for the uptake of Mn. Tf – transferrin; TfR transferrin receptor; DMT1- divalent metal transporter 1; VR – voltage-gated Ca^{2+} channel; SOC – store-operated Ca^{2+} channel; Glu Rec – glutamic acid ionotropic receptor.

(iii) store-operated Ca^{2+} channel [16] and a Tf-dependent and independent process via divalent metal transporter 1 (DMT1) [17–19]. Also indicated are members of the Slc39 gene family, ZIP 8 and 14, which have recently been identified as being involved in the transport of Mn [20–23]. Of all these proteins, DMT1 is generally considered to be the predominant transporter for Mn though this does not exclude the possibility that under varying physiological or pathological conditions and in any given cell population that the other transport processes may also contribute significantly to uptake.

2.1 *Divalent Metal Transporter 1*

As noted above, under normal conditions, transport of Mn into most cells is generally assumed to preferentially be dependent on DMT1. DMT1 has a broad substrate specificity for a variety of divalent cations including Fe^{2+}, Mn^{2+}, Cd^{2+}, Ni^{2+}, Co^{2+}, and Pb^{2+} [24–26]. The one exception to the rule is for Cu which may involve transport of the monovalent species [27]. Although DMT1 has a relatively broad substrate specificity, transport of iron is generally considered to be its principal function although, the affinity of Mn for DMT1 is similar to that for iron [17,28]. Smf1p (a homolog of DMT1) has been shown also in yeast to be a high-affinity transporter for Mn. Notably, a range of divalent metals can act as substrates for this transporter and when overexpressed in oocytes, it can increase intracellular concentrations, not only of Mn, but also of copper, cadmium, and iron.

As noted above (Figure 1), two distinct but related mechanisms are responsible for the transport of Mn^{2+} and Fe^{2+} by DMT1: a Tf-dependent and a Tf-independent pathway. In the intestines (see Figure 2), both Mn^{2+} and Fe^{2+} preferentially utilize the Tf-independent pathway, which is responsible for the direct absorption of the divalent species of both metals on the apical side of the enterocyte [29]. DMT1 is highest in the duodenum and decreases in the subsequent segments of the intestine [30] implying that transport preferentially occurs in the upper intestines. Although Mn^{2+} has a high affinity for DMT1, equivalent to that of Fe^{2+}, total uptake within the gastrointestinal tract is at best 5% of that present in ingested foods [31–34]. Once inside the enterocytes lining of the microvilli, Mn is transferred to the basolateral surface by a process which has not been adequately defined. Export of Fe^{2+} on the basal lateral side has been shown to require ferroportin (Fpn), which has recently been suggested to also play a role in the export of Mn [35–37] (see below).

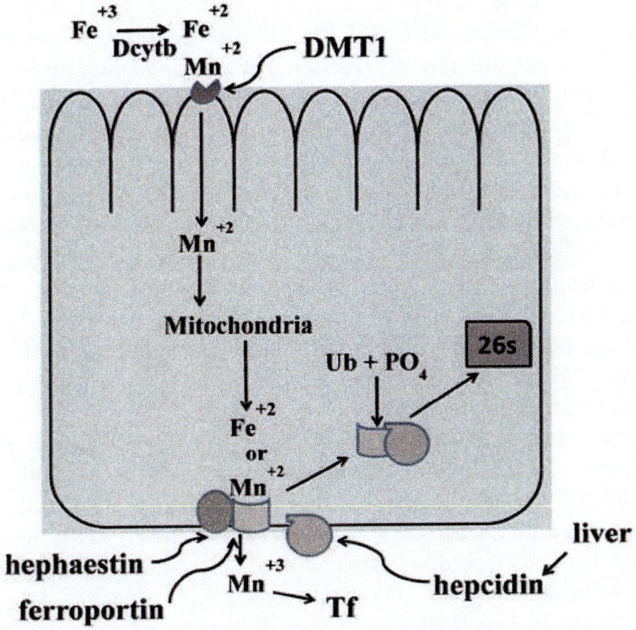

Figure 2 Mechanism for the transport of Mn across the intestine.

Levels of Fpn on the basolateral surface are controlled by the iron-regulated protein, hepcidin, produced in the liver [38,39]. Hepcidin causes internalization of Fpn, which subsequently undergoes ubiquitination before being degraded within the 26S proteasomes.

Before or during exit from the enterocyte, Fe^{2+} and presumably Mn^{2+} are oxidized to the trivalent state prior to entering the blood stream. Although this has never been directly demonstrated for Mn, it likely occurs, as a major portion of Mn entering the blood stream is tightly bound to Tf [33,40]. In the case of iron, this is catalyzed by either of the copper-containing proteins, hephaestin or ceruloplasmin [41,42], but whether either of the copper-containing ferrooxidases is also responsible for oxidation of Mn is not known, although, Cu does not have the oxidative potential to oxidize Mn from 2+ to 3+. Consistent with this is the recent study [43] suggesting that ceruloplasmin may not function in the loading of Mn onto plasma proteins or in the partitioning of Mn between the plasma and cellular fractions of blood though it does participate in tissue disposition and toxicity of Mn. Nevertheless, a major portion of Mn released from the enterocyte which enters the blood stream is bound tightly to Tf as the trivalent species [31,33,40,44]. Under normal conditions, approximately 30% of the available Tf binding sites in blood are occupied by Fe^{3+} leaving sufficient sites accessible for binding of Mn^{3+} [45]. Because Mn^{3+} forms a stable complex with Tf, hepatic removal is relatively slow and comparable to that for Tf-Fe^{3+} complex. The availability of these binding sites for Mn, however, may be compromised in several disease states such as hemochromatosis where Tf is saturated with iron. Because Fe, under these abnormal conditions, is in excess and is likely to have the higher affinity for Tf, Mn transport may be limited or carried out by other plasma components. In blood, the divalent species of Mn is preferentially bound to α2-macroglobulin though, because of the significantly greater abundance of albumin in serum, a sizable portion may also be bound to this latter protein [31,33,40,44]. The ligand interaction between Mn^{2+} and either serum proteins is relatively weak which likely accounts for the rapid hepatic elimination from blood.

In the absence of Fe, the binding sites of Tf can accommodate a number of other metals including gallium, copper, chromium, cobalt, vanadium, aluminum, terbium, and plutonium, raising the possibility that Tf functions *in vivo* as a transport agent for many of these metals. Fe is taken up by cells after Tf binds to a specific cell surface receptor and the Tf receptor complex is internalized. At its usual concentration in plasma, 3 mg/mL, and with 2 metal ion-binding sites per molecule (M_r 77000) of which only 30% are occupied by Fe^{3+}, transferrin has available 50 μ mole of unoccupied Mn^{3+} binding sites per liter.

The direct Tf-independent pathway utilizing DMT1 is also likely to be responsible for transport of Mn (in the 2+ oxidation state) into cells within the central nervous system (CNS) despite the fact that the interstitial fluid within the brain contains Tf. This is anticipated based on the fact that Fe levels in the interstitial fluid greatly exceed that of Tf resulting in saturation of the available binding sites causing the exclusion of Mn binding [46]. The actual ionic species of Mn in the extracellular fluid in the CNS and whether some of the Mn is capable of competing with Fe for binding to Tf is not known. Mn^{2+} at physiological pH is actually quite stable and because the pO_2 is low in the brain, little Mn^{3+} probably exists in the CNS. In addition, it is unknown whether Mn actually undergoes oxidation to the trivalent species upon its exit from endothelial cells comprising the blood-brain barrier (BBB). Based on a recent study by Hernández et al. [47], both Mn^{2+} or Mn^{3+} are capable of binding to citrate which is in excess in the brain interstitial fluid (225 to 573 μM) although neuronal toxicity of the Mn/citrate complex is approximately the same for all ionic species. Regardless, transport of the trivalent species of Mn, if it exists, must first be reduced to the divalent state prior to its uptake by either the Tf-dependent or independent pathway or for that matter the ZIP proteins (see below) as well. The role of citrate in this reductive process may be significant in that citrate has been reported to facilitate reduction of Fe^{3+}-bound to Tf as citrate has the ability to displace iron from holo-Tf when present in excess [48]. Reduction of Fe^{3+}- bound citrate, in fact, has been shown to be the rate limiting step in the uptake of Fe in astrocytes [49]. Whether a similar process occurs for Mn^{3+} in the CNS, however, is not known. Assuming Mn is similar to Fe, it may also bind to ascorbate in the interstitial fluid of brain [50].

The non-Tf-dependent pathway is also most likely to be accountable for the direct entry of Mn into the CNS via retrograde transport within axonal projections impinging on the nasal cavity leading from the neurons within the olfactory bulb [51–56]. Recent studies in the laboratory of Wessling-Resnick [57] have demonstrated that uptake of Mn into the presynaptic endings of these neurons is a DMT1-dependent process. The overall contribution of this pathway to Mn accumulation in the CNS in humans is not known but is likely contingent on its composition within the inspired air with the soluble forms being more efficiently taken up via this process [58]. There is evidence that extremely small ultrafine particles can also be taken up by the nerve endings and subsequently delivered to the CNS via a similar retrograde transport process [59].

For the Tf-dependent pathway, both trafficking and cellular transport of Mn^{3+} function very much like that for Fe^{3+}, indicative of the fact that a major portion of extracellular Mn^{3+} utilizes the same biochemical components as that for Fe^{3+} (Figure 1). The Tf-dependent pathway has been suggested to be the major mechanism within the choroid plexus and blood-brain barrier for transport of Mn into the brain [36,60]. Initially, the Tf-Mn complex attaches to the Tf receptors (TfR) on the cell surface, which is subsequently internalized within endosomal vesicles. Precisely how Mn is delivered to the endosomes is not understood. Furthermore, the specificity to Mn *versus* Fe has also yet to be delineated in mammalian cells. Next, these vesicles undergo acidification within the endosome via a hydrogen ion ATPase pump causing release of the metal from the Tf/TfR complex. In the case of iron, the released metal is reduced by either duodenal cytochrome *b* (Dcytb) [61] or a family of ferroreductases referred to as steap2-4 [62,63]. Whether these enzymes are also responsible for the reduction of Mn has not been investigated, although reduction must occur to enable transport via DMT1 and by the fact that the divalent species of Mn^{2+} is essentially the only form found in cells [64]. As noted above, it is questionable as to whether the Mn^{3+}-Tf complex actually is present in the interstitial fluid with the CNS.

Four different mammalian isoforms of DMT1 have been identified resulting from two autonomous start sites and alternate splicing of a single gene transcript. All are putative 12 membrane-spanning domain proteins [25,65,66] but differ both in their N- and carboxy-terminal amino acid sequences. The gene for DMT1 has two discrete promoter regions (1A and 1B) which independently regulate transcription at the two start sites. mRNA from two of the four species contain an iron response element (IRE) motif in the 3' UTR and are referred to as the +IRE species whereas those forms lacking the IRE are categorized as the –IRE species. Transcriptionally regulated splice variants, exon 1A and 1B, are present on the proximal N-terminal end for both the ±IRE forms of DMT1 mRNA. Exon 1A extends the N-terminal polypeptide 29 residues in human (31 in rats and 30 in mice) [66,67]. All four species of DMT1 distal from exon 1A share 543 residues in common but differ structurally in the last 18 (+IRE) or 25 (–IRE) carboxy terminal residues. The presence of the IRE in the message provides a site for binding of the iron response proteins, IRP1 and/or 2 [68,69], which potentially stabilize the mRNA when iron levels are low, leading to selective increases in expression of the+IRE isoforms of the transporter. This is consistent with numerous studies demonstrating that Mn disposition is critically dependent on iron levels *in vivo* [70–73] and with the +IRE isoform of the transporter being present in the apical surface of enterocytes [73]. Mn has also been reported to affect the binding of IRPs to IRE suggesting stabilization of the mRNA to DMT1 [60,74]. In addition to DMT1, the TfR and Fpn similarly contain an IRE in their 3'-UTR and like DMT1 are also regulated by iron via the IRPs [68,75].

The need for four distinct isoforms of DMT1 is not readily apparent especially since all display similar kinetic characteristics in regard to transport of both Mn and Fe [17,28,76]. In all likelihood, the physiological actions of the four isoforms of DMT1 are likely to have distinct roles dependent on the specific requirements within cells to maintain normal homeostatic levels of Fe particularly during fluctuating stress-related conditions. Because Fe is likely to be the primary operational metal transported by DMT1, alterations in expression of the different isoforms caused by changes in Fe content or other stress inducing situations will result in a corresponding adjustment in Mn transport irrespective of the actual *in vivo* requirements for Mn and thus, potentially stimulating development of Mn toxicity. This is consistent with the observation that DMT1 is elevated in the basal ganglia of iron-deficient rats and that Mn is selectively increased in the globus pallidus in iron-deficient animals exposed to Mn [77,78]. In addition to regulation of the +IRE species of DMT1 by IRPs, the N-terminal isoforms of DMT1 are also independently regulated at the level of transcription by environmental conditions which manifest in various stress-related events and inflammatory processes within the cell [79–83]. Transcription of the 1B isoforms of the transporter is upregulated by both nuclear factor kappa-light-chain-enhancer of activated B cells (NF-κB), and NF-Y (CCAAT promoter) and indirectly by nitric oxide suppression of NF-κB activity. In addition, both the 1A and 1B isoforms have also been reported to be induced by hypoxia implying the presence of an HRE within both promoter sites. Thus, expression of the different isoforms of DMT1 can be independently regulated within cells at both the level of transcription as well as translation.

In yeast and bacteria, Mn is transported from the endosomes to the Golgi apparatus and the mitochondria. Again, precise information on the process in mammalian cells is lacking, but it may involve fusion of Smf2p-containing vesicles with these organelles, as has been previously suggested for iron trafficking to mitochondria. It has also been suggested that Mn chaperones may transport the metals in the cytoplasm. The Tf transport mechanism for Fe is known to deliver both Fe^{3+} and Mn^{2+} to the mitochondria for incorporation into hemes [220–224].

2.2 *ZIP-Dependent Uptake of Manganese*

A number of studies over the past several years have demonstrated that two members of the solute-carrier-39 (SLC39) metal-transporter family, ZIP8 and ZIP14, are also capable of transporting Mn [21,23,84,85]. Unlike DMT1, both ZIP proteins are divalent cation/HCO_3^- symporters with the HCO_3^- gradient across the plasma membrane acting as the driving force for divalent metal transport. There is mounting evidence that these membrane proteins may play a significant role in Mn transport but their overall contribution relative to that of DMT1 has not been adequately assessed. Since the interstitial fluid in brain has a pH of approximately 7.3 [86], which is close to the optimum for both ZIP transporters, it is feasible their contribution to Mn uptake may be relatively more significant than anticipated in comparison to DMT1, which maximally functions at around pH 6.0. The significance of this pathway is further accentuated by the fact that Mn^{2+}, which is present in the interstitial fluid of brain, is probably not bound to Tf and thus, is available for transport by these ZIP proteins.

For the ZIP8 transporter, Cd^{2+} displays the highest affinity, exhibiting a K_m of 0.62 μM, though Mn^{2+} has been posited to be most efficiently transported, with a K_m for uptake of 2.2 μM, implying that Mn is probably the physiological substrate for ZIP8 [21]. This value is comparable to the K_m for Mn binding to DMT1 suggesting that it has the potential to significantly impact on Mn accumulation *in vivo*. Unfortunately, there have been few studies that have attempted to quantitate expression of this transporter relative to that of DMT1 and therefore the overall contribution of ZIP8 to Mn uptake in comparison to DMT1 or the other transport systems needs to be examined. One study which examined expression of ZIP8 to that of DMT1 by QPCR in the inner ear found that DMT1 levels were slightly higher than that of ZIP8 [87].

In contrast to ZIP8, its counterpart, ZIP14, has been studied more extensively in regard to both Mn and Fe transport. Of the 14-member ZIP protein family, ZIP14 is most homologous to ZIP8 [23]. Like DMT1, ZIP14 has a relatively broad substrate specificity capable of transporting a wide variety of divalent metals including, Mn, Fe, Co, Ni, Zn, Cd, Ca, and Pb [88]. Two forms of ZIP14 have been identified, A and B, having affinities for Mn^{2+} of 4.4 and 18.2 μM, respectively, with both being driven by a HCO_3^- gradient across the membrane [23]. Highest levels of ZIP14A are observed in the lung, testis, and kidney whereas the B isoform of the transporter is relatively evenly distributed. High expression of ZIP14 has also been observed in the intestines where it has the potential to play a significant role in the accumulation of Mn, although direct quantification relative to DMT1 has not been established. Several studies have indicated that like DMT1, expression of ZIP14 is capable of transporting Fe as well as being regulated by Fe levels with high iron decreasing expression whereas low Fe increasing protein levels [72,89]. Transport of Mn by ZIP14 mirrors that of Fe as both are Ca^{2+}-dependent [88].

Expression of the hemochromatosis protein, HFE, promotes a decrease in both ZIP14-dependent Tf bound and non-Tf-bound iron uptake as well as protein levels of ZIP14 with no apparent change in the mRNA level, thus suggesting that HFE decreases the stability of ZIP14 [89]. Since Tf is saturated with Fe^{3+} in the interstitial fluid in brain, appreciable amounts of both non-Tf bound Fe^{2+} and Mn^{2+} exist which may be transported by ZIP14 [90]. Like DMT1, ZIP14 also resides on the plasma membrane and within endosomal vesicles. Although greatest activity is observed at pH 7.5, ZIP14 can still function, although suboptimally, at pH 6.5, a value associated with release of divalent metals within endosomal vesicles. Nevertheless, transport of divalent metals by DMT1, which is H^+-coupled and preferentially stimulated at low pH, may serve as the predominant route of Fe^{2+} and Mn^{2+} uptake from late endosomes and lysosomes. Thus, the significance of ZIP14 in maintenance of Mn homeostasis and toxicity remains to be determined.

2.3 *Calcium(II) Channel-Dependent Transport of Manganese*

In addition to cellular uptake by the different isoforms of DMT1 and the ZIP proteins, there is also evidence for transport of Mn via several divalent metal channels which include, (i) the voltage regulated, (ii) the ionotropic glutamate receptor Ca^{2+} channels [12,13,91], (iii) TRPM7 [14,15] and (iv) the store operated Ca^{2+} channel (Figure 1) [16,92]. Evidence for the role of these ion channels in transporting Mn comes from a number of studies many of which primarily utilize Mn as a tool to measure functionality of the channel and therefore, it is not known, under normal resting conditions, what the overall contribution of each is to the total accumulation of Mn relative to the other known transport processes.

The potential significance of the voltage-gated Ca^{2+} comes from studies demonstrating that depolarization of cell membranes leads to increased uptake of Mn which can be prevented by Ca^{2+} channel blockers [12,91]. In regard to store-operated Ca^{2+} channels (SOC), Crossgrove and Yokel [93] have speculated this pathway may be responsible for the transport of Mn across the blood-brain barrier, but direct supportive evidence for this is lacking. Whether this also involves the SOC-dependent interaction of the Ca^{2+} sensor protein, STIM1, with that of calcium release-activated calcium channel protein 1 (ORAI1) at the plasma membrane is not known though the sarcoplasmic reticulum Ca-ATPase (SERCA) inhibitor, thapsigargin, has been reported to induce Mn^{2+} uptake which is inhibited by Ca^{2+} [94]. Demonstration of the direct involvement of TRPM7 in the transport of Mn was shown in studies assessing changes in transport in MCF-7 cells generated by overexpression or down-regulation of the protein [95]. Although Mn was shown to be transported by these channels, their overall contribution to neuronal toxicity within the CNS is unknown.

In contrast to the channels described above, the ionotropic glutamate receptor Ca^{2+} channel may directly impact on Mn toxicity due to the fact that neurons within the globus pallidus receive glutaminergic input from axonal projections leading from the subthalamic nuclei [96,97]. Several studies have demonstrated that Mn toxicity is attenuated by the glutamate post-synaptic ionotropic receptor inhibitor, MK801 [98,99], implying that glutamate participates in the neurotoxic actions of Mn. Similarly, the antiglutaminergic drug, riluzole, has also been shown to attenuate Mn toxicity in rats [100]. It is important to recognize that the cytotoxic events provoking Mn toxicity, to a large extent, parallel similar pathways for the excitatory neurotransmitter, glutamate, as both involve collapse of mitochondrial function initiated by excess sequestration of calcium [101,102]. The extent to which Mn uptake is increased upon glutamate binding within the γ-aminobutyric acid (GABAergic) neurons of the globus pallidus and the role this potentially plays in potentiating neurotoxicity, however, has not been directly investigated.

2.4 *Manganese Efflux*

Ferroportin (Fpn) [also referred to as IREG1 (iron-regulated protein 1) or MTP1 (metal tolerance protein 1)] is the cytoplamic Fe exporter. It is expressed in all cells, including neurons playing a key role in maintaining optimal Fe-homeostasis. Mutations in Fpn lead to type VI hemochromatosis, commonly known as Fpn disease, which is predominantly characterized by Fe accumulation in reticuloendothelial macrophages. The protein is transcriptionally, translationally, and posttranslationally regulated. Considering the shared uptake and characteristics of Mn and Fe (see above), it is surprising that only a few studies have attempted to delineate the role of Fpn in Mn efflux. Yin et al. [35] have recently reported that Fpn induction in human embryonic kidney cells (HEK293T) led to a reduction in Mn-induced toxicity, concomitant with decreased intracellular Mn accumulation. Notably, cerebellar and cortical Fpn was also increased in mice *in vivo* exposed to Mn [35]. More recently, Madejczyk and Ballatori [37] examined the role of Fpn in exporting Mn in *Xenopus laevis* oocytes demonstrating lower Mn accumulation in oocytes expressing Fpn. Furthermore, the efflux was inhibited by Fe, as well as other divalent metals. Collectively, these studies suggest Mn exposure in addition to promoting Fpn protein expression also reduces intracellular Mn levels and cytotoxicity.

3 Manganese: Toxic Mechanisms

Manganism and Parkinson's disease (PD) are two distinct neurological entities that impair basal ganglia function; the globus pallidus is predominantly damaged in the former and nigral dopaminergic (DAergic) neurons in the latter. DAergic cells from mesostriatal circuitry are also vulnerable to the toxic effects of Mn, making this metal a potential environmental risk factor in the etiology of PD. Cases of Mn poisoning have been reported in patients with chronic liver failure and long-term parenteral nutrition [103] and in individuals chronically exposed to the Mn-containing fungicide (manganese ethylene-bis-dithiocarbamate) [104] as well as other occupational cohorts (smelters, welders, etc.). Mn-induced disturbances of cellular homeostatic mechanisms include reactive oxidative species (ROS) production [105], impairment of antioxidant cellular defenses [106], mitochondrial damage [107], endoplasmatic reticulum stress [108], DNA damage [109], and inflammatory reactions [110], just to name a few. These are common features in a plethora of neurodegenerative diseases making Mn neurotoxicity studies an important tool in the understanding of the etiology of neurodegenerative processes. In this section, we discuss experimental data highlighting novel approaches to elucidate the vulnerability of DAergic nigral cells to the transition metal, Mn.

The DAergic cells that integrate the basal ganglia circuitry reside in the substantia nigra pars compacta (SNpc), a brain region with high oxidative activity. High content of oxidative enzymes and a high metabolic rate leads these cells to produce large quantities of damaging ROS [111]. Notably, catecholamine catabolism, a process which takes place in these cells, involves hydrogen peroxide production by the monoamine oxidase (MAO) and quinones by autoxidation [112]. *In vitro* assays have demonstrated that Mn increases dopamine (DA) and L-dopa autoxidation, leading to ROS and quinone production (see below); the Mn valence represents an important factor in the metal's DA oxidizing capacity, with manganic ion (Mn^{3+}) being more efficient than Mn^{2+} in potentiating DA autoxidation [113,114]. Nevertheless, whether Mn^{3+} oxidizes DA has been recently questioned, given the inability to locate Mn^{3+} within various cell types [64]. A schematic representation of Mn-induced DA autoxidation is shown in Figure 3.

3.1 *Mitochondria, Dopamine, and Manganese-Induced Neurotoxicity*

The mitochondria are critical organelles in mediating Mn-induced neurotoxicity [64]. Mn preferentially accumulates within these organelles [115] and mitochondrial superoxide dismutase 2 (SOD2) requires Mn as cofactor (there are other SODs in mitochondria, such as CuZn SOD1). Mn mitochondrial overload is toxic, secondary to its ability to impair ATP production and antioxidant defense mechanisms. Mitochondrial Mn^{2+} neurotoxicity could be related to inhibition of Ca^{2+} activation of

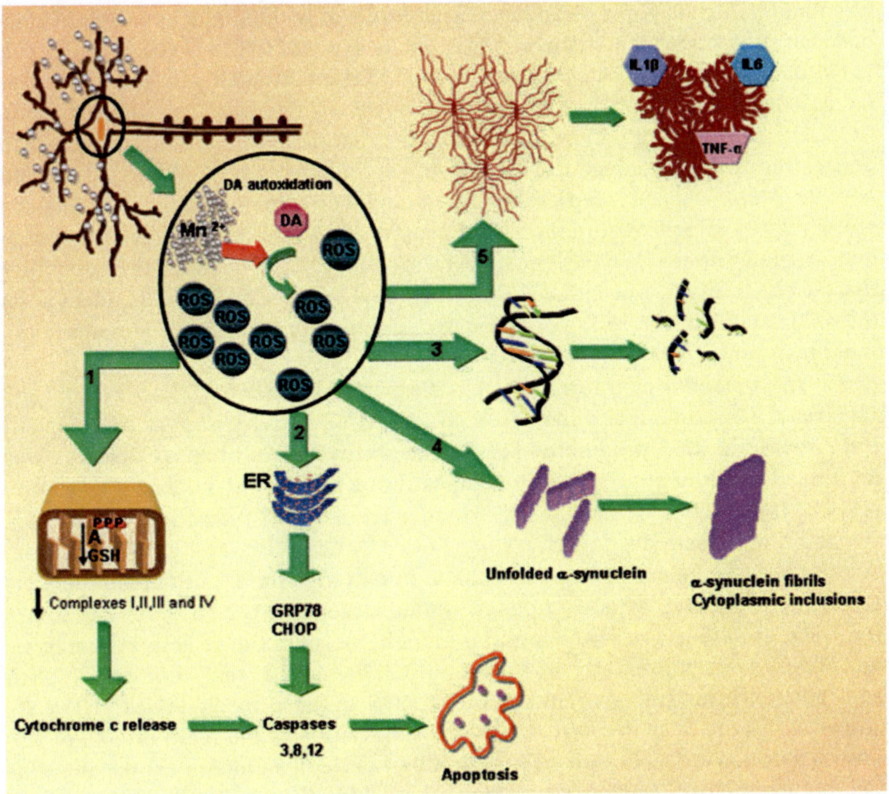

Figure 3 Schematic representation of some cellular mechanisms vulnerable to the toxic effects of Mn. Pathway 1 represents the mitochondrial toxic action with decreased ATP production and glutathione levels (GSH) leading to cytochrome *c* release and caspase activation and apoptosis. Pathway 2 represents endoplasmic reticulum stress with increased expression of glucose-regulated protein (GRP78) and pro-apoptotic protein C/EBP homologous protein (CHOP). Pathway 3 represents Mn clastogenic effect. Pathway 4 represents fibrillation of α-synuclein protein leading to the formation of cytoplasmatic inclusions. Pathway 5 represents microglial activation with the release of pro inflammatory mediators such as interleukin-6 (IL6), interleukin-1β (IL1β) and tumor necrosis factor α (TNF α).

ATP production [64]. Mn^{2+} attenuates brain mitochondrial ATP formation in two independent inhibitory sites; the primary site, complex II or fumarase, where succinate is the substrate, and the secondary site, the glutamate/aspartate exchanger, where glutamate plus malate are the substrates [107].

The ability of Mn to generate excess ROS has been observed both in cell culture and *in vivo* experiments. Cell mortality was increased when Mn was added to human fibroblast cultures, an effect that was potentiated in the presence of DA and antagonized by catalase and MnSOD2 [105]. Catechol isoquinolines, 1-methyl-6, 7-dihydroxy-1, 2, 3, 4-tetrahydroquinoline (salsolinol), and N-methyl-salsolinol induced apoptosis and increased levels of malondialdehyde (MDA) in a concentration-dependent manner in PC12 cells exposed to Mn [116]. Worley et al. [117] demonstrated in catecholaminergic CATH.a cells that Mn alone failed to increase ROS production; however, when followed by hydrogen peroxide exposure, ROS levels significantly increased, indicating that in addition to intracellular Mn concentrations, exposure duration, oxidative challenges post Mn exposure and cellular redox capacity represent important factors in mediating Mn-induced stress related effects. Mn-induced cytotoxicity in DAergic cells was also dependent on glutathione (GSH) levels in CATH.a cells [118]. This raises the possibility that susceptibility to Mn-induced ROS-mediated toxicity may be dependent upon the redox status of the cell, with those expressing high GSH levels perhaps being more resistant *versus* those with a low GSH complement.

In vivo, oral administration of Mn in the drinking water for 30 days decreased rat striatal DA content and increased striatal MAO activity, suggesting a potential increase in ROS production [119]. The ability of Mn to autoxidize DA was addressed by Sloot et al. [120] by demonstrating that intrastriatal Mn^{2+} injection resulted in depletion of striatal DA, which preceded ROS formation. In addition, reserpine pre-treatment failed to alter both DA depletion and ROS formation. Accordingly, reduced redox potential and impaired cellular antioxidant defense mechanisms likely play a key role in cellular susceptibility to develop Mn toxicity. Decreased levels of mitochondrial glutathione-peroxidase activity, catalase, and GSH were observed in striatum of Mn-treated rats, consistent with Mn-induced impairment of intracellular ROS defense mechanisms [106]. As shown in Figure 3, pathway 1 represents schematically the main effects of Mn on mitochondria. Decreased striatal and pallidal cell numbers expressing glutamine synthetase (astrocyte marker) and Mn-superoxide dismutase were observed in rats treated chronically for 13 weeks with high Mn concentrations in drinking water [121]. Aged and young rats exposed for 8 days to Mn in drinking water showed heightened susceptibility to the toxic effects of Mn concomitant with decreased GSH (also see previous paragraph) and uric acid levels in the striatum and in striatal synaptosomes [122]. Aged rats treated intraperitoneally for 30 days with Mn showed decreased mitochondrial complex II succinate dehydrogenase activity (~30%) in striatum and nigra, and increased tyrosine hydroxylase (TH) mRNA and protein levels [123]. Subcutaneous Mn administration in C57B1/C6 mice caused persistent striatal accumulation of the metal, observed at 21 days after cessation of the treatment; decreased DA release was observed at 7 and 21 days post treatment and attenuation in potassium stimulated increases in extracellular DA was noted at 1, 7, and 21 days post treatment, suggesting a long-term effect of Mn on striatal DAergic transmission [124]. Taken together these data are consistent with the preferential accumulation of Mn in basal ganglia and altered dopamine homeostasis in response to Mn exposure in various animal models.

3.2 Age-Related Effects of Manganese

The long-term effects of Mn in animals are helpful in delineating mechanisms of toxicity resulting from occupational exposures to this metal. Notably, a link between Mn and PD has been suggested, with Mn exposure exceeding twenty years, correlating with increased risk for PD [125]. Furthermore, chronic exposure to the fungicide Mn ethylene-bis-dithiocarbamate (Maneb) also increases the risk for Parkinsonism [104]. Though PD manifests in mid-life (familial PD) or senescence (idiopathic PD), it needs to be considered that its etiology evolves over decades with environmental exposures heightening the risk for the disease.

Several studies have addressed the developmental effects of Mn. For example, mice exposed to paraquat (PQ) and Maneb during the perinatal period and re-challenged as adults showed 62% reduction in striatal DA levels [126]. Juvenile mice exposed to Mn showed increased susceptibility, characterized by reduced striatal DA concentration in adulthood, induction of nitric oxide synthase (NOS), increased neuronal protein nitration and glial reactivity compared with adult mice that were not subjected to early-life Mn exposure or exposed to it only as adults [127,128]. These epidemiological and experimental findings point to an impairment in cellular antioxidant defenses induced by long-term or early-life exposure to this transition metal that diminishes the ability of the cells to cope with added oxidative challenges. The major susceptibility of aged rats to Mn neurotoxic effects likely reflects impairment in ROS defense mechanisms as a consequence of the aging process. The long-term and persistent cytotoxic effect of Mn and alterations in DAergic transmission suggest that chronic exposure to low metal concentrations in early life may facilitate the development of neurodegenerative processes later in life.

3.3 Manganese and Apoptosis

Mn-induced ROS production may also be a trigger for other processes that lead to cells demise, such as apoptosis. For example, in PC12 cells, Mn not only induces apoptosis, but also enhances L-dopa-induced apoptosis. Both effects were inhibited by antioxidants [129–131]. In the same cells, Mn was also shown to induce caspase 3-dependent proteolytic activation of protein kinase Cδ (PKCδ), which in turn contributes to apoptosis [132]. Increased intracellular calcium transient, decreased Na^+/K^+ ATPase and Ca^{2+} ATPase activities as well as increased apoptosis rates and impairment of *N*-methyl-D-aspartate (NMDA) receptor subunits synthesis were observed in Mn exposed primary neuronal cultures [133]. In SN4741 cells, a DAergic neuronal cell line, Mn-induced endoplasmic reticulum (ER) stress and activation of caspase-12 and apoptosis; the latter effect was reduced in Bcl-2-overexpressing DAergic cell lines [108].

Concentration-dependent Mn-induced ER stress response was observed in SK-N-MC human neuroblastoma cells, accompanied by increased expression of glucose-regulated protein 78 (GRP78), pro-apoptotic protein C/EBP homologous protein (CHOP) and p-eIF2α, concomitant with reduced mitochondrial complex I, II, III, and IV activities [134]. As shown in Figure 3, pathway 2 represents the ER stress response, which is also linked to pathway 1. Mn also effectively induced apoptosis in cultured mesencephalic cells, and this effect was accelerated by pre-treatment with DA. Cells were protected by blockade of NO synthesis, inhibition of NF-κB activation and pre-treatment with vitamin E. Induction of NF-κB and nitric oxide synthase activation by ROS were proposed as plausible mechanisms for Mn-induced neurotoxicity [135]. NF-κB would also lead to increased uptake of Mn as it has been reported to augment expression of DMT1, the major transport protein of Mn [80]. Treatment of neuronal stem cells with Mn was also shown to stimulate apoptosis via mitochondrial-mediated pathways, concomitant with cytochrome *c* release, caspase-3 activation, and ROS generation [136]. In immortalized DAergic cells, MES 23.5, Mn was shown to up regulate mitochondrial Bcl_2/E1B 19 KDa interacting protein (BNIP3), which correlated with mitochondrially-mediated apoptosis [137]. Finally, in rat astrocytoma C6 cells Mn led to ROS-mediated apoptosis, involving caspase-8 and mitochondrial-mediated pathways, with both the total and mitochondrial levels of Bcl-2 and Bax shifting to favor the apoptotic process [138]. Taken together, these studies establish the propensity of Mn to lead to cell demise by activation of apoptotic mechanisms.

3.4 *Effects of Manganese on DNA*

Mn also possesses clastogenic activities making DNA a potential target for its toxic action, which could result in impaired gene expression. For example, Chinese hamster V79 cells treated with L-dopa or DA showed decreased proliferative potential and elevated micronuclei frequency. The addition of Mn to the cell culture enhanced the antiproliferative and clastogenic effects of DA and its precursor [109].

Concentration-dependent DNA fragmentation was seen in Mn-treated PC12 cells; apoptosis was preceded by mitochondrial dysfunction and activation of c-Jun N-terminal kinases [139]. In striatal neurons, Mn treatment caused concentration-dependent loss of mitochondrial membrane potential, complex II activity, DNA fragmentation, and decreased microtubule-associated protein (MAP2) [140]. DA-induced oxidative DNA damage was shown to increase in Mn-treated PC12 cells, likely secondary to enhancement in DA autoxidation with semiquinone radical production. The DA-induced DNA damage in the presence of Cu^{2+} and NADH was further potentiated by Mn [141]. Mn increased the formation of mitochondrial DNA single strand breaks (SSB) probably caused by increased oxidative stress in *cultured* liver cells [142].

Mn is normally present in the heterochromatin, nucleolus, cytoplasm, and mitochondria of rat striatal astrocytes and neurons. Chronic Mn treatment leads to increased Mn levels in the mitochondria and nuclei, serving as a possible explanation for its combined ability to impair mitochondrial oxidative metabolism, increase ROS production and damage to DNA [143]. Nuclear accumulation of Mn was also observed in cultured PC12 cells [144]. The DNA damage is shown by pathway 3 in Figure 3 (see above). Taken together, research to date indicates that excessive Mn levels may disturb cellular DNA.

3.5 *Manganese and Neurotransmitter Homeostasis*

Alterations in biosynthetic neurotransmitter pathways, which lead to changes in the activity of rate-limiting enzymes or synaptic receptors may also be responsible for Mn-induced alterations in synaptic transmission and consequently behavior. Mn alters the activity of the rate-limiting enzyme in DA synthesis, tyrosine hydroxylase (TH). In DAergic neural cell lineage, N27 cells, Mn was shown to lead to distinct alterations in TH upon both acute and chronic exposures. In acute conditions TH activity increased, whereas in chronic conditions it was decreased [145]. Decreased striatal DA content was preceded by an initial phase of increased TH activity and striatal DA content in rats chronically treated with Mn in drinking water [146,147]. Manganism is characterized by two phases, commencing with a psychiatric syndrome (bizarre behavior and madness) followed by impaired motor function. Changes in TH activity inherent to the acute or chronic Mn exposure may be related to these two phases, the initial phase corresponding to increased TH activity, which is probably linked to increased DA release, and the second phase related to decrease TH activity and probably linked to reduced DA release or DA cellular content.

The psychiatric phase of manganism could be also a consequence of Mn-induced inhibition of glutamatergic corticostriatal signaling mediated by presynaptic D2-like DA receptors. For example, corticostriatal slices of rats chronically treated with Mn, showed enhanced glutamate-mediated synaptic transmission in the striatum, an effect associated with increased locomotor activity in these rats [148].

3.6 *Manganese Nanoparticles*

A recent concern on human health is associated with the potential effects associated with exposure to Mn-containing nanoparticles. Nanotechnology is focused on creation and manipulation of particles with dimensions ranging from 1 to 100 nm, leading to the production of new materials that exhibit novel physicochemical properties and functions [149]. Mn has a large industrial use involving steel and non-steel alloy production, colorants, battery manufacture, pigments, and fuel additives, among others [150]. With the availability of nanosized particles, the replacement of macro-sized Mn particles has rapidly evolved [151]. As indicated above, olfactory neuronal pathway has been shown to be efficient in translocating inhaled Mn oxide into the CNS [152]. Some experimental data obtained with nanosized Mn particles will be presented herein, focusing attention on their safety and potential health concerns.

In PC12 Mn oxide nanoparticles (40 nm) were shown to be internalized and to lead to a concentration-dependent depletion of DA and its metabolites, increased ROS production [153] and down-regulation of TH gene expression [149]. Mn nanoparticles (25–900 nm) were also shown to be internalized in N27 DAergic neuronal cells, leading to time-dependent up-regulation of the protein transporter Tf, increased ROS production, and activation of caspase-3-mediated apoptosis and pro apoptotic protein kinase Cδ (PKCδ). In addition, Mn nanoparticles caused autophagy characterized by decreased levels of the native form of Bcl-2-interacting protein-1, Beclin 1, and increased cleavage of microtubule-associated protein 1 light chain 3, LC3, both of which are associated with autophagosome formation [151].

3.7 Manganese and Cytoplasmic Inclusions

The pathological hallmark of idiopathic PD is the presence of cytoplasmatic inclusions of α-synuclein fibrils, known as Lewy bodies and Lewy neurites; their production can be affected by both genetic and environmental factors [154,155]. α-Synuclein is a heat stable protein that is soluble and natively unfolded. In the CNS it is expressed in neurons where it localizes in pre-synaptic terminals in the vicinity to synaptic vesicles. The precise function of α-synuclein is unknown; a role as modulator in synaptic and neural plasticity has been advanced [156].

In vitro assay showed that pesticides, such as rotenone, PQ and Maneb, as well as metals, such as Al^{3+}, Fe^{3+}, Co^{3+}, Cd^{2+}, and Mn^{2+}, induce a conformational change in α-synuclein accelerating its fibrillation rate; the effect of pesticides and metals when simultaneously present is synergistic [157] (see pathway 4 in Figure 3). Cultured DAergic neuronal cells, MES 23.5, exposed to the Mn-containing pesticide Maneb showed inhibition of proteasomal chymotrypsin-like and postglutamyl peptidase activities. Proteasomal dysfunction was accompanied by cytoplasmatic inclusions that were positive for α-synuclein immunostaining [158]. Metals such as arsenic, copper, zinc, mercury, cadmium, nickel, and lead as well as the metalloid selenium, have been shown to increase α-synuclein-like immunoreactivity aggregates in the CNS of white sucker fish, *Catostomus commersoni*, sampled from highly contaminated water with metal ions secondary to mining activity [159].

The propensity of Mn to increase cytoplasmic inclusions has also been shown *in vitro*. For example, Mn decreased the viability of SK-N-MC neuroblastoma cells expressing human DA transporter and α-synuclein [160], and increased expression of α-synuclein in PC12 cells, via ERK1/2 MAPK activation [161]. In SH-SY5Y neuroblastoma cells up-regulation of α-synuclein expression and α-synuclein mRNA by Mn preceded the apoptotic response [162], suggesting up-regulation of α-synuclein may be an early effect.

Similar effects have been noted *in vivo*. For example, non-human primates exposed to Mn showed up-regulation of another type of cytoplasmic inclusions, namely amyloid-β (A-β) precursor-like protein-1; its gene expression, increased along with the number of A-β diffuse plaques. In the same brains, Mn also led to α-synuclein aggregation particularly within the frontal cortex [163]. While still debatable, these findings raise the plausibility that in addition to its well-established link to PD, Mn may play a role in the etiology of Alzheimer's disease (AD).

3.8 Manganese and Dopaminergic Circuitry

DA is a universal neurotransmitter present in the animal kingdom from invertebrates to vertebrates. The interaction of Mn with DA has been noted in other species, such as mollusks where exposure to Mn was shown to impair the cilio-inhibitory system of the lateral cilia of the gill, and decreased endogenous DA levels in cerebral and visceral ganglia and gills of *Crassostrea virginica* [164].

In vertebrate animals, microinjection of Mn into the SNpc or striatum allows for the evaluation of its action on mesostriatal DAergic circuitry. For example, unilateral intra-nigral Mn microinjection in rats led to ipsilateral striatal DA loss in a dose-dependent fashion and a rotational behavior towards the lesion side in response to systemic apomorphine. Systemic administration of L-dopa plus carbidopa or pargyline increased the DAergic striatal loss in these animals [165]. Intra-nigral Mn microinjection in rats decreased nigral and ipsilateral striatal DA and TH cofactor, (6R)-L-erythro-5,6,7,8 tetrahydrobiopterin (BH_4) levels. Maximal decrease in BH_4 was observed at 60 days, with complete recovery at 90 days after the Mn microinjection [166]. Mn microinjected into rat SNpc decreased ipsilateral striatal DA concentrations, the number of TH-positive cells and dopamine- and cAMP-regulated neuronal phosphoprotein (DARPP-32) expression, and increased rotational behavior in response to systemic apomorphine administration, while systemic L-dopa plus carbidopa treatment worsened these effects [167].

Rats intra-nigrally microinjected with Mn showed rotational behavior in response to apomorphine and increased number of NADPH-d positive neurons in the ipsilateral SNpc as well as ipsi- and contralateral striatum. NO synthesis inhibition by NG-nitro-L-arginine (L-NOARG) reversed the NADPH-diaphorase increase, but worsened the rotational behavior response to systemic apomorphine, suggesting a protective role for NO in the Mn-induced neurodegenerative process [168]. Paradkar and Roth have reported that NO can inhibit NF-κB activity and the subsequent up regulation of DMT1 which may, in part, account for the protective effects of NO [80]. Mn is slowly cleared from the substantia nigra (SN), with a 50% decrease noted at 72 hours post microinjection [169]. Notably, the apomorphine-induced rotational behavior was detected at 24 hours after intra-nigral Mn microinjection reaching its maximum at 72 hours post injection. The time course of striatal TH immunostaining loss observed in Mn-treated rats followed the time course of rotational behavior, suggesting that cellular mechanisms induced by Mn, which lead to

DAergic cell death occurred shortly after injection. It also suggests that threshold metal concentration is required in order to induce neurotoxic effects [169]. The slow brain clearance of Mn could be explained by its propensity to accumulate within brain mitochondria as shown by Gavin et al. [102].

Mn-induced DAergic neurodegeneration may be a consequence of its ability to produce an indirect excitotoxic process, as demonstrated by Brouillet et al. [98]. These authors have shown that Mn microinjected into rat striatum caused a dose-dependent decrease in DA, γ-aminobutyric acid, and substance P concentrations in the striatum. Such effects were blocked by prior removal of cortico-striatal glutamatergic inputs or by treatment with MK-801, a non-competitive NMDA antagonist. Striatal slices of rats treated chronically with Mn showed an enhancement in cortical glutamate-mediated synaptic transmission in the striatum, suggesting impairment in corticostriatal glutamatergic transmission and increased excitotoxic damage [170].

3.9 *Manganese and Microglia*

Reactive microglia are present in the substantia nigra of PD patients, suggesting an inflammatory component in the neurodegenerative process. CNS inflammatory responses are predominantly mediated by microglial cells which represent the first line of defense. Microglial activation has two stages; the first one is associated with release of neurotrophins leading to increased neuronal survival, and the second phase associated with release of pro-inflammatory mediators potentially leading to impaired neuronal survival [171]. Several studies suggest a potential role for microglia in mediating Mn-induced neurodegeneration. *In vivo*, intrastriatal Mn microinjections in rats were shown (7 days post injection) to reduce the number of TH^+ cells and increase the number of activated microglia in the substantia nigra; up regulation of inducible nitric oxide synthase (iNOS) and tumor necrosis factor α gene expression was paralleled by increased protein levels of iNOS and interleukin-1β in the substantia nigra [172]. The microglial participation in Mn-induced neurodegeneration response is shown in pathway 5 in Figure 3 (see above). Non-human primates chronically treated with Mn also displayed increased microglial number in SNpc and reticulate concomitant with increased expression of iNOS, L-ferritin, and intracellular ferric ion in reactive microglial cells [173]. The decrease in TH^+ cells observed in animals treated with intrastriatal Mn microinjection could be the consequence of a microglial defense response, which depending upon the duration and intensity could lead to inflammation and cellular degeneration.

Consistent with the above, mouse N9 microglial cells treated *in vitro* with bacterial lipopolysaccharide (LPS) and challenged with Mn showed significant induction of NO production, with a concomitant increase in iNOS gene transcription [110]. The same cells, when exposed to Mn showed increased production of interleukin-6 and TNFα. In the presence of lipopolysaccharide, this response was significantly potentiated. Co-exposure to LPS and Mn also increased NO production and iNOS expression [174]. It is noteworthy that Mn regulates iNOS expression at the transcriptional level in BV2 microglial cells, increasing iNOS protein expression by the activation of JNK-ERK MAPK and PI3/Akt signaling pathways [175].

More recently, it has been shown in rat primary neuron-glia co-cultures that simultaneous treatment with Mn and LPS increased production of pro-inflammatory cytokines, including TNFα and IL-1β, ROS as well as reactive nitrogen species (RNS). Minocycline (an antibiotic) pre-treatment effectively reduced the pro-inflammatory cytokine production [176]. These experimental data suggest that Mn can potentiate the effects of bacterial toxins and that ensuing inflammatory reactions in basal ganglia thus exacerbating the neurodegenerative process.

4 Manganese and Genetics

The clinical features of the extrapyramidal symptoms of manganism can be quite diverse reflecting the complex nature of the basal ganglia which is associated with a variety of integrated inhibitory and excitatory neurochemical pathways. The degree to which Mn disrupts any of these interdependent processes promotes an imbalance of output from the basal ganglia which potentially result in capricious disparities in onset, severity, and specific symptoms expressed as well as the progression of the disorder. The source which contributes to deviations in the characteristics of the expressed symptoms and neurological lesions generated may include exposure to other noxious environmental toxins, nutrition status, state of health and, most importantly, underlying genetic variability.

All of these reasons to varying degrees have the potential to alter the biochemical processes by which Mn imbalance can occur. Of these, genetic polymorphisms have to be considered as potentially playing a dominant role in regulating alterations in specific signaling pathways controlling Mn-induced cell death. This is best exemplified by the publication by Sadek et al. [177] describing a patient who worked as a welder for a total of three years. This individual acquired the progressive symptoms of Mn toxicity within one year after beginning employment making his clinical history quite remarkable as development of manganism after only one year of welding is highly unusual as onset usually requires considerably longer exposure times. The reason for the rapid development of the neurological deficits is unknown but, most likely, reflects a genetic predisposition in this individual. In addition, within several years of the initial diagnosis, he developed a unilateral tremor in his right hand more characteristic of Parkinsonism. The prominence of a genetic component contributing to manganism is further substantiated by the reality that not all welders or Mn miners develop manganism yet exposures are comparable to their fellow workers that acquire the disorder. Clearly, this underscores the potential impact of genetic polymorphisms that likely contribute to the variability in development of Mn toxicity. It is essential to stress that although chronic exposure to Mn is not the causative agent provoking Parkinson's disease, there is compelling evidence in the literature that it may be one of the most influential metals correlating with increased susceptibility to develop this condition [178–182].

Manganism is a disorder anatomically and functionally distinct from that of Parkinson's disease, at least in regard to the critical initial site of injury, leaving us with the issue as to why many of the observed symptoms overlap between the two disorders and whether chronic exposure to Mn can eventually provoke idiopathic Parkinsonism. Some of the similarities between the two disorders simply reflect the fact that Mn can influence DAergic transmission by inhibiting dopamine release and decreasing dopamine transporter (DAT) levels in the striatum [124,183–187]. Although this may clarify why some of the symptoms of Mn toxicity overlap with that of Parkinson's disease, it fails to explain individual differences in susceptibility to develop manganism.

Over the past 15 years, a number of gene variants have been identified as being linked to early (α-synuclein, parkin, PINK1, DJ-1, and ATP13A2) and late (LRRK2) onset of Parkinson's disease [188–190]. Because many of the symptoms and neurodegenerative features between the two neurological disorders appear to be interrelated, it is feasible that polymorphisms in some of these genes may also be associated with the susceptibility to develop manganism. If a relationship does exist, then it may also explain, in a reciprocal fashion, why excess exposure to Mn may provoke early onset of Parkinson's disease. Of the six genes linked to Parkinsonism, two of these, parkin and ATP13A2, have already been reported to influence Mn toxicity in cell culture systems [191–194]. Thus, one-third of the known polymorphisms in genes linked to onset of Parkinsonism are suggested to promote increased susceptibility to acquire manganism.

The two genes, parkin and ATP13A2, are both linked to early onset of Parkinsonism but differ in their enzymatic activity though both have several features in common including alteration of α-synuclein activity and protein degradation as well as mitochondrial dysfunction [195–197]. Parkin is one of over 600 identified E3 ligases responsible for the conjugation of ubiquitin to a variety of proteins [198,199]. Of the known genes correlating with early onset of Parkinson's disease, those involving parkin are the most prevalent encompassing approximately 50% of all recessive cases [200,201]. Although considered to be an autosomal recessive gene, there is evidence in the literature suggesting that mutations in a single allele may exert sufficient imbalance in dopaminergic activity to cause subclinical features of Parkinsonism [202] and thus, even in the heterozygous state, may play a significant role in development of idiopathic Parkinsonism [203,204]. Two studies [191,205] have reported that overexpression of parkin can protect cells against Mn toxicity. Recent studies [191] have further demonstrated that parkin is responsible for the ubiquitination of DMT1, the major transport protein for Mn (see above) [191]. This process is relatively specific as it is responsible for degradation of only the 1B species of the transporter which predominates in the CNS [206,207]. Thus, mutations in parkin, which lead to its inactivation would likely increase DMT1 levels and the subsequent accumulation of Mn within the CNS.

Like parkin, ATP13A2 is also associated with the pathogenesis of both familial and sporadic Parkinson's disease as well as Kufor-Rakeb syndrome (KRS), a rare juvenile-onset autosomal recessive disease characterized by progressive Parkinsonism [190,208]. ATP13A2 encodes a large membrane-bound lysosomal P-type ATPase and when mutated acts, presumably, to disrupt its lysosomal localization, thus, linking lysosomal degradation with Parkinson's disease. A recent paper by Gitler et al. [193] demonstrated a strong genetic interaction between α-synuclein and PARK9, a yeast ortholog of human Parkinson's disease-linked gene, ATP13A2. Coexpression of PARK9 (ATP13A2) in animals overexpressing α-synuclein was shown to protect against dopaminergic neuron loss whereas knockdown of the ATP13A2 ortholog in *Caenorhabditis elegans* enhance α-synuclein misfolding. It was also reported that wild-type ATP13A2, but not the KRS pathogenic ATP13A2 mutant, protected cells from Mn-induced cell death in mammalian cell lines and primary rat neuronal cultures. In addition, wild-type ATP13A2 reduced intracellular Mn and prevented cytochrome *c* release from mitochondria when compared to the mutant cells. Based on these findings, these authors suggested that PARK9 is involved in Mn transport though direct evidence for this is lacking.

Thus, one-third of the known Parkinson-linked polymorphisms are linked to Mn toxicity leaving us with the question as to the role of the other four genes, DJ-1, PINK1, α-synuclein, and LRRK2, in the development of manganism [192,193,209, 210]. The proposed mechanisms for each of these proteins in inducing Parkinsonism are relatively complex and often overlap in that each may mutually influence the others activity. Based on the predicted mechanisms, there is compelling evidence to justify the hypothesis that mutations in these genes will augment Mn toxicity as polymorphisms produce gene products that disrupt mitochondrial function and thus, potentiate mitochondrial-induced oxidative stress which, independent of other toxic mechanisms, will facilitate Mn-induced toxicity.

Whereas the aforementioned genes are autosomal recessive, LRRK2 is an autosomal dominant gene and the only gene associated with late onset of the disorder [211]. Thus, a mutation in a single allele may be sufficient to elicit effects on Mn toxicity prior to the appearance of the symptoms of Parkinsonism. LRRK2 is a relatively complex gene which encodes a large multidomain protein that includes a Rho/Ras-like GTPase domain (termed Roc, for Ras) and a protein kinase domain. Because of this complexity, the mechanism by which mutations in LRRK2 elicits Parkinsonism is likely to be multifaceted and are anticipated to influence a number of signaling pathways [212]. In regard to dopamine function, LRRK2 can cause a decrease in the dopamine transporter, DAT [213,214], as well as impair dopamine-stimulated neurotransmission [215] both of which are also seen with overexposure to Mn [183,184,186,187]. Relevant to manganism, studies have reported that Mn can increase phosphorylation activity of the most predominant polymorphic species of LRRK2, the G2019S mutant, while inhibiting the wild-type kinase activity activated by Mg [216,217]. Thus, LRRK2 displays a number of critical functions many of which can potentially affect or be affected by Mn.

Support for a potential role for DMT1 in neurodegeneration associated with PD was recently suggested by several researchers. Salazar et al. [218] have recently shown that 1-methyl-4-phenyl-1,2,3,6-tetrahydropyridine (MPTP) intoxication in mice, a well established PD model, caused increased DMT1 expression in the ventral mesencephalon concomitant with iron accumulation, oxidative stress, and DAergic cell loss. More recently He et al. [219], direct sequencing identified two single-nucleotide polymorphisms in the DMT1 gene (CC haplotype) as a risk factor for PD in the Han Chinese population. The above studies have not directly assessed whether polymorphisms in DMT1 in PD patients also alter CNS Mn levels. Given this gene's role in regulating intracellular Mn levels (see above), it is essential that future studies in suitable cohorts also address the association between DMT1 polymorphisms and Mn accumulation.

5 Concluding Remarks

As described above, uptake of Mn in the intestines and the various processes controlling its delivery and access to the CNS represent the rate limiting steps regulating Mn levels and toxicity *in vivo*. Multiple systems are involved in the maintenance of Mn levels, all of which must function, to some extent, in the preservation of its homeostatic levels. These systems routinely work in harmony to sustain required levels of Mn even though Mn may not be an operational cation these systems were designed for. As a consequence, conditions that upset this delicate balance result in a failure of the various components to adjust to the practical needs of Mn in order to support the functional demands of the other essential cations, the most important of which is Fe. Thus, conditions of excess exposures to Mn or a variety stress inducing or inflammatory conditions are likely to provoke an imbalance in the various transport systems for Mn resulting in increased CNS toxicity.

The synopsis provided above establishes that Mn exerts neurotoxicity specifically in DAergic nigral cells. However, its aberrant effects may be much broader given recent evidence on its ability to perturb other cell types and neurotransmitters, as well as its propensity to increase cytoplasmic inclusions. Taken together, a broad array of mechanisms appears to mediate Mn-induced DAergic neurodegeneration. Considering the essentiality of Mn, future studies should continue to focus on its role both in health and disease. Given that the safety margin for Mn appears to be narrow, it is incumbent upon the research community to further delineate its toxicity, considering its ubiquitous presence in the environment and anthropogenic usage. This novel information will be highly relevant for future considerations on its safety and in delineating risk assessments for exposure and safety.

Finally, the findings to date support a possible genetic link between the manifestations of manganism with genes associated with onset of Parkinsonism. This is not totally surprising as mutations in these genes, in many respects, display similar aberrant neurological activities within the basal ganglia, which provoke the distinctive dystonic movements associated with both disorders. It remains to be determined whether the other genes linked to Parkinsonism will also be associated with the development of Mn toxicity.

ABBREVIATIONS AND DEFINITIONS

AD	Alzheimer's disease
A-β	amyloid-β
BBB	blood-brain barrier
BH_4	tetrahydrobiopterin
CHOP	C/EBP homologous protein
CNS	central nervous system
DA	dopamine
DAergic	dopaminergic
DARPP-32	dopamine- and cAMP-regulated neuronal phosphoprotein
DAT	dopamine transporter
Dcytb	duodenal cytochrome B
DMT1	divalent metal transporter 1
ER	endoplasmic reticulum
ESADDI	estimated safe and adequate dietary intake
Fpn	ferroportin
GABA	γ-aminobutyric acid
GRP78	glucose-regulated protein 78
GSH	glutathione
HEK293	human embryonic kidney cells 293
HFE	hemochromatosis protein
HRE	hypoxia response element
IL-1β	interleukin-1β
IL-6	interleukin-6
iNOS	inducible nitric oxide
IRE	iron response element
IREG1	iron-regulated protein 1
IRP	iron-responsive protein
KRS	Kufor-Rakeb syndrome
L-dopa	L-3,4-dihydroxyphenylalanine
L-NOARG	NG-nitro-L-arginine
LPS	lipopolysaccharide
Maneb	manganese ethylene-bis-dithiocarbamate
MAO	monoamine oxidase
MAP2	microtubule-associated protein
MDA	malondialdehyde
MnSOD	Mn superoxide dismutase
MPTP	1-methyl-4-phenyl-1,2,3,6-tetrahydropyridine
MTP1	metal tolerance protein 1
NADH	nicotinamide adenine dinucleotide (reduced)
NF	necrosis factor
NMDA	*N*-methyl-D-aspartate
NOS	nitric oxide synthase

NRC	National Research Council
OEC	oxygen-evolving complex
ORAI1	calcium release-activated calcium channel protein 1
PD	Parkinson's disease
PKCδ	protein kinase Cδ
PQ	paraquat
QPCR	real-time polymerase chain reaction
RDA	Recommended Dietary Allowance
RNS	reactive nitrogen species
ROS	reactive oxygen species
SERCA	sarcoplasmic reticulum Ca-ATPase
SLC39	solute-carrier-39
SNpc	substantia nigra pars compacta
SOC	store-operated Ca^{2+} channels
SOD	superoxide dismutase
SOD2	mitochondrial superoxide dismutase
SSB	single strand breaks
STIM1	Ca^{2+} sensor protein
Tf	transferrin
TfR	transferrin receptor
TH	tyrosine hydroxylase
TNFα	tumor necrosis factor α
TRPM7	transient receptor potential cation channel, subfamily M, member 7
Ub	ubiquitin
UTR	untranslated region
ZIP	zinc transporter

Acknowledgment This review was supported in part by grants from NIH, ES R01 10563 and ES P32 000267.

References

1. J. L. Greger, *J. Nutr.* **1998**, *128*, 368S–371S.
2. National Academy of Sciences (NAS). Dietary Reference Intakes for Vitamin A, Vitamin K, Arsenic, Boron, Chromium, Copper, Iodine, Iron, Manganese, Molybdenum, Nickel, Silicon, Vanadium, and Zinc. Panel on Micronutrients, Subcommittees on Upper Reference Levels of Nutrients and of Interpretation and Use of Dietary Reference Intakes, and the Standing Committee on the Scientific Evaluation of Dietary Reference Intakes. 2001, available from: <www.nap.edu/books/ 0309072794/html/>
3. P. Aisen, R. Aasa, A. G. Redfield, *J. Biol. Chem.* **1969**, *244*, 4628–4633.
4. J. W. Critchfield, C. L. Keen, *Metabolism* **1992**, *41*, 1087–1092.
5. C. L. Keen, J. G. Bell, B. Lonnerdal, *J. Nutr.* **1986**, *116*, 395–402.
6. C. L. Keen, S. Zidenberg-Cherr, *Manganese Toxicity in Humans and Experimental Animals*, in *Manganese in Health and Disease*, Ed D. J. Klimis-Tavantzis, CRC Press, Boca Raton, FL, 1994, pp 193–205.

7. C. L. Keen, B. Lonnerdal, M. Clegg, L. S. Hurley, *J. Nutr.* **1981**, *111*, 226–230.
8. C. L. Keen, J. L. Ensunsa, M. H. Watson, D. L. Baly, S. M. Donovan, M. H. Monaco, M. S. Clegg, *Neurotoxicology* **1999**, *20*, 213–223.
9. J. L. Greger, *J. Nutr.* **1998**, *128*, 368S–371S.
10. G. L. Rehnberg, J. F. Hein, S. D. Carter, R. S. Linko, J. W. Laskey, *J. Toxicol. Environ. Health* **1982**, *6*, 217–226.
11. P. S. Papavasiliou, S. T. Miller, G. C. Cotzias, *Am. J. Physiol.* **1966**, *211*, 211–216.
12. C. M. Lucaciu, C. Dragu, L. Copaescu, V. V. Morariu, *Biochim. Biophys. Acta* **1997**, *1328*, 90–98.
13. S. S. Kannurpatti, P. G. Joshi, N. B. Joshi, *Neurochem. Res.* **2000**, *25*, 1527–1536.
14. C. Grimm, R. Kraft, S. Sauerbruch, G. Schultz, C. Harteneck, *J. Biol. Chem.* **2003**, *278*, 21493–21501.
15. A. Riccio, C. Mattei, R. E. Kelsell, A. D. Medhurst, A. R. Calver, A. D. Randall, J. B. Davis, C. D. Benham, M. N. Pangalos, *J. Biol. Chem.* **2002**, *277*, 12302–12309.
16. R. A. Yokel, *Neuromol. Med.* **2009**, *11*, 297–310.
17. J. A. Roth, M. D. Garrick, *Biochem. Pharmacol.* **2003**, *66*, 1–13.
18. M. Aschner, D. C. Dorman, *Toxicol. Rev.* **2006**, *25*, 147–154.
19. M. D. Garrick, S. T. Singleton, F. Vargas, H. C. Kuo, L. Zhao, M. Knopfel, T. Davidson, M. Costa, P. Paradkar, J. A. Roth, L. M. Garrick, *Biol. Res.* **2006**, *39*, 79–85.
20. H. Fujishiro, K. Kubota, D. Inoue, A. Inoue, T. Yanagiya, S. Enomoto, S. Himeno, *Toxicology* **2011**, *280*, 118–125.
21. L. He, K. Girijashanker, T. P. Dalton, J. Reed, H. Li, M. Soleimani, D. W. Nebert, *Mol. Pharmacol.* **2006**, *70*, 171–180.
22. S. Himeno, T. Yanagiya, H. Fujishiro, *Biochimie* **2009**, *91*, 1218–1222.
23. K. Girijashanker, L. He, M. Soleimani, J. M. Reed, H. Li, Z. Liu, B. Wang, T. P. Dalton, D. W. Nebert, *Mol. Pharmacol.* **2008**, *73*, 1413–1423.
24. B. Mackenzie, H. Takanaga, N. Hubert, A. Rolfs, M. A. Hediger, *Biochem. J.* **2007**, *403*, 59–69.
25. H. Gunshin, B. Mackenzie, U. V. Berger, Y. Gunshin, M. F. Romero, W. F. Boron, S. Nussberger, J. L. Gollan, M. A. Hediger, *Nature* **1997**, *388*, 482–488.
26. J. A. Roth, *Biol. Res.* **2006**, *39*, 45–57.
27. M. Arredondo, P. Munoz, C. V. Mura, M. T. Nunez, *Am. J. Physiol. Cell Physiol.* **2003**, *284*, C1525–1530.
28. M. D. Garrick, S. T. Singleton, F. Vargas, H. C. Kuo, L. Zhao, M. Knopfel, T. Davidson, M. Costa, P. Paradkar, J. A. Roth, L. M. Garrick, *Biol. Res.* **2006**, *39*, 79–85.
29. B. Mackenzie, M. D. Garrick, *Am. J. Physiol. Gastrointest. Liver Physiol.* **2005**, *289*, G981–986.
30. J. F. Collins, *Biol. Res.* **2006**, *39*, 25–37.
31. L. Davidsson, A. Cederblad, B. Lonnerdal, B. Sandstrom, *Am. J. Dis. Child.* **1989**, *143*, 823–827.
32. L. Davidsson, A. Cederblad, B. Lonnerdal, B. Sandstrom, *Am. J. Clin. Nutr.* **1991**, *54*, 1065–1070.
33. R. A. Gibbons, S. N. Dixon, K. Hallis, A. M. Russell, B. F. Sansom, H. W. Symonds, *Biochim. Biophys. Acta* **1976**, *444*, 1–10.
34. P. E. Johnson, G. I. Lykken, E. D. Korynta, *J. Nutr.* **1991**, *121*, 711–717.
35. Z. Yin, H. Jiang, E.S. Lee, M. Ni, K. M. Erikson, D. Milatovic, A. B. Bowman, M. Aschner, *J. Neurochem.* **2010**, *112*, 1190–1198.
36. X. Wang, D. S. Miller, W. Zheng, *Toxicol. Appl. Pharmacol.* **2008**, *230*, 167–174.
37. M. S. Madejczyk, N. Ballatori, *Biochim. Biophys. Acta* **2012**, *1818*, 651–657.
38. J. F. Collins, M. Wessling-Resnick, M. D. Knutson, *J. Nutr.* **2008**, *138*, 2284–2288.
39. T. Ganz, *Blood* **2011**, *117*, 4425–4433.
40. L. Davidsson, B. Lonnerdal, B. Sandstrom, C. Kunz, C. L. Keen, *J. Nutr.* **1989**, *119*, 1461–1464.

41. G.J. Anderson, D. M. Frazer, A. T. McKie, C. D. Vulpe, *Blood Cells Mol. Dis.* **2002**, *29*, 367–375.
42. S. Cherukuri, R. Potla, J. Sarkar, S. Nurko, Z. L. Harris, P. L. Fox, *Cell Metab.* **2005**, *2*, 309–319.
43. T. Jursa, D. R. Smith, *Toxicol. Sci.* **2009**, *107*, 182–193.
44. R. A. Gibbons, S. N. Dixon, A. M. Russell, B. F. Sansom, *Biochim. Biophys. Acta* **1976**, *437*, 301–304.
45. G. J. Anderson, C. D. Vulpe, *Cell Mol. Life Sci.* **2009**, *66*, 3241–3261.
46. T. Moos, E. H. Morgan, *J. Neurosci. Res.* **1998**, *54*, 486–494.
47. R. B. Hernández, M. Farina, B. P. Esposito, N. C. Souza-Pinto, F. Barbosa, Jr., C. Sunol, *Toxicol. Sci.* **2011**, *124*, 414–423.
48. G. W. Bates, C. Billups, P. Saltman, *J. Biol. Chem.* **1967**, *242*, 2816–2821.
49. K. Tulpule, S. R. Robinson, G. M. Bishop, R. Dringen, *J. Neurosci. Res.* **2010**, *88*, 563–571.
50. Y. Ke, Z. M. Qian, *Prog. Neurobiol.* **2007**, *83*, 149–173.
51. J. Henriksson, J. Tallkvist, H. Tjalve, *Toxicol. Appl. Pharmacol.* **1999**, *156*, 119–128.
52. D. Vitarella, B. A. Wong, O. R. Moss, D. C. Dorman, *Toxicol. Appl. Pharmacol.* **2000**, *163*, 279–285.
53. L. D. Fechter, D. L. Johnson, R. A. Lynch, *Neurotoxicology* **2002**, *23*, 177–183.
54. A. W. Dobson, S. Weber, D. C. Dorman, L. K. Lash, K. M. Erikson, M. Aschner, *Biol. Trace Elem. Res.* **2003**, *93*, 113–126.
55. D. C. Dorman, B. E. McManus, C. U. Parkinson, C. A. Manuel, A. M. McElveen, J. I. Everitt, *Inhal. Toxicol.* **2004**, *16*, 481–488.
56. L. Normandin, L. Ann Beaupre, F. Salehi, A. St-Pierre, G. Kennedy, D. Mergler, R. F. Butterworth, S. Philippe, J. Zayed, *Neurotoxicology* **2004**, *25*, 433–441.
57. K. Thompson, R. M. Molina, T. Donaghey, J. E. Schwob, J. D. Brain, M. Wessling-Resnick, *FASEB J.* **2007**, *21*, 223–230.
58. D. C. Dorman, M. F. Struve, R. A. James, M. W. Marshall, C. U. Parkinson, B. A. Wong, *Toxicol. Appl. Pharmacol.* **2001**, *170*, 79–87.
59. G. Oberdorster, Z. Sharp, V. Atudorei, A. Elder, R. Gelein, W. Kreyling, C. Cox, *Inhal. Toxicol.* **2004**, *16*, 437–445.
60. X. Wang, G. J. Li, W. Zheng, *Brain Res.* **2006**, *1097*, 1–10.
61. A. T. McKie, *Biochem. Soc. Trans.* **2008**, *36*, 1239–1241.
62. R. S. Ohgami, D. R. Campagna, E. L. Greer, B. Antiochos, A. McDonald, J. Chen, J. J. Sharp, Y. Fujiwara, J. E. Barker, M. D. Fleming, *Nat. Genet.* **2005**, *37*, 1264–1269.
63. R. S. Ohgami, D. R. Campagna, A. McDonald, M. D. Fleming, *Blood* **2006**, *108*, 1388–1394.
64. T. E. Gunter, C. E. Gavin, M. Aschner, K. K. Gunter, *Neurotoxicology* **2006**, *27*, 765–776.
65. M. D. Fleming, M. A. Romano, M. A. Su, L. M. Garrick, M. D. Garrick, N. C. Andrews, *Proc. Natl. Acad. Sci. USA* **1998**, *95*, 1148–1153.
66. N. Hubert, M. W. Hentze, *Proc. Natl. Acad. Sci. USA* **2002**, *99*, 12345–12350.
67. P. L. Lee, T. Gelbart, C. West, C. Halloran, E. Beutler, *Blood Cells Mol. Dis.* **1998**, *24*, 199–215.
68. J. Kato, M. Kobune, S. Ohkubo, K. Fujikawa, M. Tanaka, R. Takimoto, K. Takada, D. Takahari, Y. Kawano, Y. Kohgo, Y. Niitsu, *Exp. Hematol.* **2007**, *35*, 879–887.
69. J. A. Roth, C. Horbinski, D. Higgins, P. Lein, M. D. Garrick, *Neurotoxicology* **2002**, *23*, 147–157.
70. S. J. Garcia, K. Gellein, T. Syversen, M. Aschner, *Toxicol. Sci.* **2007**, *95*, 205–214.
71. V. A. Fitsanakis, N. Zhang, J. G. Anderson, K. M. Erikson, M. J. Avison, J. C. Gore, M. Aschner, *Toxicol. Sci.* **2008**, *103*, 116–124.
72. S. L. Hansen, N. Trakooljul, H. C. Liu, A. J. Moeser, J. W. Spears, *J. Nutr.* **2009**, *139*, 1474–1479.
73. M. T. Nunez, V. Tapia, A. Rojas, P. Aguirre, F. Gomez, F. Nualart, *Am. J. Physiol. Cell Physiol.* **2010**, *298*, C477–485.
74. C. Kwik-Uribe, D. R. Smith, *J. Neurosci. Res.* **2006**, *83*, 1601–1610.

75. E. C. Theil, *J. Nutr.* **2011**, *141*, 724S–728S.
76. J. A. Roth, L. Feng, K. G. Dolan, A. Lis, M. D. Garrick, *J. Neurosci. Res.* **2002**, *68*, 76–83.
77. K. M. Erikson, T. Syversen, E. Steinnes, M. Aschner, *J. Nutr. Biochem.* **2004**, *15*, 335–341.
78. K. M. Erikson, Z. K. Shihabi, J. L. Aschner, M. Aschner, *Biol. Trace Elem. Res.* **2002**, *87*, 143–156.
79. P. N. Paradkar, J. A. Roth, *Biochem. J.* **2006**, *394*, 173–183.
80. P. N. Paradkar, J. A. Roth, *J. Neurochem.* **2006**, *96*, 1768–1777.
81. P. N. Paradkar, J. A. Roth, *J. Cell Physiol.* **2007**, *211*, 183–188.
82. A. Lis, P. N. Paradkar, S. Singleton, H. C. Kuo, M. D. Garrick, J.A. Roth, *Biochem. Pharmacol.* **2005**, *69*, 1647–1655.
83. D. Wang, L. H. Wang, Y. Zhao, Y. P. Lu, L. Zhu, *IUBMB Life* **2010**, *62*, 629–636.
84. H. Fujishiro, M. Doi, S. Enomoto, S. Himeno, *Metallomics* **2011**, *3*, 710–718.
85. J. P. Liuzzi, F. Aydemir, H. Nam, M. D. Knutson, R. J. Cousins, *Proc. Natl. Acad. Sci. USA* **2006**, *103*, 13612–13617.
86. M. Chesler, *Physiol. Rev.* **2003**, *83*, 1183–1221.
87. C. Ma, S. N. Schneider, M. Miller, D. W. Nebert, C. Lind, S. M. Roda, S. E. Afton, J. A. Caruso, M. B. Genter, *J. Biochem. Mol. Toxicol.* **2008**, *22*, 305–310.
88. J. J. Pinilla-Tenas, B. K. Sparkman, A. Shawki, A. C. Illing, C. J. Mitchell, N. Zhao, J. P. Liuzzi, R. J. Cousins, M. D. Knutson, B. Mackenzie, *Am. J. Physiol. Cell Physiol.* **2011**, *301*, C862–871.
89. J. Gao, N. Zhao, M. D. Knutson, C. A. Enns, *J. Biol. Chem.* **2008**, *283*, 21462–21468.
90. N. Zhao, J. Gao, C. A. Enns, M. D. Knutson, *J. Biol. Chem.* **2010**, *285*, 32141–32150.
91. K. Loutzenhiser, R. Loutzenhiser, *Circ. Res.* **2000**, *87*, 551–557.
92. R. A. Yokel, J. S. Crossgrove, *Res. Rep. Health Eff. Inst.* **2004**, 7–58; discussion 59–73.
93. J. S. Crossgrove, R. A. Yokel, *Neurotoxicology* **2005**, *26*, 297–307.
94. I. Jardin, L. J. Gomez, G. M. Salido, J. A. Rosado, *Biochem. J.* **2009**, *420*, 267–276.
95. A. Guilbert, M. Gautier, I. Dhennin-Duthille, N. Haren, H. Sevestre, H. Ouadid-Ahidouch, *Am. J. Physiol. Cell Physiol.* **2009**, *297*, C493–502.
96. D. Plenz, M. Herrera-Marschitz, S. T. Kitai, *J. Comp. Neurol.* **1998**, *397*, 437–457.
97. S. T. Rouse, M. J. Marino, S. R. Bradley, H. Awad, M. Wittmann, P. J. Conn, *Pharmacol. Ther.* **2000**, *88*, 427–435.
98. E. P. Brouillet, L. Shinobu, U. McGarvey, F. Hochberg, M. F. Beal, *Exp. Neurol.* **1993**, *120*, 89–94.
99. Z. Xu, K. Jia, B. Xu, A. He, J. Li, Y. Deng, F. Zhang, *Toxicol. Ind. Health* **2010**, *26*, 55–60.
100. Y. Deng, Z. Xu, B. Xu, Y. Tian, X. Xin, X. Deng, J. Gao, *Brain Res.* **2009**, *1289*, 106–117.
101. E. V. Stelmashook, E. R. Lozier, E. S. Goryacheva, P. Mergenthaler, S. V. Novikova, D. B. Zorov, N. K. Isaev, *Neurosci. Lett.* **2010**, *482*, 151–155.
102. C. E. Gavin, K. K. Gunter, T. E. Gunter, *Neurotoxicology* **1999**, *20*, 445–453.
103. A. Takeda, N. Sotogaku, N. Oku, *Neuroscience* **2002**, *114*, 669–674.
104. G. Meco, V. Bonifati, N. Vanacore, E. Fabrizio, *Scand. J. Work Environ. Health* **1994**, *20*, 301–305.
105. M. Parenti, L. Rusconi, V. Cappabianca, E. A. Parati, A. Groppetti, *Brain Res.* **1988**, *473*, 236–240.
106. J. J. Liccione, M. D. Maines, *J. Pharmacol. Exp. Ther.* **1988**, *247*, 156–161.
107. T. E. Gunter, B. Gerstner, T. Lester, A. P. Wojtovich, J. Malecki, S. G. Swarts, P. S. Brookes, C. E. Gavin, K. K. Gunter, *Toxicol. Appl. Pharmacol.* **2010**, *249*, 65–75.
108. H. S. Chun, H. Lee, J. H. Son, *Neurosci. Lett.* **2001**, *316*, 5–8.
109. R. D. Snyder, M. B. Friedman, *Mutat. Res.* **1998**, *405*, 1–8.
110. J. Y. Chang, L. Z. Liu, *Brain Res. Mol. Brain Res.* **1999**, *68*, 22–28.
111. A. Barbeau, *Neurotoxicology* **1984**, *5*, 13–35.
112. D. G. Graham, *Mol. Pharmacol.* **1978**, *14*, 633–643.
113. F. S. Archibald, C. Tyree, *Arch. Biochem. Biophys.* **1987**, *256*, 638–650.
114. J. Donaldson, D. McGregor, F. LaBella, *Can. J. Physiol. Pharmacol.* **1982**, *60*, 1398–1405.

115. L. S. Maynard, G. C. Cotzias, *J. Biol. Chem.* **1955**, *214*, 489–495.
116. Y. Deng, Y. Luan, H. Qing, H. Xie, J. Lu, J. Zhou, *Neurosci. Lett.* **2008**, *444*, 122–126.
117. C. G. Worley, D. Bombick, J. W. Allen, R. L. Suber, M. Aschner, *Neurotoxicology* **2002**, *23*, 159–164.
118. D. L. Stredrick, A. H. Stokes, T. J. Worst, W. M. Freeman, E. A. Johnson, L. H. Lash, M. Aschner, K. E. Vrana, *Neurotoxicology* **2004**, *25*, 543–553.
119. M. N. Subhash, T. S. Padmashree, *Food Chem. Toxicol.* **1990**, *28*, 567–570.
120. W. N. Sloot, J. Korf, J. F. Koster, L. E. De Wit, J. B. Gramsbergen, *Exp Neurol.* **1996**, *138*, 236–245.
121. M. Morello, P. Zatta, P. Zambenedetti, A. Martorana, V. D'Angelo, G. Melchiorri, G. Bernardi, G. Sancesario, *Brain Res. Bull.* **2007**, *74*, 406–415.
122. M. S. Desole, G. Esposito, R. Migheli, L. Fresu, S. Sircana, D. Zangani, M. Miele, E. Miele, *Neuropharmacology* **1994**, *34*, 289–285.
123. M. Tomas-Camardiel, A. J. Herrera, J. L. Venero, M. Cruz Sanchez-Hidalgo, J. Cano, A. Machado, *Brain Res. Mol. Brain Res.* **2002**, *103*, 116–129.
124. M. Khalid, R. A. Aoun, T. A. Mathews, *J. Neurosci. Methods* **2011**, *202*, 182–191.
125. J. M. Gorell, C. C. Johnson, B. A. Rybicki, E. L. Peterson, G. X. Kortsha, G. G. Brown, R. J. Richardson, *Neurology* **1997**, *48*, 650–658.
126. M. Thiruchelvam, E. K. Richfield, B. M. Goodman, R. B. Baggs, D. A. Cory-Slechta, *Neurotoxicology* **2002**, *23*, 621–633.
127. J. A. Moreno, K. M. Streifel, K. A. Sullivan, M. E. Legare, R. B. Tjalkens, *Toxicol. Sci.* **2009**, *112*, 405–415.
128. J. A. Moreno, E. C. Yeomans, K. M. Streifel, B. L. Brattin, R. J. Taylor, R. B. Tjalkens, *Toxicol. Sci.* **2009**, *112*, 394–404.
129. M. S. Desole, L. Sciola, M. R. Delogu, S. Sircana, R. Migheli, *Neurosci. Lett.* **1996**, *209*, 193–196.
130. M. S. Desole, L. Sciola, M. R. Delogu, S. Sircana, R. Migheli, E. Miele, *Neurochem. Int.* **1997**, *31*, 169–176.
131. R. Migheli, C. Godani, L. Sciola, M. R. Delogu, P. A. Serra, D. Zangani, G. De Natale, E. Miele, M. S. Desole, *J. Neurochem.* **1999**, *73*, 1155–1163.
132. M. Kitazawa, V. Anantharam, Y. Yang, Y. Hirata, A. Kanthasamy, A. G. Kanthasamy, *Biochem. Pharmacol.* **2005**, *69*, 133–146.
133. B. Xu, Z. F. Xu, Y. Deng, *Neurotoxicology* **2009**, *30*, 941–949.
134. H. Yoon, G. H. Lee, D. S. Kim, K. W. Kim, H. R. Kim, H. J. Chae, *Toxicol. In Vitro.* **2011**, *25*, 1259–1268.
135. K. Prabhakaran, D. Ghosh, G. D. Chapman, P. G. Gunasekar, *Brain Res. Bull.* **2008**, *76*, 361–367.
136. C. Tamm, F. Sabri, S. Ceccatelli, *Toxicol. Sci.* **2008**, *101*, 310–320.
137. K. Prabhakaran, G. D. Chapman, P. G. Gunasekar, *Neurotoxicology* **2009**, *30*, 414–422.
138. A. Alaimo, R. M. Gorojod, M. L. Kotler, *Neurochem. Int.* **2011**, *59*, 297–308.
139. Y. Hirata, K. Adachi, K. Kiuchi, *J. Neurochem.* **1998**, *71*, 1607–1615.
140. E. A. Malecki, *Brain Res. Bull.* **2001**, *55*, 225–228.
141. S. Oikawa, I. Hirosawa, S. Tada-Oikawa, A. Furukawa, K. Nishiura, S. Kawanishi, *Free Radic. Biol. Med.* **2006**, *41*, 748–756.
142. J. Jiao, Y. Qi, J. Fu, Z. Zhou, *Environ. Toxicol. Pharmacol.* **2008**, *26*, 123–127.
143. M. Morello, A. Canini, P. Mattioli, R. P. Sorge, A. Alimonti, B. Bocca, G. Forte, A. Martorana, G. Bernardi, G. Sancesario, *Neurotoxicology* **2008**, *29*, 60–72.
144. K. Kalia, W. Jiang, W. Zheng, *Neurotoxicology* **2008**, *29*, 466–470.
145. D. Zhang, A. Kanthasamy, V. Anantharam, *Toxicol. Appl. Pharmacol.* **2011**, *254*, 65–71.
146. E. Bonilla, *Neurobehav. Toxicol.* **1980**, *2*, 37–41.
147. S. V. Chandra, G. S. Shukla, *J. Neurochem.* **1981**, *36*, 683–687.
148. P. Calabresi, M. Ammassari-Teule, P. Gubellini, G. Sancesario, M. Morello, D. Centonze, G. A. Marfia, E. Saulle, E. Passino, B. Picconi, G. Bernardi, *Neurobiol. Dis.* **2001**, *8*, 419–432.

149. J. Wang, M. F. Rahman, H. M. Duhart, G. D. Newport, T. A. Patterson, R. C. Murdock, S. M. Hussain, J. J. Schlager, S. F. Ali, *Neurotoxicology* **2009**, *30*, 926–933.
150. M. J. Han, T. Ozaki, J. Yu, *J. Chem. Phys.* **2005**, *123*, 34306.
151. H. Afeseh Ngwa, A. Kanthasamy, Y. Gu, N. Fang, V. Anantharam, A. G. Kanthasamy, *Toxicol. Appl. Pharmacol.* **2011**, *256*, 227–240.
152. A. Elder, R. Gelein, V. Silva, T. Feikert, L. Opanashuk, J. Carter, R. Potter, A. Maynard, Y. Ito, J. Finkelstein, G. Oberdorster, *Environ. Health Perspect.* **2006**, *114*, 1172–1178.
153. S. M. Hussain, A. K. Javorina, A. M. Schrand, H. M. Duhart, S. F. Ali, J. J. Schlager, *Toxicol. Sci.* **2006**, *92*, 456–463.
154. A. B. Bowman, G. F. Kwakye, E. Herrero Hernandez, M. Aschner, *J. Trace Elem. Med. Biol.* **2011**, *25*, 191–203.
155. J. Lotharius, P. Brundin, *Nat. Rev. Neurosci.* **2002**, *3*, 932–942.
156. J. P. Covy, B. I. Giasson, *Neurotoxicology* **2011**, *32*, 622–629.
157. V. N. Uversky, J. Li, K. Bower, A. L. Fink, *Neurotoxicology* **2002**, *23*, 527–536.
158. Y. Zhou, F. S. Shie, P. Piccardo, T. J. Montine, J. Zhang, *Neuroscience* **2004**, *128*, 281–291.
159. H. S. Boudreau, K. M. Krol, J. K. Eibl, L. D. Williams, J. P. Rossiter, V. P. Palace, G. M. Ross, *Aquat. Toxicol.* **2009**, *92*, 258–263.
160. C. Pifl, M. Khorchide, A. Kattinger, H. Reither, J. Hardy, O. Hornykiewicz, *Neurosci. Lett.* **2004**, *354*, 34–37.
161. T. Cai, T. Yao, G. Zheng, Y. Chen, K. Du, Y. Cao, X. Shen, J. Chen, W. Luo, *Brain Res.* **2010**, *1359*, 201–207.
162. Y. Li, L. Sun, T. Cai, Y. Zhang, S. Lv, Y. Wang, L. Ye, *Brain Res Bull.* **2010**, *81*, 428–433.
163. T. R. Guilarte, *Neurotoxicology* **2010**, *31*, 572–574.
164. K. Martin, T. Huggins, C. King, M. A. Carroll, E. J. Catapane, *Comp. Biochem. Physiol. C Toxicol. Pharmacol.* **2008**, *148*, 152–159.
165. M. Parenti, C. Flauto, E. Parati, A. Vescovi, A. Groppetti, *Brain Res.* **1986**, *367*, 8–13.
166. A. J. Daniels, J. Abarca, *Neurotoxicol. Teratol.* **1991**, *13*, 483–487.
167. G. Biagini, F. Ferraguti, S. Ponzoni, M. Zoli, L. Alboni, G. Toffano, K. Fuxe, L.F. Agnati, **1994**.
168. S. Ponzoni, F. S. Guimaraes, E. A. Del Bel, N. Garcia-Cairasco, *Prog. Neuropsychopharmacol. Biol. Psychiatry* **2000**, *24*, 307–325.
169. S. Ponzoni, L. C. Gaziri, L. R. Britto, W. J. Barreto, D. Blum, *Neurosci. Lett.* **2002**, *328*, 170–174.
170. D. Centonze, P. Gubellini, G. Bernardi, P. Calabresi, *Exp. Neurol.* **2001**, *172*, 469–476.
171. E. Polazzi, B. Monti, *Prog. Neurobiol.* **2010**, *92*, 293–315.
172. F. Zhao, T. Cai, M. Liu, G. Zheng, W. Luo, J. Chen, *Toxicol. Sci.* **2009**, *107*, 156–164.
173. T. Verina, S. F. Kiihl, J. S. Schneider, T.R. Guilarte, *Neurotoxicology* **2011**, *32*, 215–226.
174. N. M. Filipov, R. F. Seegal, D. A. Lawrence, *Toxicol. Sci.* **2005**, *84*, 139–148.
175. J. H. Bae, B. C. Jang, S. I. Suh, E. Ha, H. H. Baik, S. S. Kim, M. Y. Lee, D. H. Shin, *Neurosci. Lett.* **2006**, *398*, 151–154.
176. P. Zhang, K. M. Lokuta, D. E. Turner, B. Liu, *J. Neurochem.* **2010**, *112*, 434–443.
177. A. H. Sadek, R. Rauch, P. E. Schulz, *Int. J. Toxicol.* **2003**, *22*, 393–401.
178. P. K. Pal, A. Samii, D. B. Calne, *Neurotoxicology* **1999**, *20*, 227–238.
179. J. M. Gorell, C. C. Johnson, B. A. Rybicki, E. L. Peterson, G. X. Kortsha, G. G. Brown, R. J. Richardson, *Neurotoxicology* **1999**, *20*, 239–247.
180. H. K. Hudnell, *Neurotoxicology* **1999**, *20*, 379–397.
181. Y. Kim, J. M. Kim, J. W. Kim, C. I. Yoo, C. R. Lee, J. H. Lee, H. K. Kim, S. O. Yang, H. K. Chung, D. S. Lee, B. Jeon, *Mov. Disord.* **2002**, *17*, 568–575.
182. B. A. Racette, L. McGee-Minnich, S. M. Moerlein, J. W. Mink, T. O. Videen, J. S. Perlmutter, *Neurology* **2001**, *56*, 8–13.
183. T. R. Guilarte, *Environ. Health Perspect.* **2010**, *118*, 1071–1080.
184. T. R. Guilarte, N. C. Burton, J. L. McGlothan, T. Verina, Y. Zhou, M. Alexander, L. Pham, M. Griswold, D. F. Wong, T. Syversen, J. S. Schneider, *J. Neurochem.* **2008**, *107*, 1236–1247.

185. T. R. Guilarte, M. K. Chen, J. L. McGlothan, T. Verina, D. F. Wong, Y. Zhou, M. Alexander, C. A. Rohde, T. Syversen, E. Decamp, A. J. Koser, S. Fritz, H. Gonczi, D. W. Anderson, J. S. Schneider, *Exp. Neurol.* **2006**, *202*, 381–390.
186. T. M. Peneder, P. Scholze, M. L. Berger, H. Reither, G. Heinze, J. Bertl, J. Bauer, E. K. Richfield, O. Hornykiewicz, C. Pifl, *Neuroscience* **2011**, *180*, 280–292.
187. K. Sriram, G. X. Lin, A. M. Jefferson, J. R. Roberts, R. S. Chapman, B. T. Chen, J. M. Soukup, A. J. Ghio, J. M. Antonini, *Arch. Toxicol.* **2010**, *84*, 521–540.
188. C. Wider, Z. K. Wszolek, *Parkinsonism Relat. Disord.* **2007**, *13, Suppl 3*, S229–232.
189. Y. X. Yang, N. W. Wood, D. S. Latchman, *Neuroreport* **2009**, *20*, 150–156.
190. S. Lesage, A. Brice, *Hum. Mol. Genet.* **2009**, *18*, R48–59.
191. J. A. Roth, S. Singleton, J. Feng, M. Garrick, P. N. Paradkar, *J. Neurochem.* **2010**, *113*, 454–464.
192. J. A. Roth, *Neuromol. Med.* **2009**, *11*, 281–296.
193. A. D. Gitler, A. Chesi, M. L. Geddie, K. E. Strathearn, S. Hamamichi, K. J. Hill, K. A. Caldwell, G. A. Caldwell, A. A. Cooper, J. C. Rochet, S. Lindquist, *Nat. Genet.* **2009**, *41*, 308–315.
194. J. Tan, T. Zhang, L. Jiang, J. Chi, D. Hu, Q. Pan, D. Wang, Z. Zhang, *J. Biol. Chem.* **2011**, *286*, 29654–29662.
195. T. Pan, S. Kondo, W. Le, J. Jankovic, *Brain* **2008**, *131*, 1969–1978.
196. A. Ramirez, A. Heimbach, J. Grundemann, B. Stiller, D. Hampshire, L. P. Cid, I. Goebel, A. F. Mubaidin, A. L. Wriekat, J. Roeper, A. Al-Din, A. M. Hillmer, M. Karsak, B. Liss, C. G. Woods, M. I. Behrens, C. Kubisch, *Nat. Genet.* **2006**, *38*, 1184–1191.
197. J. B. Schulz, *J. Neurol.* **2008**, *255, Suppl 5*, 3–7.
198. K. K. Dev, H. van der Putten, B. Sommer, G. Rovelli, *Neuropharmacology* **2003**, *45*, 1–13.
199. K. L. Lim, K. C. Chew, J. M. Tan, C. Wang, K. K. Chung, Y. Zhang, Y. Tanaka, W. Smith, S. Engelender, C. A. Ross, V. L. Dawson, T. M. Dawson, *J. Neurosci.* **2005**, *25*, 2002–2009.
200. Y. Mizuno, N. Hattori, H. Mori, T. Suzuki, K. Tanaka, *Curr. Opin. Neurol.* **2001**, *14*, 477–482.
201. A. B. West, D. Maraganore, J. Crook, T. Lesnick, P. J. Lockhart, K. M. Wilkes, G. Kapatos, J. A. Hardy, M. J. Farrer, *Hum. Mol. Genet.* **2002**, *11*, 2787–2792.
202. R. Hilker, C. Klein, M. Ghaemi, B. Kis, T. Strotmann, L. J. Ozelius, O. Lenz, P. Vieregge, K. Herholz, W. D. Heiss, P. P. Pramstaller, *Ann. Neurol.* **2001**, *49*, 367–376.
203. S. A. Oliveira, W. K. Scott, M. A. Nance, R. L. Watts, J. P. Hubble, W. C. Koller, K. E. Lyons, R. Pahwa, M. B. Stern, B. C. Hiner, J. Jankovic, W. G. Ondo, F. H. Allen, Jr., B. L. Scott, C. G. Goetz, G. W. Small, F. L. Mastaglia, J. M. Stajich, F. Zhang, M. W. Booze, J. A. Reaves, L.T. Middleton, J. L. Haines, M. A. Pericak-Vance, J. M. Vance, E. R. Martin, *Arch. Neurol.* **2003**, *60*, 975–980.
204. M. Periquet, M. Latouche, E. Lohmann, N. Rawal, G. De Michele, S. Ricard, H. Teive, V. Fraix, M. Vidailhet, D. Nicholl, P. Barone, N. W. Wood, S. Raskin, J. F. Deleuze, Y. Agid, A. Durr, A. Brice, *Brain* **2003**, *126*, 1271–1278.
205. Y. Higashi, M. Asanuma, I. Miyazaki, N. Hattori, Y. Mizuno, N. Ogawa, *J. Neurochem.* **2004**, *89*, 1490–1497.
206. P. Haeger, A. Alvarez, N. Leal, T. Adasme, M. T. Nunez, C. Hidalgo, *Neurotox. Res.* **2010**, *17*, 238–247.
207. I. Pelizzoni, D. Zacchetti, C. P. Smith, F. Grohovaz, F. Codazzi, *J. Neurochem.* **2012**, *120*, 269–278.
208. M. I. Behrens, N. Bruggemann, P. Chana, P. Venegas, M. Kagi, T. Parrao, P. Orellana, C. Garrido, C. V. Rojas, J. Hauke, E. Hahnen, R. Gonzalez, N. Seleme, V. Fernandez, A. Schmidt, F. Binkofski, D. Kompf, C. Kubisch, J. Hagenah, C. Klein, A. Ramirez, *Mov. Disord.* **2010**, *25*, 1929–1937.
209. F. Kamp, N. Exner, A. K. Lutz, N. Wender, J. Hegermann, B. Brunner, B. Nuscher, T. Bartels, A. Giese, K. Beyer, S. Eimer, K. F. Winklhofer, C. Haass, *EMBO J.* **2010**, *29*, 3571–3589.
210. S. J. Chung, S. M. Armasu, J. M. Biernacka, T. G. Lesnick, D. N. Rider, S. J. Lincoln, A. I. Ortolaza, M. J. Farrer, J. M. Cunningham, W. A. Rocca, D. M. Maraganore, *Mov. Disord.* **2010**, *26*, 280–288.

211. S. Lesage, C. Condroyer, A. Lannuzel, E. Lohmann, A. Troiano, F. Tison, P. Damier, S. Thobois, A. M. Ouvrard-Hernandez, S. Rivaud-Pechoux, C. Brefel-Courbon, A. Destee, C. Tranchant, M. Romana, L. Leclere, A. Durr, A. Brice, *J. Med. Genet.* **2009**, *46*, 458–464.
212. S. Saha, M. D. Guillily, A. Ferree, J. Lanceta, D. Chan, J. Ghosh, C. H. Hsu, L. Segal, K. Raghavan, K. Matsumoto, N. Hisamoto, T. Kuwahara, T. Iwatsubo, L. Moore, L. Goldstein, M. Cookson, B. Wolozin, *J. Neurosci.* **2009**, *29*, 9210–9218.
213. J. R. Adams, H. van Netten, M. Schulzer, E. Mak, J. McKenzie, A. Strongosky, V. Sossi, T. J. Ruth, C. S. Lee, M. Farrer, T. Gasser, R. J. Uitti, D. B. Calne, Z. K. Wszolek, A. J. Stoessl, *Brain* **2005**, *128*, 2777–2785.
214. R. Nandhagopal, E. Mak, M. Schulzer, J. McKenzie, S. McCormick, V. Sossi, T.J. Ruth, A. Strongosky, M. J. Farrer, Z. K. Wszolek, A.J. Stoessl, *Neurology* **2008**, *71*, 1790–1795.
215. Y. Tong, A. Pisani, G. Martella, M. Karouani, H. Yamaguchi, E. N. Pothos, J. Shen, *Proc. Natl. Acad. Sci. USA* **2009**, *106*, 14622–14627.
216. B. Lovitt, E. C. Vanderporten, Z. Sheng, H. Zhu, J. Drummond, Y. Liu, *Biochemistry* **2010**, *49*, 3092–3100.
217. J. P. Covy, B. I. Giasson, *J. Neurochem.* **2010**, *115*, 36–46.
218. J. Salazar, N. Mena, S. Hunot, A. Prigent, D. Alvarez-Fischer, M. Arredondo, C. Duyckaerts, V. Sazdovitch, L. Zhao, L. M. Garrick, M. T. Nunez, M. D. Garrick, R. Raisman-Vozari, E. C. Hirsch, *Proc. Natl. Acad. Sci. USA* **2008**, *105*, 18578–18583.
219. Q. He, T. Du, X. Yu, A. Xie, N. Song, Q. Kang, J. Yu, L. Tan, J. Xie, H. Jiang, *Neurosci. Lett.* **2011**, *501*, 128–131.
220. S. J. Lippard, J. M. Berg, *Principles of Bioinorganic Chemistry*, University Science Books, Mill Valley, Ca, 1995.
221. *The Biological Chemistry of Magnesium*, VCH Publishers, Inc., New York, 1995.
222. T. Pan, D. M. Long, O. C. Uhlenbeck, *Divalent Metal Ions in RNA Folding and Catalysis*, in *The RNA World*, Eds R. Gesteland, J. Atkins, Cold Spring Harbor Press, Cold Spring Harbor, 1993, pp. 271–302.
223. L. A. Finney, T. V. O'Halloran, *Science* **2003**, *300*, 931–936.
224. M. Pechlaner, R. K. O. Sigel, *Met. Ions. Life Sci.* **2012**, *10*, 1–42.

Chapter 7
Control of Iron Metabolism in Bacteria

Simon Andrews, Ian Norton, Arvindkumar S. Salunkhe, Helen Goodluck, Wafaa S.M. Aly, Hanna Mourad-Agha, and Pierre Cornelis

Contents

S. Andrews (✉) • I. Norton • A.S. Salunkhe • H. Goodluck
W.S.M. Aly • H. Mourad-Agha
The School of Biological Sciences, The University of Reading,
Whiteknights, Reading, RG6 6AJ, UK
e-mail: s.c.andrews@reading.ac.uk

P. Cornelis
Department of Bioengineering Sciences, Research Group Microbiology,
Flanders Interuniversity Institute of Biotechnology VIB6,
Vrije Universiteit Brussel, Brussels, Belgium

L. Banci (ed.), *Metallomics and the Cell*, Metal Ions in Life Sciences 12,
DOI 10.1007/978-94-007-5561-1_7, © Springer Science+Business Media Dordrecht 2013

Abstract Bacteria depend upon iron as a vital cofactor that enables a wide range of key metabolic activities. Bacteria must therefore ensure a balanced supply of this essential metal. To do so, they invest considerable resourse into its acquisition and employ elaborate control mechanisms to eleviate both iron-induced toxitiy as well as iron deficiency. This chapter describes the processes that bacteria engage in maintaining iron homeostasis. The focus is *Escherichia coli*, as this bacterium provides a well studied example. A summary of the current status of understanding of iron management at the 'omics' level is also presented.

Keywords ferritin • Fur • iron regulation • iron transport metallome • metalloproteome • RyhB

Please cite as: *Met. Ions Life Sci.* 12 (2013) 203–239

1 Introduction

Bacteria employ a variety of strategies to ensure effective iron management. They deposit intracellular iron stores for later utilization [1] and use high-affinity iron-transport systems to acquire iron in both complexed and free form. Bacteria also possess highly effective redox-stress resistance mechanisms providing the capacity to degrade reactive oxygen species which can be generated as a result of reactions catalyzed by free intracellular iron [2]. Bacterial species carry iron-responsive regulatory systems that are able to alter and control the homeostatic machinery in direct response to iron availability [3]. In addition, many bacterial species have the capacity to down-regulate iron-containing proteins when iron availability is restricted in the surrounding environment; this reduces the demand of the cell for iron thus ensuring that limited iron is used efficiently [4,5]. This 'iron rationing response' is supplemented by alternative, non-iron-dependent isoenzymes used under iron limitation. Thus, bacteria benefit from a sophisticated iron homeostatic machinery that counters both iron restriction and iron-induced toxicity.

This chapter describes how bacteria metabolize iron and achieve homeostatic control with respect to iron, with a major focus on the model bacterium *Escherichia coli*. The aim is to provide a general overview, rather than a comprehensive account, of the mechanisms involved.

2 Life Is Wedded to Iron – For Better or for Worse

Iron is a key transition metal, it belongs to group eight of the periodic table of elements and is an absolute requirement for almost all living organisms. It is also a highly abundant element being the second principal metal (after aluminum) within the crust of the earth. Biologically, iron is generally found in one of two redox states: the oxidized, rusty-brown, ferric form (Fe^{3+}) and the reduced, largely-colorless, ferrous form (Fe^{2+}). These alternative redox states enable iron to play a highly versatile redox role within biological systems, although it has catalytic roles also that do not depend on redox function. The ferric form is the most abundant environmentally. This is because ferrous iron is rapidly oxidized to the ferric state under aerobic conditions at neutral pH or above. Unfortunately, ferric iron exhibits very poor solubility (10^{-18} M) at neutral pH which reduces its bioavailability significantly. For this reason, iron restriction is a condition commonly encountered by living organisms and indeed the production of biomass is strictly limited by iron supply in many environments (e.g., the Southern Ocean [6]). The few bacteria that have the ability to survive without this otherwise essential element have adapted to use alternative transition metals in place of iron. One such organism is *Borrelia burgdorferi*, the causative agent of Lyme disease. The genome sequence of this organism was found to lack any apparent iron-requiring system and appears to utilize manganese instead of iron within key proteins [7].

Although such 'iron-free' organisms are the exception, some of the tricks that they employ to escape iron dependency are reiterated in other bacteria to combat iron restriction (see Section 6). The archaebacterium, *Ferroplasma acidiphilum*, represents the opposite extreme. Proteomic analysis showed that this archaebacterium possesses an unusally high number of iron-proteins (86% of its proteins!) correlating well with its preferred niche – a ferrous-iron-rich acidic pool. It is suggested that the iron atoms incorporated into its proteins act as 'iron rivets' to stabilize them against denaturation within the harsh environment that *F. acidiphilum* occupies [8].

Many important biological processes depend on iron including respiration, the tricarboxylic acid cycle, oxygen transport, nitrogen fixation, and DNA synthesis [3,9]. For iron to be utilized in biological systems it usually has to be incorporated into proteins in either a mononuclear or binuclear form, or as heme or iron sulfur clusters. The form in which iron is incorporated and the manner in which it is coordinated as a cofactor to proteins dictate its redox potential and the spin state adopted. This allows iron-containing proteins to fulfil a diverse range of biological functions.

Iron homeostasis has an important role in oxidative stress as free iron is toxic due to its tendency to stimulate the production of hydroxyl radicals [10]. The hydroxyl radicals are formed when iron reacts with peroxides through the Fenton reaction:

$$H_2O_2 + Fe^{2+} \rightarrow Fe^{3+} + OH^- + {}^{\bullet}OH$$

Superoxide participates in the process by reducing Fe^{3+} back to Fe^{2+}. Together, these two processes form the Haber-Weiss reaction in which free iron catalyzes the conversion of hydrogen peroxide and superoxide into the deadly hydroxyl radical. Hydroxyl radicals are highly reactive, highly damaging, and short lived. A major target of hydroxyl radicals is DNA which leads to mutations and eventually, once DNA damage becomes excessive, cell death. Therefore, iron management is generally a key component of redox-stress resistance [11] and free cytosolic iron levels are carefully regulated at around 10 μM to ensure there is sufficient for cellular processes but not excess that would induce toxicity [12].

3 Iron Uptake

3.1 *Siderophore-Dependent Acquisition*

3.1.1 Introduction

By far the most important process by which bacteria resist iron starvation is through the use of various types of iron uptake systems. Bacteria often possess multiple iron acquisition systems that have specificities for different forms of iron and these may operate under different environmental conditions. Iron transport pathways can be divided into two types: (i) those that target soluble, ferrous iron and (ii) those that are specific for poorly soluble, ferric iron. Ferrous iron is acquired in its free, uncomplexed form. However, due to the low solubility of ferric iron under physiological conditions, ferric iron is generally acquired in the form of a complex. Bacteria, fungi, and some plants obtain ferric iron using compounds known as 'siderophores' (or phytosiderophores for plants) [13]. Siderophores are high-affinity chelators of ferric iron which are secreted into the environment where they bind and solubilize iron to generate a ferri-siderophore complex that is subsequently internalized [14,15]. In Gram-negative bacteria, ferri-siderophore complexes are taken up via receptors in the outer membrane (OM) and transport is mediated by the energy-transducing TonB-ExbB-ExbD system located in the inner membrane (IM) and periplasm. The complexes are then delivered to the IM by periplasmic-binding proteins where they then interact with an ABC (ATP-binding cassette) transporter

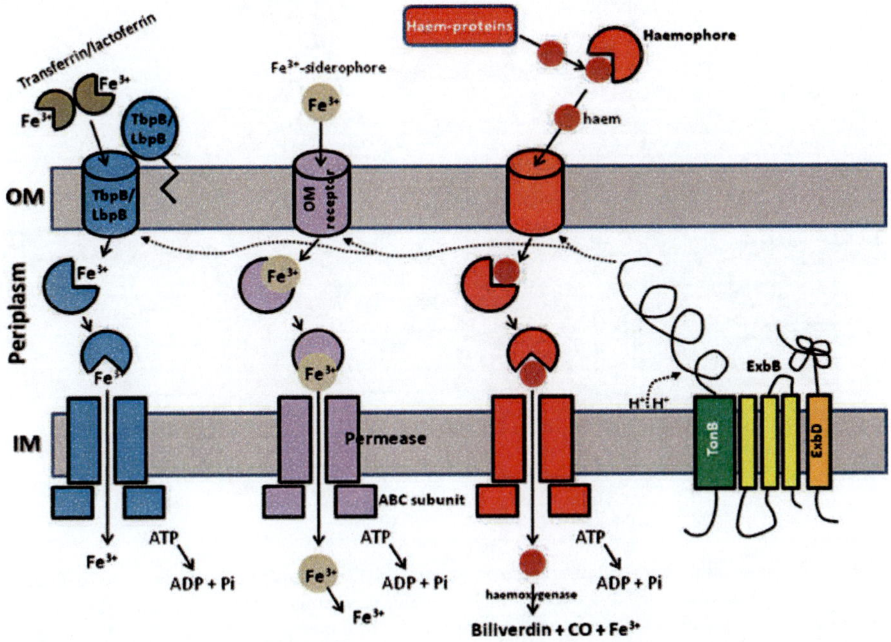

Figure 1 Schematic representation of iron uptake in Gram-negative bacteria. Gram-negative bacteria possess numerous iron uptake pathways. This schematic representation displays three uptake pathways for transferrin/lactoferrin (blue), siderophores (purple), and heme (red). All of these pathways require an outer membrane receptor, a periplasmic binding protein, and an inner membrane ABC transporter. Not all Gram-negatives have all three systems but many possess one or more of them. Uptake through the outer membrane receptors requires energy derived from the TonB-ExbB-ExbD system highlighted as a multi-coloured complex at the inner membrane.

which enables passage of the complex to the cytoplasm (Figure 1). Bacteria can be opportunistic in that they pirate the siderophores produced by other organisms [15]. Thus, bacteria can utilize siderophores that they are unable to synthesize. Gram-positive organisms lack OM receptors (due to the absence of an OM) and so only require ABC transport systems for siderophore uptake (located in the cytoplasmic membrane). These ABC permeases are usually binding protein-dependent where the binding protein is usually a tethered lipoprotein attached to the external surface of the cytoplasmic membrane (Figure 2).

Chemically, many siderophores are peptides produced by non-ribosomal peptide synthetase (NRPS) multienzyme pathways [16,17]. However, some siderophores are produced by NRPS-independent pathways [18]. Siderophores have high affinity and specificity for ferric iron and are usually produced during periods of iron restriction. They can sometimes be found at very high concentrations but this depends on the bacterial species and growth conditions. Siderophores tend to have a peptidic backbone with amino acid side chains which are often modified to generate iron-coordinating ligands such as hydroxamate, catecholate, and α-hydroxycarboxylate

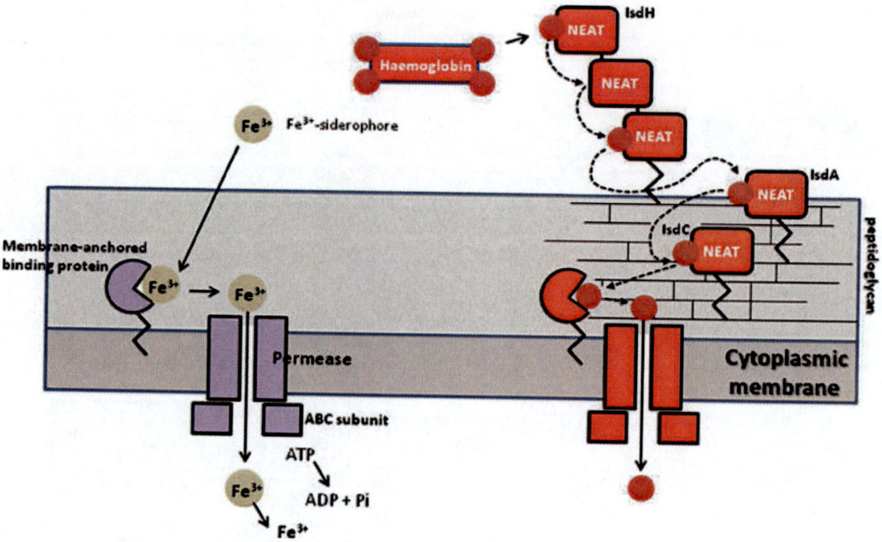

Figure 2 Schematic representation of iron uptake in Gram-positive bacteria. Unlike Gram-negative bacteria, Gram-positive bacteria lack an outer membrane, therefore when uptaking iron from heme, transferrin or siderophores the involvement of a membrane-anchored binding protein is usually required and an ABC transporter as detailed in the diagram. The diagram on the right shows a schematic representation of heme uptake using NEAT proteins.

groups [15]. More than 500 siderophores have been identified and are classified according to the functional groups which they utilize as ligands [19]. Two very well studied siderophores are enterobactin and ferrichrome, the chemical structures of these siderophores can be seen in Figure 3 along with the structure of the α-hydroxycarboxylate siderophore, rhizoferrin. After siderophores have been synthesized they are secreted, but due to the polarity of siderophores the involvement of transport proteins is necessary for this process. One such protein is known as EntS which exports enterobactin across the IM in *E. coli* [20]; export across the OM is thought to be TolC-dependent [21]. Siderophores are associated with both pathogenic and non-pathogenic bacteria [22–24].

3.1.2 Siderophore Receptors

The porins located in the OM of Gram-negative bacteria are insufficient to allow ferri-siderophore complexes to pass through as they are simply too large, however, smaller molecules such as glucose, phosphate, and amino acids can pass directly through [15]. Due to the size of ferri-siderophore complexes they require dedicated OM receptors. Most characterized siderophore receptors appear to be related [14] and in *E. coli* the structures of three have been determined, these are FepA

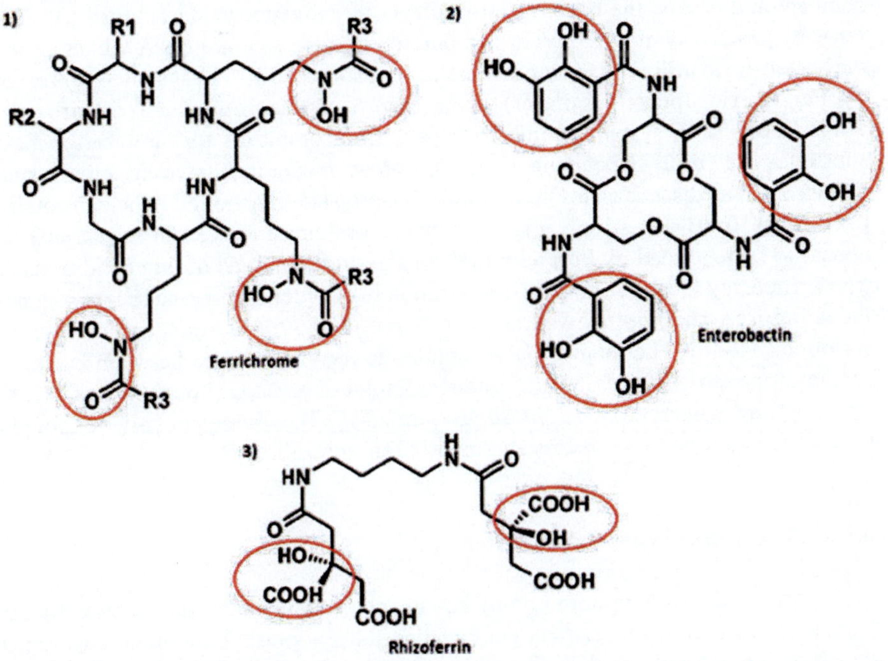

Figure 3 Siderophores and their associated functional groups. This figure shows the chemical structures of siderophores for each of the three functional group types. (**1**) Displays the chemical structure of the siderophore ferrichrome with the hydroxamate groups within the molecule highlighted; (**2**) shows the chemical structure of enterobactin with the catecholate groups highlighted in red, and (**3**) shows the chemical structure of the α-hydroxycarboxylate siderophore rhizoferrin with the associated functional groups highlighted. Image modified from [28].

(enterobactin receptor), FhuA (hydroxamate receptor), and FecA (citrate receptor) [25]. Similarly in *Pseudomonas aeruginosa* the crystal structures of FpvA (pyoverdine receptor) and FptA (pyochelin receptor) have been elucidated [26,27].

Siderophore receptors are β-barrel proteins forming a tube that completely spans the outer membrane thus allowing siderophore complexes to enter the periplasm [3]. The receptors are TonB-gated and the β-barrel structure consists of 22 antiparallel β-strands and an N-terminal globular domain often referred to as the plug domain [28]. Binding of a siderophore complex to its receptor triggers a conformational change leading to the plug domain becoming dislodged from the β-barrel pore, allowing subsequent transport to the periplasm [15]. This process requires energy generated by the TonB-ExbB-ExbD system. Many bacterial species have more than one type of siderophore receptor providing specificity for a range of different siderophores. Ligand binding sites are specific for each siderophore, in the case of FhuA it is the aromatic residues present in the receptor which bind ferrichrome [29]. Bacteria can also utilize siderophores produced by other species enabling them to compete and escape the bacteriostatic effects of exogenous siderophores [3]. The ferric dicitrate system of *E. coli* has an outermembrane receptor FecA which allows the bacterium to utilize this compound as an iron source [30]. The system is repressed by Fur (ferric uptake regulator) at high iron concentration and requires the TonB-ExbB-ExbD system for induction [31]. Iron complexes appear to be the main substrates for TonB-dependent outer-membrane receptors, however, other compounds such as disaccharides and vitamin B_{12} are also transported by such proteins [32]. Most TonB-dependent outer membrane proteins involved in iron transport appear to be regulated by Fur, with their expression repressed during conditions of iron sufficiency [33]. Fur binds to the promoters of outer membrane receptor genes including *fecA* and *fhuA* as well as directly repressing *tonB*, *exbB*, and *exbD* transcription [34–36]. The total of outer membrane receptors varies between bacterial species, for example, *E. coli* K-12 produces a total of seven OM transporters dependent on TonB whereas *P. aeruginosa* produces 34 [37], although not all are thought to be ferri-siderophore receptors and are not Fur controlled.

3.1.3 The TonB-ExbB-ExbD System

Transport of ferri-siderophore complexes is energy-dependent, this energy comes from an electrochemical gradient generated at the cytoplasmic membrane which is then transferred by the TonB-ExbB-ExbD complex. TonB which spans from the IM to the inner face of the OM, across the periplasm, and is complexed with ExbB and ExbD which are anchored in the IM. Together this complex utilizes energy from the electrochemical gradient generated across the IM [38]. TonB appears to contact an OM receptor bound to ligand, driving conformational change in the receptor which results in the ligand being transferred to the periplasm [39]. During this activation TonB appears to undergo cyclic changes of affinity for the OM [40]. The TonB system in *E. coli* has been extensively studied and it appears that ExbB and ExbD function to cycle TonB to and from its high affinity OM association. ExbB is a 26 kDa IM associated protein consisting of three transmembrane domains and ExbD is a 17 kDa protein which similarly to TonB has only one transmembrane domain and a periplasmic domain [41]. The transmembrane domain of TonB is responsible for interacting with ExbB and ExbD to form the energy transducing complex; replacement of this domain leads to loss of activity [42].

Despite many years of research on the molecular details of TonB and its action, the details of TonB interaction with OM receptors remains unclear [43,44]. The 'TonB box' is a conserved seven amino acid segment that has been found to be vital for mediated siderophore uptake. The 'TonB box' is found at the N-terminus of the plug domain and in some cases has been observed extending into the periplasm [33]. However, this is not always the case, in some instances it has not been identified or can be observed tucked up into the plug domain. It is believed that the TonB box region is what TonB contacts to induce conformational changes in the OM receptors. Instead of the mechanical model of TonB transduction, whereby TonB remains attached to the IM during the process of energy transduction, an alternative shuttling mechanism has been proposed where the TonB N-terminus leaves the IM to deliver energy to OM receptors [45,46]. However, recent studies have shown the shuttling hypothesis to be improbable by demonstrating TonB localization at the IM while carrying out its biological function [47,48]. TonB, ExbB, and ExbD are all found in a complex at the IM when in a resting state (not actively involved in energy transduction). ExbB and ExbD utilize the IM electrochemical gradient to activate TonB which then leads to receptor contact and transport of siderophore complexes to the periplasm, which in turn leads to TonB returning to its resting state [28,39]. Some bacteria possess two (or more) TonB-ExbBD systems which are coupled to distinct sets of OM receptors [49]. Due to Gram-positive organisms lacking an OM they do not appear to possess or require any OM transport genes or any of the TonB-ExbB-ExbD proteins.

3.1.4 Transport across the Periplasm and Cytoplasmic Membranes

Transport across the periplasm and cytoplasmic membranes is mediated by periplasmic binding proteins and transporters associated with the IM [3]. Each periplasmic binding protein picks up one siderophore complex at a time and functions to transport the siderophore complexes from the OM receptor to the permease associated with the IM. Siderophores are chemically distinct from one another and specific periplasmic binding proteins are responsible for shuttling different classes of siderophores [28]. In *E. coli*, for example, the periplasmic binding protein FecB is responsible for the shuttling of ferricitrate, FepB is responsible for shuttling catecholate siderophores and the best studied of the periplasmic binding proteins, FhuD, which can be found in both Gram-negative and positive species, is responsible for hydroxamate siderophore shuttling and can interact with a wide range of ferri-siderophore complexes such as ferrichrome and ferrioxamine B [50]. FhuD transfers ferrichrome to the IM-associated permease complex FhuB/C [51]. The siderophore binding site of FhuD is hydrophobic containing mainly aromatic residues.

Recent studies have shown that FhuD and TonB interact and these protein-protein interactions assist in ferrichrome binding to FhuD [51]. Crystal structure analysis of FhuD has revealed a bilobal structure, with the ligand binding site located between these two lobes [52]. Binding occurs due to specific residues in the shallow cleft between the two lobes, where the residues interact with the iron hydroxamate center of the siderophore [14]. The permease complexes located at the inner membrane contain ABC modules. There are four modules in total; 2 homologous permease modules and two ATP-binding cassette modules. The permease modules can be comprised of two different subunits, two identical subunits or even one large molecule consisting of two subunits. The ATP-binding cassette modules are almost always comprised of two identical subunits [3].

3.1.5 Fate of Internalized Siderophores

Once siderophores have entered the cell the iron must be liberated from the ferri-siderophore complex. This is achieved by reducing the iron bound to the complex, leading to dissociation due to low affinity for ferrous iron [53] or by hydrolysis of the siderophore backbone triggering iron release [54,55]. Iron reduction is achieved by ferric reductase enzymes which are present in many bacterial species, e.g. flavin reductase which reduces siderophore complexes *in vitro* although this activity may not reflect its role *in vivo* [53].

In *E. coli* before ferri-enterobactin can be utilized, an esterase protein encoded by the *fes* gene is required, this enzyme is known to hydrolyze ester bonds within the enterobactin siderophore thus weakening binding to iron [3,56]. Mechanisms for exogenous siderophores are not well understood, however, the *fhuF* gene of *E. coli* encodes a cytoplasmic ferredoxin like protein, which was shown to be responsible for reducing iron bound to hydroxamate-type siderophores, such as ferrioxamine B, coprogen, and ferrichrome [57,58].

3.1.6 The Ferric Dicitrate System of *E. coli* – an Example

A well studied example of a 'siderophore' uptake system is ferric dicitrate transport in *E. coli* (note that citrate is not strictly a siderophore). The system consists of the FecA protein located in the OM, the FecB periplasmic-binding protein, the FecCD proteins associated with the IM, and the FecE ATPase also located in the IM [59]. In this system, ferric iron is transported into the periplasm as a ferric dicitrate complex, the complex becomes dissociated and the ferric iron then crosses the cytoplasmic membrane in a process requiring energy which is derived from ATP and is driven by the FecE ATPase. This system is carefully regulated by the regulatory protein pair, FecIR, during iron limitation [59]. The transport machinery is only fully synthesized when the substrate (ferric citrate) is present in the growth medium.

This system displays a unique mechanism by which transcriptional signals begin at the cell surface, the signal is mediated by the OM transport protein FecA, which receives and transmits a signal across the OM. The TonB-ExbB-ExbD system is required for this process both for signalling and transport of ferric citrate [60]. FecR receives the signal in the periplasm and transmits the signal across the IM into the cytoplasm [61]. FecR achieves this by interaction of the N terminal region of FecA, with the C terminal region of FecR [62]. This interaction activates the FecI sigma factor which belongs to the extracytoplasmic function (ECF) sigma factor family [63]. FecI then binds RNA polymerase which initiates transcription and subsequent synthesis of the transport machinery at the *fecA* promoter [63]. The FecIR regulatory genes are repressed by the Fe^{2+}-Fur complex (Section 5.1) so the entire system remains iron-repressed [61]. Similar regulation systems have been observed for heme acquisition in *Serratia marcescens* [64] and for ferri-siderophore-mediated iron uptake in *Pseudomonas aeruginosa* and other bacteria [65,66].

3.2 *Direct Uptake Systems*

3.2.1 The FeoABC System. G-Protein-Coupled Transport

Bacteria are able to transport ferrous as well as ferric iron, but the routes used are entirely distinct for these two forms of iron [3]. The Feo transporter is the most common and best known ferrous uptake system in bacteria. It was first identified in *E. coli* and was found to be encoded by the *feoAB* genes [67]. However, more recent studies have shown there to be three closely associated genes predicted to form an operon; *feoABC* [68]. The *feoA*, *feoB*, and *feoC* genes encode proteins of 75, 773, and 78 amino acid residues. FeoB is an integral IM protein whereas FeoA and FeoC are likely to be cytoplasmic [67,69]. The *feoA* and *feoB* genes are generally found in close proximity, and in some cases the genes are fused (e.g., *Bacteroides*) indicating a close association between the corresponding proteins [69]. FeoB acts as the high-affinity ferrous iron transporter with FeoA having a more minor and undefined role in ferrous iron uptake [67]. It was initially thought that the FeoABC uptake system is dependent on ATP [67]; however, more recent studies have shown the FeoB protein to have a cytoplasmically located G-protein domain (discussed in more detail below) with GTPase activity [70] suggesting uptake may be GTP-dependent. *feoB* mutants of *E. coli* exhibit decreased gut colonization, thus indicating an important role for Feo in the anaerobic environment of the intestine and suggesting a role in pathogenesis (for intestinal pathogens [71]). A similar effect has been reported for other gut dwelling bacteria, e.g., *S. enterica* and *Helicobacter pylori* [72,73]. *Legionella pneumophilia* mutants lacking FeoB display a major decline in ferrous iron uptake indicating the importance of FeoB to this intracellular pathogen [74].

The *feoABC* operon of *E. coli* is produced under anoxic (as well as low iron) conditions. This low-oxygen induction is mediated by Fnr (a global regulator of anaerobic gene expression) [67]. This explains why the Feo ferrous iron transport system appears to be of importance during anaerobic/microaerophilic growth. This oxygen dependence is rational since (as mentioned above) ferrous iron is unstable aerobically [3]. Studies on *H. pylori* have shown that FeoB may also have some degree of ATP dependence indicating FeoB may exhibit ATPase as well as GTPase activity [73]. Facilitation of ferrous iron uptake in bacterial species has been seen to be mediated by ferric reductase activity, such enzymes act extracellularly to convert ferric iron to ferrous iron for transport. Such activity has been seen both in *Helicobacter pylori* and *E. coli* [73,75] and a corresponding diheme-cytochrome has been recently identified in *Bradyrhizobium japonicum* [76].

The precise function of FeoA is as yet unknown, although its structure is solved [77]. It has been observed to display a degree of similarity to the SH3-like C-terminal domain of the iron-dependent repressors, DtxR and IdeR [3,77] and may have a role in assisting FeoB activity. For FeoB, the importance of the G-protein region has been demonstrated in studies where mutants lacking functional G-protein regions failed to allow sufficient iron into cells to cause repression of a *fhuF-lacZ* reporter

[70]. G-proteins are involved in fundamental cellular processes such as protein synthesis and cell growth [78]. When GTP (guanosine 5′-triphosphate) is bound to a GTPase (G-protein), the corresponding system becomes activated and signalling is triggered; when GDP is bound, the system is considered inactivated and the signalling pathway is repressed [79]. G-proteins contain a highly conserved GTP-binding domain which is usually 160 residues characterized into five short motifs designated G1-G5, which appear critical in GTP and GDP binding [80]. Activation of G-proteins requires GDP dissociation resulting in a nucleotide free state. GTP binding then leads to conformational change in the protein allowing effector proteins to interact triggering signalling pathways [80]. The characteristic G1-G5 motifs found in G-protein-coupled systems appear to be slightly aberrant in *E. coli* FeoB. The G1-G4 motifs are present and identified, however, studies have failed to assign the G5 motif [70]. FeoB appears to have intrinsically slow hydrolysis of GTP which has led to the suggestion that it may operate as a nucleotide-gated Fe^{2+} channel [81].

Crystal structures of the FeoB N-terminal domain with GTP-bound and apo (GTP-free) form have been resolved. This work shows that the G-protein domain forms a trimer around a cytoplasmic pore, which appears open in the GTP-bound state and closed in the apo state [82]. The uncharacterized linker domain (between the G protein domain and the integral membrane domain) has also been proposed to act as an intrinsic effector responsible for coupling GTP binding to pore opening [82]. Rapid GDP release rates in FeoB have indicated the G5 motif has a tendency to switch to a stable out state where GDP affinity is nominal [83,84]. In the apo structure the trimer appears to be stabilized through the formation of salt bridges between Asp135 and Arg140 residues in neighboring promoters, Glu133 is pointing to the center leading to a 'closed' pore conformation [82]. With GTP bound, structural changes occur leading to a shift of the Cα atoms surrounding the center of the trimer, this enables Glu133 to form hydrogen bonds to main-chain N-atoms in Ile134 leading to a new open pore conformation [82].

FeoC is believed to be involved in regulation of the FeoABC operon [69]. It is likely to contain an iron-sulfur cluster due to the presence of four cysteine residues in close proximity and is believed to act as a transcriptional regulator as it also possesses a winged helix motif similar to that of LysR transcriptional regulators [69].

3.2.2 ZIP-Like Transporters

E. coli possesses a gene designated *zupT* which encodes a membrane protein displaying similarities to eukaryotic ZIP (ZRT, IRT-like proteins) family transporters - these are involved in metal ion transport [85]. ZRT stands for 'zinc-regulated transporters' and these are best represented by the Zrt1 and Zrt2 proteins found in yeast [86]. IRT stands for 'iron-regulated transporter' and such proteins are represented by Irt1 of *Arabidopsis* which is expressed in roots of plants deficient in iron, and by LIT1 of *Leishmania amazonensis* [87]. ZupT was shown to be responsible for zinc uptake in *E. coli* [88]; however, more recent studies have demonstrated the ability of ZupT to uptake iron and cobalt as well as zinc and manganese [85]. *zupT* in *E. coli* is constitutively expressed but at low levels and has a broad substrate specificity [85]. Mutants lacking all known iron uptake pathways in *E. coli* could still obtain iron via ZupT [85].

3.2.3 Nramp 1-Like Transporters

Natural resistance associated macrophage protein (Nramp) transporters have been identified in yeasts, mammals, and more recently, bacteria. Members of this protein family traditionally transport manganese and ferrous iron into cells, which is often coupled with hydrogen ions [89]. In *E. coli*, an Nramp-like manganese transporter designated MntH has been characterized [90,91]. The protein appears to utilize manganese as a preferred substrate and has a highly conserved DPGN motif within its transmembrane segment 1 [92]. Mutations within this region greatly reduce manganese transport indicating structural or functional importance of this motif [92].

As well as manganese transport, MntH has also been shown to uptake ferrous iron as well as other divalent cations such as cobalt and zinc. MntH transport appears dependent on the proton motive force, similar to mammalian Nramp proteins [89,90,93]. The preference of MntH for manganese indicates that this system functions predominantly as a manganese transporter [90]. Studies have characterized bacterial MntR which functions as a manganese-dependent repressor of *mntH* [94]. In *E. coli* two new MntR regulated genes *mntS* and *mntP* have been identified. *mntS* encodes for a novel 42 amino acid protein and *mntP* encodes an efflux pump which is up-regulated by MntR in the presence of manganese [95]. The *E. coli mntH* gene is also regulated by Fur (the global iron-responsive repressor), so Mn^{2+} uptake is dependent on low iron as well as manganese levels [96].

3.2.4 The EfeUOB System of *E. coli*

The EfeUOB system, formerly known as YcdNOB, has been identified as a ferrous iron transporter in *E. coli* [97,98]. The EfeUOB system (Figure 4) is very different from the other bacterial iron transporters described above. It transports ferrous iron aerobically and functions at low pH which favors ferrous iron stability against oxidation [97]. All three components of the EfeUOB system are required for function. Although the system appears functional in most bacteria where it is found (e.g., *E. coli* O157:H7 and *E. coli* Nissle 1917 [97,98]), in some cases (e.g., *E. coli* K-12 and *E. coli* O6) the system is inactive due to mutation. In the case of *E. coli* K-12, all operon components are present but, due to a single frameshift within the *efeU* gene, the EfeU translation product is disrupted leading to a non-functional system. However, the EfeO and EfeB proteins are expressed and appear functional [97,98]. The reason for the cryptic nature of the *efeUOB* system in *E. coli* is unclear. The fully functional EfeUOB in *E. coli* O157:H7 suggests a benefit to this strain, which

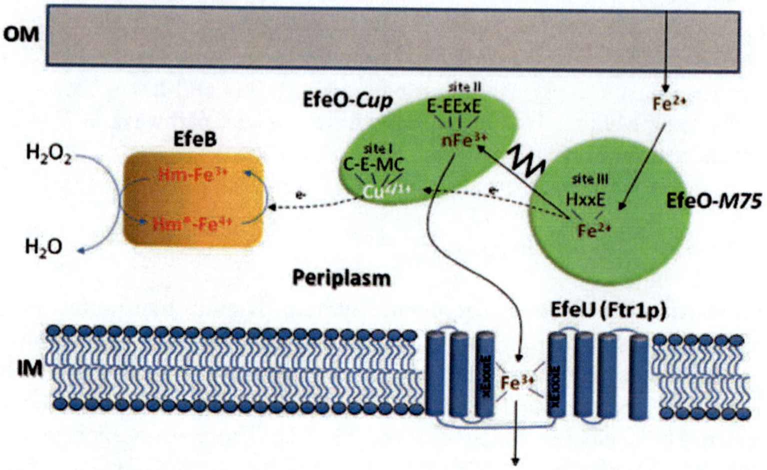

Figure 4 Schematic representation of the EfeUOB system of *E. coli.* This figure displays the components of the EfeUOB system of *E. coli.* Environmental ferrous iron is the initial substrate which is thought to be engaged in the periplasm by the iron binding protein, EfeO. The ferrous iron is then oxidised and the released electrons are delivered to an oxidant via a copper ion bound to site I and subsequently through the heme peroxidase protein EfeB. The oxidised iron is transferred to the ferric-iron permease protein EfeU located at the cytoplasmic membrane which then transports the ferric iron into the cytoplasm (adapted from [102]).

may be due to colonization or environmental survival outside hosts [97]. When the frameshift mutation is corrected, the EfeUOB system in *E. coli* K-12 is fully expressed and functional. Studies have demonstrated that in mutant strains lacking the known iron transport systems, under iron restriction a fully functional *efeUOB* operon provides a major growth advantage displaying as much as a 13-fold increase in iron uptake activity [97].

The EfeUOB system displays homology to the Ftr1p-Fet3p system found in yeast in that EfeU is homologous with the yeast Ftr1p. However, the Ftr1p system appears to be entirely dependent on oxygen whereas EfeUOB seems to rely on an alternative oxidant postulated to be peroxide [97]. The *efeUOB* operon forms an operon under the control of Fe^{2+}-Fur which matches its iron transporter role [4,97,98]. Induction of the *efeUOB* operon under acidic conditions is achieved by the two-component sensor regulator CpxAR which is known to be involved in genetic regulation in response to pH [99]. *B. subtilis* possesses an *efeUOB*-like operon designated *ywbLMN* which is also repressed by iron [100]. In *B. subtilis* the YwbLMN proteins provide an elemental iron uptake pathway in minimal media lacking citric acid [101].

EfeU, due to its homology with Ftr1p from *Saccharomyces,* is believed to be located in the IM and to function as a ferric permease [102]. Its function as a ferric permease has been observed in proteoliposomes [98]. It is believed that this protein is ultimately responsible for the transport of iron which is first oxidized by EfeO and

EfeB; this oxidation step may drive uptake and provide enhanced specificity. EfeB is a periplasmic heme-containing DyP-like peroxidase [97]. EfeB is translocated into the periplasm by the Tat pathway (twin arginine translocation) and is one of the only heme-containing proteins known to be secreted by this pathway [103]. EfeO is located in the periplasm. It consists of a C-terminal peptidase M-75 domain and an N-terminal cupredoxin domain [102]. The exact role EfeO plays is unclear, however, it is hypothesized that EfeO binds ferrous iron which is subsequently oxidized before transport to EfeU at the IM. The electrons generated from this process are then transferred to the heme group of EfeB which is hypothesized to use peroxide as an electron acceptor [102].

3.2.5 The P19 Iron Transporter of *Campylobacter jejuni*

C. jejuni expresses a member of the Ftr1 iron transporter superfamily designated cFtr1 that may form part of an iron transport system in *C. jejuni* [104]. Adjacent to the *cfrt1* gene in *C. jejuni* is a gene that encodes a protein of 19 kDa designated 'P19'; however, this protein does not appear to display homology to any of the components of the Ftr1p or EfeUOB system [104]. P19 is located in the periplasmic space [105] and is regulated by iron and Fur [106,107]. However, its precise function is unclear, although recent studies have demonstrated that P19 is crucial for growth of *C. jejuni* under conditions of iron restriction [104]. P19 is able to bind both iron and copper, and crystallization studies have identified two distinct metal binding sites [104]. The capability of this protein to bind both copper and iron suggests it may play a role in uptake of both of these metals, and it has been suggested that copper may play an important role as a cofactor for iron transport [104]. The metal binding sites appear to be in close proximity in the protein structure and Glu44 and Met88 residues may be mechanistically important as they appear to interact with both metal binding sites, however, this seems dependent on pH and metal occupancy [104]. The Glu44 residue has been hypothesized to receive iron as part of an acidic solvent channel and may mediate electron transfer during the uptake process similarly to the action of the Fet3p multi-copper oxidase in yeasts and algae [104]. Further study on this protein is needed to fully elucidate the mechanistic details of its action and what other proteins P19 interacts with.

3.2.6 Metal ABC Systems

Some ABC transport systems are utilized for the transport of free metals (uncomplexed). An example of this is the SitABCD system in *Salmonella enterica* serovar Typhimurium. This system is responsible for uptake of both Fe^{2+} and Mn^{2+}. The *sitABCD* operon appears to be regulated by Fur in response to iron availability and by the MntR protein in response to the availability of manganese [108]. SitABCD plays a role in the uptake of ferrous iron, but has preference for Mn^{2+}, and is also found in some pathogenic *E. coli* and *Shigella* strains [109]. The system is comprised of SitA which is a periplasmic-binding protein, the SitCD permease complex located at the IM and the SitB ATPase [110]. The *sit* genes are located on pathogenicity islands in *Salmonella* and other *Enterobacteriaceae* family members with the location of these genes being variable [111]. SitABCD resembles other bacterial 'metal-only' ABC transporters (e.g., FbpABC of *Neisseria gonorrhoeae*, SfuABC of *Serratia marcescens*, and YfeABCD of *Yersinia pestis*) which take up the free metal (rather than a metal complex) and do not depend on OM receptors [112–114].

4 Iron Storage and Detoxification

4.1 Introduction to Iron Storage

Once bacteria have acquired sufficient iron from the extracellular environment they can store excess in intracellular iron-storage proteins. These sources of iron can be utilized for growth when iron sources are limited. There are three main types of bacterial iron storage proteins, the bacterioferritins, ferritins, and Dps proteins. The ferritins and bacterioferritins have 24 identical (or similar) subunits and the Dps proteins have 12 identical subunits arranged to form a spherical shell which surrounds a central cavity which acts as the storage center for iron. The shell contains pores or channels for transport of iron, chelators, oxidants and reductants into and out of the molecule [115].

The bacterioferritins and ferritins have been observed to contain approximately 2000-3000 iron atoms whereas the smaller Dps proteins can only carry around 500 iron atoms [3]. The proteins contain a ferroxidase center which is located in the central portion of the individual subunits. Ferrous iron binds and forms diferric intermediates due to oxidation by oxygen, the ferric iron then migrates to the central core for storage. Oxidation of ferrous iron in Dps proteins, however, happens at a different location, at the interface between the individual subunits [116]. Ferritin and Dps proteins utilize O_2 and H_2O_2, respectively, in the process of oxidising ferrous iron. This results in the production of an insoluble iron oxide core in the oligomeric proteins [117].

4.2 Ferritins and Bacterioferritins

Ferritins accumulate iron during periods of iron sufficiency and store iron for utilization under iron limitation [1]. They also play a role in redox stress resistance by preventing iron induced toxicity. Ferritin (FtnA) in *E. coli* accumulates iron when the growth of the bacterium is favorable due to high iron availability.

Studies in *E. coli* mutants lacking functional FtnA showed reduced growth under iron restriction. This has re-enforced theories that FtnA is used for iron storage for later use [118]. If iron conditions become limited *E. coli* utilizes the bound iron as an iron source. Ferritins are able to control the reversible transition of ferrous iron to ferric iron using O_2 as oxidant [119] such that harmless water is generated rather than toxic hydrogen peroxide (along with ferric iron). Ferritins sequester iron away from superoxide and hydrogen peroxide providing a detoxification function [119].

Bacterioferritins (Bfr) are more common than ferritins in bacterial species [3]. All of these proteins are known to contain heme which is usually in the form of protoporphyrin IX. The heme groups are usually found at interfaces between the 24 subunits of the protein, so on average bacterioferritins contain 12 heme groups [3,120]. The heme groups are usually oriented with their propionate groups projecting into the center of the molecule; however, the exact purpose of these heme groups is unclear. Heme-free bacterioferritin produced in *E. coli* displayed no change in *in vitro* iron uptake indicating the heme groups are not involved in forming an iron core. However, the heme-free variants accumulated much more iron *in vivo* indicating a possible role of heme in mediating iron release [121]. Many bacterioferritin genes are associated with a gene known as *bfd* which encodes a 2Fe-2S ferredoxin known as Bfd (Bfr-associated ferredoxin) [3]. It is thought that Bfd interacts with bacterioferritin during iron starvation acting as a bacterioferritin reductase mediating iron release [122]. Other bacterial species are thought to contain electron carriers acting as reductases mediating iron release from bacterioferritin. Bacterioferritins also have a ferroxidase center and the rate of formation of the iron core and speed of iron loading is pH-dependent [123]. It has also been found that the ferroxidase center is vital throughout core formation, disruption to the core occurred in mutants lacking this ferroxidase center [123].

Despite structural similarities between ferritins in different bacterial species their function and regulation can vary [124]. This is illustrated with studies on *Salmonella* Typhimurium where bacterioferritin accounts for half of total cellular iron whereas FtnA plays a minor role in iron storage [118,125]. In contrast in *E. coli* FtnA plays a major storage role [118]. In *Salmonella* it has been observed that bacterioferritin protects the bacterial cell from cytotoxic hydroxyl radicals and reactive oxygen species by limiting the concentration of free iron within the cell [125]. Ferritin B (FtnB) characterized in *Salmonella* displays unique characteristics, on top of it providing a modest contribution to iron storage capacity, it is required for the Fe-S containing aconitase enzyme to undergo repair following oxidative damage [125]. It is expressed during iron restriction and is repressed by Fur, it is required for full virulence and appears to exacerbate oxidative stress [125]. It has therefore been suggested that FtnB may provide an iron reservoir for iron-sulfur cluster repair rather than performing an iron storage role [125]. The obligate anaerobe, *Bacteroides fragilis*, has a bacterioferritin-related gene that is significantly induced on exposure to air suggesting a protective role against reactive oxygen species [126].

4.3 Dps Proteins

Dps (DNA-binding protein from starved cells) contributes to protection against environmental stresses such as oxidative stress and nutritional limitation. Not all Dps proteins are capable of DNA binding, however, those that do, provide some degree of DNA protection by shielding DNA [127]. The extent and mechanisms by which this stress protection is provided are still not clearly understood [128]. Dps shares similar characteristics and structure to other ferritins as discussed earlier, the fully assembled protein is very stable and highly compact with the capability to bind DNA with great stability [127]. Dps forms a ferrihydrite core which binds iron tightly until required for biological processes. Dps has been described as a ferritin-like protein due to its ferroxidase activity whereby it can bind ferrous iron and oxidize it to the ferric state [129].

The 12 ferroxidase sites present in the Dps protein are located at interfaces between the subunits of the protein [130]. Each ferroxidase site oxidizes two ferrous iron ions utilising one molecule of H_2O_2 in the process. This reaction helps to avoid hydroxyl radical formation which would otherwise occur when ferrous iron and H_2O_2 react (the Fenton reaction). H_2O_2-mediated oxidation in Dps proteins differentiates them from ferritin proteins which utilize O_2 preferentially [131]. Thus, not only do Dps proteins of *E. coli* effectively store iron but they may also play a major role in protection of DNA from the damage inducing combination of hydrogen peroxide and ferrous iron [128]. Indeed, it is now hypothesized that in many cases the role of Dps proteins in iron storage may be secondary with its protection against oxidative stress being a primary role [132].

5 Global Iron Regulation

5.1 Cellular Iron and the 'Iron Proteome'

The amount of iron found within a bacterial cell is variable and depends on the species as well as the growth condition. For instance, an *E. coli* cell grown in rich broth typically carries 0.014% iron (by dry weight) in exponential phase which increases to 0.026% in stationary phase as iron stores are accumulated [118]. Mössbauer spectroscopy indicates 26% of this stationary-phase cellular iron is in the ferrous state, the remainder being ferric. Under iron-poor conditions, cellular iron levels are greatly diminished and can be as low as 0.002% (i.e., 13-fold lower than those observed under iron sufficiency!). The mechanism that enables this adjustment of cellular iron levels in response to iron availability is discussed below (see Section 5.3). A normal cellular iron content of 0.02% equates to ~1×10^6 atoms per cell (note that there are 2.4×10^6 protein molecules in an average *E. coli* cell [133]). However, it remains unclear how this iron is distributed within bacteria and what proportion of proteins are iron-containing. For *E. coli*, the majority of cellular iron is thought to be protein-associated (mainly as iron-sulfur clusters, heme, and mono- or divalent iron) with up to 50% located within storage proteins as ferric-iron clusters (see Section 4). Indeed, EPR studies using iron chelators as probes suggest that only a small fraction of cellular iron is 'free' (not incorporated into proteins). Such free iron is designated as the 'labile iron pool' and is estimated to be present at just 10 μM (~0.1% of cellular iron) [12]; this value appears realistic since it matches the affinity of Fur (the global iron regulatory protein) for its effector, ferrous iron [134] (see Section 5.2).

Cellular iron distribution has been investigated in *Bacillus anthracis* using a two-dimensional chromatography approach [135]. However, only three major pools could be identified and a comprehensive analysis of iron-protein composition has yet to be performed for any bacterium. An analysis of the proteome [136] suggested that ~85% of the metal content of the *E. coli* metalloproteome contains iron with the remainder consisting mainly of Cu, Mn, Zn, Ni, and Mo (at ~4, 4, 2, 2, and 0.8%, respectively). But, although iron would appear to dominate the metalloproteins of *E. coli* (and also *Pyrococcus furiosus*), for *Sulfolobus solfataricus* zinc and iron account for similar proportions (~37 and 60%, respectively) [136]. Thus, it seems that different bacterial species incorporate distinct sets of metals into their metalloproteomes.

Bioinformatic approaches have also been applied in an attempt to quantify and identify metalloproteins of bacteria [137]. Such studies predict that non-heme iron proteins account for 3.9% (+/–1.6%) of protein-encoding bacterial genes, of which 40% were considered to be Fe-S proteins. Needless to say, such studies are limited by an inability to account for the abundance of individual proteins or identify metalloproteins containing novel metal-binding sites.

5.2 *Fur*

5.2.1 Introduction

Iron availability usually governs how iron metabolism is regulated in most bacterial species. In *E. coli* and most other Gram-negatives, as well as some Gram-positives, iron regulation is mediated by the ferric uptake regulation protein (Fur). Fur has the ability to control the expression of numerous genes in response to iron [57]. The Fur protein forms a homodimer and consists of subunits with an approximate molecular weight of 17 kDa [138]. Fur exhibits regulatory activity once bound to its co-repressor Fe^{2+}. Under conditions of iron limitation derepression is triggered.

Fur acts as a positive repressor only repressing transcription once bound to its co-repressor. Each subunit within the Fur protein is able to bind one molecule of ferrous iron. Once metal has become bound to Fur its affinity for DNA significantly increases. The Fur subunits are divided into two distinct domains, an N-terminal domain responsible for DNA binding and a C-terminal domain that binds Fe^{2+} and assists in dimerization. Studies have also shown that zinc ions are present as a structural stabiliser of Fur and aid proper folding of the protein [139,140]. However, other members of the Fur protein family have been shown to not require zinc ions for structural stability [141].

Fur usually binds to promoters of Fur-repressed genes at sequences known as Fur boxes. These Fur boxes are usually 19 bp inverted repeat sequences. The Fur boxes are usually observed overlapping the –10 or –35 regions and block transcription of genes involved in iron metabolism. Fur-binding sites have been shown to consist of two or more Fur boxes adjacent to each other. Several Fur dimers can bind to these sites [139]. At some promoters it can be seen that Fur initially binds to a high-affinity site which leads to further Fur binding adjacent to it, leading to the appearance of Fur polymerising along a DNA duplex [142]. The Fur box was believed to consist of a set of three repeats of a six base pair motif. Multiple repeats of these motifs were seen to lead to subsequent and increased Fur binding [139]. However, this theory was not universally accepted, as when Fur regulation was studied in *Bacillus subtilis*, Fur-regulated genes were seen to contain operator sites with two of the proposed six base pair motifs embedded within a 7-1-7 motif [143]. This prompted a newly proposed model suggesting that the Fur box is recognized by two Fur dimers which seem to interact with overlapping 7-1-7 motifs from opposite faces of the helix [144].

Fur is highly abundant in *E. coli* with approximately 5000 copies per cell [145]. The reason for such a high amount of Fur per cell is unclear but suggests that it may control a larger number of genes than currently realized. As well as being able to regulate genes related to iron it is also able to regulate other genes such as those encoding hemeolysin and manganese superoxide dismutase [146]. The *fur* gene in *E. coli* is located downstream in the *fldA-fur* operon. The *fldA* gene encodes flavodoxin which is a vital protein believed to have an important redox role. Flavodoxin is believed to assist in maintaining reduced free iron in the cytosol, thus may be vital for providing iron in the reduced state for the Fur protein. The *fldA-fur* operon is induced tenfold under superoxide stress via the SoxRS system [144]. The *fur* gene also possesses an independent promoter which is induced by OxyR under hydrogen peroxide stress. Thus, Fur regulation is integrated with the redox stress response ensuring that free-iron levels are limited during oxidative challenge. This link between iron and redox stress is further illustrated by the role of OxyR as an inducer of *yaaA*. The *yaaA*-encoded YaaA protein functions to lower free cellular iron levels and thereby mediates a dampening of iron-induced redox stress [147]. In addition, Fur also contributes to the global response of *E. coli* to reactive nitrogen stress, acting alongside NorR (and NssR) as one of the major reactive nitrogen-responsive regulators [148]. This effect may be related to inhibition of Fur activity by nitrosylation [149].

5.2.2 The Fur Modulon and Global Control

The main responsibility of Fur is the repression of iron acquisition genes under iron sufficiency; this enables iron transport to be activated when access to iron becomes restricted. However, Fur not only represses iron acquisition genes, it also represses genes involved in the tricarboxylic acid (TCA) cycle, glycolysis, respiration, chemotaxis, redox stress, and DNA synthesis [150,151]. Thus, Fur can be considered as a global regulator as it adjusts many cellular processes according to iron availability. As might be expected, loss of Fur function in *E. coli* results in inability to grow on non-fermentable carbon sources, consistent with defects in respiration, and causes increased susceptibility to oxidative stress due to raised cellular free iron levels [152,153].

Transcriptomic analysis of global Fe^{2+}-Fur dependent expression in *E. coli* revealed that Fe^{2+}-Fur represses 53 genes and induces 48 [4]. These Fe^{2+}-Fur controlled genes fall into two major categories: iron metabolism and energy metabolism. Similar regulatory effects have been seen in many other bacteria (Table 1). In addition, genes involved with metabolism of other metals, stress response, phage function, and nucleic acid metabolism were also identified. The iron metabolism genes (55 in total) were largely repressed and were mostly involved in iron acquisition. The energy metabolism genes included a large number of iron-protein encoding genes that were nearly all induced by Fe^{2+}-Fur. Further analysis revealed that the total iron content and iron-proteome level of *E. coli* is greatly lowered in a *fur* mutant. This, together with the array data, indicated an 'iron-rationing' response to iron restriction. This effect was later confirmed following the discovery of the Fur-reponsive regulatory-RNA molecule, RyhB [5], and similar iron-rationing mechanisms have now been observed in other bacteria (see Section 5.3).

DNA-array based global expression studies on iron regulation have now been published for numerous bacterial species, including *Pseudomonas aeruginosa* [154], *Bacillus subtilis* [100,167], *Neisseria meningitidis* [155], *Shewanella oneidensis* [156,157,170], *Campylobacter jejuni* [106], *Helicobacter pylori* [158,161,164], *N. gonorrhoeae* [159,172], *Vibrio cholerae* [160], *Corynebacterium glutamicum* [162,163], *Bradyrhizobium japonicum* [165], *Francisella tularensis* [166], *Desulfovibrio vulgaris* [168], *Xylella fastidiosa* [169], *Pseudomonas syringae* DC3000 [171], *Listeria monocytogenes* [173], and *Leptospira interrogans* [174].

The responses reported vary between different bacteria and even for the same species in different experiments (see Table 1 for summary). However, generally there would appear to be a greater number of genes repressed by iron and/or Fur than induced. The common functional categories of iron/Fur regulated genes within different bacteria are iron metabolism, energy generation, and redox stress (Table 1) which indicates a similar overall physiological response to environmental iron.

Experiments aimed at identifying the location of Fur-binding sites across the entire chromosome have been performed for *H. pylori* using Fur-immunoprecipitation combined with DNA chip analysis [164]. This study showed that Fur binds at 200 loci with the *H. pylori* genome – this is a surprisingly large number given the relatively small size (~1560 coding genes) of the *H. pylori* genome. Another surprise was the observation that 71 of the Fur-binding loci were within cistrons or polycistronic transcriptional units. Furthermore, only about 30% of Fur-binding loci were associated with a gene that was subject to Fur-transcriptional control; the remaining 70% of Fur-binding sites were designated "orphan" Fur sites. The purpose of these orphan and intracistronic Fur sites remains obscure [164].

Table 1 Summary of transcriptomic studies on iron- and/or Fur-dependent gene control[a]

Species	Genes Fe-repressed	Genes Fe-induced	Conditions[b]	Cell functions displaying significant effects[c]	Gene no[d]
E. coli	53	48	-Fe, -Fur	Fe uptake, energy metabolism	4427
P. aeruginosa	118	87	-Fe	Fe uptake	~6000
B. subtilis	39	0	-Fe, -Fur	Fe uptake, redox-related	~4300
S. oneidensis	39	0	-Fur	Energy metabolism, redox stress	4657
C. jejuni	117	30	-Fe	Redox stress, Fe metabolism, respiration	~1700
N. meningitidis	153	80	-Fe		~2100
H. pylori	16	16	-Fe, -Fur	Metal metabolism, respiration, redox stress	~1560
N. gonorrhoeae	109	94	-Fe	Fe transport, rapid growth	~2450
V. cholerae	64	6	-Fe, -Fur	Fe transport, nitrate respiration	~3900
H. pylori	25	4	-Fur	Fe uptake, acid stress	~1560
C. glutamicum	131	126	-DtxR		~3100
C. glutamicum	51	13	-DtxR	Fe uptake, Fe-S biosynthesis	~3100
H. pylori	29	25	-Fur, ChIP		~1560
B. japonicum	134 (14)	120 (13)	-Fe (-Fur)	Fe uptake, motility, Calvin cycle	~8600
F. tularensis	80 regulated		-Fe		~1700
B. subtilis	71	71	-Fe	Fe uptake, amino acid biosynthesis, nucleoside biosynthesis, CodY, stringent response, translation	~4300
D. vulgaris	30/32 (22)	4/18 (22)	-Fur (-Fe)	Fe uptake	3531
X. fastidiosa	193/256	216/218	-Fe/+Fe	Pathogenicity, cell structure, pili	~2800
S. oneidensis	118	56	-Fur (-Fe)	Fe uptake, ribosome, (energy metabolism)	4657
P. syringae	386 regulated		-Fe	HrpL regulation, coronatine biosynthesis, Fe uptake/storage	5608
N. gonorrhoeae	300	107	-Fe	Fe metabolism, cell communication, intermediary metabolism, energy metabolism	~2450
L. monocytogenes	10 (23)	15 (0)	-Fe (-Fur)	Amino acid metabolism, Fe transport	~3000
L. interrogans	43	49	-Fe	Fe uptake	3700-4800

[a] Data is presented in chronological order. For sources of data, see references in the text.
[b] '-Fe' or '-Fur' indicate that differential expression was achieved by manipulation of iron availability or mutation of the *fur* gene (or *dtxR*), respectively; '-Fe/+Fe' indicates that differential expression was achieved by iron depletion and iron excess, respectively.
[c] Only selected, major effects are indicated (where no effects are shown this indicates that either none were reported or observed).
[d] Indicates approx. number of protein-encoded genes in corresponding genome.

Transcriptomic analysis with *E. coli* has also shown that high, toxic levels of certain metal ions (e.g., copper) can induce the Fur modulon presumably by competing with iron and thus blocking iron uptake, and that chelators utilized to lower the availability of metals other than iron can induce the Fur regulon possibly due to co-chelation of iron [175,176]. DNA-microarray studies also suggest that iron uptake systems are induced by a temperature switch from ambient (23°C) to body temperature (37°C), and that this effect is mediated by the histone-like H-NS protein [177].

5.3 Positive Regulation by Fur: RyhB

E. coli genes such as *sodB* (encoding superoxide dismutase) and *bfr* (encoding bacterioferritin) are induced by iron in a fashion that is Fur dependent. Such genes are believed to be indirectly affected by Fur as they generally do not possess a Fur box (the ferritin-encoding *ftnA* gene being a notable exception [178]). Positive regulation by Fur is now understood to be mediated by Fur-dependent expression of a small regulatory RNA (sRNA) that acts post transcriptionally to cause rapid degradation of target transcripts. In *E. coli* this regulatory RNA is called RyhB. It is 90 nucleotides in length and its expression is repressed by Fur in response to iron [179]. Under iron restriction RyhB expression is derepressed triggering decreases in target mRNA levels. As well as superoxide dismutase and bacterioferritin being regulated in this way aconitase and fumarase of the TCA cycle are other examples of RyhB targets [180]. The role of RyhB is to down regulate non-essential iron containing proteins under conditions of iron restriction to induce an 'iron sparing' or 'iron rationing' response. This characterization of RyhB has further emphasized Fur as a global iron regulator [4,5]. This iron sparing (or iron rationing) response has been observed in other bacteria such as *Pseudomonas* and *Shigella* species [181,182] and so, as indicated above, iron rationing would appear to be a common response in bacteria.

5.4 Fur Family Members

A large number of Fur homologues have been identified in bacterial species that respond to a wide variety of different metals. In *E. coli* and *B. subtilis* there is a Fur family member responsible for genetic regulation in response to zinc, designated Zur [183]. *B. subtilis* also possesses another Fur homologue known as PerR, the PerR protein is a repressor responsible for sensing peroxide stress [184]. PerR can be found associated with either manganese or iron, when bound to these metals it acts as a transcriptional repressor, but only when ferrous iron is bound can it act as a peroxide sensor [185]. PerR represses many genes including *fur* and its own expression [186]. *Rhizobium leguminosarum* contains another Fur homologue which was observed to repress the ABC-type manganese transporter SitABCD under manganese restriction, this homologue was named Mur [187] since it responds to Mn^{2+}. In *Streptomyces coelicolor* a Fur homologue responsible for responding to nickel was characterized and designated Nur [188]. Nur has been seen to regulate the nickel transport genes *nikABCDE* by binding to promoters when nickel is abundant [188]. Fur homologues are widespread within bacterial species and all are believed to play a role as metal binding dependent regulators predominantly acting in a repressive fashion. In addition, the Fur-like Irr (iron response regulatory) protein of *Bradyrhizobium japonicum* acts as a heme-responsive regulator [189].

5.5 *Alternative Iron Regulators*

5.5.1 DtxR-IdeR

In some Gram-positive organisms from the 'high GC-content' group, DtxR (diphtheria toxin regulator) plays a role in global iron regulation. The DtxR family of metalloregulators use iron or manganese as co-repressors. Recent studies on *Deinococcus radiodurans*, which is renowned for being able to resist extreme ionising and UV radiation, have characterized a DtxR metalloregulator encoded by the *dr2539* gene. This DtxR regulator has been demonstrated to negatively regulate manganese-dependent transport and act as an activator of iron transporters which differs from the activity of other DtxR homologues [190]. DtxR has been seen to regulate iron acquisition and diphtheria toxin expression in response to iron and has been identified as the master regulator of iron-dependent genes in *Corynebacteria* [191]. Under conditions of iron sufficiency DtxR in complex with iron represses genes involved in iron acquisition, such as siderophore ABC transporters [162]. Some storage genes are stimulated by DtxR such as ferritin and Dps proteins [192]. Under iron limitation Fe^{2+} dissociates from DtxR and the protein loses its DNA binding affinity in a similar mechanism of action to Fur [193]. DtxR proteins are homodimers with each subunit containing three domains: an N-terminal DNA-binding domain, a central metal/dimerization domain, and a C-terminal SH3-like domain (of unknown function). MntR found in *E. coli* (and elsewhere) which acts as a manganese regulator as mentioned previously, is also a homologue of DtxR. However, the C-terminal SH3 domain is not present in MntR proteins.

In *Staphylococcus epidermidis*, SirR is a metal-dependent repressor homologous to DtxR. Several SirR box motifs have been found within the genome and it has been hypothesized that this regulator may act as a global metal regulator like Fur, however, more work is needed to confirm this and the precise mechanisms involved [194]. IdeR is a DtxR-like, iron-dependent transcriptional regulator in *Mycobacterium tuberculosis* [195]. IdeR is essential for growth in this organism and mutations leading to a non-functional protein prove lethal [196]. IdeR acts primarily as a transcriptional repressor by binding to operator sites in close proximity to genes, although like Fur and other homologues IdeR also has the ability to enhance transcription of some genes [195].

5.5.2 RirA

RirA is an Rrf2-type transcriptional regulator found in rhizobial species such as *Rhizobium leguminosarum* and *Sinorhizobium meliloti*. Members of the Rrf2 family include the iron-sulfur cluster responsive IscR repressor as well as NsrR, an NO-responsive activator [197] from *E. coli*. RirA controls transcription of siderophore uptake genes [198–200] but appears to respond to cellular Fe-S cluster status rather than cytosolic free iron.

5.5.3 PmrBA

The two-component sensor-regulator pair, PmrA-PmrB, of *S.* Typhimurium was originally identified on the basis of its role in regulating resistance to the cationic peptide antibiotic, polymyxin B [201]. It was subsequently shown that PmrBA is also responsive to external Fe^{3+} (as well as Al^{3+} and acidity) and thus controls resistance to Fe^{3+} [202]. The PmrBA system is also induced by low Mg^{2+} levels, but this response is imposed by another two-component system, PhoPQ, in conjunction with PmrD [203]. PmrBA controls expression of lipopolysaccharide (LPS) modification factors, which at least partly explains the resistance effects observed [204]. The related BasS-BasR system of *E. coli* exhibits similar features controlling LPS modification and acid resistance in response to low pH and iron [205].

6 Iron Replacement

Another ploy used by bacteria to combat lack of iron is replacement of iron-dependent enzymes with alternative, iron-independent isoenzymes. This substitution approach allows bacteria to escape from growth restriction imposed by lack of iron [206]. Two well described cases are known for *E. coli*. One involves the down regulation of the gene specifying the Fe-containing superoxide dismutase (SodB) under low iron conditions (RyhB-dependent) with the concomitant induction of the gene specifying the 'replacement', the Mn-dependent SodA enzyme (Fur-dependent) [207]. The second example is that of the NrdEF ribonucleotide reductase (RnR). This diMn-dependent enzyme is induced by iron restriction where it functionally substitutes for the diFe-dependent NrdAB RnR isoenzyme [208].

These Mn replacement strategies are further supported by induction of Mn^{2+} uptake in response to iron starvation, which ensures that *E. coli* only employs Mn as an alternative to iron when iron is otherwise unavailable (or under redox stress) [96]. Indeed, some mononuclear iron enzymes can replace iron with manganese (e.g., ribulose-5-phosphate 3-epimerase [209]), so only the metal needs to be replaced rather than the entire isoenzyme.

7 Iron and Pathogenicity

7.1 Host Use of Iron Restriction to Combat Infection

Pathogens face the problem of acquiring sufficient iron for effective growth whilst competing with their host. The innate immune system of the host limits iron availability to prevent pathogen propagation and colonization. Mammals employ transferrin and lactoferrin which bind iron, thus reducing free extracellular iron to concentrations too low for bacterial growth [210]. Heme is a major source of iron for many pathogens and so the host usually tries to bind this molecule to proteins to limit its availability. Pathogenic bacterial species often recognise this lack of iron availability as a sign of their host localization and so trigger transcription of virulence genes in response to low environmental iron [210].

7.2 Counter-Measures. Sources and Strategies for Pathogens

7.2.1 Transferrin and Lactoferrin as Iron Sources

However, pathogens are not defenceless against iron restriction imposed by the host. Many pathogens utilize siderophores to compete with host binding proteins to acquire iron. Often receptor-mediated transport systems are used to acquire iron directly from host complexes. *Neisseria* species produce two transferrin-binding proteins which are highly conserved. Mutants lacking these proteins in *Neisseria gonorrhoeae* were avirulent demonstrating the importance of these proteins in gonococcal pathogenicity [211]. *Neisseria* species contain receptors able to bind and acquire iron from both transferrin and lactoferrin. The transferrin receptor is comprised of transferrin binding protein A (TbpA) and transferrin binding protein B (TbpB); these proteins are responsible for binding transferrin, acquiring any iron bound and then transporting it across the OM. They are usually induced under conditions of iron starvation [211].

TbpA is approximately 100 kDa and shares homology with other OM receptors responsible for siderophore uptake. Thus, it has a β-barrel structure with a plug domain – just like siderophore receptors. TbpB is a lipoprotein thought to enhance utilization but is not absolutely required [212,213]. The structure of the TbpA-TbpB-transferrin complex has recently been solved [247].

Uptake of transferrin is dependent on the ability of the TonB-ExbB-ExbD complex to utilize the proton motive force [214]. Once iron has been released from transferrin and crosses the OM it is deposited into FbpA (ferric binding protein A, a periplasmic iron-binding protein); the mechanism by which this occurs however remains elusive [215]. FbpA binds the released iron and transfers it to the IM [216]. Tight iron binding by FbpA has been shown to drive release of iron from transferrin and drive uptake by the transferrin receptor complex [217]. At the IM, FbpA docks with the receptor complex (FbpB/C) which then transports the iron to the cytoplasm. FbpABC is another example of an ABC transport system that utilizes ATP hydrolysis to drive uptake [218]. The lactoferrin receptor is closely related to the transferrin receptor but utilizes iron bound to lactoferrin as substrate. The lactoferrin receptor is comprised of two subunits just like the transferrin receptor; these are designated LbpA and LbpB [211].

7.2.2 Heme as an Iron Source

Most heme is contained within hemoglobin and free hemoglobin in serum is complexed with haptoglobin and removed by the liver. Heme in serum is bound by the host protein, hemopexin [10]. Gram-negative bacteria acquire heme through OM receptor proteins to which heme-containing compounds bind. The heme acquisition mechanism of the Gram-negative *Serratia marcescens* has been extensively studied. *S. marcescens* releases a small protein known as a 'hemophore', designated HasA, which binds heme incredibly tightly. Upon binding heme, HasA changes from an open to a closed form [219]. Heme-loaded HasA delivers heme to the OM receptor, HasR. Upon docking to the receptor one of the two axial heme ligands of the hemophore is disrupted which is followed by the breaking of the second heme coordination site of HasA through steric displacement of heme by Ile671 of HasR [220]. Following binding, heme is translocated across the OM by an energy-coupled process driven by TonB-ExbB-ExbD. TonB interacts with HasR along with another TonB-related protein, designated HasB [221]. Heme uptake via this system in *S. marcescens* only appears to occur under conditions of iron limitation [221].

Once in the periplasm, heme binds to a periplasmic-binding protein which in turn delivers heme to an ABC transporter at the IM [222]. Binding of ATP at the inside of the IM triggers a closing of the gap between the transmembrane domains of the transporter; this then causes them to flip to an outward facing conformation. ATP hydrolysis causes the transmembrane domains to form an inward facing conformation leading to substrate import [222]. In *E. coli,* NikA appears to bind heme in the periplasm and can function in heme transport although it primarily functions as a nickel-binding protein [223]. Also in *E. coli,* EfeB has been observed to remove iron from heme [224]. In *Yersinia pestis* the heme-binding protein HmuT appears to bind two heme molecules which are then subsequently transported to the cytoplasmic membrane ABC transporter HmuUV [225]. In *Haemophilus influenzae* the heme-binding protein HbpA also appears to be involved in the uptake of glutathione [226].

H. influenzae obtains heme from human hosts in a variety of ways. Utilization of hemoglobin-haptoglobin as a heme source depends on binding proteins designated HgP proteins [227]. Heme-hemopexin complexes are utilized by products of the iron-regulated *hxuCBA* gene cluster [228,229]. HxuA is secreted by *H. influenzae* and binds hemopexin which forces it to release any bound heme; the heme is then internalized via the TonB-dependent OM receptor, HxuC [230,231].

Gram-positive pathogens contain ABC transporters at the IM but, as they lack an OM, they do not possess any energy-driven systems that would usually be required for transport across this membrane. Many Gram-positive pathogens do not have well characterized heme-utilization systems; however, *Staphylococcus aureus* has an iron-regulated surface determinant (Isd) system which removes heme from heme-containing proteins such as hemoglobin and hemoglobin-hemopexin liberated from lysed erythrocytes. Heme is transferred to IsdR which is a substrate-binding lipoprotein for an ABC transporter that imports heme and leads to its degradation [232,233]. The Isd system is comprised of nine proteins some of which contain NEAr iron Transporter (NEAT) domains; these domains have been observed to bind heme and hemoglobin with high affinity [233]. IsdB, IsdH, IsdC, and IsdA have all been found to contain between one to three NEAT domains [234–237]. IsdB and IsdH are secreted into the extracellular environment and release heme from heme-containing proteins. IsdA and IsdC are found covalently anchored to peptidoglycan, IsdC is hypothesized to constitute a heme channel and IsdA appears to be multifunctional in nature (binding lactoferrin) and subsequently heme passes to IsdC and onto IsdE [238]. Import across the cytoplasmic membrane is catalyzed by an ABC system consisting of IsdF permease and IsdD ATPase. Once inside the bacterial cell iron is liberated by oxygenase enzymes IsdG and IsdI. To prevent toxicity due to an abundance of heme, excess is exported by the HtrAB proteins [239].

7.2.3 Enterobactin, Salmochelin, and Aerobactin

Many pathogens employ siderophores to sequester iron from their hosts. Enterobactin, a catechol siderophore, is produced by most enteric organisms and has the highest ferric iron-binding affinity known [240]. Many enteric pathogens also produce the siderophore aerobactin which scavanges iron during infection [24]. The genes specifying aerobactin production and utilization are often plasmid encoded [241]. Although aerobactin binds iron with less affinity than enterobactin [242], it is effective *in vivo* (enhances pathogenicity) whereas enterobactin provides little advantage under *in vivo* conditions [243,244]. For many years, the precise reason for the poor performance of enterobactin *in vivo* was unknown. However, it is now clear that enterobactin is specifically cleared from circulation by the host protein lipochalin-2, which explains its lack of effectiveness during systemic infection [245]. Indeed, some pathogenic bacteria produce a modified form of enterobactin, initially identified in *S.* Typhimurium and thus designated 'salmochelin' [246]. This siderophore is produced by C-glucosylation of enterobactin at two positions and as a result is not recognized by lipochalin-2 [246]. In this way, pathogenic bacteria such as *Salmonella* effectively counter the antibacterial activity of lipochalin-2.

8 Final Remarks

The chapter provided here highlights the multiple mechanisms used by bacteria to manage their need for iron. In particular, bacteria have a range of options for securing this biologically precious metal from their environment. But, although uptake of iron is the major process through which cellular demand for iron is met, a range of other processes combine together to ensure that a refined, integrated approach to iron metabolism and its control is practiced. Such other processes include storage, rationing, substitution, detoxification and global iron-regulation mechanisms. In addition, it is now clear that there is interaction between the iron homeostasis machinery and those processes governing the redox-stress response and manganese utilization. Thus, iron control within bacteria is a complex, multifaceted process. 'Omics' approaches have allowed the iron regulon to be defined in many bacteria and have helped to unravel a new component of iron homeostasis – iron rationing. Further work is required in order to understand the reason for the relatively high level of the Fur protein as seen in many bacterial cells, the full extent of the 'iron proteome' and whether there is a hierachical regulation of gene expression in response to changes in iron availability.

Abbreviations

ABC	ATP-binding cassette
ATP	adenosine 5′-triphosphate
Bfd	Bfr-associated ferredoxin
Bfr	bacterioferritin
Dps	DNA-binding protein from starved cells
DtxR	diphteria toxin regulator
ECF	extracytoplasmic function
EPR	electron paramagnetic resonance
Fbp	ferric binding protein
Ftn	ferritin
Fur	ferric uptake regulation protein
GDP	guanosine 5′-diphosphate
GTP	guanosine 5′-triphosphate
H-NS	histone-like nucleoid structuring protein
IM	inner membrane
IRT	iron-regulated transporter
Isd	iron-regulated surface determinant
LPS	lipopolysaccharide
NEAT	NEAr (iron) transporter
Nramp	natural resistance-associated macrophage protein
NRPS	non-ribosomal peptide synthetase
OM	intermembrane
RnR	ribonucleotide reductase
Sod	superoxide dismutase
sRNA	small regulatory RNA
Tat	twin arginine translocation
TCA	tricarboxylic acid cycle
Tbp	transferrin-binding protein
ZIP	ZRT,IRT-like proteins
ZRT	zinc-regulated transporters

Acknowledgments We thank the University of Reading, Diamond Light Source Ltd, the Felix Foundation, the Egyptian Government, and the Nigerian Institute of Medical Research for studentship funding.

References

1. S. C. Andrews, *Adv. Microb. Physiol.* **1998**, *40*, 281–351.
2. A. N. Woodmansee, J. A. Imlay, *J. Biol. Chem.* **2002**, *277*, 34055–34066.
3. S. C. Andrews, A. K. Robinson, F. Rodriguez-Quinones, *FEMS Microbiol. Rev.* **2003**, *27*, 215–237.
4. J. P. McHugh, F. Rodriguez-Quinones, H. Abdul-Tehrani, D. A. Svistunenko, R. K. Poole, C. E. Cooper, S. C. Andrews, *J. Biol. Chem.* **2003**, *278*, 29478–29486.
5. E. Masse, F. E. Escorcia, S. Gottesman, *Genes Dev.* **2003**, *17*, 2374–2383.
6. P. W. Boyd, A. J. Watson, C. S. Law, E. R. Abraham, T. Trull, R. Murdoch, D. C. Bakker, A. R. Bowie, K. O. Buesseler, H. Chang, M. Charette, P. Croot, K. Downing, R. Frew, M. Gall, M. Hadfield, J. Hall, M. Harvey, G. Jameson, J. LaRoche, M. Liddicoat, R. Ling, M. T. Maldonado, R. M. McKay, S. Nodder, S. Pickmere, R. Pridmore, S. Rintoul, K. Safi, P. Sutton, R. Strzepek, K. Tanneberger, S. Turner, A. Waite, J. Zeldis, *Nature* **2000**, *407*, 695–702.
7. J. E. Posey, F. C. Gherardini, *Science* **2000**, *288*, 1651–1653.
8. M. Ferrer, O. V. Golyshina, A. Beloqui, P. N. Golyshin, K. N. Timmis, *Nature* **2007**, *445*, 91–94.
9. P. Cornelis, Q. Wei, S. C. Andrews, T. Vinckx, *Metallomics* **2011**, *3*, 540–549.
10. V. Braun, K. Hantke, *Curr. Opin. Chem. Biol.* **2011**, *15*, 328–334.
11. D. Touati, *Arch. Biochem. Biophys.* **2000**, *373*, 1–6.
12. K. Keyer, J. A. Imlay, *Proc. Natl. Acad. Sci. USA* **1996**, *93*, 13635–13640.
13. P. Heymann, M. Gerads, M. Schaller, F. Dromer, G. Winkelmann, J. F. Ernst, *Infect. Immun.* **2002**, *70*, 5246–5255.
14. W. Koster, *Res. Microbiol.* **2001**, *152*, 291–301.
15. B. C. Chu, A. Garcia-Herrero, T. H. Johanson, K. D. Krewulak, C. K. Lau, R. S. Peacock, Z. Slavinskaya, H. J. Vogel, *Biometals*, *23*, 601–611.
16. J. H. Crosa, C. T. Walsh, *Microbiol. Mol. Biol. Rev.* **2002**, *66*, 223–249.
17. D. I. Chan, H. J. Vogel, *Biochem. J.* **2010**, *430*, 1–19.
18. G. L. Challis, *Chembiochem* **2005**, *6*, 601–611.
19. H. Drechsel, G. Winkelmann, in *Transition Metals in Microbial Metabolism*, Ed G. Winkelmann, C. J. Carrano, Harwood Academic, Amsterdam, 1997, pp. 1–49.
20. J. L. Furrer, D. N. Sanders, I. G. Hook-Barnard, M. A. McIntosh, *Mol. Microbiol.* **2002**, *44*, 1225–1234.
21. C. Bleuel, C. Grosse, N. Taudte, J. Scherer, D. Wesenberg, G. J. Krauss, D. H. Nies, G. Grass, *J. Bacteriol.* **2005**, *187*, 6701–6707.
22. T. Nishio, Y. Ishida, *Agric. Biol. Chem.* **1990**, *54*, 1837–1839.
23. T. J. Brickman, J. G. Hansel, M. J. Miller, S. K. Armstrong, *Biometals* **1996**, *9*, 191–203.
24. C. Ratledge, L. G. Dover, *Annu. Rev. Microbiol.* **2000**, *54*, 881–941.
25. A. D. Ferguson, E. Hofmann, J. W. Coulton, K. Diederichs, W. Welte, *Science* **1998**, *282*, 2215–2220.
26. D. Cobessi, H. Celia, N. Folschweiller, I. J. Schalk, M. A. Abdallah, F. Pattus, *J. Mol. Biol.* **2005**, *347*, 121–134.
27. W. W. Yue, S. Grizot, S. K. Buchanan, *J. Mol. Biol.* **2003**, *332*, 353–368.
28. K. D. Krewulak, H. J. Vogel, *Biochim. Biophys. Acta* **2008**, *1778*, 1781–1804.
29. K. P. Locher, B. Rees, R. Koebnik, A. Mitschler, L. Moulinier, J. P. Rosenbusch, D. Moras, *Cell* **1998**, *95*, 771–778.

30. S. Enz, S. Mahren, U. H. Stroeher, V. Braun, *J. Bacteriol.* **2000**, *182*, 637–646.
31. J. H. Crosa, *Microbiol. Mol. Biol. Rev.* **1997**, *61*, 319–336.
32. K. Schauer, D. A. Rodionov, H. de Reuse, *Trends Biochem. Sci.* **2008**, *33*, 330–338.
33. N. Noinaj, M. Guillier, T. J. Barnard, S. K. Buchanan, *Annu. Rev. Microbiol. 64*, 43–60.
34. A. Angerer, V. Braun, *Arch. Microbiol.* **1998**, *169*, 483–490.
35. Z. Chen, K. A. Lewis, R. K. Shultzaberger, I. G. Lyakhov, M. Zheng, B. Doan, G. Storz, T. D. Schneider, *Nucleic Acids Res.* **2007**, *35*, 6762–6777.
36. G. M. Young, K. Postle, *Mol. Microbiol.* **1994**, *11*, 943–954.
37. C. K. Stover, X. Q. Pham, A. L. Erwin, S. D. Mizoguchi, P. Warrener, M. J. Hickey, F. S. Brinkman, W. O. Hufnagle, D. J. Kowalik, M. Lagrou, R. L. Garber, L. Goltry, E. Tolentino, S. Westbrock-Wadman, Y. Yuan, L. L. Brody, S. N. Coulter, K. R. Folger, A. Kas, K. Larbig, R. Lim, K. Smith, D. Spencer, G. K. Wong, Z. Wu, I. T. Paulsen, J. Reizer, M. H. Saier, R. E. Hancock, S. Lory, M. V. Olson, *Nature* **2000**, *406*, 959–964.
38. K. Postle, R. A. Larsen, *Biometals* **2007**, *20*, 453–465.
39. K. G. Wooldridge, J. A. Morrissey, P. H. Williams, *J. Gen. Microbiol.* **1992**, *138*, 597–603.
40. P. I. Higgs, P. S. Myers, K. Postle, *J. Bacteriol.* **1998**, *180*, 6031–6038.
41. R. A. Larsen, M. G. Thomas, K. Postle, *Mol. Microbiol.* **1999**, *31*, 1809–1824.
42. J. C. Jaskula, T. E. Letain, S. K. Roof, J. T. Skare, K. Postle, *J. Bacteriol.* **1994**, *176*, 2326–2338.
43. C. S. Lopez, R. S. Peacock, J. H. Crosa, H. J. Vogel, *Biochem. J.* **2009**, *418*, 49–59.
44. A. Garcia-Herrero, R. S. Peacock, S. P. Howard, H. J. Vogel, *Mol. Microbiol.* **2007**, *66*, 872–889.
45. T. E. Letain, K. Postle, *Mol. Microbiol.* **1997**, *24*, 271–283.
46. R. A. Larsen, T. E. Letain, K. Postle, *Mol. Microbiol.* **2003**, *49*, 211–218.
47. W. A. Kaserer, X. Jiang, Q. Xiao, D. C. Scott, M. Bauler, D. Copeland, S. M. C. Newton, P. E. Klebba, *J. Bacteriol.* **2008**, *190*, 4001–4016.
48. M. G. Gresock, M. I. Savenkova, R. A. Larsen, A. A. Ollis, K. Postle, *Front. Microbiol.* **2011**, *2*, 206–206.
49. R. J. Kustusch, C. J. Kuehl, J. H. Crosa, *J. Bacteriol.*, *194*, 3250–3259.
50. T. E. Clarke, M. R. Rohrbach, L. W. Tari, H. J. Vogel, W. Koster, *Biometals* **2002**, *15*, 121–131.
51. D. M. Carter, I. R. Miousse, J. N. Gagnon, E. Martinez, A. Clements, J. Lee, M. A. Hancock, H. Gagnon, P. D. Pawelek, J. W. Coulton, *J. Biol. Chem.* **2006**, *281*, 35413–35424.
52. T. E. Clarke, S. Y. Ku, D. R. Dougan, H. J. Vogel, L. W. Tari, *Nat. Struct. Biol.* **2000**, *7*, 287–291.
53. M. Fontecave, J. Coves, J. L. Pierre, *Biometals* **1994**, *7*, 3–8.
54. I. Schroder, E. Johnson, S. de Vries, *FEMS Microbiol. Rev.* **2003**, *27*, 427–447.
55. J. M. Harrington, A. L. Crumbliss, *Biometals* **2009**, *22*, 679–689.
56. T. J. Brickman, M. A. McIntosh, *J. Biol. Chem.* **1992**, *267*, 12350–12355.
57. K. Hantke, *Curr. Opin. Microbiol.* **2001**, *4*, 172–177.
58. B. F. Matzanke, S. Anemuller, V. Schunemann, A. X. Trautwein, K. Hantke, *Biochemistry* **2004**, *43*, 1386–1392.
59. V. Braun, C. Herrmann, *J. Bacteriol.* **2007**, *189*, 6913–6918.
60. S. Mahren, H. Schnell, V. Braun, *Arch. Microbiol.* **2005**, *184*, 175–186.
61. V. Braun, F. Endriss, *Biometals* **2007**, *20*, 219–231.
62. D. Welz, V. Braun, *J. Bacteriol.* **1998**, *180*, 2387–2394.
63. A. Sauter, S. P. Howard, V. Braun, *J. Bacteriol.* **2003**, *185*, 5747–5754.
64. F. Biville, H. Cwerman, S. Letoffe, M. S. Rossi, V. Drouet, J. M. Ghigo, C. Wandersman, *Mol. Microbiol.* **2004**, *53*, 1267–1277.
65. V. Braun, S. Mahren, A. Sauter, *Biometals* **2005**, *18*, 507–517.
66. K. Poole, G. A. McKay, *Front. Biosci.* **2003**, *8*, d661–686.
67. M. Kammler, C. Schon, K. Hantke, *J. Bacteriol.* **1993**, *175*, 6212–6219.
68. K. Hantke, *Trends Microbiol.* **2003**, *11*, 192–195.

69. M. L. Cartron, S. Maddocks, P. Gillingham, C. J. Craven, S. C. Andrews, *Biometals* **2006**, *19*, 143–157.
70. T. C. Marlovits, W. Haase, C. Herrmann, S. G. Aller, V. M. Unger, *Proc. Natl. Acad. Sci. USA* **2002**, *99*, 16243–16248.
71. I. Stojiljkovic, M. Cobeljic, K. Hantke, *FEMS Microbiol. Lett.* **1993**, *108*, 111–116.
72. R. M. Tsolis, A. J. Baumler, F. Heffron, I. Stojiljkovic, *Infect. Immun.* **1996**, *64*, 4549–4556.
73. J. Velayudhan, N. J. Hughes, A. A. McColm, J. Bagshaw, C. L. Clayton, S. C. Andrews, D. J. Kelly, *Mol. Microbiol.* **2000**, *37*, 274–286.
74. M. Robey, N. P. Cianciotto, *Infect. Immun.* **2002**, *70*, 5659–5669.
75. R. E. Cowart, *Arch. Biochem. Biophys.* **2002**, *400*, 273–281.
76. S. K. Small, M. R. O'Brian, *J. Bacteriol.* **2011**, *193*, 4088–4094.
77. Y. C. Su, K. H. Chin, H. C. Hung, G. H. Shen, A. H. Wang, S. H. Chou, *Acta Crystallogr. Sect. F, Struct. Biol. Cryst. Commun.* **2011**, *66*, 636–642.
78. B. S. Finlin, R. N. Correll, C. Pang, S. M. Crump, J. Satin, D. A. Andres, *J. Biol. Chem.* **2006**, *281*, 23557–23566.
79. T. G. Cachero, A. D. Morielli, E. G. Peralta, *Cell* **1998**, *93*, 1077–1085.
80. I. R. Vetter, A. Wittinghofer, *Science* **2001**, *294*, 1299–1304.
81. D. C. Rees, E. Johnson, O. Lewinson, *Nat. Rev. Mol. Cell Biol.* **2009**, *10*, 218–227.
82. A. Guilfoyle, M. J. Maher, M. Rapp, R. Clarke, S. Harrop, M. Jormakka, *EMBO J.* **2009**, *28*, 2677–2685.
83. W. M. Oldham, H. E. Hamm, *Nat. Rev. Mol. Cell Biol.* **2008**, *9*, 60–71.
84. P. Scheerer, J. H. Park, P. W. Hildebrand, Y. J. Kim, N. Krauss, H. W. Choe, K. P. Hofmann, O. P. Ernst, *Nature* **2008**, *455*, 497–502.
85. G. Grass, S. Franke, N. Taudte, D. H. Nies, L. M. Kucharski, M. E. Maguire, C. Rensing, *J. Bacteriol.* **2005**, *187*, 1604–1611.
86. N. Zhao, J. Gao, C. A. Enns, M. D. Knutson, *J. Biol. Chem.* **2010**, *285*, 32141–32150.
87. C. Huynh, D. L. Sacks, N. W. Andrews, *J. Exp. Med.* **2006**, *203*, 2363–2375.
88. G. Grass, M. D. Wong, B. P. Rosen, R. L. Smith, C. Rensing, *J. Bacteriol.* **2002**, *184*, 864–866.
89. H. Gunshin, B. Mackenzie, U. V. Berger, Y. Gunshin, M. F. Romero, W. F. Boron, S. Nussberger, J. L. Gollan, M. A. Hediger, *Nature* **1997**, *388*, 482–488.
90. H. Makui, E. Roig, S. T. Cole, J. D. Helmann, P. Gros, M. F. M. Cellier, *Mol. Microbiol.* **2000**, *35*, 1065–1078.
91. D. G. Kehres, M. L. Zaharik, B. B. Finlay, M. E. Maguire, *Mol. Microbiol.* **2000**, *36*, 1085–1100.
92. H. A. H. Haemig, P. J. Moen, R. J. Brooker, *Biochemistry* **2010**, *49*, 4662–4671.
93. M. A. Su, C. C. Trenor, J. C. Fleming, M. D. Fleming, N. C. Andrews, *Blood* **1998**, *92*, 2157–2163.
94. S. A. Lieser, T. C. Davis, J. D. Helmann, S. M. Cohen, *Biochemistry* **2003**, *42*, 12634–12642.
95. L. S. Waters, M. Sandoval, G. Storz, *J. Bacteriol.* **2011**, *193*, 5887–5897.
96. A. Anjem, S. Varghese, J. A. Imlay, *Mol. Microbiol.* **2009**, *72*, 844–858.
97. J. Cao, M. R. Woodhall, J. Alvarez, M. L. Cartron, S. C. Andrews, *Mol. Microbiol.* **2007**, *65*, 857–875.
98. C. Grosse, J. Scherer, D. Koch, M. Otto, N. Taudte, G. Grass, *Mol. Microbiol.* **2006**, *62*, 120–131.
99. A. D. Hernday, B. A. Braaten, G. Broitman-Maduro, P. Engelberts, D. A. Low, *Mol. Cell* **2004**, *16*, 537–547.
100. N. Baichoo, T. Wang, R. Ye, J. D. Helmann, *Mol. Microbiol.* **2002**, *45*, 1613–1629.
101. J. Ollinger, K.-B. Song, H. Antelmann, M. Hecker, J. D. Helmann, *J. Bacteriol.* **2006**, *188*, 3664–3673.
102. M. B. Rajasekaran, S. Nilapwar, S. C. Andrews, K. A. Watson, *Biometals* **2010**, *23*, 1–17.
103. A. Sturm, A. Schierhorn, U. Lindenstrauss, H. Lilie, T. Bruser, *J. Biol. Chem.* **2006**, *281*, 13972–13978.

104. A. C. Chan, T. I. Doukov, M. Scofield, S. A. Tom-Yew, A. B. Ramin, J. K. Mackichan, E. C. Gaynor, M. E. Murphy, *J. Mol. Biol.* **2010**, *401*, 590–604.
105. B. Janvier, C. Constantinidou, P. Aucher, Z. V. Marshall, C. W. Penn, J. L. Fauchere, *Res. Microbiol.* **1998**, *149*, 95–107.
106. K. Holmes, F. Mulholland, B. M. Pearson, C. Pin, J. McNicholl-Kennedy, J. M. Ketley, J. M. Wells, *Microbiology* **2005**, *151*, 243–257.
107. A. H. van Vliet, K. G. Wooldridge, J. M. Ketley, *J. Bacteriol.* **1998**, *180*, 5291–5298.
108. J. S. Ikeda, A. Janakiraman, D. G. Kehres, M. E. Maguire, J. M. Slauch, *J. Bacteriol.* **2005**, *187*, 912–922.
109. D. G. Kehres, A. Janakiraman, J. M. Slauch, M. E. Maguire, *J. Bacteriol.* **2002**, *184*, 3151–3158.
110. D. G. Zhou, W. D. Hardt, J. E. Galan, *Infect. Immun.* **1999**, *67*, 1974–1981.
111. C. R. Fisher, N. M. Davies, E. E. Wyckoff, Z. Feng, E. V. Oaks, S. M. Payne, *Infect. Immun.* **2009**, *77*, 1992–1999.
112. P. Adhikari, S. A. Berish, A. J. Nowalk, K. L. Veraldi, S. A. Morse, T. A. Mietzner, *J. Bacteriol.* **1996**, *178*, 2145–2149.
113. A. Angerer, S. Gaisser, V. Braun, *J. Bacteriol.* **1990**, *172*, 572–578.
114. S. W. Bearden, R. D. Perry, *Mol. Microbiol.* **1999**, *32*, 403–414.
115. T. Tosha, H. L. Ng, O. Bhattasali, T. Alber, E. C. Theil, *J. Am. Chem. Soc.* **2010**, *132*, 14562–14569.
116. A. Ilari, S. Stefanini, E. Chiancone, D. Tsernoglou, *Nat. Struct. Biol.* **2000**, *7*, 38–43.
117. G. H. Gauss, M. A. Reott, E. R. Rocha, M. J. Young, T. Douglas, C. J. Smith, C. M. Lawrence, *J. Bacteriol.* **2012**, *194*, 15–27.
118. H. Abdul-Tehrani, A. J. Hudson, Y. S. Chang, A. R. Timms, C. Hawkins, J. M. Williams, P. M. Harrison, J. R. Guest, S. C. Andrews, *J. Bacteriol.* **1999**, *181*, 1415–1428.
119. E. C. Theil, H. Takagi, G. W. Small, L. He, A. R. Tipton, D. Danger, *Inorg. Chim. Acta* **2000**, *297*, 242–251.
120. M. A. Carrondo, *EMBO J.* **2003**, *22*, 1959–1968.
121. S. C. Andrews, N. E. Le Brun, V. Barynin, A. J. Thomson, G. R. Moore, J. R. Guest, P. M. Harrison, *J. Biol. Chem.* **1995**, *270*, 23268–23274.
122. M. A. Quail, P. Jordan, J. M. Grogan, J. N. Butt, M. Lutz, A. J. Thomson, S. C. Andrews, J. R. Guest, *Biochem. Biophys. Res. Commun.* **1996**, *229*, 635–642.
123. S. Baaghil, A. Lewin, G. R. Moore, N. E. Le Brun, *Biochemistry* **2003**, *42*, 14047–14056.
124. J. L. Smith, *Crit. Rev. Microbiol.* **2004**, *30*, 173–185.
125. J. Velayudhan, M. Castor, A. Richardson, K. L. Main-Hester, F. C. Fang, *Mol. Microbiol.* **2007**, *63*, 1495–1507.
126. C. J. Sund, E. R. Rocha, A. O. Tzianabos, W. G. Wells, J. M. Gee, M. A. Reott, D. P. O'Rourke, C. J. Smith, *Mol. Microbiol.* **2008**, *67*, 129–142.
127. T. Haikarainen, A. C. Papageorgiou, *Cell Mol. Life Sci.* **2010**, *67*, 341–351.
128. L. N. Calhoun, Y. M. Kwon, *J. Appl. Microbiol.* **2011**, *110*, 375–386.
129. S. Nair, S. E. Finkel, *J. Bacteriol.* **2004**, *186*, 4192–4198.
130. A. Ilari, P. Ceci, D. Ferrari, G. Rossi, E. Chiancone, *J. Biol. Chem.* **2002**, *5*, 5.
131. G. Zhao, P. Arosio, N. D. Chasteen, *Biochemistry* **2006**, *45*, 3429–3436.
132. X. Wei, H. Mingjia, L. Xiufeng, G. Yang, W. Qingyu, *IUBMB Life* **2007**, *59*, 675–681.
133. F. C. Neidhardt, J. L. Ingraham, M. Schaechter, *Physiology of the Bacterial Cell: A Molecular Approach*, Sinauer Ass, Inc., Sunderland, Massachusetts, 1990.
134. A. Bagg, J. B. Neilands, *Biochemistry* **1987**, *26*, 5471–5477.
135. W. Y. Tu, S. Pohl, J. Gray, N. J. Robinson, C. R. Harwood, K. J. Waldron, *J. Bacteriol.* **2012**, *194*, 932–940.
136. A. Cvetkovic, A. L. Menon, M. P. Thorgersen, J. W. Scott, F. L. Poole, 2nd, F. E. Jenney, Jr., W. A. Lancaster, J. L. Praissman, S. Shanmukh, B. J. Vaccaro, S. A. Trauger, E. Kalisiak, J. V. Apon, G. Siuzdak, S. M. Yannone, J. A. Tainer, M. W. Adams, *Nature* **2010**, *466*, 779–782.

137. C. Andreini, I. Bertini, A. Rosato, *Acc. Chem. Res.* **2009**, *42*, 1471–1479.
138. M. Coy, J. B. Neilands, *Biochemistry* **1991**, *30*, 8201–8210.
139. L. Escolar, J. Perez-Martin, V. de Lorenzo, *J. Biol. Chem.* **2000**, *275*, 24709–24714.
140. E. W. Althaus, C. E. Outten, K. E. Olson, H. Cao, T. V. O'Halloran, *Biochemistry* **1999**, *38*, 6559–6569.
141. A. C. Lewin, P. A. Doughty, L. Flegg, G. R. Moore, S. Spiro, *Microbiology* **2002**, *148*, 2449–2456.
142. V. De Lorenzo, M. Herrero, F. Giovannini, J. B. Neilands, *Eur. J. Biochem.* **1988**, *173*, 537–546.
143. N. Baichoo, J. D. Helmann, *J. Bacteriol.* **2002**, *184*, 5826–5832.
144. J. L. Lavrrar, C. A. Christoffersen, M. A. McIntosh, *J. Mol. Biol.* **2002**, *322*, 983–995.
145. M. Zheng, B. Doan, T. D. Schneider, G. Storz, *J. Bacteriol.* **1999**, *181*, 4639–4643.
146. E. C. Niederhoffer, C. M. Naranjo, K. L. Bradley, J. A. Fee, *J. Bacteriol.* **1990**, *172*, 1930–1938.
147. Y. Liu, S. C. Bauer, J. A. Imlay, *J. Bacteriol.* **2011**, *193*, 2186–2196.
148. P. Mukhopadhyay, M. Zheng, L. A. Bedzyk, R. A. LaRossa, G. Storz, *Proc. Natl. Acad. Sci. USA* **2004**, *101*, 745–750.
149. B. D'Autreaux, D. Touati, B. Bersch, J. M. Latour, I. Michaud-Soret, *Proc. Natl. Acad. Sci. USA* **2002**, *99*, 16619–16624.
150. I. Stojiljkovic, A. J. Baumler, K. Hantke, *J. Mol. Biol.* **1994**, *236*, 531–545.
151. N. Vassinova, D. Kozyrev, *Microbiology* **2000**, *146*, 3171–3182.
152. D. Touati, M. Jacques, B. Tardat, L. Bouchard, S. Despied, *J. Bacteriol.* **1995**, *177*, 2305–2314.
153. K. Hantke, *Mol. Gen. Genet.* **1987**, *210*, 135–139.
154. U. A. Ochsner, P. J. Wilderman, A. I. Vasil, M. L. Vasil, *Mol. Microbiol.* **2002**, *45*, 1277–1287.
155. R. Grifantini, S. Sebastian, E. Frigimelica, M. Draghi, E. Bartolini, A. Muzzi, R. Rappuoli, G. Grandi, C. A. Genco, *Proc. Natl. Acad. Sci. USA* **2003**, *100*, 9542–9547.
156. X. F. Wan, N. C. Verberkmoes, L. A. McCue, D. Stanek, H. Connelly, L. J. Hauser, L. Wu, X. Liu, T. Yan, A. Leaphart, R. L. Hettich, J. Zhou, D. K. Thompson, *J. Bacteriol.* **2004**, *186*, 8385–8400.
157. D. K. Thompson, A. S. Beliaev, C. S. Giometti, S. L. Tollaksen, T. Khare, D. P. Lies, K. H. Nealson, H. Lim, J. Yates, C. C. Brandt, J. M. Tiedje, J. Zhou, *Appl. Environ. Microbiol.* **2002**, *68*, 881–892.
158. F. D. Ernst, S. Bereswill, B. Waidner, J. Stoof, U. Mader, J. G. Kusters, E. J. Kuipers, M. Kist, A. H. M. van Vliet, G. Homuth, *Microbiology* **2005**, *151*, 533–546.
159. T. F. Ducey, M. B. Carson, J. Orvis, A. P. Stintzi, D. W. Dyer, *J. Bacteriol.* **2005**, *187*, 4865–4874.
160. A. R. Mey, E. E. Wyckoff, V. Kanukurthy, C. R. Fisher, S. M. Payne, *Infect. Immun.* **2005**, *73*, 8167–8178.
161. H. Gancz, S. Censini, D. S. Merrell, *Infect. Immun.* **2006**, *74*, 602–614.
162. I. Brune, H. Werner, A. T. Huser, J. Kalinowski, A. Puhler, A. Tauch, *BMC Genomics* **2006**, *7*, 21–21.
163. J. Wennerhold, M. Bott, *J. Bacteriol.* **2006**, *188*, 2907–2918.
164. A. Danielli, D. Roncarati, I. Delany, V. Chiarini, R. Rappuoli, V. Scarlato, *J. Bacteriol.* **2006**, *188*, 4654–4662.
165. J. Yang, I. Sangwan, M. R. O'Brian, *Mol. Genet. Genomics* **2006**, *276*, 555–564.
166. K. Deng, R. J. Blick, W. Liu, E. J. Hansen, *Infect. Immun.* **2006**, *74*, 4224–4236.
167. M. Miethke, H. Westers, E. J. Blom, O. P. Kuipers, M. A. Marahiel, *J. Bacteriol.* **2006**, *188*, 8655–8657.
168. K. S. Bender, H.-C. B. Yen, C. L. Hemme, Z. Yang, Z. He, Q. He, J. Zhou, K. H. Huang, E. J. Alm, T. C. Hazen, A. P. Arkin, J. D. Wall, *Appl. Environ. Microbiol.* **2007**, *73*, 5389–5400.

169. P. A. Zaini, A. C. Fogaca, F. G. N. Lupo, H. I. Nakaya, R. Z. N. Vencio, A. M. da Silva, *J. Bacteriol.* **2008**, *190*, 2368–2378.
170. Y. Yang, D. P. Harris, F. Luo, L. Wu, A. B. Parsons, A. V. Palumbo, J. Zhou, *BMC Genomics* **2008**, *9, Suppl 1*, S11.
171. P. A. Bronstein, M. J. Filiatrault, C. R. Myers, M. Rutzke, D. J. Schneider, S. W. Cartinhour, *BMC Microbiol.* **2008**, *8*, 209–209.
172. L. A. Jackson, T. F. Ducey, M. W. Day, J. B. Zaitshik, J. Orvis, D. W. Dyer, *J. Bacteriol.* **2010**, *192*, 77–85.
173. N. Ledala, M. Sengupta, A. Muthaiyan, B. J. Wilkinson, R. K. Jayaswal, *Appl. Environ. Microbiol.* **2010**, *76*, 406–416.
174. M. Lo, G. L. Murray, C. A. Khoo, D. A. Haake, R. L. Zuerner, B. Adler, *Infect. Immun.* **2010**, *78*, 4850–4859.
175. C. J. Kershaw, N. L. Brown, C. Constantinidou, M. D. Patel, J. L. Hobman, *Microbiology* **2005**, *151*, 1187–1198.
176. T. K. Sigdel, J. A. Easton, M. W. Crowder, *J. Bacteriol.* **2006**, *188*, 6709–6713.
177. C. A. White-Ziegler, A. J. Malhowski, S. Young, *J. Bacteriol.* **2007**, *189*, 5429–5440.
178. A. Nandal, C. C. Huggins, M. R. Woodhall, J. McHugh, F. Rodriguez-Quinones, M. A. Quail, J. R. Guest, S. C. Andrews, *Mol. Microbiol.* **2010**, *75*, 637–657.
179. E. Masse, M. Arguin, *Trends Biochem. Sci.* **2005**, *30*, 462–468.
180. J. W. Lee, J. D. Helmann, *Biometals* **2007**, *20*, 485–499.
181. S. M. Payne, E. E. Wyckoff, E. R. Murphy, A. G. Oglesby, M. L. Boulette, N. M. Davies, *Biometals* **2006**, *19*, 173–180.
182. P. J. Wilderman, N. A. Sowa, D. J. FitzGerald, P. C. FitzGerald, S. Gottesman, U. A. Ochsner, M. L. Vasil, *Proc. Natl. Acad. Sci. USA* **2004**, *101*, 9792–9797.
183. A. Gaballa, J. D. Helmann, *J. Bacteriol.* **1998**, *180*, 5815–5821.
184. V. Duarte, J. M. Latour, *Mol. Biosyst.* **2010**, *6*, 316–323.
185. Z. Ma, J. W. Lee, J. D. Helmann, *Nucleic Acids Res.* **2011**, *39*, 5036–5044.
186. J. D. Helmann, M. F. W. Wu, A. Gaballa, P. A. Kobel, M. M. Morshedi, P. Fawcett, C. Paddon, *J. Bacteriol.* **2003**, *185*, 243–253.
187. E. Diaz-Mireles, M. Wexler, G. Sawers, D. Bellini, J. D. Todd, A. W. Johnston, *Microbiology* **2004**, *150*, 1447–1456.
188. B.-E. Ahn, J. Cha, E.-J. Lee, A.-R. Han, C. J. Thompson, J.-H. Roe, *Mol. Microbiol.* **2006**, *59*, 1848–1858.
189. Z. Qi, I. Hamza, M. R. O'Brian, *Proc. Natl. Acad. Sci. USA* **1999**, *96*, 13056–13061.
190. H. Chen, R. Wu, G. Xu, X. Fang, X. Qiu, H. Guo, B. Tian, Y. Hua, *Biochem. Biophys. Res. Commun.* **2010**, *396*, 413–418.
191. M. P. Schmitt, B. G. Talley, R. K. Holmes, *Infect. Immun.* **1997**, *65*, 5364–5367.
192. J. Frunzke, C. Gatgens, M. Brocker, M. Bott, *J. Bacteriol.* **2011**, *193*, 1212–1221.
193. J. A. D'Aquino, J. Tetenbaum-Novatt, A. White, F. Berkovitch, D. Ringe, *Proc. Natl. Acad. Sci. USA* **2005**, *102*, 18408–18413.
194. C. Massonet, V. Pintens, R. Merckx, J. Anne, E. Lammertyn, J. Van Eldere, *BMC Microbiol.* **2006**, *6*, 103.
195. B. Gold, G. M. Rodriguez, S. A. Marras, M. Pentecost, I. Smith, *Mol. Microbiol.* **2001**, *42*, 851–865.
196. G. M. Rodriguez, M. I. Voskuil, B. Gold, G. K. Schoolnik, I. Smith, *Infect. Immun.* **2002**, *70*, 3371–3381.
197. D. M. Bodenmiller, S. Spiro, *J. Bacteriol.* **2006**, *188*, 874–881.
198. T.-C. Chao, J. Buhrmester, N. Hansmeier, A. Puhler, S. Weidner, *Appl. Environ. Microbiol.* **2005**, *71*, 5969–5982.
199. K. H. Yeoman, A. R. J. Curson, J. D. Todd, G. Sawers, A. W. B. Johnston, *Microbiology* **2004**, *150*, 4065–4074.
200. J. D. Todd, M. Wexler, G. Sawers, K. H. Yeoman, P. S. Poole, A. W. B. Johnston, *Microbiology* **2002**, *148*, 4059–4071.

201. J. S. Gunn, *Trends Microbiol.* **2008**, *16*, 284–290.
202. K. Nishino, F.-F. Hsu, J. Turk, M. J. Cromie, M. M. S. M. Wosten, E. A. Groisman, *Mol. Microbiol.* **2006**, *61*, 645–654.
203. A. Kato, T. Latifi, E. A. Groisman, *Proc. Natl. Acad. Sci. USA* **2003**, *100*, 4706–4711.
204. C. M. Herrera, J. V. Hankins, M. S. Trent, *Mol. Microbiol.* **2010**, *76*, 1444–1460.
205. D. Hagiwara, T. Yamashino, T. Mizuno, *Biosci. Biotechnol. Biochem.* **2004**, *68*, 1758–1767.
206. S. C. Andrews, *Mol. Microbiol.* **2011**, *80*, 286–289.
207. I. Compan, D. Touati, *J. Bacteriol.* **1993**, *175*, 1687–1696.
208. J. E. Martin, J. A. Imlay, *Mol. Microbiol.* **2011** *80*, 319–334.
209. J. M. Sobota, J. A. Imlay, *Proc. Natl. Acad. Sci. USA* **2011** *108*, 5402–5407.
210. C. M. Litwin, S. B. Calderwood, *Clin. Microbiol. Rev.* **1993**, *6*, 137–149.
211. C. N. Cornelissen, P. F. Sparling, *Mol. Microbiol.* **1994**, *14*, 843–850.
212. C. Ronpirin, A. E. Jerse, C. N. Cornelissen, *Infect. Immun.* **2001**, *69*, 6336–6347.
213. A. J. DeRocco, M. K. Yost-Daljev, C. D. Kenney, C. N. Cornelissen, *Biometals* **2009**, *22*, 439–451.
214. C. D. Kenney, C. N. Cornelissen, *J. Bacteriol.* **2002**, *184*, 6138–6145.
215. C. J. Siburt, P. L. Roulhac, K. D. Weaver, J. M. Noto, T. A. Mietzner, C. N. Cornelissen, M. C. Fitzgerald, A. L. Crumbliss, *Metallomics* **2009**, *1*, 249–255.
216. P. L. Roulhac, K. D. Weaver, P. Adhikari, D. S. Anderson, P. D. DeArmond, T. A. Mietzner, A. L. Crumbliss, M. C. Fitzgerald, *Biochemistry* **2008**, *47*, 4298–4305.
217. A. G. Khan, S. R. Shouldice, S. D. Kirby, R.-h. Yu, L. W. Tari, A. B. Schryvers, *Biochem. J.* **2007**, *404*, 217–225.
218. H. R. Strange, T. A. Zola, C. N. Cornelissen, *Infect. Immun.* **2011**, *79*, 267–278.
219. N. Wolff, N. Izadi-Pruneyre, J. Couprie, M. Habeck, J. Linge, W. Rieping, C. Wandersman, M. Nilges, M. Delepierre, A. Lecroisey, *J. Mol. Biol.* **2008**, *376*, 517–525.
220. S. Krieg, F. Huche, K. Diederichs, N. Izadi-Pruneyre, A. Lecroisey, C. Wandersman, P. Delepelaire, W. Welte, *Proc. Natl. Acad. Sci. USA* **2009**, *106*, 1045–1050.
221. N. Benevides-Matos, C. Wandersman, F. Biville, *J. Bacteriol.* **2008**, *190*, 21–27.
222. N. Benevides-Matos, F. Biville, *Microbiology* **2010**, *156*, 1749–1757.
223. M. Shepherd, M. D. Heath, R. K. Poole, *Biochemistry* **2007**, *46*, 5030–5037.
224. S. Letoffe, G. Heuck, P. Delepelaire, N. Lange, C. Wandersman, *Proc. Natl. Acad. Sci. USA* **2009**, *106*, 11719–11724.
225. D. Mattle, A. Zeltina, J.-S. Woo, B. A. Goetz, K. P. Locher, *J. Mol. Biol.* **2010**, *404*, 220–231.
226. B. Vergauwen, J. Elegheert, A. Dansercoer, B. Devreese, S. N. Savvides, *Proc. Natl. Acad. Sci. USA* **2010**, *107*, 13270–13275.
227. D. J. Morton, P. W. Whitby, H. Jin, Z. Ren, T. L. Stull, *Infect. Immun.* **1999**, *67*, 2729–2739.
228. L. D. Cope, R. Yogev, U. Muller-Eberhard, E. J. Hansen, *J. Bacteriol.* **1995**, *177*, 2644–2653.
229. P. W. Whitby, T. M. Vanwagoner, T. W. Seale, D. J. Morton, T. L. Stull, *J. Bacteriol.* **2006**, *188*, 5640–5645.
230. C. Fournier, A. Smith, P. Delepelaire, *Mol. Microbiol.* **2011**, *80*, 133–148.
231. L. D. Cope, S. E. Thomas, Z. Hrkal, E. J. Hansen, *Infect. Immun.* **1998**, *66*, 4511–4516.
232. J. C. Grigg, C. L. Vermeiren, D. E. Heinrichs, M. E. P. Murphy, *Mol. Microbiol.* **2007**, *63*, 139–149.
233. S. K. Mazmanian, E. P. Skaar, A. H. Gaspar, M. Humayun, P. Gornicki, J. Jelenska, A. Joachmiak, D. M. Missiakas, O. Schneewind, *Science* **2003**, *299*, 906–909.
234. R. M. Pilpa, S. A. Robson, V. A. Villareal, M. L. Wong, M. Phillips, R. T. Clubb, *J. Biol. Chem.* **2009**, *284*, 1166–1176.
235. M. Watanabe, Y. Tanaka, A. Suenaga, M. Kuroda, M. Yao, N. Watanabe, F. Arisaka, T. Ohta, I. Tanaka, K. Tsumoto, *J. Biol. Chem.* **2008**, *283*, 28649–28659.
236. M. Pluym, N. Muryoi, D. E. Heinrichs, M. J. Stillman, *J. Inorg. Biochem.* **2008**, *102*, 480–488.

237. V. J. Torres, G. Pishchany, M. Humayun, O. Schneewind, E. P. Skaar, *J. Bacteriol.* **2006**, *188*, 8421–8429.
238. V. A. Villareal, R. M. Pilpa, S. A. Robson, E. A. Fadeev, R. T. Clubb, *J. Biol. Chem.* **2008**, *283*, 31591–31600.
239. D. B. Friedman, D. L. Stauff, G. Pishchany, C. W. Whitwell, V. J. Torres, E. P. Skaar, *PLoS Pathog* **2006**, *2*, e87.
240. B. R. Byers, J. E. L. Arceneaux, in *Metal Ions in Biological Systems: Iron Transport and Storage in Microorganisms, Plants, and Animals*, Eds A. Sigel, H. Sigel, Marcel Dekker, Inc., New York, NY, 1998, Vol. 35, pp. 37–66.
241. P. J. Warner, P. H. Williams, A. Bindereif, J. B. Neilands, *Infect. Immun.* **1981**, *33*, 540–545.
242. F. Gibson, D. I. Magrath, *Biochim. Biophys. Acta* **1969**, *192*, 175–184.
243. J. B. Neilands, *Microbiol. Sci.* **1984**, *1*, 9–14.
244. P. H. Williams, N. H. Carbonetti, *Infect. Immun.* **1986**, *51*, 942–947.
245. S. I. Muller, M. Valdebenito, K. Hantke, *Biometals* **2009**, *22*, 691–695.
246. K. Hantke, G. Nicholson, W. Rabsch, G. Winkelmann, *Proc. Natl. Acad. Sci USA* **2003**, *100*, 3677–3682.
247. N. Noinaj, N. C. Easley, M. Oke, N. Mizuno, J. Gumbart, E. Boura, A.N. Steere, O. Zak, P. Aisen, E. Tajkhorshid, R.W. Evans, A.R. Gorringe, A.B. Mason, A.C. Steven, S.K. Buchanan, *Nature* **2012**, *483*, 53–58.

Chapter 8
The Iron Metallome in Eukaryotic Organisms

Adrienne C. Dlouhy and Caryn E. Outten

Contents

A.C. Dlouhy • C.E. Outten (✉)
Department of Chemistry and Biochemistry,
University of South Carolina, Columbia, SC 29208, USA
e-mail: outten@mailbox.sc.edu

L. Banci (ed.), *Metallomics and the Cell*, Metal Ions in Life Sciences 12,
DOI 10.1007/978-94-007-5561-1_8, © Springer Science+Business Media Dordrecht 2013

Abstract This chapter is focused on the iron metallome in eukaryotes at the cellular and subcellular level, including properties, utilization in metalloproteins, trafficking, storage, and regulation of these processes. Studies in the model eukaryote *Saccharomyces cerevisiae* and mammalian cells will be highlighted. The discussion of iron properties will center on the speciation and localization of intracellular iron as well as the cellular and molecular mechanisms for coping with both low iron bioavailability and iron toxicity. The section on iron metalloproteins will emphasize heme, iron-sulfur cluster, and non-heme iron centers, particularly their cellular roles and mechanisms of assembly. The section on iron uptake, trafficking, and storage will compare methods used by yeast and mammalian cells to import iron, how this iron is brought into various organelles, and types of iron storage proteins. Regulation of these processes will be compared between yeast and mammalian cells at the transcriptional, post-transcriptional, and post-translational levels.

Keywords eukaryote • heme • iron metallome • iron-sulfur cluster • iron trafficking • metal homeostasis

Please cite as: *Met. Ions Life Sci.* 12 (2013) 241–278

1 Introduction

Organisms use a variety of transition metals as catalytic centers in proteins, including iron, copper, manganese, and zinc. Iron is well suited to redox reactions due to its capability to act as both an electron donor and acceptor. In eukaryotic cells, iron is a cofactor for a wide variety of metalloproteins involved in energy metabolism, oxygen binding, DNA biosynthesis and repair, synthesis of biopolymers, cofactors, and vitamins, drug metabolism, antioxidant function, and many others. Because iron is so important for survival, organisms utilize several techniques to optimize uptake and storage to ensure maintenance of sufficient levels for cellular requirements. However, the redox properties of iron also make it extremely toxic if cells have excessive amounts. Free iron can catalyze the formation of reactive oxygen species such as the hydroxyl radical, which in turn can damage proteins, lipids, membranes, and DNA. Cells must maintain a delicate balance between iron deficiency and iron overload that involves coordinated control at the transcriptional, post-transcriptional, and post-translational levels to help fine tune iron utilization and iron trafficking.

This chapter will focus on the iron metallome in eukaryotes from uptake to regulation. The physicochemical properties will be investigated, including bioavailability for uptake, chemical forms and subcellular locations of intracellular iron, and mechanisms of toxicity. In addition, the function and formation of iron metalloproteins such as heme and iron-sulfur clusters will be considered. Uptake systems, iron chaperones, and storage systems in yeast and mammalian cells will be compared. Regulation of these processes at the gene, mRNA, and proteins levels will be examined, again comparing yeast and mammalian cells.

2 Physicochemical Properties of the Iron Metallome

2.1 Intracellular Concentrations, Oxidation State, and Speciation

Iron is the most abundant metal on the Planet Earth as a whole, thus it is not surprising that almost all organisms have evolved to exploit the unique chemical properties of this ubiquitous transition metal. Iron primarily exists in either the ferrous (Fe^{2+}) or ferric (Fe^{3+}) oxidation state in biological systems. Due to its critical role in cell metabolism, iron constitutes a significant portion of the eukaryotic metallome [1]. Intracellular iron concentrations vary with cell type, environmental conditions, and disease state. The iron concentration of human erythroid cells was measured at 300–400 μM [2], while isolated rat hepatocytes maintain iron concentrations close to 1 mM [3]. Iron overload diseases caused by mutations in iron handling proteins can lead to 10- to 20-fold increases in these intracellular iron levels in specific tissues [4–6]. The local bioavailability of iron also strongly influences intracellular concentrations. For example, analysis of the single-celled model eukaryote *S. cerevisiae* demonstrated intracellular iron concentrations ranging from 250 μM to 600 μM depending on the iron content of the growth medium [1,7].

To better study the iron metallome, biophysical probes such as Mössbauer and electron paramagnetic resonance (EPR) have been recently employed to measure not only the absolute iron concentration, but also the types of iron and how this varies within specific organelles [8]. Lindahl and colleagues have used an integrated biophysical approach to characterize the iron speciation in *S. cerevisiae* whole cells and organelles under several growth conditions [7,9–11]. These studies clearly demonstrate that the mitochondria and vacuole are the two central hubs of iron metabolism in this organism. In general, yeast mitochondria contain 700–800 μM Fe. In respiring cells, most of this mitochondrial iron is present as the prosthetic groups of the respi-

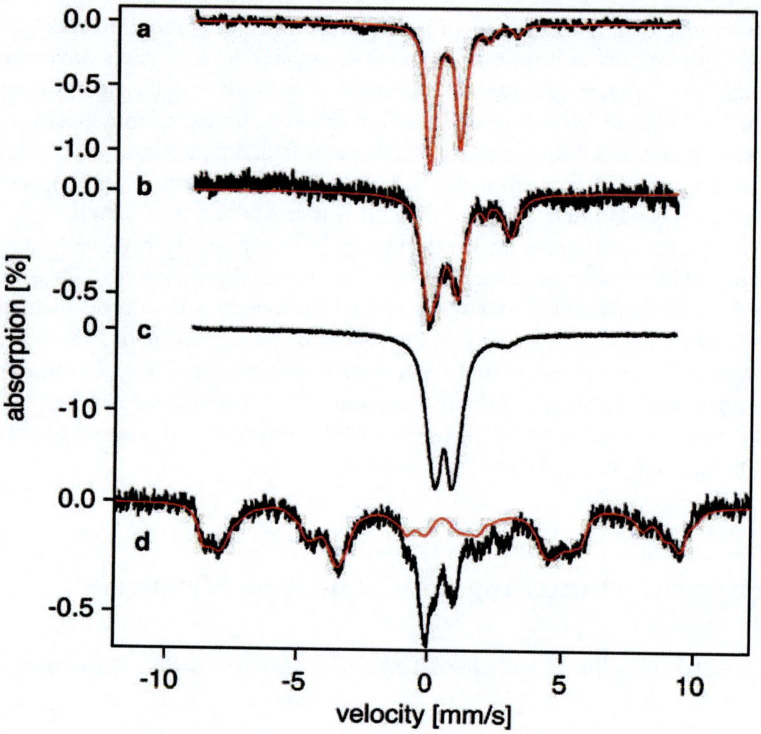

Figure 1 Mössbauer spectra comparing differing iron species found in (**a**) respiring mitochondria, (**b**) fermenting mitochondria, (**c**) Atm1-depleted mitochondria, and (**d**) whole fermenting yeast cells. Reprinted from ref [8] with permission from Elsevier; copyright 2011.

ratory complexes (~70% $[4Fe\text{-}4S]^{2+}$ clusters and heme centers), with the remaining iron present as $[2Fe\text{-}2S]^{1+}$ clusters in enzymes and as non-heme, high spin Fe^{2+} ions. Conversely, in fermenting cells the iron from respiratory complexes is reduced to ~30% of the total iron, there is an increase in non-heme iron, and the appearance of ferric phosphate nanoparticles. Mutations in Fe-S cluster assembly and trafficking proteins leads to increased concentration of these nanoparticles with a concomitant rise in reactive oxygen species [7,11]. The differences in iron speciation in each of these different situations are readily apparent from the Mössbauer spectra (Figure 1). The other major iron repository in yeast is the vacuole. Vacuoles isolated from fermenting yeast contain an average of 220 μM Fe as ferric iron, which is expected given the acidic environment of this organelle (pH ~5). Vacuolar iron is present in both a soluble Fe(III) complex and insoluble, magnetically-interacting Fe(III) nanoparticles, which are proposed to interconvert based on changes in pH [9].

Characterization of iron speciation in other organelles and organisms is still in the initial stages as most published studies are based on a single technique, instead of verification by several methods. In addition, various techniques have been used to study different organisms, making comparison of results challenging. Iron can be found in a variety of different forms based on location (and thus pH and redox potential), available ligands, and cellular need. The integrated approach described above is one of the most promising for studying the iron metallome: by combining Mössbauer, EPR, X-ray absorption spectroscopy (XAS), electronic absorption spectroscopy, and electron microscopy, one can resolve different groups of iron species (such as Fe-S clusters, hemes, and nanoparticles) at relatively low concentrations. In particular, XAS techniques such as X-ray absorption near edge structure (XANES) and extended X-ray absorption fine structure (EXAFS) give information about oxidation state, geometry, and ligation. All forms of iron can be detected and quantified, so changing levels of species can also be monitored [12].

Another promising method for future studies is X-ray fluorescence microscopy (XRFM), which provides information about metal distribution, oxidation state, and coordination. XRFM offers high spatial resolution of biological samples by detection of emitted X-rays from the sample after irradiation. Pairing this with XANES can provide more information about iron speciation and subcellular distribution. A combination of XRFM and XAS studies on brain tissue from Alzheimer's disease patients showed an increased concentration of iron found mainly in the oxidized form [13]. Fluorescence intensity from XRFM studies is directly proportional to the element concentration, providing some quantitative analysis of samples. Quantification of transition metals in cells and organelles can also be accomplished using particle-induced X-ray emission (PIXE). Like XRFM, PIXE analysis detects X-rays that are emitted which are characteristic of elements in the sample. This technique is capable of detecting and quantifying trace elements (including Fe, Mn, Zn, and Cu) in the μg/g range [12].

2.2 *Subcellular Distribution*

As mentioned above, the majority of cellular iron is found in the mitochondria and the cytosol for utilization in iron-dependent proteins. While yeast stores excess iron in the vacuole, mammals express iron storage proteins such as ferritin and mitochondrial ferritin for this purpose. In addition, iron is recycled in lysosomes after iron-containing proteins are degraded. For example, human liver and spleen cells from patients with hemochromatosis (an iron overload disease) were found to contain iron-loaded lysosomes (siderosomes), hemosiderin (a degradation product of ferritin), and ferritin [14]. There has not been a significant amount of research focused on the concentration and chemical nature of iron in the endoplasmic reticulum (ER). However, the existence of iron pools in this organelle is likely since a number of heme and non-heme iron proteins are located in the ER. For more information on iron-containing organelles and iron storage proteins, refer to Sections 4.2 and 4.3.

2.3 Bioavailability, Intracellular Labile Iron Pools

2.3.1 Bioavailability

Although iron is one of the most abundant elements on Earth, the environment is usually oxygenated, non-acidic, and aqueous. Under these conditions, extracellular iron is predominantly found in the poorly soluble ferric (Fe^{3+}) state. One way that organisms such as yeast improve iron bioavailability is by acidifying the local environment. The solubility of ferric iron is pH-dependent, changing from 10^{-18} M at pH 7 to 10^{-3} M at pH 2. By lowering the pH of the surrounding environment, organisms facilitate solubilization and uptake of iron. ATP-driven proton transporters move H^+ ions from the cytosol across the plasma membrane to control the pH at the cell surface [15]. Humans also use an acidic environment to facilitate uptake of dietary iron. Uptake mainly occurs though enterocytes in the duodenum, which receives the acidic contents of the stomach. Iron can then be absorbed for storage in intestinal cells or delivery to other cells [15,16].

Many microorganisms, including some fungi, also secrete low molecular weight compounds known as siderophores into their surroundings, which form high-affinity (~10^{-33} M) complexes with ferric iron to make it bioavailable for uptake. Transporters on the cell surface then recapture the Fe^{3+}-siderophores complexes. For infectious microorganisms, these molecules help the invading pathogen acquire iron from the host for survival. Interestingly, two recent reports suggest that mammalian cells may also synthesize their own siderophores [17,18]. In both cases, the siderophore-like compound was isolated by screening for molecules that bound to siderocalins, a class of lipocalins that specifically bind exogenous siderophores. Siderocalins are weapons in the immune system arsenal, designed to prevent the invading organisms from acquiring iron by sequestering Fe^{3+}-bound siderophores [19]. However, these new studies suggest that siderocalins may also bind Fe^{3+} complexed with endogenous siderophores to facilitate iron trafficking. The candidate endogenous siderophore-like compounds isolated include catechol and catechol-like compounds [17], as well as a molecule with a 2,5-dihydroxybenzoic acid (2,5-DHBA) iron-binding moiety, which is isomeric to 2,3-DHBA found in the bacterial siderophore enterobactin (Figure 2) [18]. BDH2, a homologue of bacterial EntA (which catalyzes

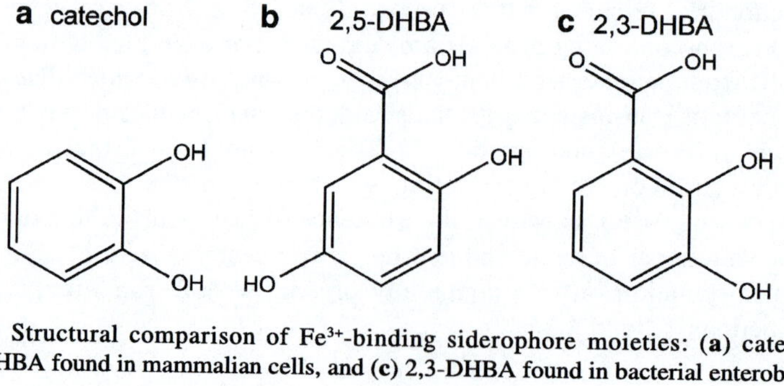

Figure 2 Structural comparison of Fe^{3+}-binding siderophore moieties: (**a**) catechol and (**b**) 2,5-DHBA found in mammalian cells, and (**c**) 2,3-DHBA found in bacterial enterobactin.

2,3-DHBA production), was found to be responsible for 2,5-DHBA production. Knockdowns of BDH2 suggested that the 2,5-DHBA-containing mammalian siderophore is involved in regulating both cytosolic and mitochondrial iron levels [18].

2.3.2 Intracellular Labile Iron Pools

The vast majority of iron is bound to proteins and enzymes for use as a cofactor (for more information, see Section 3) and stored in ferritin, vacuoles, and lysosomes. The remaining iron in the cell is proposed to be part of a labile iron pool (LIP), also known as chelatable or free iron, which is most likely present as ferrous complexes given the neutral pH and reducing conditions inside the cell. The LIP is thought to constitute only 0.2–3% of total cellular iron [20]. Studies suggest that the LIP acts as a crossroad in iron trafficking, providing iron for incorporation into metalloenzymes, feeding pathways for heme and Fe-S biosynthesis, and directing excess iron towards storage or export proteins [21]. The LIP is also assumed to be dynamic in nature, shrinking and growing in response to the needs of the cell. Only recently has research focused on uncovering the chemical nature of iron in the cytosolic LIP [22]. At physiological conditions, small molecular weight ligands such as phosphates, citrate, cysteine, and glutathione (GSH) are available to bind Fe(II) [20]. While the predominant ligand for the LIP remains an open question, potentiometric and binding affinity studies suggest that only GSH is present at high enough concentrations with sufficient binding affinity to buffer the labile Fe(II) pool, forming the proposed pentaaqua-Fe(II)-GSH complex shown in Figure 3. This iron-GSH complex is suggested to be a way for cells to distinguish between Fe(II) and Mn(II), which have similar intracellular concentrations. In addition, the Fe(II)-GSH complex may play a role in iron

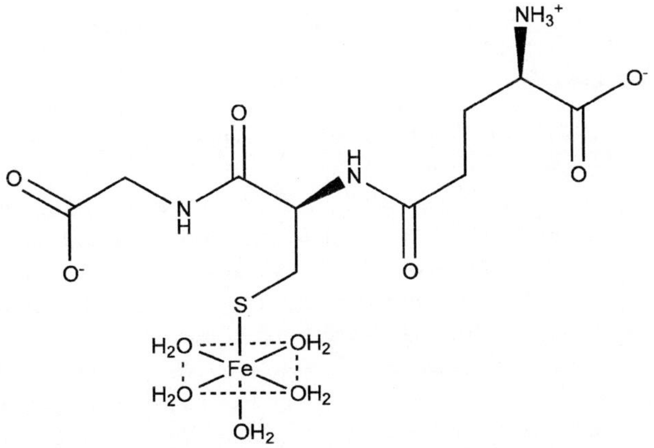

Figure 3 Proposed structure of the glutathione-coordinating complex with Fe(II) found in the labile iron pool.

trafficking based on the interaction of GSH with monothiol glutaredoxins, which are essential for iron regulation and trafficking (see Section 4.1) [23].

There is some evidence that a labile iron pool exists in individual organelles as well as the cytosol. Using fluorescent indicators, the LIP in mammalian mitochondria was measured between 1 and 16 μM depending on the cell type, which constitutes <0.4% of total mitochondrial iron [24–26]. Studies on mitochondrial iron speciation in yeast suggest that a somewhat larger pool of non-heme, high spin Fe^{2+} is used in assemblies of hemes and Fe-S clusters. The exact nature of this iron is still unknown, although it is presumably loosely bound by low-molecular-weight ligands similar to the cytosol. In actively respiring mitochondria, this pool constitutes ~2% of total mitochondrial iron, but grows to 20% during fermentation when the rate of Fe-S cluster and heme biosynthesis decreases. Since the total mitochondrial iron in yeast is nearly identical in respiring *versus* fermenting cells, these results demonstrate dynamic shifts in subcellular iron speciation rather than mitochondrial iron import in response to changes in energy metabolism [10].

2.4 Mechanisms of Iron Toxicity

2.4.1 Oxidative Stress and Formation of Reactive Oxygen Species

Although iron is required for many cellular processes, excess iron levels can be toxic to cells. Iron has a central role in the production of one of the most reactive oxygen species (ROS) found in the cell, namely the hydroxyl radical ($HO^{\bullet}$). Intracellular iron catalyzes formation of $HO^{\bullet}$ non-enzymatically by reacting with the superoxide anion ($O_2^{\bullet-}$) (Eq. 1) and hydrogen peroxide (H_2O_2) (Eq. 2) [27].

$$Fe^{3+} + O_2^{\bullet-} \rightarrow Fe^{2+} + O_2 \tag{1}$$

$$Fe^{2+} + H_2O_2 \rightarrow Fe^{3+} + HO^- + HO^{\bullet} \quad \text{(Fenton reaction)} \tag{2}$$

Net reaction (1) + (2):

$$O_2^{\bullet-} + H_2O_2 \rightarrow O_2 + HO^- + HO^{\bullet} \quad \text{(Haber-Weiss reaction)} \tag{3}$$

ROS such as H_2O_2 and $O_2^{\bullet-}$ are produced naturally *in vivo* through enzymatic reactions and auto-oxidation from endogenous compounds and have well-documented roles in signal transduction pathways and immune cell response [28]. However, when left unchecked, these molecules together with $HO^{\bullet}$ have the ability to initiate oxidative damage to DNA, lipids, and proteins, all of which contribute to cell death, aging, and various diseases. Thus, iron overload diseases are often characterized by elevated levels of biomarkers for oxidative stress, including protein carbonyls, DNA oxidation products, lipid peroxidation, advanced glycation end products, and malondialdehyde formation. Accumulation of iron in the brain coupled with oxidative stress is also a common feature of neurodegenerative diseases such as Alzheimer's and Parkinson's disease [29].

2.4.2 Iron Interference in Other Metal Trafficking Pathways

Recent studies also suggest that iron toxicity may not be solely due to iron-catalyzed ROS formation. Kaplan and coworkers demonstrated that toxicity is not dependent on the presence of oxygen, since iron is toxic to yeast even under anaerobic growth conditions [30]. Alternatively iron toxicity may stem from the interference of excess iron in other metal trafficking pathways. Strong evidence for this hypothesis is the effect of excess iron on manganese trafficking to the antioxidant enzyme superoxide dismutase (SOD). Eukaryotic cells express an SOD in the mitochondria, SOD2, which preferentially binds manganese over iron under normal conditions. SOD2 is an essential antioxidant enzyme since deletion of the *SOD2* gene leads to neonatal lethality in mouse models [31,32]. SOD2 activity requires the correct insertion of manganese into the enzyme, while misincorporation of iron renders it inactive [33,34]. Studies of SOD2 mismetallation in yeast revealed the presence of two distinct iron pools in the mitochondria, one being "SOD2-inert" and the other "SOD2-reactive". Disruption of iron homeostasis increases the reactive pool (without significantly affecting total mitochondrial iron), allowing for iron incorporation into SOD2. In particular, disruptions in the late stages of mitochondrial Fe-S cluster biogenesis led to diversion of iron to SOD2 (see Section 3.2.2). A somewhat similar situation was observed in a mouse model of the iron overload disease hereditary hemochromatosis, albeit via a different mechanism. In this case, cytosolic iron overaccumulation was found to disrupt trafficking of copper, zinc, and manganese to mitochondria, leading to deficiencies of these essential metals in this organelle. Consequently, Mn-SOD2 activity was significantly reduced leading to lower respiratory activity and increased lipid peroxidation [35].

3 Iron Metalloproteins

3.1 Types of Iron Metalloproteins

3.1.1 Heme-Containing Proteins

Organisms utilize heme-containing proteins for a variety of processes, including sensing and transport of oxygen, energy metabolism, transcriptional regulation, and protein stability. Heme consists of iron bound to a porphyrin ring (shown in Figure 4), where the iron can act as an electron source or sink for redox and electron transfer processes. In mammals, heme is one of the most important iron cofactors. It is best known as an oxygen carrier when bound to hemoglobin in red blood cells. Cytochromes are hemoproteins involved in electron transfer reactions. For example, cytochrome *c* transfers electrons between Complexes III and IV in the electron transport chain of the mitochondria. The cytochrome P450 enzyme family catalyzes the oxidation of many organic compounds, including lipids, hormones, and xenobiotics. Catalases and peroxidases are both hemoprotein families that protect against peroxide damage. Catalase prevents H_2O_2 damage by catalyzing its decomposition to H_2O and O_2. Peroxidases use heme to convert peroxides into alcohols using electron

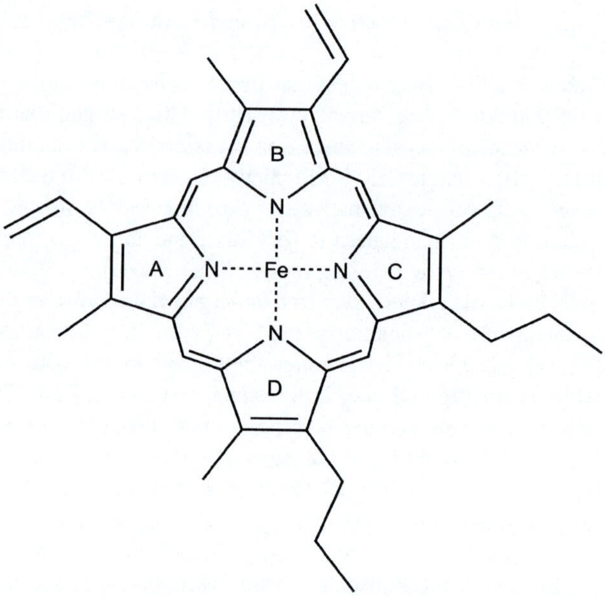

Figure 4 Heme cofactor. Structure of protoporphyrin IX with ferrous iron inserted.

donors and protons, again to prevent damage caused by reactive peroxides. Recently, it was shown that the nuclear receptor Rev-erbα binds heme and regulates circadian rhythmicity as well as other metabolic pathways [36].

3.1.2 Iron-Sulfur Cluster-Containing Proteins

Similar to the heme cofactor, organisms employ iron in the iron-sulfur cluster cofactor for its versatility in electron transfer reactions, with redox potentials ranging from –500 to +300 mV. Iron sulfur clusters form as complexes of iron (Fe^{2+} or Fe^{3+}) and inorganic sulfide (S^{2-}) in various arrangements, with two of the most common being [2Fe-2S] and [4Fe-4S] clusters (Figure 5). Certain prokaryotes and cyanobacteria also utilize larger, more complex clusters that incorporate additional metals such as molybdenum, nickel, and vanadium. For example, the nitrogenase enzyme contains a FeMoco cofactor cluster ($MoFe_7S_9$) and a P cluster (Fe_8S_7) in addition to a [4Fe-4S] cluster in the heterodimer form [37]. Evidence for these types of complex clusters in higher eukaryotes is lacking.

Fe-S proteins can also convert between cluster forms. The enzyme aconitase is active in the [4Fe-4S] cluster form, while partial disassembly to a [3Fe-4S] cluster renders it reversibly inactive. Fe-S clusters are usually bound to proteins via coordination of the iron to sulfur from cysteine and nitrogen from histidine, although serine and arginine have also been shown to ligate Fe-S clusters. In addition to binding to protein residues, the iron can bind other small molecules such as GSH, homocitrate, CO, and CN^-. Recent studies also demonstrate that a carbon atom coordinates all 6 Fe at the core of the FeMoco cluster in nitrogenase from

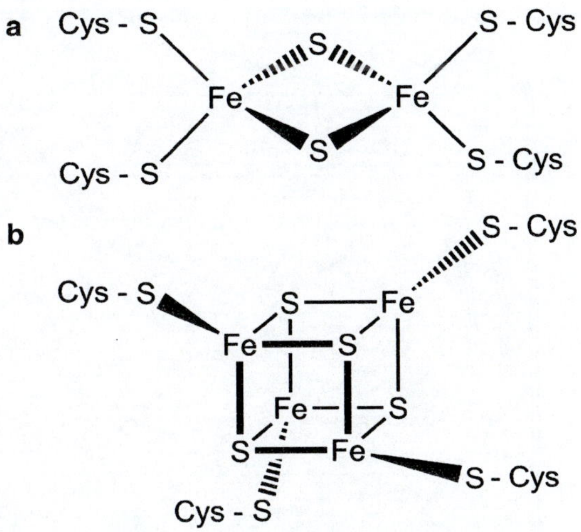

Figure 5 Common forms of iron-sulfur cluster cofactors: (**a**) [2Fe-2S] and (**b**) [4Fe-4S].

nitrogen-fixing bacteria [38,39]. In addition to their well-known redox function in electron transfer reactions, Fe-S clusters are also involved in heme biosynthesis, DNA synthesis and repair, ribosome assembly, tRNA modification, nucleotide and amino acid metabolism, and biogenesis of Fe-S proteins [40].

3.1.3 Mono- and Dinuclear Non-Heme Iron Proteins

While heme iron and Fe-S clusters are two of the most common ways that proteins use iron as a cofactor, it is found in other forms. Non-heme iron cofactors can be bound directly to proteins as mononuclear and dinuclear iron centers with a variety of amino acid ligands and bridging atoms adapted for specific roles. Many non-heme iron centers catalyze similar reactions to heme enzymes. For example, both heme and non-heme diiron enzymes can act as monooxygenases that insert oxygen atoms into substrate molecules. Non-heme iron proteins catalyze a wide array of reactions, such as converting nucleoside diphosphates (NDP) to deoxyNDPs (ribonucleotide reductase), catalyzing biomineralization of iron for intracellular storage (ferritin), sensing oxygen (prolyl hydroxylases), synthesizing eicosanoids (lipoxygenases), and modifying histones (lysine demethylases).

3.2 Assembly and Insertion of Iron Cofactors

3.2.1 Assembly and Insertion of Heme Cofactors

The steps involved in the synthesis of heme are well conserved from prokaryotes to eukaryotes (Figure 6). As mentioned previously, free iron is toxic to cells due to generation of ROS. Both porphyrin and heme are also toxic, generating oxygen radicals

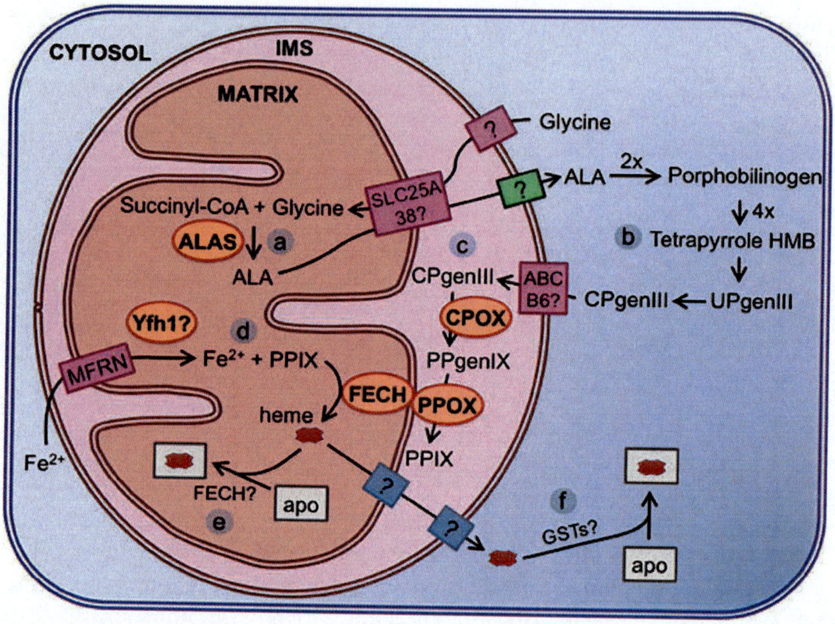

Figure 6 Heme biosynthesis pathway in eukaryotes. (**a**) Glycine is transported into the mitochondrial matrix via an unknown mechanism where it is combined with succinyl-CoA by ALA synthase (ALAS) to form ALA. (**b**) ALA is transported out to the cytosol where it is converted to CPgenIII through four conserved steps. (**c**) ABCB6 is the transporter proposed to import CPgenIII to the intermembrane space (IMS) where it is converted first to PPgenIX by CPOX, then to protoporphyrin IX (PPIX) by PPOX. (**d**) PPIX is transported to the matrix where iron is inserted by ferrochelatase (FECH). The proposed Fe(II) importers are Mrs3/4 (Mfrn1/2 in mammalian cells). (**e**) Assembled heme is inserted into target apo proteins to form hemoproteins in the mitochondria, possibly aided by FECH. (**f**) Heme is inserted into target apo proteins in the cytosol, possibly aided by GSTs.

and peroxidase activity, respectively. To reduce the risk of these potentially toxic molecules, heme biogenesis is linked to intracellular iron concentrations and synthesis of hemoprotein precursors. The heme synthesis machinery is distributed in both the cytosol and the mitochondria, requiring intermediates in this pathway to be shuttled across membranes. Transport of these porphyrin intermediates must be tightly regulated, again to reduce the risk of toxic components accumulating in the cell [41].

The first phase in heme biosynthesis is formation of the pyrrole. Initially, ALA synthase (ALAS) catalyzes this condensation reaction between succinyl-CoA and glycine to form δ-aminolevulinic acid (ALA) in the mitochondrial matrix. ALA is then transported to the cytosol, possibly via exchange for glycine by the mitochondrial carrier protein SLC25A38, where aminolevulinate dehydratase (ALAD) catalyzes the condensation of two ALA molecules to form the monopyrrole porphobilinogen. After formation of the monopyrrole is complete, porphobilinogen deaminase (PBGD) catalyzes assembly of the unstable tetrapyrrole hydroxymethylbilane (HMB) from four molecules of porphobilinogen. Formation of the tetrapyrrole macrocycle is completed by uroporphyrinogen III synthase (URO3S), which catalyzes the ring inversion and closure of HMB to make uroporphyrinogen III (UPgenIII) [42,43]. Once the tetrapyrrole is formed, the side chains need to be modified to form the correct porphyrin before insertion of iron. Uroporphyrinogen decarboxylase (UROD) catalyzes the removal of carboxyl groups from the acetic acid side chains of UPgenIII to form coproporphyrinogen III (CPgenIII). Coproporphyrinogen oxidase (CPOX) catalyzes the conversion of CPgenIII to protoporphyrinogen IX (PPgenIX) via oxidative decarboxylation of the pyrrole ring propionate groups to vinyl groups. CPOX is cytosolic in yeast, and located in the mitochondrial intermembrane space (IMS) in higher eukaryotes. Several studies suggest that the ATP-binding cassette transporter protein ABCB6 is either the CPgenIII transporter, or is somehow involved in the transport of CPgenIII to the IMS in mammals [44]. Protoporphyrinogen oxidase (PPOX) located on the outer surface of the mitochondrial inner membrane catalyzes oxidation of PPgenIX to protoporphyrin IX (PPIX) in the IMS [42].

The final step in forming heme is the insertion of ferrous iron into PPIX by ferrochelatase (FECH) in the mitochondrial matrix. There is some evidence that FECH physically interacts with PPOX across the mitochondrial inner membrane to allow substrate channeling of PPIX between these two enzymes [45]. Human and *S. pombe* forms of FECH contain a structural [2Fe-2S] cluster that is sensitive to nitric oxide, while *S. cerevisiae* and bacterial ferrochelatases lack the Fe-S cluster [46]. Studies indicate that ferrous iron may be imported into the matrix by Mrs3/4 importers (Mfrn1/2 or mitoferrin in mammalian cells). The yeast homologue of human frataxin, Yfh1, is proposed to act as an iron chaperone and donate Fe(II) to ferrochelatase for heme biosynthesis [47]. However, human frataxin does not seem to be involved in heme biosynthesis, although it may have a role in Fe-S protein assembly [40,48].

Once heme is fully assembled, it must be transported from FECH in the matrix across one or more membranes to target hemoproteins found in various organelles, such as the IMS, cytosol, nucleus, ER, and lysosomes [49]. FECH may act as a heme shuttle for proteins in the matrix that are in the same vicinity, such as cytochrome P450. For proteins outside the matrix, there is no known heme chaperone in mammals, although heme chaperones for cytochrome *c* have been identified in plants and bacteria [50]. Some cytosolic heme-binding proteins have also been suggested to have a role in heme transport, including glutathione S-transferases (GSTs) from liver and red blood cells [51]. Hemoproteins found in the secretory pathway may obtain heme in the ER, indicating a role for the ER in heme delivery. The ER and mitochondria have been shown to physically interact via a tethering complex that may provide a path for heme transport from the mitochondria [45].

3.2.2 Assembly and Insertion of Iron-Sulfur Clusters

There are two identified systems in eukaryotes for assembly of iron-sulfur clusters: the mitochondrial iron-sulfur cluster (ISC) assembly machinery and the cytosolic Fe-S protein assembly (CIA) machinery (Figure 7). In the mitochondria, Fe-S protein

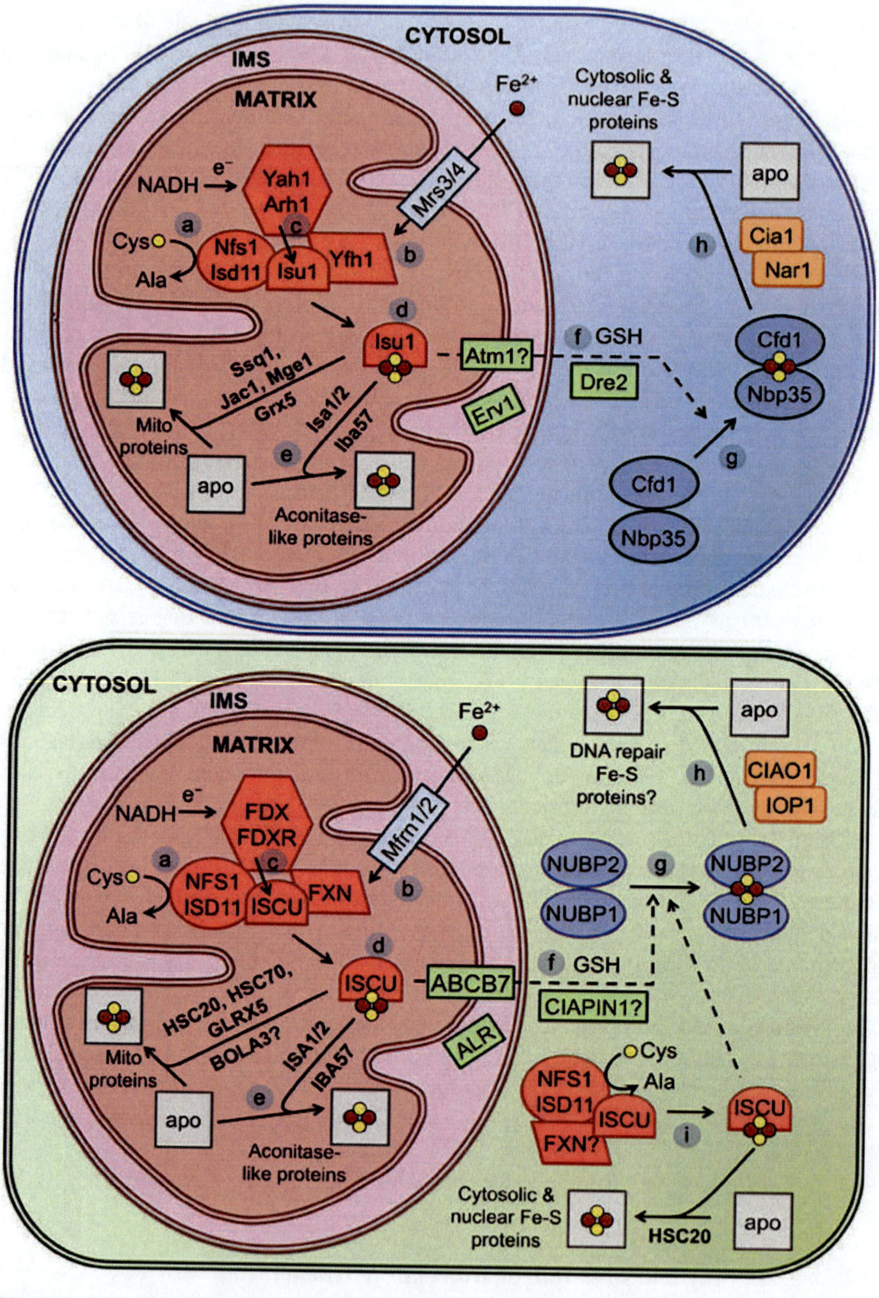

Figure 7 Mitochondrial and cytosolic Fe-S cluster assembly in yeast (top panel) and mammalian cells (bottom panel). (**a**) In the mitochondria, sulfur is obtained from the cysteine desulfurase Nfs1, interacting with Isd11. Nfs1 transfers sulfur as a persulfide to Isu1/2 (ISCU in humans). (**b**) Iron is imported to the mitochondria by the transporters Mrs3/4 (mitoferrins or Mfrn1/2 in humans) and

maturation occurs via two distinct stages. First, Fe(II) and sulfur are combined on a homodimeric scaffold protein (Isu) to form a labile Fe-S cluster. Once formed, the nascent Fe-S cluster is then transferred to its ultimate target protein via additional accessory proteins. In the initial stage, sulfur is obtained from free cysteine via the pyridoxal phosphate-dependent cysteine desulfurase Nfs1. Both yeast and human Nfs1 are mainly mitochondrial, although a portion of Nfs1 localizes to the nucleus and cytosol in both systems. In yeast, the mitochondrial form is essential for both mitochondrial and cytosolic cluster assembly, while the cytosolic form may act as a sulfur donor for other pathways requiring mobilized sulfur, such as cytosolic tRNA thio-modification [52]. Nfs1 function is dependent on formation of a stable complex with a partner protein named Isd11, although the specific role of Isd11 is not clear. The Nfs1-Isd11 complex produces sulfane sulfur (S^0) via persulfide formation, thus a source of electrons is required to reduce this molecule to sulfide (S^{2-}) for Fe-S cluster synthesis. This electron transfer is most likely performed by the ferredoxin-ferredoxin reductase pair Yah1 and Arh1 in yeast (FDX1/2 and FDXR in humans) using electrons from NADH [48,53].

It is clear that the Nfs1-Isd11 complex transfers sulfur to the Isu scaffold proteins (Isu1 and Isu2 in yeast, ISCU in humans). Similar to Nfs1, a small amount of human ISCU exhibits cytosolic localization where it may function as a scaffold for extramitochondrial Fe-S cluster biogenesis [48,53]. The mechanisms of cluster assembly on scaffold proteins, including the order of Fe and S binding and the sources of iron, remain open questions. As far as iron delivery, one proposal is that Yfh1, the yeast homologue of human frataxin, delivers iron to the Isu proteins. Yfh1 was shown to bind iron *in vitro*, and to bind Isu1/Nfs1 in an iron-dependent manner in the mitochondria, possibly enabling iron transfer [54]. More recently, frataxin was proposed to stabilize the active form of the Nfs1-Isd11-Isu complex thereby promoting sulfur transfer and enhancing Fe-S cluster formation [55].

Once the cluster is assembled, it must be transferred from the scaffold to the target apoprotein. Bacterial U-type scaffolds (IscU, NifU) are capable of making and transferring both [2Fe-2S] and [4Fe-4S] clusters, thus eukaryotic Isu proteins may also be able to assemble both types of clusters. The cluster transfer system is formed by the chaperone Ssq1 (an Hsp70 chaperone), a J-type accessory chaperone,

Figure 7 (continued) possibly donated to Isu1/2 through frataxin (FXN, or Yfh1 in yeast). (**c**) Electrons are donated by NADH through the ferredoxin-ferredoxin reductase pair Yah1-Arh1 to reduce S^0 to S^{2-}. (**d**) The assembled Fe-S cluster is transferred to target proteins by a chaperone system consisting of Ssq1, Jac1, Mge1, and Grx5 (HSC20, HSC70, and GLRX5 in humans). (**e**) Isa1/2 specifically delivers clusters to aconitase-like proteins. (**f**) An unknown substrate produced by the ISC machinery is exported out to the cytosol by the transporter Atm1 (ABCB7 in humans). This process may also include the sulfhydryl oxidase Erv1 (ALR in humans), GSH, and Dre2 (CIAPIN1). (**g**) Fe-S clusters are assembled in the cytosol on the scaffold complex formed by Cfd1 and Nbp35 (NUBP2 and NUBP1 in humans, respectively). (**h**) The assembled cluster is transferred to target cytosolic and nuclear proteins by Nar1 and Cia1 (IOP1 and CIAO1 in humans). (**i**) Mammalian cells also express cytosolic versions of NFS1, ISD11, ISCU, HSC20, and possibly FXN that may also facilitate *de novo* assembly of Fe-S clusters outside the mitochondria.

Jac1, and the chaperone/nucleotide release factor, Mge1. While not required for cluster assembly on Isu proteins, depletion of these chaperones results in cluster accumulation on Isu1. This system most likely takes the transiently bound cluster from the Isu scaffold and inserts it into the target protein. The monothiol glutaredoxin Grx5 also may play a role in cluster transfer, as Grx5 depletion results in cluster accumulation on Isu1, although a specific role has not yet been determined. Studies on zebrafish and human forms of Grx5 show that it is important for cytosolic Fe-S assembly, and thus may regulate heme synthesis by facilitating Fe-S cluster assembly on IRP1 (see Section 5.2.2) [48,53]. Additional mitochondrial Fe-S biogenesis assembly factors (e.g., Isa1, Isa2, Iba57, BolA3, Nfu1, Ind1) are proposed to function as intermediate scaffold/delivery proteins between ISCU and specific subsets of target proteins [40].

In addition to the mitochondrial ISC system, yeast and mammalian cells have cytosolic iron-sulfur assembly (CIA) components that build clusters for cytosolic and nuclear proteins. Since the mitochondrial form of yeast Nfs1 is essential for both mitochondrial and cytosolic Fe-S cluster assembly, one theory is that the mitochondria exports a sulfur-containing substrate produced by the ISC machinery that is used by the CIA machinery to build and/or insert Fe-S clusters [40]. In yeast, the ABC transporter Atm1 (ABCB7 in humans) is proposed to export the unidentified sulfur-containing substrate from the mitochondria to the cytosol since depletion of Atm1 inhibits cytosolic Fe-S assembly. A recent study also implicates the mammalian ABC transporter ABCB8 in export of mitochondrial iron for cytosolic Fe-S cluster biogenesis [56]. The mitochondrial IMS-localized sulfhydryl oxidase Erv1 and the ubiquitous tripeptide GSH are also implicated in export of the sulfur-containing substrate from the mitochondria for cytosolic Fe-S cluster assembly, although their specific roles are nebulous [40]. However, a recent study reported that the cytosolic Fe-S assembly defect reported for Atm1-depleted strains was an artifact of the strain background used in the initial report, indicating that Atm1 activity is not required for cytosolic Fe-S cluster assembly in yeast [57]. In addition, studies in human cells suggest that the cytosolic Fe-S cluster assembly machinery is independent of the mitochondrial system [48].

Regardless of the specific details regarding initial Fe-S cluster assembly in the cytosol and the requirement of the mitochondrial ISC system, it is clear that a number of additional proteins are essential for assembling Fe-S clusters for cytosolic/nuclear proteins. Two potential scaffold proteins are the P-loop NTPases Cfd1 and Nbp35 that form a [4Fe-4S]-bridged heterotetramer. Yeast Nar1 (IOP1 and IOP2 in humans) is similar to bacterial iron hydrogenases, although it has no hydrogenase activity. Nar1 has two Fe-S clusters whose assembly requires the mitochondrial ISC systems, as well as Cfd1 and Nbp35. Depletion of mammalian IOP1 disrupts cytosolic Fe-S cluster assembly, and homology of Nar1/IOP1/2 to other hydrogenases has led to the suggestion that these proteins may act as electron donors in the CIA system, either in assembly or transfer of a cluster. A fourth member of the CIA machinery is Cia1, which might play a role late in biogenesis after Nbp35 and Nar1, possibly in cluster transfer to target proteins. In addition, Dre2, which is localized to both the cytosol and the mitochondrial intermembrane space, was recently found to be a possible link between the mitochondrial ISC and cytosolic CIA systems. Dre2 can be purified with both a [4Fe-4S] and a [2Fe-2S] cluster that seem to play catalytic or structural roles based on cluster stability. Since depletion of Dre2 impairs cytosolic Fe-S cluster assembly, it may work early in the CIA pathway and deliver the ISC system substrate necessary for cytosolic cluster assembly. The human homologue of Dre2, anamorsin (or CIAPIN1) is proposed to function in electron transfer in the CIA system, similar to Dre2. While the function appears conserved between yeast and humans, anamorsin only binds a [2Fe-2S] cluster [58]. Currently, the only proteins known to require the CIA system bind [4Fe-4S] clusters [40,59], thus there may exist a parallel pathway for assembly of cytosolic [2Fe-2S] clusters such as those found on the Grx3/4 proteins (more in Section 5.3) [60].

3.2.3 Assembly and Insertion of Non-Heme Iron Cofactors

Due to the multitude of different non-heme iron centers, there is not a singular system for assembly and insertion of these cofactors. In most cases, a specific set of proteins are required: a chaperone for iron delivery, redox proteins to maintain the oxidation state of iron, and enzymes involved in protein folding that allow for proper insertion of the metal. For metalloproteins that merely need their cofactor inserted (such as mononuclear iron), cells can minimize metal misincorporation by compartmentalizing proteins and using metal chaperones. The metal concentrations in different subcellular compartments can vary, and a metalloprotein's metal affinity is usually tailored to these specific ranges. Here, we focus on one of the better-characterized assembly systems for incorporation and maturation of the dinuclear iron center of class Ia ribonucleotide reductases (RNRs).

Class I RNRs are divided into subgroups based on their metallocofactor. Class Ia RNRs contain a diferric-tyrosyl radical ($Fe^{3+}Fe^{3+}$-$Y^{\bullet}$) that is involved in the oxidation of a conserved active site cysteine to a thiyl radical ($S^{\bullet}$), which in turn reduces the substrate. Fe^{2+} first needs access to the metal binding site of the RNR that is buried in the protein. One hypothesis is that one subunit of the RNR acts as an iron chaperone to another subunit, then becomes part of the functional enzyme. Once the protein contains a diferrous center ($Fe^{2+}Fe^{2+}$), this can react with O_2 to form a diferric peroxide. A conserved Trp residue reduces the diiron center to an $Fe^{4+}Fe^{3+}$ intermediate, producing a Trp cation radical ($Trp^{+\bullet}$). The intermediate can then oxidize the active site Tyr (Y), forming $Fe^{3+}Fe^{3+}$ and the Tyr radical ($Y^{\bullet}$) needed for catalysis [61]. Figure 8 shows diferric-tyrosyl radical cofactor formation (a) and proteins involved in the biosynthesis pathway (b).

Recent studies on the cytosolic monothiol glutaredoxins, Grx3 and Grx4, and cytosolic Fe-S biogenesis factor Dre2 have revealed a possible role for these proteins in cofactor assembly on yeast RNR2. Grx3/4 is involved in intracellular iron trafficking for the assembly of many iron cofactors, including Fe-S loading onto Dre2. Dre2 and the reductase Tah18 are postulated to act as an electron source for cytosolic Fe-S cluster assembly (analogous to the Yah1-Arh1 pair in the mitochondrial ISC system). Grx3/4 was shown to be the source of iron in RNR cofactor synthesis, either directly donating the iron, or indirectly by making it bioavailable

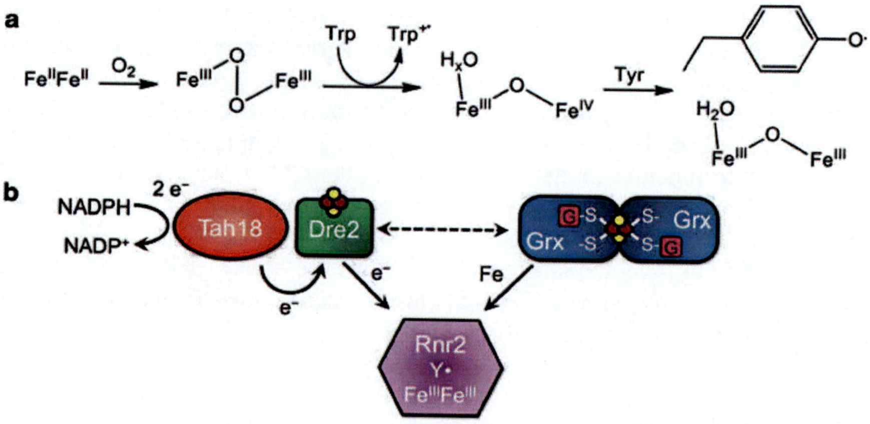

Figure 8 Formation of the class Ia RNR diferric-tyrosyl radical cofactor. (**a**) Formation reaction for the diferric-tyrosyl radical cofactor. (**b**) Model for the biosynthesis pathway of active Rnr2. Tah18 and Dre2 are proposed to donate electrons for diferric-tyrosyl radical formation on Rnr2. The [2Fe-2S] Grx3/4 homodimer most likely donates iron for the cofactor. Reducing equivalents from Dre2 may also be required for iron delivery (dashed line).

for assembly. Dre2 also plays a role in RNR cofactor formation that seems to be separate from the role of Grx3/4. Although this specific function has not yet been determined, the Dre2-Tah18 pair may provide reducing equivalents to maintain the iron in its reduced state for delivery [62].

4 Iron Uptake, Trafficking, and Storage

4.1 Iron Uptake and Transport

4.1.1 Iron Uptake and Transport in *S. cerevisiae*

The yeast *S. cerevisiae* expresses an extensive system of membrane transporters for uptake of environmental iron (Figure 9). As previously mentioned, yeasts have the ability to scavenge iron-loaded siderophores from their surroundings. Arn1, Taf1, Sit1, and Enb1 (also known as Arn1-4, respectively) are transporters specific for Fe^{3+}-siderophore complexes. Environmental Fe^{3+} is reduced to the more soluble Fe^{2+} by the ferrireductases Fre1 and Fre2, which are also capable of reducing Cu^{2+}. Once reduced, Fe^{2+} can be imported by the high-affinity uptake system encoded by *FET3* and *FTR1*. Fet3 is a multicopper oxidase that oxidizes the Fre1/2-produced Fe^{2+} to Fe^{3+}, which is then transferred across the plasma membrane by the transmembrane permease Ftr1. In addition, *S. cerevisiae* possesses low-affinity iron transporters (Fet4 and Smf1) that can also transport other transition metals. Fet4, in addition to

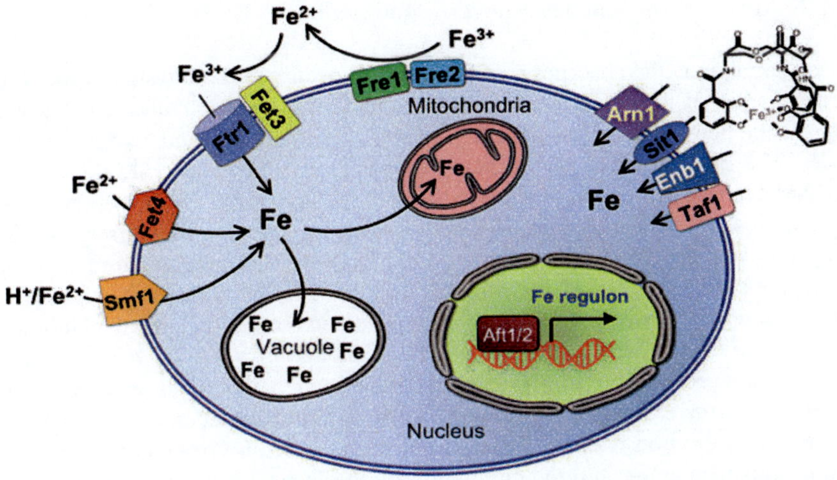

Figure 9 Iron uptake systems in *S. cerevisiae*. Arn1, Sit1, Enb1, and Taf1 are transporters for Fe^{3+}-siderophore complexes. The ferrireductases Fre1/2 reduce environmental Fe^{3+} to Fe^{2+}. Fet3 and Ftr1 form the high-affinity iron uptake system. Fet3 oxidizes Fe^{2+} to Fe^{3+} and Ftr1 transports this across the plasma membrane. Fet4 is a low-affinity transporter responsible for uptake under iron-replete conditions. Smf1, a member of the NRAMP family of transporters, is a H^+/M^+ symporter for some transition metals like Fe^{2+}, Mn^{2+}, and Zn^{2+}. Aft1 and Aft2 are transcriptional regulators that activate expression of the iron regulon, including these iron uptake systems, under low iron conditions.

being a low-affinity iron transporter, can import copper and zinc, and is responsible for most of the uptake under iron-replete conditions. Smf1, a member of the NRAMP transporter family found in both prokaryotes and eukaryotes, is a H^+/M^+ symporter that uses a pH gradient to import transition metals such as Fe^{2+}, Mn^{2+}, and Zn^{2+}. Yeasts express two other NRAMP proteins, Smf2 and Smf3, although they do not seem to play a significant role in environmental iron uptake. In addition to these uptake systems, the yeast genome encodes transmembrane ATP-driven H^+ transporters to acidify the environment, which increases the solubility of Fe^{3+} [15,63].

After iron is imported into cells, it needs to be transported either for storage or utilization. Since free iron is redox-active and can damage proteins and membranes, it seems likely that cells would use a chaperone to bind and transport iron through the cell. In yeast, the cytosolic monothiol Grxs with a signature CGFS active site, Grx3 and Grx4, are suggested to have an essential role in cellular iron trafficking. Depletion of Grx3/4 leads to defects in iron-dependent proteins, independent of induction of the Aft1 iron uptake system (see Regulation in Section 5.3 for more). Although these cells had excess cytosolic iron, iron delivery is impaired. Grx3/4 have previously been shown to bind a [2Fe-2S] cluster in a homodimeric complex. Not only is this Fe-S cluster essential for their role in iron trafficking, it is also required in iron sensing and regulation of the transcription factors Aft1 and Aft2 [23,64,65].

4.1.2 Iron Uptake and Transport in Mammalian Cells

In mammalian cells, absorption of dietary iron occurs in the intestine through the brush border of duodenal enterocytes. Inorganic iron mainly comes from vegetables, while heme iron comes from degradation of hemoglobin and myoglobin in red meat. The divalent metal transporter 1 (DMT1) also known as SLC11A1, is an NRAMP family protein that carries inorganic iron across the apical membrane into enterocytes. The ferric reductase Dcytb (duodenal cytochrome *b*) is required to reduce Fe^{3+} to Fe^{2+} prior to uptake by DMT1 since dietary inorganic iron is primarily in the oxidized form in the acidic environment of the duodenum. The mechanisms for uptake of dietary heme, which is absorbed more efficiently than inorganic iron, are somewhat nebulous. One possible candidate for dietary heme uptake is heme carrier protein HCP1 [66]. After internalization, heme iron is released into the enterocyte cytosol by heme oxygenase-1 (HO-1) via degradation of the heme molecule. Cytosolic iron is either stored in ferritin (see below) or exported across the basolateral membrane into the circulation by the ferrous iron exporter ferroportin (FPN). In enterocytes, FPN functions in concert with the multicopper oxidase hephaestin, which oxidizes exported Fe^{2+} to Fe^{3+} to facilitate iron loading onto the plasma iron carrier transferrin [15]. Many other mammalian cell types also have the ability to export iron via FPN, including macrophages and hepatocytes. The plasma multicopper oxidase ceruloplasmin functions with FPN in these cell types in an analogous manner to hephaestin [67].

Transferrin (Tf) binds two ferric ions with high affinity, providing iron for most human cell types (Figure 10). Iron-bound Tf is imported into the cell through the cell surface Tf receptor 1 (TfR1), forming a complex that is internalized by endocytosis. ATP-dependent proton pumps acidify the resulting endosome so iron is released from Tf, although this release likely also requires reduction of the iron to Fe^{2+}. Members of the STEAP (six-transmembrane epithelial antigen of the prostate) protein family are ferric reductases that catalyze this reaction. Apo-Tf is reutilized after delivery to the plasma membrane and released back into the circulation. The freed ferrous iron is then transported from the endosome to the cytoplasm via DMT1 [67]. In addition to Tf-iron, some mammalian cells can utilize hemoglobin iron from degraded erythrocytes. Senescent red blood cells (RBCs) are phagocytosed by macrophages in order to recycle body iron. Hemoglobin binds to the plasma protein haptoglobin, which then binds to CD163 on the surface of monocytes and macrophages. Similar to uptake of Tf-iron, this complex undergoes endocytosis before being broken down for iron release. The released heme iron can then bind to another plasma protein, hemopexin, which is endocytosed following binding to the CD91 receptor of certain cells [45]. The transmembrane protein HRG-1 has also been implicated in heme import [68].

Once iron is delivered to the cytoplasm, it must be transferred to various sites for utilization and storage. When iron concentrations exceed cellular need, iron is deposited in the iron storage protein ferritin (see below). Iron is most likely trafficked via a chaperone so that it cannot prematurely react with cellular components. While no general iron chaperone has been confirmed yet, several possible candidates have been identified. Poly (rC) binding protein 1 (PCBP1) was shown to facilitate iron

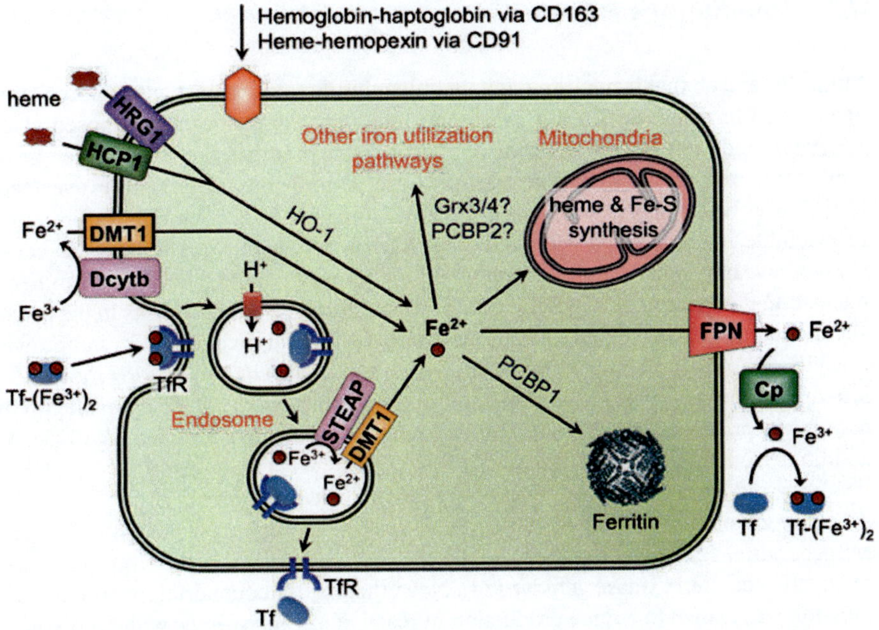

Figure 10 Iron import and export in a generic mammalian cell. The plasma protein transferrin (Tf) binds 2 ferric ions for uptake by the transferrin receptors (TfR1 and TfR2). TfR1 is found in all cell types, while TfR2 is limited to liver, intestinal, and red blood cells. $Tf(Fe^{3+})_2$-bound TfR is internalized by endocytosis and ferric iron is released in the acidic environment of the endosome. Tf and TfR are recycled to the plasma membrane. DMT1 is involved in release of Tf iron from the endosome following reduction of Fe^{3+} to Fe^{2+} by STEAP ferrireductases. Dcytb (cytochrome b-like ferrireductase) reduces dietary Fe^{3+} to Fe^{2+}, which is imported by DMT1 at the plasma membrane. The plasma proteins haptoglobin and hemopexin bind hemoglobin and free heme, respectively, produced by erythrocyte destruction. Haptoglobin-hemoglobin and heme-hemopexin complexes are recognized by CD163 and CD91 receptors, respectively, for subsequent endocytosis. Heme is also imported via HCP1 and HRG1. HO-1 (heme oxygenase-1) catalyzes degradation of the heme to remove iron. Imported iron can be stored in ferritin or trafficked to the mitochondria for synthesis of heme and Fe-S clusters. FPN is the iron exporter, transporting Fe^{2+} out of the cell. Ceruloplasmin (Cp) oxidizes this Fe^{2+} to Fe^{3+} for binding to Tf.

loading to ferritin; however, the exact mechanism of delivery is not yet characterized [54]. In addition, PCBP1 and its homologue PCBP2 were shown to act as iron chaperones to the iron-dependent enzymes HIF prolyl hydroxylases and factor inhibiting HIF (FIH1). As with the ferritin-PCBP1 interaction, the mechanism of iron delivery is not known; however, it seems likely that the PCBPs interact with their targets post-translationally for iron incorporation. Whether this takes place in the nucleus or the cytosol remains to be seen [69]. Cellular iron trafficking in yeast by [2Fe-2S]-bridged Grx3/4 homodimers has already been described (see above) [23]. The human ortholog of yeast Grx3/4, namely GLRX3, shares high sequence similarity with its yeast counterparts and forms analogous [2Fe-2S] bridged homodimers [70,71], thus human GLRX3 may play a similar role as an iron chaperone.

4.2 *Iron Storage Proteins*

Intracellular iron that is not immediately utilized is directed to the cytosolic iron storage protein ferritin. Ferritin self-assembles into a protein nanocage composed of 24 heavy (H) and light (L) subunits housing ferric oxide biominerals, with an average of 1000-1500 iron atoms stored per ferritin cage. During intracellular iron overload, ferritin iron levels can reach 3000-4000 iron atoms/nanocage, leading to formation of insoluble material from damaged ferritin known as hemosiderin. Several recent studies have unveiled the unique structural properties of ferritin that facilitate mineral nucleation during iron entry [72]. Each H subunit binds two ferrous ions in the active site that combine with O_2 to form the di-Fe(III)O mineral precursor. Crystal growth is achieved as the mineral precursors from each active site coalesce during movement into the central cavity. Release of ferritin iron back into bioavailable cytosolic pools occurs via proteasomal or lysosomal degradation of the ferritin nanocage [73]. In addition, iron may also be released from ferritin via a non-destructive pathway involving reduction and dissolution of the ferric oxide biomineral [72].

A mitochondrial version of ferritin is also expressed in some human cell types. Mitochondrial ferritin is similar in sequence and structure to ferritin, but possesses relatively weak ferroxidase activity [74]. Nevertheless, mitochondrial ferritin overexpression was shown to reduce production of reactive oxygen species while promoting mitochondrial iron loading and cytosolic iron depletion [75]. However, the increased mitochondrial iron in cells overexpressing mitochondrial ferritin is less bioavailable for heme and Fe-S cluster biogenesis [76]. Interestingly, mitochondrial ferritin transcription does not seem to be regulated by iron, but rather by the oxidative metabolic activity demand. Taken together, these studies thus suggest that mitochondrial ferritin functions to sequester redox-active iron in this organelle and highlight the importance of the mitochondrion in regulation of whole cell iron metabolism [76,77].

4.3 *Iron Entry and Exit into Extramitochondrial Organelles*

4.3.1 Yeast Vacuoles

As mentioned earlier, in addition to mitochondria, the vacuole is the other major hub of iron metabolism in *S. cerevisiae*. As yeasts do not possess ferritin for iron storage, the vacuole is designed to serve this purpose. The acidic pH in vacuoles, which ranges from 4.5-5.5, permits increased iron solubility, and is regulated by ATP-driven proton pumps. Iron is imported from the cytosol to the vacuole through the transporter Ccc1, found in the vacuolar membrane. In addition to being an iron (and manganese) storage organelle, vacuoles help protect cells from the toxic effects of excess iron by sequestering it in an unreactive form. The protein complexes involved in iron export from the vacuole are comparable to the cell surface iron import machinery. A member of the FRE family, Fre6, is localized to the vacuole where it acts as a ferrireductase in exporting iron from the vacuole. Fre6 works in combination with the Fet5-Fth1 ferrous iron transport complex, which is analogous to the cell surface Fet3-Ftr1 import complex. Smf3, another member of the NRAMP family and homologous to Smf1, is also found in the vacuolar membrane. Smf3 and Fet5-Fth1 all work downstream from Fre6 in iron export from the vacuole [63,78–80].

4.3.2 Lysosomes

In mammalian cells, lysosomes play a significant role in iron recycling as the primary site for degradation of cytosolic ferritin. These organelles contain acidic hydrolases that store and degrade biological waste produced by the cell, including damaged proteins and organelles. Much of the material taken up by lysosomes contains iron, mainly derived from ferritin and respiratory complexes. Iron concentrations in lysosomes can vary greatly depending on how much material has been taken up, and if iron-rich compounds have recently been degraded. Lysosomes with higher iron content tend to be more susceptible to oxidative stress and destabilization. Fenton chemistry becomes more likely under the favorable lysosomal conditions of an acidic pH and the availability of reducing equivalents. Oxidative damage to membranes can cause leaky lysosomes, which exposes other cellular components to the harmful materials contained in this organelle [81].

4.3.3 Endoplasmic Reticulum

As mentioned earlier, studies suggest that the ER may acquire heme via direct contact with the mitochondrion [45]. The existence of non-heme iron enzymes in the mammalian ER, such as prolyl and lysyl hydroxylases, suggests that ionic iron is also transported into the ER and/or Golgi. Although the yeast genome does not encode homologues for these enzymes, prolyl hydroxylases heterologously expressed in *S. cerevisiae* are correctly targeted and enzymatically active, demonstrating iron delivery to the secretory pathway [82]. In addition, iron-free ferritin monomers are translocated into the ER in yeast and mammalian cells [83], and iron-loaded ferritin is detected in the secretory pathway of *Drosophila* cells [84]. Taken together, these studies provide evidence that ionic iron is transported into the ER/Golgi compartment in both lower and higher eukaryotes [83]. However, specific ER iron transporters for either yeast or mammalian cells have not yet been identified.

4.3.4 Nucleus

Numerous nuclear proteins contain heme, non-heme iron, and Fe-S cluster cofactors that are necessary for their activity. Examples of heme-binding nuclear proteins include the nuclear receptor Rev-ervα and the transcriptional repressor Bach1 [49], while diiron-binding nuclear proteins include RNRs and HIF prolyl hydroxylases.

A growing list of Fe-S proteins involved in DNA repair, DNA replication, and transcriptional elongation are also localized to the nucleus [40]. In each of these cases, it is not known whether iron cofactor acquisition and assembly occurs before or after entry in the nucleus. Nevertheless, a chelatable, labile iron pool similar in concentration to the cytosolic labile iron pool has been detected in mammalian nuclei [20], thus nuclear iron is bioavailable for insertion into iron metalloproteins if necessary. Nuclear ferritin has also been detected in some mammalian cells types, demonstrating the capacity of this organelle to store and sequester DNA-damaging, redox-active iron [85]. The specific mechanisms by which iron enters the nucleus are unclear. Labile iron bound to low molecular weight ligands may freely diffuse through nuclear pores. Alternatively, there is evidence that iron is actively transported across the nuclear membrane via an ATP-dependent transport system [86].

5 Regulation of the Iron Metallome

5.1 Regulation at the Transcriptional Level

5.1.1 Yeast Iron-Responsive Transcription Factors

5.1.1.1 *S. cerevisiae* Hap1:

S. cerevisiae regulate iron homeostasis via transcriptional, post-transcriptional, and post-translational mechanisms. At the transcriptional level, yeast utilize several regulatory factors that respond not only to cellular iron status, but also oxygen levels. Yeasts use heme as an oxygen sensor to differentiate between aerobic and anaerobic growth conditions. Heme is well-suited to this role since it not only binds to oxygen, but the synthesis of porphyrin for heme depends on oxygen availability. Lower oxygen levels lead to decreased heme synthesis, which in turn decreases synthesis of respiratory proteins. The transcriptional activator Hap1 controls this process in *S. cerevisiae*. Hap1 is a heme-binding protein that activates transcription of respiratory proteins under normal oxygen levels, as well as the repressor Rox1. Rox1 then represses transcription of genes involved in anaerobic growth. Under low or no oxygen conditions when heme synthesis is attenuated, Hap1 is inactive, thus aerobic metabolism genes are turned off and anaerobic genes are turned on. Neither Hap1 nor Rox1 has any effect on transcription of the iron regulon (more below) [87].

5.1.1.2 *S. cerevisiae* Aft1/Aft2:

In *Saccharomyces cerevisiae,* expression of high affinity ionic iron and siderophore transporters is primarily controlled by the transcriptional activator Aft1 and to a lesser extent, its paralog Aft2. In addition to these transporters, Aft1 and Aft2 control several other genes that together comprise the iron regulon. The iron regulon includes genes encoding the high affinity iron transport system (*FET3*, *FTR1*, *CCC2*, *ATX1*, *FRE1-FRE6*), the siderophore transport system (*ARN1-4*, *FIT1-3*), and vacuolar iron export systems (*SMF3*, *FET5*, *FTH1*) [15]. Aft1/2 nuclear localization is controlled by the cellular iron status: in iron-starved cells, Aft1/2 accumulates in the nucleus, while under iron-replete conditions, Aft1/2 is shuttled to the cytosol by the exportin Msn5. Aft1/2 localization is regulated by a complex formed of Grx3/4, Fra1, and Fra2 proteins, which transmits an inhibitory signal that is dependent on the synthesis of mitochondrial Fe-S clusters (for more, see Section 5.3.1) [87]. Aft mutants are unable to grow on iron-poor media, although this is not primarily due to a malfunctioning iron uptake system, but rather the lack of control of intracellular iron use. This misuse of iron also renders cells more sensitive to oxidative stress, most likely due to metal toxicity and formation of ROS [88–90].

Aft1 and Aft2 have both overlapping and independent functions. Many iron regulon genes are induced by both Aft1 and Aft2; although in most cases, Aft1 activation elicits a stronger response. Aft1 and Aft2 regulate gene expression by binding to iron-responsive elements (FeREs) with the conserved sequence 5′-CACCC-3′. However, Aft1 binds more selectively than Aft2 since it preferentially binds 5′-TGCACC-3′, while Aft2 prefers 5′-G/ACACC-3′. Transcriptional analysis of Aft1 and Aft2 target genes suggest that Aft1 is primarily involved in cellular iron uptake, while Aft2 specifically regulates intracellular trafficking to vacuoles and mitochondria [90,91]. However, biophysical analysis of iron speciation in yeast mutants that express constitutively active forms of Aft1 or Aft2 suggests that the Aft proteins do not regulate trafficking of cytosolic iron into mitochondria and vacuoles [7].

5.1.1.3 *S. cerevisiae* Yap5:

The regulatory functions of the low iron-sensing transcriptional activators Aft1/Aft2 are also complemented by a high iron-sensing transcriptional activator named Yap5. Under high iron conditions, Yap5 activates expression of the vacuolar iron transporter Ccc1 resulting in increased iron transport into the vacuole, which effectively lowers cytosolic iron levels [92]. Yap5 binds to the *CCC1* promoter independent of iron levels, but only induces expression under high-iron conditions via formation of an intramolecular disulfide bond [92,93]. A recent microarray study identified a handful of Yap5-regulated genes, including *TYW1*, *GRX4*, and *CUP1*. Tyw1 is an Fe-S enzyme involved in modification of tRNA bases; however, the catalytic function of Tyw1 is not implicated in iron response. Instead, the data suggests that Yap5 protects the cell from excess iron by increasing Tyw1 levels, which sequesters iron into Fe-S clusters. It is possible that upregulation of Grx4 expression has a similar effect since this [2Fe-2S]-binding protein is implicated in iron trafficking [23]. *CUP1* encodes a Cu-binding metallothionein that is important for resistance to copper toxicity [94]. Interestingly, biophysical studies indicate that Cup1 also binds four Fe^{2+} atoms/monomer *in vitro* [95], and thus may also play a role in iron sequestration under toxic iron conditions. Taken together, these studies suggest that Yap5 responds to high cellular iron levels by decreasing free cytosolic iron through sequestration in vacuoles or incorporation into Fe- or Fe-S cluster-binding proteins [92].

5.1.2 Mammalian HIF-2α

Unlike the transcriptional regulation of iron uptake and storage genes in yeast, studies suggest that mammalian cellular iron homeostasis is primarily regulated by post-transcriptional control of mRNA translation and stability (see Section 5.2.2). However, iron-dependent regulation at the transcriptional level was recently implicated in control of intestinal iron absorption. As described in Section 4.2.1, mammalian enterocytes control iron import and export by regulating levels of the ferric reductase DcytB and iron importer DMT1 located at the apical membrane, as well as the basolateral iron exporter FPN. While the peptide hormone hepcidin is the main regulator of FPN (see Section 5.3.2.1), the hypoxia-inducible factor HIF-2α (also known as EPAS1) controls expression of DctyB and DMT1. Under iron deficiency or hypoxic conditions, HIF-2α forms a heterodimeric complex with HIF-1β, also known as aryl hydrocarbon nuclear receptor (ARNT), that induces transcription of both DcytB and DMT1. When iron levels are sufficient, HIF-2α is degraded (see Section 5.3.2.2) and does not activate expression of the iron absorption genes [96,97].

5.2 *Regulation at the Post-Transcriptional Level*

5.2.1 Iron-Responsive mRNA-Binding Proteins in Yeast

In addition to activation of iron uptake and transport genes by Aft1 and Aft2, iron deficiency in *S. cerevisiae* also leads to reprogramming of iron-dependent metabolic pathways in order to preserve limited iron pools for essential pathways [15,98]. This metabolic reprogramming is primarily accomplished by two mRNA binding proteins, Cth1 and Cth2, whose expression is upregulated by Aft1/2 under low iron conditions. Both proteins bind to the 3′ untranslated regions of mRNAs encoding iron-utilizing proteins in non-essential pathways, thereby facilitating their degradation [99,100]. Cth1 and its paralog Cth2 are tandem zinc finger (TZF) proteins related to the mammalian protein tristetraprolin (TTP). Similar to TTP, Cth1/2 binds to AU-rich elements (ARE) in target mRNAs. Cth2 binds to mRNAs that encode enzymes in heme biosynthesis, Fe-S cluster biosynthesis, the TCA cycle, the electron transport chain, and components of fatty acid metabolism pathways. Gene targets of Cth1 are mainly mitochondrial proteins involved in respiration and amino acid synthesis. Cth1 shares a subset of target genes with Cth2, including some involved in oxidative phosphorylation [99–101].

Cth1 and Cth2 regulation helps to redistribute iron to vital processes under iron starvation. One of the enzymes that receives the limited iron is ribonucleotide reductase (RNR). RNR is composed of a large R1 subunit that houses the catalytic site and a small R2 subunit that contains the diferric tyrosyl radical cofactor. One facet of RNR regulation in yeast involves subcellular localization, since the R1 subunit is primarily localized to the cytosol while the R2 subunit resides in the nucleus under normal conditions. During genotoxic stress, the iron-containing R2 subunit is exported to the cytosol allowing assembly of the active, holo enzyme [102]. Nuclear localization of the R2 subunit depends on an interaction with the nuclear protein Wtm1. Interestingly, the R2 subunit is also shuttled to the cytoplasm under iron-deficient conditions, suggesting that the limited available iron is partly funneled to this essential enzyme. Iron-responsive transport of R2 to the cytoplasm is dependent on Cth1 and Cth2, which are responsible for binding to AREs in the *WTM1* mRNA and degrading it in response to low iron. Cth1 and Cth2 enhance RNR activity not only by facilitating relocalization, but also by degrading mRNA of nonessential iron-dependent pathways to increase available iron. In addition, Cth1 and Cth2 degrade new RNR transcripts to optimize the use of the limited iron available in the redistributed RNR enzymes [103].

5.2.2 Iron-Responsive mRNA-Binding Proteins in Mammalian Cells

In contrast to yeast, iron uptake and storage in mammalian cells is mainly regulated at the post-transcriptional level through two iron regulatory proteins, IRP1 and IRP2 [73,104,105]. Under iron-depleted conditions, IRP1 and IRP2 bind to iron response elements (IREs) that form hairpin structures in the 5′ or 3′ untranslated regions (UTRs) of their target mRNAs. IREs are found in mRNAs encoding a variety of proteins involved in iron metabolism, including ferritin, TfR, heme synthesis, DMT1, and FPN. IRP binding to mRNA either stabilizes it or sterically blocks its translation depending on the location of the IRE. When IREs are found in the 3′ UTR, such as in TfR, IRP binding protects the mRNA from degradation, allowing transcription and subsequent iron uptake. When IREs are found in the 5′ UTR, such as with ferritin mRNA, IRP binding blocks translation and thus decreases iron storage. Under iron replete conditions when IRP1 and IRP2 lose their IRE-binding activity, mRNAs with IREs in the 3′ UTR are degraded while mRNAs with IREs in the 5′ UTR are stable and freely translated [105]. Figure 11 illustrates how various target mRNAs are regulated by IRPs.

The IRPs are cytoplasmic proteins that belong to the aconitase family of isomerases. IRP1 is capable of switching from mRNA-binding activity to aconitase activity by binding a [4Fe-4S] cluster that prevents IRE binding. When cells are iron-depleted, the cluster is degraded and IRP1 gains mRNA-binding activity. In addition to being regulated by iron availability, IRP1 activity is influenced by Fe-S cluster biogenesis: IRE-binding activity is increased when the Fe-S assembly machinery is impaired. This cluster can also be degraded by exposure to hydrogen peroxide or nitric oxide. Although IRP1 and IRP2 are 57% identical, IRP2 does not bind an Fe-S cluster. Instead, it is synthesized and stable under low iron conditions, and degraded under high iron conditions. In addition to regulation by iron, IRP2 is also inhibited by ROS, RNS, and phosphorylation. IRP2 appears to be the main regulator of iron homeostasis as it can compensate for a loss of IRP1, while IRP1 cannot necessarily compensate for loss of IRP2 [105,106].

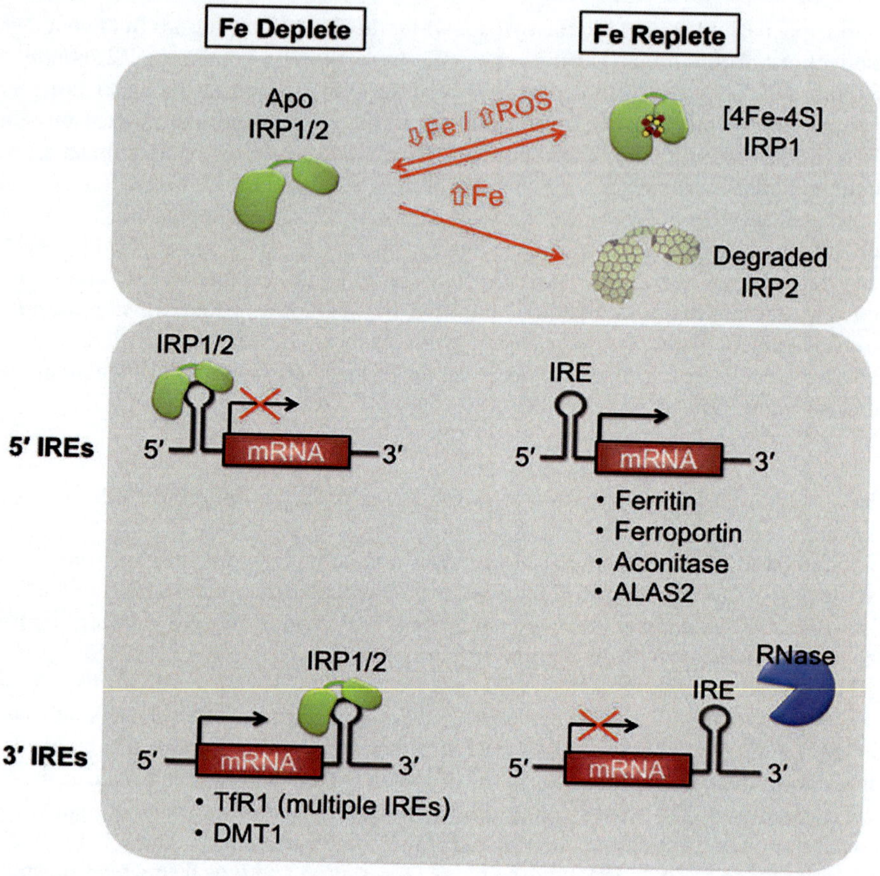

Figure 11 mRNA regulation of iron trafficking and utilization factors by the IRE-IRP system. In iron-deplete cells, IRP1 lacks the [4Fe-4S] and IRP2 is stabilized, thus both IRPs are able to bind the target IREs. Translation of mRNAs with 5′ IREs (e.g., ferritin) is blocked, while mRNAs with 3′ IREs (e.g., TfR1) are stabilized and translated. In iron-replete cells, IRP1 binds a [4Fe-4S] cluster that precludes IRE binding, while IRP2 is degraded by the proteasome. IREs of the target RNAs are unoccupied. mRNAs with IREs in the 5′ UTR are translated, while RNase degrades mRNAs with IREs in the 3′ UTR.

5.3 *Regulation at the Post-Translational Level*

5.3.1 Post-Translational Iron Regulation in *S. cerevisiae*

Recent studies have started to unveil the post-translational mechanisms that govern the iron-dependent activity of the *S. cerevisiae* transcription factors Aft1 and Aft2 (see Section 5.1.1.2 for more on Aft1/2). Genetic studies suggest that Aft1/2 does not directly sense iron, but rather responds indirectly to mitochondrial Fe-S cluster assembly [107]. Interpretation of the mitochondrial Fe-S signal is dependent on a signaling pathway involving four cytosolic proteins: Grx3, Grx4, Fra1, and Fra2, which all have homologues in mammalian cells. Grx3 and Grx4 are multidomain CGFS monothiol glutaredoxins that form [2Fe-2S] cluster-bound homodimers with a role in intracellular iron trafficking (see Section 4.1.1) [23,65]. Grx3 and Grx4 perform redundant functions since deletion of *GRX3* or *GRX4* singly has little or no phenotypic consequence, while *grx3Δgrx4Δ* double mutants are severely growth impaired or inviable [23,108]. Fra1 is an aminopeptidase P-like protein that is also implicated in regulation of vacuolar iron uptake [109]. Fra2 is a member of the BolA protein family of unknown function, although recent work has linked both prokaryotic and eukaryotic BolA homologues to Fe-S cluster biogenesis [110]. Iron-dependent regulation of Aft1/2 occurs at the protein level since Aft1/2 consistently localizes to the nucleus and binds its target promoters in the absence of the Fra1/Fra2/Grx3/Grx4 signaling pathway [64].

A variety of protein-protein interactions control signaling in this pathway. Grx3/4 interacts with Aft1 and Aft2 via a conserved CDC motif in Aft1/2. In addition, Grx3/4 and Fra1 both bind to Fra2. Biochemical and spectroscopic studies revealed that Grx3/4 and Fra2 form [2Fe-2S]-bridged heterodimers, while mutagenesis studies confirmed that the amino acids necessary for cluster binding *in vitro* were also required for inhibition of Aft1/2 activity *in vivo* [60,65]. Thus, Aft1/2 is proposed to sense the cellular iron status based on the ability of Grx3/4 and Fra2 to bind a [2Fe-2S] cluster [110]. Interestingly, Fe-S binding by Grx3/4 *in vivo* does not require the CIA machinery (see Section 3.2.2), indicating that a parallel pathway exists for insertion of the [2Fe-2S] cluster on Grx3/4. Under Fe replete conditions when Fra2-Grx3/4 binds an Fe-S cluster, the Fra1/Fra2/Grx3/4 signaling pathway is proposed to induce multimerization of Aft1/2. This conformational change in turn facilitates interaction with the exportin Msn5, leading to cytosolic localization of Aft1/2 and deactivation of the iron regulon. If the Fe-S signal is not received (iron-deplete conditions), the complex is unable to inhibit Aft1/2 activity, and these transcription factors move to the nucleus and induce the iron regulon to increase cellular iron levels [60,64]. However, the iron regulon is not fully activated in iron-sufficient medium in *fra1Δ* or *fra2Δ* mutants or upon disruption of mitochondrial Fe-S assembly pathways, suggesting that a separate iron signal may partially inhibit Aft1/2 activity in these mutants. The specific molecular mechanisms for inhibiting Aft1/2 activity via the Grx3/Grx4/Fra1/Fra2 signaling pathway or an alternate iron signal are still elusive. Figure 12 shows a model for regulation of Aft1 and Aft2 based on the available information for this system.

5.3.2 Post-Translational Iron Regulation in Mammalian Cells

5.3.2.1 Hepcidin:

Systemic iron homeostasis in mammals requires communication between cells that acquire (enterocytes), recycle (macrophages), store (hepatocytes), and utilize iron (developing erythrocytes). The coordination of systemic iron acquisition and usage

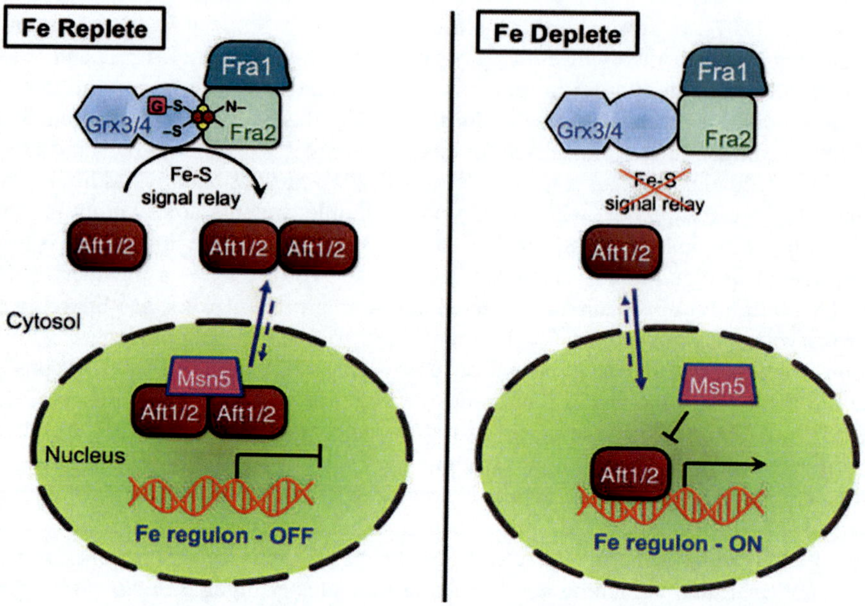

Figure 12 Proposed model for regulation of Aft1/2 in *S. cerevisiae*. Under iron replete conditions (left panel), the [2Fe-2S]-bridged Fra2-Grx3/4 complex (and possibly Fra1) relays the cellular iron status to Aft1/2, leading to Aft1/2 oligomerization. Oligomeric Aft1/2 favors cytosolic localization, possibly via nuclear export by the exportin Msn5, deactivating the iron regulon. Under low iron conditions (right panel), Grx3/4 and Fra2 cannot bind an Fe-S cluster, and Aft1/2 does not receive the Fe-S signal. Msn5 does not recognize Aft1/2 and Aft1/2 accumulates in the nucleus where it activates the iron regulon.

is mainly regulated post-translationally by the peptide hormone hepcidin. Hepcidin is a cysteine-rich, 25-amino acid peptide secreted by the liver and its only known target is the protein FPN. Hepcidin circulates in the plasma and binds to FPN found on the surface of intestinal duodenal cells and macrophages, thereby promoting FPN phosphorylation and subsequent degradation. Since FPN is the only known mammalian iron exporter, its degradation leads to decreased iron absorption from the intestine and disruption of iron recycling from macrophages [111].

Hepcidin expression is regulated at the transcriptional level by several stimuli, including iron availability, inflammation, and hypoxia. Basal transcriptional expression of the hepcidin gene requires C/EBPα (CCAAT enhancer-binding protein α) binding to a CCAAT sequence in the hepcidin promoter. Iron-dependent regulation of hepcidin expression occurs in response to two factors: iron stores and circulating iron. Hepatic iron stores activate hepcidin expression through bone morphogenic protein (BMP) signaling via an unknown mechanism. BMP binding to the BMP receptor activates a cascade that leads to hepcidin transcription. While several BMPs can induce hepcidin expression, BMP6 seems to be the most relevant as its expression is regulated by hepatic iron stores. Plasma iron levels can also regulate hepcidin expression through the major histocompatibility complex class 1-like protein HFE and TfR. HFE is a membrane protein that interacts with TfR1 and TfR2. When plasma iron levels are low, TfR1 is proposed to bind HFE. When iron-bound Tf increases, it displaces HFE from TfR1, allowing HFE to interact with TfR2, which activates a signaling pathway leading to hepcidin expression. Inflammatory response and ER stress can also induce hepcidin expression through the BMP and C/EBPα pathways [73,112].

5.3.2.2 Prolyl Hydroxylase Regulation of HIF-2α:

As previously mentioned, HIF-2α controls iron absorption in enterocytes via transcriptional activation of iron uptake genes. The mechanism of activation is indirectly dependent on intracellular iron levels: HIF-2α protein levels are controlled by a prolyl hydroxylase (PHD) that requires O_2 and Fe^{2+} for activity. Under normoxia conditions with sufficient iron, PHD is active and hydroxylates HIF-2α, resulting in ubiquitination by the VHL (von Hippel-Lindau) ubiquitin ligase. Ubiquitinated HIF-2α is subsequently targeted for degradation by the proteasome. In hypoxic conditions and/or under iron deficiency, PHD is inactive, thus HIF-2α is not hydroxylated and degraded, and instead forms a heterodimeric complex with HIF-1β that induces transcription of DcytB and DMT1 [96]. Thus, HIF-2α directly links iron absorption with both oxygen and iron metabolism. The mRNA for HIF-2α also contains an IRE in the 5′ UTR that is a target for IRP1. HIF-2α is therefore post-transcriptionally regulated in an iron-dependent fashion. The relationship between PHDs, HIF-2α, and IRP1 demonstrates feedback regulation between iron and oxygen metabolism that involves transcriptional, post-transcriptional, and post-translational control mechanisms [113].

5.3.2.3 FBXL5 Regulation of IRP2:

IRP2 plays a role in iron homeostasis by controlling the stability of mRNAs that encode proteins involved in iron trafficking and iron utilization. While it is clear that IRP2 is degraded under high iron conditions by a ubiquitin ligase, the identity of this enzyme was only recently uncovered. The SCF (SKP1-CUL1-F-box) ubiquitin ligases were found to be specific for IRP2 when customized by the F-box protein FBXL5. FBXL5 forms an SCF complex that physically interacts with IRP2 in an iron-dependent fashion. This regulation does not occur at the translational level, as FBXL5 mRNA levels were not affected by changes in iron. Rather, the FBXL5 protein is targeted for degradation based on iron availability. FBXL5 stability also depends on intracellular oxygen concentrations since exposure to oxygen leads to degradation. These observations are explained by the fact that FBXL5 contains a hemerythrin domain that binds iron and oxygen. Many iron-bound hemerythrin domains function as oxygen sensors and metal storage sites. Binding of iron to the FBXL5 hemerythrin domain influences the protein's stability. If an iron center cannot be bound either due to insufficient iron or hypoxic conditions, the protein is targeted for degradation. The stability of FBXL5 in turn regulates IRP2. Degradation of FBXL5 prevents formation of the SCF ubiquitin ligase, thus IRP2 is not targeted for degradation, and can bind IREs. Under iron-replete conditions, the F-box domain of FBXL5 can bind its iron cofactor and form the SCF ligase complex, which then promotes degradation of IRP2 by ubiquitination [114,115].

6 Concluding Remarks and Future Directions

While a great deal of progress has been made in characterizing the iron metallome of eukaryotes, there is still more left to uncover. Recent progress in spectroscopic techniques has aided in revealing the various forms of cellular iron. Future studies will focus on the nature of labile iron pools, including location, iron ligands, and cellular concentrations. How does iron speciation vary among eukaryotes, and under different growth conditions?

The overall assembly of heme and Fe-S cluster cofactors is well-established, but there are some holes in our knowledge here as well. Even though some common iron chaperones have been identified, these need to be studied in more detail. Future work in this area will determine how iron is delivered for cofactor assembly, and how the assembled cofactors are trafficked to target proteins. One intriguing area of study is the identity of the product exported from the mitochondrial ISC machinery for the cytosolic CIA machinery.

In addition, assembly of non-heme iron cofactors demands more investigation, as it seems this process is unique to the target protein. Several iron transporters have been identified and categorized to different organelles, particularly the mitochondria and vacuoles. However, just as chaperones to transfer iron for cofactor assembly are not well-characterized, neither are chaperones for intracellular trafficking. In addition, while transport proteins have been identified for iron import, the mechanisms of import must be further characterized. Iron storage proteins are another focus for future studies, especially the mechanisms for releasing iron once cellular need increases. Are there specific enzymes that target storage proteins for degradation or facilitate dissolution of ferritin biominerals? How is this regulated?

One of the most important topics for future study is regulation of the iron metallome. From yeast to humans, eukaryotes have several complex mechanisms to maintain iron levels, from transcriptional to post-transcriptional to post-translational control. There appears to be connections not only between these different regulatory levels, but also to oxygen metabolism. For example, hypoxia plays a regulatory role in controlling the expression and stability of several proteins: the hypoxia-inducible factors control iron uptake, and hypoxic conditions may induce expression of the hormone hepcidin. Several forms of regulation depend on indirect sensing of cellular iron levels, such as yeast Aft1 and Aft2. What is the mechanism for interpreting these cellular iron signals and how are Fe-S clusters involved? Furthermore, failures in regulation have been linked to many iron-related diseases, including hemochromatosis and anemia. A deeper understanding of the regulatory intricacies of iron metabolism will help in the treatment of these diseases.

Abbreviations

ABC	ATP-binding cassette
ALA	δ-aminolevulinic acid
ALAD	aminolevulinate dehydratase
ALAS	ALA synthase
ARE	AU-rich elements
ATP	adenosine 5'-triphosphate
BDH2	3-hydroxybutyrate dehydrogenase, type 2
BMP	bone morphogenic protein
C/EBPα	CCAAT enhancer-binding protein α
CIA	cytosolic Fe-S protein assembly
CoA	coenzyme A
Cp	ceruloplasmin
CPgenIII	coproporphyrinogen III
CPOX	coproporphyrinogen oxidase
DHBA	dihydroxybenzoic acid
DMTI	divalent metal transporter 1 (= SCL11A1)
dNDP	deoxynucleoside diphosphate
dNTP	deoxynucleoside triphosphate
EPR	electron paramagnetic resonance
ER	endoplasmic reticulum
EXAFS	extended X-ray absorption fine structure
FECH	ferrochelatase
FeRE	iron-responsive element
FIH1	factor inhibiting HIF
FPN	ferroportin
Grx	glutaredoxin
GSH	glutathione
GST	glutathione S-transferase
HCP1	heme carrier protein 1
HIF	hypoxia-inducible factor
HMB	hydroxymethylbilane
HO-1	heme oxygenase-1
HRG	heme responsive gene
IMS	intermembrane space
IRE	iron regulatory element
IRP	iron regulatory protein
ISC	iron-sulfur cluster
LIP	labile iron pool
Mfrn	mitoferrin
NADH	nicotinamide adenine dinucleotide reduced
NDP	nucleoside 5'-diphosphate
NRAMP	natural resistance-associated macrophage protein
NTP	nucleoside 5'-triphosphate

PBGD	porphobilinogen deaminase
PCBP	poly (rC) binding protein
PHD	prolyl hydroxylase
PIXE	particle-induced X-ray emission
PPgenIX	protoporphyrinogen IX
PPIX	protoporphyrin IX
PPOX	protoporphyrinogen oxidase
RBC	red blood cells
RNR	ribonucleotide reductase
ROS	reactive oxygen species
RNS	reactive nitrogen species
SCF	SKP1-CUL1-F-box
SOD	superoxide dismutase
STEAP	six-transmembrane epithelial antigen of the prostate
TCA	tricarboxylic acid
Tf	transferrin
TfR	Tf receptor
TTP	tristetraprolin
TZF	tandem zinc finger
UPgenIII	uroporphyrinogen III
URO3S	uroporphyrinogen III synthase
UROD	uroporphyrinogen decarboxylase
UTR	untranslated region
VHL	von Hippel-Lindau
XANES	X-ray absorption near edge structure
XAS	X-ray absorption spectroscopy
XRFM	X-ray fluorescence microscopy

Acknowledgments This work was funded by the National Institutes of Health Grants ES013780 and GM086619 and the South Carolina Research Foundation. The authors would like to thank Dr. L. Celeste for helpful discussions.

References

1. D. J. Eide, S. Clark, T. M. Nair, M. Gehl, M. Gribskov, M. L. Guerinot, J. F. Harper, *Genome Biol.* **2005**, *6*, R77.
2. S. Epsztejn, H. Glickstein, V. Picard, I. N. Slotki, W. Breuer, C. Beaumont, Z. I. Cabantchik, *Blood* **1999**, *94*, 3593–3603.
3. F. Petrat, U. Rauen, H. de Groot, *Hepatology* **1999**, *29*, 1171–1179.
4. D. Ceccarelli, D. Gallesi, F. Giovannini, M. Ferrali, A. Masini, *Biochem. Biophys. Res. Commun.* **1995**, *209*, 53–59.
5. X. Gao, M. Qian, J. L. Campian, J. Marshall, Z. Zhou, A. M. Roberts, Y. J. Kang, S. D. Prabhu, X. F. Sun, J. W. Eaton, *Free Radic. Biol. Med.* **2010**, *49*, 401–407.
6. J. Petrak, D. Myslivcova, P. Man, R. Cmejla, J. Cmejlova, D. Vyoral, *Am. J. Physiol. Gastrointest. Liver Physiol.* **2006**, *290*, G1059–1066.

7. R. Miao, G. P. Holmes-Hampton, P. A. Lindahl, *Biochemistry* **2011**, *50*, 2660–2671.
8. P. A. Lindahl, G. P. Holmes-Hampton, *Curr. Opin. Chem. Biol.* **2011**, *15*, 342–346.
9. A. L. Cockrell, G. P. Holmes-Hampton, S. P. McCormick, M. Chakrabarti, P. A. Lindahl, *Biochemistry* **2011**, *50*, 10275–10283.
10. G. P. Holmes-Hampton, R. Miao, J. Garber Morales, Y. Guo, E. Munck, P. A. Lindahl, *Biochemistry* **2010**, *49*, 4227–4234.
11. R. Miao, H. Kim, U. M. Koppolu, E. A. Ellis, R. A. Scott, P. A. Lindahl, *Biochemistry* **2009**, *48*, 9556–9568.
12. R. Ortega, G. Deves, A. Carmona, *J. R. Soc. Interface* **2009**, *6 Suppl 5*, S649–658.
13. C. J. Fahrni, *Curr. Opin. Chem. Biol.* **2007**, *11*, 121–127.
14. T. C. Iancu, Y. Deugnier, J. W. Halliday, L. W. Powell, P. Brissot, *J. Hepatol.* **1997**, *27*, 628–638.
15. C. D. Kaplan, J. Kaplan, *Chem. Rev.* **2009**, *109*, 4536–4552.
16. M. W. Hentze, M. U. Muckenthaler, N. C. Andrews, *Cell* **2004**, *117*, 285–297.
17. G. Bao, M. Clifton, T. M. Hoette, K. Mori, S. X. Deng, A. Qiu, M. Viltard, D. Williams, N. Paragas, T. Leete, R. Kulkarni, X. Li, B. Lee, A. Kalandadze, A. J. Ratner, J. C. Pizarro, K. M. Schmidt-Ott, D. W. Landry, K. N. Raymond, R. K. Strong, J. Barasch, *Nat. Chem. Biol.* **2010**, *6*, 602–609.
18. L. R. Devireddy, D. O. Hart, D. H. Goetz, M. R. Green, *Cell* **2010**, *141*, 1006–1017.
19. C. Correnti, R. K. Strong, *J. Biol. Chem.* **2012**, *287*, 13524–13531.
20. F. Petrat, H. de Groot, U. Rauen, *Biochem. J.* **2001**, *356*, 61–69.
21. M. W. Hentze, M. U. Muckenthaler, B. Galy, C. Camaschella, *Cell* **2010**, *142*, 24–38.
22. R. C. Hider, X. L. Kong, *Biometals* **2011**, *24*, 1179–1187.
23. U. Muhlenhoff, S. Molik, J. R. Godoy, M. A. Uzarska, N. Richter, A. Seubert, Y. Zhang, J. Stubbe, F. Pierrel, E. Herrero, C. H. Lillig, R. Lill, *Cell Metab.* **2010**, *12*, 373–385.
24. F. Petrat, D. Weisheit, M. Lensen, H. de Groot, R. Sustmann, U. Rauen, *Biochem. J.* **2002**, *362*, 137–147.
25. U. Rauen, A. Springer, D. Weisheit, F. Petrat, H. G. Korth, H. de Groot, R. Sustmann, *Chembiochem* **2007**, *8*, 341–352.
26. B. Sturm, U. Bistrich, M. Schranzhofer, J. P. Sarsero, U. Rauen, B. Scheiber-Mojdehkar, H. de Groot, P. Ioannou, F. Petrat, *J. Biol. Chem.* **2005**, *280*, 6701–6708.
27. J. P. Kehrer, *Toxicology* **2000**, *149*, 43–50.
28. I. I. Rovira, T. Finkel, B. S. Masters, M. B. Dickman, J. Lee, S. W. Ragsdale, C. C. Lee, in *Redox Biochemistry*, Eds R. Banerjee, D. F. Becker, M. B. Dickman, V. N. Gladyshev, S. W. Ragsdale, John Wiley & Sons, Inc., Hoboken, NJ, USA, 2007.
29. K. Jomova, M. Valko, *Toxicology* **2011**, *283*, 65–87.
30. H. Lin, L. Li, X. Jia, D. M. Ward, J. Kaplan, *J. Biol. Chem.* **2011**, *286*, 3851–3862.
31. R. M. Lebovitz, H. Zhang, H. Vogel, J. Cartwright, Jr., L. Dionne, N. Lu, S. Huang, M. M. Matzuk, *Proc. Natl. Acad. Sci. USA* **1996**, *93*, 9782–9787.
32. Y. Li, T. T. Huang, E. J. Carlson, S. Melov, P. C. Ursell, J. L. Olson, L. J. Noble, M. P. Yoshimura, C. Berger, P. H. Chan, D. C. Wallace, C. J. Epstein, *Nat. Genet.* **1995**, *11*, 376–381.
33. A. Naranuntarat, L. T. Jensen, S. Pazicni, J. E. Penner-Hahn, V. C. Culotta, *J. Biol. Chem.* **2009**, *284*, 22633–22640.
34. M. Yang, P. A. Cobine, S. Molik, A. Naranuntarat, R. Lill, D. R. Winge, V. C. Culotta, *EMBO J.* **2006**, *25*, 1775–1783.
35. H. A. Jouihan, P. A. Cobine, R. C. Cooksey, E. A. Hoagland, S. Boudina, E. D. Abel, D. R. Winge, D. A. McClain, *Mol. Med.* **2008**, *14*, 98–108.
36. L. Yin, N. Wu, J. C. Curtin, M. Qatanani, N. R. Szwergold, R. A. Reid, G. M. Waitt, D. J. Parks, K. H. Pearce, G. B. Wisely, M. A. Lazar, *Science* **2007**, *318*, 1786–1789.
37. M. K. Johnson, *Curr. Opin. Chem. Biol.* **1998**, *2*, 173–181.
38. K. M. Lancaster, M. Roemelt, P. Ettenhuber, Y. Hu, M. W. Ribbe, F. Neese, U. Bergmann, S. DeBeer, *Science* **2011**, *334*, 974–977.

39. T. Spatzal, M. Aksoyoglu, L. Zhang, S. L. Andrade, E. Schleicher, S. Weber, D. C. Rees, O. Einsle, *Science* **2011**, *334*, 940.
40. A. Sheftel, O. Stehling, R. Lill, *Trends Endocrinol. Metab.* **2010**, *21*, 302–314.
41. I. Hamza, *ACS Chem. Biol.* **2006**, *1*, 627–629.
42. R. S. Ajioka, J. D. Phillips, J. P. Kushner, *Biochim. Biophys. Acta* **2006**, *1763*, 723–736.
43. P. M. Shoolingin-Jordan, A. Al-Dbass, L. A. McNeill, M. Sarwar, D. Butler, *Biochem. Soc. Trans.* **2003**, *31*, 731–735.
44. P. C. Krishnamurthy, G. Du, Y. Fukuda, D. Sun, J. Sampath, K. E. Mercer, J. Wang, B. Sosa-Pineda, K. G. Murti, J. D. Schuetz, *Nature* **2006**, *443*, 586–589.
45. I. J. Schultz, C. Chen, B. H. Paw, I. Hamza, *J. Biol. Chem.* **2010**, *285*, 26753–26759.
46. H. A. Dailey, *Biochem. Soc. Trans.* **2002**, *30*, 590–595.
47. S. Park, O. Gakh, H. A. O'Neill, A. Mangravita, H. Nichol, G. C. Ferreira, G. Isaya, *J. Biol. Chem.* **2003**, *278*, 31340–31351.
48. T. A. Rouault, *Dis. Model. Mech.* **2012**, *5*, 155–164.
49. S. Severance, I. Hamza, *Chem. Rev.* **2009**, *109*, 4596–4616.
50. N. Spielewoy, H. Schulz, J. M. Grienenberger, L. Thony-Meyer, G. Bonnard, *J. Biol. Chem.* **2001**, *276*, 5491–5497.
51. J. W. Harvey, E. Beutler, *Blood* **1982**, *60*, 1227–1230.
52. Y. Nakai, N. Umeda, T. Suzuki, M. Nakai, H. Hayashi, K. Watanabe, H. Kagamiyama, *J. Biol. Chem.* **2004**, *279*, 12363–12368.
53. R. Lill, U. Muhlenhoff, *Annu. Rev. Biochem.* **2008**, *77*, 669–700.
54. P. Subramanian, A. V. Rodrigues, S. Ghimire-Rijal, T. L. Stemmler, *Curr. Opin. Chem. Biol.* **2011**, *15*, 312–318.
55. J. Bridwell-Rabb, C. Iannuzzi, A. Pastore, D. P. Barondeau, *Biochemistry* **2012**, *51*, 2506–2514.
56. Y. Ichikawa, M. Bayeva, M. Ghanefar, V. Potini, L. Sun, R. K. Mutharasan, R. Wu, A. Khechaduri, T. Jairaj Naik, H. Ardehali, *Proc. Natl. Acad. Sci. USA* **2012**, *109*, 4152–4157.
57. T. Bedekovics, H. Li, G. B. Gajdos, G. Isaya, *J. Biol. Chem.* **2011**, *286*, 40878–40888.
58. L. Banci, I. Bertini, S. Ciofi-Baffoni, F. Boscaro, A. Chatzi, M. Mikolajczyk, K. Tokatlidis, J. Winkelmann, *Chem. Biol.* **2011**, *18*, 794–804.
59. A. K. Sharma, L. J. Pallesen, R. J. Spang, W. E. Walden, *J. Biol. Chem.* **2010**, *285*, 26745–26751.
60. H. Li, D. T. Mapolelo, N. N. Dingra, G. Keller, P. J. Riggs-Gelasco, D. R. Winge, M. K. Johnson, C. E. Outten, *J. Biol. Chem.* **2011**, *286*, 867–876.
61. J. A. Cotruvo, J. Stubbe, *Annu. Rev. Biochem.* **2011**, *80*, 733–767.
62. Y. Zhang, L. Liu, X. Wu, X. An, J. Stubbe, M. Huang, *J. Biol. Chem.* **2011**, *286*, 41499–41509.
63. M. R. Bleackley, R. T. Macgillivray, *Biometals* **2011**, *24*, 785–809.
64. A. Kumanovics, O. S. Chen, L. Li, D. Bagley, E. M. Adkins, H. Lin, N. N. Dingra, C. E. Outten, G. Keller, D. Winge, D. M. Ward, J. Kaplan, *J. Biol. Chem.* **2008**, *283*, 10276–10286.
65. H. Li, D. T. Mapolelo, N. N. Dingra, S. G. Naik, N. S. Lees, B. M. Hoffman, P. J. Riggs-Gelasco, B. H. Huynh, M. K. Johnson, C. E. Outten, *Biochemistry* **2009**, *48*, 9569–9581.
66. M. Shayeghi, G. O. Latunde-Dada, J. S. Oakhill, A. H. Laftah, K. Takeuchi, N. Halliday, Y. Khan, A. Warley, F. E. McCann, R. C. Hider, D. M. Frazer, G. J. Anderson, C. D. Vulpe, R. J. Simpson, A. T. McKie, *Cell* **2005**, *122*, 789–801.
67. G. J. Anderson, C. D. Vulpe, *Cell. Mol. Life Sci.* **2009**, *66*, 3241–3261.
68. A. Rajagopal, A. U. Rao, J. Amigo, M. Tian, S. K. Upadhyay, C. Hall, S. Uhm, M. K. Mathew, M. D. Fleming, B. H. Paw, M. Krause, I. Hamza, *Nature* **2008**, *453*, 1127–1131.
69. A. Nandal, J. C. Ruiz, P. Subramanian, S. Ghimire-Rijal, R. A. Sinnamon, T. L. Stemmler, R. K. Bruick, C. C. Philpott, *Cell Metab.* **2011**, *14*, 647–657.
70. P. Haunhorst, C. Berndt, S. Eitner, J. R. Godoy, C. H. Lillig, *Biochem. Biophys. Res. Commun.* **2010**, *394*, 372–376.

71. H. Li, D. T. Mapolelo, S. Randeniya, M. K. Johnson, C. E. Outten, *Biochemistry* **2012**, *51*, 1687–1696.
72. E. C. Theil, *Curr. Opin. Chem. Biol.* **2011**, *15*, 304–311.
73. J. Wang, K. Pantopoulos, *Biochem. J.* **2011**, *434*, 365–381.
74. F. Bou-Abdallah, P. Santambrogio, S. Levi, P. Arosio, N. D. Chasteen, *J. Mol. Biol.* **2005**, *347*, 543–554.
75. P. Santambrogio, B. G. Erba, A. Campanella, A. Cozzi, V. Causarano, L. Cremonesi, A. Galli, M. G. Della Porta, R. Invernizzi, S. Levi, *Haematologica* **2011**, *96*, 1424–1432.
76. D. R. Richardson, D. J. Lane, E. M. Becker, M. L. Huang, M. Whitnall, Y. Suryo Rahmanto, A. D. Sheftel, P. Ponka, *Proc. Natl. Acad. Sci. USA* **2010**, *107*, 10775–10782.
77. M. L. Huang, D. J. Lane, D. R. Richardson, *Antioxid. Redox Signal.* **2011**, *15*, 3003–3019.
78. L. Li, O. S. Chen, D. McVey Ward, J. Kaplan, *J. Biol. Chem.* **2001**, *276*, 29515–29519.
79. C. C. Philpott, O. Protchenko, *Eukaryot. Cell* **2008**, *7*, 20–27.
80. A. Singh, N. Kaur, D. J. Kosman, *J. Biol. Chem.* **2007**, *282*, 28619–28626.
81. T. Kurz, J. W. Eaton, U. T. Brunk, *Int. J. Biochem. Cell Biol.* **2011**, *43*, 1686–1697.
82. P. D. Toman, G. Chisholm, H. McMullin, L. M. Giere, D. R. Olsen, R. J. Kovach, S. D. Leigh, B. E. Fong, R. Chang, G. A. Daniels, R. A. Berg, R. A. Hitzeman, *J. Biol. Chem.* **2000**, *275*, 23303–23309.
83. I. De Domenico, M. B. Vaughn, P. N. Paradkar, E. Lo, D. M. Ward, J. Kaplan, *Cell Metab.* **2011**, *13*, 57–67.
84. F. Missirlis, S. Kosmidis, T. Brody, M. Mavrakis, S. Holmberg, W. F. Odenwald, E. M. Skoulakis, T. A. Rouault, *Genetics* **2007**, *177*, 89–100.
85. A. A. Alkhateeb, J. R. Connor, *Biochim. Biophys. Acta* **2010**, *1800*, 793–797.
86. S. A. Gurgueira, R. Meneghini, *J. Biol. Chem.* **1996**, *271*, 13616–13620.
87. J. Kaplan, D. McVey Ward, R. J. Crisp, C. C. Philpott, *Biochim. Biophys. Acta* **2006**, *1763*, 646–651.
88. P. L. Blaiseau, E. Lesuisse, J. M. Camadro, *J. Biol. Chem.* **2001**, *276*, 34221–34226.
89. J. C. Rutherford, A. J. Bird, *Eukaryot. Cell* **2004**, *3*, 1–13.
90. J. C. Rutherford, S. Jaron, D. R. Winge, *J. Biol. Chem.* **2003**, *278*, 27636–27643.
91. M. Courel, S. Lallet, J. M. Camadro, P. L. Blaiseau, *Mol. Cell. Biol.* **2005**, *25*, 6760–6771.
92. L. Li, X. Jia, D. M. Ward, J. Kaplan, *J. Biol. Chem.* **2011**, *286*, 38488–38497.
93. L. Li, D. Bagley, D. M. Ward, J. Kaplan, *Mol. Cell. Biol.* **2008**, *28*, 1326–1337.
94. D. R. Winge, K. B. Nielson, W. R. Gray, D. H. Hamer, *J. Biol. Chem.* **1985**, *260*, 14464–14470.
95. X. Q. Ding, E. Bill, A. X. Trautwein, H. J. Hartmann, U. Weser, *Eur. J. Biochem.* **1994**, *223*, 841–845.
96. M. Mastrogiannaki, P. Matak, B. Keith, M. C. Simon, S. Vaulont, C. Peyssonnaux, *J. Clin. Invest.* **2009**, *119*, 1159–1166.
97. Y. M. Shah, T. Matsubara, S. Ito, S. H. Yim, F. J. Gonzalez, *Cell Metab.* **2009**, *9*, 152–164.
98. C. C. Philpott, S. Leidgens, A. G. Frey, *Biochim. Biophys. Acta* **2012**, *1823*, 1509–1520.
99. S. Puig, S. V. Vergara, D. J. Thiele, *Cell Metab.* **2008**, *7*, 555–564.
100. S. Puig, E. Askeland, D. J. Thiele, *Cell* **2005**, *120*, 99–110.
101. C. D. Kaplan, J. Kaplan, *Cell Metab.* **2005**, *2*, 4–6.
102. R. Yao, Z. Zhang, X. An, B. Bucci, D. L. Perlstein, J. Stubbe, M. Huang, *Proc. Natl. Acad. Sci. USA* **2003**, *100*, 6628–6633.
103. N. Sanvisens, M. C. Bano, M. Huang, S. Puig, *Mol. Cell* **2011**, *44*, 759–769.
104. S. Recalcati, G. Minotti, G. Cairo, *Antioxid. Redox Signal.* **2010**, *13*, 1593–1616.
105. T. A. Rouault, *Nat. Chem. Biol.* **2006**, *2*, 406–414.
106. K. Pantopoulos, *Ann. N. Y. Acad. Sci.* **2004**, *1012*, 1–13.
107. J. C. Rutherford, L. Ojeda, J. Balk, U. Muhlenhoff, R. Lill, D. R. Winge, *J. Biol. Chem.* **2005**, *280*, 10135–10140.
108. N. Pujol-Carrion, G. Bellí, E. Herrero, A. Nogues, M. A. de la Torre-Ruiz, *J. Cell Sci.* **2006**, *119*, 4554–4564.

109. L. Li, G. Murdock, D. Bagley, X. Jia, D. M. Ward, J. Kaplan, *J. Biol. Chem.* **2010**, *285*, 10232–10242.
110. H. Li, C. E. Outten, *Biochemistry* **2012**, *51*, 4377–4389.
111. J. Kaplan, D. M. Ward, I. De Domenico, *Int. J. Hematol.* **2011**, *93*, 14–20.
112. K. Gkouvatsos, G. Papanikolaou, K. Pantopoulos, *Biochim. Biophys. Acta* **2012**, *1820*, 188–202.
113. M. Sanchez, B. Galy, M. U. Muckenthaler, M. W. Hentze, *Nat. Struct. Mol. Biol.* **2007**, *14*, 420–426.
114. A. A. Salahudeen, J. W. Thompson, J. C. Ruiz, H. W. Ma, L. N. Kinch, Q. Li, N. V. Grishin, R. K. Bruick, *Science* **2009**, *326*, 722–726.
115. A. A. Vashisht, K. B. Zumbrennen, X. Huang, D. N. Powers, A. Durazo, D. Sun, N. Bhaskaran, A. Persson, M. Uhlen, O. Sangfelt, C. Spruck, E. A. Leibold, J. A. Wohlschlegel, *Science* **2009**, *326*, 718–721.

Chapter 9
Heme Uptake and Metabolism in Bacteria

David R. Benson and Mario Rivera

Contents

D.R. Benson • M. Rivera (✉)
Department of Chemistry, University of Kansas, Multidisciplinary Research Building, 2030 Becker Dr., Lawrence, KS 66047, USA
e-mail: drb@ku.edu; mrivera@ku.edu

L. Banci (ed.), *Metallomics and the Cell*, Metal Ions in Life Sciences 12, 279
DOI 10.1007/978-94-007-5561-1_9, © Springer Science+Business Media Dordrecht 2013

Abstract All but a few bacterial species have an absolute need for heme, and most are able to synthesize it via a pathway that is highly conserved among all life domains. Because heme is a rich source for iron, many pathogenic bacteria have also evolved processes for sequestering heme from their hosts. The heme biosynthesis pathways are well understood at the genetic and structural biology levels. In comparison, much less is known about the heme acquisition, trafficking, and degradation processes in bacteria. Gram-positive and Gram-negative bacteria have evolved similar strategies but different tactics for importing and degrading heme, likely as a consequence of their different cellular architectures. The differences are manifested in distinct structures for molecules that perform similar functions. Consequently, the aim of this chapter is to provide an overview of the structural biology of proteins and protein-protein interactions that enable Gram-positive and Gram-negative bacteria to sequester heme from the extracellular milieu, import it to the cytosol, and degrade it to mine iron.

Keywords bacterial heme metabolism • bacterial heme transport • bacterial heme uptake • heme-binding proteins • heme-degrading enzymes • hemophore

Please cite as: *Met. Ions Life Sci.* 12 (2013) 279–332

1 Introduction

Heme is a ubiquitous cofactor needed for the proper function of a large number of proteins and enzymes. It is therefore not surprising that bacteria developed a heme biosynthetic machinery to meet the demands of the many metabolic paths where heme is required. Pathogenic bacteria seem to be unique in that they possess, in addition to heme biosynthesis, the ability to sequester heme from their mammalian hosts in order to obtain essential iron. Likely as a consequence of depending upon heme as a source of iron, some pathogenic bacteria have lost the ability to synthesize heme. For these organisms exogenous heme is not only a source of iron but is also directly used as a cofactor in heme containing enzymes and proteins. Bacteria in this category include the human pathogens *Porphyromonas gingivalis* [1] and *Haemophilus influenzae* [2,3].

Iron is an essential nutrient for all but a few organisms. Bacteria have evolved a variety of mechanisms for obtaining iron from their surroundings, for example by secreting molecules with very high affinity for the ferric (Fe^{3+}) ion known as siderophores [4,5]. Siderophores transport the captured ferric ion to the cell surface from where the complex is transported into the cytosol. Pathogenic bacteria face the unique problem that the vast majority of iron in their hosts is sequestered in enzyme active sites, or in the storage molecules ferritin and transferrin. Pathogenic bacteria have generally adapted to these low-iron conditions by developing additional iron acquisition pathways that target the proteins harboring the nutrient. Approximately 95% of all iron in mammals is complexed in the heme prosthetic group, primarily in hemoglobin (Hb) [6]. As a consequence, hemoglobin is one of the major iron sources for pathogenic bacteria. Host proteins that scavenge heme or hemoglobin from blood plasma for eventual recycling, such as serum albumin, haptoglobin, and hemopexin (HPX), serve as additional targets [7].

In this chapter our objective is to present a summary of the immense literature describing heme uptake and metabolism, with emphasis on the structural biology of the proteins that enable bacteria to capture heme from a mammalian host and transport it to the cytosol for utilization or degradation. Although we focus primarily on systems which have been characterized at the atomic level, we want to emphasize that the structural knowledge summarized here has been largely enabled by previous extensive genetic analysis. The scope of the chapter is captured in Schemes 1a and 1b, which summarize the main aspects of heme uptake and metabolism that occur in Gram-negative and in Gram-positive bacteria, respectively.

2 Heme Biosynthesis

Heme is the ferrous (Fe^{2+}) complex of protoporphyrin IX (PPIX). The corresponding ferric (Fe^{3+}) complex is referred to as hemin. When referring generally to iron PPIX we will use the term heme, but we will use hemin when the oxidized form is specifically implicated.

2.1 δ-Aminolevulinic Acid, the First Committed Precursor for Tetrapyrroles

The steps involved in the biosynthesis of heme are well-established, and have been detailed in several reviews [8–10]. The key intermediates in the process are shown in Scheme 2, and their transformations are summarized below.

The first committed precursor for tetrapyrrole synthesis in all organisms is δ-aminolevulinic acid (ALA). Although two pathways have been identified for ALA synthesis in bacteria, the pathogenic bacteria at the center of the present review all utilize the "C_5-pathway." This pathway begins with glutamyl-tRNA, which is reduced to glutamate-1-semialdehyde (GSA) by the NADPH-dependent glutamyl-tRNA reductase. GSA is subsequently converted to ALA via a transamination reaction, catalyzed by the pyridoxal phosphate (PLP)-dependent enzyme glutamate-1-semialdehyde-2,1-aminomutase. Condensation of two molecules of ALA by δ-aminolevulinate dehydratase yields the substituted pyrrole porphobilinogen, which bears acetate (A) and propionate (P) substituents on the β-positions. The next series of reactions is catalyzed by porphobilinogen deaminase, which harbors a dipyrromethane unit (two pyrroles connected by a methylene group) covalently attached via a thioether linkage with a cysteine side chain. Four molecules of porphobilinogen are sequentially added to this dipyrromethane unit, with release of an ammonium ion in each step, to give an enzyme-linked hexapyrrole. Cleavage of the hexapyrrole releases a linear tetrapyrrole which, following reaction with water,

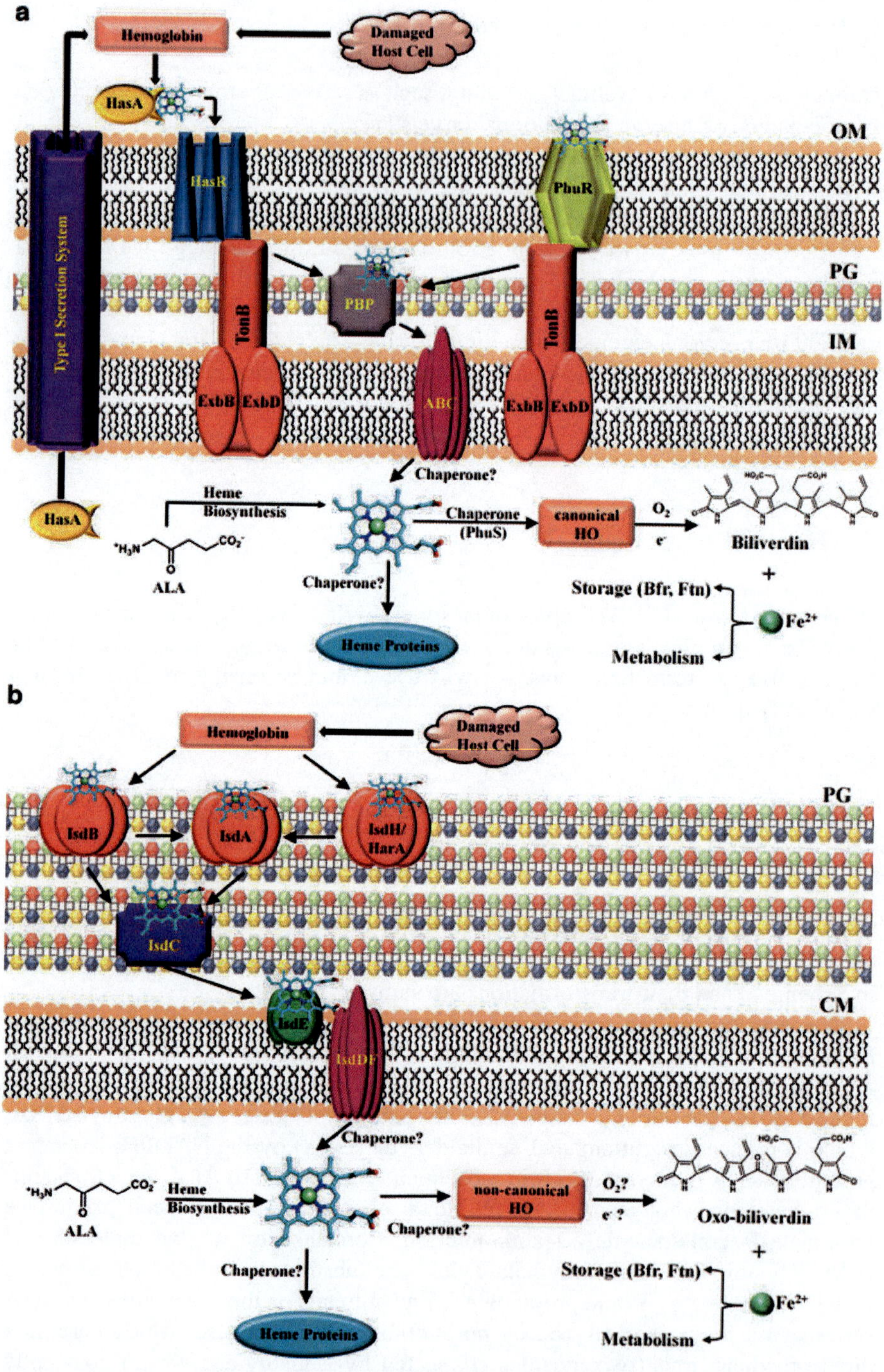

Scheme 1 (**a**) Cartoon representation of heme uptake, biosynthesis, and degradation in Gram-negative bacteria. The scheme is based largely on information obtained from *Serratia marcescens* and *Pseudomonas aeruginosa*. ALA is the first committed precursor for heme biosynthesis (see Scheme 2). Heme in the cytosol is likely trafficked by heme chaperones, which deliver it for incorporation into heme proteins or for degradation by heme oxygenase (HO) enzymes. Heme uptake across the outer membrane can be redundant, and involves TonB-dependent cell surface

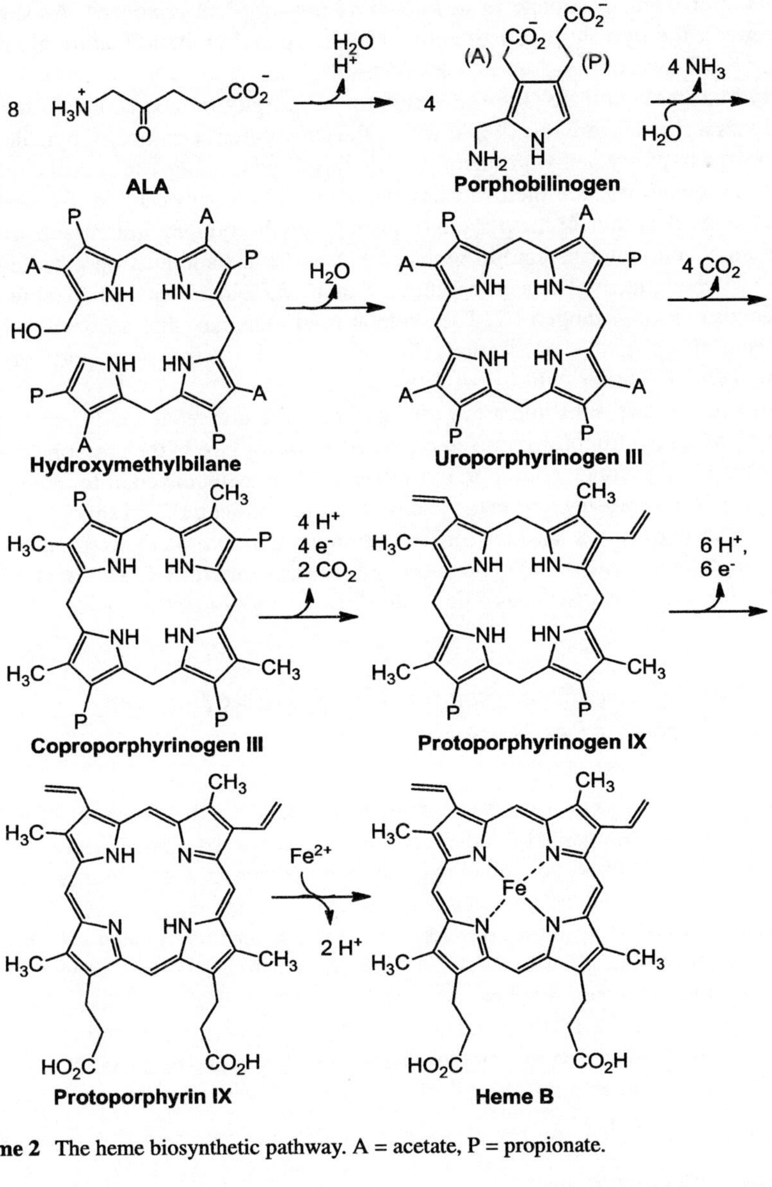

Scheme 2 The heme biosynthetic pathway. A = acetate, P = propionate.

Scheme 1 (continued) receptors with or without the aid of secreted hemophores (e.g., HasA). Heme is transported across the periplasmic space by heme-binding periplasmic binding proteins (PBPs) which deliver their cargo to heme-specific ABC transporters for import into the cytosol. OM (outer membrane), IM (inner membrane). **(b)** Analogous representation in Gram positive bacteria as exemplified by *Staphylococcus aureus*. The IsdEDF complex is an ABC transporter to import heme into the cytosol. Some Gram-positives utilize hemophores (e.g., *Bacillus anthracis*) in addition to cell surface receptors. It is also interesting to note that heme degradation by non-canonical HO enzymes in Gram-positives produces oxo-biliverdin, whereas the canonical enzymes in Gram-negatives produce biliverdin. CM (cell membrane), PG (peptidoglycan).

yields hydroxymethylbilane (also known as pre-uroporphyrinogen). As shown in Scheme 2, the pyrrole β-substituents in hydroxymethylbilane feature alternating A and P substituents (A-P-A-P-A-P-A-P).

Hydroxymethylbilane is the substrate for uroporphyrinogen III synthase, which catalyzes a ring closure reaction in which the substituted α-carbon of pyrrole ring 4 attacks the hydroxymethyl group of pyrrole ring 1, generating a spirocyclic intermediate. This intermediate then reopens between pyrrole rings 3 and 4 (numbering refers to the positions of the pyrrole rings in the hydroxymethylbilane substrate), to yield an acyclic intermediate in which the A and P substituents on pyrrole ring 4 have switched places. The acylic intermediate then recloses via a Michael reaction, yielding uroporphyrinogen III. The skeletal rearrangement that occurs as a consequence of this sequence of reactions results in the substituents in uroporphyrinogen III following the order A-P-A-P-A-P-P-A.

All four acetate substituents in this intermediate are sequentially converted to methyl groups by uroporphyrinogen III decarboxylase. Two of the propionate groups in the resulting product, coproporphyrinogen III, are then converted to vinyl groups by coproporphyrinogen oxidase to yield protoporphyrinogen IX. Protoporphyrinogen IX oxidase performs a six electron oxidation on this intermediate, producing the fully conjugated immediate heme precursor, protoporphyrin IX. The final stage of heme biosynthesis is insertion of Fe^{2+} into PPIX by ferrochelatase.

2.2 *Heme B Is a Precursor for Tetrapyrroles Involved in Bacterial Respiration*

The respiratory chains of many bacteria are branched and utilize a variety of terminal oxidases. The most extensively studied are the heme-copper oxygen reductases, typified by cytochrome *c* oxidase, but quinol oxidases are common as well. These terminal oxidases utilize one of two tetrapyrrole cofactors derived from heme B known as heme O and heme A. Heme O synthase converts heme B into heme O by attaching a hydrophobic hydroxethylfarnesyl tail to the vinyl group on position 3. Heme A synthase then converts heme O to heme A by oxidation of the 8-methyl group to an aldehyde.

Cytochromes *bd* are an additional class of oxygen reductases in bacteria [11]. As the name implies, these reductases employ heme B, but also utilize heme D, which is a heme B derivative containing a spirolactone functionality and two hydroxyl groups.

3 Heme Acquisition

3.1 *Heme Sources*

The vast majority of heme is present in the blood, where it serves as the cofactor in the oxygen transport protein hemoglobin. Proper functioning of Hb requires the heme iron atoms to be in the ferrous (Fe^{2+}) oxidation state. Autoxidation, resulting from transfer of an electron from heme to molecular oxygen, generates methemoglobin (metHb) in which iron is in the ferric (Fe^{3+}) state. A very efficient reductase system returns most methemoglobin molecules to the functional ferrous form as long as they remain within erythrocytes [12]. However, erythrocytes occasionally undergo spontaneous lysis, resulting in release of their contents into the plasma. Once in the plasma, the tetrameric Hb molecules enter into equilibrium with the α/β-dimeric form, and become more susceptible to autoxidation [13]. Ferric heme is bound much less strongly than ferrous heme, and therefore is more readily released into the plasma.

Mammals possess an efficient system for recycling of the iron atoms from Hb following hemolysis [7], culminating in degradation of the PPIX ring by heme oxygenase in the liver and other organs [14]. Pathogenic bacteria take advantage of this system to acquire heme [4,5,7,15], which may be incorporated into their own hemoproteins [16] or enzymatically degraded to release the iron [17,18]. Many pathogenic bacteria further aid their survival by promoting hemolysis, through the agency of multimeric pore forming proteins known as hemolysins [19,20].

The dimeric form of Hb in the plasma is efficiently captured by the homo-dimeric protein haptoglobin (Hp) [7], thereby preventing unwanted redox chemistry from occurring [21]. The Hp-Hb complex is recognized by a specific receptor protein (hemoglobin scavenger receptor), which mediates endocytosis of the complex [22]. Endocytosis occurs primarily in the liver and spleen, but also in macrophages.

Free heme is also scavenged rapidly, primarily by high- and low-density lipoproteins, with the remainder being taken up by human serum albumin (HSA) and hemopexin. Hemin bound to the lipoproteins and HSA is subsequently transferred to HPX [7]. HSA represents about half of the protein present in blood serum, and is a promiscuous binder of hydrophobic compounds including fatty acids and hemin. Crystal structure analysis of the hemin-HSA complex has shown that the hemin is bound specifically, in a site normally occupied by a fatty acid [23]. Heme binding specificity is aided by salt bridges with the two heme propionate groups, and coordination of hemin iron by the side chain of tyrosine residue Tyr-36. The complex is quite strong, with $K_d \sim 10$ nM [24].

HPX is a glycoprotein that is present at low abundance in plasma but has one of the highest known affinities for heme (K_d below 1 pM) [25,26]. HPX is therefore highly efficient at scavenging heme that is released into the plasma, and at sequestering it from lipoproteins and from heme-HSA. HPX delivers its hemin cargo to a specific cell surface receptor. The primary uptake occurs in the liver, but macrophages and other cell types also express the receptor [27].

Free heme, Hb, the Hb-Hp complex, and the heme-bound forms of HSA and HPX all serve as sources of heme for pathogenic bacteria.

3.2 *Heme Scavenging from the Extracellular Milieu*

Gram-positive bacteria feature cell walls with a thick peptidoglycan layer, whereas Gram-negative bacteria possess two lipid bilayer membranes, outer and inner, separated by periplasm. This difference in cellular architecture has necessitated the evolution of distinctly different systems for heme thievery by Gram-positive and Gram-negative bacteria. In pathogenic bacteria, one or more cell surface receptor proteins receive heme directly from a hemoprotein target. In some cases, however, the bacteria release soluble proteins (called hemophores) that either retrieve heme from target proteins and deliver them to the surface-localized receptors, or deliver the hemoprotein itself to the cell surface.

In this section we will first focus on scavenging of heme from the extracellular milieu by hemophores and/or surface proteins of pathogenic bacteria, and their subsequent transport into the cytoplasm. Our principal focus will be heme acquisition systems that have been extensively characterized in terms of structures of the proteins involved, and the mechanisms by which they function. We will emphasize those systems that have been characterized since the mini-review published by Tong and Guo in early 2009 [28]. We will first discuss Gram-negative systems, and then turn our attention to the Gram-positives.

3.2.1 Gram-Negative Bacterial Heme Uptake Systems

3.2.1.1 Classification:

In a review published in 2000, Wandersman and Stojiljkovic classified two types of heme uptake systems in Gram-negative bacteria [29]. In one class, heme uptake involves direct binding of heme or heme-containing proteins to specific receptors localized on the outer membrane. These include ShuA from *Shigella dysenteriae* [30], ChuA from *E. coli* O157:H7 [31], and PhuR from *Pseudomonas aeruginosa* [32]. The second class has the same type of surface receptors as above, but heme is delivered to them by secreted proteins known as hemophores. The heme acquisition system (Has) from *S. marcescens* [33,34] is the most thoroughly characterized one in this category, followed closely by the corresponding system from *P. aeruginosa* [35]. Other bacteria with a related hemophore/receptor pair include *Pseudomonas fluorescens* [36] and *Yersinia pestis* [37]. Distinctly different hemophores have been identified in the Gram-negative bacteria *H. influenzae* (HuxA) [38,39] and in *P. gingivalis* (HusA) [40].

3.2.1.2 Outer Membrane Receptor Proteins:

All known outer membrane (OM) heme uptake receptors are members of the TonB-dependent outer transporter (TBDT) family. TonB, together with the proteins ExbB and ExbD constitute an energy-transducing complex, which uses the proton motive force to drive substrate transport [41]. Structures of several members of the TBDT protein family have been determined by X-ray crystallography [42], including the heme uptake protein ShuA from *Shigella* (PDB code 3FHH) [43], which is shown in Figure 1. The TBDT fold comprises two unique domains: (i) a β-barrel with 22 antiparallel strands, having an internal diameter of 35–40 Å; and (ii) an N-terminal

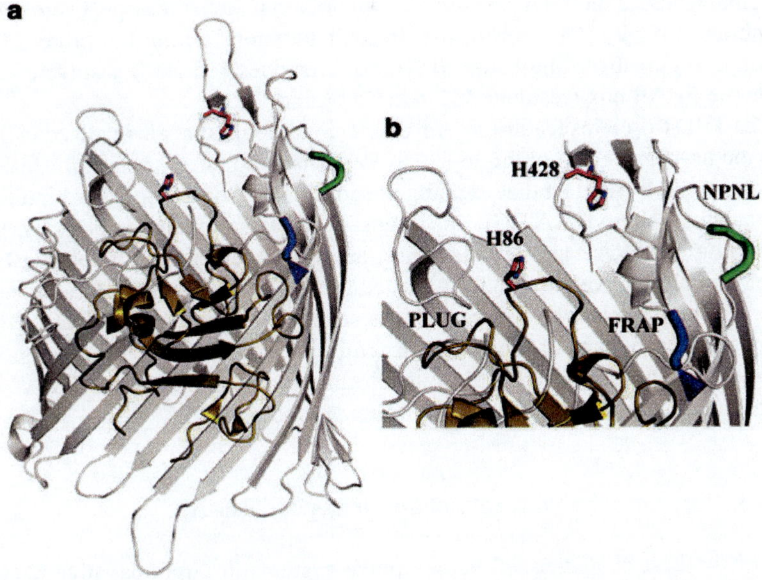

Figure 1 Cartoon representation of the X-ray crystal structure of ShuA from *Shigella dysenteriae* (PDB code 3FHH). (**a**) Full structure, with the 22-stranded β-barrel domain shown at 50% transparency to better reveal the structure of the central plug domain, represented in olive-green. (**b**) Enlarged view of the heme-binding region of the protein. In both panels the locations of the FRAP and NPNL boxes are indicated in blue and green, respectively, and essential His86 and His428 are shown in sticks.

"hatch" or "cork" domain that partially fills the internal cavity and prevents passive diffusion through the barrel. TBDT proteins also contain a sequence near their N-terminus, known as the TonB box, which undergoes a conformational transition following ligand binding which allows the TonB box to interact with TonB [44,45].

Two highly conserved His residues have been identified among TBDT proteins involved in heme transport [46]. As highlighted for ShuA in Figure 1, one is located in an extracellular loop (His428), the other on the extracellular face of the cork (His86) [43,47]. These conserved His residues have been shown to be essential for heme uptake *in vivo* in ShuA [48], HmuR from *P. gingivalis* [49], HemR from *Y. enterocolitica* [48], and HasR from *S. marcescens* [50]. The recently characterized heme receptor (Hma) from a uropathogenic *E. coli* strain lacks both of the conserved His residues, but instead has a Tyr residue (Tyr126) that is essential for Hma-mediated heme utilization [51].

Two conserved amino acid motifs (the FRAP and NPNL boxes) have also been identified, which mark the beginning and end, respectively, of the extracellular loop containing one of the conserved His residues [46]. Mutagenesis studies have indicated that the NPNL box plays a role in binding of heme and heme proteins, whereas the FRAP box is likely involved in heme transport across the cell membrane [49,52]. This is reflected in their locations within the protein fold: as shown in Figure 1 for ShuA, the NPNL box (residues 414–417) is solvent-exposed while the FRAP box (residues 442-445) is buried.

The TBDT proteins feature several extracellular loops that are involved in docking with heme-bearing donor proteins, as demonstrated in the case of HasR from *S. marcescens* by the fact that excising portions of two loops abolishes binding of its cognate hemophore HasA [53]. The largest of the extracellular loops (L7) is the one containing the FRAP and NPNL boxes, along with essential His residue His420, which is well-resolved in the ShuA structure (Figure 1). Several of the other loops could not be modeled, however, due to absence of corresponding electron density. An interesting feature of the ShuA structure in Figure 1 is that the essential His residues His86 and His420 are quite distant from one another (9.86 Å), which suggested an induced fit mechanism for heme binding [43].

3.2.1.3 Structure and Properties of the Hemophore HasA:

One of the best characterized heme uptake systems in Gram-negative bacteria is produced by the Has operon. The system includes the TBDT surface receptor, HasR, which is able to obtain heme directly from Hb, but efficiency of the process is increased by 100-fold through the agency of a secreted hemophore, HasA [33]. The Has system of *S. marcescens* was the first to be discovered, and remains the most thoroughly characterized in terms of both structure and function. The analogous system in *P. aeruginosa* has been the subject of more recent structural and biophysical studies, particularly the hemophore component. In the following discussion, HasA from *S. marcescens* and from *P. aeruginosa* will be referred to as HasAs and HasAp, respectively.

HasAs contains 188 amino acids, and folds stably in the absence of heme. In fact, CD data have revealed that heme binding by apo-HasAs exerted little effect on polypeptide conformation [54]. This observation was largely confirmed in subsequent X-ray crystallography studies, discussed below. Initial biophysical studies suggested that hemin is bound in the low-spin state, but subsequent direct-detected ^{13}C NMR studies revealed a high-spin low-spin thermal equilibrium [55]. The initial biophysical studies also revealed that HasAs exhibits one of the most negative redox potentials ever recorded for a hemoprotein (–550 mV) [54]. X-ray crystal structures have been obtained for the wild-type holo forms of this hemophore, initially for HasAs (PDB code 1B2V) [56], and later for HasAp (PDB code 3ELL) [57]. The crystal structure of HasAs, shown in Figure 2, revealed a protein fold that is novel in several respects. For example, the bound heme is surrounded almost completely by loops, whereas in most heme proteins the heme binding pocket is dominated by helices. Moreover, the overall fold was new, resembling a fish, with the mouth of the fish representing the heme binding site. The body of the fish can be thought of as having two structurally distinct sides that are packed against one another. One side comprises a 7-stranded β-sheet with a strong twist, while the other contains four

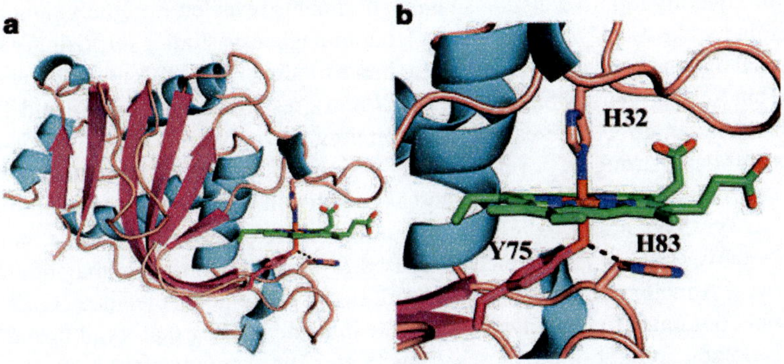

Figure 2 Cartoon representation of the X-ray crystal structure of the hemophore HasA from *Serratia marcescens* (PDB code 1B2V) (**a**) Full structure showing overall arrangement of helices and strands; (**b**) Enlarged view of the heme-binding pocket, showing ligation of hemin by His32 and Tyr75, and the hydrogen-bonding interaction between Tyr75 and His83.

α-helices. Perhaps the most surprising aspect of the protein structure was the unusual heme ligation. Prior to the structure determination, EPR data had been interpreted to indicate that the heme was bis-histidine ligated, but with one of the ligands in the conjugate base form (His^-) [54]. The crystal structure demonstrated that ligation was instead provided by one His side chain (His32) and one Tyr side chain (Tyr75). The only previously reported example of His/Tyr ligation in a natural heme protein was heme d_1 in cytochrome cd_1 nitrite reductase [58]. Catalases were the only other proteins known to have tyrosine as a hemin axial ligand, but with the sixth coordination site unoccupied [59]. The observation of Tyr ligation in HasAs provided a ready explanation for its very negative redox potential, as Tyr75 most likely coordinates in its conjugate base form ($TyrO^-$) which would strongly favor the ferric oxidation state for heme. Another possible factor stabilizing Fe^{3+} relative to Fe^{2+}, suggested by the crystal structure, is fairly extensive solvent exposure of the heme (186 $Å^2$).

Each heme ligand in HasA was shown to reside in a loop. The loop containing Tyr75 separates two strands of the β-sheet portion of the main fold, while the loop containing His83 bridges a β-sheet strand on one side of the body and an α-helix on the other side. An interesting feature in the second coordination sphere is the side chain of His83, which forms a tight hydrogen bond with the Tyr75 hydroxyl group via its Nδ1 NH group. It has been proposed that this hydrogen bonding interaction increases the tyrosinate character of Tyr75, thereby strengthening it as a ligand [55,60]. This is supported by NMR data revealing that the high-spin/low-spin equilibrium exhibited by HasA does not involve a change in heme ligation, but rather relates to modulation of the strength of the His83-Tyr75 H-bond [61]. The observation from the crystal structure that the Nε1 nitrogen of His83 is solvent-exposed suggested a model for heme release. Specifically, the His83 side chain might serve as a trigger for heme release, because it could relay a proton from bulk solvent to the Tyr75 phenoxide group. Protonation of the Tyr75 phenoxide group would weaken it as a ligand, favoring release of the heme.

Subsequently reported displacement experiments revealed extremely tight binding of hemin by HasAs ($K_a = 5.3 \times 10^{10}$ M^{-1}), but mutagenesis studies surprisingly showed that axial ligand His32 is not essential for heme binding (the H32A mutation decreased heme binding strength by only about 1 kcal/mol) [62]. Mutating heme ligand Tyr75 to Ala had a larger effect, decreasing heme binding strength 400-fold, but residual binding is still quite strong ($K_a = 1.3 \times 10^8$ M^{-1}). Similar decreases accompanied mutation of the non-ligating His83 with Ala, and also with Gln which can maintain H-bonding with the Tyr75 phenolate oxygen. Larger effects were observed for double mutations, most notably H32A/Y75A, which decreased K_a nearly six orders of magnitude to 5.9 $\times 10^4$ M^{-1}. A major conclusion of this work is that heme pocket residue His83, which does not coordinate to Fe, plays a larger role in heme binding than axial ligand His32. This is notable in light of the fact that His83 and Tyr75 are invariant in known examples of HasA, whereas His32 is not. For example, position 32 is occupied by Gln in HasA from *Y. pseudotuberculosis* and *Y. pestis* [63].

In a follow-up spectroscopic and crystallographic study, it was shown that the H83A mutation caused dissociation of the Tyr75 ligand [61]. The mutant exhibits equilibrium involving a high-spin five-coordinate form and six-coordinate high- and low-spin forms. The fifth ligand in all three species appears to be His32, while the sixth ligand is thought to be water, possibly with some hydroxide character due to H-bonding with the Tyr75 side chain. It was proposed that this species represents an intermediate in the HasA heme transfer mechanism.

The solution structure of apo-HasAs from *S. marcescens* was solved by 3D NMR methods [64], and the results confirmed that apo- and holo-HasAs have largely identical folds. The loop containing Tyr75 and His83 was shown to be virtually identical in the apo and holo structures. Nonetheless, the apoprotein was observed to exhibit an "open" structure because reorientation of the loop containing heme ligand His32 moves this residue ~30 Å from its location in the holoprotein [64]. More recent X-ray crystallographic studies comparing the apo and holo forms of HasAp have confirmed this observation (see Figure 3 below) [63].

3.2.1.4 The Heme Loading Mechanism of HasA:

At present, little is known about the mechanism by which HasA obtains its hemin cargo from metHb. Analytical ultracentrifugation experiments involving mixtures of apo-HasAs and Hb yielded no evidence for stable complex formation [65]. However, results of studies in which human metHb was titrated into a sample of ^{15}N-labelled HasAp, monitored by ^{1}H-^{15}N-HSQC NMR spectra, indicated a specific recognition involving part of the HasAp loop containing ligand His32 [57]. This transient complex formation, which may be involved in the hemin transfer reaction, appears not to accelerate heme transfer. Indeed, rate constants for hemin transfer from the α- and β-subunits of metHb to apo-HasAp are virtually the same as literature values for dissociation of hemin from those subunits [66]. Thus, hemin uptake by HasA from metHb appears to be a passive process.

Despite the paucity of evidence regarding acquisition of hemin from metHb by apo-HasA, substantial insight into the mechanism by which apo-HasA binds free hemin has been obtained. A key question in the mechanism of this reaction is the order in which the ligands coordinate to heme iron. Initial evidence supporting Tyr75 ligation followed by His32 ligation was obtained by resonance Raman (RR) studies by Lukat-Rodgers et al., who demonstrated that reduction of Fe^{3+} to Fe^{2+} in holo-HasAs in the presence of CO led to a complex exhibiting TyrO-Fe(II)-CO ligation, suggesting that His32 is the more readily displaced ligand [67]. Studies with the H32A mutant of HasAp have shown this to be the case for the ferric oxidation state as well [63]. The apo form of the H32A mutant is monomeric, a characteristic it shares with the wild-type apoprotein. In contrast to the wild-type apoprotein, however, reconstitution of the apo H32A mutant with hemin yields monomeric and dimeric holoproteins which are not in dynamic equilibrium and thus could be separated. Interestingly, crystallization experiments with the monomeric holoprotein yielded a dimer exhibiting a co-facial interaction between the two bound hemin molecules. Each hemin molecule in the dimer is ligated by the Tyr75 side chain of one of the protein components, while the conformation of the loop normally containing His32 (referred to as the Ala32 loop) is virtually identical to that of the corresponding loop in wild-type apo-HasAp. This was taken as additional evidence that Tyr75 ligates to hemin iron first, followed by ligation of His32 which is accompanied by a large conformational reorganization. This observation also suggested that dimer formation may be physiologically relevant for pathogenic bacteria in which residue 32 of HasA is not occupied by His (e.g., *Y. pestis*).

Follow-up investigations carried out with stopped-flow methods coupled to electronic absorption spectroscopy and rapid-freeze quench methods coupled to resonance Raman spectroscopy provided experimental support for the induced-fit mechanism of heme binding described above [66]. In addition, these investigations showed that hemin loads onto HasAp within 4 ms but surprisingly, closing of the His32 loop, monitored by the spectroscopic changes associated with the formation of the His32-Fe bond, occurs in approximately one second. Thus, formation of the His-Fe bond occurs orders of magnitude more slowly than formation of the Tyr-Fe bond, as illustrated in Figure 3 [66].

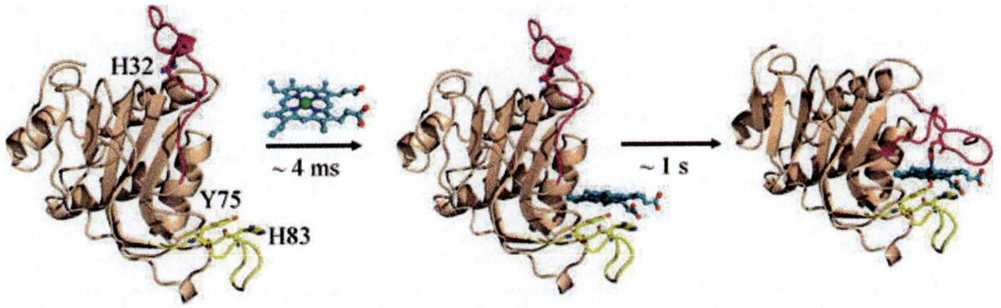

Figure 3 A model of the heme-loading mechanism of HasA, as determined from studies with the *P. aeruginosa* protein. On the left is the X-ray crystal crystal structure of apo-HasAp, with the His32 loop shown in yellow and the Tyr75 loop shown in magenta. Initial binding of hemin is complete within ~4 ms and involves only interactions with the Tyr75 loop. Hemin binding triggers closure of the His32 loop and ligation of hemin iron by the His32 side chain, a process that requires approximately one second.

Efficient capture of hemin on the Tyr75 loop of HasAp was attributed to a combination of factors [63]. Perhaps most importantly, the Tyr75 loop structure is virtually identical in the apo and holo states, and features several exposed hydrophobic residues. It therefore appears to serve as a "sticky" platform for hemin docking. In addition, the structural organization of the Tyr75 loop favors H-bonding between the side chains of His83 and Tyr75. This is likely to facilitate hemin docking by allowing coordination of Fe^{3+} by a phenolate-like ligand. Data from NMR experiments and targeted molecular dynamics studies suggest that subsequent closure of the His32 loop is due to changes in polypeptide dynamic behavior arising from heme docking [63].

3.2.1.5 *S. marcescens* Secretes a Dimeric Form of HasAs:

It has been demonstrated that, under low iron conditions, HasAs is secreted by *S. marcescens* in a dimeric form as well as in the monomeric form described above [68]. This dimer binds two equivalents of hemin, but is unable to deliver hemin to HasR. The dimer-loaded HasAs can relinquish its cargo to monomeric HasAs, however, which led to the suggestion that it might serve as a heme reservoir [68]. A crystal structure (PDB code 2CN4) revealed the complex to be a domain-swapped dimer. The structure of each protein component is largely identical to that of monomeric HasAs. However, each heme iron is ligated by Tyr75 of one monomer and His32 of the other monomer. The hemin molecules in the dimer do not make direct contact with one another. Binding is thus distinct from that observed for the H32A mutant of HasAp described above, in which bound hemin itself is dimeric and each hemin molecule is ligated by Tyr75 from one protein component.

3.2.1.6 A Model for Delivery of Heme from HasAs to HasR:

HasR is able to sequester both free heme, and to obtain it from holo-HasA [33]. Although HasR has weaker affinity for heme than does HasA, the transfer occurs spontaneously *in vivo* [50], using energy from the strong binding between the hemophore and the receptor (binding is strong whether or not heme is bound in HasAs; K_d=7 nM in both cases [65]). While a crystal structure of free HasR is not yet available, a structure of the homologous heme TBDT protein ShuA has been described (see above). In addition, structures have been reported for complexes in which holo-HasAs and apo-HasAs are bound to wild-type HasR (PDB codes 3CSL and 3CSN, respectively) and holo-HasA is bound to the I671G mutant of HasR (PDB code 3DDR) [47]. The structures have led to a model for heme delivery from holo-HasA to HasR that was originally suggested by results of solution NMR studies [69]. The model is illustrated in Figure 4, and described below. In Figure 4 and accompanying description, heme receptor residues are numbered according to their locations in HasR.

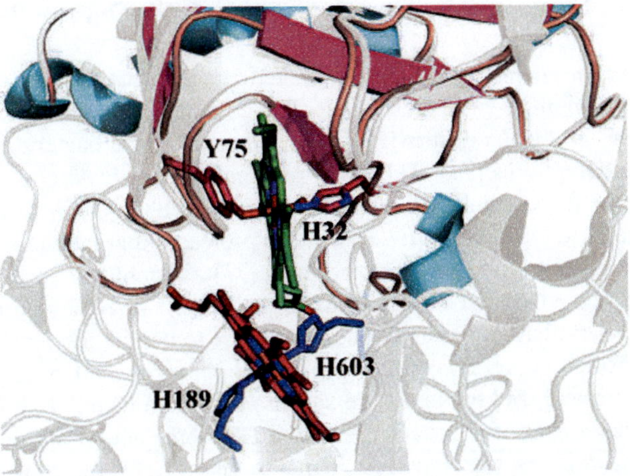

Figure 4 Superimposed crystal structures of holo-HasA (PDB ID 1B2V) and of the HasA-hemin-HasR complex (PDB ID 3CSL) from *Serratia marcescens*. The structure of the HasA-hemin-HasR complex is shown in grey at 80% transparency. The overlay highlights the change in hemin ligation and location that accompany its transfer from HasA to HasR. The HasR heme ligands His189 and His603 are shown in blue. These correspond to essential His86 and His428, respectively, in the structure of ShuA shown in Figure 1. In the HasA-hemin-HasR complex, the His32 loop could not be modeled due to absence of corresponding electron density, suggesting it has been displaced in a manner analogous to that observed in the structure of apo-HasAp in Figure 3.

As noted above, the two heme-binding His residues in the ShuA structure are 9.86 Å apart, and it is reasonable to assume that the corresponding residues in HasR are similarly displaced (His603 in the solvent exposed loop L7; His189 on the cytosolic face of the central plug). Prior mutagenesis studies had identified both of these His residues as being involved in heme acquisition by *S. marcescens*, with His189 playing the dominant role [50]. The structure of the complex between holo-HasA and the I671G mutant of HasR is thought to represent the wild-type complex before heme transfer. In this complex, heme is still ligated by Tyr75 of HasA, but His32 has dissociated and the His32 loop has pivoted away (the loop could not be modeled due to the absence of electron density). It is therefore consistent with the structures of apo-HasAs obtained in the absence of HasR, which as noted previously revealed an open conformation resulting from His32 loop movement. In addition, loop L7 of HasR has undergone a large conformational change relative to its position in the homologous ShuA structure, bringing ligand His603 to within 7 Å of His189. In the complex between holo-HasA and wild-type HasR, the heme has moved 9 Å relative to its location in the complex with HasR (I671G), and is coordinated by HasR residues His189 and His603. This observation provided support to a hypothesis that Ile176 aids in heme delivery from HasA to HasR by means of a steric push. Another factor contributing to the push may be changes in the dynamic and conformational properties of the Tyr75 loop that occurs in the HasA/HasR complex, leading to weakening of the H-bond between His83 and Tyr75 [64].

3.2.1.7 Heme Uptake by *Porphyromonas gingivalis* May Involve Hemophores:

Porphyromonas gingivalis is a black-pigmented Gram-negative anaerobic bacterium that is an early colonizer of dental plaque, and a leading pathogen in chronic periodontitis [70]. *P. gingivalis* differs from other Gram-negative bacteria in that it does not produce siderophores [71]. The black pigmentation of *P. gingivalis* arises from deposits of heme μ-oxo dimers on the cell surface, which may serve as a rudimentary antioxidant defense [72]. *P. gingivalis* secretes proteases that facilitate heme release from Hb, as well as from other heme proteins [73]. Current evidence indicates that acquisition of heme from Hb involves initial proteolysis of Hb by arginine-specific gingipain RgpA, which results in oxidation of heme to hemin, thereby facilitating its dissociation. Following release of hemin from the damaged metHb, the Hb polypeptide is further proteolyzed by the lysine-specific gingipain Kgp [74–76].

The major recipient of heme released from Hb by the action of the gingipains is HmuY, which delivers the captured heme to the TBDT receptor HmuR [71]. Although HmuY is an outer-membrane associated lipoprotein, its expression is increased under iron starvation conditions and some of the protein is released into the culture medium within vesicles as well as in soluble form [77]. Under these growth conditions, HmuY may therefore function as a hemophore. Partitioning of a heme-acquisition protein between the cell envelope and the medium has also been observed in *Corynebacterium diphtheriae*, the causative agent of diphtheria [78], as well as in *Haemophilus influenzae*, described in the next section.

HmuY binds hemin in the low-spin state, which is readily reduced to the low-spin ferrous state ($E^{0'} = 136$ mV) [79]. Mutagenesis studies revealed that heme is ligated by the side chains of His134 and His166 [79]. In addition, CD data indicated that HmuY adopts a largely β-structure [79,80] that is not altered by heme binding [81]. Both apo- and holo-HmuY are resistant to trypsin and to the major gingipains released by *P. gingivalis*, suggesting a compact fold with or without heme present. A crystal structure of the heme-bound protein (PDB code 3H8T) confirmed the largely β-structure predicted by CD, revealing a novel fold comprising 15 β-strands that resembles a right hand [81] (see Figure 5). Heme is bound between the fingers of the hand (a four-stranded anti-parallel β-sheet) and the thumb (a two-stranded antiparallel β-sheet). Heme is best described as lying between two anti-parallel β-strand pairs (i.e., two fingers and the thumb), each pair of strands separated by a short loop. Heme ligands His134 and His166 are each located near the terminus of one of the β-strand pairs. The authors of the report note that hemopexin is the only other known heme-binding protein involved in heme-transport that has an all β-structure.

An interesting feature of the HmuY structure is that the heme propionate groups on pyrrole rings A and D project toward the palm of the hand, where they engage in hydrogen bonding interactions. This places the methyl and vinyl substituents of heme pyrrole rings B and C in a location where they would be solvent-exposed if the protein were a monomer, as it likely is prior to heme transfer to HmuR. Notably, however, the quaternary structure of HmuY in the crystal is a tetramer, in which the tips of the thumb and fingers in the heme-binding motifs meet in a cross-like arrangement. In this arrangement, the bound heme molecules make van der Waals contact via the methyl and vinyl substituents on rings B and C. This is consistent with analytical

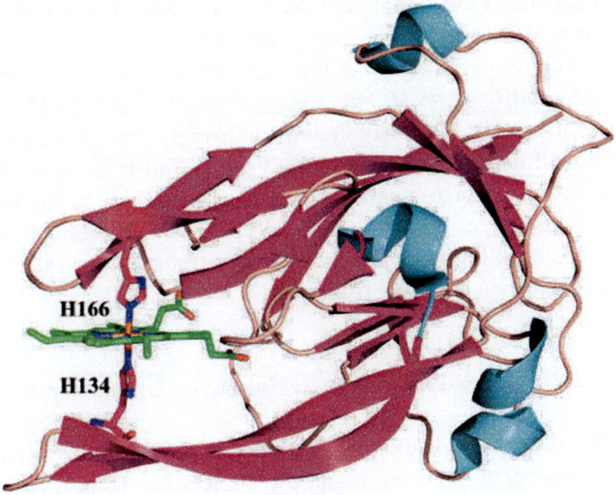

Figure 5 Cartoon representation of the X-ray crystal structure of the hemophore HmuY from *P. gingivalis,* in complex with hemin (PDB code 3H8T). Hemin ligands His134 and His166 are shown in sticks.

size-exclusion chromatography data which indicate that HmuY forms a tetramer upon binding heme [80].

A paper describing another possible *P. gingivalis* hemophore has recently been published. HusA (heme uptake system protein A) is a heme-binding protein that is essential for growth of the organism under heme-limited conditions [40]. It has no significant homology to proteins of known function. As in the case of HmuY, HusA is primarily associated with the outer cellular membrane but under heme-limiting conditions is also released into the growth medium [40]. Heme titration experiments at pH 8.0 indicated formation of a 1:1 heme-protein complex, with $K_a \sim 7 \times 10^{-10}$ M. However, spectroscopic changes upon incubation of HusA with an aqueous solution of hemin (which exhibits a dimer-monomer equilibrium) suggested preference for binding to the μ-oxo dimer form of hemin. Binding studies with the black μ-oxo dimer-containing pigment from *P. gingivalis* revealed a binding constant for dimeric heme at pH 6.5 of $K_d \sim 2 \times 10^{-9}$ M, stronger than binding of monomeric heme at the same pH ($K_d \sim 1 \times 10^{-8}$ M). Size exclusion chromatography data indicated that binding of dimeric hemin caused dimerization of the HusA protein, analogous to the observation for the H32A mutant of HasAp described above.

3.2.1.8 HxuA from *Haemophilus influenzae* Is also a Hemophore:

Relatively little is known about the mechanism of heme acquisition by *H. influenzae*, but it is of interest as it is one of few pathogenic bacteria that can utilize the heme-hemopexin complex as a source of heme. As noted above, *H. influenzae* is a heme auxotroph because it lacks the heme biosynthesis pathway. HxuA is known to be anchored to the outer membrane by a C-terminal signal peptide. The Hxu operon has been reconstituted in *E. coli*. In studies with this strain a small amount of HxuA was recovered from the extracellular medium in a truncated form lacking the 21 residues of the signal peptide [82]. It is not clear whether *H. influenzae* releases the truncated form as well, so it may or may not qualify as a hemophore. Nonetheless, a recombinant version of the secreted form was shown to reconstitute heme acquisition ability of an *H. influenzae* mutant lacking the HxuA gene [82]. The recombinant protein was also shown to strongly bind hemopexin, both with and without bound heme, but had no significant affinity for heme itself. However, transfer of heme from hemopexin to OM receptor HxuC in the HxuA mutant only occurred when the recombinant excreted HxuA was present. HxuA thus appears to mediate transfer of heme from heme-hemopexin to HxuC, rather than to deliver heme to the receptor via complex formation.

3.2.2 Gram-Positive Bacterial Heme Uptake Systems

Relatively fewer examples of heme acquisition systems have been identified in Gram-positive bacteria, and the subject has been recently reviewed [83]. The best characterized system, from *Staphylococcus aureus* [18] and *Bacillus anthracis* [84], is referred to as the iron-responsive determinant (Isd) system. Each gene in the Isd operon is under the control of the Fur repressor; repression is relieved when iron is limiting [85]. The Isd surface proteins (and in *B. anthracis*, the hemophore) contain one to five domains known as NEAT (NEAr iron transporter) domains [86]. A notably different system, which involves two membrane-anchored heme-binding proteins that also contain NEAT domains (Shr and Shp) has been described in *Streptococcus pyogenes* [87]. A heme acquisition system from the Gram-positive species *Mycobacterium tuberculosis*, which includes a hemophore with a unique fold, has been recently discovered as well [88,89].

3.2.2.1 *Staphylococcus aureus*:

There are at least 9 proteins in the *S. aureus* Isd system, four of which are cell-wall-anchored binding proteins (IsdA, IsdB, IsdC, and IsdH) [90]. IsdA, IsdB and IsdH are covalently attached to the peptidoglycan layer by the action of sortase A, while attachment of IsdC is accomplished by sortase B [91]. IsdH (also known as HarA; **ha**ptoglobin **r**eceptor A) is not part of the same operon in *S. aureus*. IsdA and IsdC (the central conduit of the Isd heme uptake system) each contain one NEAT domain, whereas IsdB contains two ($IsdB_{N1}$ and $IsdB_{N2}$) and IsdH contains 3 ($IsdH_{N1}$, $IsdH_{N2}$, and $IsdH_{N3}$). The various domains exhibit widely divergent amino acid sequences. For example, the two IsdB NEAT domains share only 12% pair-wise sequence identity [92] while IsdA and IsdC are only 19% identical [93]. The low homology likely reflects the fact that the various NEAT domains play diverse roles.

IsdB scavenges heme from Hb, and is essential for Hb binding to the cell [94]. Of the two NEAT domains in IsdB, only N1 binds Hb [95] and only N2 binds heme [96,97]. IsdH is able to obtain hemin from Hb as well as from the Hp-Hb complex. Available evidence indicates that $IsdH_{N1}$ and $IsdH_{N2}$ bind Hb, while $IsdH_{N3}$ binds the extracted hemin molecule [98].

3.2.2.2 The Direction of Hemin Transport in *S. aureus*:

Muryoi et al. investigated the directionality of heme transfer in the *S. aureus* Isd system using proteins representing the individual NEAT domains known to bind heme [96]. Utilizing absorption and MCD spectroscopic data, combined with electrospray mass spectrometric data, it was shown that heme flows unidirectionally from IsdA to IsdC to IsdE. They also showed that IsdA is able to obtain heme from both $IsdB_{N2}$ and $IsdH_{N3}$. These findings confirmed a slightly earlier report by Lei and coworkers that IsdA transfers hemin directly to apo-IsdC [99]. However, Lei and coworkers subsequently demonstrated that IsdB can transfer heme to apo-IsdA as well as to apo-IsdC [100], an interesting observation in light of studies showing that IsdA is not required for growth of *S. aureus* when metHb is the iron source [101]. This suggests a possible redundant role for IsdA in heme acquisition from Hb [100]. The last protein in this chain, IsdE, is a lipoprotein and is a component of the IsdEDF complex that transports heme into the cytosol.

Hemin is transferred from IsdA to IsdC more than 70,000 times faster than it is released from IsdA into solution as assessed using the H64Y/V68F double mutant of apoMb as a non-specific receptor [99]. This indicates that the hemin transfer occurs via an activated complex. The directionality of transfer is favored by the fact that IsdC binds hemin about ten times more strongly than does IsdA. Transfer of hemin from IsdB to the stronger binder IsdC also involves an activated complex, occurring 11,500 times faster than its transfer to H64Y/V68F apoMb [100].

It has been reported that binding of heme by IsdC in *S. aureus* [102] is weak (on the order of 1 μM), thereby favoring quick capture and release. This stands in contrast to results of Lei and co-workers, which have shown that IsdC binds hemin more strongly than the H64Y/V68F double mutant of apoMb [99]. Those studies suggest that apo-IsdC binds hemin from solution in two stages, the first involving formation of an intermediate which reorganizes to the properly ligated complex. The first step does involve weak binding ($K_d \sim 3 \times 10^{-6}$ M), whereas overall complexation is much stronger ($K_d \sim 2.6 \times 10^{-12}$ M).

3.2.2.3 Structural Characterization of *S. aureus* NEAT Domains:

The first structure published of a NEAT domain was for $IsdH_{N1}$ (PDB code 2H3K) [103], which binds Hb but not heme. Structures of the apo- and heme-bound forms of the *S. aureus* IsdA NEAT domain, determined by X-ray crystallography (PDB codes 2ITE and 2ITF, respectively), soon followed [104]. The holo-IsdA structure, shown in Figure 6, shows that the NEAT domain fold comprises a β-sandwich having two five-stranded β-sheets, and resembles immunoglobulin. In the holoprotein, bound hemin is ligated by a "proximal" Tyr ligand, with a second Tyr residue four residues away that makes van der Waals contact with the heme while forming an H-bond with the ligating Tyr. The YXXXY sequence containing these Tyr residues is located in a β-hairpin comprising strands β7 and β8, which serves as a structural platform for the heme. Substantial hemin surface area (278 Å^2; ~35%) is exposed to

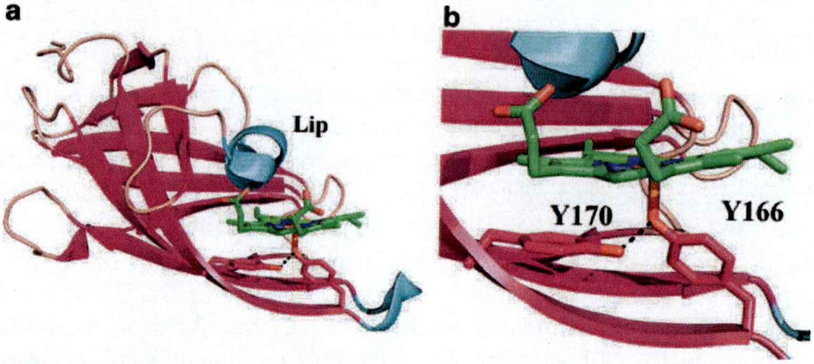

Figure 6 Cartoon representation of the X-ray crystal structure of the IsdA NEAT domain from *S. aureus*, with hemin bound (PDB code 2ITF). (**a**) Full structure illustrating the immunoglobulin-like fold, and indicating the location of the "lip". (**b**) Enlarged view of the hemin binding pocket, highlighting coordination by Tyr166 and the H-bonding interaction between Tyr166 and Tyr170.

solvent, particularly on the "distal" side of the heme pocket. The extent of solvent exposure is even greater than in the Gram-negative hemophore HasA (186 Å^2; ~25%). The heme coordination is strikingly similar to that described above for HasA, in which a His side chain forms an H-bond with the ligating Tyr residue, and may represent an example of convergent evolution. As in the HasA case, this arrangement has been proposed to modulate the Tyr ligand strength [105]. Notably, all of the heme-binding NEAT domains feature a YXXXY motif, whereas those that do not bind heme (IsdB$_{N1}$, IsdH$_{N1}$ and IsdH$_{N2}$) lack this signature [103].

An X-ray crystal structure of the heme-bound form of the IsdC NEAT domain (PDB code 2O6P) [105] was published shortly after the paper reporting the IsdA structures. These authors also noted that the distal side of heme is significantly solvent exposed, with about half of the porphyrin plane covered by a short 3_{10} helix (residues 48–52) described as acting as a "lip". The lip and associated heme exposure are characteristic of NEAT domains, and likely explains the fact that in the crystal structure of IsdH$_{N3}$ (PDB code 2Z6F) [106], two crystallographically related proteins are oriented such that their bound hemes lie parallel to one another. The parallel hemes are in van der Waals contact, with their iron atoms 8.7 Å apart.

In all but one heme-binding NEAT domain from *S. aureus* the heme is penta-coordinate, with the ferric iron lying out of the average porphyrin plane in the direction of the Tyr ligand. The only exception is the IsdB$_{N2}$ domain (PDB code 3RTL), in which the side chain of a Met residue serves as a distal ligand [92]. No other heme-binding protein exhibiting Tyr/Met ligation is known. Ligation by the distal Met ligand appears to be rather weak, however, as indicated by the fact that its side chain adopted two similarly populated conformations in the structure, only one of which places the sulfur atom close enough to the iron atom to allow ligation.

3.2.2.4 On the Mechanism of Hemin Acquisition from Methemoglobin by IsdH:

Recently reported size-exclusion chromatography and isothermal titration calorimetry (ITC) data have shown that the individual $IsdH_{N1}$ and $IsdH_{N2}$ NEAT domains bind Hb, with strong preference in each case for the α-subunit over the β-subunit [107]. Binding was shown to be independent of oxidation or ligation state of Hb. Furthermore, spectroscopic studies indicated that complexation does not induce conformational changes in metHb, and does not promote hemin release. In fact, complexation appears to stabilize metHb against hemin loss relative to a control solution lacking $IsdH_{N1}$. This suggests that release of hemin from metHb in the complex must be facilitated by interactions with the heme-binding NEAT domain, $IsdH_{N3}$.

Insight into the mechanism of acquisition of heme from metHb by IsdH has been obtained through a combination of NMR solution studies of $IsdH_{N1}$ (PDB code 2H3K) [103] and an X-ray crystal structure of the $IsdH_{N1}$-metHb complex (PDB code 3SZK) [107]. Both structures revealed that $IsdH_{N1}$ adopts a fold similar to those of the other Isd proteins that have been characterized, with the exception that it includes an extra helix-loop at the N-terminus. In the crystal of the complex (Figure 7), Hb is present as the αHb-βHb dimer, a species it is likely to encounter in blood serum *in vivo*. $IsdH_{N1}$ interacts most extensively with the A helix of the α-Hb subunit, far from the heme pocket, consistent with the biophysical data summarized above. Docking of $IsdH_{N1}$ to metHb occurs at the same face of the molecule that represents the heme pocket in other Isd proteins. Loops between β-sheet strands represent the predominant structural features that interact with the bound Hb. A particularly important loop (L2) contains five consecutive aromatic residues (Tyr125, Tyr126, His127, Phe128, and Phe129). This loop is disordered in the apo-$IsdH_{N1}$ NMR structure, but becomes helical in the complex with Hb. Many of the side chains of these aromatic residues pack against non-polar residues on α-Hb. Loop 8 also undergoes a major conformation change upon complex formation, involving movements of 8–10 Å. Helix A in the α-Hb subunit contains a Lys residue (Lys11) that is conserved in all known mammalian hemoglobins, while the corresponding conserved residue in the β-subunit is Thr. In the crystal structure of the complex, the Lys11 side chain projects into a pocket in $IsdH_{N1}$ containing multiple carbonyl and hydroxyl groups. This is a key binding determinant, as evidenced by the fact that mutating Lys11 to Thr abolished binding of Hb by $IsdH_{N1}$ and by $IsdH_{N2}$.

The authors suggest that a major role of NEAT domains N1 and N2 in IsdH is to orient Hb to favor efficient transfer of bound hemin from both the α- and β-subunits to $IsdH_{N3}$. They also note that knowledge of the interface between the receptor and its target may aid development of new anti-virulence agents.

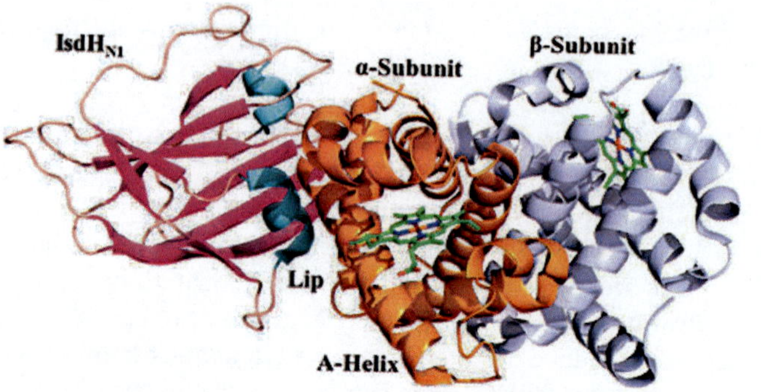

Figure 7 Cartoon representation of the X-ray crystal structure of IsdH$_{N1}$ from *S. aureus*, complexed with the α/β dimer of Hb. The Hb α-subunit is rendered in orange, and the β-subunit is rendered in gray. The major interaction is with the A helix of the α-subunit.

3.2.2.5 The Mechanism of Hemin Transfer:

The authors of the paper reporting the holo-IsdC structure proposed that opening of the 3_{10} helix of the lip, to a more flexible open structure, might aid in hemin association and dissociation and therefore transfer [105]. Consistent with this argument, a subsequently published solution structure of apo-IsdC revealed that it exhibits a flexible binding pocket [102]. In contrast, comparison of the apo-IsdA and holo-IsdA crystal structures revealed the presence of a largely pre-formed cavity in the former, with the 3_{10} helix on the distal side of the heme pocket intact [104]. While this may be a consequence of more efficient packing of the "closed" form of the apoprotein relative to a more disorganized "open" form, it suggests that the direction of transfer of hemin from holo-IsdA to apo-IsdC is not predicated on movement from a less pre-organized site to a more pre-organized site.

Insight into the atomic-level mechanism of hemin transfer from holo-IsdA to apo-IsdC has been provided by recent solution NMR experiments, utilizing paramagnetic relaxation enhancement (PRE) methods [108]. The experiments revealed absence of interactions between apo-IsdA and apo-IsdC, but that specific and transient "ultra-weak" interactions occurred in the presence of the diamagnetic heme analogue Zn PPIX. Utilizing the PRE data and rigid-body simulated annealing refinement, the authors generated a model for the transient complex. In the model, the ZnPPIX-bound forms of the two proteins are related by a pseudosymmetric two-fold axis, with the 3_{10} helix of each protein packing against the loop connecting strands β7 and β8 in the other protein. In other words, the interactions are analogous to those crystal packing interactions described above in crystal structures of heme-bound NEAT domains. In their proposed mechanism for hemin transfer, hemin must move ~18 Å from the holo-IsdA heme pocket to the apo-IsdC heme pocket, which would not require significant conformational reorganization of either protein. A presumption of the model is that a distortion of the IsdC heme pocket occurs which weakens the Tyr166-Fe bond. Support for the mechanism was provided by targeted mutagenesis of residues at the predicted protein-protein interface, which interrupted hemin transfer as evidenced in stopped-flow kinetics measurements.

Structural and kinetic studies with IsdA mutants led Grigg et al. to propose an analogous mechanism for hemin transfer from $IsdB_{N2}$ to IsdA, and from IsdA to IsdC [109]. For example, it was shown that mutating hemin ligand Tyr166 in IsdA to Phe or to Ala led to enhanced rates of hemin release to apoMb in comparison to the wild-type IsdA, whereas the mutants were impaired with respect to hemin transfer to apo-IsdC. This suggested that the Tyr axial ligand is a key determinant of NEAT domain hemin transfer. Their proposed mechanism for hemin transfer was motivated by the observation that heme-bound Isd proteins tend to crystallize to give a packing motif containing two protein molecules that are related by a 180° rotation, and with the hemes in contact with one another. The protein complex modeling server ClusPro was used to predict that an analogous arrangement of proteins, but with only one heme between them, would lead to a complex favorable for hemin transfer. The IsdA-IsdC complex obtained in this modeling study (shown in Figure 8) is effectively identical to the handclasp complex independently proposed by Villareal et al. on the basis of NMR studies [108]. The authors propose that hemin is shuttled from the Tyr ligand of the donor protein directly to the Tyr ligand of the acceptor protein. Available data did not allow determination of whether both Tyr ligands were coordinated simultaneously at any stage of the transfer.

In independent studies with $IsdH_{N3}$, it was shown that replacing hemin with the Ga(III) and Mn(III) analogues caused no changes in complex structure, little change in affinity, and little change in rate of transfer to IsdA [110]. In contrast, $IsdH_{N3}$ did not form stable complexes with the Zn(II) and Cu(II) complexes of

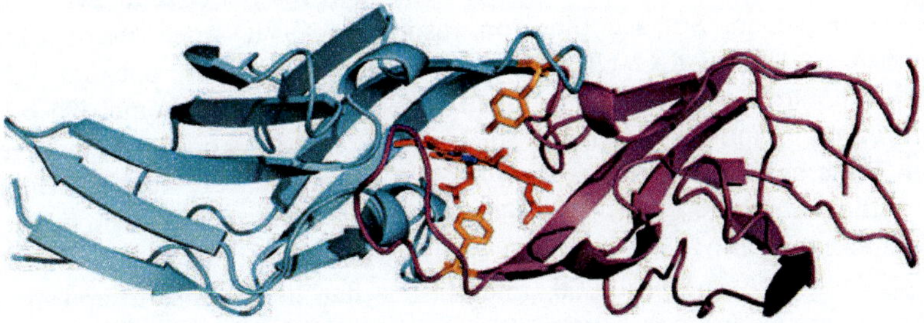

Figure 8 Model proposed by Grigg et al. for transfer of hemin from the NEAT domain in IsdA (rendered in cyan) to the NEAT domain in apo-IsdC (rendered in purple). Heme is shown in red, and the Tyr ligand in each protein is shown in orange. Adapted from Figure 7 in reference [109], with permission from the Publisher; copyright 2011.

PPIX. This suggests that electrostatic complementarity between the tyrosinate ligand and bound cargo is important for efficient hemin transfer. Tyr is the only ligand other than Cys with a sufficiently acidic side chain to allow ligation to hemin iron in an anionic form.

3.2.2.6 *Bacillus anthracis*:

B. anthracis also utilizes an Isd system for heme uptake, which is similar to that described above for *S. aureus* but with several key differences. One difference is absence of cell wall-anchored proteins corresponding to IsdA, IsdB, and IsdH [84]. Another is that *B. anthracis* secretes two hemophores, IsdX1 and IsdX2, which promote growth when Hb is the sole source of iron [84,111,112]. The lone IsdX1 NEAT domain is the only NEAT domain known to date that performs all functions exhibited by hemophores: Hb and heme binding, heme extraction from Hb, and transfer of heme to receptors (specifically IsdC, the central conduit for hemin uptake in the *B. anthracis* Isd heme uptake system as it is in *S. aureus*) [111,112]. The structure of IsdX1 has recently been elucidated by X-ray crystallography, both in the holo and apo forms (PDB codes 3SIK and 3SZ6, respectively) [113]. Heme ligation involves a YXXXY motif, analogous to those described above for the heme-binding *S. aureus* NEAT domains. As in the case of apo-IsdA from *S. aureus*, the apo form of IsdX1 has a largely pre-formed heme pocket.

In contrast to IsdX1, IsdX2 contains five NEAT domains, which together perform all the functions needed to obtain hemin from metHb and deliver it to IsdC [112]. Employing a reductionist approach, Honsa and Maresso showed that four of the five IsdX2 NEAT domains (N1, N3, N4 and N5) are able to bind hemin, with N5 exhibiting the strongest affinity [84]. Notably, N1, N3, N4, and N5 all contain a YXXXY sequence analogous to those for IsdX1 and the heme-binding *S. aureus* NEAT domains discussed above. In the sole NEAT domain that does not bind heme (N2), a His residue replaces the second Tyr to give a YXXXH sequence. Surface plasmon resonance (SPR) spectroscopy studies showed that all five IsdX2 NEAT domains interact with Hb, but that only the two with the strongest affinity (N1 and N5) can extract the heme. No binding was observed to apo-Hb, however. The hemin extraction appears to involve formation of an activated complex. Only N1, N3 and N4 were able to efficiently transfer their hemin cargo to apo-IsdC, but at the same rates at which hemin is lost to apomyoglobin. This stands in contrast to hemin delivery from IsdX1 to apo-IsdC, which occurs ~10,000 times faster than hemin loss to apoMb, suggesting an active process involving specific holoprotein-protein interactions [112]. It was suggested that IsdX2 acts as a "heme sponge", obtaining heme from multiple sources (including IsdX1) and thereby enhancing *B. anthracis* survival under iron starvation conditions.

3.2.2.7 *Streptococcus pyogenes*:

Heme acquisition by the Group A Streptococcus bacterium *Streptococcus pyogenes* involves the heme binding outer membrane proteins Shr (**S**treptococcal **h**emoprotein **r**eceptor) and Shp (**S**treptococcal **h**eme-associated **p**rotein), and an ATP-binding cassette (ABC) transporter (HtsABC or SiaABC) [87,114]. Shr has not been structurally characterized, but is known to contain two NEAT domains, albeit with considerably different domain architecture than exhibited by the NEAT receptors from *S. aureus* and *B. anthracis* [115]. Shr is anchored to the cell membrane, but reaches the cell surface [116]. Current evidence indicates that Shr obtains hemin from metHb (or from the metHb-haptoglobin complex) and transfers it to Shp, which is associated with the cell surface. Shp then delivers the heme to SiaA (also known as HtsA), the lipoprotein component of the ABC transporter [117–119]. Shp shows little discrimination for binding of ferric *versus* ferrous heme, suggesting that both forms can be scavenged by *S. pyogenes* [120]. This is noteworthy in light of a recent report that Shr can reversibly reduce bound hemin to heme [121]. Shp delivers its heme or hemin cargo to HtsA in an affinity-driven mechanism [118,120]. In the case of hemin, transfer occurs ~100,000-fold more rapidly than simple dissociation of hemin from Shp [118].

A 2.1 Å structure of the Shp heme-binding domain (PDB code 2Q7A) revealed a fold similar to the NEAT domains described above, with one equivalent of hemin bound per protein [93]. Thus, although Shp was not originally believed to belong to the NEAT family [86], it appears to be at least distantly related [93]. In fact, the ABC transporter lipoprotein HtsA from *S. pyogenes* is nearly 40% identical to IsdE from *S. aureus* suggesting that they play similar roles, and that heme uptake in the two organisms follows similar mechanisms [93]. There are two molecules of Shp in the asymmetric unit, and these are separated by two stacked molecules of hemin which resemble the dimeric form of hemin that forms in solution. These exogenous hemin molecules are not likely to be physiologically relevant, however, as ultracentrifugation and gel filtration experiments indicate that full-length Shp and its recombinant heme-binding domain exist as monomers in solution. The Shp structure revealed that the hemin cargo is ligated by the thioether side chains of two methionine residues, Met66 and Met153, resulting in a mostly low-spin state. Only bacterioferritin had previously been shown to exhibit bis-Met heme ligation, but in that case the Met ligands come from different protein subunits [122]. The Met153 ligand is located in a loop between strands β7 and β8, corresponding to the β-hairpin platform that houses the Tyr ligands in the hemin-binding NEAT domains from *S. aureus* and *B. anthracis*. Ligand Met66 resides in a region corresponding to the distal "lip" in the previously described NEAT domains, which in Shp is a 3-turn α-helix rather than a short 3_{10} helix.

Knowledge of the heme ligation mode in holo-Shp [93], together with kinetic data obtained with axial ligand mutants M66A and M153A [120], has added to previous understanding of the mechanism of hemin transfer from holo-Shp to apo-SiaA [117,118]. Both of the axial ligand mutants with bound hemin are high spin, indicative of ligation by one Met. The M66A mutation was found to increase hemin binding strength by approximately 3-fold, whereas the M153A mutation decreased hemin-binding by 138-fold. Hemin bound more rapidly to the M66A and M153A mutants than to wild-type Shp, and the rate of hemin loss to the H64Y/V68RF mutant of apoMb was accelerated 700-fold by the M153A mutation. These results were interpreted to indicate that the role of Met153 is to provide high affinity binding of heme, while the role of Met66 is to weaken heme binding and thereby facilitate its transfer to HtsA. Whereas hemin transfer from wild-type Shp to HtsA exhibits simple first order kinetics, the corresponding reactions with the axial ligand mutants occurred in two distinct steps. The authors proposed the first step is formation of an intermediate species in which hemin is bound to one ligand from the Shp mutant and one ligand from HtsA. Hemin transfer then proceeds from this ternary complex in a first order process. This is fundamentally analogous to the proposed mechanism of hemin transfer between NEAT domains in *S. aureus* and *B. anthracis*.

3.3 *Heme in the Periplasmic Space of Gram-Negative Bacteria*

3.3.1 Periplasmic-Binding Proteins

The environment of the periplasm responds to growth conditions and nutrient availability, thus it is a continuously changing and likely crowded environment. Periplasmic-binding proteins (PBP) therefore serve the crucial function of capturing specific solute (cargo) from a cognate outer membrane importer for subsequent transport across the periplasmic space. A number of PBPs also function in signal transduction by detecting signals in the exterior environment and then transmitting information to the cytoplasm via their cognate transmembrane transport partners. Components in the cytoplasm (typically proteins) respond by altering rates of gene expression or protein degradation [123].

Carbohydrates, amino acids, peptides, metal ions, and other nutrients are transported selectively across the periplasmic space. Historically, Tam and Saier developed the first classification scheme to organize the vast number of Gram-negative and Gram-positive bacterial proteins belonging to the PBP family [124]. These were grouped into 8 clusters or subfamilies, according to the chemical nature of their cargo and the molecular size of the PBPs. Although members of the PBP family share relatively small sequence conservation, almost all PBPs fold into a similar structural motif consisting of two globular domains made of α/β secondary structure separated by a shallow or deep binding cleft [125]. A significant number of PBP structures in the Protein Data Bank enabled structural classification of the PBP family, which was initially subdivided into 2 classes (I and II). Class I PBPs have three short connecting segments or hinges between the N- and C-terminal globular domains, whereas class II have two [125,126]. Subsequently, structural elucidation of the Mn^{2+}-binding protein from *Streptococcus pneumoniae* (PSaA) [127], the

Zn^{2+}-binding protein from *Treponema pallidum* (TroA) [128], and the ferrichrome-binding protein FhuD from *E. coli* [129], revealed a new PBP fold (class III). In class III PBPs the N- and C-terminal domains are linked by a long (thought to be rigid) α-helical linker and cargo is loaded onto a cleft between the two domains.

The distinct architecture of class III relative to class I and II PBPs appears to exert important influence on their mechanisms of ligand binding and release: Class I and II proteins consistently undergo large hinge-bending motions between the two domains. In the apo-proteins the two domains are further apart, exposing a ligand binding cleft to the aqueous environment [125]. Ligand binding is accompanied by large domain rotation about the hinge, enabling capture of the ligand by both domains with simultaneous extrusion of water molecules associated with the ligand and the binding cleft. This type of motion, which was termed "Venus flytrap", was firmly established upon elucidation of crystal structures corresponding to the ligand-bound and ligand-free forms of the maltose binding protein (MBP) [130, 131], and the leucine/isoleucine/valine binding protein (LivJ) [132]. The hinge-bending motions are illustrated in Figure 9 (a and b) with the apo- and valine-bound forms of LivJ.

The structure of TroA bound to Zn^{2+} revealed that, unlike class I and II PBPs, the two domains are linked by a long and seemingly inflexible α-helix that interacts along its entire length with both the N- and C-terminal domains of this Zn^{2+} binding PBP [128]. Subsequent structural elucidation of the apo-TroA structure showed a small hinge bending/rotation of the two domains (< 4°) upon ligand binding/release [133], compared to >35° rotation observed with class I and II PBPs. Similarly, the structures of apo- and ligand-bound BtuF, a vitamin B_{12}-binding protein from

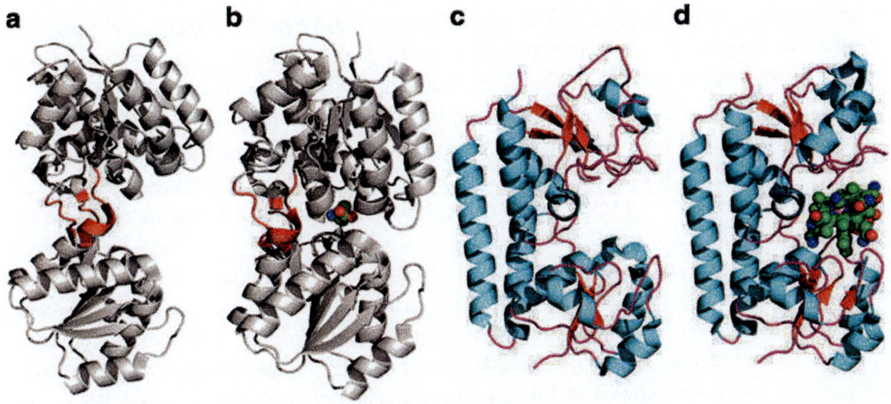

Figure 9 Contrasting behavior in domain motions observed between Class I/II and Class III PBPS. The "Venus flytrap" motion accompanying ligand binding/egress from class I and II PBPs is illustrated with the structure of LivJ in its apo-form (**a**) and valine-bound (**b**) conformations. For comparison the small conformational rearrangements accompanying binding/release of vitamin B_{12} from BtuF are illustrated in the structures of apo-BtuF (**c**) and vitamin B_{12} bound BtuF (**d**). The Protein Data Bank entries for apo- and Val-bound LivJ are 1Z15 and 1Z18, respectively, and entries for apo- and ligand-bound BtuF are 1N4D and 1N2Z, respectively.

E. coli, provided additional evidence suggesting that the long α-helix linking the two domains prevents domain rotation upon ligand binding and egress [134]. This is illustrated in Figure 9 (c and d) with the structures of apo- and vitamin B_{12}-bound BtuF, in order to allow ready visualization of the striking contrast in structural reorganization that differentiates class III from class I and II PBPs, at least as can be inferred from comparing crystal structures in the ligand free and bound forms. Molecular dynamics simulations, however, suggest that class III PBPs may be more dynamic than can be inferred from the crystal structures available thus far (see below).

3.3.1.1 Structurally Characterized Heme-Binding Periplasmic-Binding Proteins Belong to Class III:

As of this writing, the only heme-binding PBPs that have been structurally characterized are PhuT from *Pseudomonas aeruginosa*, ShuT from *Shigella dysenteriae*, [135] and HmuT from *Yersinia pestis* [136]. The three proteins have N- and C-terminal domains bridged by a long α-helix, thus they belong to class III PBPs. The PhuT and ShuT structures are very similar (Figure 10 a and 10 b). The heme loads onto a cleft formed by the N- and C-terminal domains, where the heme iron is coordinated by a conserved Tyr residue. In the distal side there is no protein provided ligand, which is in agreement with conclusions drawn from spectroscopic investigation suggesting that the heme in ShuT and PhuT is high-spin and five-coordinate, with an anionic proximal ligand [137,138]. In PhuT the coordinating phenol oxygen of axial ligand Tyr71 is within hydrogen bonding distance of Arg73, which also forms a hydrogen bond with one of the heme propionates; in the distal side the side chain of Arg228 forms a hydrogen bond with the second heme propionate. In ShuT, the heme is coordinated by Tyr67 in the proximal side but in contrast to the situation in PhuT, the coordinating phenol oxygen of Tyr67 does not engage in hydrogen bonding. The O-Fe bond distance in PhuT and ShuT is 2.3 Å, which is longer than equivalent O-Fe bonds in HasAp (2.0 Å) [57] and in beef liver catalase (2.17 Å) [139]. The relatively long O-Fe distance in PhuT and ShuT was proposed to suggest that the heme is not held in the binding pocket as tightly as in catalase, which may be functionally important to the bind and release function of PhuT/ShuT [135].

In this context, it is interesting that the O-Fe distance in HasAs and HasAp is the shortest, despite the fact that the hemophores also function by binding heme and releasing it to a cognate receptor. It is also interesting that, despite the similarity in the PhuT and ShuT structures, heme sits differently in each of the two PBPs. In PhuT the heme propionates are exposed to the aqueous environment, whereas in ShuT the heme propionates are buried within the heme binding cleft. The distal sites of PhuT and ShuT are clearly distinct in their organization and contacts with the macrocycle. It has been pointed out that, although these structural differences would likely affect the function of heme proteins and enzymes, they do not have a pronounced effect on the heme shuttling function of PhuT and ShuT [135]. It is possible that the differences reflect structural/dynamic properties that allow each of the two PBPs to efficiently interact with their cognate partners to load and then release heme.

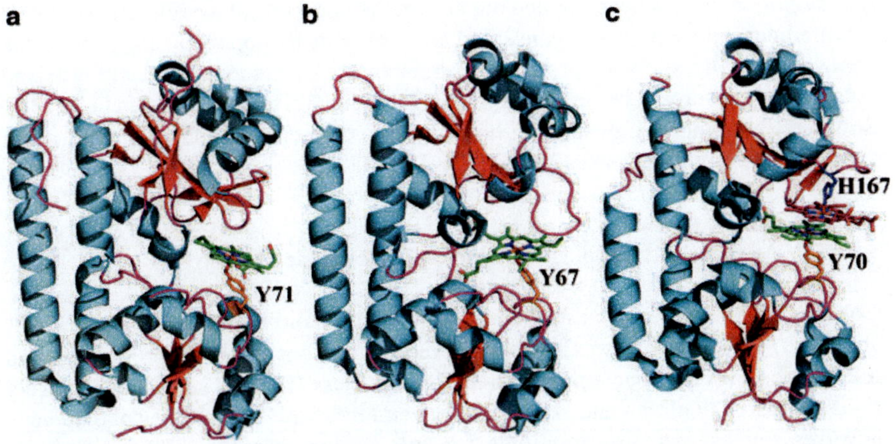

Figure 10 The structures of heme binding PBPs are very similar and use a conserved Tyr ligand to coordinate heme iron. Comparing the structures of PhuT (**a**) and ShuT (**b**) suggests that specific heme-protein interactions are not conserved, despite a very similar fold. The structure of HmuT (**c**) has two heme molecules within a fold similar that of PhuT and ShuT. A conserved Tyr functions as ligand to one of the heme molecules, whereas the other heme is coordinated by a non-conserved His residue. The Protein Data Bank entries for PhuT, ShuT and HmuT are 2R79, 2RG7 and 3NU1, respectively.

More recently, the X-ray crystal structure of the periplasmic heme-binding protein HmuT from *Yersinia pestis* has been reported [136]. The structure shows that two heme molecules are bound in the HmuT structure, with their heme propionate groups pointing in opposite directions, thus producing a situation where a pair of heme propionates is exposed to the aqueous environment, whereas the other pair is sequestered in the interior of the heme binding cleft (Figure 10 c). One of the heme molecules is coordinated by conserved Tyr70, whereas the second heme molecule is stacked parallel to the first and is coordinated by His167. The latter residue is not conserved but is consistently present in a group of homologues, so it has been proposed that its presence in a particular sequence may be used to predict that the corresponding homologue may bind two heme molecules per PBP molecule [136]. The prediction awaits validation from structural and biochemical characterization of additional heme binding PBPs. Along the same vein, if dimeric heme is indeed transported across the periplasmic space, the mechanism and order of heme delivery to the cognate receptor (ABC transporter) in the inner membrane will be an interesting challenge to biochemists and structural biologists. In this context, note that a similarly stacked-heme dimer has been observed in two other proteins that function in the capture and transport of heme. In contrast to HmuT, however, the stacked-heme dimer in these other structures causes dimerization of two monomeric proteins. The first example was observed in the structure of ChaN, a lipoprotein thought to be associated with the heme outer membrane receptor of *Campylobacter jejuni* [140] and the second is the H32A mutant of HasAp, the hemophore secreted by

P. aeruginosa. As has been pointed out above, the latter may have relevance to heme capture because His32 is not conserved amongst Has hemophores, suggesting that heme-stacked dimers may form upon heme capture by hemophores lacking His32 [141]. The structures of heme free ShuT and PhuT have also been reported [135,136]. Structural alignments indicate that heme bound and heme free PhuT do not differ significantly, as may be expected for class III PBPs.

3.3.1.2 Are Class III Periplasmic-Binding Proteins Flexible?

Comparing the X-ray crystal structures of apo- and ligand-bound BtuF, ShuT, PhuT, and HmuT suggest only small structural reorganizations upon ligand binding and release, observations that support the notion that class III PBPs are likely relatively rigid compared to class I and II proteins. Several relatively recent lines of evidence, however, suggest that large amplitude dynamic motions may play an important role in the function of class III PBPs:

(i) The crystal structure of apo-FitE [142], a siderophore binding PBP found in pathogenic strains of *E. coli* [143], showed that the protein in the crystal can be trapped in several conformations differing in the orientation of its N- and C-terminal domains about the intervening α-helix. These predicted motions are thought to be representative of open-to-closed domain transition associated with ligand binding.

(ii) The structures of two lipoprotein surface receptors that capture ferrated siderophores for transport across heterodimeric permeases in the Gram-positive bacterium *Staphylococcus aureus* have been reported [144,145]. *S. aureus* produces at least two siderophores, staphyloferrin A (SA) and staphyloferrin B (SB) [146,147]. Their ferric complexes, Fe^{3+}-SA and Fe^{3+}-SB, are sensed and imported by lipoprotein surface receptors HtsA and SirA, respectively, which transfer them to the heterodimeric permeases HtBC and SirBC [148,149]. The crystal structure of apo-HtsA devoid of its N-terminal secretion signal revealed a fold consisting of two domains connected by a single α-helix, akin to that characteristic of class III PBPs [148]. Subsequent determination of the apo-HTsA structure followed by structural alignment showed that HtsA undergoes large conformational changes in specific regions of its C-terminal domain upon Fe^{3+}-SA binding/release, while maintaining its interdomain orientations nearly unchanged [145]. Similarly, alignment of the apo-SirA and SirA-SB structures shows that binding of the Fe^{3+}-siderophore is accompanied by relatively large structural rearrangements in specific loops of the N-terminal domain but minimal relative change of the two domains [144]. The structure of FeuA, a siderophore-binding protein attached to the membrane of *B. subtilis* by a lipid anchor also shows N- and C-terminal domains separated by a long α-helix. When FeuA binds its ferrated siderophore ligand, a 20° motion of the two domains is observed, which though not as large as the ~35° hinge motions observed in class I and II PBPs, is significantly larger than similar motions observed when comparing apo- and ligand-bound class III PBPs [150].

(iii) Molecular dynamics simulations of apo- and heme-bound PhuT, ShuT, and BtuF also suggest that these PBPs may be more dynamic than can be inferred from aligning the corresponding X-ray structures. The motions observed with these calculations are proposed to be intrinsic to the apo-proteins and suggestive of Venus flytrap motions in class III PBPs [151–153].

The issue of flexibility in PBPs is important because the distinct conformational and dynamic states of class I and II PBPs are thought to provide a means for distinguishing between loaded and empty PBPs, and for directing appropriate interactions with their cognate ABC transporters, which are required to move molecules from the periplasmic space into the cytosol (see below). Consequently, future investigations directed at experimentally testing the notions derived from molecular dynamics simulations will be important to understand the mechanism of binding, release and recognition of heme-binding and other class III PBPs.

3.3.1.3 Is HbpA a Heme-Binding Periplasmic-Binding Protein?

H. influenzae lacks all enzymes in the biosynthetic pathway of protoporphyrin IX [154,155] although most strains possess a gene coding for ferrochelatase [155,156]. Consequently, *H. influenzae* requires heme or protoporphyrin IX for growth. HbpA is a periplasmic protein that has been implicated in host heme utilization [157–160]. Recently, however, HbpA has been identified as a protein that binds reduced and oxidized glutathione (GSSH and GSSG, respectively) with high affinity (K_d~10 μM) and with 1:1 stoichiometry. In comparison, although heme was found to bind HbpA it does so with significantly lower affinity (K_d~650 μM) [161]. The authors also determined the structure of HbpA from *Haemophilus parasuis* in complex with GSSG and observed a structure consistent with a ligand-bound class I/II PBP.

3.3.2 Across the Inner Membrane in Gram-Negative Bacteria

Translocation of heme into the cytoplasm is carried out by specific ABC transporters, which consist of transmembrane domains harboring a specific path for substrate translocation across the membrane and two cytosolic **a**denosine triphosphate (ATP) **b**inding **c**assettes, hence the term ABC. ABC transporters constitute one of the largest protein super families, which are involved in a variety of cellular processes, such as nutrient uptake, cell division, and maintenance of osmotic pressure [162]. The prototypical ABC transporter consists of four subunits; two of them, the ABCs, or nucleotide binding domain (NBD), are located in the cytoplasm and are associated with the other two subunits, which form the α-helical transmembrane domain (TMD) [134,163]. The ABCs harbor conserved amino acid sequence motifs for binding and hydrolyzing ATP, which provides the energy required to translocate cargo against cellular concentration gradients [164]. ABC transporters are thought to be substrate specific; their specificity is defined by the ligand selection of the PBP and by interactions with cognate ligand-bound PBPs [164,165].

The following structural conditions have been proposed for ABC importers and exporters to function as pumps [162,166]: (i) their structures must harbor a cavity of sufficient dimensions to admit a ligand (cargo), (ii) their transmembrane domain structures should enable inward- and outward-facing configurations of the cavity such that it can open to the cytoplasm or periplasm side of the membrane, respectively, and (iii) their structures should contain a binding site for cognate cargo within the cavity with different affinity in the inward- and outward-configurations of the cavity. Within the frame of these structural properties translocation across the membrane is thought to be initiated by specific interactions with a cognate, ligand-bound PBP (Figure 11). These interactions may trigger closing of the NBD interface and opening of the periplasmic side (outward-facing configuration), in turn causing a conformational change on the PBP that reduces its affinity for the bound ligand. ATP hydrolysis destabilizes the NBD dimer, causes opening of the cavity toward the cytosol (inward-facing configuration) and facilitates release of the PBP [162,166–169].

Although at present there are no published structures of heme ABC transporters, a few ABC transporters with homology to heme importers have been characterized structurally, including the vitamin B_{12} ABC transporter BtuCD. Hence, the B_{12} ABC importer from *E. coli* will be summarized here in an attempt to provide a general idea of the molecular and structural requirements for importing heme from the periplasm into the cytoplasm. In this context, the *Shigella dysenteriae* heme ABC transporter, which shares homology with the B_{12} ABC importer has been functionally characterized *in vitro* and is thought to be assembled and function in a manner similar to that summarized below for the B_{12} ABC importer [169].

The BtuCD transporter is a heterotetramer ($BtuC_2D_2$). The transmembrane domain (TMD) is composed of the $BtuC_2$ dimer and the ABC cassettes or nucleotide binding domain (NBD) is composed of the $BtuD_2$ dimer [163,170]. Each of the BtuC subunits traverses the membrane 10 times for a total of 20 TM helices. The interface between subunits creates a relatively large cavity (9 Å maximum diameter) that opens to the periplasmic space, outward-facing conformation (Figure 12a). The cavity is closed to the cytoplasm by residues in the loops connecting TM4 to TM5,

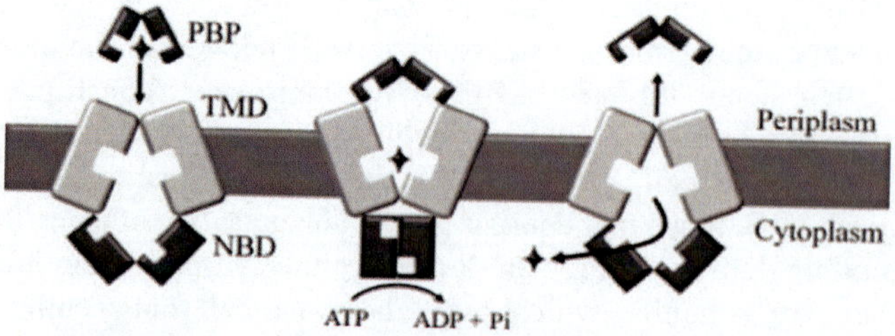

Figure 11 Schematic representation of the putative mechanism of ligand translocation across the inner membrane of Gram-negative bacteria. PBP, TMD, and NBD represent the periplasmic-binding protein, transmembrane domain and nucleotide binding domain, respectively. The figure was adapted from references [162] and [169].

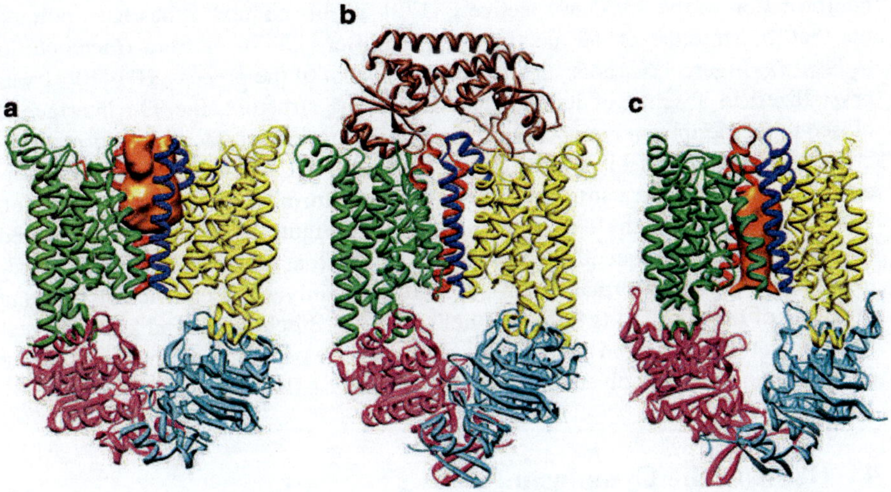

Figure 12 Cartoon rendering of (**a**) the B_{12} importer BtuCD from *E. coli* (PBD 1L7V), (**b**) BtuCD-F, the BtuCD importer in complex with its cognate PBP, BtuF (PDB 2QI9), and (**c**) HI1470-71, the metal chelate importer from *Haemophilus influenzae* (PDB 2NQ2). The TM subunit dimers are shown in green and yellow, the NBD domains in magenta and cyan and BtuF is shown in brown. To illustrate the model discussed in the text, TM helix 5 in each of the TM subunit dimers has been colored red and blue to illustrate that a switch in the cavity from the outward-facing (a) to the closed (b) to the inward-facing (c) conformations is likely a consequence of a different tilt of pore lining TM helix 5. The outward- and inward-facing cavities rendered in orange were calculated with ^{3}V: cavity, channel and volume calculator and extractor (http://3vee.molmovdb.org). The figure was adapted from references [162] and [173].

which are thought to function as a gate. The BtuD (NBD) domains are in the cytoplasm and their fold consists of a six-stranded β-sheet surrounded by nine α-helices and a peripheral β-sheet. The interface between the NBDs and the TMDs transmits the conformational changes caused by the binding and hydrolysis of ATP from the BtuD to the BtuC subunits.

The structure of BtuCD in complex with its PBP (BtuF) has also been determined [171] (Figure 12b). In the BtuCD-BtuF structure the PBP is bound to the periplasmic face of BtuCD and is devoid of its cargo, despite the fact that B_{12} binds to BtuF with $K_d \sim 15$ nM [172]. Note that each of the two non-identical lobes (domains) of the PBP contacts primarily one of the BtuC subunits. Interestingly, the structure reveals a significant opening of the N- and C-terminal domains of BtuF relative to its stand-alone structure, suggesting that the structure captured the PBP in its ligand-releasing conformation. It is also noteworthy that one of the BtuC subunits is in a different conformation relative to its position in the BtuCD structure, creating asymmetry at the TM interface and abolishing the outward-facing conformation (cavity). Hence it has been suggested that the BtuCD and BtuCD-BtuF structures may represent snapshots of intermediate conformations in the translocation process, representing an outward-facing conformation and a closed conformation of the TMD, respectively [173]. In this context, it has been pointed out that the structure of the metal chelate importer HI1470/71 from *Haemophilus influenzae* (Figure 12C) adds an additional snapshot of the process. HI1470/71 was crystallized in absence of its cognate PBP. In the structure, the TM interface is closed to the periplasm but is open to the cytoplasm (inward-facing conformation), creating a cavity with 11 Å maximum diameter [173]. The main structural differences amongst the three structures are on the pore-forming helices, particularly on TM helix 5 in each of the TMD subunits, which in Figure 12 have been highlighted in blue and red. Consequently, it has been suggested that the transition from inward- to outward-facing conformations of the cavity is mainly a consequence of a change in the tilt of the pore-lining helix TM helix 5 [162]. Biochemical and kinetic investigations capitalizing from the available structures are beginning to shed light on the complicated and likely diverse mechanisms used by ABC transporters [174].

4 Heme in the Cytoplasm

4.1 Heme Degradation and Iron Release

Pathogenic bacteria attempting to colonize a host are confronted with severely limiting concentrations of essential iron. One way to overcome this limitation is to utilize the rich source of iron presented by its host's heme. The intrinsic redox chemistry of heme and the presence of dioxygen (O_2) rapidly engage the macrocycle in Fenton-type reactions which result in the formation of superoxide, peroxide and hydroxyl radicals. Consequently, heme concentrations in bacteria, as in any other cell living in a micro-aerobic to aerobic habitat, are tightly regulated; for recent reviews, see [175,176]. Mechanisms to decrease heme toxicity in bacteria include heme sequestration by cytoplasmatic heme chaperones, heme utilization by its incorporation into functional heme containing proteins and enzymes, and heme degradation.

4.1.1 Heme Degradation in Gram-Negative Bacteria

Heme degradation in Gram-negative bacteria is carried out by the enzyme heme oxygenase (HO). HO enzymes have been found in several pathogenic strains, including *P. aeruginosa* (*pa*-HO), *N. meningitidis* (*nm*-HO) and *C. diphtheria* (*cd*-HO). The enzymes have been characterized extensively by biochemical, structural and spectroscopic investigations [177–180]. The X-ray crystal structures of *nm*-HO [181], *cd*-HO [182], and *pa*-HO [183] revealed a conserved α-helical fold, which is very similar to the fold of mammalian heme oxygenase enzymes [184,185], despite low sequence identity (~25%) between the bacterial and mammalian enzymes. As will be discussed below, HO enzymes from Gram-positive bacteria fold differently, and it is becoming common to refer to the fold characteristic of HOs from mammals and Gram-negative bacteria as the canonical HO fold. In Figure 13 the structure of

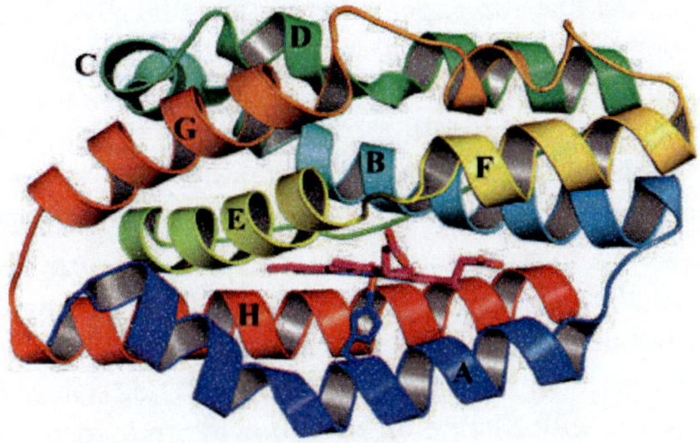

Figure 13 The canonical HO fold is represented by the crystal structure of *pa*-HO (PDB 1SK7) which shows a completely α-helical fold where heme is sandwiched between the helices and the heme-iron coordinated by conserved His-26.

pa-HO is shown to illustrate the canonical HO fold, which is comprised of 8 α-helices harboring a heme molecule between helices A and F; the heme-iron is coordinated by a proximal His residue and by a distal H_2O or OH^-. A unique feature of HO enzymes that distinguishes them from the globins, cytochromes, and peroxidases is the proximity of the distal helix to the face of the heme and the absence of polar residues that can stabilize a distal ligand.

4.1.1.1 The Catalytic Cycle of Heme Degradation by Canonical Heme Oxygenase:

In contrast to the function of the hemophores and transporters discussed in the preceding sections which function by binding and then releasing heme, HOs are enzymes that carry out a complex set of reactions involving electrons, oxygen and heme. The catalytic cycle of heme degradation (Scheme 3) accomplishes the feat of breaking down the highly conjugated heme macrocycle. In this complex cycle HOs utilize heme in a dual role of substrate and cofactor [177], thus in essence, heme bound to heme oxygenase catalyzes its own degradation.

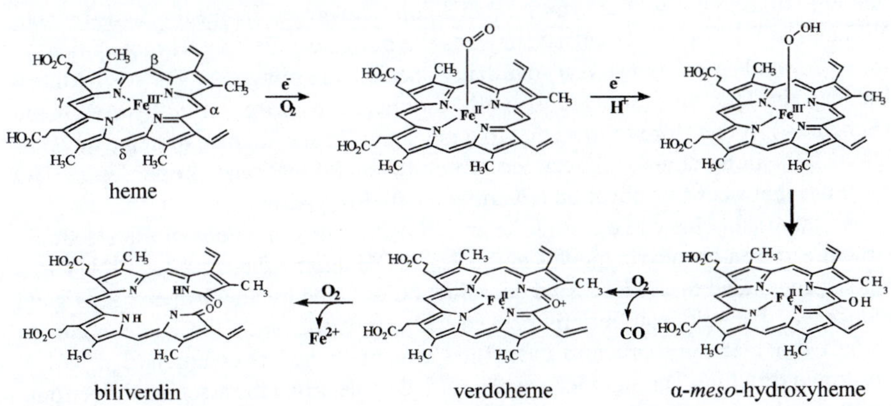

Scheme 3 Schematic representation of the heme oxidation reaction carried by canonical HOs, leading to the formation of CO, iron and biliverdin.

As may be expected from their nearly identical folds, the mechanism of heme degradation observed with the Gram-negative HO enzymes is the same as that elucidated for mammalian HO enzymes. What is known about the mechanism can be summarized as follows: Resting state (ferric) enzyme is reduced to the ferrous state, enabling the binding of an O_2 molecule in the distal site to form an oxyferrous complex (Fe^{2+}-O_2), which, upon acceptance of a second electron and a proton, is converted into an activated ferric hydroperoxy (Fe^{3+}-OOH) oxidizing species [186] (Scheme 3). Whereas this part of the mechanism is similar to that followed by monooxygenases such as cytochrome P450 to activate oxygen at a heme center, the subsequent reaction is unique to HO enzymes. The Fe^{3+}-OOH intermediate in HO reacts with a heme-*meso* carbon to produce *meso*-hydroxyheme, a very reactive species that undergoes a rapid O_2-dependent elimination of the hydroxylated *meso* carbon as carbon monoxide (CO) to form verdoheme [186,187]; note that in verdoheme an O atom replaces a *meso* carbon in the macrocycle. Verdoheme is subsequently oxidized to Fe^{3+}-biliverdin in a reaction that requires electrons and O_2 [188]. The 7 electrons necessary to break down the heme are provided to mammalian HOs by cytochrome P450 reductase [189].

In bacteria the nature of the cognate electron donor is less clear, although a relatively recent report indicates that in *Pseudomonas aeruginosa*, a ferredoxin $NADP^+$ reductase (*pa*-FPR) supports the degradation of heme to Fe^{3+}-biliverdin by *pa*-HO [190]. The structural and biochemical characterization of *pa*-FPR revealed that the enzyme is an NADPH-dependent flavoenzyme [190,191], thus suggesting that flavoenzymes function as physiological electron donors supporting the catalytic activity of canonical HOs in eukaryotes and in Gram-negative bacteria.

4.1.1.2 Activation of Heme Reactivity by Canonical Heme Oxygenase:

The first steps in the mechanism of heme oxygenation leading the formation of the obligatory Fe^{3+}-OOH oxidizing species are similar in the mechanism of O_2 activation followed by heme monooxygenases, such as the well-known cytochrome P450 and peroxidase family of enzymes. At this point, however, the mechanisms branch: Whereas in P450 and peroxidase enzymes the obligatory Fe^{3+}-OOH moiety decays into a formally Fe^{5+} oxo (ferryl) oxidizing species, the obligatory Fe^{3+}-OOH intermediate in HO enzymes *reacts* with the heme to form *meso*-hydroxyheme (Scheme 4). It therefore follows that to understand the mechanism of heme oxygenation, it is important to dissect the structural and functional properties of HO enzymes that steer O_2 activation toward heme hydroxylation.

An important clue came from EPR and ^{13}C NMR investigations of the electronic structure of iron-porphyrin models of the Fe^{3+}-OOH intermediate in HO [192]. These studies concluded that axial ligand coordination of heme by an alkylperoxide ligand induces a dynamic equilibrium between planar heme with the more common $(d_{xy})^2(d_{xz}d_{yz})^3$ electronic structure and ruffled heme with the less common $(d_{xz}d_{yz})^4(d_{xy})^1$ electronic structure. The significance of a ruffled heme is that the associated electronic structure allows significant unpaired electron (spin) delocalization at the *meso*-carbons

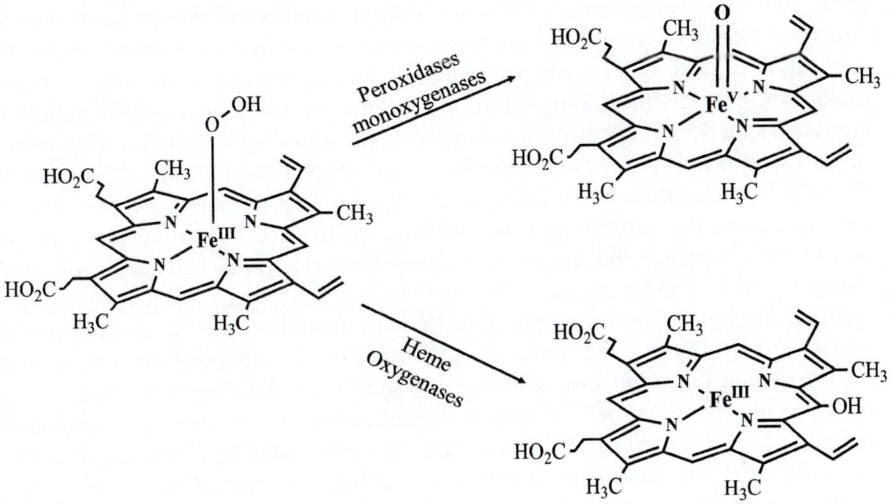

Scheme 4 Distinct reactivity of the obligatory Fe^{3+}-OOH intermediate in canonical HO enzymes and in monooxygenases and peroxidases.

[193–195]. Hence, it was proposed that if these two electronic structures and porphyrin ring conformations exist in equilibrium in the HO system, the ruffled heme ring would aid the attack of the terminal oxygen of the Fe^{3+}-OOH intermediate on the *meso* carbon, and the large spin density at the meso carbons would suggest the possibility of a radical mechanism for heme hydroxylation in HO [192].

In a subsequent study ^{13}C NMR spectroscopy was used to demonstrate that in the alkaline form of *pa*-HO (Fe^{3+}-OH) the heme is in dynamic equilibrium between at least three populations of Fe^{3+}-OH complexes with distinct electronic structures and non-planar porphyrin ring distortions that are in slow exchange relative to the NMR time scale [196]. In this dynamic equilibrium approximately 2% of the total population access the $(d_{xz}d_{yz})^4(d_{xy})^1$ electronic structure with ruffled heme, which was interpreted to suggest that if the coordinated OOH^- in the Fe^{3+}-OOH intermediate causes a similar dynamic equilibrium, the resultant large spin density at the *meso*-carbons and non-planar deformations of the heme ring would prime the macrocycle to actively participate in its own hydroxylation [196]. In this context, it is interesting that a theoretical study has suggested that heme hydroxylation by HO is more likely to occur via a stepwise mechanism in which hemolytic cleavage of the O-O bond is followed by trapping of the OH radical at the porphyrin ring [197].

The above discussion regarding electronic activation of heme reactivity is not complete without consideration of another crucial aspect of the HO structure, the presence of a conserved network of hydrogen bonded water molecules that communicate the heme distal ligand with the protein surface [183–185,198,199]. This network functions to deliver a proton to the nascent Fe^{3+}-OO^- [200,201] and formation of the Fe^{3+}-OOH is thought to weaken the O-O bond to facilitate its cleavage

[202,203]. The presence of this network of hydrogen bonded waters in the crystal structures of HO enzymes has been corroborated by NMR measurements to persist in solution [204,205]. The role played by the hydrogen bonding network of water molecules in promoting the $(d_{xz}d_{yz})^4(d_{xy})^1$ electronic structure associated with ruffled heme and spin density at the *meso*-carbons was uncovered by ^{13}C NMR investigations of *pa*-HO coordinated by hydroxide (OH^-), azide (N_3^-) and cyanide (CN^-) ligands [196,206]. It was determined that ligands capable of accepting a hydrogen bond at the same atom that coordinates the heme iron (OH^- and N_3^-) promote porphyrin non-planar distortions which are associated with electronic structures not commonly found in heme proteins, including the $(d_{xz}d_{yz})^4(d_{xy})^1$ which is associated with ruffling. This phenomenon stems from the coordinating atom in the axial ligand accepting a hydrogen bond from the distal network of waters or from a conserved Gly residue in the distal site, which lowers the ligand field strength. This in turn increases the equatorial field (porphyrin ligand), which results in non-planar distortions of the macrocycle and concomitant rearrangements in electronic structure [177,178,196,206]. As will be shown below, it has been recently suggested that a similar electronic activation takes place in HO enzymes from Gram-negative bacteria, despite their very different fold and absence of a network of hydrogen bonded waters in the distal cavity [207].

4.1.1.3 The Different Nature of the Coupled Oxidation *versus* Heme Oxygenation Reactions:

Free heme as well as heme contained in heme-proteins and enzymes can be oxidatively degraded by addition of a sacrificial electron donor (i.e., ascorbate) in the presence of O_2. This non-enzymatic process has been termed coupled oxidation [208–210]. The coupled oxidation reaction produces the same *meso*-hydroxyheme and verdoheme intermediates and releases CO, thus it is similar to the heme oxygenation reaction carried out by HO. The fact that both reactions require reducing equivalents and O_2 and proceed via formation of the same intermediates suggested that their mechanisms may be similar. Consequently, the coupled oxidation reaction has been used to model the heme oxygenation reaction carried out by HO [209,211–217], work that produced important insights into the mechanism of heme degradation. It is noteworthy, however, that subsequent work demonstrated that the coupled oxidation and heme oxygenation reactions differ fundamentally in their mechanisms leading to the formation of *meso*-hydroxyheme [218,219].

Not appreciating these differences has profound connotations in the assignment of heme oxygenase activity to newly isolated heme-binding proteins and enzymes. Therefore, it is important to appreciate that in the *heme oxygenation reaction* O_2 coordinated to the heme-iron in HO (Fe^{2+}-O_2 in Scheme 3) is reduced to coordinated hydroperoxide (Fe^{3+}-OOH), which then adds a hydroxyl group to a *meso*-carbon to produce *meso*-hydroxyheme. In contrast, in the *coupled oxidation reaction* non-coordinated O_2 is reduced by the sacrificial electron donor (i.e., ascorbate) to non-coordinated H_2O_2, which then reacts with Fe^{2+}-heme to produce *meso*-hydroxyheme [219]. It therefore follows that whenever a heme containing protein and enzyme is treated with a reducing agent in the presence of oxygen, the Fe^{2+}-heme and H_2O_2 that form independently will react to produce the highly reactive *meso*-hydroxyheme, which commits the macrocycle to release CO and form verdoheme. Six-coordinate heme-proteins with strong bis-axial ligation are an exception because the reaction between Fe^{2+}-heme and H_2O_2 requires that one axial ligand be labile or absent [215–217]. We will come back to this issue below when discussing cytoplasmic heme binding proteins.

4.1.1.4 The Regioselectivity of Heme Oxidation:

The activity of mammalian HO enzymes oxidizes exclusively the heme α-*meso* carbon (see Scheme 3), thus producing only α-biliverdin [220]. Although most bacterial HOs also produce α-biliverdin, the bacterium *P. aeruginosa* is thus far unique in that: (i) it expresses two distinct HOs, namely, *pa*-HO [221] and BphO [222]. The gene coding for the *pa*-HO enzyme (*pigA*), is part of a five-gene operon (*pigA-E*) expressed under iron limitation conditions [223]. In comparison, *bphO* occurs in an operon with the gene *bphP*, which encodes a bacterial phytochrome, and its expression is not regulated by iron concentrations [222]. (ii) Unlike all other known canonical HOs, *pa*-HO produces δ-biliverdin [221,224], whereas BphO, like most other canonical HO enzymes, produces only α-biliverdin [222]. Efforts aimed at understanding the unusual regioselectivity of heme degradation exhibited by *pa*-HO contributed significantly to understand how polypeptide-heme propionate interactions control the regioselectivity of heme oxidation. Before these investigations are summarized, however, it is interesting to ponder why the two HOs in *P. aeruginosa* oxidize heme with distinct regioselectivity. Although the reasons are not yet clear, it is known that α-biliverdin is a substrate for the bacterial phytochrome (BphP), which generates a photoactive phytochrome. In contrast, the product of *pa*-HO, δ-biliverdin is not [222], which suggests that each of the HOs in *P. aeruginosa* evolved to make distinct biliverdin isomers in order to maintain the selectivity of downstream signals.

The regioselectivity of heme oxidation is dictated by two important structural properties of HO enzymes:

(i) Steric constraints in the distal pocket steer the terminal OH moiety of the Fe^{3+}-OOH oxidizing species toward the heme-*meso* carbon to be hydroxylated. Because the Fe^{3+}-OOH intermediate is too reactive to be characterized structurally, evidence supporting this notion stems from the crystal structure of the Fe^{2+}-O_2 complex in *cd*-HO [225] which shows that specific interactions in the distal pocket direct the terminal O atom toward the heme α-*meso* carbon. Since the Fe^{2+}-O_2 complex is the immediate precursor of the Fe^{3+}-OOH oxidizing species, it is reasonable to assume that the terminal OH hydroxylates the α-*meso* carbon and therefore commits heme degradation toward the formation of α-biliverdin, the product of heme oxidation by *cd*-HO.

(ii) Specific interactions between the heme propionates and the HO polypeptide place the appropriate *meso*-carbon where it can be hydroxylated by the sterically constrained Fe^{3+}-OOH oxidizing species. Realization of the importance played by the heme propionates in directing the regioselectivity of heme oxidation was initially obtained in NMR studies of *pa*-HO site-directed mutants [224] and then confirmed in the crystal structure of *pa*-HO [183].

In brief: In the structure of *pa*-HO, which is identical to the structure of *nm*-HO or any other α-hydroxylating HO, interactions between the heme propionate and the *pa*-HO polypeptide seat the heme so that the δ-*meso* carbon is susceptible to oxidation by the Fe^{3+}-OOH moiety (Figure 14, left). In comparison, the heme-polypeptide interactions in the structure of *nm*-HO seat the heme with the α-*meso* carbon in a position equivalent to that occupied by the δ-*meso* carbon in *pa*-HO, so that the α-*meso*-carbon is hydroxylated (Figure 14, right). Additional evidence for the crucial role played by the heme propionate-polypeptide interactions was obtained by eliminating hydrogen-bonding or ionic interactions between heme propionates and key side chains in the HO structure using site-directed mutagenesis. Elimination of these key interactions causes large (~90°) in-plane rotations of the heme that exchange the positions (seating) of *meso*-carbons within the structure [224]; these large in-plane heme rotations have thus far been observed only in HO enzymes. As a consequence, these mutants produce a mixture of different biliverdin isomers whose concentrations could be predicted from the NMR spectra of the heme resonances [224,226–228].

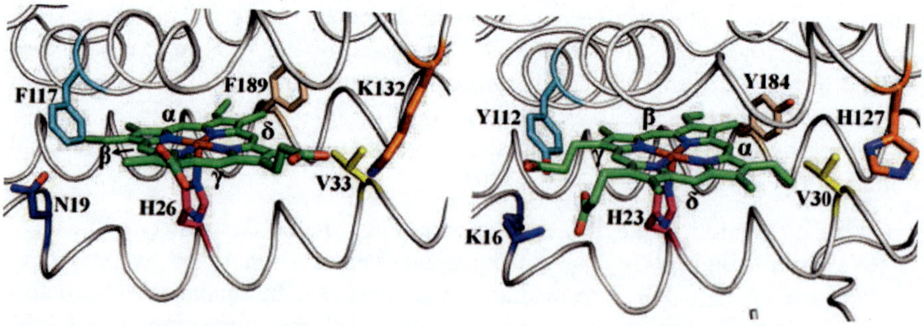

Figure 14 Structures of *pa*-HO (PDB 1SK7) and *nm*-HO (1J77) illustrating the fact that the δ-*meso* carbon in *pa*-HO and α-*meso* carbon in *nm*-HO share the same location in the two structures. The isostructural nature of these *meso*-carbons is accomplished by ~90° in-plane rotation of the heme in *pa*-HO relative to *nm*-HO, and the resultant heme seatings in each of the structures are stabilized by unique polypeptide-heme propionate interactions.

4.1.2 Heme Degradation in Gram-Positive Bacteria

Heme degrading enzymes were not identified in Gram-positive bacteria until 1994 when two cytosolic homolog proteins from *Staphylococcus aureus*, IsdG and IsdI, were shown to cleave the heme macrocycle in the presence of NADPH and cytochrome P450 reductase [229]. Following this report several orthologs have been identified in pathogenic strains of bacteria, including, *Bacillus anthracis*, and *Listeria* [230–232]. The first structural reports concerned the apo-forms of IsdG and IsdI, which revealed a dimer with a fold reminiscent of the ferredoxin architecture ($\alpha+\beta$ sandwich) in which the β-barrel is formed at the dimer interface [233]. Clearly, this fold is entirely different from the all α-helical fold characteristic of canonical HO enzymes. The biochemical characterization of IsdG and IsdI led to the identification of a triad of residues critical to heme degradation catalytic activity, Asn7, Trp 67, and His77 [233]. More recently, the structure of an Asn7Ala mutant IsdG in complex with heme showed that the protein fold remains nearly unchanged upon heme binding, and that one heme binds to each monomer in the asymmetric unit [234] (Figure 15). As is the case in canonical HOs, the heme-iron in IsdG is coordinated by a conserved His proximal ligand (His77). However, unlike the canonical enzymes, the heme-propionates in IsdG are buried deep within the heme binding cavity, where they interact with two Arg side chains. A striking feature in the structure is the very large deformation from planarity (ruffling) exhibited by the heme, with distortions as large as 2.1 Å out of plane. This distortion of the porphyrin macrocycle, also observed in the structure of IsdG bound by cobalt protoporphyrin IX, was suspected to be at the center of heme degradation, although the distortion by itself is not sufficient, because only heme (Fe-protoporphyn IX) is degraded by IsdG and IsdI [234,235].

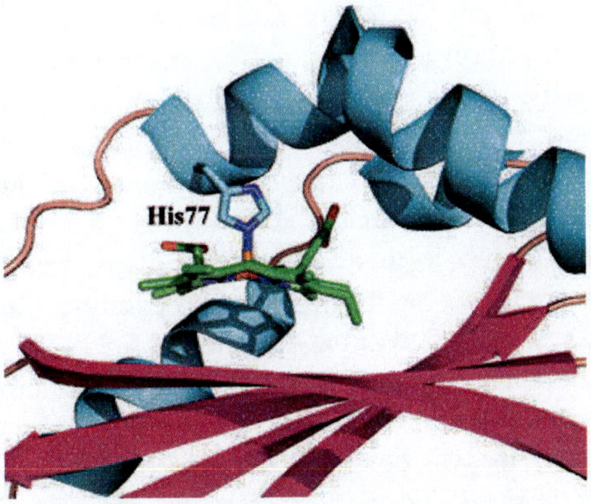

Figure 15 Crystal structure of the IsdG-heme complex (PDB 2ZDO). The biological molecule is a dimer with the β-barrel (magenta) at the dimer interface. In the figure, a monomer is shown to provide a zoomed-in view of the highly distorted (ruffled) heme, which is coordinated by His77. The figure was adapted from reference [234].

Recently, NMR spectroscopy and functional studies were used to determine that the extreme ruffling of heme bound to IsdI stabilizes the less common $(d_{xz}d_{yz})^4(d_{xy})^1$ electronic structure [207], which as mentioned above activates the reactivity of the heme ring by placing significant spin delocalization to the heme-*meso* carbons. As in the case in canonical HO enzymes, the heme in IsdI is in dynamic equilibrium between planar heme with the more common $(d_{xy})^2(d_{xz}d_{yz})^3$ electronic structure and ruffled heme with the less common $(d_{xz}d_{yz})^4(d_{xy})^1$ electronic structure. However, unlike the small population accessing the reactive $(d_{xz}d_{yz})^4(d_{xy})^1$ electronic configuration in *pa*-HO [196], a relatively large population of IsdI molecules have the reactive electronic configuration. Hence, it has been proposed that the increased stabilization of the $(d_{xz}d_{yz})^4(d_{xy})^1$ electronic structure by IsdI is necessary to compensate for the absence of a hydrogen bonding network in its distal site [207]. It is thus remarkable that, in spite of the very different folds, canonical HOs and the more recently discovered IsdG enzymes, appear to activate O_2 via a similar stabilization of a reactive heme electronic structure.

The mechanism that allows IsdG and IsdI enzymes to degrade heme remains largely unknown. Nevertheless an interesting recent discovery is that the final product of degradation is not biliverdin, as is the case with the canonical HO enzymes. With a combination of biochemical, NMR and mass spectrometric studies, it was determined that IsdG breaks down heme to a novel chromophore (oxo-biliverdin), which has been termed staphylobilin [236]. It therefore remains to be discovered whether heme degradation proceeds via verdoheme and if so, at what point does the mechanism branch from that of canonical HOs, so as to produce staphylobilin.

4.1.3 Does a Bacterial Dechelatase exist?

A relatively recent report by Letoffé et al. [237] suggested that two paralogs from *E. coli*, YfeX and EfeB, carry out the *in vivo* dechelation of heme to release iron while maintaining the protoporphyrin IX ring intact. Given that YfeX and EfeB are highly conserved in bacteria, the authors proposed that the physiological function of these enzymes is to retrieve iron from heme. These surprising observations would place YfeX and EfeB in a unique category of enzymes that procure iron from heme without breaking the macrocycle, a process that is very different from that observed with the HO enzymes discussed above. In this context, it is also interesting to consider that although porcine ferrochelatase has been shown to catalyze the dechelation of iron from heme, the reaction was carried out at 45° and pH 5.5 [238], conditions that have been reported to cause denaturation of mammalian ferrochelatases [239]. Hence, the dechelatase activity reported for YfeX and EfeB is indeed unique. It is important to note that dechelatase activity was observed only in cell extracts obtained from cells overexpressing YfeX or EfeB. Formation of protoporphyrin IX was detected by fluorescence, HPLC and mass spectrometry. *In vitro*, however, neither of the enzymes carried out the heme dechelation reaction, and the authors suggest that the isolated enzyme may be missing a co-factor which is present in the cell extracts [237].

A more recent report suggests that *E. coli* YfeX is not a heme dechelatase, but rather a porphyrinogen oxidase [240]. The authors point out that in their hands dechelatase activity was not observed in cell extracts or *in vitro*. Instead, they found that holo-YfeX oxidizes protoporphyrinogen IX and corproporphyrinogen III to their corresponding porphyrins, and suggest that the enhanced fluorescence observed by Letoffé et al. could have resulted from oxidation of endogenously produced porphyrinogens that would have accumulated in the absence of available iron for ferrochelatase [240]. These authors also argue that in bacteria where global iron-dependent gene regulation has been studied in detail, *E. coli*, *V. cholerae*, and *P. aeruginosa*, YfeX homologs do not appear to be regulated by iron availability and that this observation would appear to contradict the expectation for a protein predicted to function in iron retrieval from heme. The structure of EfeB bound to heme has been determined [241] and its fold was found to be identical that of dye-decolorizing peroxidases [242]. Although the report speculates regarding a channel to enable access of a yet unknown cofactor that donates protons for the putative dechelation reaction of EfeB, the structure does not reveal how EfeB may function as a dechelatase, if indeed this is a function endowed to this enzyme.

4.2 Cytosolic Heme Chaperones

A number of Gram-negative bacteria have cytosolic proteins encoded by genes located in heme uptake operons. Examples are HemS (*Yersinia enterolitica*), PhuS (*Pseudomonas aeruginosa*), ChuS (*Escherichia coli*), and ShuS (*Shigella dysenteriae*). As will be discussed below, these proteins were initially thought to be heme degrading enzymes. Upon further scrutiny, however, it is becoming clear that these proteins do not function in heme degradation, but rather as heme storage or heme chaperoning proteins. In an early study the relevance of these proteins to heme metabolism was brought to the front by showing that HemS is a cytoplasmic protein required to prevent heme-induced toxicity in an *E. coli* strain engineered to overexpress an outer membrane heme receptor [243]. These observations led to the proposal that HemS is a heme oxygenase, although its sequence lacks homology with known HO enzymes. A subsequent investigation was carried out with ShuS [244], which has approximately 64% sequence homology to HemS. It was shown that ShuS binds heme with 1:1 stoichiometry and that the heme-ShuS complex is six-coordinate and low-spin. It was noted that this spectroscopic signature is in stark contrast to the characteristic high-spin signature of HO enzymes, and therefore not consistent with a heme degradation function for ShuS. The author also observed that ShuS binds DNA, which led to the proposal that ShuS may function as a heme storage protein when heme uptake is active and as DNA binding protein to protect DNA from possible oxidative stress [244].

A few years later the structure of a homologous protein from *E. coli* (ChuS) was determined in its apo form [245]. The structure revealed a novel fold consisting of nine antiparallel β-strands that form a saddle-shape motif flanked by α-helices, which bears no resemblance to canonical or the more recently characterized Isd heme degrading enzymes. It was argued that ChuS is a heme oxygenase because when the ChuS-heme complex was incubated with NADPH and cytochrome P450 reductase heme was degraded to Fe^{3+}-biliverdin and CO was released [245]. This report was followed by the structural elucidation of the ChuS-heme complex, which showed that heme is bound in a cleft region delineated by the C-terminal and N-terminal domains and that the heme-iron is coordinated by a His ligand (His193) located in the C-terminal domain [246]. Shortly after this structure was published a new study was reported with the homolog PhuS [247]. This report demonstrated that PhuS binds heme with a 1:1 stoichiometry but that PhuS is not a heme degrading enzyme. To reach this conclusion the authors reacted the PhuS-heme complex with a sacrificial electron donor (ascorbate) in the presence and in the absence of catalase. As has been indicated above, heme degradation occurs via the coupled oxidation reaction because the sacrificial electron donor reduces O_2 to H_2O_2, which in turn reacts with the Fe^{2+}-heme [219]. Hence, in the presence of catalase H_2O_2 is efficiently removed from solution and heme degradation by coupled oxidation is prevented. In contrast, in the heme oxygenation reaction, the Fe^{3+}-OOH intermediate is generated at the heme active site of HO and therefore is not susceptible to catalase [219]. Consequently, the observation that the heme in PhuS is not degraded in the presence of ascorbate and catalase strongly suggests that PhuS and its homologs are not heme degrading enzymes [247].

Moreover, it has also been determined that protein-protein interactions drive the transfer of heme from PhuS to *pa*-HO, one of the heme degrading enzymes in *P. aeruginosa*, thus strongly suggesting a heme chaperoning function [247,248]. Recently, a model for heme transfer from PhuS to *pa*-HO has been proposed in which heme is loaded into apo-PhuS and initially coordinated by His209. These heme-protein interactions trigger a conformational change in PhuS that enables it to interact specifically with *pa*-HO. In turn, the energy released from the protein-protein interactions drives a change in heme-iron coordination from His209 to His212 in PhuS, which renders the latter competent for heme transfer [249].

The structure of HemS in its apo and heme bound states has also been determined [250]. The structures reveal that binding of heme causes interdomain motions that effectively clamp the heme between the N- and C-terminal domains, with the heme-iron coordinated by His196 (Figure 16a). The effect of heme binding to HemS, which is accompanied by main chain displacements > 4 Å and burial of ~ 350 $Å^2$ of solvent-accessible surface area, has been described as induced-fit binding [250]. Superposing the heme-bound structures of HemS and ChuS (Figure 16b) reveals that the two structures are nearly identical (backbone RMSD = 1.1 Å). Hence, taken together, the evidence thus far strongly suggest that HemS, PhuS and ChuS are cytoplasmic heme binding proteins that may function in heme storage and heme trafficking.

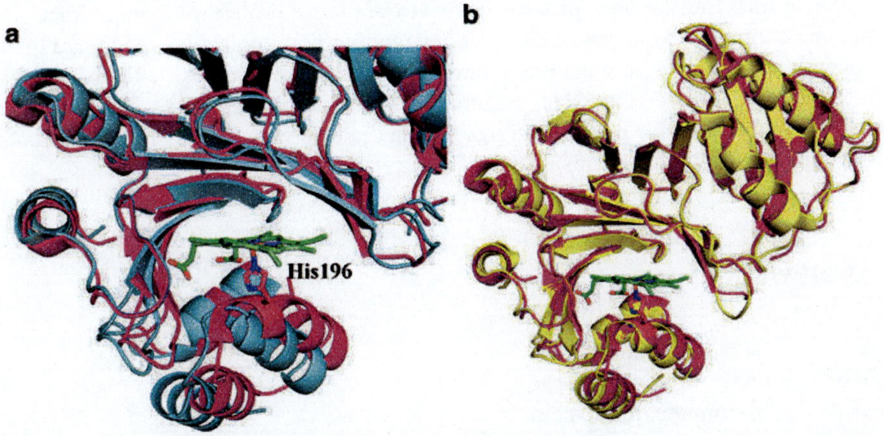

Figure 16 (**a**) Superposition of the apo- (cyan) and heme-bound (magenta) structures of HemS, illustrating the conformational rearrangements caused by heme binding/release. The PDB codes are 2J0R and 2J0P for the apo- and heme-bound HemS structures. (**b**) Structure of the heme chaperone HemS (magenta) superimposed to the structure of ChuS (yellow), showing the nearly identical fold (backbone RMSD 1.1 Å) of these enzymes, and suggesting that ChuS may be a heme-binding protein and not a heme degrading enzyme. PDB codes are: HemS (2J0P) and ChuS (2HQ2).

5 Conclusions and Outlook

The structural biology efforts summarized in this chapter largely follow a road map sketched by prior and ongoing genetic approaches. Although significant inroads have been made with atomic level description of proteins and enzymes involved in all aspects of bacterial heme uptake and metabolism, the large majority of systems are not completely structurally characterized. The available structures have largely expanded the repertoire of folds and heme coordination motifs that characterize the classical (eukaryotic) heme proteins and enzymes. They are also beginning to illuminate the structural features that enable tight binding of heme while also allowing rapid and directional transfer of the cargo. In this context, a surprising observation is the high representation of Tyr heme ligation in heme transport proteins, including hemophores. This may be due to the fact that Tyr can bind in its conjugate base form, thereby balancing the positive charge on ferric heme iron and also endowing it with a very negative redox potential, thus effectively preventing heme reduction and consequent oxidative stress. An interesting example highlighting the similar strategies but different tactics used by Gram-positive and Gram-negative bacteria are the heme-degrading enzymes. While Gram-negative bacteria use canonical HO enzymes to convert heme to biliverdin, Gram-positive bacteria have evolved an enzyme with a different fold which generates a similar product (oxo-biliverdin), likely via a similar mechanism.

The efforts thus far have provided a few examples of protein-protein interactions that are central to heme trafficking. Understanding how molecules recognize their cognate partners and how the recognition facilitates heme transfer is likely an area that will see continued intensive scrutiny. The information that will result from these endeavors may provide possible targets for future development of therapeutic agents to combat bacterial infections.

Abbreviations

A	acetate substituent
ABC	ATP -binding cassette
ALA	δ-aminolevulinic acid
ATP	adenosine 5'-triphosphate
BphP	bacterial phytochrome
CD	circular dichroism
EPR	electron paramagnetic resonance
GSA	glutamate-1-semialdehyde
HarA	haptoglobin receptor A = IsdH
Has	heme acquisition system
Hb	hemoglobin
heme	Fe^{2+} complex of protoporphyrin IX
hemin	Fe^{3+} complex of protoporphyrin IX
Hma	heme receptor
HO	heme oxygenase
Hp	haptoglobin
HPX	hemopexin
HSA	human serum albumin
HSQC	heteronuclear single quantum coherence spectroscopy
Isd	iron-responsive determinant
ITC	isothermal titration colorimetry
K_a	affinity constant
K_d	dissociation constant
MBP	maltose binding protein
MCD	magnetic circular dichroism
metHB	methemoglobin
NADPH	nicotinamide adenine dinucleotide phosphate (reduced)
NBD	nucleotide binding domain
NEAT	near iron transporter
NMR	nuclear magnetic resonance
OM	outer membrane
P	propionate substituent
PBP	periplasmic binding proteins
PDB	Protein Data Bank

PMP	pyridoxamine-5'-monophosphate
PPIX	protoporphyrin IX
PRE	paramagnetic relaxation enhancement
RR	resonance Raman
Shp	Streptococcal heme-associated protein
Shr	Streptococcal hemoprotein receptor
SPR	surface plasmon resonance
TBDT	TonB-dependent outer transportet
TM	transmembrane
TMD	transmembrane domain

Acknowledgments The authors are indebted to Mr. Ritesh Kumar for his assistance in the preparation of figures. The authors' work reported in this review was carried out with support from grants from the National Institutes of Health (GM 50503) and the National Science Foundation (MCB-0818488).

References

1. A. Kusaba, T. Ansai, S. Akifusa, K. Nakahigashi, S. Taketani, H. Inokuchi, T. Takehara, *Oral Microbiol. Immunol.* **2002**, *17*, 290–295.
2. D. C. White, S. Granick, *J. Bacteriol.* **1963**, *85*, 842–850.
3. M. R. Loeb, *J. Bacteriol.* **1995**, *177*, 3613–3615.
4. C. Wandersman, P. Delepelaire, *Annu. Rev. Microbiol.* **2004**, *58*, 611–647.
5. V. Braun, K. Handtke, *Microbiol. Monogr.* **2007**, *6*, 190–219.
6. B. R. Otto, A. M. J. J. Verweij-van Vught, D. M. Maclaren, *Crit. Rev. Microbiol.* **1992**, *18*, 217–233.
7. P. Ascenzi, A. Bocedi, P. Visca, F. Altruda, E. Tolosano, T. Beringhelli, M. Fasano, *IUBMB Life* **2005**, *57*, 749–759.
8. N. Frankenberg, J. Moser, D. Jahn, *Appl. Microbiol. Biotechnol.* **2003**, *63*, 115–127.
9. G. Layer, J. Reichelt, D. Jahn, D. W. Heinz, *Protein Sci.* **2010**, *19*, 1137–1161.
10. I. U. Heinemann, M. Jahn, D. Jahn, *Arch. Biochem. Biophys.* **2008**, *474*, 238–251.
11. V. B. Borisov, R. B. Gennis, J. Hemp, M. I. Verkhovsky, *Biochim. Biophys. Acta - Bioenergetics* **2011**, *1807*, 1398–1413.
12. D. E. Hultquist, P. G. Passon, *Nature New Biology* **1971**, *275*, 51–61.
13. R. L. Nagel, Q. H. Gibson, *J. Biol. Chem.* **1971**, *246*, 69–73.
14. G. Kikuchi, T. Yoshida, M. Noguchi, *Biochem. Biophys. Res. Comm.* **2005**, *338*, 558–567.
15. I. Stojiljkovic, D. Perkins-Balding, *DNA & Cell Biology* **2002**, *21*, 281–295.
16. F. Tiburzi, F. Imperi, P. Visca, *IUBMB Life* **2009**, *61*, 80–83.
17. A. Wilks, K. A. Burkhard, *Natural Product Reports* **2007**, *24*, 511–522.
18. M. Reniere, V. Torres, E. Skaar, *BioMetals* **2007**, *20*, 333–345.
19. E. Gouaux, *J. Struct. Biol.* **1998**, *121*, 110–122.
20. M. L. Vasil, R. M. Berka, G. L. Gray, H. Nakai, *J. Bacteriol.* **1982**, *152*, 431–440.
21. F. S. Boretti, P. W. Buehler, F. D'Agnillo, K. Kluge, T. Glaus, O. I. Butt, Y. Jia, J. Goede, C. P. Pereira, M. Maggiorini, G. Schoedon, A. I. Alayash, D. J. Schaer, *J. Clin. Invest.* **2009**, *119*, 2271–2280.
22. M. Kristiansen, J. H. Graversen, C. Jacobsen, O. Sonne, H.-J. Hoffman, S. K. A. Law, S. K. Moestrup, *Nature* **2001**, *409*, 198–201.
23. P. Zunszain, J. Ghuman, T. Komatsu, E. Tsuchida, S. Curry, *BMC Struct. Biol.* **2003**, *3*, 6.
24. P. A. Adams, M. C. Berman, *Biochem. J.* **1980**, *191*, 95–102.

25. Z. Hrkal, U. Mueller-Eberhard, *Biochemistry* **1971**, *10*, 1746–1750.
26. W. T. Morgan, U. Muller-Eberhard, *J. Biol. Chem.* **1972**, *247*, 7181–7187.
27. V. Hvidberg, M. B. Maniecki, C. Jacobsen, P. Højrup, H. J. Møller, S. K. Moestrup, *Blood* **2005**, *106*, 2572–2579.
28. Y. Tong, M. Guo, *Arch. Biochem. Biophys.* **2009**, *481*, 1–15.
29. C. Wandersman, I. Stojiljkovic, *Curr. Opin. Microbiol.* **2000**, *3*, 215–220.
30. M. Mills, S. M. Payne, *Infection Immunity* **1997**, *65*, 5358–5363.
31. A. G. Torres, S. M. Payne, *Mol. Microbiol.* **1997**, *23*, 825–833.
32. U. A. Ochsner, Z. Johnson, M. L. Vasil, *Microbiolog* **2000**, *146*, 185–198.
33. J. M. Ghigo, S. Létoffé, C. Wandersman, *J. Bacteriol.* **1997**, *179*, 3572–3579.
34. S. Létoffé, J. M. Ghigo, C. Wandersman, *J. Bacteriol.* **1994**, *176*, 5372–5377.
35. S. Létoffé, V. Redeker, C. Wandersman, *Mol. Microbiol.* **1998**, *28*, 1223–1234.
36. S. Létoffé, K. Omori, C. Wandersman, *J. Bacteriol.* **2000**, *182*, 4401–4405.
37. M.-S. Rossi, J. D. Fetherston, S. Létoffé, E. Carniel, R. D. Perry, J.-M. Ghigo, *Infection Immunity* **2001**, *69*, 6707–6717.
38. M. S. Hanson, S. E. Pelzel, J. Latimer, U. Muller-Eberhard, E. J. Hansen, *Proc. Natl. Acad. Sci. USA* **1992**, *89*, 1973–1977.
39. L. D. Cope, S. E. Thomas, Z. Hrkal, E. J. Hansen, *Infection Immunity* **1998**, *66*, 4511–4516.
40. J.-L. Gao, K.-A. Nguyen, N. Hunter, *J. Biol. Chem.* **2010**, *285*, 40028–40038.
41. C. Bradbeer, *J. Bacteriol.* **1993**, *175*, 3146–3150.
42. D. P. Chimento, R. J. Kadner, M. C. Wiener, *Proteins: Structure, Function, Bioinformatic,* **2005**, *59*, 240–251.
43. D. Cobessi, A. Meksem, K. Brillet, *Proteins: Structure, Function, Bioinformatics* **2010**, *78*, 286–294.
44. A. D. Ferguson, J. Deisenhofer, *Biochim. Biophys Acta - Biomembranes* **2002**, *1565*, 318–332.
45. A. D. Ferguson, C. A. Amezcua, N. M. Halabi, Y. Chelliah, M. K. Rosen, R. Ranganathan, J. Deisenhofer, *Proc. Natl. Acad. Sci. USA* **2007**, *104*, 513–518.
46. C. S. Bracken, M. T. Baer, A. Abdur-Rashid, W. Helms, I. Stojiljkovic, *J. Bacteriol.* **1999**, *181*, 6063–6072.
47. S. Krieg, F. Huché, K. Diederichs, N. Izadi-Pruneyre, A. Lecroisey, C. Wandersman, P. Delepelaire, W. Welte, J. Deisenhofer, *Proc. Natl. Acad. Sci. USA* **2009**, *106*, 1045–1050.
48. K. A. Burkhard, A. Wilks, *J. Biol. Chem.* **2007**, *282*, 15126–15136.
49. X. Liu, T. Olczak, H.-C. Guo, D. W. Dixon, C. A. Genco, *Infection Immunity* **2006**, *74*, 1222–1232.
50. N. Izadi-Pruneyre, F. Huché, G. S. Lukat-Rodgers, A. Lecroisey, R. Gilli, K. R. Rodgers, C. Wandersman, P. Delepelaire, *J. Biol. Chem.* **2006**, *281*, 25541–25550.
51. E. C. Hagan, H. L. T. Mobley, *Mol. Microbiol.* **2009**, *71*, 79–91.
52. D. Perkins-Balding, M. T. Baer, I. Stojiljkovic, *Microbiology* **2003**, *149*, 3423–3435.
53. C. Barjon, K. Wecker, N. Izadi-Pruneyre, P. Delepelaire, *J. Bacteriol.* **2007**, *189*, 5379–5382.
54. N. Izadi, Y. Henry, J. Haladjian, M. E. Goldberg, C. Wandersman, M. Delepierre, A. Lecroisey, *Biochemistry* **1997**, *36*, 7050–7057.
55. C. Caillet-Saguy, M. Delepierre, A. Lecroisey, I. Bertini, M. Piccioli, P. Turano, *J. Am. Chem. Soc.* **2006**, *128*, 150–158.
56. P. Arnoux, R. Haser, N. Izadi, A. Lecroisey, M. Delepelaire, C. Wandersman, M. Czjzek, *Nature Struct. Biol.* **1999**, *6*, 516–520.
57. A. Y. Alontaga, J. C. Rodríguez, E. Schönbrunn, A. Becker, T. Funke, E. T. Yukl, T. Hayashi, J. Stobaugh, P. Moënne-Loccoz, M. Rivera, *Biochemistry* **2009**, *48*, 96–109.
58. P. A. Williams, V. Fulop, E. F. Garman, N. F. W. Saunders, S. J. Ferguson, J. Hajdu, *Nature* **1997**, *389*, 406–412.
59. T. J. Reid, M. R. Murthy, A. Sicignano, N. Tanaka, W. D. Musick, M. G. Rossmann, *Proc. Natl. Acad. Sci. USA* **1981**, *78*, 4767–4771.
60. N. Wolff, C. Deniau, S. Létoffé, C. Simenel, V. Kumar, I. Stojiljkovic, C. Wandersman, M. Delepierre, A. Lecroisey, *Protein Sci.* **2002**, *11*, 757–765.

61. C. Caillet-Saguy, P. Turano, M. Piccioli, G. S. Lukat-Rodgers, M. Czjzek, B. Guigliarelli, N. Izadi-Pruneyre, K. R. Rodgers, M. Delepierre, A. Lecroisey, *J. Biol. Chem.* **2008**, *283*, 5960–5970.
62. C. Deniau, R. Gilli, N. Izadi-Pruneyre, S. Létoffé, M. Delepierre, C. Wandersman, C. Briand, A. Lecroisey, *Biochemistry* **2003**, *42*, 10627–10633.
63. G. Jepkorir, J. C. Rodríguez, H. Rui, W. Im, S. Lovell, K. P. Battaile, A. Y. Alontaga, E. T. Yukl, P. Moënne-Loccoz, M. Rivera, *J. Am. Chem. Soc.* **2010**, *132*, 9857–9872.
64. N. Wolff, N. Izadi-Pruneyre, J. Couprie, M. Habeck, J. Linge, W. Rieping, C. Wandersman, M. Nilges, M. Delepierre, A. Lecroisey, *J. Mol. Biol.* **2008**, *376*, 517–525.
65. S. Létoffé, F. Nato, M. E. Goldberg, C. Wandersman, *Mol. Microbiol.* **1999**, *33*, 546–555.
66. E. T. Yukl, G. Jepkorir, A. Y. Alontaga, L. Pautsch, J. C. Rodriguez, M. Rivera, P. Moënne-Loccoz, *Biochemistry* **2010**, *49*, 6646–6654.
67. G. S. Lukat-Rodgers, K. R. Rodgers, C. Caillet-Saguy, N. Izadi-Pruneyre, A. Lecroisey, *Biochemistry* **2008**, *47*, 2087–2098.
68. M. Czjzek, S. Létoffé, C. Wandersman, M. Delepierre, A. Lecroisey, N. Izadi-Pruneyre, *J. Mol. Biol.* **2007**, *365*, 1176–1186.
69. C. l. Caillet-Saguy, M. Piccioli, P. Turano, N. Izadi-Pruneyre, M. Delepierre, I. Bertini, A. Lecroisey, *J. Am. Chem. Soc.* **2009**, *131*, 1736–1744.
70. S. C. Holt, L. Kesavalu, S. Walker, C. A. Genco, *Periodontology 2000,* **1999**, *20*, 168–238.
71. T. Olczak, W. Simpson, X. Liu, C. A. Genco, *FEMS Microbiol. Rev.* **2005**, *29*, 119–144.
72. J. W. Smalley, J. Silver, P. J. Marsh, A. J. Birss, *Biochem. J.* **1998**, *331*, 681–685.
73. A. Sroka, M. Sztukowska, J. Potempa, J. Travis, C. A. Genco, *J. Bacteriol.* **2001**, *183*, 5609–5616.
74. J. W. Smalley, A. J. Birss, B. Szmigielski, J. Potempa, *Biol. Chem.* **2008**, *389*, 1235–1238.
75. J. W. Smalley, A. J. Birss, B. Szmigielski, J. Potempa, *Arch. Biochem. Biophys.* **2007**, *465*, 44–49.
76. J. W. Smalley, D. P. Byrne, A. J. Birss, H. Wojtowicz, A. Sroka, J. Potempa, T. Olczak, *PLoS ONE* **2011**, *6*, e17182.
77. T. Olczak, H. Wojtowicz, J. Ciuraszkiewicz, M. Olczak, *BMC Microbiol.* 2010, *10*, 134.
78. C. E. Allen, M. P. Schmitt, *J. Bacteriol.* **2009**, *191*, 2638–2648.
79. H. Wójtowicz, J. Wojaczyński, M. Olczak, J. Króliczewski, L. Latos-Grażyński, T. Olczak, *Biochem. Biophys. Res. Comm.* **2009**, *383*, 178–182.
80. T. Olczak, A. Sroka, J. Potempa, M. Olczak, *Arch. Microbiol.* **2008**, *189*, 197–210.
81. H. Wójtowicz, T. Guevara, C. Tallant, M. Olczak, A. Sroka, J. Potempa, M. Solà, T. Olczak, F. X. Gomis-Rüth, *PLoS Pathog.* **2009**, *5*, e1000419.
82. C. Fournier, A. Smith, P. Delepelaire, *Mol. Microbiol.* **2011**, *80*, 133–148.
83. C. L. Nobles, A. W. Maresso, *Metallomics* **2011**, *3*, 788–796.
84. E. Honsa, A. Maresso, *BioMetals* **2011**, *24*, 533–545.
85. E. P. Skaar, O. Schneewind, *Microbes Infection.* **2004**, *6*, 390–397.
86. M. Andrade, F. Ciccarelli, C. Perez-Iratxeta, P. Bork, *Genome Biology* **2002**, *3*, research0047.0041 - research0047.0045.
87. C. S. Bates, G. E. Montañez, C. R. Woods, R. M. Vincent, Z. Eichenbaum, *Infection Immunity,* **2003**, *71*, 1042–1055.
88. M. V. Tullius, C. A. Harmston, C. P. Owens, N. Chim, R. P. Morse, L. M. McMath, A. Iniguez, J. M. Kimmey, M. R. Sawaya, J. P. Whitelegge, M. A. Horwitz, C. W. Goulding, *Proc. Natl. Acad. Sci. USA* **2011**, *108*, 5051–5056.
89. C. P. Owens, J. Du, J. H. Dawson, C. W. Goulding, *Biochemistry* **2012**, *51*, 1518–1531.
90. S. K. Mazmanian, E. P. Skaar, A. H. Gaspar, M. Humayun, P. Gornicki, J. Jelenska, A. Joachmiak, D. M. Missiakas, O. Schneewind, *Science* **2003**, *299*, 906–909.
91. L. A. Marraffini, A. C. Dedent, O. Schneewind, *Microbiol. Mol. Biol. Rev.* **2006**, *70*, 192–221.
92. C. F. M. Gaudin, J. C. Grigg, A. L. Arrieta, M. E. P. Murphy, *Biochemistry* **2011**, *50*, 5443–5452.

93. R. Aranda IV, C. E. Worley, M. Liu, E. Bitto, M. S. Cates, J. S. Olson, B. Lei, G. N. Phillips Jr, *J. Mol. Biol.* **2007**, *374*, 374–383.
94. V. J. Torres, G. Pishchany, M. Humayun, O. Schneewind, E. P. Skaar, *J. Bacteriol.* 2006, *188*, 8421–8429.
95. H. K. Kim, A. DeDent, A. G. Cheng, M. McAdow, F. Bagnoli, D. M. Missiakas, O. Schneewind, *Vaccine* **2010**, *28*, 6382–6392.
96. N. Muryoi, M. T. Tiedemann, M. Pluym, J. Cheung, D. E. Heinrichs, M. J. Stillman, *J. Biol. Chem.* **2008**, *283*, 28125–28136.
97. M. T. Tiedemann, N. Muryoi, D. E. Heinrichs, M. J. Stillman, *Biochem. Soc. Trans.* **2008**, *36*, 1138–1143.
98. R. M. Pilpa, S. A. Robson, V. A. Villareal, M. L. Wong, M. Phillips, R. T. Clubb, *J. Biol. Chem.* **2009**, *284*, 1166–1176.
99. M. Liu, W. N. Tanaka, H. Zhu, G. Xie, D. M. Dooley, B. Lei, *J. Biol.Chem.* **2008**, *283*, 6668–6676.
100. H. Zhu, G. Xie, M. Liu, J. S. Olson, M. Fabian, D. M. Dooley, B. Lei, *J. Biol. Chem.* **2008**, *283*, 18450–18460.
101. S. K. Mazmanian, E. P. Skaar, A. H. Gaspar, M. Humayun, P. Gornicki, J. Jelenska, A. Joachmiak, D. M. Missiakas, O. Schneewind, *Science* **2003**, *299*, 906–909.
102. V. A. Villareal, R. M. Pilpa, S. A. Robson, E. A. Fadeev, R. T. Clubb, *J. Biol. Chem.* **2008**, *283*, 31591–31600.
103. R. M. Pilpa, E. A. Fadeev, V. A. Villareal, M. L. Wong, M. Phillips, R. T. Clubb, *J. Mol. Biol.* **2006**, *360*, 435–447.
104. J. C. Grigg, C. L. Vermeiren, D. E. Heinrichs, M. E. P. Murphy, *Mol. Microbiol.* **2007**, *63*, 139–149.
105. K. H. Sharp, S. Schneider, A. Cockayne, M. Paoli, *J. Biol. Chem.* **2007**, *282*, 10625–10631.
106. M. Watanabe, Y. Tanaka, A. Suenaga, M. Kuroda, M. Yao, N. Watanabe, F. Arisaka, T. Ohta, I. Tanaka, K. Tsumoto, *J. Biol. Chem.* **2008**, *283*, 28649–28659.
107. K. Krishna Kumar, D. A. Jacques, G. Pishchany, T. Caradoc-Davies, T. Spirig, G. R. Malmirchegini, D. B. Langley, C. F. Dickson, J. P. Mackay, R. T. Clubb, E. P. Skaar, J. M. Guss, D. A. Gell, *J. Biol. Chem.* **2011**, *286*, 38439–38447.
108. V. A. Villareal, T. Spirig, S. A. Robson, M. Liu, B. Lei, R. T. Clubb, *J. Am. Chem. Soc.* **2011**, *133*, 14176–14179.
109. J. C. Grigg, C. X. Mao, M. E. P. Murphy, *J. Mol. Biol.* **2011**, *413*, 684–698.
110. Y. Moriwaki, J. M. M. Caaveiro, Y. Tanaka, H. Tsutsumi, I. Hamachi, K. Tsumoto, *Biochemistry* **2011**, *50*, 7311–7320.
111. A. W. Maresso, G. Garufi, O. Schneewind, *PLoS Pathog.* **2008**, *4*, e1000132.
112. M. Fabian, E. Solomaha, J. S. Olson, A. W. Maresso, *J. Biol. Chem.* **2009**, *284*, 32138–32146.
113. M. T. Ekworomadu, C. B. Poor, C. P. Owens, M. A. Balderas, M. Fabian, J. S. Olson, F. Murphy, E. Balkabasi, E. S. Honsa, C. He, C. W. Goulding, A. W. Maresso, *PLoS Pathog.* **2012**, *8*, e1002559.
114. B. Lei, L. M. Smoot, H. M. Menning, J. M. Voyich, S. V. Kala, F. R. Deleo, S. D. Reid, J. M. Musser, *Infection Immunity* **2002**, *70*, 4494–4500.
115. M. Fisher, Y.-S. Huang, X. Li, K. S. McIver, C. Toukoki, Z. Eichenbaum, *Infection Immunity* **2008**, *76*, 5006–5015.
116. M. Fisher, Y. S. Huang, X. Li, K. S. McIver, C. Toukoki, Z. Eichenbaum, *Infection Immunity* **2008**, *76*, 5006–5015.
117. M. Liu, B. Lei, *Infection Immunity* **2005**, *73*, 5086–5092.
118. T. K. Nygaard, G. C. Blouin, M. Liu, M. Fukumura, J. S. Olson, M. Fabian, D. M. Dooley, B. Lei, *J. Biol. Chem.* **2006**, *281*, 20761–20771.
119. H. Zhu, M. Liu, B. Lei, *BMC Microbiol.* **2008**, *8*, 15.
120. Y. Ran, H. Zhu, M. Liu, M. Fabian, J. S. Olson, R. Aranda, G. N. Phillips, D. M. Dooley, B. Lei, *J. Biol. Chem.* **2007**, *282*, 31380–31388.

121. M. Ouattara, E. Bentley Cunha, X. Li, Y.-S. Huang, D. Dixon, Z. Eichenbaum, *Mol. Microbiol.* **2010**, *78*, 739–756.
122. F. Frolow, A. J. Kalb, J. Yariv, *Nature Struct. Biol.* **1994**, *1*, 453–460.
123. B. C. Chu, H. J. Vogel, *Biol. Chem.* **2011**, *392*, 39–52.
124. R. Tam, M. H. Saier, Jr., *Microbiol. Rev.* **1993**, *57*, 320–346.
125. F. A. Quiocho, P. S. Ledvina, *Mol. Microbiol.* **1996**, *20*, 17–25.
126. N. Yao, S. Trakhanov, F. A. Quiocho, *Biochemistry* **1994**, *33*, 4769–4779.
127. M. C. Lawrence, P. A. Pilling, V. C. Epa, A. M. Berry, A. D. Ogunniyi, J. C. Paton, *Structure* **1998**, *6*, 1553–1561.
128. Y. H. Lee, R. K. Deka, M. V. Norgard, J. D. Radolf, C. A. Hasemann, *Nat. Struct. Biol.* **1999**, *6*, 628–633.
129. T. E. Clarke, S. Y. Ku, D. R. Dougan, H. J. Vogel, L. W. Tari, *Nat. Struct. Biol.* **2000**, *7*, 287–291.
130. J. C. Spurlino, G. Y. Lu, F. A. Quiocho, *J. Biol. Chem.* **1991**, *266*, 5202–5219.
131. A. J. Sharff, L. E. Rodseth, J. C. Spurlino, F. A. Quiocho, *Biochemistry* **1992**, *31*, 10657–10663.
132. S. Trakhanov, N. K. Vyas, H. Luecke, D. M. Kristensen, J. Ma, F. A. Quiocho, *Biochemistry* **2005**, *44*, 6597–6608.
133. Y. H. Lee, M. R. Dorwart, K. R. Hazlett, R. K. Deka, M. V. Norgard, J. D. Radolf, C. A. Hasemann, *J. Bacteriol.* **2002**, *184*, 2300–2304.
134. N. K. Karpowich, H. H. Huang, P. C. Smith, J. F. Hunt, *J. Biol. Chem.* **2003**, *278*, 8429–8434.
135. W. W. Ho, H. Li, S. Eakanunkul, Y. Tong, A. Wilks, M. Guo, T. L. Poulos, *J. Biol. Chem.* **2007**, *282*, 35796–35802.
136. D. Mattle, A. Zeltina, J. S. Woo, B. A. Goetz, K. P. Locher, *J. Mol. Biol.* **2010**, *404*, 220–231.
137. S. Eakanunkul, G. S. Lukat-Rodgers, S. Sumithran, A. Ghosh, K. R. Rodgers, J. H. Dawson, A. Wilks, *Biochemistry* **2005**, *44*, 13179–13191.
138. Y. Tong, M. Guo, *J. Biol. Inorg. Chem.* **2007**, *12*, 735–750.
139. I. Fita, M. G. Rossmann, *Proc. Natl. Acad. Sci. USA* **1985**, *82*, 1604–1608.
140. A. C. K. Chan, B. Lelj-Garolla, F. I. Rosell, K. A. Pedersen, A. G. Mauk, M. E. P. Murphy, *J. Mol. Biol.* **2006**, *362*, 1108–1119.
141. G. Jepkorir, J. C. Rodríguez, H. Rui, W. Im, S. Lovell, K. P. Battaille, A. Y. Alontaga, E. T. Yukl, P. Moënne-Loccoz, M. Rivera, *J. Am. Chem. Soc.* **2010**, *132*, 9857-9872.
142. R. Shi, A. Proteau, J. Wagner, Q. Cui, E. O. Purisima, A. Matte, M. Cygler, *Proteins* **2009**, *75*, 598–609.
143. Z. Ouyang, R. Isaacson, *Infection Immunity* **2006**, *74*, 6949–6956.
144. J. C. Grigg, J. Cheung, D. E. Heinrichs, M. E. Murphy, *J. Biol. Chem.* **2010**, *285*, 34579–34588.
145. J. C. Grigg, J. D. Cooper, J. Cheung, D. E. Heinrichs, M. E. Murphy, *J. Biol. Chem.* **2010**, *285*, 11162–11171.
146. S. Konetschny-Rapp, G. Jung, J. Meiwes, H. Zahner, *Eur. J. Biochem.* **1990**, *191*, 65–74.
147. H. Drechsel, S. Freund, G. Nicholson, H. Haag, O. Jung, H. Zahner, G. Jung, *Biometals* **1993**, *6*, 185–192.
148. F. C. Beasley, E. D. Vines, J. C. Grigg, Q. Zheng, S. Liu, G. A. Lajoie, M. E. Murphy, D. E. Heinrichs, *Mol. Microbiol.* **2009**, *72*, 947–963.
149. S. E. Dale, M. T. Sebulsky, D. E. Heinrichs, *J. Bacteriol.* **2004**, *186*, 8356–8362.
150. F. Peuckert, M. Miethke, A. G. Albrecht, L. O. Essen, M. A. Marahiel, *Angew. Chem. Int. Ed. Engl.* **2009**, *48*, 7924–7927.
151. C. Kandt, Z. Xu, D. P. Tieleman, *Biochemistry* **2006**, *45*, 13284–13292.
152. M. Liu, J. G. Su, R. Kong, T. G. Sun, J. J. Tan, W. Z. Chen, C. X. Wang, *Biophys. Chem.* **2008**, *138*, 42–49.
153. M. Liu, T. Sun, J. Hu, W. Chen, C. Wang, *Biophys. Chem.* **2008**, *135*, 19–24.
154. D. C. White, S. Granick, *J. Bacteriol.* **1963**, *85*, 842–850.

155. H. Panek, M. R. O'Brian, *Microbiology* **2002**, *148*, 2273–2282.
156. M. R. Loeb, *J. Bacteriol.* **1995**, *177*, 3613–3615.
157. M. S. Hanson, E. J. Hansen, *Mol. Microbiol.* **1991**, *5*, 267–278.
158. M. S. Hanson, C. Slaughter, E. J. Hansen, *Infection Immunity* **1992**, *60*, 2257–2266.
159. D. J. Morton, L. L. Madore, A. Smith, T. M. Vanwagoner, T. W. Seale, P. W. Whitby, T. L. Stull, *FEMS Microbiol. Lett.* **2005**, *253*, 193–199.
160. D. J. Morton, T. W. Seale, L. O. Bakaletz, J. A. Jurcisek, A. Smith, T. M. VanWagoner, P. W. Whitby, T. L. Stull, *Int. J. Med. Microbiol.* **2009**, *299*, 479–488.
161. B. Vergauwen, J. Elegheert, A. Dansercoer, B. Devreese, S. N. Savvides, *Proc. Natl. Acad. Sci. USA* **2010**, *107*, 13270–13275.
162. P. M. Jones, M. L. O'Mara, A. M. George, *Trends Biochem. Sci.* **2009**, *34*, 520–531.
163. K. P. Locher, E. Borths, *FEBS Lett.* **2004**, *564*, 264–268.
164. C. F. Higgins, *Annu. Rev. Cell Biol.* **1992**, *8*, 67–113.
165. A. Wilks, K. A. Burkhard, *Nat. Prod. Rep.* **2007**, *24*, 511–522.
166. K. Hollenstein, R. J. Dawson, K. P. Locher, *Curr. Opin. Struct. Biol.* **2007**, *17*, 412–418.
167. R. J. Dawson, K. Hollenstein, K. P. Locher, *Mol. Microbiol.* **2007**, *65*, 250–257.
168. C. F. Higgins, K. J. Linton, *Nat. Struct. Mol. Biol.* **2004**, *11*, 918–926.
169. K. A. Burkhard, A. Wilks, *Biochemistry* **2008**, *47*, 7977–7979.
170. K. P. Locher, A. T. Lee, D. C. Rees, *Science* **2002**, *296*, 1091–1098.
171. R. N. Hvorup, B. A. Goetz, M. Niederer, K. Hollenstein, E. Perozo, K. P. Locher, *Science* **2007**, *317*, 1387–1390.
172. N. Cadieux, C. Bradbeer, E. Reeger-Schneider, W. Koster, A. K. Mohanty, M. C. Wiener, R. J. Kadner, *J. Bacteriol.* **2002**, *184*, 706–717.
173. H. W. Pinkett, A. T. Lee, P. Lum, K. P. Locher, D. C. Rees, *Science* **2007**, *315*, 373–377.
174. O. Lewinson, A. T. Lee, K. P. Locher, D. C. Rees, *Nat. Struct. Mol. Biol.* **2010**, *17*, 332–338.
175. L. L. Anzaldi, E. P. Skaar, *Infection Immunity* 2010, *78*, 4977–4989.
176. S. Severance, I. Hamza, *Chem. Rev.* **2009**, *109*, 4596–4616.
177. M. Rivera, J. C. Rodriguez, *The Dual Role of Heme as Cofactor and Substrate in the Biosynthesis of Carbon Monoxide*, in *Metal Ions in the Life Sciences,* Vol. 6, Eds A. Sigel, H. Sigel, R. K. O. Sigel, RSC Publishing, Cambridge, UK, **2009**, pp 241–293,.
178. M. Rivera, Y. Zeng, *J. Inorg. Biochem.* **2005**, *99*, 337–354.
179. N. Frankenberg-Dinkel, *Antioxid. Redox Signa.* **2004**, *6*, 825–834.
180. M. Ratliff-Griffin, A. Wilks, I. Stojiljkovic, *Iron Transport in Bacteria: Molecular Genetics, Biochemistry, Microbial Pathogenesis and Ecology*, in *Bacterial Heme Oxygenases,* Eds J. H. Crosa, A. R. Mey, S. M. Payne, ASM Press, Washington, D.C. **2004**, pp. 86-94.
181. D. J. Schuller, W. Zhu, I. Stojiljkovic, A. Wilks, T. L. Poulos, *Biochemistry* **2001**, *40*, 11552–11558.
182. S. Hirotsu, G. C. Chu, M. Unno, D.-S. lee, T. Yoshida, S.-Y. Park, Y. Shiro, M. Ikeda-Saito, *J. Biol. Chem.* **2004**, *279*, 11937–11947.
183. J. Friedman, L. Lad, H. Li, A. Wilks, T. L. Poulos, *Biochemistry* **2004**, *43*, 5239–5245.
184. D. J. Schuller, A. Wilks, P. R. Ortiz de Montellano, T. L. Poulos, *Nature Struct. Biol.* **1999**, *6*, 860–867.
185. M. Sugishima, Y. Omata, Y. Kakuta, H. Sakamoto, M. Noguchi, K. Fukuyama, *FEBS Lett.* **2000**, *471*, 61–66.
186. T. Yoshida, M. Noguchi, G. Kikuchi, *J. Biol. Chem.* **1980**, *255*, 4418–4420.
187. T. Yoshida, M. Noguchi, G. Kikuchi, *J. Biochem.* **1980**, *88*, 557–563.
188. W. Lai, H. Chen, T. Matsui, K. Omori, M. Unno, M. Ikeda-Saito, S. Shaik, *J. Am. Chem. Soc.* **2010**, *132*, 12960–12970.
189. Y. Liu, P. R. Ortiz de Montellano, *J. Biol. Chem.* **2000**, *275*, 5297–5307.
190. A. Wang, Y. Zeng, H. Han, S. Weeratunga, B. N. Morgan, P. Moënne-Loccoz, E. Schönbrunn, M. Rivera, *Biochemistry* **2007**, *46*, 12198–12211.

191. A. Wang, J. C. Rodríguez, H. Han, E. Schönbrunn, M. Rivera, *Biochemistry* **2008**, *47*, 8080–8093.
192. M. Rivera, G. A. Caignan, A. V. Astashkin, A. M. Raitsimring, T. K. Shokhireva, F. A. Walker, *J. Am. Chem. Soc.* **2002**, *124*, 6077–6089.
193. M. K. Safo, F. A. Walker, A. M. Raitsimring, W. P. Walters, D. P. Dolata, P. G. Debrunner, W. R. Scheidt, *J. Am. Chem. Soc.* **1994**, *116*, 7760–7770.
194. F. A. Walker, H. Nasri, I. Torowska-Tyrk, K. Mohanrao, C. T. Watson, N. V. Shkhirev, P. G. Debrunner, W. R. Scheidt, *J. Am. Chem. Soc.* **1996**, *118*, 12109–12118.
195. G. Simonneaux, V. Schünemann, C. Morice, L. Carel, L. Toupet, H. Winkler, A. X. Trautwein, F. A. Walker, *J. Am. Chem. Soc.* **2000**, *122*, 4366–4377.
196. G. A. Caignan, R. Deshmukh, Y. Zeng, A. Wilks, R. A. Bunce, M. Rivera, *J. Am. Chem. Soc.* **2003**, *125*, 11842–11852.
197. P. K. Sharma, R. Kevorkiants, S. P. de Visser, D. Kumar, S. Shaik, *Angew. Chem. Int. Ed.* **2004**, *43*, 1129–1132.
198. L. Koenigs Lightning, H. Huang, P. Moënne-Loccoz, T. M. Loehr, D. J. Schuller, T. L. Poulos, P. R. Ortiz de Montellano, *J. Biol. Chem.* **2001**, *276*, 10612–10619.
199. H. Fujii, X. Zhang, T. Tomita, M. Ikeda-Saito, T. Yoshida, *J. Am. Chem. Soc.* **2001**, *123*, 6475–6484.
200. R. Davydov, J. D. Satterlee, H. Fujii, A. Sauer-Masarwa, D. H. Busch, B. M. Hoffman, *J. Am. Chem. Soc.* **2003**, *125*, 16340–16346.
201. R. Davydov, T. M. Makris, V. Kofman, D. E. Werst, S. G. Sligar, B. M. Hoffman, *J. Am. Chem. Soc.* **2001**, *123*, 1403–1415.
202. D. L. Harris, G. H. Loew, *J. Am. Chem. Soc.* **1998**, *120*, 8941–8948.
203. J. Zheng, d. Wang, W. Thiel, S. Shaik, *J. Am. Chem. Soc.* **2006**, *128*, 13204–13215.
204. Y. Li, R. T. Syvitski, K. Auclair, A. Wilks, P. R. Ortiz de Montellano, G. N. La Mar, *J. Biol. Chem.* **2002**, *277*, 33018–33031.
205. Y. Li, R. T. Syvitski, K. Auclair, P. R. Ortiz de Montellano, G. N. La Mar, *J. Am. Chem. Soc.* **2003**, *125*, 13392–13403.
206. Y. Zeng, G. A. Caignan, R. A. Bunce, J. C. Rodriguez, A. Wilks, M. Rivera, *J. Am. Chem. Soc.* **2005**, *127*, 9794–9807.
207. S. J. Takayama, G. Ukpabi, M. E. Murphy, A. G. Mauk, *Proc. Natl. Acad. Sci. USA* **2011**, *108*, 13071–13076.
208. R. Lemberg, *Rev. Pure Appl. Chem.* **1956**, *6*, 1–23.
209. P. O'Carra, E. Colleran, *FEBS Lett.* **1969**, *5*, 295–298.
210. S. B. Brown, A. A. Chabot, E. A. Enderby, A. C. T. North, *Nature* **1981**, *289*, 93–95.
211. S. Sano, T. Sano, I. Morishima, Y. Shiro, Y. Maeda, *Proc. Natl. Acad. Sci. USA* **1986**, *83*, 531–535.
212. J. Crusats, A. Suzuki, T. Mizutani, H. Ogoshi, *J. Org. Chem.* **1998**, *63*, 602–607.
213. D. P. Hildebrand, H.-L. Tang, Y. Luo, C. L. Hunter, M. S. Hunter, G. D. Brayer, A. G. Mauk, *J. Am. Chem. Soc.* **1996**, *118*, 12909–12915.
214. T. Murakami, I. Morishima, M. Toshitaka, S.-i. Ozaki, I. Hara, H.-J. Yang, Y. Watanabe, *J. Am. Chem. Soc.* **1999**, *121*, 2007–2011.
215. J. C. Rodriguez, M. Rivera, *Biochemistry* **1998**, *37*, 13082–13090.
216. L. Avila, H.-w. Huang, J. C. Rodríguez, P. Moënne-Loccoz, M. Rivera, *J. Am. Chem. Soc.* **2000**, *122*, 7618–7619.
217. J. K. Rice, I. M. Fearnley, P. D. Barker, *Biochemistry* **1999**, 16847–16856.
218. J. A. Sigman, X. Wang, Y. Lu, *J. Am. Chem. Soc.* **2001**, *123*, 6945–6946.
219. L. Avila, H. Huang, C. O. Damaso, S. Lu, P. Moënne-Loccoz, M. Rivera, *J. Am. Chem. Soc.* **2003**, *125*, 4103–4110.
220. R. Tenhunen, H. S. Marver, R. Schmid, *J. Biol. Chem.* **1969**, *244*, 6388–6394.
221. M. Ratliff, W. Zhu, R. Deshmukh, A. Wilks, I. Stojiljkovic, *J. Bacteriol.* **2001**, *183*, 6394–6403.

222. R. Wegele, R. Tasler, Y. Zeng, M. Rivera, N. Frankenberg-Dinkel, *J. Biol. Chem.* **2004**, *279*, 45791–45802.
223. M. L. Vasil, U. A. Ochsner, *Mol. Microbiol.* **1999**, *34*, 399–413.
224. G. A. Caignan, R. Deshmukh, A. Wilks, Y. Zeng, H. Huang, P. Moënne-Loccoz, R. A. Bunce, M. A. Eastman, M. Rivera, *J. Am. Chem. Soc.* **2002**, *124*, 14879–14892.
225. M. Unno, T. Matsui, G. C. Chu, M. Couture, T. Yoshida, D. L. Rousseau, J. S. Olson, M. Ikeda-Saito, *J. Biol. Chem.* **2004**, *279*, 21055–21061.
226. Y. Zeng, R. Deshmukh, G. A. Caignan, R. A. Bunce, M. Rivera, A. Wilks, *Biochemistry* **2004**, *43*, 5222–5238.
227. R. Deshmukh, Y. Zeng, L. M. Furci, H.-w. Huang, B. N. Morgan, S. Sander, A. Alontaga, R. A. Bunce, P. Moënne-Loccoz, M. Rivera, A. Wilks, *Biochemistry* **2005**, *44*, 13713–13723.
228. J. Wang, J. P. Evans, H. Ogura, G. N. La Mar, P. R. Ortiz de Montellano, *Biochemistry* **2006**, *45*, 61–73.
229. E. P. Skaar, A. H. Gaspar, O. Schneewind, *J. Biol. Chem.* **2004**, *279*, 436–443.
230. E. P. Skaar, A. H. Gaspar, O. Schneewind, *J. Bacteriol.* **2006**, *188*, 1071–1080.
231. S. Puri, M. R. O'Brian, *J. Bacteriol.* **2006**, *188*, 6476–6482.
232. N. Chim, A. Iniguez, T. Q. Nguyen, C. W. Goulding, *J. Mol. Biol.* **2010**, *395*, 595–608.
233. R. Wu, E. P. Skaar, R. Zhang, G. Joachimiak, P. Gornicki, O. Schneewind, A. Joachimiak, *J. Biol. Chem.* **2005**, *280*, 2840–2846.
234. W. C. Lee, M. L. Reniere, E. P. Skaar, M. E. Murphy, *J. Biol. Chem.* **2008**, *283*, 30957–30963.
235. J. C. Grigg, G. Ukpabi, C. F. Gaudin, M. E. Murphy, *J. Inorg. Biochem.* **2010**, *104*, 341–348.
236. M. L. Reniere, G. N. Ukpabi, S. R. Harry, D. F. Stec, R. Krull, D. W. Wright, B. O. Bachmann, M. E. Murphy, E. P. Skaar, *Mol. Microbiol.* **2010**, *75*, 1529–1538.
237. S. Letoffe, G. Heuck, P. Delepelaire, N. Lange, C. Wandersman, *Proc. Natl. Acad. Sci. USA* **2009**, *106*, 11719–11724.
238. T. T. Chau, M. Ishigaki, T. Kataoka, S. Taketani, *Biosci. Biotechnol. Biochem.* **2010**, *74*, 1415–1420.
239. H. A. Dailey, V. M. Sellers, T. A. Dailey, *J. Biol. Chem.* **1994**, *269*, 390–395.
240. H. A. Dailey, A. N. Septer, L. Daugherty, D. Thames, S. Gerdes, E. V. Stabb, A. K. Dunn, T. A. Dailey, J. D. Phillips, *MBio* **2011**, *2*, e00248–00211.
241. X. Liu, Q. Du, Z. Wang, D. Zhu, Y. Huang, N. Li, T. Wei, S. Xu, L. Gu, *J. Biol. Chem.* **2011**, *286*, 14922–14931.
242. C. Zubieta, R. Joseph, S. S. Krishna, D. McMullan, M. Kapoor, H. L. Axelrod, M. D. Miller, P. Abdubek, C. Acosta, T. Astakhova, D. Carlton, H. J. Chiu, T. Clayton, M. C. Deller, L. Duan, Y. Elias, M. A. Elsliger, J. Feuerhelm, S. K. Grzechnik, J. Hale, G. W. Han, L. Jaroszewski, K. K. Jin, H. E. Klock, M. W. Knuth, P. Kozbial, A. Kumar, D. Marciano, A. T. Morse, K. D. Murphy, E. Nigoghossian, L. Okach, S. Oommachen, R. Reyes, C. L. Rife, P. Schimmel, C. V. Trout, H. van den Bedem, D. Weekes, A. White, Q. Xu, K. O. Hodgson, J. Wooley, A. M. Deacon, A. Godzik, S. A. Lesley, I. A. Wilson, *Proteins* **2007**, *69*, 234–243.
243. I. Stojiljkovic, K. Hantke, *Mol. Microbiol.* **1994**, *13*, 719–732.
244. A. Wilks, *Arch. Biochem. Biophys.* **2001**, *387*, 137–142.
245. M. D. Suits, G. P. Pal, K. Nakatsu, A. Matte, M. Cygler, Z. Jia, *Proc. Natl. Acad. Sci. USA* **2005**, *102*, 16955–16960.
246. M. D. Suits, N. Jaffer, Z. Jia, *J. Biol. Chem.* **2006**, *281*, 36776–36782.
247. I. B. Lansky, G. S. Lukat-Rodgers, D. Block, K. R. Rodgers, M. Ratliff, A. Wilks, *J. Biol. Chem.* **2006**, *281*, 13652–13662.
248. M. N. Bhakta, A. Wilks, *Biochemistry* **2006**, *45*, 11642–11649.
249. M. J. O'Neill, M. N. Bhakta, K. G. Fleming, A. Wilks, *Proc. Natl. Acad. Sci. USA* **2012**, 109, 5639–5644.
250. S. Schneider, K. H. Sharp, P. D. Barker, M. Paoli, *J. Biol. Chem.* **2006**, *281*, 32606–32610.

Chapter 10
Cobalt and Corrinoid Transport and Biochemistry

Valentin Cracan and Ruma Banerjee

Contents

V. Cracan • R. Banerjee (✉)
Department of Biological Chemistry, University of Michigan,
4220C MSRB III, 1150 W. Medical Center Dr., Ann Arbor, MI 48109-0600, USA
e-mail: vfcracan@umich.edu; rbanerje@umich.edu

L. Banci (ed.), *Metallomics and the Cell*, Metal Ions in Life Sciences 12,
DOI 10.1007/978-94-007-5561-1_10, © Springer Science+Business Media Dordrecht 2013

Abstract In this chapter, we focus on the biochemistry of non-corrin cobalt and on a subset of corrinoid-containing enzymes. We review the import of cobalt in prokaryotes and discuss two members of the non-corrin cobalt-dependent enzymes, nitrile hydratase and methionine aminopeptidase. Cobalt is best known for its central role in alkylcorrinoid cofactors, where the unique properties of the cobalt-carbon bond are exploited to catalyze chemically challenging biotransformations. We discuss the import of corrinoids and the reactions catalyzed by the acyl-CoA mutases, the fastest-growing subfamily of adenosylcobalamin (AdoCbl)-dependent enzymes. AdoCbl is used as a radical reservoir to catalyze 1,2 rearrangement reactions. The loading of AdoCbl-dependent enzymes with the correct cofactor form is critically important for their functions and is gated by chaperones that use the chemical energy of GTP hydrolysis to ensure the fidelity of the process. Recent insights into the organization and editing functions of G-protein chaperones in the context of AdoCbl-dependent enzymes that they support, are discussed.

Keywords cobalamin • cobalt • corrin • metallochaperones • radical • transport

Please cite as: *Met. Ions Life Sci.* 12 (2013) 333–374

1 Introduction

In contrast to other transition metals, cobalt is generally used as a cofactor in the corrinoid form. Only eight non-corrin cobalt-containing enzymes are known so far of which nitrile hydratase and methionine aminopeptidase are the best studied [1,2]. Many reviews covering selected topics on corrinoid biochemistry have appeared recently including the organometallic chemistry of cobalamins [3], biochemistry of cobalamin- and corrinoid-dependent enzymes [4–6] and cobalamin transport and trafficking in humans [7,8]. An exhaustive review of the bacterial and Archaeal corrinoid-dependent methyltransferases has also been reported recently [5].

In this review, we focus on the import of cobalt and corrinoids in prokaryotes, much of which has been informed by comparative genomics studies. We also discuss two members of the non-corrin cobalt-dependent enzymes, nitrile hydratase and methionine aminopeptidase and some of the more recently described 5'-deoxyadenosylcobalamin (AdoCbl)-dependent acyl-CoA mutases [9–11] and their G-protein chaperones [12].

2 Cobalt and Cobalamin Transport

2.1 Cobalt Transporters

Corrinoids and the non-corrin cobalt serve as cofactors for a number of enzymes [1,2,5,13]. Organisms that utilize non-corrin cobalt or synthesize corrinoids are dependent on the efficient uptake of cobalt ions from the environment. Clues about cobalt transport in prokaryotes came from careful bioinformatic analyses of gene clusters involved in the biosynthesis of vitamin B_{12} and gene clusters encoding non-corrin cobalt-dependent enzymes [14–16]. NhlF is a permease that is found in the gene cluster of the Co-type nitrile hydratase (NHase) in *Rhodococcus rhodochrous* J1 (see Figures 1 and 7 below) and shares high sequence similarity with other permeases involved in nickel uptake: HoxN from *Ralstonia eutropha* H16, HupN from *Bradyrhizobium japonicum*, NixA from *Helicobacter pylori*, and UreH from *Bacillus sp.* TB-90 [17–20]. Members of the nickel-cobalt (NiCoT) permease family contain eight transmembrane domains and are found in gram-negative and gram-positive bacteria, Archaea and some fungi including *Schizosaccharomyces pombe* and *Neurospora crassa* [15,21]. NiCoT-like proteins with substrate specificity for nickel are also sometimes found adjacent to Ni^{2+}-dependent enzymes, e.g., HoxN is located adjacent to the operon encoding two [NiFe] hydrogenases in *Ralstonia eutropha* H16 [22].

When overexpressed in *E.coli*, NhlF facilitates transport of both $^{57}Co^{2+}$ and $^{63}Ni^{2+}$ with cobalt being the preferred substrate [23]. Transport studies with chimeric proteins constructed using HoxN and NhlF [24] suggested that the transmembrane domains I and II play a role in ion specificity [24]. Both HoxN and NhlF are conserved at the primary sequence level and the signature motif, HA(**V**/**F**)DADHI in transmembrane domain II, contains residues that are important for ion specificity [24]. In this motif, phenylalanine is conserved in NhlF and other verified Co^{2+} transporters, whereas in HoxN and other Ni^{2+} transporters, valine is present [24]. When the first histidine in the motif is replaced by Ala, Ile or Ser in HoxN and NhlF, it reduces the affinity for Ni^{2+} and Co^{2+}, respectively [20,25]. In contrast, mutation of the second histidine residue abolishes transport completely. The Phe→Val mutation in NhlF results in significantly lower Co^{2+} transport. Interestingly, when the valine residue in HoxN is replaced by phenylalanine (as found in NhlF), Ni^{2+} transport is enhanced and some specificity for Co^{2+} is detected [24].

Overexpression of proteins belonging to the NiCoT family in *E. coli* led to the identification of two additional Co^{2+} transporters (from *Novosphingobium aromaticivorans* and *Rhodopseudomonas palustris*) [21]. In both organisms, the cobalt permeases are adjacent to the cluster of genes involved in B_{12} biosynthesis [21]. This confirms that gene neighborhood analyses of putative cobalt permeases and associations with genes encoding cobalamin biosynthesis, non-corrin cobalt-dependent enzymes or B_{12} riboswitches serves as an important guide for discovering Co^{2+}-specific transporters [14,15].

The *Bacillus pallidus* NHase gene cluster contains the *cbi*MNQ genes instead of the NiCoT-type permease [26]. These genes encode components of an ATP-binding cassette (ABC-type) Co^{2+} transporter, representing a second system for Co^{2+} transport (Figure 1) [14]. This Co^{2+} uptake system was initially found in the gene cluster encoding B_{12} biosynthesis in *Salmonella typhimurium* and belongs to the family of energy-coupling factor (ECF) transporters [27]. The Co^{2+} transporter comprises the following components: (i) CbiMN, a substrate-binding transmembrane protein or the S component, (ii) CbiO, an ATPase or A component, and (iii) CbiQ, a conserved transmembrane protein or T component (Figure 1) [27,28]. In general, different ECF transporters in prokaryotes are used for the uptake of water-soluble vitamins,

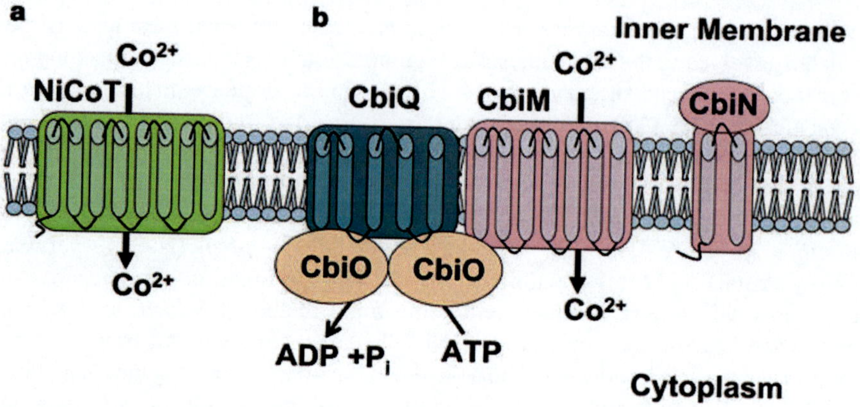

Figure 1 Cobalt transport systems. Cobalt is transported across the inner membrane through a nickel-cobalt (NiCoT) permease (**a**) and an ECF CbiMNQO transporter (**b**). TheCbiMNQO transporter comprises the following components: (i) CbiMN or the S component (shown in pink), (ii) CbiO or the A component (shown in light orange) and (iii) CbiQ or the T component (shown in cyan). Membrane-bound components are shown by rectangles, the transmembrane domains by cylinders and the ATPase component of the ECF transporter is shown as an oval.

cofactors, and their precursors, queuosine, tryptophan and the two transition metals, Ni^{2+} and Co^{2+} [28]. Based on their genomic context, the Ni^{2+}- and Co^{2+}-specific ECFs are distinguished by *nik* and *cbi* prefixes, respectively (*nik*ABCDE and *cbi*MNQO) [14]. These transporters differ from canonical ABC importers since they are not dependent on extracytoplasmic solute-binding proteins and have a different modular architecture. ECF transporters contain A, T, and S components whereas canonical ABC transporters contain two transmembrane domains and two nucleotide-binding domains (Figures 1 and 3, below). The transporter function of the CbiMNQO complex has been verified by overexpression of the *S. enterica* and *Rhodobacter capsulatus* genes in *E. coli* [14]. Interestingly, when either the CbiMNQ or the CbiMN components of the transporter are expressed, significant Co^{2+} uptake is detected, but the transport activity is lower when all four components were present [14]. This result suggests that the CbiMN (S component of the transporter) might be the minimal module required for Co^{2+} transport and that the S component recruits an ATPase component of another ABC transporter from *E. coli* (see also Section 2.2) [14,28].

2.2 *Corrinoid Transport and Metabolism in Bacteria*

Cobalamin and related derivatives are essential cofactors for many enzymes that catalyze numerous transmethylation, rearrangement, and dehalogenation reactions in bacteria, Archaea, algae, and animals [4–6,13]. Only in a limited number of bacteria and Archaea, the complete pathway for cobalamin biosynthesis is present.

The remaining corrinoid-utilizing organisms import the cofactor or its precursor and subsequently converts it to the active form in a process described as salvaging [29,30].

Recent bioinformatic analyses had predicted that members of the family of ECF transporters are involved in the import of both Co^{2+} and cobalamin (Section 2.1) [14,31]. In numerous organisms, two S components of ECF transporters, CbrT and CblT are postulated to be specific for cobalamin and dimethylbenzimidazole (DMB), respectively (Figure 2) [31,32]. Frequently, various S components of ECF transporters are scattered in the genome, which suggests that S components are interchangeable [28]. In fact, in *Lactococcus lactis*, eight S components interact with a single AT module, giving rise to functional transporters of vitamins and related compounds [33].

It should be stressed that functional predictions of CbrT and CblT are based solely on their genomic context (i.e., close proximity to cobalamin biosynthesis genes, vitamin B_{12}-dependent enzymes or regulation by B_{12} riboswitches) and have not been experimentally validated [31]. In *Lactobacillus casei*, CbrT is found in the gene cluster with the PduO-type adenosyltransferase and AdoCbl-dependent ribonucleotide reductase [31]. Alternatively, in some organisms CbrT is found in the operon with an AT component of the transporter (CbrTUV) (Figure 2b). This organization is found in some Fermicutes, Actinibacteria, and in the archaeon, *Methanosphaera stadtmanae*. In certain organisms, CbrTUV is accompanied by other components, CbrX as in *Bacillus cereus* and CbrY-Z as in *Moorella thermoacetica*, which might also be associated with the transporter [31] (Figure 2b).

The BtuCD-F system (3.A.1.13.1) is the cobalamin transporter in *E.coli* and one of the best-studied ABC transporters (Figure 3). In contrast to the ECF transporters discussed above, the BtuCD-F system is widespread in bacteria and

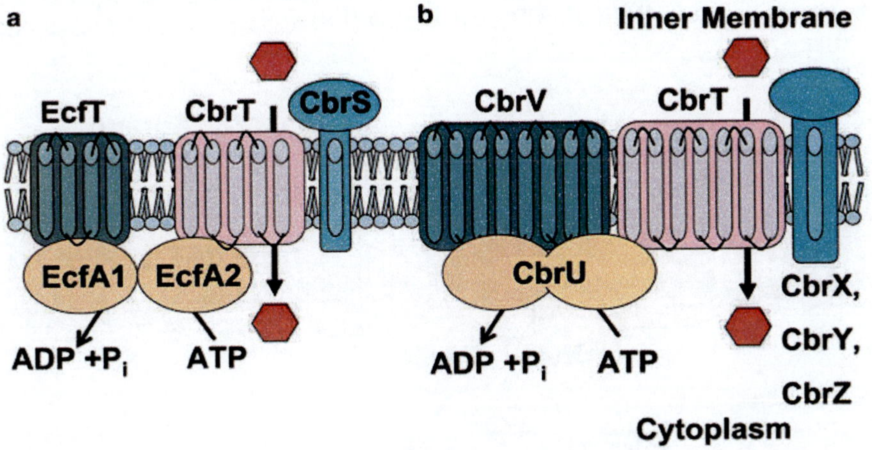

Figure 2 ECF transporters for cobalamin. Cobalamin is transported across the inner membrane via two types of ECF transporters: the S component CbrT (shown in pink) interacts with common A and T modules (shown in light orange and cyan) (**a**) and a dedicated CbrUVT transporter (**b**). Both types of transporters are accompanied by one or more transmembrane components (shown in pale cyan). Hexagons represent cobalamin molecules.

Archaea and its function is well established [32,34]. The BtuCD-F cobalamin transporter is very similar to iron-chelate ABC transporters and is frequently missannotated in bacterial genomes as FepBCD enterobactin transporter, a high affinity iron siderophore [35,36].

In *E. coli* and other Gram-negative bacteria, translocation of cobalamins across the outer membrane involves the action of the TonB-dependent transporter, BtuB [37–39]. BtuB is located in the outer membrane and interacts with three inner membrane proteins, TonB-ExbB-ExbD, which are shared between various TonB-dependent transporters [39–41]. In this process, the energy derived from the proton motive force across the inner membrane is used by BtuB to import cobalamins into the periplasmic space. The expression of BtuB is tightly regulated in many bacteria. Frequently, the *btu*B gene is found in the same operon as the cobalamin biosynthesis genes, and is controlled by a B_{12} riboswitch [42].

In the periplasmic space, cobalamin is bound by BtuF, which is a component of the BtuCD-F transporter (Figure 3). The functional form of the cobalamin transporter is the $BtuC_2D_2$-F complex, where BtuC is the membrane spanning subunit, BtuD is an ATPase and BtuF, a periplasmic component (Figure 3). Cyanocobalamin (CNCbl or vitamin B_{12}) has been used in most studies on the *E. coli* BtuCD-F transporter, although other forms of cobalamin or its precursors including adenosylcobalamin (AdoCbl), cyanocobinamide (CNCbi), and adenosylcobinamide (AdoCbi) are also transported [43].

While the genes encoding components of the BtuCD-F transporter and cobalamin biosynthesis are co-localized and are often regulated by a B_{12} riboswitch, some components of the transporter appear to be missing in some organisms [32]. In these cases, the missing components of the B_{12} transporter might be shared with other ABC transporters as it was shown for ECF transporters [32].

Great strides have been made on structural studies on the B_{12} transporter and structures of BtuB, BtuF, BtuCD, and BtuCD-F have been reported [38,44–46].

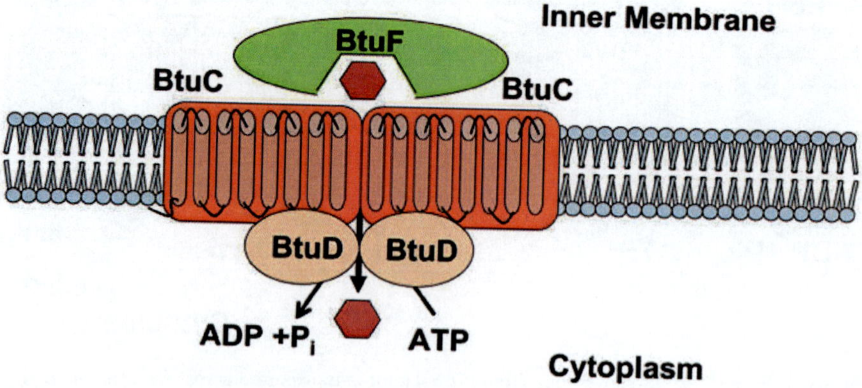

Figure 3 The ABC-type BtuCD-F cobalamin transporter. The symbols used are as described in the legend of Figure 1 and hexagons represent cobalamin molecules.

BtuB binds CNCbl very tightly and the affinity increases in the presence of Ca^{2+} ($K_D \approx 0.3$ nM) [47]. The periplasmic binding protein BtuF contains two domains or lobes connected by an α helix which is very similar to the organization of other periplasmic binding proteins [48,49]. Cobalamin is bound between both lobes of BtuF with very high affinity ($K_D \approx 15$nM) [50]. The two BtuD subunits in the transporter are found in a "head-to-tail" organization [44]. Hence, the active site of $BtuD_2$ is formed by the P-loop of one monomer and the conserved C-motif (LSGGQ in *E. coli*) of the other. The C-motif is located just upstream of the Walker B motif and was shown to be important for ATP binding and catalysis [51]. The BtuCD complex has a very high basal ATP hydrolysis rate, which is stimulated by the presence of BtuF both in proteoliposomes and in solution [52]. It was shown that two ATP molecules are hydrolyzed per molecule of B_{12} transported [53]. Nucleotides drive the dimerization of $BtuD_2$ and in the absence of nucleotides, recombinant BtuD is a monomer [34]. BtuF forms a very stable complex with BtuCD which can be purified by gel-filtration chromatography [54]. Extensive crystallographic, surface plasmon resonance, and EPR techniques using spin-labeled cysteines have been very informative for elucidating the effect of nucleotides and cobalamin on the BtuCD-BtuF complex association and dissociation [53,54].

Multiple lines of evidence suggest a "two-state, alternating access" model [53,54] (Figure 4). According to this model, the transporter in its resting state exists preferentially in the apo- or ADP-bound form and the translocation pore is open towards the periplasmic space (Figure 4). Subsequently, cobalamin-bound BtuF docks to

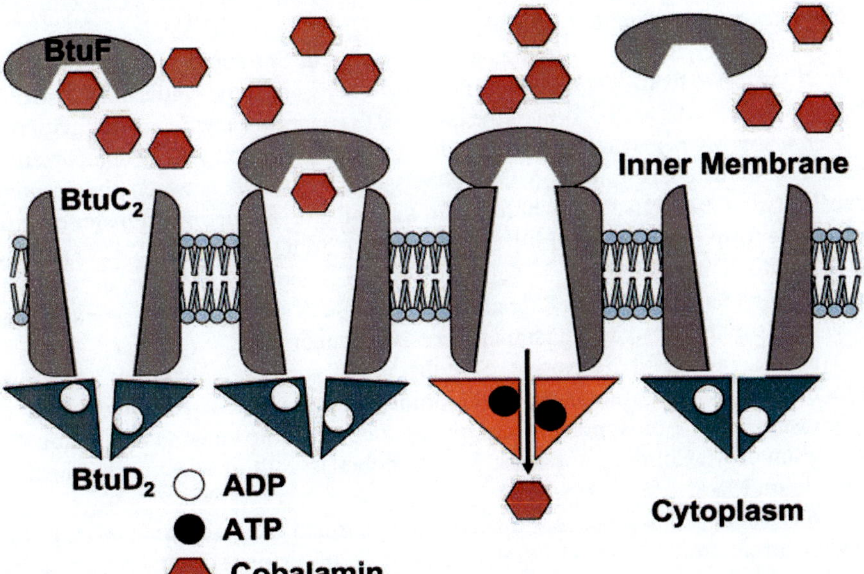

Figure 4 Model of the BtuCD-F transporter cycle. The black and white circles depict ADP and ATP.

BtuCD, forming the BtuCD-F complex. ATP binding to $BtuD_2$ leads to a conformational change inducing an inward-facing conformation and allows escape of the cobalamin into the cytoplasm. EPR spectroscopy has been used to measure distances between the spin-labeled cysteines with different combinations of transporter components and ligand. In the absence of BtuF and in apo-, ADP- or ATP-bound form, BtuC is always in an inward-facing conformation [53]. In the presence of high concentration of ATP and CNCbl, BtuF dissociates from the BtuCD complex so that the next round of transport can follow [53,54] (Figure 4). This "two-state, alternating access" model might be applicable to other ABC transporters as well [28].

Cobalamin salvage is an economical strategy since only processing enzymes that convert cobalamin precursors obtained via import to the active cofactor forms rather than the entire pathway are needed [32]. ATR or ATP:cob(I)alamin adenosyltransferase is needed for synthesis of the AdoCbl form of the cofactor (Figure 5). There are three families of ATRs that are unrelated at the amino acid sequence level: CobA-type, PduO-type, and EutT-type. CobA (in some organisms designated as BtuR or CobO) is involved in *de novo* cobalamin biosynthesis in both the aerobic and anaerobic pathways and preferentially catalyzes the adenosylation of cobyrinic acid a,c-diamide (Figure 5) [32]. In contrast, PduO-like and EutT-like ATRs are involved in assimilation of corrinoids and adenosylate both cobinamide (Cbi) and cobalamin (Cbl) [55,56] (Figure 5). Similarly, the PduO-type ATR from *Lactobacillus reuteri* adenosylates both Cbi ($k_{cat} = (2.4 \pm 0.1) \times 10^{-2}\ min^{-1}$)) and Cbl ($k_{cat} = (2.0 \pm 0.2) \times 10^{-2}\ min^{-1}$)). Since the K_M values for both substrates are similar $K_{Cbl} = (13 \pm 1) \times 10^{-2}\ \mu M$ and $K_{Cbi} = (9.6 \pm 1.4) \times 10^{-2}\ \mu M$, the catalytic efficiencies are comparable. In contrast, the EutT-like ATR from *Salmonella enterica* adenosylates Cbi with only 21% of the activity reported for Cbl.

The PduO-type ATR is often associated with AdoCbl-dependent mutases. The PduO-type ATR from *Methylobacterium extorquens* not only synthesizes AdoCbl, but also delivers the active form of the cofactor to its target enzyme, methylmalonyl-CoA mutase (MCM), without releasing it into solution [57]. Remarkably, complexation of a G-protein chaperone to MCM allows discrimination between the inactive cob(II)alamin and active AdoCbl forms of the cofactor and permits transfer only of the active form (Section 5.4) [58]. In humans, the PduO form of ATR is expressed in the mitochondrion, where human MCM is also found. Mutations in the *cblB* locus, which encodes this ATR, lead to methylmalonic aciduria, a rare genetic disease that is inherited in an autosomal recessive manner [7].

Another important derivative of cobalamin is methylcobalamin (MeCbl), which serves as a cofactor for methionine synthase (MS). Although ATR is responsible for the biosynthesis of AdoCbl, a comparable enzyme catalyzing the formation of MeCbl does not exist. Rather, cob(II)alamin is converted to MeCbl directly on MS.

MS is a cobalamin-dependent methyltransferase and catalyzes the transfer of the methyl group from 5-methyl tetrahydrofolate (CH_3-H_4folate) to L-homocysteine leading to the formation of tetrahydrofolate (H_4folate) and L-methionine [5]. During the reaction cycle, the methyl group is initially transferred to the cob(I)alamin form of the cofactor to form MeCbl and then subsequently donated to the thiolate of

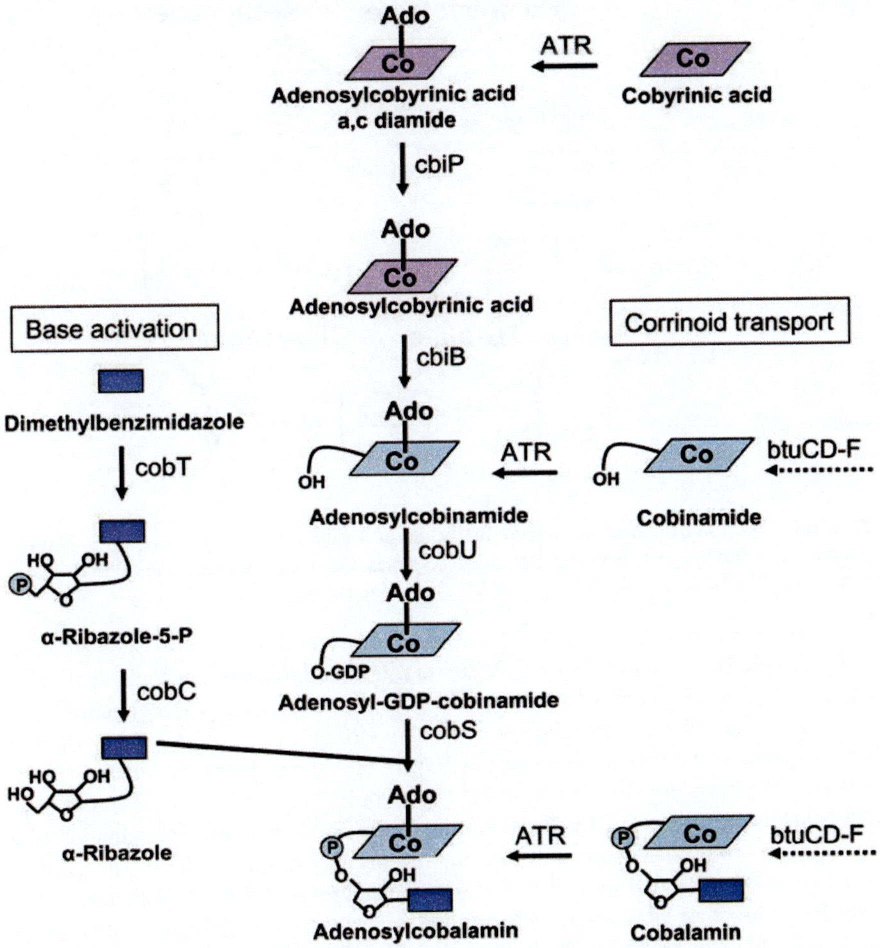

Figure 5 Late steps in cobalamin biosynthesis *versus* the salvage pathway. Schematic representation of reactions catalyzed by enzymes involved in cobalamin scavenging. CobA-, EutT- and PduO-like ATRs all synthesize AdoCbl albeit in different metabolic contexts (see text). Cobyrinic acid and its derivatives, which are part of *de novo* cobalamin synthesis, are shown in purple. Cobinamide and cobalamin and their derivatives, which can be imported in the cell by the BtuCD-F transport system are shown in light blue. Dimethylbenzimidazole is shown in dark blue box and Ado is 5'-deoxyadenosine.

homocysteine (Figure 6). The cob(I)alamin form of the cofactor is highly reactive and susceptible to oxidative inactivation [59]. The C-terminal domain of MS or the reactivating module, binds the methyl donor, S-adenosylmethionine (AdoMet), which during the reductive methylation reaction, donates its methyl group to cob(I)alamin, generated from cob(II)alamin by electron transfer from an oxidoreductase (Figure 6). In bacteria, the reducing system comprises of flavodoxin, flavodoxin reductase, and NADPH [60,61].

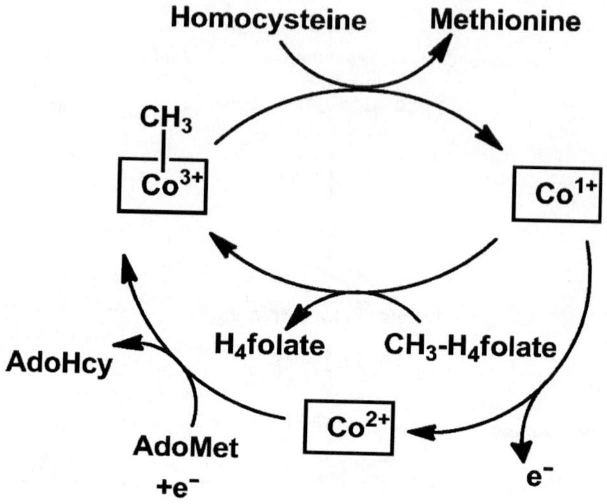

Figure 6 Reaction catalyzed by methionine synthase. In the reaction catalyzed by methionine synthase, cobalamin cycles between the +1 and +3 states. Occasional oxidation leads to formation of inactive cob(II)alamin, which is rescued by a reductive methylation.

Mammals lack a stand-alone flavodoxin and flavodoxin reductase. Instead, they employ a fusion protein, methionine synthase reductase (MSR) containing an N-terminal domain homologous to flavodoxin and a C-terminal domain homologous to flavodoxin reductase [62,63]. In both the prokaryotic and eukaryotic systems, electrons are shuttled from NADPH via FAD and FMN cofactors to inactive cob(II)alamin-bound MS. In mammals, apo-MS is postulated to be loaded with the cob(II)alamin form of the cofactor that is a product of the trafficking chaperone, CblC encoded by *cblC* locus [7,64]. It has been suggested that MSR is needed not only for reductive reactivation of MS but also serves as a chaperone, which stabilizes MS and converts aquacobalamin (OH_2Cbl) to cob(II)alamin [62,65].

2.3 Cobalamin Transport in Eukaryotes

Mammals are not able to synthesize vitamin B_{12} and depend on cobalamin uptake from the diet [4,8]. Vitamin B_{12} and its derivatives are bulky molecules and cannot be absorbed directly from the intestine. Rather, a complex assimilation system exists in mammals, which allows an efficient mechanism for capturing ingested B_{12} and delivering it to cells. This system includes three cobalamin-binding proteins: haptocorrin (HC), intrinsic factor (IF), and transcobalamin (TC) [8]. Interestingly, all three proteins are related at their amino acid sequence level and have molecular masses of ~45 kDa each. IF and HC are glycoproteins and glycosylation is postulated to be important for receptor recognition and for protecting against proteolysis in the intestine [8,66–68].

Briefly, assimilation starts when vitamin B_{12} and its derivatives are bound by HC in saliva and stomach. When the complex of vitamin B_{12} and HC reaches the duodenum, the cofactor is released and subsequently binds IF. The latter allows corrinoids to cross intestinal cells by receptor-mediated endocytosis. Finally, in the plasma, cobalamins are passed to TC, which is the main protein vehicle for distribution of cobalamins to the cells via bloodstream. TC is very specific for intact cobalamin, i.e., it binds only B_{12} derivatives with the intact lower axial ligand, i.e., the dimethylbenzimidazole (DMB) moiety [69]. In contrast, HC binds cobalamin derivatives lacking the DMB moiety, suggesting that HC might play a role in clearing inactive corrinoids from circulation [70].

Cobalamin-loaded TC interacts with TC receptor on the surface of the cell, and following internalization into the lysosome, is degraded, while the receptor is recycled [8]. Subsequently, cobalamin is transported from lysosomes to the cytoplasm by the recently discovered CblF transporter [7]. It is estimated that the distribution of various cobalamin forms in the body is: ~60% AdoCbl, ~20% MeCbl and ~20% OH_2Cbl [8].

3 Non-Corrin Cobalt Enzymes

3.1 Nitrile Hydratase

Nitrile hydratase is a cobalt-containing enzyme that catalyzes the conversion of nitriles (organic cyanides) to amides as shown in equation (1):

$$R\text{–}C{\equiv}N + H_2O \rightarrow R\text{–}(C{=}O)\text{–}NH_2 \qquad (1)$$

The breadth of structural diversity of NHase substrates is truly remarkable [71]. Studies on the genetics and structural enzymology of NHases have led to the development of several biotechnological applications [72]. For example, NHases are used in the production of acrylamide, nicotinamide, and other commodity chemicals from the corresponding nitriles on an industrial scale [73–76].

NHase activity was discovered in *Rhodococcus rhodochrous* J1 (formerly known as *Arthrobacter sp.* J-1) [72,77,78]. Subsequently, NHase was found in many bacteria, particularly in the proteobacteria, actinobacteria, cyanobacteria and firmicutes phyla. Based on the metal cofactor, NHases can be divided in two major groups, cobalt-dependent (Co-type) and iron-dependent. Co-type NHases possess a non-corrin Co^{3+} center whereas the iron-dependent enzymes contain a non-heme Fe^{3+} center [71,74]. Interestingly, the *R. rhodochrous* J1 strain contains two forms of the Co-type NHases that are transcribed from two gene clusters (Figure 7) [71,79]. In addition to genes encoding subunits of NHases, both clusters encode transcriptional regulators, cobalt transporters, and NHase chaperones (Figure 7). NHases are comprised of α and β subunits with molecular masses ranging from 22–63 kDa

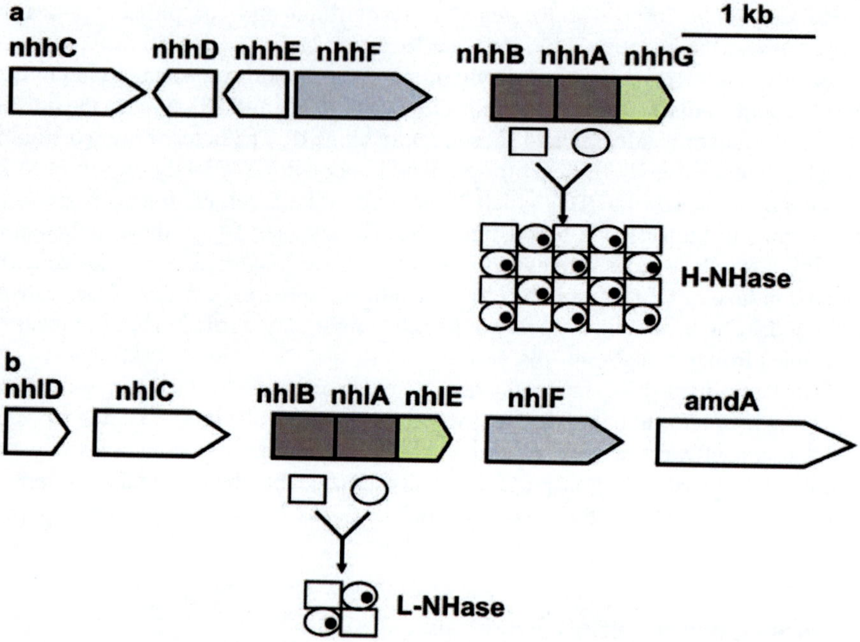

Figure 7 Organization of the H-NHase and L-NHase gene clusters in *R. rhodochrous* J1. The gene clusters for H-NHase (top) and L-NHase (bottom) are shown with the individual genes denoted as arrows. The genes encode the following functions: *nhh*C, *nhh*D, transcriptional regulators, *nhh*E, homolog of the β subunit of H-NHase, *nhh*F, cobalt transporter, *nhh*B, β-subunit of H-NHase, *nhh*A, α subunit of H-NHase, *nhh*G, swapping chaperone; *nhl*D, *nhl*C, transcriptional regulators, *nhl*B, β subunit of L-NHase, *nhl*A, α subunit of L-NHase, *nhl*E-swapping chaperone, *nhl*F-cobalt transporter. The difference between the oligomeric organization of the H- and L-NHase is shown. The cobalt ion is shown as a black dot.

(Table 4 in [74]). Low-molecular-weight NHase (L-NHase) is an $\alpha_2\beta_2$ heterotetramer with a molecular mass of ~130 kDa (Figure 7). High-molecular-weight NHase (H-NHase) is a multimer of αβ heterodimers with a molecular mass reaching ~520 kDa (Figure 7). The expression of L- and H-NHases depends on different inducers in the culture medium [74]. Thus, when the culture media is supplemented with urea, *R. rhodochrous* J1 produces H-NHase whereas supplementation with cyclohexanecarboxamide results in induction of L-NHase. Aromatic or heterocyclic nitriles (3-cyanopyridine, benzonitrile, and cyanopyrazine) are preferential substrates for L-NHase whereas H-NHase exhibits higher activity with aliphatic nitriles (acrylonitrile) [71,72,74]. In contrast to *R. rhodochrous* J1, in other organisms only one type of NHase is found and different substrates/products or their analogs can act as inducers. Several NHases (e.g., from *Bacillus pallidus* Dac 521, *Brevibacterium sp.* R312, *Rhodococcus sp.* N771) are constitutively expressed [80–82].

Regardless of the oligomerization state of NHase, only the α subunit contains a conserved metal-binding **VC(T/S)LCSC(Y/T)** motif and binds non-corrin cobalt

[74]. Importantly, the second and third cysteines in this motif are posttranslationally modified to cysteine-sulfinic (Cys-SO_2H) and cysteine-sulfenic (Cys-SOH) acid, respectively. Co-type NHase can be easily distinguished from the Fe-dependent NHase, based on the amino acid substitutions in the metal-binding motif. Thus, all Co-type NHases have a threonine and tyrosine residue in the motif, whereas in the Fe-dependent enzymes, these amino acids are substituted by serine and threonine, respectively [74]. Three cysteines from the motif, two amide nitrogens of the peptide backbone, and a water molecule or hydroxide serve as ligands for the six-coordinate cobalt ion. This metal coordination geometry was termed "claw setting" [83] and most importantly, posttranslational modification of cysteines in the active site was shown to be crucial for NHase activity [84–86]. NHases have an absorption maximum at 410 nm and are EPR silent, which is expected for low-spin Co^{3+}. Despite the fact that the structure around the metal is very similar and the ionic radii of the trivalent Co^{3+} and Fe^{3+} are similar, in both Co-type and Fe-dependent NHases, the identity of metal is crucial for activity [83,87–89]. When the Fe-dependent NHase from *Rhodococcus sp.* N-771 was reconstituted with cobalt and purified, the specific activity with methacrylonitrile was only 5.9% of the activity of the recombinant native enzyme and 9.9% of the activity of Co-type NHase from *Klebsiella sp.* AM1 [90].

X-ray crystallographic studies revealed that the Co-type and Fe-dependent NHases have very similar overall structures and that the active site is located at the interface between the α and β subunits [83,87–89]. The metal-binding site is deeply buried and the distance from the surface to the Co^{3+} center is ~15 Å [74]. It was proposed that the substrate accesses the active site via the channel formed at the interface between the α and β subunits [74]. Since NHases from different organisms contain different numbers of αβ modules, it suggests that these enzymes can tune their catalytic efficiency and substrate specificity based not only on the amino acids lining the active site but also on the oligomeric state of the enzyme (Figure 7) [91]. NHase from *Rhodococcus sp.* N771 exists in solution in equilibrium between a heterodimer and heterotetramer [92]. Notably, two arginines (Arg52 and Arg157) in the β subunit (*Rhodococcus sp.* N771 numbering) form hydrogen bonds with the oxygen atoms of Cys-SO_2H 111 and Cys-SOH 113 and are important for stabilizing the "claw-setting" orientation of ligands for metal-binding [74].

In the reaction mechanism of NHase, it is generally accepted that the cobalt ion functions as a Lewis acid, assisting hydrolysis of the metal-bound nitrile [93–96]. The mechanism involves the initial binding of the nitrile moiety of the substrate to the cobalt atom, displacing the water/hydroxide ligand. The imidate intermediate rearranges to give the final product (Figure 8). It was originally proposed that the nucleophilic water is activated by an active site base and that catalysis proceeds without direct coordination of substrate to the metal [2,75]. However, a strictly conserved **YYX(H/R)(W/Y)** motif has been identified in the β subunit of NHases and βTyr68 and βTrp72 are proposed to form a catalytic triad with αSer112 (*Pseudonocardia thermophila* JCM3095 numbering) with one of these residues functioning as a general base (Figure 8). Substitution of βTyr68 by phenylalanine leads to a 17-fold decrease in k_{cat} and a 10-fold increase in K_M [97].

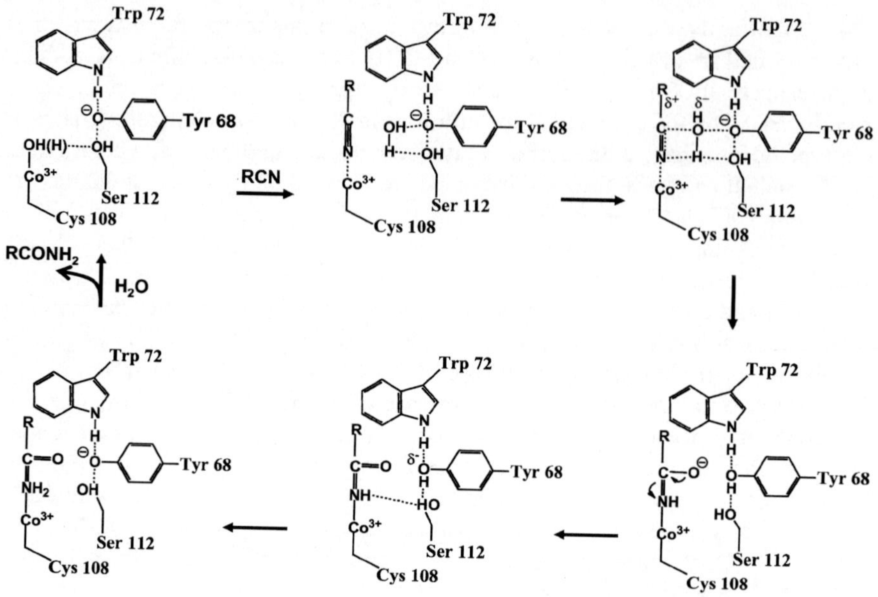

Figure 8 Proposed reaction mechanism for Co-type NHase. Adapted from [221].

Unsuccessful attempts to express *Rhodococcus sp.* N-774 NHase in *E. coli* [98,99] led to the recognition that genes flanking the α and β subunits encode auxiliary proteins that are important for production of active enzyme (Figure 7) [84,100–102]. NhlF was shown to be a Co^{2+} transporter, but the functions of NhhG and NhlE, which have sequence similarity with the NHase subunits, remained a mystery for a while [74].

Both L- and H-NHases in *R. rhodochrous* J1 are matured by a "self-subunit swapping" mechanism [103,104]. A generalized scheme for this process is depicted in Figure 9. The swapping chaperones are encoded by *nhl*E for L-NHase and *nhh*G for H-NHase. The swapping chaperones interact exclusively with the respective apo-α subunits of L- and H-NHases forming mediator complexes NhlAE and NhhAG (designated as αe_2 or αg_2) (Figure 9). The swapping chaperones facilitate the posttranslational modifications of cysteines to Cys-SO_2H and Cys-SOH and insertion of cobalt in the active site. Subsequently, the cobalt-loaded mediator complexes, holo-αe_2 or holo-αg_2, interact with the respective apo-NHases, which results in replacement of apo-α subunits with holo-α subunits and the dissociation of chaperones (Figure 9). AnhE, the chaperone for a newly discovered Co-type NHase from *R. jostii* RHA1, has no detectable sequence similarity to NhlE or NhhG but nevertheless participates in a similar cobalt insertion process [105].

Two other enzymes have been discovered that are very similar in their operonic organization to NHases: thiocyanate hydrolase (SCNase) and toyocamycin nitrile hydratase (TNHase) [106,107]. SCNase from *Thiobacillus thioparus* THI 115 catalyzes the hydration of thiocyanate to carbonyl sulfide and ammonia according to equation (2) [106,108]:

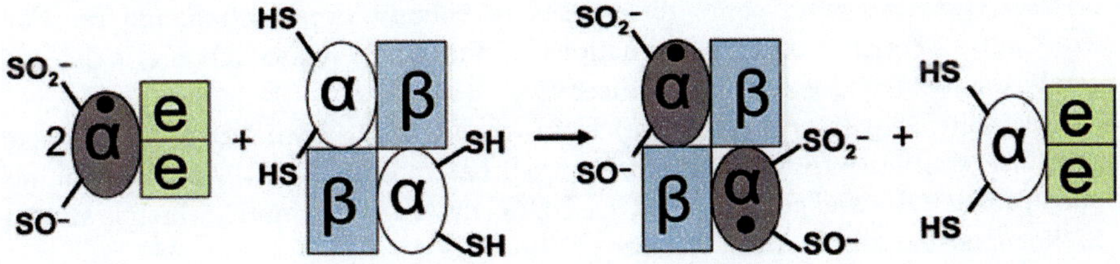

Figure 9 Generalized scheme for the "self-subunit swapping" maturation mechanism for L-NHase. The cobalt ion is shown as black dot. The two cysteines in the metal-binding motif are denoted as –SH, or –SO⁻ and –SO²⁻. The holo-α-subunit is shown in gray and the NHase chaperone is shown in light green.

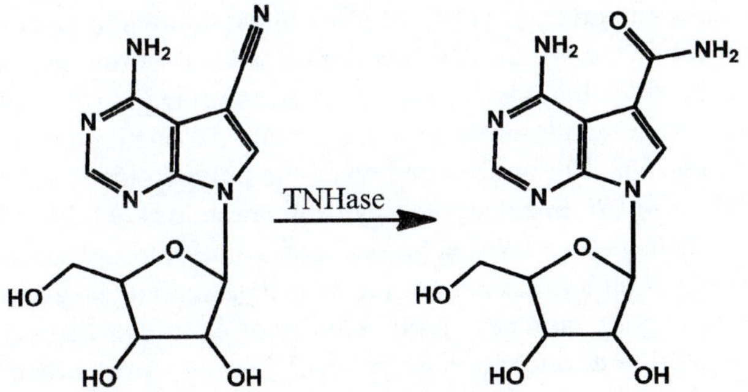

Figure 10 Toyocamycin nitrile hydratase (TNHase) catalyzes the formation of toyocamycin to sangivamycin.

$$SCN^- + 2H_2O \rightarrow COS + NH_3 + OH^- \quad (2)$$

SCNase contains three subunits (α-15 kDa, β-18 , and γ-28 kDa), which are homologous to the subunits of NHase, with the γ subunit possessing the metal-binding VCTLCSC motif [109]. SCNase is a heterododecamer $(\alpha\beta\gamma)_4$ with a molecular mass of 240 kDa [110]. Interestingly, the α and β subunits of SCNase might have arisen via gene fission of the β subunit of NHase while the γ subunit of SCNase is homologous to the α subunit of NHase. The crystal structure of SCNase from *T. thioparus* revealed that it contains cobalt in the active site and that the two cysteines in the metal-binding motif are modified as seen in NHases [109,110]. Active holo-enzyme was recovered only when SCNase was co-expressed in the presence of the swapping chaperone, P15K [106,110].

TNHase catalyzes the formation of toyocamycin (containing a cyanide moiety attached to a deazapurine) to sangivamycin (Figure 10) [107,111]. TNHase from *Streptomyces rimosus* was recently cloned and purified and contains three subunits (ToyJ of 21 kDa, ToyK of 10 kDa, and ToyL of 11 kDa) that are homologous to the three subunits of SCNase [112]. ToyJ is homologous to the γ subunit of SCNase and contains the metal-binding motif, VCTLCSC, which is identical to the one found in SCNase [112]. Both SCNase and TNHase belong to the family of Co-type

NHases, however with some differences in subunit organization and reaction mechanism. Perhaps other combinations of the core NHase subunits exist and catalyze novel reactions that await discovery.

Acetonitrile hydratase (ANHase) from *Rhodococcus jostii* has no amino acid sequence similarity to NHases and is an αβ heterodimer [113]. ANhase requires cobalt for activity and contains the CLLGCAC cysteine-rich motif, which is similar to the metal-binding motif in NHases [113].

3.2 Methionine Aminopeptidase

Aminopeptidases are exopeptidases that cleave the N-terminal amino acid residues from polypeptides and proteins [114]. Methionine aminopeptidases or MetAPs catalyze the removal of the N-terminal methionine and constitute the only family of aminopeptidases, which historically, were viewed as being Co^{2+}-dependent. MetAPs cleave the N-terminal methionine only if the second residue is small and uncharged (Gly, Ala, Ser, Thr, Val, Pro or Cys) and when the polypeptide is at least three residues in length [114–116]. In eukaryotes, cytosolic proteins are initiated by methionine, whereas proteins in prokaryotes, mitochondria, and chloroplasts are initiated by formylmethionine, which is a substrate for deformylase that leaves a free methionine at the N-terminus. In both cases, the majority of mature proteins do not contain an N-terminal methionine and cleavage by MetAPs occurs primarily co-translationally. The reaction catalyzed by MetAPs is crucial for maturation, subcellular localization and degradation of many proteins [114,117–121]. Inactivation of MetAP genes is lethal as demonstrated in *E. coli*, *S. typhimurium* and *Saccharomyces cerevisiae* [122–125].

Multiple crystal structures of MetAPs from different organisms reveal that these proteins are monomeric and possess the "pita-bread" fold (Figure 11b). The two halves of the monomer are referred to as pseudo domains, and each contains two α-helices and two antiparallel β-strands. Both domains contribute five conserved residues to the dinuclear metal center [126–128] providing monodentate (His171 and Glu204 *E. coli* numbering) and bidentate (Asp97, Asp108, and Glu235) coordination to the divalent metals, which are bridged by water or a hydroxide ion [114,129]. In contrast to NHases, posttranslational modifications of amino acids in the metal-binding residues are not seen in MetAPs. Several related aminopeptidases contain the "pita-bread" fold including aminopeptidase P, creatine amidinohydrolase and prolidase [114,127] (Figure 11b).

MetAPs are classified into the type I and type II classes based on the presence of the insert sequence (Figure 11a). Interestingly, eukaryotic organisms always express both types of MetAPs [129,130]. Humans contain two cytosolic methionine aminopeptidases, MetAP-1 and MetAP-2, and the mitochondrial MetAP-1D, type I enzyme [131–134]. MetAP from *E. coli* belongs to the type Ia class and contains an intact aminopeptidase domain (Figure 11a). Types Ib and Ic enzymes contain N-terminal domain extensions, including a zinc finger domain (e.g., human MetAP-1). The type II MetAPs are distinguished from type I by an ~60 amino acid

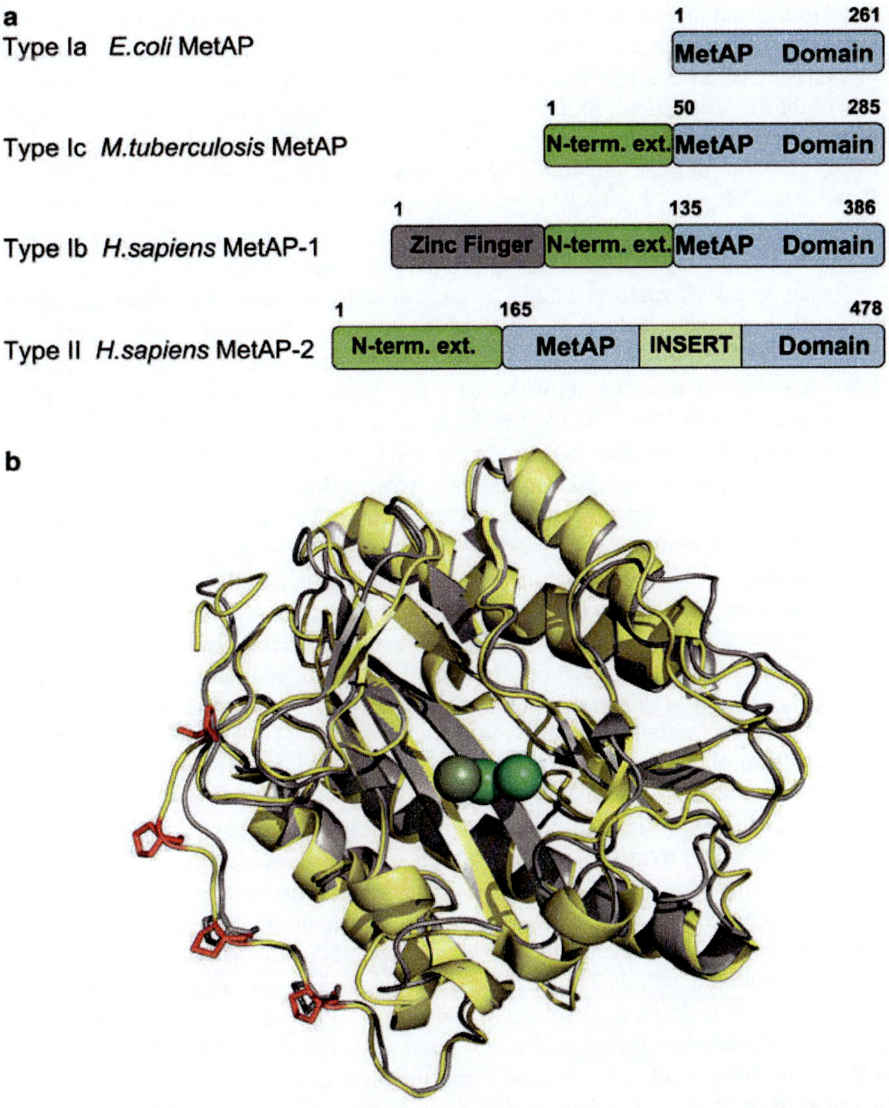

Figure 11 Domain organization and structures of MetAPs. (**a**) Organization of type I and type II MetAPs. (**b**) An overlay of the truncated type I *H. sapiens* MetAP (2B3H, yellow) and *M. tuberculosis* (1Y1N, grey) structures. The prolines in the **PXXP** motifs in both structures are shown in stick representation (*H. sapiens* in red, *M.tuberculosis* in grey). Two cobalt ions are shown in light green and a third cobalt is shown in olive.

insertion in the middle of the aminopeptidase domain (human MetAP-2) (Figure 11a). The additional N-terminal extensions are proposed to be important for protein-protein and protein-nucleic acid interactions. A slow-growth phenotype of *S. cerevisiae* was observed upon deletion of the zinc finger domain of MetAP [129,135]. The

N-terminal extension in type Ic MetAP from *Mycobacterium tuberculosis* is suggested to be important for ribosomal interaction [132]. The structure of truncated human MetAP-1 (lacking the zinc finger domain) superimposes well with the structure of the *M. tuberculosis* MetAP (Figure 11b). The N-terminal extension of MetAP from *M. tuberculosis*, which contains the $\mathbf{P}_{14}\mathbf{XXP}_{17}$ motif is homologous to a similar region in the human enzyme, which contains three tandem repeats of the $\mathbf{P}_{98}\mathbf{XXP}_{101}\mathbf{XXP}_{104}\mathbf{XXP}_{107}$ motif (Figure 11b, shown in red). Since the **PXXP** motif binds the SH3 domains of ribosomal proteins, the N-terminal extension in MetAP-1 is believed to be important for binding to the ribosome [132].

Based on early structural studies, it was generally accepted that MetAPs contain two Co^{2+} ions in the active site [126,129]. However, the number and identity of divalent metals present in the active site of MetAPs is still controversial [131,136–138]. Apo-MetAPs can be activated by a number of divalent cations: Mn^{2+}, Fe^{2+}, Ni^{2+}, and Zn^{2+} [129,139,140]. Co^{2+} is the best activator for apo-MetAP1D and other divalent metal ions are less effective [131]. While human MetAP-1D and MetAP from *E. coli* require two metal ions, human cytosolic MetAP-1 requires three, which is consistent with structural studies on the human MetAP-1 and *E. coli* MetAP revealing the presence of three and two Co^{2+} ions, respectively in the active site [126,132]. MetAPs bind two metals between two double β-antiparallel sheets [132]. Both metals have octahedral geometry and are bridged by a water molecule. It was demonstrated that in MetAP-1 containing three metals, His212 plays an important role in binding a third Co^{2+} ion [132]. In the His212Ala and His212Lys mutants of MetAP-1, a decrease in metal-binding stoichiometry from 3 to 2.2 and 1.4, respectively, was observed but no changes in K_M were seen [131]. It was suggested that a third Co^{2+} ion plays a role in activation of enzyme activity exclusively for human MetAP-1. Interestingly, in several crystal structures with recently reported MetAP-1 inhibitors, a third Co^{2+} is in direct contact with an inhibitor [131].

Since MetAP activity is associated with many crucial cellular processes, the development of specific MetAP inhibitors has attracted significant attention [114,129]. The fungal metabolite fumagillin is a covalent inhibitor of MetAP and is antiangiogenic, i.e., it prevents the growth of new blood vessels [141,142]. MetAPs represent potential targets for antifungal, antibacterial, and anticancer therapies. Fumagillin and its derivatives (ovalicin and TNP-470) inhibit only human MetAP-2 [143,144]. Fumagillin forms a covalent bond with His231 in MetAP-2, a residue that is conserved in all MetAPs [145], and prevents binding of substrates by interacting with residues in the methionine-binding pocket. In addition, the oxygen of the ring-opened epoxide interacts with Co^{2+} and the bridging water [145]. The structure of human MetAP-1 with a truncated N-terminal zinc finger domain revealed that residues 298-303 are pushed into the active site restricting access of ovalicin [132]. On the other hand, the structure of MetAP-2 is more open and does not block access of the inhibitor. Non-covalent inhibitors and transition state analogs of MetAP including amastatin and bestatin derivatives, methionine phosphonate, norleucine phosphonate, and methionine phosphinate have been used to study the reaction mechanism of MetAP [114]. All these compounds exhibit a common mode of binding [128].

The identity of the metal in the active site of MetAP also adds a layer of complexity to the inhibitor studies. The affinity of inhibitors obtained from a screening study is influenced by whether the *E. coli* MetAP is loaded with Co^{2+}, Mn^{2+} or Fe^{2+} [146]. Since there are no major differences between the structures of MetAP with different metals and inhibitors, it was proposed that the coordination chemistry between the compounds and metal ions plays a major role in the selectivity. Using metal-selective inhibitors, the identity of the metal bound to *E. coli* MetAP was investigated. The growth of *E. coli* strains was not inhibited at 1 mM concentrations of the Co^{2+}- and Mn^{2+}-specific inhibitors in contrast to Fe^{2+}-specific inhibitors suggesting that Fe^{2+} might represent the metal used *in vivo* [146]. However, these results are not unequivocal leaving open the question of metal identity *in vivo*.

4 Coenzyme B_{12}-Dependent Isomerases

4.1 Methylmalonyl-Coenzyme A Mutase

Methylmalonyl-CoA mutase (EC 5.4.99.2) is a coenzyme B_{12} or AdoCbl-dependent enzyme that catalyzes the reversible interconversion between (2*R*)-methylmalonyl-CoA and succinyl-CoA [5,147,148] (Figure 12). In fact, this is the only AdoCbl-dependent mutase that is found in bacteria, Archaea and also in animals, where it is located in the mitochondrion.

Different aspects of the MCM-catalyzed reaction mechanism have been reviewed previously [5,13,147,149]. The reaction catalyzed by MCM starts with the homolytic cleavage of the cobalt-carbon bond of AdoCbl, which generates cob(II)alamin and a 5'-deoxyadenosyl radical (Figure 13). Following this, the highly reactive 5'-deoxyadenosyl radical abstracts a H atom from the substrate, which leads to the formation of 5'-deoxyadenosine and a substrate radical. In the last part of the reaction, rearrangement of the substrate radical via a putative cyclopropylcarbinyl intermediate leads to formation of a product radical, which after H atom re-abstraction from 5'-deoxyadenosine, yields product. Finally, the product is released and the two cofactor radicals, cob(II)alamin and the 5'-deoxyadenosyl, recombine to give AdoCbl, completing the turnover cycle [147,150]. The dramatic acceleration ($\sim 10^{12}$-fold) of the homolysis rate of the cobalt-carbon bond is induced upon substrate binding [150]. During the reaction catalyzed by acyl-CoA mutases and other AdoCbl-dependent enzymes, like aminomutases and diol dehydratases, the reactive radicals that are formed can undergo unproductive side reactions precluding re-formation of AdoCbl at the end of the catalytic cycle and leading to enzyme inactivation [9,151–155].

MCM plays an indispensable role for the metabolism of propionate in mammals, where propionyl-CoA, derived from catabolism of branched-chain amino acids, odd-chain fatty acids and cholesterol, is converted to succinyl-CoA [6,156]. In humans, genetic defects in MCM cause methylmalonic aciduria, an autosomal

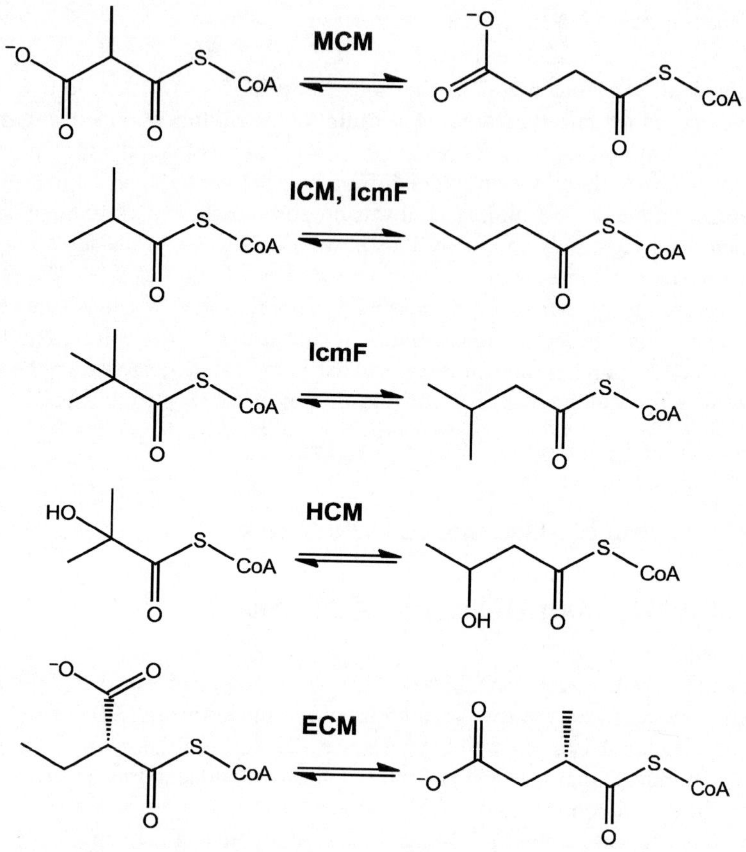

Figure 12 Reactions catalyzed by methylmalonyl-coenzyme A mutase (MCM), isobutyryl-coenzyme A mutase (ICM), isobutyryl-coenzyme A mutase fused (IcmF), 2-hydroxyisobutyryl-coenzyme A mutase (HCM), and ethylmalonyl-coenzyme A mutase (ECM).

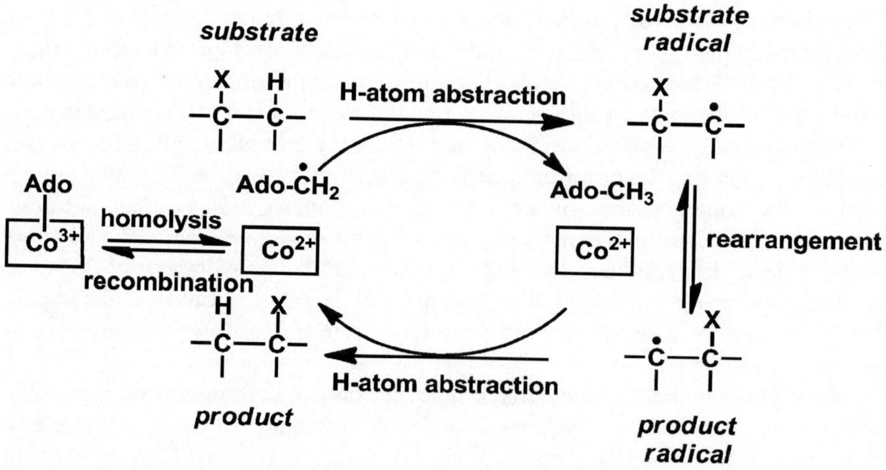

Figure 13 Generalized scheme for the 1,2-rearrangement reaction catalyzed by AdoCbl-dependent mutases.

recessive metabolic disorder [7,157,158]. Methylmalonic aciduria also can result from mutations in genes that impair B_{12} trafficking, i.e. the assimilation and delivery of cofactor [7].

In bacteria, the reaction catalyzed by MCM is important in the reverse metabolic direction, i.e. in the production of propionate [159]. In the genus *Streptomyces*, the MCM-catalyzed reaction is important for the formation of methylmalonyl-CoA, a building block for polyketide biosynthesis [160]. The vast majority of bacterial MCMs are αβ heterodimers, where the α catalytic subunit binds both the cofactor and the substrate (Figure 14). The two subunits of bacterial MCM are related in sequence suggesting that they are products of a gene duplication event [5]. However, only the α subunit contains an intact C-terminal domain which exhibits the typical Rossman-like fold with the canonical "DxHxxG" cofactor-binding motif [161,162]. The N-terminal part of both subunits exhibits a triose phosphate isomerase (TIM)

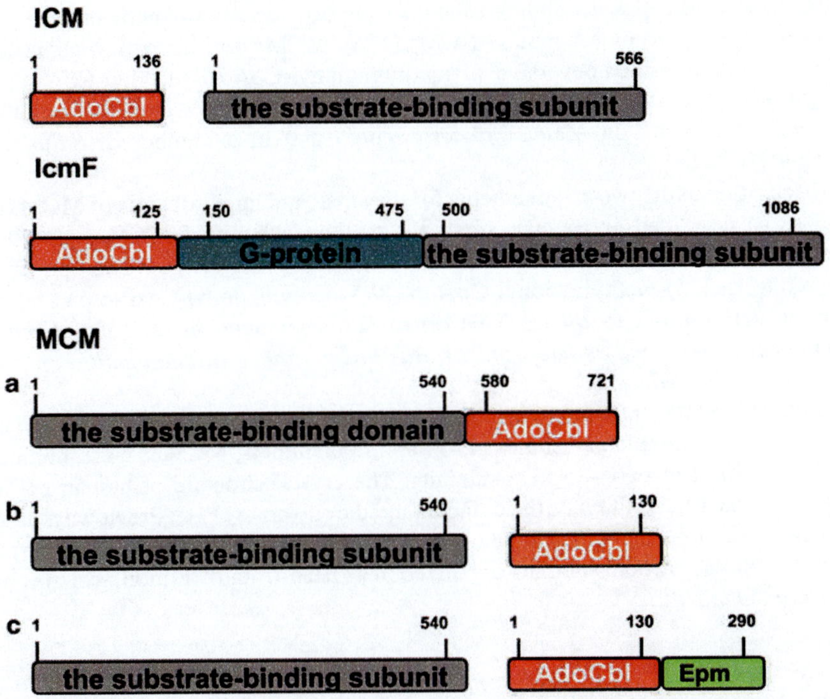

Figure 14 Domain organization of the genes encoding isobutyryl-coenzyme A mutase (ICM), isobutyryl-coenzyme A mutase fused (IcmF), and methylmalonyl-coenzyme A mutase (MCM). In ICMs, the AdoCbl-binding and the substrate-binding domains are always encoded as two separate polypeptides (IcmB and IcmA). IcmF comprises three domains: the N-terminal AdoCbl-binding module that is homologous to the small subunit of ICM, a middle P-loop GTPase domain, and a C-terminal part that is homologous to the large subunit of ICM. In the vast majority of MCMs, the AdoCbl-binding and the substrate-binding domains are found on a single subunit (**a**). In several organisms, MCMs are organized similarly to ICM (**b**), where the AdoCbl-binding and the substrate-binding polypeptides are separate and sometimes, the AdoCbl-binding domain is fused to methylmalonyl-CoA epimerase (Epm) (**c**).

$\alpha_8\beta_8$ barrel fold, and based on the crystal structure of the $\alpha\beta$ heterodimer of MCM, the inactive β subunit appears to serve a structural role [161].

Genomic analyses show that heterodimeric MCMs are found in the vast majority but not in all bacteria and usually, both MCM subunits are found in the same operon. In *E. coli*, the *sbm* (sleeping beauty mutase) gene encodes MCM, which is a functional homodimeric enzyme [163,164]. *Sbm* is not the only example of a bacterial MCM with homodimeric organization. The enzyme from *Sinorhizobium meliloti* is also a homodimer [165]. Notably, in some Archaea (for example in *Metallosphaera sedula, Sulfolobus solfataricus, S. tokodaii, S. islandicus*) and in some bacteria (e.g., in *Petrotoga mobilis, Thermoanaerobacter sp., Ilyobacter polytropus, Thermoanaerobacter tengcongensis, Clostridium sp. OhILAs, Veillonella parvula, Selenomonas sputigen*) the substrate-binding and the AdoCbl-binding domains of MCM are not found in a single polypeptide, but rather, reside on separate subunits of ~63 kDa and ~15 kDa molecular mass (Figure 14). The corresponding MCMs are expected to be heterotetramers comprised of two substrate-binding subunits and two AdoCbl-binding subunits in an $\alpha_2\beta_2$-heterotetrameric organization as seen in isobutyryl-CoA mutase (ICM) [166]. MCMs with an $\alpha_2\beta_2$ organization also catalyze the interconversion of methylmalonyl-CoA to succinyl-CoA in the 3-hydroxypropionate/4-hydroxybutyrate autotrophic carbon dioxide assimilation pathway discovered in *Metallosphaera sedula* and in a number of *Sulfolobus* species [167,168].

In addition to the two arrangements for the small and large subunits of MCM ($\alpha\beta$ and $\alpha_2\beta_2$) described above, the small B_{12}-binding subunit of MCM is fused to methylmalonyl-CoA epimerase in several organisms (Figure 14). The epimerase interconverts (2*S*)-methylmalonyl-CoA to (2*R*)-methylmalonyl-CoA to present the correct stereoisomer to MCM. This organization is found in *Caldalkalibacillus thermarum, Brevibacillus laterosporus, Bacillus tusciae, Ornithinibacillus sp.*, and *Bacillus selenitireducens*.

In the photosynthetic coccolithophorid alga *Pleurochrysis carterae* [169], in the round worm *Ascaris lumbricoides* [170], and in mammals, MCM is an α_2 homodimer, comprised of two ~75 kDa subunits. The crystal structure of human MCM, shows that each subunit binds the cofactor and the substrate [171]. Previously, based on genetic complementation studies, it was suggested that each active site of human MCM is formed by both subunits organized in a "head-to-tail" fashion, which is not the case [172].

4.2 Isobutyryl-Coenzyme A Mutase

Isobutyryl-CoA mutase (ICM) (EC 5.4.99.13) is a coenzyme B_{12}-dependent enzyme, which catalyzes the reversible interconversion of isobutyryl-CoA and *n*-butyryl-CoA [147,166,173] (Figure 12). ICM activity was discovered in cell extracts of the gram-positive, filamentous soil bacterium, *Streptomyces cinnamonensis* [174]. The reaction catalyzed by ICM is very similar to that catalyzed by MCM, which is better studied and more widely distributed in nature. In both reactions, a 1,2-rearrangement of the carbon skeleton takes place where the –(C=O)S–CoA substituent and a H atom on vicinal carbons exchange positions (Section 4.1) (Figure 13) [13,147].

Although the role of ICM in the synthesis of polyketides in several bacteria belonging to the *Streptomyces* genus was well established [160], the presence of ICM-like proteins in other bacteria was neither biochemically nor genetically confirmed. Interestingly, ICM-like activity appears to be involved in the conversion of isobutyrate to butyrate in numerous anaerobic bacteria and cultures enriched in methanogenic bacteria [175–178]. Using ^{13}C-labelled butyrate, several sulfate reducers (*Desulforhabdus amnigenus*, *Desulfobacterium vacuolatum*, *Desulfoarculus baarsii*, *Desulfotomaculum sp.*) have been shown to be capable of isomerization of butyrate to isobutyrate [176]. Also, isomerization of butyrate and isobutyrate has been reported in the anaerobe, *Pelospora glutarica* WoG13 [179,180], and in a thermophilic bacterium, *Synthrophothermus lipocalidus* [181]. We have examined the available bacterial genomic sequences for MCM-like proteins where the interconversion of isobutyrate to butyrate has been documented and found "stand-alone" ICMs in several bacteria (see also Section 4.3) (Figure 15), thereby expanding their known distribution beyond the genus *Streptomyces*.

The genes encoding ICM were first cloned and sequenced from *S. cinnamonensis* [166,173]. ICM is an $\alpha_2\beta_2$-heterotetramer composed of two large subunits (IcmA) of 62.5 kDa and two small subunits (IcmB) of 14.3 kDa (Figure 14). Remarkably, the α and β subunits of ICM have very low affinity for each other in the absence of AdoCbl. The IcmA subunit can be viewed as a truncated form of the large α subunit of MCM lacking the C-terminal AdoCbl-binding domain (Figure 14). Since IcmA does not have a C-terminal AdoCbl-binding domain with the canonical DxHxxG

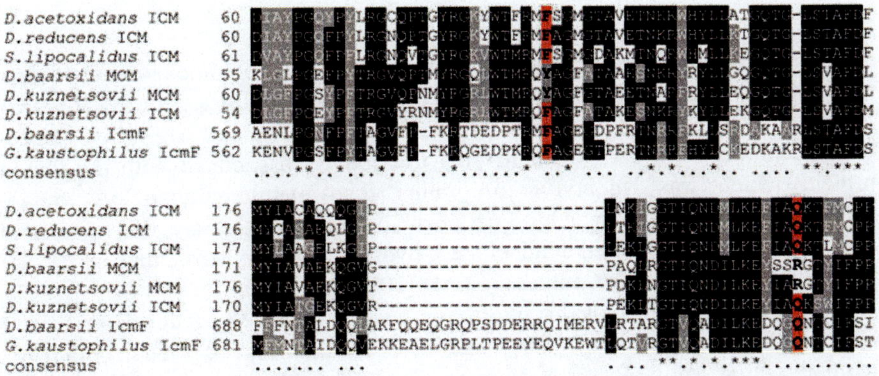

Figure 15 "Stand-alone" ICMs identified in bacteria for which the interconversion of isobutyrate to butyrate was detected. Multiple sequence alignment of IcmF from *G. kaustophilus* and *Desulfarculus baarsii*, MCMs from *Desulfarculus baarsii* and *Desulfotomaculum kuznetsovii*, and "stand-alone" ICMs from *Syntrophothermus lipocalidus*, *D. acetoxidans*, *D. kuznetsovii*, and *D. reducens*. Two residues in ICM and IcmF, Phe and Gln are substituted by Tyr and Arg in MCM.

motif, it requires a separate small IcmB subunit to bind the cofactor (Figure 14). In this respect, ICM resembles other AdoCbl-dependent mutases that exhibit a similar organization. For example, glutamate mutase is also composed of two subunits: the large subunit MutE, which contains the substrate binding site and a small subunit MutS, with the AdoCbl-binding site [182]. The small subunit, IcmB, has high sequence similarity to the AdoCbl-binding domains of MCM, glutamate mutase and 2-methyleneglutarate mutase [166].

Substantial effort has been dedicated to elucidation of the pathways involved in the biosynthesis of polyketide monensin A and B in the genus *Streptomyces* [160]. These studies highlighted the importance of both MCM and ICM-catalyzed reactions for supplying methylmalonyl-CoA to polyketide synthases (PKSs) (Section 4.3) [160].

Recently, in our laboratory, we have discovered and characterized IcmF (isobutyryl-CoA mutase fused), a fusion protein between the two subunits of ICM and its G-protein chaperone, which is found in >80 bacterial species (Figures 12 and 14) [9,12]. This discovery expands the known distribution of ICM activity well beyond the genus *Streptomyces* where it is involved in polyketide biosynthesis. We have also discovered that IcmF catalyzes a novel coenzyme B_{12}-dependent 1,2-rearrangement of isovaleryl-CoA and pivalyl-CoA (2,2-dimethylpropionyl-CoA) (Figure 12) [9]. IcmF variants in which the base specificity loop motif, NKxD is modified to NKxE, catalyze the hydrolysis of both GTP and ATP rather than GTP alone [9]. IcmF is susceptible to more rapid inactivation than other AdoCbl-dependent enzymes. GTP confers some protection from inactivation, albeit only during turnover of isovaleryl-CoA/pivalyl-CoA [9].

4.3 *Novel Activities Catalyzed by Coenzyme B_{12}-Dependent Isomerases*

Two substitutions in the active site of IcmF are key for accommodating the difference between a carboxylate *versus* a methyl group in the substrates of MCM *versus* ICM/IcmF, respectively. These substitutions are Tyr→Phe and Arg→Gln, respectively, and were critical for identifying the activity associated with the fusion protein which was isobutyryl-CoA rather than methylmalonyl-CoA mutase (Figure 15) [12]. Recently, additional acyl-CoA mutases that are very similar to MCM have been discovered and include 2-hydroxyisobutyryl-CoA mutase (HCM) [11] and ethylmalonyl-CoA mutase (ECM) [10] (Figure 12).

In HCM, the amino acids that interact with substrate-specific features are Ile and Gln, respectively (Figure 16). Thus while the Gln residue is conserved in both ICM/IcmF and in HCM, the Phe in ICM/IcmF is substituted by Ile to accommodate the bulkier 2-hydroxyisobutyryl-CoA substrate. HCM was identified in strains that are able to grow on 2-hydroxyisobutyrate (2-HIBA) (Figure 12) [11]. In the pathway for conversion of 2-HIBA to 3-hydroxybutyrate, a putative AdoCbl-dependent mutase with high sequence similarity to ICM was shown to be involved [11].

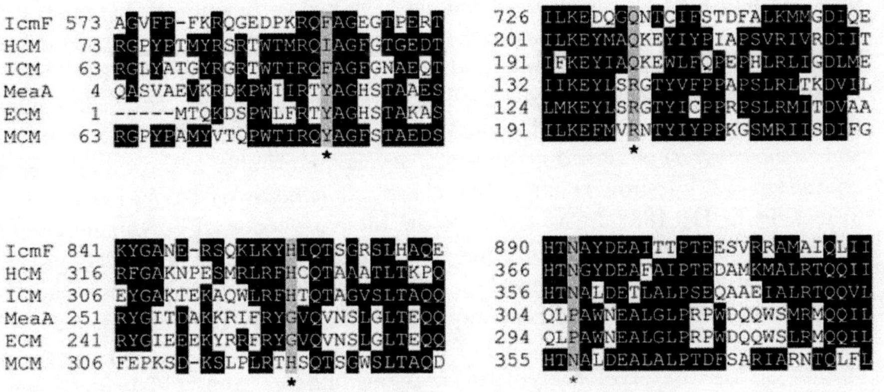

Figure 16 Multiple sequence alignment of the substrate-binding domain of different AdoCbl-dependent mutases: IcmF from *G. kaustophilus*, ICM from *S. cinnamonensis*, MCM from *M. extorquens*, MeaA from *M. extorquens*, ECM from *R. sphaeroides* and HCM from *M. petroleiphilum*. Four residues, which were recognized to be important for substrate binding are highlighted in gray and indicated with asterisks. Two residues Phe and Gln, found in ICM and IcmF are substituted by Tyr and Arg in MCM and ECM or Ile and Gln in HCM. ECM is different from other acyl-CoA mutases by the substitution of His and Asp to Gly and Pro, respectively. ECM was previously known as MeaA in *M. extorquens*.

In ECM, which catalyzes the interconversion of ethylmalonyl-CoA and methylsuccinyl-CoA, the corresponding two amino acids are identical to those found in MCM. The altered substrate specificity can be rationalized on the basis of two key substitutions in ECM to accommodate the bulkier ethylmalonyl-CoA substrate: a His and an Asn residue in MCM are replaced by Gly and Pro, respectively [10] (Figure 16). Interestingly, MCM can catalyze a novel double rearrangement reaction, albeit very inefficiently, of glutaryl-CoA, with a three-carbon template, to methylsuccinyl-CoA and ethylmalonyl-CoA [183]. In addition it was shown that MCM also accepts ethylmalonyl-CoA which is converted mainly to (2*R*)-methylsuccinyl-CoA and some (2*S*)-diastereoisomer [184].

Additionally, as discussed above, IcmF catalyzes the interconversion of pivalyl-CoA and isovaleryl-CoA [9], a reaction that was predicted to be catalyzed by an AdoCbl-dependent mutase [185,186]. Our discovery of the isovaleryl-CoA/pivalyl-CoA mutase activity adds to the growing list of carbon-skeleton rearrangements catalyzed by AdoCbl-dependent isomerases. Based on their high amino acid sequence similarity, the newly-described acyl-CoA mutases are expected to resemble the overall structure of MCM [161]. The "mutase core" thus appears to be quite versatile for the following reasons: (i) substitutions of a limited number of key active site residues alter substrate specificity and (ii) relaxed substrate specificity allows alternative reactions to be catalyzed by the same active site as exemplified by IcmF and MCM.

In early gene inactivation studies in *S. cinnamonensis*, it was shown that this organism possesses another route for making methylmalonyl-CoA that is independent of ICM and MCM [160]. MeaA, a protein of unknown function at that time, with high sequence similarity to MCM, was proposed as the enzyme that allowed formation of methylmalonyl-CoA independently of ICM and MCM [160,187]. When both the ICM and MeaA genes were inactivated in *S. cinnamonensis*, the mutant was unable to produce monensin A and B, whereas the single mutants (Δ*mut::hygB* and Δ*icmA::hygB*) produced monensin at levels comparable to those in the wild-type strain [187]. It turned out that MeaA in *S. cinnamonensis* and *M. extorquens* is in fact, ECM [10,188]. ECM catalyzes the interconversion of ethylmalonyl-CoA and methylsuccinyl-CoA and operates in a recently-discovered ethylmalonyl-CoA pathway for acetate assimilation [10]. In this pathway, three molecules of acetyl-CoA, one molecule of CO_2 and one molecule of bicarbonate form malate and succinyl-CoA. Since the ethylmalonyl-CoA pathway allows the formation of propionyl-CoA from acetoacetyl-CoA, previous findings from ^{13}C-labeling experiments in *S. cinnamonensis* can now be elegantly explained. Moreover, ECMs are found in other bacteria belonging to the genus *Streptomyces* including *S. avermitilis* and *S. coelicolor.*

5 G-Protein Chaperones for Coenzyme B_{12}-Dependent Isomerases

5.1 Organization and Common Features of G-Proteins Belonging to the G3E Family

GTPases, proteins that hydrolyze GTP, are crucial for many aspects of cellular life [189-191]. The main feature of P-loop GTPases and related ATPases is the presence of the so-called "G domain". It is estimated that 10–18% of all gene products in most organisms contain a G domain [190,191]. G domains are found fused to various domains in different proteins, making members of this diverse group of proteins key players in countless cellular processes [190,192]. According to the classification developed by Koonin and colleagues, P-loop GTPases and related ATPases can be divided into two large classes: TRAFAC (named after **tra**nslation **fac**tor-related) and SIMIBI (named after its three largest subgroups, the s**i**gnal recognition particle, **Mi**nD and **Bi**oD superfamilies) [190].

In the past few decades, a number of P-loop GTPases from the SIMIBI class were described as chaperones involved in the assembly of target metalloenzymes. Most of these chaperones belong to two families within the SIMIBI class: the Mrp/MinD and the G3E. The Mrp/MinD family is characterized by the consensus GKGGxGK[ST] Walker A motif [190]. The lysine residue in the KGG portion of the motif is crucial for catalysis [190]. Interestingly, the conserved nucleotide specificity NKxD motif (or G4 motif) found in many GTPases, is not found in members of the Mrp/MinD family (only asparagine (N) is conserved). The aspartate residue in the NKxD motif is a determinant for guanine (*versus* adenine) binding specificity. The best-studied chaperones in the Mrp/MinD family are NifH and CooC. NifH is involved in an ATP-dependent maturation of the iron-molybdenum cofactor of nitrogenase [193,194]. CooC is an ATPase that participates in Ni^{2+} incorporation into the active site of Ni,Fe-carbon monoxide dehydrogenase [195–197].

The G3E family is so named because members of this family have a glutamic acid (E) residue in the Walker B (or G3) motif. Most members of the G3E family have GxxGxGK[ST] Walker A motif and an intact NKxD G4 motif. The G3E family contains four fairly well-studied subfamilies: ArgK, HypB, UreG, and CobW [190]. Recently, studies in our laboratory have demonstrated that MeaB, a GTPase that belongs to the ArgK subfamily, is involved in AdoCbl docking to MCM. MeaB also protects MCM from inactivation during catalytic turnover [58,198-200].

HypB (hydrogenase pleiotropic B) is a metal-binding G-protein that regulates, together with HypA and its homolog HybF, nickel incorporation into the [NiFe]-hydrogenase [201–203]. UreG is a metallochaperone that is highly homologous in sequence and function to HypB and is involved in nickel incorporation into urease [204,205]. Formation of a heterotrimeric complex of the accessory proteins UreG-UreF-UreH, is proposed to be crucial for urease maturation [206].

In thinking about the role of the G3E family of proteins in maturation of metalloenzymes, one must keep in mind that while some chaperones bind metal and deliver it to the active site of the target enzyme, others like MeaB, do not bind metal cofactors and instead, regulate holoenzyme formation in addition to performing auxiliary functions, e.g., protection of target enzyme from inactivation. Based on this dichotomy of function, two general roles for members of the G3E family have been proposed: (i) facilitation of cofactor incorporation in an energy-dependent manner into target active sites (i.e., an insertase role) and (ii) storage and delivery of a metal/cofactor to a target metalloprotein. In light of our findings on MeaB, the G-protein chaperone for MCM [58], it is necessary to expand the suggested categories to include, (iii) protection of the catalytically active form of the enzyme and/or other auxiliary functions.

While MeaB, HypB, and UreG have target metalloenzymes for which they serve as chaperones, not much is known about a large group of proteins that are classified as CobW or COG0523 [207]. CobW from *Pseudomonas denitrificans* was the first member of this group to be described, and it was shown that disruption of the corresponding gene disabled cobalamin synthesis [208]. Only ~12% of COG0523 are orthologs of CobW from *Pseudomonas denitrificans* [207]. Clearly, there is a substantial diversity among COG0523 members, which suggests that these chaperones support functions of several distinct metalloenzymes. In *Rhodococcus sp.* RHA1, *Rhodococcus sp.* N-771, and *R. erythropolis*, a CobW-like protein is found downstream of the β subunit of the iron-dependent NHase. This protein was characterized as an NHase activator and is not related at an amino acid sequence level to the swapping chaperones needed for NHase assembly (Section 3.1) [209]. CobW-like proteins are found in all kingdoms of life [207]. In humans, four copies of CobW-like proteins are found and their function is unknown. Interestingly, proteins belonging to COG0523 were linked to the ability of pathogens to overcome zinc deficiency induced by the host organism [207].

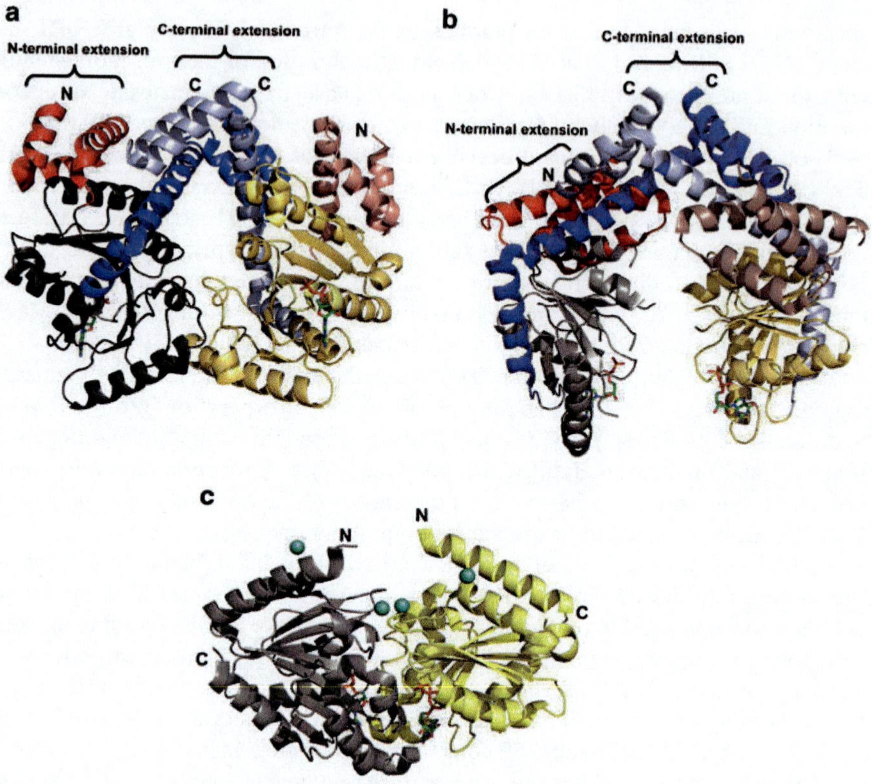

Figure 17 Dimer assembly in the G3E family. Crystal structures of (**a**) MeaB (2QM7A), (**b**) MMAA (2WWW), and (**c**) HypB (2HF8) in complex with nucleotides (MeaB and MMAA with GDP, HypB with GTPγS). Nucleotides are indicated by stick representation. Subunits are shown in grey and yellow. The N- and C-terminal extensions are indicated in red and blue or in pale colors for the second subunit. Zinc atoms bound to HypB are shown as green spheres. All three G3E family GTPases exhibit a modified Rossman fold with a central parallel β-sheet flanked by α-helices, but they differ significantly in the arrangement of the dimer interface.

Proteins from the G3E family can be seen as comprising a core G domain with α-helical N- and C-terminal extensions (Figure 17). These extensions, which sometimes contain histidine-rich regions (or "histidine stretches"), are implicated in metal binding and storage. Most HypBs have "histidine stretches" at the N terminus. On the other hand, HypB from *E. coli* and some other bacteria have no "histidine stretches" at the N terminus but still bind two Ni^{2+} ions per monomer [201]. In HypB, metal binding is achieved via a CxxCGC motif at the N terminus in addition to another metal binding site located within the G-domain [210].

Interestingly, the absence of "histidine stretches" in UreGs in certain organisms is compensated by the presence of another protein, UreE, which is capable of binding metals [207]. On the other hand, in some bacteria and plants where UreEs are absent,

UreG-like proteins always contain "histidine stretches" of different length at the N-terminus [207,211].

CobW-like proteins have "histidine stretches" exclusively at the C termini, where the minimal motif is HxHxHxH (where x represents 0-4 residues). However, some CobW-like proteins have many more histidines in this region, for example, a protein from *Anabaena variabilis* has a "histidine stretch" containing 29 histidine residues [207]! In summary, the variety in the length of the histidine stretche in HypB, UreG and CobW members of the G3E family of G-proteins highlights their importance in metal storage and transport to target enzymes. MeaB from *M. extorquens* and the human ortholog MMAA, encoded by the *cbl*A locus, possess α-helical extensions at both the N- and C-termini but are devoid of metal or cofactor binding motifs (Figure 17). Based on X-ray crystallographic studies, the C-terminal extension appears to be essential for dimerization of MeaB and MMAA [171,212].

Another interesting feature of the G3E family is the connection between their function and oligomerization (Figure 17). Recently, a new classification for specific P-loop NTPases has been proposed, which is based on their dimerization mode in the presence of ligands [213]. GADs (G-proteins activated by dimerization) are G-proteins that have a relatively weak affinity for nucleotides and therefore do not require an external GEF (guanine nucleotide exchange factor). Their dimerization is nucleotide-dependent. Following dimerization, they can bind their corresponding GAPs (GTPase activating proteins) and perform their biological functions [213]. All G-protein chaperones from the G3E family have very low or undetectable GTPase activity, which is enhanced in the presence of cognate GAPs. Although nucleotide-dependent dimerization was proposed to be a common feature of the SIMIBI class of G-proteins or at least, those that bind nucleotides at the dimer interface [213], MeaB and MMAA exist as dimers in the absence of nucleotides (Figure 17) [171,212], challenging the model that all G-proteins in the SIMBI class are GADs. On the other hand, CooC and HypB are monomers in solution and undergo nucleotide-dependent dimerization [196,201].

The structure of HypB from *Archeoglobus fulgidus* has been reported recently and is very similar to the structure reported for HypB from *Methanocaldococcus jannaschii* (Figure 17c) [214]. HypB also dimerizes upon GTP binding and Asp72 forms an intermolecular salt-bridge with Lys148. Interestingly, in the Asp72Ala mutant, GTP-dependent dimerization is abolished, while nucleotide binding and the intrinsic GTPase activity are unaffected. It was proposed that upon GTP binding, HypB dimerizes and an additional Ni^{2+}-binding site at the dimer interface is formed. When nickel-loaded HypB docks with hydrogenase, its GTPase activity is enhanced, Ni^{2+} is delivered to the hydrogenase active site and the HypB dimer dissociates.

Another mutation that disrupts dimerization has been described in the *E. coli* HypB [202]. The monomeric form of the Leu242Ala/Leu246Ala double mutant has both intrinsic GTPase activity and metal-binding behavior (two binding sites) similar to wild-type protein. Leu242Ala/Leu246Ala double mutant supports maturation of the [NiFe]-hydrogenase both *in vitro* and *in vivo* albeit only half of the activity is reconstituted as compared to wild-type HypB [202].

5.2 MeaB Is a Chaperone for Methylmalonyl-Coenzyme A Mutase

MeaB was recognized as an auxiliary protein for MCM in studies on the glyoxylate regeneration pathway in *M. extrorquens* [198,215]. MCM activity was completely lost when the *mea*B gene was disrupted (Δ*mea*B::*kan*) and the bacterium was unable to grow on C_1 and C_2 compounds [215]. In cell extracts of the strain where only the *meaB* gene (Δ*mea*B::*kan*) was disrupted, MCM activity could not be restored by addition of exogenous AdoCbl. Interestingly, in strains in which *meaB* and *meaD* (PduO-type **a**denosyl**t**ransfe**r**ase or ATR) or *epm* (methylmalonyl-CoA epimerase) genes were inactivated (*ΔmeaDΔmeaB::kan, ΔepmΔmeaB::kan,*), MCM activity was restored by addition of exogenous AdoCbl to cell extracts [198].

While ATR catalyzes the formation of AdoCbl from cob(I)alamin and ATP, methylmalonyl-CoA epimerase catalyzes the conversion of *S*-methylmalonyl-CoA to *R*-methylmalonyl-CoA, i.e., it provides the correct stereoisomer of the substrate used by MCM. The phenotype of the Δ*mea*B::*kan* strain suggests that MCM is inactivated during catalysis and that MCM activity can not be recovered simply by addition of exogenous AdoCbl. Knowing the interplay between and functions of the MeaB, ATR, and MCM trio of proteins, the results from the various gene disruption strains can now be explained (Section 5.4) [58].

Bioinformatics pattern searches reveal that MeaB-like proteins are strongly associated with AdoCbl-dependent mutases [207,216]. In fact 63% of all MeaB-like proteins found in prokaryotic genomes are adjacent to or reside within the same operon as MCMs or the B_{12}-binding domains of other AdoCbl-dependent mutases [207]. In fact, the strong operonic association between MeaB and MCM, led to the identification of the corresponding human gene, MMAA (for **m**ethyl**m**alonic **a**ciduria linked to the *cbl***A** complementation group) [216]. MMAA is the human ortholog of MeaB and functions as a chaperone for human MCM. The gene maps to human chromosome 4q31 [216].

The crystal structures of MeaB [212] and MMAA [171] show that their G domains are organized as in other members of the G3E family (Figures 17a and b). MeaB possesses the signature G1-G4 motifs of the G3E family: it has the GxxGxGK[ST] Walker A or G1 and the DxxxxExxG Walker B or G2 motifs, a [V/I]xxD Mg^{2+}-binding or G3 motif and an NKxD GTP-binding or G4 motif. MeaB and MMAA are homodimers with a subunit molecular weight of ~34-46 kDa. The overall structure of the MeaB dimer resembles a starfish, with a central G-domain comprised of parallel β-strands surrounded by α-helices (Figure 17a). In both MeaB and MMAA, the C-terminal extension, is hook-like and forms a "dimerization arm". In MeaB, the dimerization arms together with the G domains in both subunits form the dimer interface. In contrast, only by the dimerization arm in MMAA forms the dimer interface causing the protein to adopt an open "U-shaped" configuration (Figure 17b). The N-terminal extensions in MeaB are proposed to be important for interacting with MCM. Unfortunately, numerous attempts to co-crystallize MeaB and MCM have so far been unsuccessful [171].

MeaB possesses a low intrinsic GTPase activity (k_{cat} ~0.04 min^{-1}) that is enhanced ~100 fold in the presence of MCM. Thus, MCM acts as a GAP for MeaB [200]. MeaB exhibits a weak affinity for GMPPNP and GDP (7.3 ± 1.9 μM and 6.2 ± 0.7 μM, respectively), which is in agreement with previous findings that members of the G3E family do not require GEF proteins to cycle between the GDP- and GTP-bound forms [213].

Native-PAGE can be employed to monitor direct protein-protein interaction between MeaB and MCM. The thermodynamic parameters associated with complex formation have been obtained by isothermal titration calorimetry [200]. Homodimeric MeaB and heterodimeric bacterial MCM from *M. extorquens* form a tight 1:1 complex [200]. In contrast, homodimeric human MMAA and the homodimeric human MCM form a 2:1 complex, in agreement with MCM having two active sites, each of which needs to be loaded with a cofactor [171]. Immunoprecipitation was used to demonstrate that the *E. coli* ortholog of MeaB, YgfD, interacts with MCM in cell lysates [163].

The K_D for binding of the *M. extorquens* MCM and MeaB ranges from 34 ± 4 to 524 ± 66 nM depending on the presence or absence of AdoCbl and the presence/absence and type of nucleotide present. The highest affinity (K_D of 34 ± 4 nM) was determined for binding of MCM•AdoCbl and MeaB•GMPPNP, while the weakest affinity (K_D of 524 ± 66 nM) was obtained for holo-MCM and MeaB/GDP [200].

The presence of MeaB and nucleotides affects the activity of MCM. MeaB/GDP increases k_{cat} for the mutase reaction 1.8-fold (from 132 ± 16 min^{-1} to 237 ± 12 min^{-1}) while also increasing the K_M for methylmalonyl-CoA by 1.7-fold [199]. As a result, the MeaB•GDP complex did not increase the catalytic efficiency (k_{cat}/K_M) of the mutase. Notably, MeaB•GDP decreased the K_{act} for AdoCbl ~4.6-fold and AdoCbl binding to MCM was not observed in the presence of MeaB•GMPPMP, indicating that cofactor binding requires GTP hydrolysis.

5.3 *Biochemical Properties of MMAA, a Human Ortholog of MeaB*

Until recently, most of the biochemical data on MCM and its G-protein chaperone was available only for the bacterial duo of proteins from *M. extorquens* since the human proteins were not available in sufficient quantities. More recently, the two human proteins have been successfully overexpressed and purified in quantities sufficient for crystallography and biochemical studies [171]. Both proteins were overexpressed as N-terminally truncated variants (73–424 for MMAA and 12–750 for MCM). As with the bacterial proteins, the kinetic parameters of MMAA are modulated by the presence of MCM. In the presence of apo-MCM, the GTPase activity of MMAA increased 5-fold, whereas the K_M for GTP decreased 16-fold (from 1210 ± 330 μM to 74 ± 8 μM). Consequently, the k_{cat}/K_M for the MMAA-catalyzed reaction increased 82-fold. The presence of holo-MCM decreased the K_M for GTP and the catalytic efficiency of MMAA was increased only 8-fold [171].

The crystal structures of MMAA and MCM together with the structures of their bacterial counterparts set an important framework for understanding how AdoCbl-dependent mutases interact with their cognate G-protein chaperones and modulate each other's activities [171].

5.4 *The Chaperoning Role of MeaB: The Interplay between Three Proteins*

Based on the functional classification of metallochaperones in the G3E family, MeaB plays a role both as an insertase and a chaperone (Section 5.1). The role of MeaB in the interplay with two other proteins, MCM and ATR, has been elucidated [58]. ATR catalyzes the transfer of the 5'-deoxyadenosyl moiety from ATP to the cobalt atom of cob(I)alamin to form AdoCbl [217]. This reaction is important not only in the *de novo* biosynthesis of AdoCbl in bacteria but also in higher animals in the assimilation of coenzyme B_{12} precursors (Section 2.2).

Human and the *M. extorquens* ATR belong to the PduO family [218]. It has been shown with the *M. extorquens* ATR that it functions not only in the enzymatic synthesis of AdoCbl, but also as a chaperone in the direct delivery of AdoCbl to MCM [57]. This strategy is similar to substrate channeling when one enzyme transfers the product of its reaction to the next enzyme in the pathway or to another active site in a multifunctional protein, without releasing the product into solution. The participation of these "molecular tunnels" helps to minimize unwanted side reactions of unstable intermediates and averts dilution of reactants in the cytosol. Substrate channeling is employed by several enzymes including the pyruvate dehydrogenase complex, tryptophan synthase and dihydrofolate reductase-thymidylate synthase [219,220]. Although coenzyme B_{12} can be synthesized only by some bacteria, many organisms depend on cobalamin salvage and transport to meet their needs for this rare cofactor. Thus, the strategy for sequestering AdoCbl following its synthesis by ATR and its direct transfer to MCM, ensures efficient reconstitution of MCM [57].

Once the role of ATR in cofactor transfer to MCM was established, it immediately raised questions about the role of MeaB in the transfer process [57]. Interestingly, when apo-MCM is complexed with MeaB•GMPPNP, cofactor transfer from AdoCbl-loaded ATR is not seen (Figure 18) [58]. Remarkably, an inactive cofactor form, cob(II)alamin, can be transferred from ATR to MCM that is not complexed to MeaB. In other words, the presence of MeaB prevents loading of MCM with cob(II)alamin which would lead to reconstitution of inactive MCM [199]. This editing function of MeaB uses the energy of GTP binding rather than hydrolysis (Figure 18).

Occasionally during catalysis by MCM, the 5'-deoxyadenosine moiety is lost from the active site. When this happens, AdoCbl can not be reformed and cob(II)alamin can be oxidized to OH_2Cbl [199]. MCM binds OH_2Cbl tightly and renders the mutase inactive. Formation of OH_2Cbl can be monitored by UV-visible spectroscopy by following the increase in absorption of the characteristic absorption

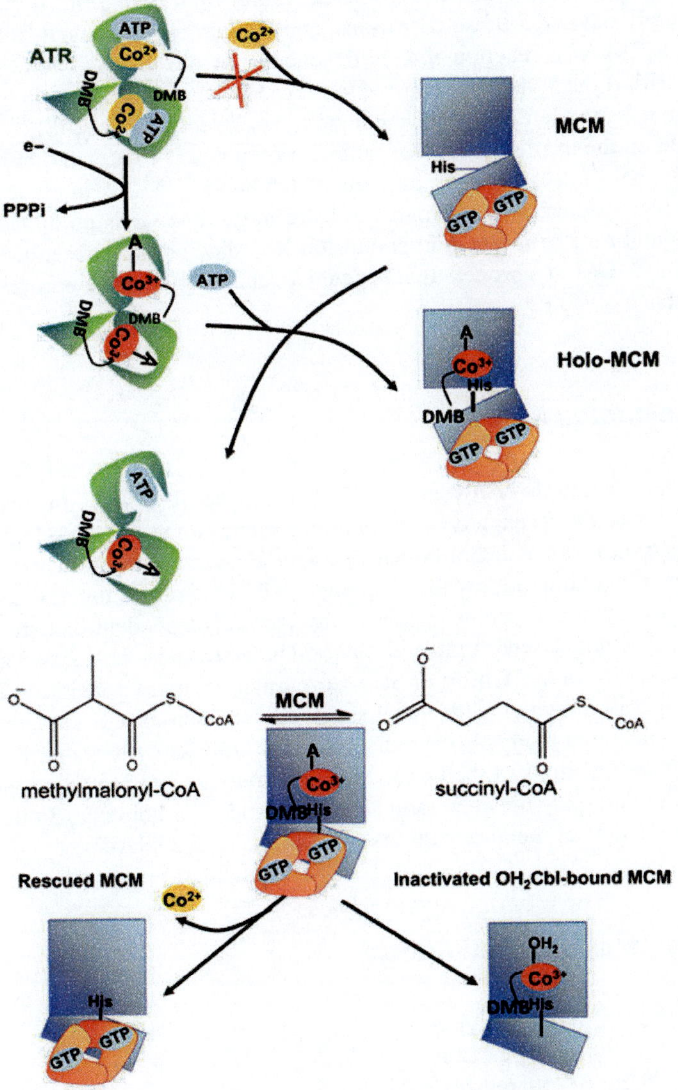

Figure 18 Schematic model for the editing, gating and repair functions of MeaB. Adapted from [58].

peak at 350 nm [199]. The inactivation rate of the *M. extorquens* MCM is 0.0072 min^{-1}, and is decreased 12-fold in the presence of MeaB•GTP [199]. When OH_2Cbl bound to MCM is mixed with MeaB•GTP, the damaged cofactor is not released, unlike the situation with ATP-dependent chaperones described for other AdoCbl-dependent enzymes, e.g., diol dehydratase [152,153]. In the MeaB•MCM complex, loss of the 5'-deoxyadenosine moiety is sensed (rather than oxidation of cob(II)alamin to OH_2Cbl) and under these conditions, cob(II)alamin is expunged from the active site of MCM in a reaction that is dependent on the GTPase activity of MeaB (Figure 18). If the 5'-deoxyadenosine moiety is lost and cob(II)alamin is oxidized to OH_2Cbl, it is "too late" for the chaperone to rescue the enzyme [58].

Based on the available data for human proteins, it is expected that in mammals MMAB (the ATR ortholog in humans), MMAA, and MCM interact in a similar way as their bacterial counterparts. Accordingly, MMAB is predicted to convert cob(II)alamin entering the mitochondrion, to AdoCbl and subsequently transfer AdoCbl to MCM in a process that is gated and edited by the chaperone, MMAA in the presence of GTP.

6 Concluding Remarks

In this chapter, the diverse uses of cobalt in biochemical reactions are discussed with a focus on the recent advances on transporter systems and on select non-corrin cobalt enzymes and AdoCbl-dependent acyl-CoA mutases. Research in several laboratories including our own are leading to a fuller appreciation of the diversity of bacterial metabolic functions supported by AdoCbl-dependent enzymes and their prevalence in unique environmental niches. Understanding the subtle basis of substrate specificity in the family of acyl-CoA mutases could potentially lead to the development of enzymes with novel substrate specificities and chemical reactivities, with biotechnological potential. Studies on the chaperones that load AdoCbl and modulate functions of the target acyl-CoA mutases, are providing insights into the complex interplay between radical enzyme and G-protein chaperones with relevance to the human metabolic disease, methylmalonic aciduria.

Abbreviations and Definitions

2-HIBA	2-hydroxyisobutyrate
ABC	ATP-binding cassette
AdoCbi	5'-deoxy-5'-adenosylcobinamide
AdoCbl	5'-deoxy-5'-adenosylcobalamin, adenosylcobalamin
AdoMet	S-adenosylmethionine
ADP	adenosine 5'-diphosphate
ANHase	acetonitrile hydratase
ATP	adenosine 5'-triphosphate
ATR	ATP:cob(I)alamin adenosyltransferase
Cbi	cobinamide
Cbl	cobalamin

CblC	cobalamin chaperone
CNCbi	cyanocobinamide
CNCbl	cyanocobalamin
CoA	coenzyme A
Cys-SO_2H	cysteine-sulfinic acid
Cys-SOH	cysteine-sulfenic acid
DMB	dimethylbenzimidazole
ECF	energy-coupling factor
ECM	ethylmalonyl-CoA mutase
Epm	methylmalonyl-coenzyme A epimerase
FAD	flavin adenine dinucleotide
FMN	flavin mononucleotide
GAD	G-protein activated by dimerization
GAP	GTPase activating protein
GDP	guanosine 5'-diphosphate
GEF	guanine nucleotide exchange factor
GMPPNP	guanosine 5'-(β,γ-imido)triphosphate
GTP	guanosine 5'-triphosphate
HC	haptocorrin
HCM	2-hydroxyisobutyryl-CoA mutase
H-NHase	high-molecular-weight nitrile hydratase
ICM	isobutyryl-CoA mutase
IcmF	isobutyryl-CoA mutase fused
IF	intrinsic factor
L-NHase	low-molecular-weight nitrile hydratase
MCM	methylmalonyl-CoA mutase
MeCbl	methylcobalamin
MetAP	methionine aminopeptidase
MMAA	methylmalonic aciduria linked to the cblA complementation group
MS	methionine synthase
MSR	methionine synthase reductase
NADPH	nicotinamide adenine dinucleotide phosphate (reduced)
NHase	nitrile hydratase
NiCoT	nickel-cobalt permease
OH_2Cbl	aquacobalamin
PAGE	polyacrylamide gel electrophoresis
SCNase	thiocyanate hydrolase
TC	transcobalamin
TIM	triose phosphate isomerase
TNHase	toyocamycin nitrile hydratase

Acknowledgment This work was supported in part by a grant from the National Institutes of Health (DK45776).

References

1. S. Okamoto, L. D. Eltis, *Metallomics* **2011**, *3*, 963–970.
2. M. Kobayashi, S. Shimizu, *Eur. J. Biochem.* **1999**, *261*, 1–9.
3. B. Kräutler, *Met. Ions Life Sci.* **2009**, *6*, 1–51.
4. B. Kräutler, *Subcell. Biochem.* **2012**, *56*, 323–346.
5. R. G. Matthews, *Met. Ions Life Sci.* **2009**, *6*, 53–114.
6. K. Gruber, B. Puffer, B. Kräutler, *Chem. Soc. Rev.* **2011**, *40*, 4346–4363.
7. R. Banerjee, C. Gherasim, D. Padovani, *Curr. Opin. Chem. Biol.* **2009**, *13*, 484–491.
8. S. N. Fedosov, *Subcell. Biochem.* **2012**, *56*, 347–367.
9. V. Cracan, R. Banerjee, *J. Biol. Chem.* **2012**, *287*, 3723–3732.
10. T. J. Erb, J. Retey, G. Fuchs, B. E. Alber, *J. Biol. Chem.* **2008**, *283*, 32283–32293.
11. T. Rohwerder, U. Breuer, D. Benndorf, U. Lechner, R. H. Muller, *Appl. Environ. Microbiol.* **2006**, *72*, 4128–4135.
12. V. Cracan, D. Padovani, R. Banerjee, *J. Biol. Chem.* **2010**, *285*, 655–666.
13. R. Banerjee, S. W. Ragsdale, *Annu. Rev. Biochem.* **2003**, *72*, 209–247.
14. D. A. Rodionov, P. Hebbeln, M. S. Gelfand, T. Eitinger, *J. Bacteriol.* **2006**, *188*, 317–327.
15. Y. Zhang, V. N. Gladyshev, *J. Biol. Chem.* **2011**, *285*, 3393–3405.
16. Y. Zhang, V. N. Gladyshev, *J. Biol. Chem.* **2011**, *286*, 23623–23629.
17. H. Komeda, M. Kobayashi, S. Shimizu, *Proc. Natl. Acad. Sci. USA* **1997**, *94*, 36–41.
18. L. Wolfram, B. Friedrich, T. Eitinger, *J. Bacteriol.* **1995**, *177*, 1840–1843.
19. J. F. Fulkerson, Jr., H. L. Mobley, *J. Bacteriol.* **2000**, *182*, 1722–1730.
20. T. Eitinger, J. Suhr, L. Moore, J. A. Smith, *Biometals* **2005**, *18*, 399–405.
21. P. Hebbeln, T. Eitinger, *FEMS Microbiol. Lett.* **2004**, *230*, 129–135.
22. E. Schwartz, A. Henne, R. Cramm, T. Eitinger, B. Friedrich, G. Gottschalk, *J. Mol. Biol.* **2003**, *332*, 369–383.
23. O. Degen, M. Kobayashi, S. Shimizu, T. Eitinger, *Arch. Microbio.l* **1999**, *171*, 139–145.
24. O. Degen, T. Eitinger, *J. Bacteriol.* **2002**, *184*, 3569–3577.
25. T. Eitinger, L. Wolfram, O. Degen, C. Anthon, *J. Biol. Chem.* **1997**, *272*, 17139–17144.
26. R. A. Cameron, M. Sayed, D. A. Cowan, *Biochim. Biophys. Acta* **2005**, *1725*, 35–46.
27. J. R. Roth, J. G. Lawrence, M. Rubenfield, S. Kieffer-Higgins, G. M. Church, *J. Bacteriol.* **1993**, *175*, 3303–3316.
28. T. Eitinger, D. A. Rodionov, M. Grote, E. Schneider, *FEMS Microbiol. Rev.* **2011**, *35*, 3–67.
29. M. J. Gray, N. K. Tavares, J. C. Escalante-Semerena, *Mol. Microbiol.* **2008**, *70*, 824–836.
30. J. D. Woodson, J. C. Escalante-Semerena, *Proc. Natl. Acad. Sci. USA* **2004**, *101*, 3591–3596.
31. D. A. Rodionov, P. Hebbeln, A. Eudes, J. ter Beek, I. A. Rodionova, G. B. Erkens, D. J. Slotboom, M. S. Gelfand, A. L. Osterman, A. D. Hanson, T. Eitinger, *J. Bacteriol.* **2009**, *191*, 42–51.
32. D. A. Rodionov, A. G. Vitreschak, A. A. Mironov, M. S. Gelfand, *J. Biol. Chem.* **2003**, *278*, 41148–41159.
33. J. ter Beek, R. H. Duurkens, G. B. Erkens, D. J. Slotboom, *J. Biol. Chem.* **2011**, *286*, 5471–5475.
34. K. P. Locher, *Curr. Opin. Struct. Biol.* **2004**, *14*, 426–431.
35. B. A. Ozenberger, M. S. Nahlik, M. A. McIntosh, *J. Bacteriol.* **1987**, *169*, 3638–3646.
36. E. E. Wyckoff, A. M. Valle, S. L. Smith, S. M. Payne, *J. Bacteriol.* **1999**, *181*, 7588–7596.
37. N. Noinaj, M. Guillier, T. J. Barnard, S. K. Buchanan, *Annu. Rev. Microbiol.* **2010**, *64*, 43–60.
38. D. P. Chimento, A. K. Mohanty, R. J. Kadner, M. C. Wiener, *Nat. Struct. Biol.* **2003**, *10*, 394–401.
39. D. P. Chimento, R. J. Kadner, M. C. Wiener, *J. Mol. Biol.* **2003**, *332*, 999–1014.
40. C. Bradbeer, *J. Bacteriol.* **1993**, *175*, 3146–3150.
41. M. C. Wiener, *Curr. Opin. Struct. Biol.* **2005**, *15*, 394–400.

42. M. D. Lundrigan, W. Koster, R. J. Kadner, *Proc. Natl. Acad. Sci. USA* **1991**, *88*, 1479–1483.
43. C. Bradbeer, J. S. Kenley, D. R. Di Masi, M. Leighton, *J. Biol. Chem.* **1978**, *253*, 1347–1352.
44. K. P. Locher, A. T. Lee, D. C. Rees, *Science* **2002**, *296*, 1091–1098.
45. R. N. Hvorup, B. A. Goetz, M. Niederer, K. Hollenstein, E. Perozo, K. P. Locher, *Science* **2007**, *317*, 1387–1390.
46. D. P. Chimento, A. K. Mohanty, R. J. Kadner, M. C. Wiener, *Acta Crystallogr. D Biol. Crystallogr.* **2003**, *59*, 509–511.
47. C. Bradbeer, P. R. Reynolds, G. M. Bauler, M. T. Fernandez, *J. Biol. Chem.* **1986**, *261*, 2520–2523.
48. E. L. Borths, K. P. Locher, A. T. Lee, D. C. Rees, *Proc. Natl. Acad. Sci. USA* **2002**, *99*, 16642–16647.
49. N. K. Karpowich, H. H. Huang, P. C. Smith, J. F. Hunt, *J. Biol. Chem.* **2003**, *278*, 8429–8434.
50. N. Cadieux, C. Bradbeer, E. Reeger-Schneider, W. Koster, A. K. Mohanty, M. C. Wiener, R. J. Kadner, *J. Bacteriol.* **2002**, *184*, 706–717.
51. P. M. Jones, A. M. George, *FEMS Microbiol. Lett.* **1999**, *179*, 187–202.
52. E. L. Borths, B. Poolman, R. N. Hvorup, K. P. Locher, D. C. Rees, *Biochemistry* **2005**, *44*, 16301–16309.
53. B. Joseph, G. Jeschke, B. A. Goetz, K. P. Locher, E. Bordignon, *J. Biol. Chem.* **2011**, *286*, 41008–41017.
54. O. Lewinson, A. T. Lee, K. P. Locher, D. C. Rees, *Nat. Struct. Mol. Biol.* **2010**, *17*, 332–338.
55. C. L. Johnson, E. Pechonick, S. D. Park, G. D. Havemann, N. A. Leal, T. A. Bobik, *J. Bacteriol.* **2001**, *183*, 1577–1584.
56. K. Park, P. E. Mera, J. C. Escalante-Semerena, T. C. Brunold, *Biochemistry* **2008**, *47*, 9007–9015.
57. D. Padovani, T. Labunska, B. A. Palfey, D. P. Ballou, R. Banerjee, *Nat. Chem. Biol.* **2008**, *4*, 194–196.
58. D. Padovani, R. Banerjee, *Proc. Natl. Acad. Sci. USA* **2009**, *106*, 21567–21572.
59. J. T. Drummond, S. Huang, R. M. Blumenthal, R. G. Matthews, *Biochemistry* **1993**, *32*, 9290–9295.
60. K. Fujii, F. M. Huennekens, *J. Biol. Chem.* **1974**, *249*, 6745–6753.
61. D. A. Hall, T. C. Jordan-Starck, R. O. Loo, M. L. Ludwig, R. G. Matthews, *Biochemistry* **2000**, *39*, 10711–10719.
62. K. Yamada, R. A. Gravel, T. Toraya, R. G. Matthews, *Proc. Natl. Acad. Sci. USA* **2006**, *103*, 9476–9481.
63. C. E. Meints, F. S. Gustafsson, N. S. Scrutton, K. R. Wolthers, *Biochemistry* **2011**, *50*, 11131–11142.
64. J. Kim, C. Gherasim, R. Banerjee, *Proc. Natl. Acad. Sci. USA* **2008**, *105*, 14551–14554.
65. K. R. Wolthers, N. S. Scrutton, *FEBS J.* **2009**, *276*, 1942–1951.
66. J. E. Hewitt, M. M. Gordon, R. T. Taggart, T. K. Mohandas, D. H. Alpers, *Genomics* **1991**, *10*, 432–440.
67. J. Gliemann, E. Nexo, *Ugeskr. Laeger* **1998**, *160*, 5666–5667.
68. O. Platica, R. Janeczko, E. V. Quadros, A. Regec, R. Romain, S. P. Rothenberg, *J. Biol. Chem.* **1991**, *266*, 7860–7863.
69. S. N. Fedosov, L. Berglund, N. U. Fedosova, E. Nexo, T. E. Petersen, *J. Biol. Chem.* **2002**, *277*, 9989–9996.
70. S. N. Fedosov, N. U. Fedosova, B. Kräutler, E. Nexo, T. E. Petersen, *Biochemistry* **2007**, *46*, 6446–6458.
71. H. Yamada, S. Shimizu, M. Kobayashi, *Chem. Record* **2001**, *1*, 152–161.
72. H. Alvarez, *Biology of Rhodococcus*, Ed A. Steinbuchel, Springer-Verlag Berlin, Heidelberg, 2010, pp. 365.

73. M. Kobayashi, T. Nagasawa, H. Yamada, *Trends Biotechnol.* **1992**, *10*, 402–408.
74. S. Prasad, T. C. Bhalla, *Biotechnol. Adv.* **2010**, *28*, 725–741.
75. M. Kobayashi, S. Shimizu, *Nat. Biotechnol.* **1998**, *16*, 733–736.
76. C. D. Mathew, T. Nagasawa, M. Kobayashi, H. Yamada, *Appl. Environ. Microbiol.* **1988**, *54*, 1030–1032.
77. Y. Asano, K. Fujishiro, Y. Tani, H. Yamada, *Agri. Biol. Chem.* **1982**, *46*, 1165–1174.
78. Y. Asano, Y. Tani, H. Yamada, *Agri. Biol. Chem.* **1980**, *44*, 2251–2252.
79. M. Kobayashi, Y. Fujiwara, M. Goda, H. Komeda, S. Shimizu, *Proc. Natl. Acad. Sci. USA* **1997**, *94*, 11986–11991.
80. R. A. Cramp, D. A. Cowan, *Biochim. Biophys. Acta* **1999**, *1431*, 249–260.
81. T. Nagasawa, K. Ryuno, H. Yamada, *Biochem. Biophys. Res. Commun.* **1986**, *139*, 1305–1312.
82. H. Yamada, M. Kobayashi, *Biosci. Biotechnol. Biochem.* **1996**, *60*, 1391–1400.
83. S. Nagashima, M. Nakasako, N. Dohmae, M. Tsujimura, K. Takio, M. Odaka, M. Yohda, N. Kamiya, I. Endo, *Nat. Struct. Biol.* **1998**, *5*, 347–351.
84. M. Nojiri, M. Yohda, M. Odaka, Y. Matsushita, M. Tsujimura, T. Yoshida, N. Dohmae, K. Takio, I. Endo, *J. Biochem.* **1999**, *125*, 696–704.
85. T. Murakami, M. Nojiri, H. Nakayama, M. Odaka, M. Yohda, N. Dohmae, K. Takio, T. Nagamune, I. Endo, *Protein Sci.* **2000**, *9*, 1024–1030.
86. L. Heinrich, A. Mary-Verla, Y. Li, J. Vassermann, J. Chottard, *Eur. J. Inorg. Chem.* **2001**, *9*, 2203–2206.
87. W. Huang, J. Jia, J. Cummings, M. Nelson, G. Schneider, Y. Lindqvist, *Structure* **1997**, *5*, 691–699.
88. S. Hourai, M. Miki, Y. Takashima, S. Mitsuda, K. Yanagi, *Biochem. Biophys. Res. Commun.* **2003**, *312*, 340–345.
89. A. Miyanaga, S. Fushinobu, K. Ito, T. Wakagi, *Biochem. Biophys. Res. Commun.* **2001**, *288*, 1169–1174.
90. M. Nojiri, H. Nakayama, M. Odaka, M. Yohda, K. Takio, I. Endo, *FEBS Lett.* **2000**, *465*, 173–177.
91. A. E. Blackwell, E. D. Dodds, V. Bandarian, V. H. Wysocki, *Anal. Chem.* **2011**, *83*, 2862–2865.
92. M. Nakasako, M. Odaka, M. Yohda, N. Dohmae, K. Takio, N. Kamiya, I. Endo, *Biochemistry* **1999**, *38*, 9887–9898.
93. J. A. Kovacs, *Chem. Rev.* **2004**, *104*, 825–848.
94. T. C. Harrop, P. K. Mascharak, *Acc. Chem. Res.* **2004**, *37*, 253–260.
95. M. Tsujimura, M. Odaka, H. Nakayama, N. Dohmae, H. Koshino, T. Asami, M. Hoshino, K. Takio, S. Yoshida, M. Maeda, I. Endo, *J. Am. Chem. Soc.* **2003**, *125*, 11532–11538.
96. A. Dey, M. Chow, K. Taniguchi, P. Lugo-Mas, S. Davin, M. Maeda, J. A. Kovacs, M. Odaka, K. O. Hodgson, B. Hedman, E. I. Solomon, *J. Am. Chem. Soc.* **2006**, *128*, 533–541.
97. A. Miyanaga, S. Fushinobu, K. Ito, H. Shoun, T. Wakagi, *Eur. J. Biochem.* **2004**, *271*, 429–438.
98. O. Ikehata, M. Nishiyama, S. Horinouchi, T. Beppu, *Eur. J. Biochem.* **1989**, *181*, 563–570.
99. M. Kobayashi, M. Nishiyama, T. Nagasawa, S. Horinouchi, T. Beppu, H. Yamada, *Biochim. Biophys. Acta* **1991**, *1129*, 23–33.
100. Y. Hashimoto, M. Nishiyama, S. Horinouchi, T. Beppu, *Biosci. Biotechnol. Biochem.* **1994**, *58*, 1859–1865.
101. H. Komeda, M. Kobayashi, S. Shimizu, *J. Biol. Chem.* **1996**, *271*, 15796–15802.
102. M. Nishiyama, S. Horinouchi, M. Kobayashi, T. Nagasawa, H. Yamada, T. Beppu, *J. Bacteriol.* **1991**, *173*, 2465–2472.
103. Z. Zhou, Y. Hashimoto, K. Shiraki, M. Kobayashi, *Proc. Natl. Acad. Sci. USA* **2008**, *105*, 14849–14854.
104. Z. Zhou, Y. Hashimoto, T. Cui, Y. Washizawa, H. Mino, M. Kobayashi, *Biochemistry* **2010**, *49*, 9638–9648.

105. S. Okamoto, F. Van Petegem, M. A. Patrauchan, L. D. Eltis, *J. Biol. Chem.* **2010**, *285*, 25126–25133.
106. Y. Katayama, Y. Narahara, Y. Inoue, F. Amano, T. Kanagawa, H. Kuraishi, *J. Biol. Chem.* **1992**, *267*, 9170–9175.
107. T. Uematsu, R. J. Suhadolnik, *Arch. Biochem. Biophy.s* **1974**, *162*, 614–619.
108. Y. Katayama, H. Kuraishi, *Can. J. Microbiol.* **1978**, *24*, 804–810.
109. Y. Katayama, K. Hashimoto, H. Nakayama, H. Mino, M. Nojiri, T. A. Ono, H. Nyunoya, M. Yohda, K. Takio, M. Odaka, *J. Am. Chem. Soc.* **2006**, *128*, 728–729.
110. T. Arakawa, Y. Kawano, S. Kataoka, Y. Katayama, N. Kamiya, M. Yohda, M. Odaka, *J. Mol. Biol.* **2007**, *366*, 1497–1509.
111. T. Uematsu, R. J. Suhadolnik, *Methods Enzymol.* **1975**, *43*, 759–762.
112. R. M. McCarty, V. Bandarian, *Chem. Biol.* **2008**, *15*, 790–798.
113. S. Okamoto, L. D. Eltis, *Mol. Microbiol.* **2007**, *65*, 828–838.
114. W. T. Lowther, B. W. Matthews, *Chem. Rev.* **2002**, *102*, 4581–4608.
115. S. Tsunasawa, J. W. Stewart, F. Sherman, *J. Biol. Chem.* **1985**, *260*, 5382–5391.
116. P. H. Hirel, M. J. Schmitter, P. Dessen, G. Fayat, S. Blanquet, *Proc. Natl. Acad. Sci. USA* **1989**, *86*, 8247–8251.
117. R. Jackson, T. Hunter, *Nature* **1970**, *227*, 672–676.
118. A. Varshavsky, *Genes Cells* **1997**, *2*, 13–28.
119. J. T. Prchal, D. P. Cashman, Y. W. Kan, *Proc. Natl. Acad. Sci. USA* **1986**, *83*, 24–27.
120. J. A. Boutin, *Cell Signal* **1997**, *9*, 15–35.
121. J. Y. Tso, M. A. Hermodson, H. Zalkin, *J. Biol. Chem.* **1982**, *257*, 3532–3536.
122. S. Y. Chang, E. C. McGary, S. Chang, *J. Bacteriol.* **1989**, *171*, 4071–4072.
123. Y. H. Chang, U. Teichert, J. A. Smith, *J. Biol. Chem.* **1992**, *267*, 8007–8011.
124. X. Li, Y. H. Chang, *Proc. Natl. Acad. Sci. USA* **1995**, *92*, 12357–12361.
125. C. G. Miller, A. M. Kukral, J. L. Miller, N. R. Movva, *J. Bacteriol.* **1989**, *171*, 5215–5217.
126. S. L. Roderick, B. W. Matthews, *Biochemistry* **1993**, *32*, 3907–3912.
127. J. F. Bazan, L. H. Weaver, S. L. Roderick, R. Huber, B. W. Matthews, *Proc. Natl. Acad. Sci. USA* **1994**, *91*, 2473–2477.
128. W. T. Lowther, A. M. Orville, D. T. Madden, S. Lim, D. H. Rich, B. W. Matthews, *Biochemistry* **1999**, *38*, 7678–7688.
129. W. T. Lowther, B. W. Matthews, *Biochim. Biophys. Acta* **2000**, *1477*, 157–167.
130. S. M. Arfin, R. L. Kendall, L. Hall, L. H. Weaver, A. E. Stewart, B. W. Matthews, R. A. Bradshaw, *Proc. Natl. Acad. Sci. USA* **1995**, *92*, 7714–7718.
131. X. V. Hu, X. Chen, K. C. Han, A. S. Mildvan, J. O. Liu, *Biochemistry* **2007**, *46*, 12833–12843.
132. A. Addlagatta, X. Hu, J. O. Liu, B. W. Matthews, *Biochemistry* **2005**, *44*, 14741–14749.
133. A. Serero, C. Giglione, A. Sardini, J. Martinez-Sanz, T. Meinnel, *J. Biol. Chem.* **2003**, *278*, 52953–52963.
134. M. Leszczyniecka, U. Bhatia, M. Cueto, N. R. Nirmala, H. Towbin, A. Vattay, B. Wang, S. Zabludoff, P. E. Phillips, *Oncogene* **2006**, *25*, 3471–3478.
135. S. Zuo, Q. Guo, C. Ling, Y. H. Chang, *Mol. Gen. Genet.* **1995**, *246*, 247–253.
136. S. Mitra, K. M. Job, L. Meng, B. Bennett, R. C. Holz, *FEBS J.* **2008**, *275*, 6248–6259.
137. S. C. Chai, Q. Z. Ye, *BMC Biochem.* **2009**, *10*, 32.
138. S. Mitra, G. Sheppard, J. Wang, B. Bennett, R. C. Holz, *J. Biol. Inorg. Chem.* **2009**, *14*, 573–585.
139. M. D'Souza V, R. C. Holz, *Biochemistry* **1999**, *38*, 11079–11085.
140. J. Y. Li, L. L. Chen, Y. M. Cui, Q. L. Luo, J. Li, F. J. Nan, Q. Z. Ye, *Biochem. Biophys. Res. Commun.* **2003**, *307*, 172–179.
141. D. Ingber, T. Fujita, S. Kishimoto, K. Sudo, T. Kanamaru, H. Brem, J. Folkman, *Nature* **1990**, *348*, 555–557.
142. E. Corey, A. Guzman-Perez, M. Noe, *J. Am. Chem. Soc.* **1994**, *116*, 12109–12110.
143. N. Sin, L. Meng, M. Q. Wang, J. J. Wen, W. G. Bornmann, C. M. Crews, *Proc. Natl. Acad. Sci. USA* **1997**, *94*, 6099–6103.

144. E. C. Griffith, Z. Su, B. E. Turk, S. Chen, Y. H. Chang, Z. Wu, K. Biemann, J. O. Liu, *Chem. Biol.* **1997**, *4*, 461–471.
145. S. Liu, J. Widom, C. W. Kemp, C. M. Crews, J. Clardy, *Science* **1998**, *282*, 1324–1327.
146. W. L. Wang, S. C. Chai, M. Huang, H. Z. He, T. D. Hurley, Q. Z. Ye, *J. Med. Chem.* **2008**, *51*, 6110–6120.
147. R. Banerjee, *Chem. Rev.* **2003**, *103*, 2083–2094.
148. M. L. Ludwig, R. G. Matthews, *Annu. Rev. Biochem.* **1997**, *66*, 269–313.
149. D. Bucher, G. M. Sandala, B. Durbeej, L. Radom, D. M. Smith, *J. Am. Chem. Soc.* **2011**, *134*, 1591–1599.
150. R. Padmakumar, R. Padmakumar, R. Banerjee, *Biochemistry* **1997**, *36*, 3713–3718.
151. D. Padovani, R. Banerjee, *Biochemistry* **2006**, *45*, 2951–2959.
152. T. Toraya, *Cell Mol. Life Sci.* **2000**, *57*, 106–127.
153. T. Toraya, N. Tamura, T. Watanabe, M. Yamanishi, N. Hieda, K. Mori, *J. Biochem.* **2008**, *144*, 437–446.
154. K. H. Tang, C. H. Chang, P. A. Frey, *Biochemistry* **2001**, *40*, 5190–5199.
155. P. A. Frey, G. H. Reed, *Biochim. Biophys. Acta* **2011**, *1814*, 1548–1557.
156. H. F. Willard, L. E. Rosenberg, *J. Clin. Invest.* **1980**, *65*, 690–698.
157. F. Deodato, S. Boenzi, F. M. Santorelli, C. Dionisi-Vici, *Am. J. Med. Genet. C Semin. Med. Genet.* **2006**, *142C*, 104–112.
158. H. F. Willard, L. E. Rosenberg, *Biochem. Biophys. Res. Commun.* **1977**, *78*, 927–934.
159. S. H. Allen, R. W. Kellermeyer, R. L. Stjernholm, H. G. Wood, *J. Bacteriol.* **1964**, *87*, 171–187.
160. J. W. Vrijbloed, K. Zerbe-Burkhardt, A. Ratnatilleke, A. Grubelnik-Leiser, J. A. Robinson, *J. Bacteriol.* **1999**, *181*, 5600-5605.
161. F. Mancia, N. H. Keep, A. Nakagawa, P. F. Leadlay, S. McSweeney, B. Rasmussen, P. Bosecke, O. Diat, P. R. Evans, *Structure* **1996**, *4*, 339-350.
162. C. L. Drennan, S. Huang, J. T. Drummond, R. G. Matthews, M. L. Lidwig, *Science* **1994**, *266*, 1669-1674.
163. D. S. Froese, C. M. Dobson, A. P. White, X. Wu, D. Padovani, R. Banerjee, T. Haller, J. A. Gerlt, M. G. Surette, R. A. Gravel, *Microbiol. Res.* **2009**, *164*, 1-8.
164. T. Haller, T. Buckel, J. Retey, J. A. Gerlt, *Biochemistry* **2000**, *39*, 4622-4629.
165. E. Miyamoto, F. Watanabe, T. C. Charles, R. Yamaji, H. Inui, Y. Nakano, *Arch. Microbiol.* **2003**, *180*, 151–154.
166. A. Ratnatilleke, J. W. Vrijbloed, J. A. Robinson, *J. Biol. Chem.* **1999**, *274*, 31679–31685.
167. W. H. Ramos-Vera, M. Weiss, E. Strittmatter, D. Kockelkorn, G. Fuchs, *J. Bacteriol.* **2011**, *193*, 1201–1211.
168. I. A. Berg, D. Kockelkorn, W. Buckel, G. Fuchs, *Science* **2007**, *318*, 1782–1786.
169. E. Miyamoto, F. Watanabe, Y. Yamaguchi, H. Takenaka, Y. Nakano, *Comp. Biochem. Physiol. B Biochem. Mol. Biol.* **2004**, *138*, 163–167.
170. Y. S. Han, J. M. Bratt, H. P. Hogenkamp, *Comp. Biochem. Physiol. B* **1984**, *78*, 41–45.
171. D. S. Froese, G. Kochan, J. R. Muniz, X. Wu, C. Gileadi, E. Ugochukwu, E. Krysztofinska, R. A. Gravel, U. Oppermann, W. W. Yue, *J. Biol. Chem.* **2010**, *285*, 38204–38213.
172. C. L. Drennan, R. G. Matthews, D. S. Rosenblatt, F. D. Ledley, W. A. Fenton, M. L. Ludwig, *Proc. Natl. Acad. Sci. USA* **1996**, *93*, 5550–5555.
173. K. Zerbe-Burkhardt, A. Ratnatilleke, N. Philippon, A. Birch, A. Leiser, J. W. Vrijbloed, D. Hess, P. Hunziker, J. A. Robinson, *J. Biol. Chem.* **1998**, *273*, 6508–6517.
174. J. R. e. G. Brendelberger, D. M. Ashworth, K. Reynolds, F. Willenbrock, J. A. Robinson, *Angew. Chem. Int. Ed. Engl.* **1988**, *27*, 1089–1091.
175. I. Angelidaki, B. K. Ahring, *Antonie Van Leeuwenhoek* **1995**, *68*, 285–291.
176. S. J. Oude Elferink, P. N. Lens, C. Dijkema, and A. J. Stams, *FEMS Microbiol. Lett.* **1996**, *142*, 237–241.
177. J.-L. Tholozan, E. Samain, and J.-P. Grivet, *FEMS Microbiol. Lett.* **1988**, *53*, 187–191.
178. W. M. Wu, M. K. Jain, J. G. Zeikus, *Appl. Environ. Microbiol.* **1994**, *60*, 2220–2226.

179. C. Matthies, N. Springer, W. Ludwig, B. Schink, *Int. J. Syst. Evol. Microbiol.* **2000**, *50 Pt 2*, 645–648.
180. C. Matthies, B. Schink, *Appl. Environ. Microbiol.* **1992**, *58*, 1435–1439.
181. Y. Sekiguchi, Y. Kamagata, K. Nakamura, A. Ohashi, H. Harada, *Int. J. Syst. Evol. Microbiol.* **2000**, *50 Pt 2*, 771–779.
182. E. N. Marsh, *Bioorg. Chem.* **2000**, *28*, 176–189.
183. R. Padmakumar, R. Banerjee, *Biofactors* **1995**, *5*, 83–86.
184. J. Retey, E. H. Smith, B. Zagalak, *Eur. J. Biochem.* **1978**, *83*, 437–451.
185. C. Probian, A. Wulfing, J. Harder, *Appl. Environ. Microbiol.* **2003**, *69*, 1866–1870.
186. T. Rohwerder, H. Müller, in *Vitamin B: New Research*, Ed C. M. Elliot, Nova Science Publishers, Hauppage, N.Y., 2007, pp. 81–98.
187. W. Zhang, K. A. Reynolds, *J. Bacteriol.* **2001**, *183*, 2071–2080.
188. L. M. Smith, W. G. Meijer, L. Dijkhuizen, P. M. Goodwin, *Microbiology* **1996**, *142*, 675–684.
189. N. Verstraeten, M. Fauvart, W. Versees, J. Michiels, *Microbiol. Mol. Biol. Rev.* **2011**, *75*, 507–542, second and third pages of table of contents.
190. D. D. Leipe, Y. I. Wolf, E. V. Koonin, L. Aravind, *J. Mol. Biol.* **2002**, *317*, 41–72.
191. A. Wittinghofer, I. R. Vetter, *Annu. Rev. Biochem.* **2011**, *80*, 943–971.
192. I. R. Vetter, A. Wittinghofer, *Science* **2001**, *294*, 1299–1304.
193. P. Rangaraj, V. K. Shah, P. W. Ludden, *Proc. Natl. Acad. Sci. USA* **1997**, *94*, 11250–11255.
194. B. Schmid, O. Einsle, H. J. Chiu, A. Willing, M. Yoshida, J. B. Howard, D. C. Rees, *Biochemistry* **2002**, *41*, 15557–15565.
195. W. B. Jeon, J. Cheng, P. W. Ludden, *J. Biol. Chem.* **2001**, *276*, 38602–38609.
196. J. H. Jeoung, T. Giese, M. Grunwald, H. Dobbek, *J. Mol. Biol.* **2010**, *396*, 1165–1179.
197. J. H. Jeoung, T. Giese, M. Grunwald, H. Dobbek, *Biochemistry* **2009**, *48*, 11505–11513.
198. N. Korotkova, M. E. Lidstrom, *J. Biol. Chem.* **2004**, *279*, 13652–13658.
199. D. Padovani, R. Banerjee, *Biochemistry* **2006**, *45*, 9300–9306.
200. D. Padovani, T. Labunska, R. Banerjee, *J. Biol. Chem.* **2006**, *281*, 17838–17844.
201. R. Gasper, A. Scrima, A. Wittinghofer, *J. Biol. Chem.* **2006**, *281*, 27492–27502.
202. F. Cai, T. T. Ngu, H. Kaluarachchi, D. B. Zamble, *J. Biol. Inorg. Chem.* **2011**, *16*, 857–868.
203. H. Kaluarachchi, K. C. Chan Chung, D. B. Zamble, *Nat. Prod. Rep.* **2011**, *27*, 681–694.
204. B. Zambelli, F. Musiani, S. Benini, S. Ciurli, *Acc. Chem. Re.s* **2011**, *44*, 520–530.
205. J. L. Boer, S. Quiroz-Valenzuela, K. L. Anderson, R. P. Hausinger, *Biochemistry* **2010**, *49*, 5859–5869.
206. Y. H. Fong, H. C. Wong, C. P. Chuck, Y. W. Chen, H. Sun, K. B. Wong, *J. Biol. Chem.* **2011**, *286*, 43241–43249.
207. C. E. Haas, D. A. Rodionov, J. Kropat, D. Malasarn, S. S. Merchant, V. de Crecy-Lagard, *BMC Genomics* **2009**, *10*:470, 1–21.
208. J. Crouzet, S. Levy-Schil, B. Cameron, L. Cauchois, S. Rigault, M. C. Rouyez, F. Blanche, L. Debussche, D. Thibaut, *J. Bacteriol.* **1991**, *173*, 6074–6087.
209. J. Lu, Y. Zheng, H. Yamagishi, M. Odaka, M. Tsujimura, M. Maeda, I. Endo, *FEBS Lett.* **2003**, *553*, 391–396.
210. M. R. Leach, S. Sandal, H. Sun, D. B. Zamble, *Biochemistry* **2005**, *44*, 12229–12238.
211. R. Real-Guerra, F. Staniscuaski, B. Zambelli, F. Musiani, S. Ciurli, C. R. Carlini, *Plant Mol. Biol.* **2012**, *78*, 461–475.
212. P. A. Hubbard, D. Padovani, T. Labunska, S. A. Mahlstedt, R. Banerjee, C. L. Drennan, *J. Biol. Chem.* **2007**, *282*, 31308–31316.
213. R. Gasper, S. Meyer, K. Gotthardt, M. Sirajuddin, A. Wittinghofer, *Nat. Rev. Mol. Cell Biol.* **2009**, *10*, 423–429.
214. K. H. Chan, T. Li, C. O. Wong, K. B. Wong, *PLoS One* **2012**, *7*, e30547.
215. N. Korotkova, L. Chistoserdova, V. Kuksa, M. E. Lidstrom, *J. Bacteriol.* **2002**, *184*, 1750–1758.
216. C. M. Dobson, T. Wai, D. Leclerc, A. Wilson, X. Wu, C. Dore, T. Hudson, D. S. Rosenblatt, R. A. Gravel, *Proc. Natl. Acad. Sci. USA* **2002**, *99*, 15554–15559.

217. M. St Maurice, P. E. Mera, M. P. Taranto, F. Sesma, J. C. Escalante-Semerena, I. Rayment, *J. Biol. Chem.* **2007**, *282*, 2596–2605.
218. K. E. Erger, T. J. Conlon, N. A. Leal, R. Zori, T. A. Bobik, T. R. Flotte, *J. Gene Med.* **2007**, *9*, 462–469.
219. E. W. Miles, S. Rhee, D. R. Davies, *J. Biol. Chem.* **1999**, *274*, 12193–12196.
220. X. Huang, H. M. Holden, F. M. Raushel, *Annu. Rev. Biochem.* **2001**, *70*, 149–180.
221. S. Mitra, R. C. Holz, *J. Biol. Chem.* **2007**, *282*, 7397–7404.

Chapter 11
Nickel Metallomics: General Themes Guiding Nickel Homeostasis

Andrew M. Sydor and Deborah B. Zamble

Contents

A.M. Sydor • D.B. Zamble (✉)
Department of Chemistry, University of Toronto,
80 St. George Street, Toronto, ON M5S 3H6, Canada
e-mail: dzamble@chem.utoronto.ca

L. Banci (ed.), *Metallomics and the Cell*, Metal Ions in Life Sciences 12,
DOI 10.1007/978-94-007-5561-1_11, © Springer Science+Business Media Dordrecht 2013

Abstract The nickel metallome describes the distribution and speciation of nickel within the cells of organisms that utilize this element. This distribution is a consequence of nickel homeostasis, which includes import, storage, and export of nickel, incorporation into metalloenzymes, and the modulation of these and associated cellular systems through nickel-regulated transcription. In this chapter, we review the current knowledge of the most common nickel proteins in prokaryotic organisms with a focus on their coordination environments. Several underlying themes emerge upon review of these nickel systems, which illustrate the common principles applied by nature to shape the nickel metallome of the cell.

Keywords accessory proteins • nickel homeostasis • nickel metallochaperones • nickel metalloenzymes • transcription factors • transporters

Please cite as: *Met. Ions Life Sci.* 12 (2013) 375–416

1 Introduction

Metallomics examines the speciation of metal ions within biological systems [1], whether bound to proteins or small molecule ligands. As for other transition metals, this field illuminates many aspects of nickel bioinorganic chemistry, including how nickel is distributed within the cell and the coordination environments, and it also provides hints about metal concentrations [2]. In order to understand the biochemical mechanisms at work within the metallome that collectively maintain nickel homeostasis, we need to define the distribution of this element and the chemical basis for the selected ligands and metal-binding sites (Figure 1). Several recent reviews focused on the nickel proteome [3–7], highlighting the diversity of nickel proteins in Nature.

In this chapter, the focus will be on the underlying themes in nickel homeostasis, with specific examples highlighted, in order to gain an appreciation for the processes that shape the nickel metallome. This chapter includes a description of some of the key players that contribute to prokaryotic membrane transport, enzyme assembly and activity, storage, and regulation, followed by a "view from the top" in which the current knowledge is placed into a cellular context. From this context, some overarching themes and principles emerge, such as the roles of thermodynamics

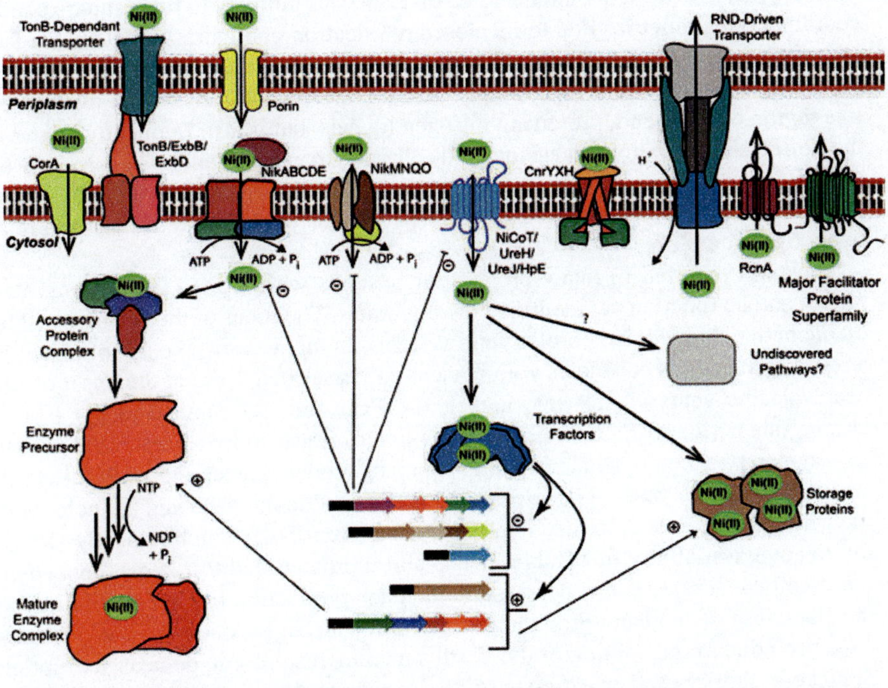

Figure 1 Summary of the protein systems involved in bacterial nickel homeostasis. This figure is a composite of systems from multiple species and is not meant to represent any one particular organism. Cellular small molecule nickel complexes are uncharacterized and are not included in the diagram. Despite current knowledge, many questions remain, including how the nickel ions are transferred from the transporters to the cytosolic proteins, the role of thermodynamics *versus* kinetics in driving Ni(II) transfer, and the coordination spheres in several of the Ni(II)-binding sites.

and kinetics, graded responses to metal concentrations by transcription factors, the generation of different metal pools in the cell, nickel-specific allosteric responses in proteins, and the requirement of metallochaperones for metalloenzyme assembly.

1.1 The Chemistry of Nickel

Nickel is a first row transition metal found primarily in the +2 oxidation state, but it also occurs in formal 0, +1 and +3 oxidation states in biological systems [8,9]. Four-coordinate Ni(II) complexes are typically square planar, with tetrahedral geometries less common, whereas trigonal-bipyramidal and square pyramidal geometries are frequently associated with five-coordinate Ni(II) and six-coordinate Ni(II) is octahedral [9]. The square planar geometry, which is only shared with Cu(II) in biological systems, is a result of the d^8 electron configuration of Ni(II), wherein the planar tetra-ligand coordination causes the $d_{x^2-y^2}$ orbital to be uniquely high in energy and thus left vacant [10]. It should be noted that Zn(II) is unlikely to form square planar coordination geometries due to its closed d^{10} electron configuration [10,11], a fact important in metal selectivity. Moving from left to right across the first row, the transition metals become increasingly softer. Nickel is situated within the borderline region of this trend, preferring nitrogen ligands, but is also frequently bound to the softer sulfur and/or occasionally the harder oxygen ligands [12]. Moreover, given its position in the Irving-William series of small molecule complex stabilities [13], Mn(II) < Fe(II) < Co(II) < Ni(II) < Cu(II) > Zn(II), nickel(II) tends to form relatively tight complexes.

Nickel is required in many prokaryotic systems with limited use in fungi and plants, and no function yet identified in vertebrates. The focus of this chapter will be on the better characterized prokaryotic systems. In many cases, the use of nickel in enzymatic reactions correlates with ancient processes that arose at the onset of life near volcanic vents where iron, nickel, H_2, CO_2, and CO were abundant [9,12]. During this period of Earth's history, soluble nickel was more accessible than zinc and copper, two metals frequently found in many modern bacterial metallomes [12]. Around 2.4 billion years ago, a decrease in atmospheric methane and an increase in oxygen production lead to the Great Oxidation Event [14], which radically altered the relative availability of soluble metals and ultimately led to oxygen-dependent multi-cellular life [15]. The decrease in methane production has been attributed to the starvation of methane-producing bacteria for nickel (Section 3.5), which may have been due to the cooling of the Earth's mantle, resulting in decreased eruption of nickel-rich rocks and an estimated 50% decrease in oceanic nickel concentrations [14]. Nonetheless, there remain a multitude of systems for which Ni(II) is the ideal metal, many of which will be discussed throughout this chapter.

2 Nickel Transport

In order for cellular nickel demands to be met, the cytosol must effectively accumulate the metal to acceptable levels. However, a buildup of excess nickel must be prevented because it can be toxic for a variety of reasons, including the replacement of the cognate metals in iron or zinc enzymes, nickel-induced allosteric inhibition of enzymes, and the indirect activation of oxidative stress [16–18]. Thus, nickel transport across the cell membrane must balance uptake and export with the nutritional requirements of the organism.

To drive nickel transport, there are a variety of known nickel proteins that harness cellular energy sources such as ATP or potentiometric gradients. Some nickel may also enter the cell in a non-specific manner via other uptake systems such as the inner membrane CorA that mediates the transport of divalent metal ions such as Mg(II) [19–21]. Although Ni(II) can pass through this system, the K_a is ~ 300 μM, calling into question the physiological relevance of this uptake mechanism [22,23], as environmental nickel concentrations tend to be in the nM range, as is the case of sea water [24] and human blood serum [25]. Of the various specific nickel transporters discussed below, only a subset may be found in any particular organism.

2.1 Nickel Importers

2.1.1 ABC Ni(II) Transporters

The ATP-binding cassette (ABC) transporters are the most widely used pathways for moving Ni(II) across the cytoplasmic membrane, and they are represented by the NikABCDE and the NikMNQO proteins [26]. These multicomponent systems generally consist of integral membrane channels and associated ATP-hydrolyzing proteins to provide energy for transport, with a soluble periplasmic Ni(II)-binding protein as an additional component of NikABCDE [27]. *Escherichia coli* NikABCDE is the best characterized Ni(II) transporter to date, serving as the main source of nickel for several [NiFe]-hydrogenases [28–32]. In this system, NikA is the periplasmic binding protein (PBP), which presumably picks up nickel in the periplasm and delivers it to NikB and NikC, which form the heterodimeric transmembrane core and are coupled to the ATP-hydrolyzing NikD and NikE proteins [28].

E. coli NikA is a mixed α/β PBP capable of binding a single nickel ion per monomer in a cleft between two globular domains (Figure 2) [33,34]. Crystallographic and spectroscopic evidence suggest that nickel is bound as a complex with a small organic molecule contributing several carboxylate ligands, currently modeled as butane-1,2,4-tricarboxylate (BTC) [34–36]. However, definitive assignment of the molecule as BTC is suspect, given that no biosynthetic pathway for this molecule is known in *E. coli*. An essential histidine ligand (His416) also coordinates the Ni(II) [35,37]. Given the clear parallels with bacterial siderophores used to assimilate Fe(III) [38], it is reasonable that *E. coli* and other bacteria utilize a small-molecule chelator to compete for low abundance nickel in the environment. It should be noted that the NikABCDE transporter is phylogenetically more closely related to oligopeptide transporters than to other metal transporters [39], supporting the hypothesis that NikABCDE transports a nickel complex rather than a "naked" nickel ion.

2.1.2 Secondary Nickel Transporters

The secondary nickel transporters, which can be divided into the NiCoT, HupE/UreJ, and UreH families, are high-affinity permeases that allow passive transport of nickel across the cytoplasmic membrane. As opposed to the multi-component ABC systems, these transporters consist of a single polypeptide chain.

The nickel/cobalt transporter (NiCoT) family is widely distributed, spanning all three domains of life [40], and is characterized by an eight transmembrane-domain (TMD) architecture and a conserved HX_4DH motif located in the second transmembrane segment [41]. NiCoTs work efficiently when their substrates are present in

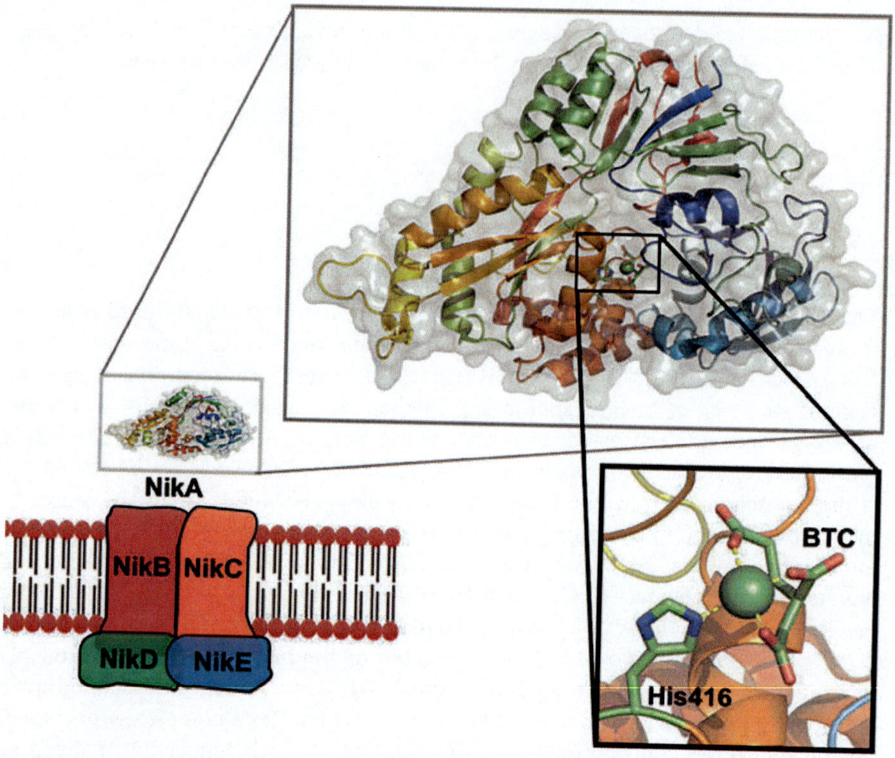

Figure 2 Structure of the *E. coli* NikA periplasmic-binding protein (PDB 3DP8). NikA, the periplasmic protein component of the ATP-dependent nickel transporter, NikABCDE, is a mixed α/β protein capable of binding a single nickel ion. The protein is believed to bind a small-molecule nickel complex, the identity of which is unknown, modeled here as butane-1,2,4-tricarboxylate (BTC). In the crystal structure, BTC contributes three O ligands. This complex is coordinated to the protein by His416 from NikA. The color scheme is nickel, green sphere; oxygen, red and nitrogen, blue. The images were generated using PyMol.

the low nM range [42]. The two best described members of the NiCoT family are the NixA and HoxN transporters. *Alcaligenes eutrophus* HoxN, the first NiCoT transporter identified, has a transport equilibrium constant (K_T) for nickel of 20 nM and is highly specific for Ni(II) over Co(II), Zn(II), and Mn(II) [43–46]. Central to the transport of metals ions are the His and Asp residues in the conserved HX_4DH motif [47]. In contrast, although the *Helicobacter pylori* HoxN homolog NixA also facilitates high-affinity Ni(II) transport (K_T = 11.3 nM), it is inhibited by Co(II), Cu(II), and Zn(II) [41,48].

The UreH proteins also feature 8 TMDs and the conserved HX_4DH motif on the second TMD [49,50] and *Y. pseudotuberculosis* UreH is a selective Ni(II) transporter with a K_T of 50 nM [50]. The HupE/UreJ transporters are more distantly related to the NiCoTs than UreH, featuring only 6 TMDs and a slightly altered signature motif (HX_5DH) located on the first TMD [42,51]. In addition, they also have a FHGX[AV]HGXE motif found only in the HupE/UreJ family. Mutation of either His residue in the HX_5DH signature motif or the second His in the FHGX[AV]HGXE motif leads to diminished transport [51].

The ability of the Ni(II)-permeases to facilitate high-affinity and specific nickel uptake is not well understood. Aside from the identity of the conserved motifs in these transporters, very little is known about the coordination environment of Ni(II). It is interesting to note that the Co(II)-transporting NhlF features the same overall topology and conserved motifs as HoxN, yet can transport both Ni(II) and Co(II) ions [46]. Mutation experiments swapping residues in HoxN with the corresponding NhlF residues demonstrated the role of non-metal-binding ligands in dictating the complicated process of transport kinetics and specificity [52].

2.1.3 Outer Membrane Transporters

In gram-negative bacteria, cellular nutrients must pass through an outer membrane before reaching the cytoplasmic membrane. One mechanism for Ni(II) transport across the outer membrane is through trimeric β-barrel proteins known as porins, which allow for the non-selective passive diffusion of metal ions [53]. However, Ni(II) may cross the outer membrane through other mechanisms as well. In the case of complexes, such as the putative small molecule Ni(II) complex bound by NikA, a TonB-dependent transporter (TBDT) system is needed [54]. TonB spans the periplasm and couples the transport of larger species across the outer membrane to the activity of the inner membrane proteins ExbB and ExbD, which access the local proton gradient [54,55]. Nickel-transporting TBDTs have been characterized in *Helicobacter* species, where a TonB ortholog contributes to Ni(II) acquisition [56] and is proposed to energize NikH [56,57], FecA3, and FrpB4 [58–60], all of which are outer membrane Ni(II) transporters. In particular, the TBDT-FrpB4 system is dedicated to nickel accumulation at low pH, thereby allowing the pathogenic *H. pylori* to colonize the acidic human stomach by activation of apo-urease with an influx of Ni(II) [60] (Section 3.2). Details of the transport mechanisms of these proteins and the identity of their substrates are unknown.

2.2 Nickel Export

One mechanism that organisms use to avoid nickel toxicity is nickel efflux. To date, nickel efflux proteins can be broadly classified into three major categories: RcnA, Major Facilitator Protein Superfamily (MFS), and the Resistance-Nodulation-Cell Division (RND)-driven exporters. Cation diffusion facilitator (CDF) proteins also transport Ni(II) [61], but have not been demonstrated to be a major Ni(II) efflux pathway. A putative P_1-type ATPase efflux pump for nickel has also been noted in *Mycobacterium tuberculosis*, a human pathogen responsible for most cases of tuberculosis, but has not been characterized (Section 5.3) [62].

Resistance to cobalt and nickel protein A (RcnA) is the only nickel efflux protein identified in *E. coli* [63]. The *rcnA* gene product includes a His-rich loop consisting of 17 His, 3 Asp and 3 Glu in a 26-residue stretch [63]. Upon inactivation of the *rcnA* gene *E. coli* become more sensitive to Ni(II) and Co(II) and accumulate more Ni(II) [63,64]. The proposed topology of RcnA includes 6 TMDs and does not bear significant homology with other transporters [63]. The *rcnB* gene located directly downstream of *rcnA* is co-expressed with *rcnA* and encodes a soluble periplasmic protein [65]. Deletion of *rcnB* results in increased Ni(II) and Co(II) resistance and a decrease in the overall cellular levels of the two metals, suggesting that this protein may regulate the efflux activity of RcnA through an unknown mechanism [65].

The MFS is a diverse group of transporters that includes several Ni(II)-transporting members. These types of transporters feature multiple components, including a membrane-spanning protein exemplified by NreB from *Ralstonia metallidurans* strain 31A [66]. This protein is proposed to consist of 12 TMDs and contains a histidine-rich C-terminus believed to be involved in nickel storage [61,66]. NreB-like proteins are widely distributed in nature, with several different homologs (NirA, NcrA, and NrsD) annotated in bacterial genomes [67–69]. When these MFS genes are deleted, an increase in nickel sensitivity or cytosolic nickel concentrations is observed [66,67,69]. A second component of these MFSs is a cytosolic histidine-rich transcription factor, responsible for repression of this efflux system in the absence of nickel [70]. No direct evidence for nickel binding to this transcription factor has been yet reported.

The RND-driven exporters are found in all major domains of life and are responsible for transporting a wide variety of substrates including heavy metals, hydrophobic compounds, and proteins [61]. RND proteins consist of 12 transmembrane helices with the metal-transporting members containing the consensus sequence DFGX$_3$DGAX$_3$VEN in TMD IV [61]. It is believed that a cycle of protonation/deprotonation of the acidic residues in this sequence is responsible for driving the translocation of the substrate out of the cell [61]. Several major Ni(II)-transporting variants of the RND-driven exporters are known, such as CznABC in *H. pylori* [71] and NccCBA in *R. metallidurans* strain 31A [72], none of which are specific for just Ni(II). One representative member of the RND-driven exporters is the CnrCBA system, responsible for a high level of cobalt and nickel resistance [73–75]. Although little is known about the mechanism of transport by the CnrCBA system, significant work has been conducted on the regulation of this Ni(II) exporter, discussed in Section 5.5.

3 Nickel Enzymes: Structure and Assembly

3.1 *[NiFe]-Hydrogenase*

Hydrogenase enzymes catalyze the reversible oxidation of hydrogen gas to protons and electrons (Figure 3), a process central to microbial energy metabolism [76,77]. [NiFe]-hydrogenases are found mainly in archaea and bacteria [78,79] and, in a few

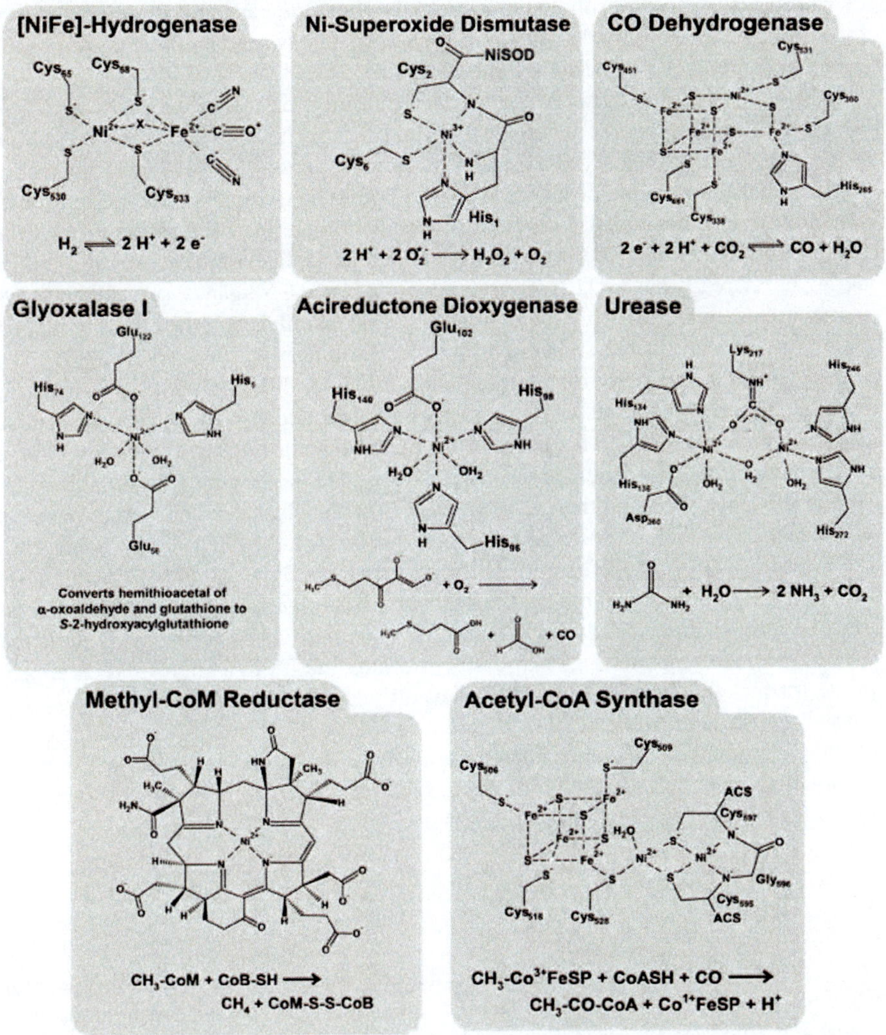

Figure 3 The active sites of nickel-containing enzymes and the reactions catalyzed. The structures shown are: [NiFe]-hydrogenase, *Desulfovibrio gigas* (PDB 1FRV), oxidized inactive form where X is a small molecule (O in the *D. gigas* crystal structure); nickel-superoxide dismutase, *Streptomyces coelicolor* (PDB 1T6V), oxidized form; CO dehydrogenase, *Rhodospirillum rubrum* (PDB 1JQK); glyoxalase I, *Escherichia coli* (PDB 1F9Z); acireductone dioxygenase, *Klebsiella oxytoca* (PDB 1ZRR); urease, *Klebsiella aerogenes* (PDB 1FWE); methyl-CoM reductase, *Methanothermobacter marburgensis* (PDB 3POT), structure is of the isolated F430 cofactor; acetyl-CoA synthase, *Moorella thermoacetica* (PDB 1OAO); active form of the catalytic site in an open conformation.

notable cases such as pathogenic *H. pylori*, are critical for colonization [77]. [NiFe]-hydrogenases are heterodimers with the [NiFe] active site in a large subunit (LS) and a series of iron-sulfur clusters responsible for electron transfer in the small subunit (SS) [80]. In the active site, nickel is coordinated to four Cys thiolates, two of which bridge to the iron center which also has one carbon monoxide and two cyanide ligands (Figure 3) [80]. In some organisms, one of the non-bridging cysteines is replaced by a selenocysteine [79,81].

Although phylogenetically distinct, [NiFe]- and [FeFe]-hydrogenases appear to be a case of evolutionary convergence as the two enzymes have remarkably similar active sites and both catalyze the same chemical reaction [78,82]. There is no consensus regarding the role of nickel in [NiFe]-hydrogenase, making the reason for its selection over other metals a source of speculation. It has been proposed that the presence of nickel in [NiFe]-hydrogenase is due to relatively high nickel availability during Earth's early history [12]. The presence of CO and CN ligands, as well as putative inorganic molecules bridging the two metals, draw parallels to those naturally occurring inorganic compounds found in primordial Earth near volcanic vents [8,83].

Due to the complexity of the metal site, the assembly of [NiFe]-hydrogenases requires an array of accessory proteins, most encoded by the *hyp* (*hydrogenase pleiotropic*) operon [84]. This process has been best characterized in *E. coli*, where at least 3 hydrogenase isoenzymes are expressed [5]. Following ribosomal translation of the gene encoding the LS, biosynthesis of the cluster is believed to occur in two stages, with the iron inserted first, followed by nickel [85]. The iron is incorporated into the LS (encoded by *hycE* for hydrogenase 3) by HypCDEF, which are also thought to be responsible for the synthesis and incorporation for at least one of the diatomic ligands [81,85–87]. HypC remains associated with HycE until the metal center is complete [88,89], possibly holding the LS in a conformation appropriate for nickel insertion, which is achieved by HypA, HypB and SlyD and is discussed in detail below. HypC then dissociates from the LS and proteolytic cleavage of the C-terminal tail by an isoenzyme-specific protease occurs [81,90]. A conformational change then yields the mature LS that can associate with the SS, prepared by a parallel biosynthetic pathway [79,81].

3.1.1 HypA

HypA is believed to serve as the scaffold for assembly of the nickel insertion proteins with the [NiFe]-hydrogenase large subunit [91]. HypA can interact with HypB and HycE, and there is evidence that HypA forms a complex with HypB and SlyD in the absence of HycE, suggesting that the nickel-insertion complex can preassemble prior to association with the hydrogenase precursor protein [92-94].

HypA consists of 2 metal-binding domains [95,96]. Both HypA and its homolog HybF, which substitutes for HypA in the biosynthesis of *E. coli* hydrogenases 1 and 2 [97], bind stoichiometric nickel with micromolar affinity [93,94,98]. This N-terminal Ni(II) site includes the conserved His2, which is essential for hydrogenase activity [94,98], but the rest of the site is under debate. Spectroscopic analysis of *H. pylori* HypA with extra N-terminal residues supports a square planar coordination including the terminal amine and the backbone nitrogens of two nearby residues [96], whereas XAS of unmodified protein suggested that the site may instead be 5–6 coordinate, with a coordination sphere composed of 3–4 (N/O) and 1–2 His ligands [99,100].

HypA and HybF also bind stoichiometric zinc (*E. coli* HypA $K_{D\text{-}Zn(II)}$=0.9 nM [93]) in a site composed of cysteines from two conserved CXXC motifs [95,96,98,101], similar to the coordination sphere found in structural zinc sites [102]. In *H. pylori* HypA (HpHypA), this Zn(II) site is modulated by both nickel loading and pH changes [100]. At the diminished pH of 6.3, consistent with the cytoplasm of bacteria exposed to acid shock, the zinc site undergoes ligand substitution to convert the Zn(II)Cys_4 site to a Zn(II)HisCys_3 site. If nickel is also present at these lower pH values, further rearrangement of the Zn(II) site occurs to yield a Zn(II)His_2Cys_2 coordination. Furthermore, the decrease in pH also reduces the nickel stoichiometry from one ion per monomer to one ion per dimer [100]. In *H. pylori*, HypA also contributes to urease biosynthesis (Section 3.2.2) and these pH- and Ni(II)-dependent changes to the Zn(II) coordination sphere of HypA are proposed to direct Ni(II) to the different enzyme assembly pathways by favoring or disfavoring interaction with HypB or the urease accessory protein UreE [100].

3.1.2 HypB

HypB is a GTPase essential for the synthesis of [NiFe]-hydrogenases [103,104] and it is thought to contribute to nickel delivery in cooperation with HypA [105–107]. All HypB homologs possess a low-affinity metal-binding site located between two of the conical GTPase motifs in the C-terminal part of the protein [6,108], which is required for hydrogenase biosynthesis in *E. coli* [109]. This site binds one nickel ion with a K_D=12 μM or K_D=150 nM for the *E. coli* (EcHypB) and *H. pylori* (HpHypB) homologs, respectively, with the coordination environment involving a conserved CH motif [110,111]. However, Zn(II) binds 10–100 times more tightly, prompting speculation regarding the identity of the physiologically relevant metal [110,111]. In the crystal structure of HypB from *Methanocaldococcus jannaschii*, two Zn(II) are bound at an asymmetric site at the dimer interface with the first ion bound by water and three Cys residues, one of which bridges to a second Zn(II) bound by another Cys, one His, and one water [108]. However, the relevance of this coordination environment *in vivo* is not clear as a mutant EcHypB lacking the ability to dimerize can still bind metal in the low-affinity site and function in hydrogenase biosynthesis [112].

Isolated HypB displays low levels of GTPase activity that is believed to serve a regulatory role in hydrogenase maturation [104], and disruption of this activity abolishes hydrogenase production due to lack of nickel incorporation into the large subunit [104]. In HpHypB, Zn(II) binding to the low-affinity site abolishes GTPase activity whereas nickel had little effect on catalysis [111]. In contrast, both metals impair GTPase activity in EcHypB, with Zn(II) imparting the more inhibitory effect [112].

In addition to the invariant metal site in the GTPase domain, some HypB homologs possess additional N-terminal metal motifs. *Bradyrhizobium japonicum* HypB bears 24 histidines within the first 39 amino acids and is capable of binding 9 Ni(II) ions per monomer with an average apparent K_D=2.3 μM [113]. Deletion of this His-rich region abolishes nearly all Ni(II) binding, as the truncated protein binds a single Ni(II) with a K_D=14.8 μM [114]. This His-rich sequence is proposed

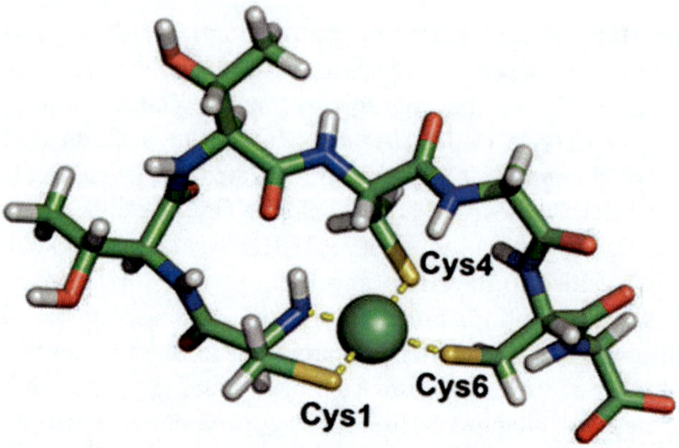

Figure 4 Model of the high-affinity nickel-binding site of *E. coli* HypB determined from density functional theory (DFT) calculations of the N-terminal 7-residues. The nickel ion is coordinated to three cysteine residues and the N-terminal amine of the CXXCGC motif in a square planar geometry. This site, critical for hydrogenase maturation in *E. coli*, binds Ni(II) with a $K_D \approx 0.1$ pM. The color scheme is nickel, green sphere; oxygen, red; sulfur, yellow and; nitrogen, blue. The image was generated using PyMol.

to serve in nickel storage, although it may also play a role in optimal hydrogenase expression [114,115]. EcHypB lacks this His-rich region, but instead possesses a high-affinity nickel-binding sequence CXXCGC at the N-terminus of the protein (Figure 4) [110]. The Ni(II) binds in a square planar site with three Cys and the N-terminal amine serving as ligands with a $K_D \approx 0.1$ pM [109,110,116]. Although not conserved in all HypB homologs, this site is critical for hydrogenase maturation in *E. coli* [109].

3.1.3 SlyD

SlyD was isolated with hydrogenase biosynthetic factors in pull down experiments in *E. coli* and *H. pylori* [117,118]. Deletion of *E. coli slyD* results in reduced hydrogenase activity, a phenotype suppressed by the addition of excess nickel in the growth media [117]. The available structures of SlyD reveal a multi-domain protein consisting of a peptidyl-prolyl isomerase (PPIase) domain and a chaperone domain as well as an unstructured C-terminal metal binding region [119–121]. The C-terminal sequence is rich in potential metal-binding residues, including 15 histidines, 6 cysteines, and a dozen carboxylate amino acids [122–124], and is required for hydrogenase maturation [125], unlike the PPIase activity [126]. SlyD is proposed to contribute to several aspects of hydrogenase maturation. SlyD may function as a general chaperone given that it interacts with HycE prior to iron and nickel insertion [92]. In addition, SlyD can act as a source of nickel for HypB and modulate both the high-affinity nickel-binding site of HypB and GTP hydrolysis, revealing a link between nickel release from HypB and the GTPase activity of the protein [125,127].

In addition to its role in hydrogenase maturation, SlyD acts as a nickel storage protein in *E. coli*, as deletion of *slyD* results in lower nickel accumulation *in vivo*

[117,127], and it also impacts the activity of the nickel responsive transcription factor NikR [128]. The full-length protein can bind up to seven nickel ions in a non-cooperative manner with an overall apparent $K_D < 0.1$ nM [128,129]. The Ni(II) is bound in several different geometries with His and Cys residues from the C-terminal tail serving as ligands [129]. The protein also binds a variety of other metals with a trend similar to the Irving-Williams series, but only Ni(II) binding appears to be relevant *in vivo* [128]. Given that the metal-binding domains (MBDs) of SlyD homologs vary considerably, it is unclear how conserved these roles of SlyD are.

3.2 *Urease*

Urease, the only nickel enzyme found in eukaryotes [40], catalyzes urea hydrolysis to yield ammonia and carbamate, which further decomposes into a second molecule of ammonia as well as carbon dioxide (see Figure 3). This reaction is a key component of the global nitrogen cycle and is critical for many pathogenic bacteria, which utilize the ammonia as a means to buffer environmental pH changes [4,130–132]. The catalytic center of urease contains two nickel ions: one is bound in a square pyramidal environment to two histidines and one water molecule, as well as a carbamylated lysine and a second water serving as bridging ligands. The pseudo-octahedral coordination sphere of the second nickel is completed by two histidines, an aspartic acid, and a water molecule (Figure 3) [4,130,133]. The ligands are pre-organized in the apo-enzyme [134], likely contributing to the high affinity and specificity of the enzyme for nickel ions [132]. The choice of Ni(II) for urea hydrolysis is not well understood, but may be related to the high affinity of the urea-Ni(II) complex and/or the stability of the Ni(II)-bound active site as compared to similarly coordinated dinuclear Zn(II) centers [9]. Furthermore, the recently discovered iron-containing urease applies the same ligand set as nickel-urease, but is less efficient, suggesting that the chemistry of nickel plays a vital role in catalysis [135].

Much like [NiFe]-hydrogenase, the assembly of the urease active site requires several accessory proteins, UreDEFG [7,130]. Following translation, apo-urease may be held in an open conformation by GroEL/GroES in order to allow the accessory proteins to carbamylate the active site Lys and to insert the metals [136,137]. The main function of the accessory proteins appears to be to speed up a very slow process, as apo-urease can be partially activated *in vitro* in the absence of UreDEFG if supplemented with nickel and a source of CO_2 [138]. The functional unit responsible for urease activation is UreDFG [139–141]. In this complex UreG is a GTPase [140,142,143], UreD (called UreH in *H. pylori*) interacts directly with urease [144–146] and UreF modulates a conformational change around the urease active site [146]. UreG is capable of binding nickel but biochemical evidence suggests that Zn(II) may be the biologically relevant metal [147,148]. The proposed Zn(II)-binding site is in the same location relative to the GTPase motifs as the low-affinity site in homologous HypB [148]. Metal binding enhances the transition from a partially disordered to structured state in some UreG proteins, the significance of which is currently not understood [147–149]. Furthermore, Zn(II) stabilizes the interaction between *H. pylori* UreG and a fourth accessory protein, UreE [150]. The presence of UreE, which is believed to be the source of nickel for urease, decreases the GTP concentration required for urease activation by UreDFG [140].

3.2.1 UreE

Whether or not UreE is critical for urease maturation depends on the organism. In the case of *Klebsiella aerogenes* deletion of *ureE* results in a urease-deficient phenotype that can be complemented by nickel, suggesting a nickel metallochaperone function [151]. In contrast, in several *Helicobacter* species *ureE* mutations are not complemented by nickel [152,153]. The crystal structures of *Bacillus pasteurii* [154], *H. pylori* [155,156], and truncated *K. aerogenes* UreE [157] feature a symmetrical homodimer with a metal-binding site at the dimerization interface involving a conserved histidine residue (Figure 5). In *H. pylori* UreE (HpUreE), this site has a K_D=0.15 μM for nickel [150,158]. Although the crystal structures of HpUreE and *B. pasteurii* UreE (BpUreE) revealed tetramers, this oligomeric form may not be relevant *in vivo* [150,156,159]. During urease maturation, the position of one UreE dimer in these structures may be replaced by an UreG dimer to form the UreE-UreG complex discussed earlier [130,150].

In some organisms, UreE features additional metal-binding capabilities beyond the conserved site at the dimer interface. *K. aerogenes* UreE (KaUreE) possesses a His-rich C-terminus and can bind 6 nickel ions per dimer in pseudo-octahedral sites composed of O/N donors including multiple histidines with an average apparent K_D=9.6 μM [151,160,161]. Co(II), Zn(II), and Cu(II) can compete with Ni(II) bind-

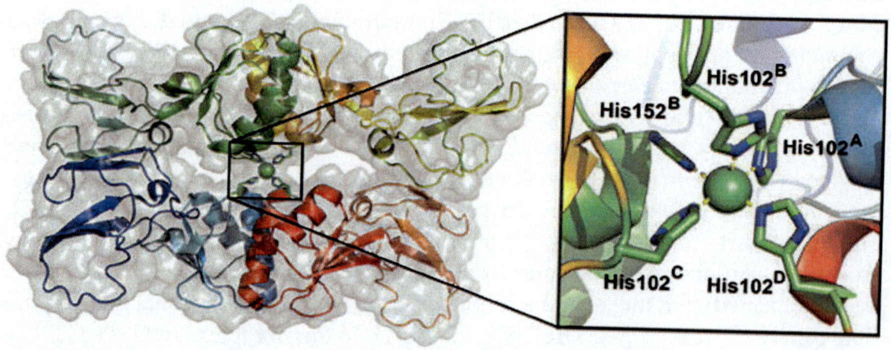

Figure 5 Structure of *H. pylori* urease accessory protein UreE (PDB 3NYQ). HpUreE lacks the His-rich sequence found on the C-termini of many UreE homologs, but it possesses the conserved metal site at the oligomeric interface of the protein. This Ni(II)-binding site is composed of a His102 ligand from each monomer, in addition to a single His152 ligand. Metal binding is critical for oligomerization of the protein. Although the protein crystallized as a tetramer, it is believed to function as a dimer during urease maturation, with the position of one UreE dimer in the structures replaced by a UreG dimer. The color scheme is nickel, green sphere; oxygen, red; sulfur, yellow and; nitrogen, blue. The images were generated using PyMol.

ing, but they have distinct coordination environments, suggesting that nickel may have a selective allosteric effect on KaUreE [151,160–162]. Upon deletion of the His-rich C-terminus, the protein binds two Ni(II) per dimer and this truncated protein is still able to function in urease maturation, so the His-rich motif might instead contribute to metal sequestration [151]. BpUreE lacks this His-rich C-terminus but can bind additional nickel ions in octahedral sites of micromolar affinity within the C-terminal region of the protein [159,163]. Although truncated KaUreE and BpUreE are structurally similar, isothermal titration calorimetry suggests significant differences between the Ni(II)-binding activities of the two proteins, as Ni(II) binding to KaUreE is enthalpically favored as opposed to the entropically driven process in BpUreE [162]. These experiments also indicate that the number and types of metal-binding sites vary depending on UreE concentration, likely due to the influence on oligomeric structure [162]. Based on the data, it is proposed that Ni(II) ions are bound to the periphery of UreE in the resting state and only upon nickel delivery to urease does the conserved histidine at the dimer interface become involved [162].

3.2.2 Connection to [NiFe]-Hydrogenase Maturation

In *H. pylori*, in addition to impacting hydrogenase maturation, *hypA* and *hypB* mutations produce a urease deficiency that is nearly fully complemented by nickel supplementation of the growth media [105,06]. This double duty by HpHypA and HpHypB may be common to the *Helicobacter* species, as it has also been noted in *H. hepaticus* [153]. An interaction between HpHypA and HpUreE is critical for urease maturation, although no direct nickel transfer from HpHypA to HpUreE was detected [152]. Furthermore, introduction of an additional chromosomal copy of *hpUreE* partially complements mutations in *hypA* and *hypB* and engineered versions of HpUreE with His-rich C-terminal sequences yield the greatest urease activity in the *hypA* and *hypB* mutants, demonstrating that there is a correlation between *H. pylori* urease activity and the nickel sequestering ability of UreE [158]. These experiments suggest that even though wild type HpUreE lacks the His-rich C-terminus found in homologous proteins, *H. pylori* has an alternative way of directing nickel to urease by recruiting the hydrogenase proteins HypA and HypB [158].

SlyD also plays an important role in urease maturation in *H. pylori*, which overlaps with that of HypA [165]. In contrast to EcSlyD discussed earlier, the *H. pylori* homolog (HpSlyD) binds 2 Ni(II) per monomer with an apparent $K_D = 2.74$ μM [119]. The significance of this lower Ni(II)-binding ability is unknown and further characterization of the function of HpSlyD in urease maturation is needed.

3.3 *Carbon Monoxide Dehydrogenase/Acetyl-CoA Synthase*

Carbon monoxide dehydrogenase (CODH) catalyzes the reversible oxidation of CO to CO_2, which allows organisms to use CO or CO_2 as a carbon source [4,166]. The CODH dimer bears three [Fe_4S_4] clusters and two nickel-containing C-clusters in

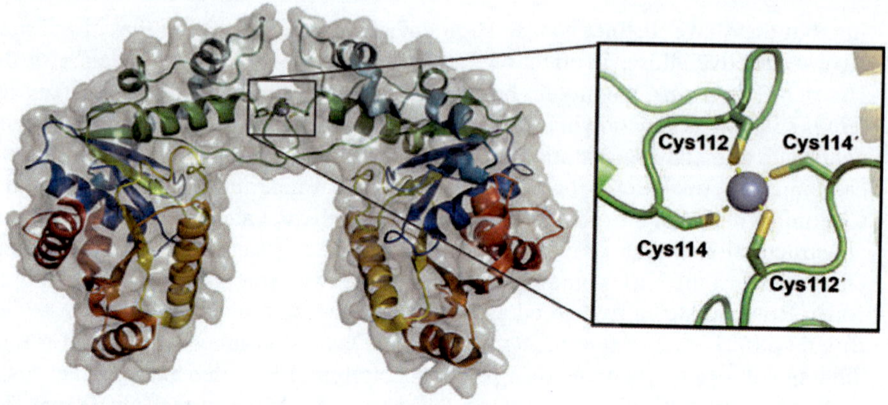

Figure 6 Structure of *Carboxydothermus hydrogenoformans* CooC (PDB 3KJI). CooC is the ATPase required for the maturation of carbon monoxide dehydrogenase. Only the apo and Zn(II)-loaded crystal structures have been determined, of which the Zn(II)-bound structure is shown here. The protein is a metal-dependent dimer, with a metal coordinated at the dimer interface by the four Cys residues from two CXC motifs. The color scheme is zinc, grey sphere; oxygen, red; sulfur, yellow and; nitrogen, blue. The images were generated using PyMol.

which a nickel ion is connected to a $[Fe_3S_4]$ cluster through a bridging Cys ligand (Figure 3) [4,166,167]. Maturation of CODH in *Rhodospirillum rubrum* involves several accessory proteins encoded by the *cooCTJ* operon, which are required for inserting Ni(II) into CODH [168]. CooT has a weak homology with HypC, but its role remains unclear [168]. CooJ may be a nickel metallochaperone, because it contains 16 histidines in the final 34 amino acids and can bind four Ni(II) per monomer with an apparent $K_D=4.3$ μM [169].

CooC, homologous to HypB and UreG, is the NTPase required for the CODH maturation pathway but in contrast to HypB and UreG, CooC is likely an ATPase [170]. Nickel impairs the ATPase activity of CooC [170], induces dimerization and is bound with a $K_D=0.41$ μM [171]. The nickel-binding site consists of four Cys residues, two each from a CXC motif on each monomer (Figure 6) [171,172]. CooC is believed to be allosterically modulated by both nickel and nucleotide, but how these factors come into play during CODH assembly is unknown [171,172].

In some cases, CODHs are components of a larger enzyme complex where they are found with acetyl-CoA synthetase or acetyl-CoA decarbonylase (Figure 3) [4,166,173,174]. In these enzymes, the active site contains two nickel ions, one coordinated by two bridging Cys residues and two backbone amide nitrogens in a square planar site [175]. The second Ni(II) is coordinated by the bridging cysteines, a third bridging Cys linked to an $[Fe_4S_4]$ cluster, with a remaining open coordination site for the substrate [4,166,176]. Very little is known about the assembly of this site, although an NTPase of unknown function, AcsF, has been implicated in its assembly [177].

The choice of nickel in CODH and acetyl-CoA synthetase or acetyl-CoA decarbonylase is obscured by the lack of consensus regarding the mechanistic role of nickel in these enzymes [4,166]. Considering the mineral availability of primordial Earth as noted earlier, the inclusion of nickel in these enzymes may be mostly historical, rather than chemical [14,178].

3.4 *Superoxide Dismutase*

Superoxide dismutase (SOD) catalyzes the disproportionation of the toxic superoxide radical to hydrogen peroxide and molecular oxygen. There are several variations of SOD with different metals at the active sites [179], one of which contains Ni(II) (Figure 3) [180,181]. Found in cyanobacteria and *Streptomyces* [182,183], NiSOD is encoded by the *sodN* gene and does not resemble other non-nickel SODs [184]. *Streptomyces* NiSOD is a hexameric protein [185] and the active site is proposed to switch between a square planar N_2S_2 Ni(II) state and a square pyramidal N_3S_2 Ni(III) state during the catalytic cycle [186]. This coordination site is provided by an N-terminal "hook" motif with the consensus sequence HCXXPCGXY, in which the His1 imidazole cycles on and off the nickel center and the constant ligands are the side chains of Cys2 and Cys6 and the backbone nitrogens of His1 and Cys2 [186].

It is possible that NiSOD is expressed instead of enzymes using other metals so that organisms can take advantage of the higher nickel concentrations in their environments [182]. The presence of the NiSOD gene, *sodN*, frequently correlates with the genomic absence of the gene encoding for FeSOD (a notable exception is in the *Streptomyces* species which possess Ni-, Fe- and ZnSODs; see Sections 5.4 and 5.5), suggesting that the use of NiSOD instead of FeSOD may have been an evolutionary adaptation to reduce bacterial iron requirements [182]. However, it is more than a simple matter of swapping metal ions – in order to perform the dismutation chemistry on superoxide, nickel must cycle between the Ni(II) and Ni(III) oxidation states with a redox potential within range of the ($O_2^{\bullet-}/O_2$) and ($O_2^{\bullet-}/H_2O_2$) couples [185,187]. Aqueous nickel does not have such a redox potential [187] and Ni(III) is an unstable oxidation state that tends to be reduced to Ni(II) upon thiolate coordination [188]. To overcome these limitations, the protein provides two backbone amides (His1 and Cys2) and two thiolates (Cys2 and Cys6) with an axial His1 in the oxidized state [185,186]. Such a coordination environment supplies sufficient electron density to retard autoreduction of the Ni(III) center [188]. The thiolate coordination sphere is also responsible for lowering the reduction potential to a suitable range for superoxide dismutation [189–191]. In this way, nature tweaks the chemistry of nickel to perform otherwise difficult chemistry.

With respect to maturation, NiSOD is expressed *in vivo* with an N-terminal leader sequence that is cleaved in a nickel-dependent manner, yielding the active enzyme [184]. The gene *sodX* is adjacent to *sodN* and encodes a putative peptidase required for this N-terminal cleavage reaction [192]. NiSOD maturation is proposed to begin with hexamer formation, followed by N-terminal proteolytic cleavage coupled to nickel loading [186], analogous to the nickel-dependent processing observed in [NiFe]-hydrogenase maturation. A putative nickel chelatase, CbiXhp, may serve a role in nickel delivery [193]. Following the proteolytic cleavage, the N-terminal "hook" is completed upon *trans*-to-*cis* isomerization of Pro5, possibly catalyzed by a PPIase accessory protein [186,192].

3.5 Methyl-Coenzyme M Reductase

Methyl-coenzyme M reductase is the final enzyme in the methane biosynthetic pathway within methanogenic archaea and is responsible for all biologically produced methane, thus playing a central role in the global carbon cycle [166,194]. The enzyme catalyzes the reduction of methyl-coenzyme M to yield methane with coenzyme B serving as an electron donor [166,194], and also catalyzes the reverse reaction to oxidize methane in methanotrophs [195]. The key to this catalysis is the nickel-containing cofactor F430 (Figure 3) [196–198]. In the proposed mechanisms [166,194], nickel varies between the +1, +2 and +3 oxidation states. These oxidation states are associated with substantial changes in coordination geometry around the metal center, dictating the need for a flexible biological ligand [8]. In relation to other scaffolds, the corphin ring of F430 is very reduced and the lack of conjugation allows F430 to accommodate these changes in coordination geometry. Furthermore, the corphin ring is believed to stabilize the Ni(III) oxidation state [199].

The biosynthesis of coenzyme F430 proceeds through 5-aminolevulinic acid and uroporphyrinogen III, in a manner similar to other cyclic tetrapyrroles such as vitamin B12 and heme [197]. The existence of a specific chelatase responsible for nickel insertion into the corphin has been proposed, similar to the chelatases involved in heme and vitamin B12 biosynthesis [200], and additional accessory proteins may be required to introduce five post-translational modifications near the active site [201].

3.6 Glyoxalase I

The glyoxalase system, composed of glyoxalase I (GlxI) and glyoxalase II (GlxII), catalyzes the conversion of toxic methylglyoxal to lactate [202]. Methylglyoxal reacts with glutathione to produce the hemithioacetal substrate of GlxI, which then catalyzes an isomerization reaction to yield S-D-lactoylglutathione, the substrate for GlxII (Figure 3) [202,203]. GlxI in yeast and humans is a zinc-dependent enzyme, but the *E. coli* homolog is inactive with Zn(II) and maximally activated with Ni(II) [204]. The active site consists of Ni(II) coordinated in an octahedral geometry by two His, two Glu, and two water molecules (Figure 3) [205,206]. This coordination geometry is in contrast to the Zn(II)-bound *E. coli* GlxI which utilizes the same protein ligands but binds the metal in a trigonal bipyramidal geometry with only one water molecule ligand [205–207]. The enzyme is also activated by other divalent metals, such as Co(II) and Cd(II) [204,208], which bind in an octahedral geometry, suggesting that this type of geometry at the active site is critical for enzymatic activity [202]. GlxI binds Zn(II) more tightly than Ni(II) [208], suggesting that a mechanism for nickel delivery exists, but no information regarding such a system is known.

3.7 Acireductone Dioxygenase

Acireductone dioxygenase (ARD) is a key enzyme in the methionine salvage pathway [209]. This enzyme is an example of a naturally occurring metalloenzyme with distinct enzymatic activities solely determined by the bound metal [209,210]. The iron-bound *Klebsiella oxytoca* ARD converts acireductone to the α-ketoacid precursor of methionine and formate, whereas the nickel-bound ARD produces methylthiopropionate, carbon monoxide, and formate [210,211]. The nickel-loaded enzyme features an octahedral Ni(II) center with His96, His98, Glu102, His140, and two water molecules serving as ligands (Figure 3) [209,212]. Surprisingly, iron uses the same ligand set, but conformational changes in the protein are proposed to yield distinct enzymology [213]. The use of the same protein scaffold but different metals to yield two distinct chemistries highlights the diverse roles the first row transition metals can play, but also raises the issue of how the organism can control incorporation of the different metals. This issue is complicated by the fact that ARD was found to bind nickel more tightly than iron [210], suggesting the possibility of unknown metallochaperones responsible for metal delivery to ARD.

4 Nickel Storage Proteins

Nickel sequestration can be achieved by proteins with other functions, as in the case of the His-rich domains of some HypB and UreE homologs and the MBD of EcSlyD. Additional proteins that contribute to nickel storage are found in bacteria if the aforementioned domains are missing or the bacteria have a greater need of nickel. For example, in order to maintain sufficient levels of urease and [NiFe]-hydrogenase activities, *H. pylori* accumulates significantly more nickel than *E. coli* [214], in part due to the presence of the unique TonB-dependent outer membrane nickel uptake system [60] and the high-affinity nickel transporter NixA [48]. While the majority of this metal is incorporated into the urease and [NiFe]-hydrogenase enzymes, additional nickel can be stored in the HspA, Hpn and Hpn-like nickel sequestering proteins for times of nickel starvation and/or to deal with short-term overflow [215–219].

HspA is a unique homolog of the GroES family of heat-shock proteins, with an extra C-terminal domain rich in potential metal-binding residues (8 His and 4 Cys in a 27-residue domain) [136]. The protein is a heptamer in solution and binds two nickel ions per monomer with an apparent $K_D = 1–2$ μM [137,220], in a coordination environment consisting of nitrogen and thiolate ligands [220]. One nickel is bound in a square planar Cys_2His_2 site, with the second site formed by a HX_4DH motif, analogous to NiCoTs [217,221,222].

The two-domain structure of HspA suggests that the protein contributes to both classical GroES chaperonin activity as well as nickel-carrier functions [137]. The C-terminal MBD is involved in protecting the cell from nickel toxicity [217,220]. Heterologous expression of *H. pylori* urease in *E. coli* suggested a requirement for HspA in urease maturation [136,137], but subsequent experiments in *H. pylori* did not yield the same results [217]. Instead, HspA mutants in *H. pylori* impacted hydrogenase maturation, suggesting that HspA substitutes for the lack of the poly-His region in HpHypB or the extended MBD in HpSlyD [217].

Two other nickel-sequestering proteins from the *Helicobacter* species are the Hpn and Hpn-like proteins [218,219]. Hpn contains 28 histidines out of 60 residues [219] and binds 5 Ni(II) per monomer with an apparent $K_D = 7.1$ μM [215]. Hpn-like contains 18 His and 30 Gln in a short 72 amino acid sequence [218], and is capable of binding 2 Ni(II) with an apparent $K_D = 3.8$ μM [223]. *In vitro* the metal affinities of Hpn and Hpn-like follow the Irving-Williams series [219,223,224], but *in vivo* the mutation of *hpn* and *hpn-like* affects sensitivity to nickel more significantly than other metals [216,218,223,224], suggesting a role in nickel sequestration [215,216,218]. Protease digestion and chemical denaturation experiments support this *in vivo* selectivity of Hpn-like, as Ni(II) confers structural stability to the protein that is not observed with other metals [225]. Double deletion of *hpn* and *hspA* leads to a more pronounced phenotypic response to nickel compared to single mutants, suggesting overlapping and redundant roles in nickel detoxification [217]. It has been proposed that the His-rich proteins may provide a rapid detoxification response to transient nickel fluctuations before the transcriptional control of transporter expression can manifest [218].

5 Nickel-Based Genetic Regulation

Metalloregulatory proteins offer a unique glimpse into nickel homeostasis and the nickel metallome. Homologous proteins permit a better understanding of the structural determinants for metal selectivity [62]. Furthermore, by monitoring the activity from a target promoter *in vivo*, it is possible to infer the metal occupancy of the metalloregulator within a cellular context [62]. Despite the wide variety of known Ni(II)-responsive transcription factors, they all appear to share several common characteristics with other metalloregulators: they function as homo-oligomers; the metal-binding ligands are found at subunit interfaces, driving secondary or quaternary structural changes that allosterically affect DNA affinity and; metal ion selectivity is dictated by the coordination geometry imposed by the chemical preferences of the metal [226–229].

5.1 NikR

NikR is the only known metal-regulated member of the ribbon-helix-helix family of DNA-binding proteins [230–232], characterized by intertwining N-terminal sequences of two monomers to produce an antiparallel β-strand that binds to the major groove of the DNA recognition sequence [232]. The protein is functional as a tetramer [233–236], although there may be exceptions [237], with a central C-terminal MBD core connected to two flanking DNA-binding domains (DBD) through a flexible linker [234–236]. NikR has been found in a wide range of archaea and eubacteria species [230], and *E. coli* and *H. pylori* NikR are the best characterized homologs.

5.1.1 *E. coli* NikR

NikR was first described in *E. coli* where it acts as a nickel-dependent repressor of the *nikABCDE* operon [28,238]. *E. coli* NikR (EcNikR, Figure 7) can bind one Ni(II) per monomer with a $K_D = 0.9$ pM [233,239] (termed the high-affinity site) and additional Ni(II) with a $K_D \approx 30$ nM (termed the low-affinity site) [240]. The two sites allow EcNikR to bind to the palindromic operator sequence (GTATGA-N_{16}-TCATAC) in the *nikABCDE* promoter (P_{nik}) with two different affinities. When the high-affinity site is occupied by Ni(II), EcNikR binds DNA with a $K_D \sim 5$ nM [233,240,241] and this complex tightens to a $K_D \sim 20$ pM upon binding of additional Ni(II) [240,241].

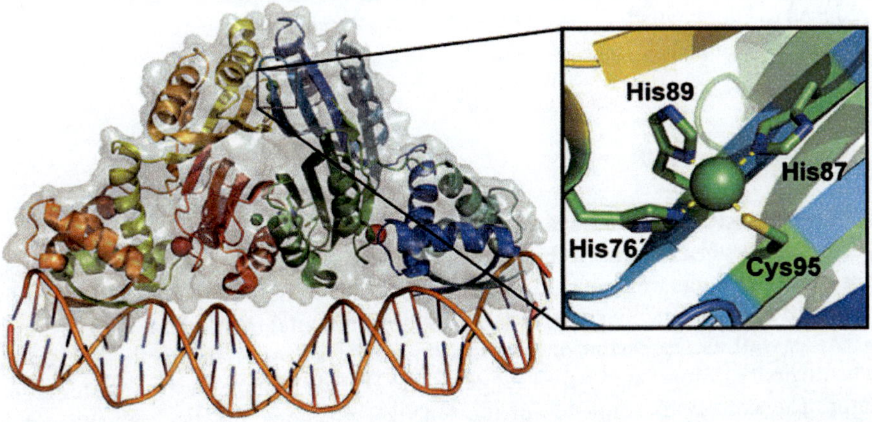

Figure 7 Structure of the nickel-responsive transcription factor *E. coli* NikR in complex with DNA (PDB 2HZV). NikR is a member of the ribbon-helix-helix family of DNA-binding proteins and features a central MBD flanked by two DNA-binding domains. The tetrameric protein binds four Ni(II) ions in a square planar site with Cys95, His87, His89 contributed from one monomer and His76′ contributed by a second monomer. Upon Ni(II) binding, organization of the α3-helix allows for non-specific DNA contacts to be made, thereby localizing the protein to the DNA. The color scheme is nickel, green sphere; potassium, red sphere; oxygen, red; sulfur, yellow and; nitrogen, blue. The images were generated using PyMol.

The high-affinity site of EcNikR is square planar with His87, His89, Cys95, and His76′ from an adjacent monomer as ligands (Figure 7) [234,242]. Upon nickel binding, the secondary structure of EcNikR changes, producing an increase in α-helix content [233]. The crystal structure of EcNikR in the apo- and holo-forms (high-affinity site occupied) demonstrates that nickel binding allows resolution of the previously unstructured α3-helix [234,243]. The stabilization of this helix is proposed to allow several non-specific protein-DNA contacts [243], thereby localizing the protein to DNA where the protein can undergo a one-dimensional search for the *nik* promoter [244]. Potassium ions bridge the DBD and the MBD, and are critical for Ni(II)-responsive DNA binding [245,246]. The induction of tighter DNA binding occurs upon coordination of Ni(II) to the low-affinity site, proposed to consist of His48 and His110 and possibly nearby carboxylate ligands, all located between the MBD and the DBD [247,248], but some controversy exists regarding the exact location of this site [249]. The mechanism through which this second site induces tighter DNA binding is unknown.

Despite the high-affinity site following the Irving-Williams series in terms of metal binding, nickel provides the most protection in denaturation [239], protease degradation [250], and H/D exchange experiments [248]. This selective allosteric effect by nickel is also observed in relation to DNA-binding affinity, as only Ni(II) can induce the formation of the tightest DNA-NikR complex [240]. The basis for this observed selectivity was proposed to be the square planar coordination geometry, which only Ni(II) and Cu(II) adopt [226,242,248,251]. The formation of the proper coordination geometry allows for the formation of the α3-helix and subsequent DNA contacts [244,251]. *In vivo* experiments demonstrate that NikR responds only to Ni(II) [248], consistent with the fact that in the cell Cu(II) is reduced to Cu(I), which is incapable of forming the square planar site and triggering the required allosteric change [226,242,248].

5.1.2 *H. pylori* NikR

H. pylori NikR (HpNikR) shares 30% identity and 68% similarity with EcNikR [252], but in stark contrast to EcNikR, HpNikR acts as both an activator or repressor of a wide range of genes [253,254]. Regulation by holo-HpNikR includes activation of *ureA-B* (encodes the urease structural subunits), *hpn*, *hpn-like*, *hspA*, and *nixA*, as well as repression of *frpB4*, *exbB*/*exbD*, *fur*, *nikR* and *fecA3* [253,254]. The diversity of regulated genes suggests that HpNikR functions as a master regulator. The consensus sequence of the HpNikR promoter has been elusive but a study examining the NikR operators in the closely related ferret pathogen, *Helicobacter mustelae*, lead to the prediction of a putative consensus sequence, TRWYA-N_{15}-TRWYA [57]. Despite this progress, recent studies suggest that HpNikR binding to DNA is not as simple as EcNikR, as the protein exhibits distinct conformations when bound to different promoters, a property that may be involved in DNA recognition [255]. Furthermore, HpNikR exhibits two tiers of DNA-binding affinities for its target sequences *in vitro* [256] and distinct temporal control of activated *versus* repressed genes *in vivo* [253]. Responsiveness to pH has also been described, suggesting that HpNikR may link nickel homeostasis to acid adaptation in this organism [257–260].

Multiple crystal structures of HpNikR generated under subtly different conditions have been reported, revealing a variety of combinations of nickel sites [261–264]. All of the structures have some nickel located in a square planar site similar to EcNikR, named the final (F) site, consisting of Cys107, His99, His101, and His88′ from an adjacent monomer [261–264]. The second site, called the exit (X) site, consists of His74, His75, Glu104, and solvent molecules, arranged in an octahedral geometry [261]. The third site is termed the intermediate (I) site because it contains residues from the F- and X-site [261]. The I site is also octahedral and features His74 and His101 from one monomer, His88′ from the adjacent monomer, as well as three water ligands, although slight variations have been observed [263]. Mutation of the residues from any of these three sites disrupts the nickel response of NikR in *H. pylori*, suggesting that all three sites may be physiologically relevant [261,262], but whether differential occupancy of these sites can modulate DNA binding to the various target sequences is not clear.

The affinity with which HpNikR binds stoichiometric nickel is also a source of debate. An original $K_D = 3.5 \times 10^{-12}$ M was determined using competition experiments [265], yielding a number comparable to that of EcNikR. However, more recent ITC data suggest that tetrameric HpNikR binds two nickel ions with a $K_D \approx 10$ nM and two additional ions with a $K_D \approx 100$ nM, with additional metal binding with micromolar affinities [266]. Unlike EcNikR, there is no evidence of a DNA-binding response to excess Ni(II). Instead, HpNikR requires an additional metal ion such as magnesium, calcium, or manganese, for *ureA* promoter binding [267,268].

5.2 *RcnR*

The RcnA Ni(II)/Co(II) exporter is controlled at the transcriptional level by the Ni(II)/Co(II)-responsive metalloregulatory protein RcnR [269]. In the apo form, RcnR represses *rcnA* and *rcnR* and dissociates from the promoters upon binding nickel or cobalt [64,269–271], whereas Cu(II) and Zn(II) elicit a response only at much higher concentrations [271]. RcnR functions as a tetramer and binds to the promoter sequence ($K_D = 255$ nM) with a large Hill coefficient of 3.3, indicating positive cooperativity for DNA binding [270].

RcnR binds one equivalent of nickel per monomer with a $K_D < 25$ nM [272]. Both Ni(II) and Co(II) bind to the protein in an octahedral geometry with $His_2Cys(N/O)_3$ ligands including the side chains of Cys35, His3, His64, and the N-terminal amine [272]. In addition, His60 has been proposed to act as a ligand, along with the amide backbone of the second residue [272]. Although Cys35 is conserved in all RcnR homologs, this residue is dispensable for Ni(II)-responsiveness *in vivo* [272]. Neither Ni(II) nor Co(II) appear to change the secondary structure of the protein upon binding, but both do increase protein stability [272]. How this plays into the allosteric mechanism remains to be elucidated.

5.3 *Members of the ArsR/SmtB Family*

The ArsR/SmtB family of transcriptional repressors regulate genes encoding metal efflux and sequestration proteins [273]. *M. tuberculosis* possesses twelve ArsR/SmtB transcription factors [273], including the nickel-responsive NmtR and KmtR.

NmtR represses transcription of the *M. tuberculosis nmtA* gene, which encodes a putative P_1-type ATPase efflux pump for Ni(II) and Co(II) [62]. Upon binding Ni(II), NmtR dissociates from the promoter, allowing transcription [62]. The homodimeric NmtR binds either a single Ni(II) per monomer with a $K_D \approx 10^{-10}$ M or a single Co(II) with a K_D roughly 10^4 times weaker than that of Ni(II) [62,274]. The isolated protein has a binding selectivity of Zn(II) > Ni(II) >> Co(II), but despite Zn(II) binding with the tightest affinity, it does not elicit a response [62]. Ni(II) is bound to NmtR in an octahedral environment with three histidines and three N/O ligands [275]. Mutagenesis identified 1 Asp and 3 His ligands on the predicted α5-helix, a location where other members of the ArsR/SmtB family also bind metal [62,273,274]. The remaining ligands are a source of debate [62,274]. In contrast to Ni(II), Zn(II) binds in a tetrahedral environment to three His and one unidentified N/O ligand on the α5-helix [275]. This tetrahedral geometry prevents Zn(II) from forming all of the needed protein contacts to activate protein dissociation from the DNA [62,275]. In this way, nature appears to have adapted a protein family to sense nickel through the introduction of two Ni(II) ligands onto a common ArsR/SmtB scaffold, utilizing nickel's ability to bind in an octahedral coordination geometry to elicit the proper biochemical response.

In addition to NmtR, *M. tuberculosis* also expresses the nickel-responsive transcription factor KmtR, responsible for repressing the *cdf* (encoding a CDF-family metal exporter) and *kmtR* genes in the absence of Ni(II) [276]. Although the Ni(II) site in KmtR is located on the predicted α5 helix, the exact location is not shared with any other ArsR/SmtB homologs [276]. Four His and one Glu are critical for the Ni(II) response, with an Asp residue predicted to serve as a sixth ligand, forming an octahedral site [276]. Competition experiments between NmtR and KmtR indicate that KmtR can bind Ni(II) with more than 100-fold tighter affinity, suggesting a KmtR Ni(II) $K_D \leq 10^{-12}$ M [274,276]. The presence of both NmtR and KmtR in *M. tuberculosis* allows for a graded response to Ni(II) concentrations. As the metal accumulates in the cell, KmtR responds to Ni(II) first, resulting in derepression of the CDF nickel exporter. As nickel levels continue to rise, NmtR then binds Ni(II) and allows expression of the NmtA efflux pump. In this way, the cell can protect itself from increasing nickel levels, and the use of the ATP-dependent P_1-type pump is restricted to circumstances when the CDF pump is insufficient [276].

5.4 *Nur*

In *Streptomyces coelicolor*, nickel represses the transcription of the FeSOD gene while inducing production of NiSOD [277]. The transcription factor responsible for this behavior is Nur, a Fur (ferric-uptake regulator) homolog [278]. Although

Ni(II)-Nur binds to the FeSOD promoter, it does not directly bind to the *sodN* promoter, prompting speculation that it instead represses the expression of another transcription factor responsible for the repression of *sodN* [278]. In addition to regulating expression of Fe- and NiSODs in *S. coelicolor*, Nur regulates the putative NikABCDE and NikMNOQ transporters as well, repressing these genes in the presence of nickel [278,279].

Nur is highly selective for Ni(II) both *in vitro* and *in vivo*, with no DNA binding activity observed in the presence of any other first row transition metal [278]. Nur is a homodimer with a modular architecture consisting of two winged-helix DBDs attached to two dimerization domains that form the core of the protein (Figure 8) [279]. Each monomer contains two types of metal sites, termed the M-site and the N-site (Figure 8). The M-site is located at the domain interface and involves four His residues from a single monomer forming a square planar geometry around

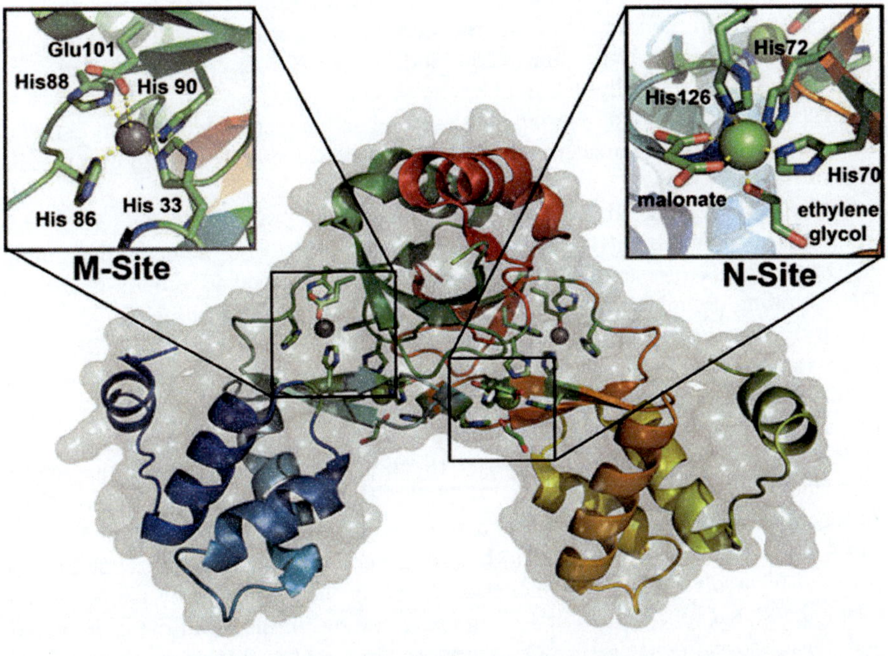

Figure 8 Structure of the nickel-responsive transcription factor Nur from *Streptomyces coelicolor* (PDB 3EYY). Nur is responsible for the regulation of a NiSOD in *S. coelicolor* and has been putatively assigned a role in repressing the nickel transporters NikABCDE and NikMNOQ. The protein is a dimer, with each monomer composed of a central dimerization domain flanked by a DNA-binding domain. The protein bears two metal-binding sites: the M-site features 4 His ligands forming a square planar geometry around Ni(II). If Zn(II) binds to this site, an additional Glu residue can replace one of the coordinated His ligands. The N-site is solvent exposed and the Ni(II) is coordinated in an octahedral geometry by 3 His ligands and 3 solvent molecules (depicted here by malonate and ethylene glycol present in the crystallization solvent). The color scheme is nickel, green sphere; oxygen, red; sulfur, yellow; nitrogen, blue; zinc, grey sphere. The images were generated using PyMol.

Ni(II) [279]. This site is also able to accommodate Zn(II) upon substitution of one of the His ligands with a nearby Glu. In contrast, the N-site is unique among Fur family members, and contains a Ni(II) coordinated in an octahedral geometry with three His ligands and three solvent molecules [279]. Both sites influence DNA binding, but the N-site appears to influence the secondary structure of the dimerization domain and the DNA contacts made by the protein, suggesting that this site is responsible for the nickel-responsive activation of Nur [279].

5.5 *Additional Nickel-Responsive Genetic Regulators*

Reports of multiple new Ni(II)-responsive transcription factors have recently emerged, and several of those with unusual features are described here.

The YqjI protein in *E. coli* represses the *yqjH* gene, which encodes a putative ferric siderophore reductase [280]. YqjI releases the promoter when exposed to elevated nickel levels, which may help prevent disruption of iron homeostasis by nickel. An uncommon aspect of this regulator is the presence of a SlyD-like metal-binding domain on the N-terminus. YqjI binds two Ni(II) ions per monomer, through a mixture of His or Cys ligands, and Zn(II) was unable to displace any of the bound Ni(II) [280]. The extent of overlap between iron and nickel homeostasis in *E. coli* is not yet clear, but is reminiscent of the cross-talk between HpNikR and iron regulation in *H. pylori* [254].

The SrnRQ regulator is composed of two proteins, SrnR and SrnQ, which, in a 1:1 ratio, mediate the Ni(II)-dependent repression of the iron- and zinc-containing SOD in *Streptomyces griseus* [281]. SrnR possesses the putative helix-turn-helix DNA-binding domain while SrnQ binds stoichiometric Ni(II) with a $K_D = 0.65\ \mu M$ [281]. Upon nickel binding, a conformational change in SrnQ is relayed to SrnR and increases the DNA-binding ability of the complex [281]. This is the first example of a two-component metalloregulatory of the ArsR/SmtB family as well as the first two-component Ni(II)-dependent transcription factor complex [281].

Regulation of the CnrCBA exporter (Section 2.2) is the responsibility of the multicomponent CnrYXH system [74,282,283]. CnrH is a sigma factor of the extracellular function family, whose activity is governed by the membrane-bound CnrY and CnrX proteins [282,283]. The general mechanism for regulation is believed to involve sensing of nickel or cobalt in the periplasm by CnrX, which relays the signal to a CnrY/CnrH complex, resulting in release of cytosolic CnrH and transcription of *cnrCBA*. CnrY is proposed to have two TMDs and serves as an anti-sigma factor against CnrH [282], whereas CnrX (Figure 9) possesses a C-terminal periplasmic domain [284] that contains six His residues [74,284]. The isolated periplasmic domain of *Cupriavidus metallidurans* CH34 CnrX binds metal and mutation of the conserved His residues results in decreased nickel binding by the protein [284]. The crystal structure of the periplasmic domain of CnrX revealed a 1:1 stoichiometry for both Ni(II) and Co(II) with almost superimposable coordination spheres [285]. The protein ligands are His42, His46, His119, and one O atom

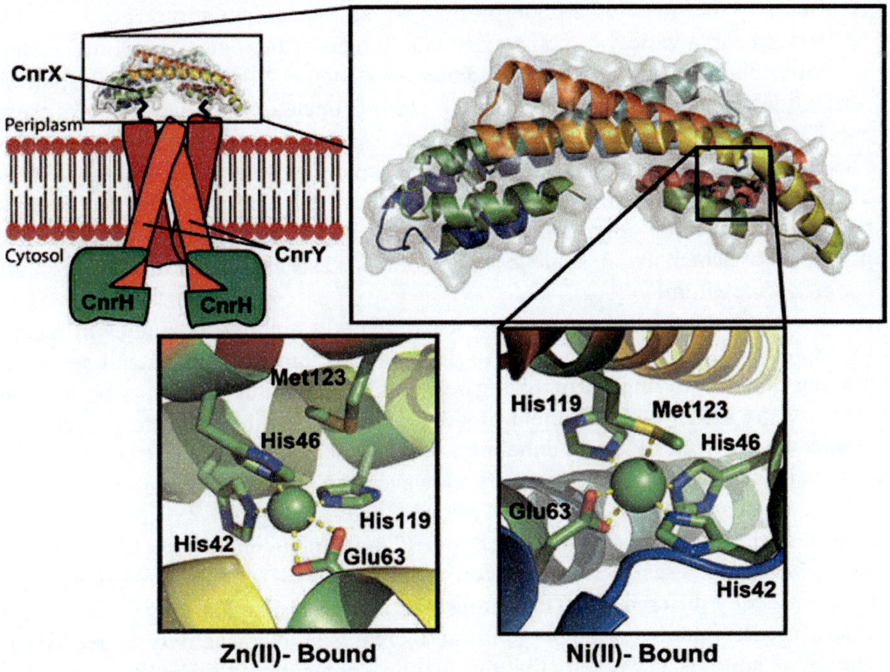

Figure 9 Structures of the periplasmic domain of *Cupriavidus metallidurans* CH34 CnrX (Ni(II)-bound: PDB 2Y39; Zn(II)-bound: PDB 2Y3D). The CnrYXH complex regulates the transcription of the Ni(II)-efflux pump CnrCBA. Ni(II) binding by CnrX is transmitted through the complex to the sigma factor CnrH. The nickel ion is bound to the protein in an octahedral geometry with a His_3GluMet ligand set. In contrast, Zn(II), which does not elicit a biological response from CnrX, binds in a tetrahedral environment to His_3Glu. The color scheme is nickel, green sphere; zinc, grey sphere; oxygen, red; sulfur, yellow and; nitrogen, blue. The images were generated using PyMol.

from the carboxylate side chain of Glu63 defining the equatorial plane of an octahedral geometry and the second O atom from Glu63 and the thioether sulfur of Met123 serving as axial ligands (Figure 9). Upon Ni(II) or Co(II) binding, a remodeling of the structure of CnrX is observed, with the recruitment of Met123 critical for this responsiveness [285]. In the Zn(II)-bound structure, Met123 does not serve as a ligand and, as a result, the corresponding conformational change and subsequent signal transduction is not triggered (Figure 9) [285].

6 The View From The Top

To date, the entire nickel metallome of a cell has not been comprehensively described, although a handful of studies examined portions of it in various organisms [286–289]. One outstanding issue is defining which nickel-containing species contribute to the metallome under normal physiological conditions. For example, Ni(II) can form a variety of complexes with numerous biological small molecules, including amino acids such as histidine and cysteine [290], sugars, phosphates, nucleotides [291], and glutathione [292], but it is unclear if and when any of these species are relevant *in vivo*. Nickel homeostasis is a complex process that includes nickel transporters, storage proteins, metallochaperones, and nickel-responsive regulatory proteins. However, given the current knowledge about individual components, such as those described above, we can begin to understand the common principles by which nature achieves nickel homeostasis and which ultimately shape the nickel metallome.

Thermodynamic binding constants are frequently reported in studies of metalloproteins because they provide a useful starting point for understanding metal binding and identifying participating residues. These values can also be used to compare the activities of different proteins, and it is feasible that thermodynamic gradients are responsible for funneling metals to their final cellular destinations, as observed for copper trafficking [293]. However, the current information regarding nickel proteins in *E. coli* and *H. pylori* (Tables 1 and 2) suggests that the thermodynamics of the isolated proteins does not yield a complete picture of nickel homeostasis. In several cases, intracellular competition for Ni(II) occurs between proteins with drastically different Ni(II) affinities. For example, deletion of any component of the *H. pylori* urease assembly pathway increases HpNikR activity under Ni(II)-limiting conditions, suggesting that this pathway competes with HpNikR for nickel ions [165]. This competition is in spite of the large difference between the nickel affinities of the urease accessory proteins and HpNikR, in favor of HpNikR (Table 2), and is perhaps achieved through the nickel sink of urease. Similarly, Hpn and Hpn-like also compete with the Ni(II)-dependent urease maturation pathway under low nickel conditions [218], despite seemingly weaker affinities. One aspect that will clearly impact nickel distribution in conjunction with the thermodynamic affinities is the relative expression levels of all the components, as well as rates of turnover, but this has been examined for only a few nickel metallochaperones [103,127].

Equilibrium titration experiments have also been used to define how selective nickel homeostasis proteins are for nickel ions relative to other biologically relevant transition metals. However, it appears that the binding affinities of many of the proteins examined follow the Irving-Williams series of small molecule chelators for divalent metal ions: Mn(II) < Fe(II) < Co(II) < Ni(II) < Cu(II) > Zn(II) [13]. Although nickel is near the top of this order, if proteins were binding metal from a pool of free metal ions then they would all be copper- or perhaps zinc-bound. It is now clear that this situation does not exist, and that one strategy used by Nature to overcome this problem is to rigorously control metal availability. First of all, despite the presence of Cu(II) at the top of the Irving-Williams series, Cu(I) is the predominant copper species in the reducing cytosol [294]. Thus, even if Cu(II) is capable of binding to a protein and eliciting a response *in vitro*, as in the case of NikR, it would not be sufficiently available *in vivo* [226,242,248]. Furthermore, the transport of copper and zinc, as well as intracellular buffering, is so tightly regulated that there is essentially no metal available for binding to adventitious sites [295–299]. Thus the metal

Table 1 Summary of Ni(II)-binding proteins in *Escherichia coli*.

Protein	Ni(II) per monomer	Apparent Ni(II) K_D (M)	Nickel Coordination Geometry and Ligands	Function	Reference
NikA	1	1×10^{-7} [a]	O-ligands on butane-1,2,4-tricarboxylate; His416	Nickel transporter	[33–35,37]
HypA	1	6×10^{-5}	Square planar or octahedral; His2, possibly backbone N's and N-terminal amine	Enzyme accessory protein	[93,99,100]
HybF	1	1.9×10^{-6}	Square planar (predicted); His2, possibly backbone N's and N-terminal amine	Enzyme accessory protein	[98]
SlyD	≤7	$<1\times10^{-10}$	Mixed coordination geometries; His, Cys ligands	Enzyme accessory protein, Ni(II) storage	[128,129]
HypB	2	LA[b]: 1.2×10^{-5} HA[b]: 1×10^{-13}	Unknown geometry; Cys166, His167, Cys198 Square planar; Cys in CXXCGC motif and N-terminal amine	Enzyme accessory protein	[108–110,116]
[NiFe]-Hydrogenase[c]	1	Unknown	Square pyramidal; 4 Cys residues and a small unknown molecule	Enzyme	[7]
Glyoxalase I	1	$<1\times10^{-8}$	Octahedral; 2 His, 2 Glu, 2 water	Enzyme	[205,206,208]
NikR	2	HA[b]: 9×10^{-13} LA[b]: 3×10^{-8}	Square planar; His87, His89, Cys95, Cys76′ Octahedral; His48, His110	Transcription factor	[233,234,239,240,247
RcnR	1	$<2.5\times10^{-8}$	Octahedral; His3, His64, possibly His60, Cys 35 and N-terminal amine	Transcription factor	[269,270,272]

[a] Value is for aqueous Ni(II) ion binding to NikA. [b] HA: high affinity Ni(II) site; LA: low affinity Ni(II) site. [c] Data for [NiFe]-hydrogenase was determined in other organisms.

Table 2 Summary of Ni(II)-binding proteins in *Helicobacter pylori*.

Protein	Ni(II) per monomer	Apparent Ni(II) K_D (M)	Nickel Coordination Geometry and Ligands	Function	Reference
NixA	Unknown	1.1×10^{-8} [a]	Unknown geometry; HX_4DH motif and additional Asp, Glu, His residues	Nickel transporter	[41,48]
Hpn	5	7.1×10^{-6}	Unknown geometry; most likely His residues	Storage protein	[215,218,224]
Hpn-like	2	3.8×10^{-6}	Unknown geometry; most likely His residues (His29, His31, His32)	Storage protein	[218,223,225]
HspA	2	$1\text{-}2 \times 10^{-6}$	Unknown geometry; Cys and nitrogen ligands; Possibly HX_4DH site	Storage protein	[137,220]
SlyD	2.4	2.4×10^{-6}	Unknown geometry; His residues	Enzyme accessory protein, Ni(II) storage (?)	[119]
UreE	0.5	1.5×10^{-7}	Square planar/ trigonal bipyramidal; His residues	Enzyme accessory protein	[150,155,156,158]
HypA	1	$0.13\text{–}5.8 \times 10^{-5}$	Unknown geometry; His2, possibly N-terminal amines and backbone N's	Enzyme accessory protein	[94,96]
HypB	1	1.5×10^{-7}	Unknown geometry; Cys and His ligands	Enzyme accessory protein	[111]
[NiFe] Hydrogenase [c]	1	Unknown	Square pyramidal; 4 Cys residues and a small unknown molecule	Enzyme	[7]
Urease [c]	2	$<4.0 \times 10^{-19}$	Trigonal bipyramidal; His, Asp and bridging carbamylated Lys	Enzyme	[7,306]
NikR	1	3.5×10^{-12} or $0.1\text{–}1 \times 10^{-7}$ [b]	Square planar; Cys107, His99, His101, His88′ [b]	Transcription factor	[264–266]

[a] Value is the measured K_T for the transporter, not the measured K_D. [b] There is a current debate in the literature regarding the apparent Ni(II) K_D and the coordination environment for HpNikR. [c] Data for [NiFe]-hydrogenase and urease were determined in other organisms.

selectivity of a protein's activity is determined not only by its own intrinsic properties but by the extrinsic influences of the cellular environment.

If non-cognate metals do bind to metal homeostasis proteins, in many cases the discriminating response to the cognate metal can still be achieved through selective allosteric conformational changes. In particular, there are several examples of systems where the correct biological response is achieved only if a specific set of ligands coordinate to the metal. For example, Ni(II), but not Zn(II), is capable of eliciting a response from CnrX because Zn(II) does not bind to the required Met residue (Figure 9) [285]. Similar types of selective effects based on distinct coordination spheres were noted for several transcription factors, including NmtR, RcnR, and NikR [62,239,240,248,271,275]. In this way, nature can utilize the preferred octahedral and square planar coordination geometries of nickel to be selective.

How much intracellular nickel is available and in what form(s) is not yet clear. Evidence is mounting that, as with copper and zinc, there is little, if any, loosely-bound nickel on hand under healthy conditions. The fact that teams of accessory proteins have been identified for the biosynthesis of many of the nickel-containing enzymes is perhaps a testament to the intricate nature of the prosthetic groups. However, at least one accessory protein in each system has been assigned a role in nickel delivery because the enzyme deficiency caused by mutation of that accessory protein is complemented by growing the mutant bacteria in media supplemented with excess nickel [103,105,151,164,168]. Thus it appears that one of the functions of the metallocenter assembly pathway is to ensure that nickel is delivered to the enzyme site under nickel limiting conditions, which may be imposed by the strong chelating capacity of the cytoplasm, and that this pathway can be bypassed by flooding the cells with extra nickel. Furthermore, many of the nickel metalloregulators examined respond to nickel at concentrations that are at or below what is produced by even just a few "free" nickel ions in the cytoplasm, suggesting that the metal is not allowed to accumulate past these levels in a healthy cell. Finally, it is likely that the amount and speciation of available nickel depends on the type of cell system in question. An examination of NmtR in two different bacterial hosts revealed that the protein sensed nickel as predicted from *in vitro* experiments in its native *M. tuberculosis*, but was unable to respond to Ni(II) in cyanobacteria [62]. Subsequent metal analysis revealed substantially higher total nickel in the mycobacterial cells, suggesting that the cyanobacteria did not accumulate sufficient nickel to bind to NmtR, even upon supplementations of the growth media.

Graded responses to nickel appear to be a common mechanism of restricting the levels of available metal in the cytosol. One instance of this graded response is found in *M. tuberculosis* with the stepwise activation of KmtR and the NmtR (discussed in Section 5.3). Another example is found in *E. coli*, where nickel homeostasis is controlled by the pair of regulators NikR (high affinity site $K_D \approx 10$ pM; low affinity site $K_D \approx 30$ nM) and RcnR ($K_D < 25$ nM) [229,230,239,240,272,300]. Thus, a suggested stepwise regulatory response may consist of partial repression of nickel uptake as the high-affinity site on NikR is saturated, followed by full repression if the nickel concentration continues to increase concurrent with Ni(II) binding to RcnR and expression of the RcnA nickel exporter [300]. In this manner, the overlap of these nickel affinities appears to achieve seamless control of nickel uptake and efflux. A related form of graded response has also been proposed for HpNikR given the distinct DNA affinities observed for subsets of target promoters *in vitro* (Section 5.1.2) [256], although how this relates to the temporal control of the various genes is not yet clear [253].

The issue of cytosolic metal availability is further complicated by the evidence for segregated intracellular metal pools. For example, the nickel imported via the NikABCDE transporter in *E. coli* appears to be destined for hydrogenase maturation and is not sensed directly by NikR [32]. This separation may be explained by the recently reported observation that *E. coli* HypA is localized near the cell membrane [91], where the hydrogenase maturation process is believed to occur, which leads to the proposal that the metal exiting the NikABCDE transporter is shuttled directly to the nascent hydrogenase precursor proteins, via protein-protein interactions that include HypA. Communication between hydrogenase maturation and nickel uptake may be through proteins such as SlyD, which contributes to hydrogenase biosynthesis and modulates NikR activity, but does not influence other aspects of nickel homeostasis in *E. coli* such as RcnR activity or sensitivity of the bacteria to nickel toxicity [128]. These types of studies support the compartmentalization of nickel in the cytoplasm of *E. coli* despite the absence of organelles. Spatial separation of nickel systems could also explain how the nickel accessory proteins with weaker nickel binding activity can function in the face of much stronger binding by a metalloregulator such as NikR [165].

An additional layer of control of nickel homeostasis that also has to be considered is that of kinetics. Metal ions trapped in tight binding sites are unlikely to be replaced under a physiological time frame and can be defined as kinetically stable, whereas some metal sites are more kinetically labile [295,301]. Only a few examples of kinetic analysis on various nickel metallochaperones such as UreE, Hpn and HypB/SlyD are present in the literature [127,224,302]. In these cases, the rates measured *in vitro* with the isolated factors appear to be very slow, especially if one considers the generation time of the bacteria involved. Some aspect of the kinetic behavior of these enzymes is thus eluding researchers and it is too early to tell what role kinetics plays on nickel homeostasis.

Another common theme in nickel homeostasis is the use of poly-His regions in nickel storage, as found in some homologs of nickel proteins with a variety of functions. The use of His is consistent with elementary nickel chemistry (Section 1.1). By placing several His ligands together in a protein, nature can take advantage of the polychelate effect to produce stronger Ni(II) binding than that of the free histidine present in the cytosol. The distribution of these poly-His regions is not readily explained, as they are not conserved among homologs, so the presence of the extra nickel sequestering ability must reflect on the nickel homeostasis requirements of individual organisms.

A final common theme pertains to metallocenter biosynthesis, where several general mechanisms work synergistically to allow for nickel incorporation along with the rest of the cofactor [303]. The use of metallochaperones ensures that metal delivery is tightly controlled and efficient. Post-translational modifications of the enzyme after or during metal insertion may serve as an important check-point to ensure proper metal loading [81,90]. In addition, the use of nucleotide triphosphate hydrolysis for the metallocenter assembly is common [303]. The exact role hydrolysis plays is not yet known, but could include a switch function that contributes to the insertion of a positively-charged metal ion into the hydrophobic interior of an enzyme, or a more regulatory responsibility.

7 Concluding Remarks

It is now clear that nickel enzymes are critical for various biological processes including the global carbon cycle [166,194] and bacterial pathogenesis [4,77,130,131]. With the discovery of additional nickel enzymes, such as methylene-urease from the *Burkholderia* species [203], the *B. subtilis* glycerol-1-phosphate dehydrogenase AraM [304], and the monocupin dioxygenase quercetinase from *Streptomyces* sp. FLA [305], it is clear that the breadth of nickel-biological chemistry must be more expansive.

In this chapter we have reviewed much of our current knowledge pertaining to cellular nickel systems. From this, several common themes can be defined that impact the mechanisms of nickel homeostasis and ultimately shape the nickel metallome. It should be noted that these principles are not limited to nickel homeostasis, but are generally applicable to all metal homeostasis pathways. Despite our current knowledge, many questions such as those discussed above still remain, and we are only beginning to understand the guiding principles of nickel homeostasis.

Abbreviations and Definitions

ABC	ATP-binding cassette
ADP	adenosine 5'-diphosphate
ARD	acireductone dioxygenase
ATP	adenosine 5'-triphosphate
BTC	butane-1,2,4-tricarboxylate
CDF	cation diffusion facilitator
CODH	carbon monoxide dehydrogenase
DBD	DNA-binding domain
DFT	density functional theory
Fur	ferric uptake regulator
Glx	glyoxalase
GTP	guanosine 5'-triphosphate
ITC	isothermal titration calorimetry
LS	large subunit
MBD	metal-binding domain

MFS	major facilitator protein superfamily
NiCoT	nickel/cobalt transporter
NikR	nickel-responsive transcription factor
PBP	periplasmic-binding protein
PDB	Protein Data Bank
P_i	inorganic phosphate
PPIase	peptidyl-prolyl isomerase
RND	resistance-nodulation-cell division
SOD	superoxide dismutase
SS	small subunit
TBDT	TonB-dependent transporter
XAS	X-ray absorption spectroscopy
TMD	transmembrane-domain

Acknowledgments The authors would like to thank the members of the Zamble laboratory for their helpful discussions and Harini Kaluarachchi, Sandra Krecisz, Colin Douglas, and Thanh Ngu for editing. The authors would also like to acknowledge the Canadian Institute for Health Research and the Canadian Research Chairs program for funding.

References

1. J. Szpunar, *Anal. Bioanal. Chem.* **2004**, *378*, 54–56.
2. J. Szpunar, *The Analyst* **2005**, *130*, 442–465.
3. R. J. Maier, S. L. Benoit, S. Seshadri, *Biometals* **2007**, *20*, 655–664.
4. S. W. Ragsdale, *J. Biol. Chem.* **2009**, *284*, 18571–18575.
5. R. Hausinger, D. Zamble, in *Molecular Microbiology of Heavy Metals*, Eds D. Nies, S. Silver, Springer, Berlin, Heidelberg, 2007, Vol. 6, pp. 287–320.
6. H. Kaluarachchi, K. C. C. Chung, D. B. Zamble, *Nat. Prod. Rep.* **2010**, *27*, 681–694.
7. Y. Li, D. B. Zamble, *Chem. Rev.* **2009**, *109*, 4617–4643.
8. H. Küpper, P. M. H. Kroneck, in *Nickel and Its Surprising Impact in Nature*, Vol. 2 of *Metal Ions in Life Sciences*, Eds A. Sigel, H. Sigel, R. K. O. Sigel, John Wiley & Sons Ltd., Chichester, UK, 2007, pp. 31–62.
9. J. Fontecilla-Camps, *Struct. Bonding* **1998**, *91*, 1–30.
10. F. A. Cotton, G. Wilkinson, *Advanced Inorganic Chemistry*, John Wiley & Sons, New York, NY, 1988.
11. L. Rulísek, J. Vondrásek, *J. Inorg. Biochem.* **1998**, *71*, 115–127.
12. J. J. R. Frausto da Silva, R. J. P. Williams, *The Biological Chemistry of the Elements*, Oxford University Press, Oxford, 2001.
13. H. Irving, R. J. P. Williams, *Nature* **1948**, *162*, 746–747.
14. K. O. Konhauser, E. Pecoits, S. V. Lalonde, D. Papineau, E. G. Nisbet, M. E. Barley, N. T. Arndt, K. Zahnle, B. S. Kamber, *Nature* **2009**, *458*, 750–753.
15. M. A. Saito, *Nature* **2009**, *458*, 714–715.
16. L. Macomber, R. P. Hausinger, *Metallomics* **2011**, *3*, 1153–1162.
17. L. Macomber, S. P. Elsey, R. P. Hausinger, *Mol. Microbiol.* **2011**, *82*, 1291–1300.
18. H. Chen, N. C. Giri, R. Zhang, K. Yamane, Y. Zhang, M. Maroney, M. Costa, *J. Biol. Chem.* **2010**, *285*, 7374–7383.
19. L. Li, A. F. Tutone, R. S. M. Drummond, R. C. Gardner, S. Luan, *The Plant Cell* **2001**, *13*, 2761–2775.

20. V. N. Tripathi, S. Srivastav, *J. Biosci.* **2006**, *31*, 61–66.
21. D. Niegowski, S. Eshaghi, *Cell. Mol. Life Sci.* **2007**, *64*, 2564–2574.
22. M. B. C. Moncrief, M. E. Maguire, *J. Biol. Inorg. Chem.* **1999**, *4*, 523–527.
23. R. L. Smith, M. E. Maguire, *Mol. Microbiol.* **1998**, *28*, 217–226.
24. T. M. Nieminen, L. Ukonmaanaho, N. Rausch, W. Shotyk, in *Nickel and Its Surprising Impact in Nature*, Vol. 2 of *Metal Ions in Life Sciences*, Eds A. Sigel, H. Sigel, R. K. O. Sigel, John Wiley & Sons Ltd., Chichester, UK, 2007, pp. 1–30.
25. F. W. Sunderman, *Scand. J. Work Environ. Health* **1993**, *19 (Suppl. 1)*, 34–38.
26. D. A. Rodionov, P. Hebbeln, M. S. Gelfand, T. Eitinger, *J. Bacteriol.* **2006**, *188*, 317–327.
27. Y. Zhang, V. N. Gladyshev, *Chem. Rev.* **2009**, *109*, 4828–4861.
28. C. Navarro, L. F. Wu, M. A. Mandrand-Berthelot, *Mol. Microbiol.* **1993**, *9*, 1181–1191.
29. L. F. Wu, M. A. Mandrand-Berthelot, *Biochimie* **1986**, *68*, 167–179.
30. L. F. Wu, M. A. Mandrand-Berthelot, R. Waugh, C. J. Edmonds, S. E. Holt, D. H. Boxer, *Mol. Microbiol.* **1989**, *3*, 1709–1718.
31. L. F. Wu, C. Navarro, M. A. Mandrand-Berthelot, *Gene* **1991**, *107*, 37–42.
32. J. L. Rowe, G. L. Starnes, P. T. Chivers, *J. Bacteriol.* **2005**, *187*, 6317–6323.
33. K. De Pina, C. Navarro, L. McWalter, D. H. Boxer, N. C. Price, S. M. Kelly, M. A. Mandrand-Berthelot, L. F. Wu, *Eur. J. Biochem.* **1995**, *227*, 857–865.
34. J. Heddle, D. J. Scott, S. Unzai, S. Y. Park, J. R. H. Tame, *J. Biol. Chem.* **2003**, *278*, 50322–50329.
35. M. V. Cherrier, C. Cavazza, C. Bochot, D. Lemaire, J. C. Fontecilla-Camps, *Biochemistry* **2008**, *47*, 9937–9943.
36. M. V. Cherrier, L. Martin, C. Cavazza, L. Jacquamet, D. Lemaire, J. Gaillard, J. C. Fontecilla-Camps, *J. Am. Chem. Soc.* **2005**, *127*, 10075–10082.
37. C. Cavazza, L. Martin, E. Laffly, H. Lebrette, M. V. Cherrier, L. Zeppieri, P. Richaud, M. Carrière, J. C. Fontecilla-Camps, *FEBS Lett.* **2011**, *585*, 711–715.
38. R. C. Hider, X. Kong, *Nat. Prod. Rep.* **2010**, *27*, 637–657.
39. R. Tam, J. M. H. Saier, *Microbiol. Rev.* **1993**, *57*, 320–346.
40. Y. Zhang, D. A. Rodionov, M. S. Gelfand, V. N. Gladyshev, *BMC Genomics* **2009**, *10*, 78–104.
41. J. F. Fulkerson, R. M. Garner, H. L. T. Mobley, *J. Biol. Chem.* **1998**, *273*, 235–241.
42. T. Eitinger, J. Suhr, L. Moore, J. A. C. Smith, *Biometals* **2005**, *18*, 399–405.
43. G. Eberz, T. Eitinger, B. Friedrich, *J. Bacteriol.* **1989**, *171*, 1340–1345.
44. T. Eitinger, B. Friedrich, *J. Biol. Chem.* **1991**, *266*, 3222–3227.
45. P. Hebbeln, T. Eitinger, *FEMS Microbiol. Lett.* **2004**, *230*, 129–135.
46. O. Degen, M. Kobayashi, S. Shimizu, T. Eitinger, *Arch. Microbiol.* **1999**, *171*, 139–145.
47. T. Eitinger, L. Wolfram, O. Degen, C. Anthon, *J. Biol. Chem.* **1997**, *272*, 17139–17144.
48. H. L. T. Mobley, R. M. Garner, P. Bauerfeind, *Mol. Microbiol.* **1995**, *16*, 97–109.
49. M. Maeda, M. Hidaka, A. Nakamura, H. Masaki, T. Uozumi, *J. Bacteriol.* **1994**, *176*, 432–442.
50. F. Sebbane, M. A. Mandrand-Berthelot, M. Simonet, *J. Bacteriol.* **2002**, *184*, 5706–5713.
51. B. Brito, R. I. Prieto, E. Cabrera, M. A. Mandrand-Berthelot, J. Imperial, T. Ruiz-Argueso, J. M. Palacios, *J. Bacteriol.* **2010**, *192*, 925–935.
52. O. Degen, T. Eitinger, *J. Bacteriol.* **2002**, *184*, 3569–3577.
53. Z. Ma, F. E. Jacobsen, D. P. Giedroc, *Chem. Rev.* **2009**, *109*, 4644–4681.
54. N. Noinaj, M. Guillier, Barnard, Travis J., S. K. Buchanan, *Annu. Rev. Microbiol.* **2010**, *64*, 43–60.
55. M. C. Wiener, *Curr. Opin. Struct. Biol.* **2005**, *15*, 394–400.
56. J. Stoof, E. J. Kuipers, G. Klaver, A. H. M. van Vliet, *Infect. Immun.* **2010**, *78*, 4261–4267.
57. J. Stoof, E. Kuipers, A. van Vliet, *Biometals* **2010**, *23*, 145–159.
58. G. S. Davis, E. L. Flannery, H. L. T. Mobley, *Infect. Immun.* **2006**, *74*, 6811–6820.
59. F. D. Ernst, J. Stoof, W. M. Horrevoets, E. J. Kuipers, J. G. Kusters, A. H. M. van Vliet, *Infect. Immun.* **2006**, *74*, 6821–6828.

60. K. Schauer, B. Gouget, M. Carrière, A. Labigne, H. De Reuse, *Mol. Microbiol.* **2007**, *63*, 1054–1068.
61. D. H. Nies, *FEMS Microbiol. Rev.* **2003**, *27*, 313–339.
62. J. S. Cavet, W. Meng, M. A. Pennella, R. J. Appelhoff, D. P. Giedroc, N. J. Robinson, *J. Biol. Chem.* **2002**, *277*, 38441–38448.
63. A. Rodrigue, G. Effantin, M. A. Mandrand-Berthelot, *J. Bacteriol.* **2005**, *187*, 2912–2916.
64. D. Koch, D. Nies, G. Grass, *Biometals* **2007**, *20*, 759–771.
65. C. Bleriot, G. Effantin, F. Lagarde, M. A. Mandrand-Berthelot, A. Rodrigue, *J. Bacteriol.* **2011**, *193*, 3785–3793.
66. G. Grass, B. Fan, B. P. Rosen, K. Lemke, H. G. Schlegel, C. Rensing, *J. Bacteriol.* **2001**, *183*, 2803–2807.
67. M. Garcia-Dominguez, L. Lopez-Maury, F. J. Florencio, J. C. Reyes, *J. Bacteriol.* **2000**, *182*, 1507–1514.
68. J. E. Park, H. G. Schlegel, H. G. Rhie, H. S. Lee, *Int. Microbiol.* **2004**, *7*, 27–34.
69. J. S. Park, S. J. Lee, H. G. Rhie, H. S. Lee, *J. Microbiol. Biotechnol.* **2008**, *18*, 1040–1043.
70. T. Zhu, J. Tian, S. Zhang, N. Wu, Y. Fan, *PLoS One* **2011**, *6*, e17367.
71. F. N. Stahler, S. Odenbreit, R. Haas, J. Wilrich, A. H. M. V. Vliet, J. G. Kusters, M. Kist, S. Bereswill, *Infect. Immun.* **2006**, *74*, 3845–3852.
72. T. Schmidt, H. G. Schlegel, *J. Bacteriol.* **1994**, *176*, 7045–7054.
73. Q. Dong, M. Mergeay, *Mol. Microbiol.* **1994**, *14*, 185–187.
74. H. Liesegang, K. Lemke, R. A. Siddiqui, H. G. Schlegel, *J. Bacteriol.* **1993**, *175*, 767–778.
75. C. Chaintreuil, F. Rigault, L. Moulin, T. Jaffre, J. Fardoux, E. Giraud, B. Dreyfus, X. Bailly, *Appl. Environ. Microbiol.* **2007**, *73*, 8018–8022.
76. R. J. Maier, *Biochem. Soc. Trans.* **2005**, *33*, 83–85.
77. J. W. Olson, R. J. Maier, *Science* **2002**, *298*, 1788–1790.
78. P. M. Vignais, B. Billoud, J. Meyer, *FEMS Microbiol. Rev.* **2001**, *25*, 455–501.
79. P. M. Vignais, B. Billoud, *Chem. Rev.* **2007**, *107*, 4206–4272.
80. A. Volbeda, M. H. Charon, C. Piras, E. C. Hatchikian, M. Frey, J. C. Fontecilla-Camps, *Nature* **1995**, *373*, 580–587.
81. A. Böck, P. W. King, M. Blokesch, M. C. Posewitz, in *Adv. Microb. Physiol.*, Ed K. P. Robert, Academic Press, 2006, *51*, pp. 1–72.
82. S. Shima, O. Pilak, S. Vogt, M. Schick, M. S. Stagni, W. Meyer-Klaucke, E. Warkentin, R. K. Thauer, U. Ermler, *Science* **2008**, *321*, 572–575.
83. A. Volbeda, J. C. Fontecilla-Camps, *Coord. Chem. Rev.* **2005**, *249*, 1609–1619.
84. A. Jacobi, R. Rossmann, A. Böck, *Arch. Microbiol.* **1992**, *158*, 444–451.
85. M. R. Leach, D. B. Zamble, *Curr. Opin. Chem. Biol.* **2007**, *11*, 159–165.
86. J. C. Fontecilla-Camps, A. Volbeda, C. Cavazza, Y. Nicolet, *Chem. Rev.* **2007**, *107*, 4273–4303.
87. L. Forzi, R. Sawers, *Biometals* **2007**, *20*, 565–578.
88. M. Blokesch, A. Böck, *J. Mol. Biol.* **2002**, *324*, 287–296.
89. N. Drapal, A. Böck, *Biochemistry* **1998**, *37*, 2941–2948.
90. E. Theodoratou, R. Huber, A. Böck, *Biochem. Soc. Trans.* **2005**, *33*, 108–111.
91. K. C. Chan Chung, D. B. Zamble, *J. Biol. Chem.* **2011**, *286*, 43081–43090.
92. K. C. Chan Chung, D. B. Zamble, *FEBS Lett.* **2011**, *585*, 291–294.
93. A. Atanassova, D. B. Zamble, *J. Bacteriol.* **2005**, *187*, 4689–4697.
94. N. Mehta, J. W. Olson, R. J. Maier, *J. Bacteriol.* **2003**, *185*, 726–734.
95. S. Watanabe, T. Arai, R. Matsumi, H. Atomi, T. Imanaka, K. Miki, *J. Mol. Biol.* **2009**, *394*, 448–459.
96. W. Xia, H. Li, K. H. Sze, H. Sun, *J. Am. Chem. Soc.* **2009**, *131*, 10031–10040.
97. M. Hube, M. Blokesch, A. Böck, *J. Bacteriol.* **2002**, *184*, 3879–3885.
98. M. Blokesch, M. Rohrmoser, S. Rode, A. Böck, *J. Bacteriol.* **2004**, *186*, 2603–2611.
99. D. C. Kennedy, R. W. Herbst, J. S. Iwig, P. T. Chivers, M. J. Maroney, *J. Am. Chem. Soc.* **2007**, *129*, 16–17.

100. R. W. Herbst, I. Perovic, V. Martin-Diaconescu, K. O'Brien, P. T. Chivers, S. S. Pochapsky, T. C. Pochapsky, M. J. Maroney, *J. Am. Chem. Soc.* **2010**, *132*, 10338–10351.
101. C. Fu, R. J. Maier, *Biochim. Biophys. Acta* **1994**, *1184*, 135–138.
102. W. Maret, Y. Li, *Chem. Rev.* **2009**, *109*, 4682–4707.
103. T. Maier, A. Jacobi, M. Sauter, A. Böck, *J. Bacteriol.* **1993**, *175*, 630–635.
104. T. Maier, F. Lottspeich, A. Böck, *Eur. J. Biochem.* **1995**, *230*, 133–138.
105. J. W. Olson, N. S. Mehta, R. J. Maier, *Mol. Microbiol.* **2001**, *39*, 176–182.
106. L. Rey, J. Imperial, J. M. Palacios, T. Ruiz-Argueso, *J. Bacteriol.* **1994**, *176*, 6066–6073.
107. R. Waugh, D. H. Boxer, *Biochimie* **1986**, *68*, 157–166.
108. R. Gasper, A. Scrima, A. Wittinghofer, *J. Biol. Chem.* **2006**, *281*, 27492–27502.
109. A. V. Dias, C. M. Mulvihill, M. R. Leach, I. J. Pickering, G. N. George, D. B. Zamble, *Biochemistry* **2008**, *47*, 11981–11991.
110. M. R. Leach, S. Sandal, H. Sun, D. B. Zamble, *Biochemistry* **2005**, *44*, 12229–12238.
111. A. M. Sydor, J. Liu, D. B. Zamble, *J. Bacteriol.* **2011**, *193*, 1359–1368.
112. F. Cai, T. Ngu, H. Kaluarachchi, D. Zamble, *J. Biol. Inorg. Chem.* **2011**, *16*, 857–868.
113. C. Fu, J. W. Olson, R. J. Maier, *Proc. Natl. Acad. Sci. USA* **1995**, *92*, 2333–2337.
114. J. W. Olson, R. J. Maier, *J. Bacteriol.* **2000**, *182*, 1702–1705.
115. J. W. Olson, C. Fu, R. J. Maier, *Mol. Microbiol.* **1997**, *24*, 119–128.
116. K. C. Chan Chung, L. Cao, A. V. Dias, I. J. Pickering, G. N. George, D. B. Zamble, *J. Am. Chem. Soc.* **2008**, *130*, 14056–14057.
117. J. W. Zhang, G. Butland, J. F. Greenblatt, A. Emili, D. B. Zamble, *J. Biol. Chem.* **2005**, *280*, 4360–4366.
118. K. Stingl, K. Schauer, C. Ecobichon, A. Labigne, P. Lenormand, J. C. Rousselle, A. Namane, H. de Reuse, *Mol. Cell. Proteomics* **2008**, *7*, 2429–2441.
119. T. Cheng, H. Li, W. Xia, H. Sun, *J. Biol. Inorg. Chem.* **2011**, *17*, 331–343.
120. C. Löw, P. Neumann, H. Tidow, U. Weininger, C. Haupt, B. Friedrich-Epler, C. Scholz, M. T. Stubbs, J. Balbach, *J. Mol. Biol.* **2010**, *398*, 375–390.
121. L. Martino, Y. He, K. L. D. Hands-Taylor, E. R. Valentine, G. Kelly, C. Giancola, M. R. Conte, *FEBS J.* **2009**, *276*, 4529–4544.
122. C. Wülfing, J. Lombardero, A. Plückthun, *J. Biol. Chem.* **1994**, *269*, 2895–2901.
123. W. D. Roof, S. M. Horne, K. D. Young, R. Young, *J. Biol. Chem.* **1994**, *269*, 2902–2910.
124. S. Hottenrott, T. Schumann, A. Plückthun, G. Fischer, J. U. Rahfeld, *J. Biol. Chem.* **1997**, *272*, 15697–15701.
125. M. R. Leach, J. W. Zhang, D. B. Zamble, *J. Biol. Chem.* **2007**, *282*, 16177–16186.
126. J. W. Zhang, M. R. Leach, D. B. Zamble, *J. Bacteriol.* **2007**, *189*, 7942–7944.
127. H. Kaluarachchi, J. W. Zhang, D. B. Zamble, *Biochemistry* **2011**, *50*, 10761–10763.
128. H. Kaluarachchi, J. F. Siebel, S. Kaluarachchi-Duffy, S. Krecisz, D. E. K. Sutherland, M. J. Stillman, D. B. Zamble, *Biochemistry* **2011**, *50*, 10666–10677.
129. H. Kaluarachchi, D. E. K. Sutherland, A. Young, I. J. Pickering, M. J. Stillman, D. B. Zamble, *J. Am. Chem. Soc.* **2009**, *131*, 18489–18500.
130. B. Zambelli, F. Musiani, S. Benini, S. Ciurli, *Acc. Chem. Res.* **2011**, *44*, 520–530.
131. K. A. Eaton, C. L. Brooks, D. R. Morgan, S. Krakowka, *Infect. Immun.* **1991**, *59*, 2470–2475.
132. E. L. Carter, N. Flugga, J. L. Boer, S. B. Mulrooney, R. P. Hausinger, *Metallomics* **2009**, *1*, 207–221.
133. E. Jabri, M. Carr, R. Hausinger, P. Karplus, *Science* **1995**, *268*, 998–1004.
134. E. Jabri, P. A. Karplus, *Biochemistry* **1996**, *35*, 10616–10626.
135. E. L. Carter, D. E. Tronrud, S. R. Taber, P. A. Karplus, R. P. Hausinger, *Proc. Natl. Acad. Sci. USA* **2011**, *108*, 13095–13099.
136. S. Suerbaum, J. M. Thiberge, I. Kansau, R. L. Ferrero, A. Labigne, *Mol. Microbiol.* **1994**, *14*, 959–974.
137. I. Kansau, F. Guillain, J. M. Thiberge, A. Labigne, *Mol. Microbiol.* **1996**, *22*, 1013–1023.
138. I. Park, R. Hausinger, *Science* **1995**, *267*, 1156–1158.
139. I. S. Park, R. P. Hausinger, *J. Bacteriol.* **1995**, *177*, 1947–1951.

140. A. Soriano, G. J. Colpas, R. P. Hausinger, *Biochemistry* **2000**, *39*, 12435–12440.
141. J. U. Park, J. Y. Song, Y. C. Kwon, M. J. Chung, J. S. Jun, J. W. Park, S. G. Park, H. R. Hwang, S. H. Choi, S. C. Baik, H. L. Kang, H. S. Youn, W. K. Lee, M. J. Cho, K. H. Rhee, *Mol. Cells* **2005**, *20*, 371–377.
142. A. Soriano, R. P. Hausinger, *Proc. Natl. Acad. Sci.* **1999**, *96*, 11140–11144.
143. M. B. Moncrief, R. P. Hausinger, *J. Bacteriol.* **1997**, *179*, 4081–4086.
144. I. Park, M. B. Carr, R. P. Hausinger, *Proc. Natl. Acad. Sci.* **1994**, *91*, 3233–3237.
145. I. S. Park, R. P. Hausinger, *Biochemistry* **1996**, *35*, 5345–5352.
146. Z. Chang, J. Kuchar, R. P. Hausinger, *J. Biol. Chem.* **2004**, *279*, 15305–15313.
147. B. Zambelli, M. Stola, F. Musiani, K. De Vriendt, B. Samyn, B. Devreese, J. Van Beeumen, P. Turano, A. Dikiy, D. A. Bryant, S. Ciurli, *J. Biol. Chem.* **2005**, *280*, 4684–4695.
148. B. Zambelli, P. Turano, F. Musiani, P. Neyroz, S. Ciurli, *Proteins: Struct., Funct., Bioinf.* **2009**, *74*, 222–239.
149. B. Zambelli, F. Musiani, M. Savini, P. Tucker, S. Ciurli, *Biochemistry* **2007**, *46*, 3171–3182.
150. M. Bellucci, B. Zambelli, F. Musiani, P. Turano, S. Ciurli, *Biochem. J.* **2009**, *422*, 91–100.
151. T. G. Brayman, R. P. Hausinger, *J. Bacteriol.* **1996**, *178*, 5410–5416.
152. S. L. Benoit, N. Mehta, M. V. Weinberg, C. Maier, R. J. Maier, *Microbiology* **2007**, *153*, 1474–1482.
153. S. L. Benoit, A. L. Zbell, R. J. Maier, *Microbiology* **2007**, *153*, 3748–3756.
154. H. Remaut, N. Safarov, S. Ciurli, J. Van Beeumen, *J. Biol. Chem.* **2001**, *276*, 49365–49370.
155. K. Banaszak, V. Martin-Diaconescu, B. Z. Matteo Bellucci, W. Rypniewski, M. J. Maroney, S. Ciurli, *Biochem. J.* **2012**, *441*, 1017–1026.
156. R. Shi, C. Munger, A. Asinas, S. L. Benoit, E. Miller, A. Matte, R. J. Maier, M. Cygler, *Biochemistry* **2010**, *49*, 7080–7088.
157. H. K. Song, S. B. Mulrooney, R. Huber, R. P. Hausinger, *J. Biol. Chem.* **2001**, *276*, 49359–49364.
158. S. Benoit, R. J. Maier, *J. Bacteriol.* **2003**, *185*, 4787–4795.
159. M. Stola, F. Musiani, S. Mangani, P. Turano, N. Safarov, B. Zambelli, S. Ciurli, *Biochemistry* **2006**, *45*, 6495–6509.
160. G. J. Colpas, T. G. Brayman, J. McCracken, M. A. Pressler, G. T. Babcock, L. J. Ming, C. M. Colangelo, R. A. Scott, R. P. Hausinger, *J. Biol. Inorg. Chem.* **1998**, *3*, 150–160.
161. M. H. Lee, H. S. Pankratz, S. Wang, R. A. Scott, M. G. Finnegan, M. K. Johnson, J. A. Ippolito, D. W. Christianson, R. P. Hausinger, *Protein Sci.* **1993**, *2*, 1042–1052.
162. N. E. Grossoehme, S. B. Mulrooney, R. P. Hausinger, D. E. Wilcox, *Biochemistry* **2007**, *46*, 10506–10516.
163. H. S. Won, Y. H. Lee, J. H. Kim, I. S. Shin, M. H. Lee, B. J. Lee, *J. Biol. Chem.* **2004**, *279*, 17466–17472.
164. N. Mehta, S. Benoit, R. J. Maier, *Microb. Pathog.* **2003**, *35*, 229–234.
165. E. L. Benanti, P. T. Chivers, *J. Bacteriol.* **2009**, *191*, 2405–2408.
166. S. W. Ragsdale, *J. Inorg. Biochem.* **2007**, *101*, 1657–1666.
167. C. L. Drennan, J. Heo, M. D. Sintchak, E. Schreiter, P. W. Ludden, *Proc. Natl. Acad. Sci. USA* **2001**, *98*, 11973–11978.
168. R. L. Kerby, P. W. Ludden, G. P. Roberts, *J. Bacteriol.* **1997**, *179*, 2259–2266.
169. R. K. Watt, P. W. Ludden, *J. Biol. Chem.* **1998**, *273*, 10019–10025.
170. W. B. Jeon, J. Cheng, P. W. Ludden, *J. Biol. Chem.* **2001**, *276*, 38602–38609.
171. J. H. Jeoung, T. Giese, M. Grünwald, H. Dobbek, *Biochemistry* **2009**, *48*, 11505–11513.
172. J. H. Jeoung, T. Giese, M. Grünwald, H. Dobbek, *J. Mol. Biol.* **2009**, *396*, 1165–1179.
173. P. A. Lindahl, *Biochemistry* **2002**, *41*, 2097–2105.
174. S. W. Ragsdale, E. Pierce, *Biochim. Biophys. Acta* **2008**, *1784*, 1873–1898.
175. T. I. Doukov, T. M. Iverson, J. Seravalli, S. W. Ragsdale, C. L. Drennan, *Science* **2002**, *298*, 567–572.
176. C. Darnault, A. Volbeda, E. J. Kim, P. Legrand, X. Vernede, P. A. Lindahl, J. C. Fontecilla-Camps, *Nat. Struct. Mol. Biol.* **2003**, *10*, 271–279.

177. H. K. Loke, P. A. Lindahl, *J. Inorg. Biochem.* **2003**, *93*, 33–40.
178. D. J. Evans, *Coord. Chem. Rev.* **2005**, *249*, 1582–1595.
179. J. J. P. Perry, D. S. Shin, E. D. Getzoff, J. A. Tainer, *Biochim. Biophys. Acta* **2010**, *1804*, 245–262.
180. H. D. Youn, E. J. Kim, J. H. Roe, Y. C. Hah, S. O. Kang, *Biochem. J.* **1996**, *318*, 889–896.
181. H. D. Youn, H. Youn, J. W. Lee, Y. I. Yim, J. K. Lee, Y. C. Hah, S. O. Kang, *Arch. Biochem. Biophys.* **1996**, *334*, 341–348.
182. C. L. Dupont, K. Neupane, J. Shearer, B. Palenik, *Environ. Microbiol.* **2008**, *10*, 1831–1843.
183. V. Leclere, P. Boiron, R. Blondeau, *Curr. Microbiol.* **1999**, *39*, 365–368.
184. E. J. Kim, H. J. Chung, B. Suh, Y. C. Hah, J. H. Roe, *Mol. Microbiol.* **1998**, *27*, 187–195.
185. J. Wuerges, J. W. Lee, Y. I. Yim, H. S. Yim, S. O. Kang, K. D. Carugo, *Proc. Natl. Acad. Sci. USA* **2004**, *101*, 8569–8574.
186. D. P. Barondeau, C. J. Kassmann, C. K. Bruns, J. A. Tainer, E. D. Getzoff, *Biochemistry* **2004**, *43*, 8038–8047.
187. P. A. Bryngelson, S. E. Arobo, J. L. Pinkham, D. E. Cabelli, M. J. Maroney, *J. Am. Chem. Soc.* **2003**, *126*, 460–461.
188. J. Hanss, H.-J. Krüger, *Angew. Chem.* **1998**, *37*, 360–363.
189. R. W. Herbst, A. Guce, P. A. Bryngelson, K. A. Higgins, K. C. Ryan, D. E. Cabelli, S. C. Garman, M. J. Maroney, *Biochemistry* **2009**, *48*, 3354–3369.
190. O. Johnson, K. Ryan, M. Maroney, T. Brunold, *J. Biol. Inorg. Chem.* **2010**, *15*, 777–793.
191. K. Ryan, O. Johnson, D. Cabelli, T. Brunold, M. Maroney, *J. Biol. Inorg. Chem.* **2010**, *15*, 795–807.
192. T. Eitinger, *J. Bacteriol.* **2004**, *186*, 7821–7825.
193. I. K. Kim, Y. I. Yim, Y. M. Kim, J. W. Lee, H. S. Yim, S. O. Kang, *FEMS Microbiol. Lett.* **2003**, *228*, 21–26.
194. B. Jaun, R. K. Thauer, in *Nickel and Its Surprising Impact in Nature*, Vol. 2 of *Metal Ions in Life Sciences*, Eds A. Sigel, H. Sigel, R. K. O. Sigel, John Wiley & Sons Ltd., Chichester, UK, 2007, pp. 323–356.
195. S. Scheller, M. Goenrich, R. Boecher, R. K. Thauer, B. Jaun, *Nature* **2010**, *465*, 606–608.
196. P. E. Cedervall, M. Dey, X. Li, R. Sarangi, B. Hedman, S. W. Ragsdale, C. M. Wilmot, *J. Am. Chem. Soc.* **2011**, *133*, 5626–5628.
197. H. C. Friedmann, A. Klein, R. K. Thauer, *FEMS Microbiol. Lett.* **1990**, *87*, 339–348.
198. S. Shima, M. Krueger, T. Weinert, U. Demmer, J. Kahnt, R. K. Thauer, U. Ermler, *Nature* **2011**, *481*, 98–101.
199. M. D. Liptak, K. M. V. Heuvelen, T. C. Brunold, in *Metal-Carbon Bonds in Enzymes and Cofactors*, Vol. 6 of *Metal Ions in Life Sciences*, Eds A. Sigel, H. Sigel, R. K. O. Sigel, Royal Society of Chemistry, Cambridge, 2009, pp. 417–460.
200. H. L. Schubert, E. Raux, M. A. A. Matthews, J. D. Phillips, K. S. Wilson, C. P. Hill, M. J. Warren, *Biochem. Soc. Trans.* **2002**, *30*, 595–600.
201. T. Selmer, J. Kahnt, M. Goubeaud, S. Shima, W. Grabarse, U. Ermler, R. K. Thauer, *J. Biol. Chem.* **2000**, *275*, 3755–3760.
202. U. Suttisansanee, J. F. Honek, *Semin. Cell Dev. Biol.* **2011**, *22*, 285–292.
203. S. B. Mulrooney, R. P. Hausinger, *FEMS Microbiol. Rev.* **2003**, *27*, 239–261.
204. S. L. Clugston, J. F. J. Barnard, R. Kinach, D. Miedema, R. Ruman, E. Daub, J. F. Honek, *Biochemistry* **1998**, *37*, 8754–8763.
205. G. Davidson, S. L. Clugston, J. F. Honek, M. J. Maroney, *Inorg. Chem.* **2000**, *39*, 2962–2963.
206. M. M. He, S. L. Clugston, J. F. Honek, B. W. Matthews, *Biochemistry* **2000**, *39*, 8719–8727.
207. U. Suttisansanee, K. Lau, S. Lagishetty, K. N. Rao, S. Swaminathan, J. M. Sauder, S. K. Burley, J. F. Honek, *J. Biol. Chem.* **2011**, *286*, 38367–38374.
208. S. L. Clugston, R. Yajima, J. F. Honek, *Biochem. J.* **2004**, *377*, 309–316.

209. T. C. Pochapsky, T. Ju, M. Dang, R. Beaulieu, G. M. Pagani, B. OuYang, in *Nickel and Its Surprising Impact in Nature*, Vol. 2 of *Metal Ions in Life Sciences*, Eds A. Sigel, H. Sigel, R. K. O. Sigel, John Wiley & Sons Ltd., Chichester, UK, 2007, pp. 473–500.
210. Y. Dai, P. C. Wensink, R. H. Abeles, *J. Biol. Chem.* **1999**, *274*, 1193–1195.
211. J. W. Wray, R. H. Abeles, *J. Biol. Chem.* **1993**, *268*, 21466–21469.
212. T. Pochapsky, S. Pochapsky, T. Ju, C. Hoefler, J. Liang, *J. Biomol. NMR* **2006**, *34*, 117–127.
213. T. Ju, R. B. Goldsmith, S. C. Chai, M. J. Maroney, S. S. Pochapsky, T. C. Pochapsky, *J. Mol. Biol.* **2006**, *363*, 823–834.
214. P. Bauerfeind, R. M. Garner, L. T. Mobley, *Infect. Immun.* **1996**, *64*, 2877–2880.
215. R. Ge, R. M. Watt, X. Sun, J. A. Tanner, Q. Y. He, J. D. Huang, H. Sun, *Biochem. J.* **2006**, *393*, 285–293.
216. H. L. T. Mobley, R. M. Garner, G. R. Chippendale, J. V. Gilbert, A. V. Kane, A. G. Plaut, *Helicobacter* **1999**, *4*, 162–169.
217. K. Schauer, C. Muller, M. Carriere, A. Labigne, C. Cavazza, H. De Reuse, *J. Bacteriol.* **2010**, *192*, 1231–1237.
218. S. Seshadri, S. L. Benoit, R. J. Maier, *J. Bacteriol.* **2007**, *189*, 4120–4126.
219. J. Gilbert, J. Ramakrishna, F. Sunderman, Jr, A. Wright, A. Plaut, *Infect. Immun.* **1995**, *63*, 2682–2688.
220. S. Cun, H. Li, R. Ge, M. C. M. Lin, H. Sun, *J. Biol. Chem.* **2008**, *283*, 15142–15151.
221. M. Rowinska-Zyrek, D. Witkowska, S. Bielinska, W. Kamysz, H. Kozlowski, *Dalton Trans.* **2011**, *40*, 5604–5610.
222. M. Rowinska-Zyrek, D. Witkowska, D. Valensin, W. Kamysz, H. Kozlowski, *Dalton Trans.* **2010**, *39*, 5814–5826.
223. Y.-B. Zeng, D.-M. Zhang, H. Li, H. Sun, *J. Biol. Inorg. Chem.* **2008**, *13*, 1121–1131.
224. R. Ge, Y. Zhang, X. Sun, R. M. Watt, Q.-Y. He, J.-D. Huang, D. E. Wilcox, H. Sun, *J. Am. Chem. Soc.* **2006**, *128*, 11330–11331.
225. Y. B. Zeng, N. Yang, H. Sun, *Chemistry – A European Journal* **2011**, *17*, 5852–5860.
226. S. C. Wang, A. V. Dias, D. B. Zamble, *Dalton Trans.* **2009**, 2459–2466.
227. D. P. Giedroc, A. I. Arunkumar, *Dalton Trans.* **2007**, 3107–3120.
228. M. A. Pennella, D. P. Giedroc, *Biometals* **2005**, *18*, 413–428.
229. H. Reyes-Caballero, G. C. Campanello, D. P. Giedroc, *Biophys. Chem.* **2011**, *156*, 103–114.
230. P. T. Chivers, R. T. Sauer, *Protein Sci.* **1999**, *8*, 2494–2500.
231. N. S. Dosanjh, S. L. J. Michel, *Curr. Opin. Chem. Biol.* **2006**, *10*, 123–130.
232. E. R. Schreiter, C. L. Drennan, *Nat. Rev. Microbiol.* **2007**, *5*, 710–720.
233. P. T. Chivers, R. T. Sauer, *Chem. Biol.* **2002**, *9*, 1141–1148.
234. E. R. Schreiter, M. D. Sintchak, Y. Guo, P. T. Chivers, R. T. Sauer, C. L. Drennan, *Nat. Struct. Mol. Biol.* **2003**, *10*, 794–799.
235. P. T. Chivers, T. H. Tahirov, *J. Mol. Biol.* **2005**, *348*, 597–607.
236. I. Delany, R. Ieva, A. Soragni, M. Hilleringmann, R. Rappuoli, V. Scarlato, *J. Bacteriol.* **2005**, *187*, 7703–7715.
237. E. L. Benanti, P. T. Chivers, *J. Bacteriol.* **2010**, *192*, 4327–4336.
238. K. De Pina, V. Desjardin, M. A. Mandrand-Berthelot, G. Giordano, L. F. Wu, *J. Bacteriol.* **1999**, *181*, 670–674.
239. S. C. Wang, A. V. Dias, S. L. Bloom, D. B. Zamble, *Biochemistry* **2004**, *43*, 10018–10028.
240. S. L. Bloom, D. B. Zamble, *Biochemistry* **2004**, *43*, 10029–10038.
241. P. T. Chivers, R. T. Sauer, *J. Biol. Chem.* **2000**, *275*, 19735–19741.
242. P. E. Carrington, P. T. Chivers, F. Al-Mjeni, R. T. Sauer, M. J. Maroney, *Nat. Struct. Mol. Biol.* **2003**, *10*, 126–130.
243. E. R. Schreiter, S. C. Wang, D. B. Zamble, C. L. Drennan, *Proc. Natl. Acad. Sci. USA* **2006**, *103*, 13676–13681.
244. C. M. Phillips, C. M. Stultz, C. L. Drennan, *Biochemistry* **2010**, *49*, 7757–7763.
245. S. C. Wang, Y. Li, C. V. Robinson, D. B. Zamble, *J. Am. Chem. Soc.* **2010**, *132*, 1506–1507.

246. C. M. Phillips, P. S. Nerenberg, C. L. Drennan, C. M. Stultz, *J. Am. Chem. Soc.* **2009**, *131*, 10220–10228.
247. S. C. Wang, Y. Li, M. Ho, M. E. Bernal, A. M. Sydor, W. R. Kagzi, D. B. Zamble, *Biochemistry* **2010**, *49*, 6635–6645.
248. S. Leitch, M. J. Bradley, J. L. Rowe, P. T. Chivers, M. J. Maroney, *J. Am. Chem. Soc.* **2007**, *129*, 5085–5095.
249. C. M. Phillips, E. R. Schreiter, C. M. Stultz, C. L. Drennan, *Biochemistry* **2010**, *49*, 7830–7838.
250. A. Dias, D. Zamble, *J. Biol. Inorg. Chem.* **2005**, *10*, 605–612.
251. C. M. Phillips, E. R. Schreiter, Y. Guo, S. C. Wang, D. B. Zamble, C. L. Drennan, *Biochemistry* **2008**, *47*, 1938–1946.
252. A. H. M. van Vliet, S. W. Poppelaars, B. J. Davies, J. Stoof, S. Bereswill, M. Kist, C. W. Penn, E. J. Kuipers, J. G. Kusters, *Infect. Immun.* **2002**, *70*, 2846–2852.
253. C. Muller, C. Bahlawane, S. Aubert, C. M. Delay, K. Schauer, I. Michaud-Soret, H. De Reuse, *Nucleic Acids Res.* **2011**, *39*, 7564–7575.
254. M. Contreras, J.-M. Thiberge, M.-A. Mandrand-Berthelot, A. Labigne, *Mol. Microbiol.* **2003**, *49*, 947–963.
255. E. L. Benanti, P. T. Chivers, *J. Biol. Chem.* **2011**, *286*, 15728–15737.
256. N. S. Dosanjh, A. L. West, S. L. J. Michel, *Biochemistry* **2009**, *48*, 527–536.
257. A. H. M. van Vliet, E. J. Kuipers, J. Stoof, S. W. Poppelaars, J. G. Kusters, *Infect. Immun.* **2004**, *72*, 766–773.
258. S. Bury-Moné, J. M. Thiberge, M. Contreras, A. Maitournam, A. Labigne, H. D. Reuse, *Mol. Microbiol.* **2004**, *53*, 623–638.
259. A. H. M. van Vliet, F. D. Ernst, J. G. Kusters, *Trends Microbiol.* **2004**, *12*, 489–494.
260. Y. Li, D. B. Zamble, *Biochemistry* **2009**, *48*, 2486–2496.
261. C. Dian, K. Schauer, U. Kapp, S. M. McSweeney, A. Labigne, L. Terradot, *J. Mol. Biol.* **2006**, *361*, 715–730.
262. C. Bahlawane, C. Dian, C. Muller, A. Round, C. Fauquant, K. Schauer, H. de Reuse, L. Terradot, I. Michaud-Soret, *Nucleic Acids Res.* **2010**, *38*, 3106–3118.
263. A. L. West, F. St. John, P. E. M. Lopes, A. D. MacKerell, E. Pozharski, S. L. J. Michel, *J. Am. Chem. Soc.* **2010**, *132*, 14447–14456.
264. S. Benini, M. Cianci, S. Ciurli, *Dalton Trans.* **2011**, *40*, 7831–7833.
265. L. O. Abraham, Y. Li, D. B. Zamble, *J. Inorg. Biochem.* **2006**, *100*, 1005–1014.
266. B. Zambelli, M. Bellucci, A. Danielli, V. Scarlato, S. Ciurli, *Chem. Commun.* **2007**, 3649–3651.
267. E. L. Benanti, P. T. Chivers, *J. Biol. Chem.* **2007**, *282*, 20365–20375.
268. N. S. Dosanjh, N. A. Hammerbacher, S. L. J. Michel, *Biochemistry* **2007**, *46*, 2520–2529.
269. J. S. Iwig, J. L. Rowe, P. T. Chivers, *Mol. Microbiol.* **2006**, *62*, 252–262.
270. J. S. Iwig, P. T. Chivers, *J. Mol. Biol.* **2009**, *393*, 514–526.
271. D. Blaha, S. Arous, C. Blériot, C. Dorel, M. A. Mandrand-Berthelot, A. Rodrigue, *Biochimie* **2011**, *93*, 434–439.
272. J. S. Iwig, S. Leitch, R. W. Herbst, M. J. Maroney, P. T. Chivers, *J. Am. Chem. Soc.* **2008**, *130*, 7592–7606.
273. D. Osman, J. S. Cavet, *Nat. Prod. Rep.* **2010**, *27*, 668–680.
274. H. Reyes-Caballero, C. W. Lee, D. P. Giedroc, *Biochemistry* **2011**, *50*, 7941–7952.
275. M. A. Pennella, J. E. Shokes, N. J. Cosper, R. A. Scott, D. P. Giedroc, *Proc. Natl. Acad. Sci. USA* **2003**, *100*, 3713–3718.
276. D. R. Campbell, K. E. Chapman, K. J. Waldron, S. Tottey, S. Kendall, G. Cavallaro, C. Andreini, J. Hinds, N. G. Stoker, N. J. Robinson, J. S. Cavet, *J. Biol. Chem.* **2007**, *282*, 32298–32310.
277. E. J. Kim, H. P. Kim, Y. C. Hah, J. H. Roe, *Eur. J. Biochem.* **1996**, *241*, 178–185.
278. B. E. Ahn, J. Cha, E. J. Lee, A. R. Han, C. J. Thompson, J. H. Roe, *Mol. Microbiol.* **2006**, *59*, 1848–1858.

279. Y. J. An, B. E. Ahn, A. R. Han, H. M. Kim, K. M. Chung, J. H. Shin, Y. B. Cho, J. H. Roe, S. S. Cha, *Nucleic Acids Res.* **2009**, *37*, 3442–3451.
280. S. Wang, Y. Wu, F. W. Outten, *J. Bacteriol.* **2011**, *193*, 563–574.
281. J. S. Kim, S. O. Kang, J. K. Lee, *J. Biol. Chem.* **2003**, *278*, 18455–18463.
282. C. Tibazarwa, S. Wuertz, M. Mergeay, L. Wyns, D. van der Lelie, *J. Bacteriol.* **2000**, *182*, 1399–1409.
283. G. Grass, C. Groe, D. H. Nies, *J. Bacteriol.* **2000**, *182*, 1390–1398.
284. G. Grass, B. Fricke, D. H. Nies, *Biometals* **2005**, *18*, 437–448.
285. J. Trepreau, E. Girard, A. P. Maillard, E. de Rosny, I. Petit-Haertlein, R. Kahn, J. Covès, *J. Mol. Biol.* **2011**, *408*, 766–779.
286. X. Sun, R. Ge, J. F. Chiu, H. Sun, Q. Y. He, *Metal-Based Drugs* **2008**, 1–6.
287. D. M. Templeton, B. Sarkar, *Biochem. J.* **1985**, *230*, 35–42.
288. M. Lucassen, B. Sarkar, *J. Toxicol. Environ. Health Part A* **1979**, *5*, 897–905.
289. L. Ouerdane, S. Mari, P. Czernic, M. Lebrun, R. Lobinski, *J. Anal. At. Spectrom.* **2006**, *21*.
290. T. Kowalik-Jankowska, H. Kozlowski, E. Farkas, I. Sóvágó, in *Nickel and Its Surprising Impact in Nature*, Vol. 2 of *Metal Ions in Life Sciences*, Eds A. Sigel, H. Sigel, R. K. O. Sigel, John Wiley & Sons Ltd., Chichester, UK, 2007, pp. 63–108.
291. R. K. O. Sigel, H. Sigel, in *Nickel and Its Surprising Impact in Nature*, Vol. 2 of *Metal Ions in Life Sciences*, Eds A. Sigel, H. Sigel, R. K. O. Sigel, John Wiley & Sons Ltd., Chichester, UK, 2007, pp. 109–180.
292. A. Krezel, W. Bal, *Bioinorg. Chem. Appl.* **2004**, *2*, 293–305.
293. L. Banci, I. Bertini, S. Ciofi-Baffoni, T. Kozyreva, K. Zovo, P. Palumaa, *Nature* **2010**, *465*, 645–648.
294. L. Banci, I. Bertini, K. S. McGreevy, A. Rosato, *Nat. Prod. Rep.* **2010**, *27*, 695–710.
295. S. Tottey, D. Harvie, N. Robinson, in *Molecular Microbiology of Heavy Metals*, Eds. D. Nies, S. Silver, Vol. 6, Springer, Berlin, Heidelberg, 2007, pp. 3–35.
296. K. J. Waldron, N. J. Robinson, *Nat. Rev. Microbiol.* **2009**, *6*, 25–35.
297. D. L. Huffman, T. V. O'Halloran, *Annu. Rev. Biochem.* **2001**, *70*, 677–701.
298. T. V. O'Halloran, V. C. Culotta, *J. Biol. Chem.* **2000**, *275*, 25057–25060.
299. C. E. Outten, T. V. O'Halloran, *Science* **2001**, *292*, 2488–2492.
300. J. S. Iwig, P. T. Chivers, *Nat. Prod. Rep.* **2010**, *27*, 658–667.
301. L. A. Ba, M. Doering, T. Burkholz, C. Jacob, *Metallomics* **2009**, *1*, 292–311.
302. G. J. Colpas, R. P. Hausinger, *J. Biol. Chem.* **2000**, *275*, 10731–10737.
303. J. Kuchar, R. P. Hausinger, *Chem. Rev.* **2004**, *104*, 509–526.
304. H. Guldan, R. Sterner, P. Babinger, *Biochemistry* **2008**, *47*, 7376–7384.
305. H. Merkens, R. Kappl, R. P. Jakob, F. X. Schmid, S. Fetzner, *Biochemistry* **2008**, *47*, 12185–12196.
306. N. E. Dixon, R. L. Blakeley, B. Zerner, *Can. J. Biochem.* **1980**, *58*, 469–473.

Chapter 12
The Copper Metallome in Prokaryotic Cells

Christopher Rensing and Sylvia Franke McDevitt

Contents

C. Rensing
Department of Plant and Environmental Sciences,
University of Copenhagen, Thorvaldsensvej 40,
DK-1870 Frederiksberg C, Denmark

S. Franke McDevitt (✉)
Department of Biology, Skidmore College,
815 North Broadway, Saratoga Springs, NY 12866, USA
e-mail: sfranke@skidmore.edu

L. Banci (ed.), *Metallomics and the Cell*, Metal Ions in Life Sciences 12,
DOI 10.1007/978-94-007-5561-1_12, © Springer Science+Business Media Dordrecht 2013

Abstract As a trace element copper has an important role in cellular function like many other transition metals. Its ability to undergo redox changes [Cu(I) ↔ Cu(II)] makes copper an ideal cofactor in enzymes catalyzing electron transfers. However, this redox change makes copper dangerous for a cell since it is able to be involved in Fenton-like reactions creating reactive oxygen species (ROS). Cu(I) also is a strong soft metal and can attack and destroy iron-sulfur clusters thereby releasing iron which can in turn cause oxidative stress. Therefore, copper homeostasis has to be highly balanced to ensure proper cellular function while avoiding cell damage.

Throughout evolution bacteria and archaea have developed a highly regulated balance in copper metabolism. While for many prokaryotes copper uptake seems to be unspecific, others have developed highly sophisticated uptake mechanisms to ensure the availability of sufficient amounts of copper. Within the cytoplasm copper is sequestered by various proteins and molecules, including specific copper chaperones, to prevent cellular damage. Copper-containing proteins are usually located in the cytoplasmic membrane with the catalytic domain facing the periplasm, in the periplasm of Gram-negative bacteria, or they are secreted, limiting the necessity of copper to accumulate in the cytoplasm. To prevent cellular damage due to excess copper, bacteria and archaea have developed various copper detoxification strategies. In this chapter we attempt to give an overview of the mechanisms employed by bacteria and archaea to handle copper and the importance of the metal for cellular function as well as in the global nutrient cycle.

Keywords copper-binding sites • copper cofactor • copper resistance mechanisms • copper toxicity • copper uptake

Please cite as: *Met. Ions Life Sci.* 12 (2013) 417–450

1 Evolution of Copper Use and Homeostasis in Prokaryotes

The most dramatic geochemical event in Earth's history was the production of oxygen as part of photosynthesis by cyanobacteria and later plants. An oxidative atmosphere changed the bioavailability of many metals dramatically. Soft metals such as Cu(I) and Cd(II) were bound to sulfides and when these were oxidized became accessible. In addition, Cu(I) is hardly soluble whereas the oxidized form, Cu(II), is very soluble. Since bacteria had developed under conditions where iron was readily available, many enzymes of their core metabolic pathways contained iron, often in the form of iron-sulfur clusters, as part of their catalytic center [1]. Soft metals have a high affinity to thiolates and other sulfides and are able to degrade iron-sulfur clusters. This still is the main cause of toxicity due to soft metals [2,3]. Undoubtedly, organisms had to immediately find ways to protect themselves from the toxic effect of soft metals such as Cu(I) even as the desirable traits of some soft metals were put into enzymatic use [4].

Copper proteins account for less than 1% of the proteome of an organism, with about 45% of the copper proteins involved in homeostasis and storage of this trace element. Copper proteins in archaea and bacteria usually contain homologues in the other domains, while eukaryotes also contain unique copper proteins [5].

Bacteria and archaea already had a long evolutionary experience to deal with toxic metals so the blueprint for transporters handling copper was already there and merely had to be tweaked to specifically recognize and transport copper. Since copper in the cytoplasm is in the reduced Cu(I) form, P_{1B}-type ATPases responsible for Cu efflux or incorporation of Cu into cytochrome oxidase or nitrous oxide reductase exclusively transport monovalent cations such as Cu(I) and Ag(I) [6,7]. Another mechanism to manage copper homeostasis is to restrict the use of copper containing enzymes in the cytoplasm. Most copper-containing enzymes are periplasmic, extracellular or embedded in the cytoplasmic membrane facing the periplasm. This allows the concentrations of copper to be extremely low in the cytoplasm thereby avoiding any damage from this metal. The only exception being enzymes involved in photosynthesis located in intracellular membrane vesicles. In order to safely guide Cu(I) through the cytoplasm it is bound by chaperones [8,9]. To ensure proper acquisition of copper by enzyme complexes such as cbb_3-type cytochrome c oxidase or nitrous oxide reductase, copper is taken up by transporters such as CcoA in *Rhodobacter capsulatus*, a member of the major facilitator superfamily [10], or YcnJ/CopCD in *Bacillus subtilis* and *Pseudomonas syringae*, respectively. Once inside the cytoplasm, Cu(I) is transported into the periplasm by special Cu(I)-transporting P_{1B}-type ATPases that pump Cu(I) much slower than the related efflux P_{1B}-type ATPases. The presence of the two Cu(I)-transporting P_{1B}-type ATPases ensures that the right metal is inserted into the nascent enzyme system even under conditions where copper concentrations are low and the concentrations of other metals high.

The most common mechanism bacteria and archaea use to relieve the cell of excess metals is efflux. In the case of copper as Cu(I), this is in (almost) all cases a P_{1B}-type ATPase [11–13]. There are only two known functions for Cu(I)-translocating P_{1B}-type ATPases. First, as an efflux pump to pump Cu(I) from the cytoplasm across the cytoplasmic membrane into the periplasm. Second, a much slower transport of Cu(I) across the cytoplasmic membrane to insert copper into cytochrome oxidase or nitrous oxide reductase [7]. There is no known uptake Cu(I) P_{1B}-type ATPase. A possible exception might be HmtA from *Pseudomonas aeruginosa* [14]. Other mechanisms include sequestration by bacterial metallothioneins [8] or binding by polyphosphate [15]. In both cases this would bind Cu(I) more tightly than glutathione (GSH) or other cellular components would have. There basically are hardly any free copper ions in the cytoplasm as most is more or less tightly bound to different cellular molecules. However, when Cu(I) is bound to GSH it can still be reactive, being bound to metallothionein or a metallochaperone would reduce reactivity. This can in principle reduce toxicity.

A summary of the identified copper uptake and detoxification mechanisms is provided in Figure 1.

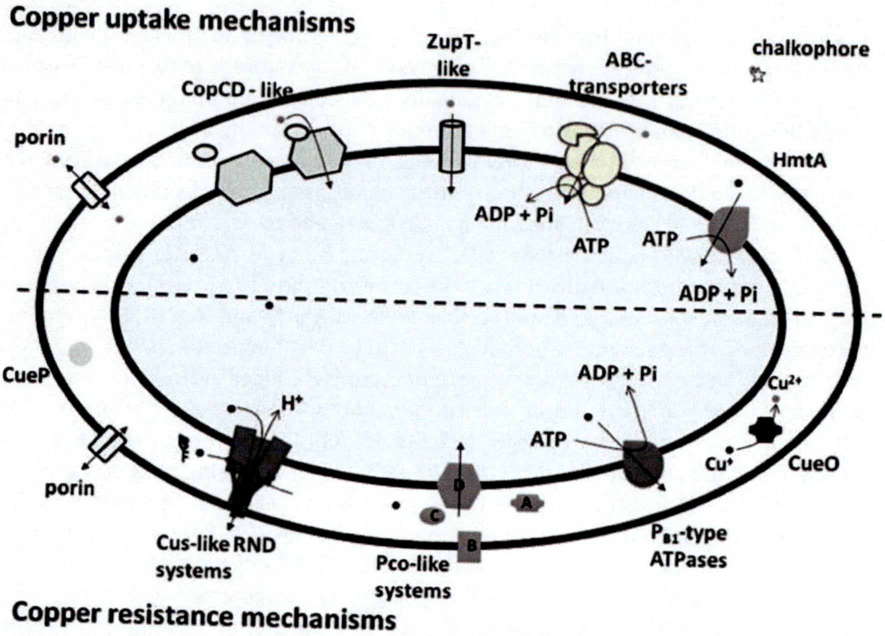

Figure 1 Overview over prokaryotic copper uptake and resistance mechanisms. Copper uptake has been described via broad substrate range transporters (ZupT, ABC-transporters), CopCD (separate proteins or one fusion protein), chelators (copper-specific chalkophore or Fe-uptake siderophores), ATPases (HmtA), and porins. Copper resistance mechanisms include transport (Cus-like RND systems, P_{B1}-type ATPases) and periplasmic (CueP and others, Pco-like, CueO) systems.

2 Prokaryotic Copper Uptake

In order for bacteria and archaea to ensure sufficient copper to be available as cofactor for essential enzymes, copper ions, as Cu(I) and/or Cu(II), have to be taken up. Whereas copper-resistance mechanisms (see Section 4 of this chapter) have been studied in great detail, our knowledge of how Cu(II) and Cu(I) enter the cell is much more limited. In Gram-negative bacteria the copper ions have to first enter the periplasm before they can be taken up into the cytoplasm. Passage of ions and molecules across the outer membrane can be part of an active, energy-dependent transport or via porins. In 1977 Lutkenhaus published his finding on the role of *Escherichia coli*'s outer membrane proteins. Strains lacking protein b (OmpF) were copper resistant, indicating the entrance of Cu(II) via this porin [16]. The importance of porins in copper uptake have also been described for methanotrophic bacteria, *Methylosinus trichosporum* OB3b [17], and a recent study identified an additional outer membrane protein, ComC, in *E. coli* which might be involved in Cu(II) entering the periplasm [18].

Several mechanisms have been described for Cu(II)/Cu(I) to enter the cytoplasm. Several transport proteins involved in uptake of other transition metal ions have been shown to have a broad substrate spectrum and also being able to take up copper. In *E. coli* ZupT, a member of the ZIP family, has a broad substrate spectrum [19] and overexpression of *zupT* increased copper sensitivity of the bacterium, indicating that Cu(II) can enter the cells via ZupT [20]. In *Streptococcus pyogenes* MstABC, encoding a member of the ABC transporter family, showed affinity for Zn(II), Fe(III) and Cu(II) indicating that more than one type of metal ions could enter the cell via this protein complex [21].

Early reports described the P_{1B}-type ATPases CopA from *Enterococcus hirae* and CtaA from *Synechocystis* PCC 6803 involved in Cu(I) uptake [22,23]. However, later studies showed that both proteins are involved in Cu(I) export [7]. In 2009 a new P-type ATPase from *P. aeruginosa*, HmtA, was described. Expression of *hmtA* in *E. coli* rendered cells sensitive to Cu(II), Ag(I), Zn(II), and Cd(II) suggesting that the protein has a broad substrate spectrum and is indeed an uptake system [14].

The CopCD uptake system is homologous to PcoCD and consists of cytoplasmic membrane proteins CopD/PcoD and CopC/PcoC, small, soluble periplasmic protein that contain two separate binding sites, one for Cu(II) (consisting of two histidine, one aspartate, and one glutamate residue) and one for Cu(I) (consisting of one histidine and three methionine residues) [24–27]. Copper in CopC can undergo the redox change from Cu(II) to Cu(I) with the different copper ions bound at different sites of the protein [26]. YcnJ of *B. subtilis* and other homologous proteins are CopCD fusion proteins. In contrast to CopC, YcnJ only contains a Cu(II)-binding site at the N-terminus and no Cu(I)-binding site [28]. Expression of CopCD alone makes cells copper-hypersensitive indicating it is involved in uptake [29]. Expression of CopBCD and CopACD also made cells more sensitive but how either interacts with CopCD is not known. In any case, since YcnJ only contains a Cu(II)-binding site it might be tempting to speculate that YcnJ transports Cu(II) and that Cu(II) is only reduced to Cu(I) in the cytoplasm [28]. However, metal binding sites in transport proteins often have a regulatory function and are not involved in the transport process. Therefore, it is possible that the actual transported species is Cu(I). Expression of the fusion protein YcnJ is upregulated under copper-limiting conditions and a disruption reduces the copper content of cells and leads to slower growth in copper-depleted medium [28]. Comparison of the sequences of YcnJ and CopCD with the genomic database using BLAST analysis [30] indicate the presence of homologous ORFs in members of Halobacteriaceae (archaea); actinobacteria, cyanobacteria, firmicutes (only Bacillales) and proteobacteria (bacteria).

Interestingly, in *Pseudomonas fluorescens* CopCD seems to be involved in copper tolerance, since deletion of *copC*, disruption of *copD* or deletion of *copR*, encoding the regulator of a two-component regulatory system, results in increased copper sensitivity [31].

The first indication of the involvement of secreted Cu chelators (chalkophore), similar to siderophores used for Fe(III), date back to 1993 [32]. A copper-binding compound was isolated from *Methanococcus capsulatus* Bath [33], and a similar compound was also found to be secreted when *M. trichosporium* OB3b was grown at very low copper concentrations [34]. Methanobactin ($C_{45}N_{12}O_{14}H_{62}$), as this compound was named, binds mainly Cu(I), using two N and two S atoms as ligands [35]. New evidence indicates that methanotrophic bacteria use a porin-dependent as well as a chalkophore (methanobactin) dependent copper uptake system [17].

In *Paracoccus denitrificans*, coproporphyrin III has been shown to be involved in copper acquisition. This bacterium requires copper for respiration and denitrification. Under copper limiting conditions, *P. denitrificans* secrets Zn-coproporphyrin III, a red pigment, which could be purified via a copper column, indicating Zn(II) could be replaced by Cu(II). In copper-free medium, growth of *P. denitrificans* was limited to 70-80% of the control and addition of Cu-coproporphyrin could restore cell yields to 90–95% indicating that *P. denitrificans* can use Cu-coproporphyrin as a copper source [36]. The cyanobacterium *Anabaena* sp. PCC 7120 uses siderophores for both, copper and iron, acquisition [37]. Chelators are widely used for iron uptake. How many of them are also used to ensure sufficient copper supply as shown for *Anabaena* sp. PCC7120 or the occurrence of copper-specific chalkophores is currently unknown.

In *Rhodobacter capsulatus* a novel transporter, CcoA (RCAP_rcc02192), has been identified, required for synthesis of *cbb*$_3$-type cytochrome *c* oxidase [10]. CcoA is a member of the major facilitator superfamily (MFS) and its presence is essential for proper *cbb*$_3$-COX assembly. Deletion of *ccoA* decreased the intracellular copper concentration and drastically decreased *cbb*$_3$-COX activity. Enzyme activity could be restored to wild-type levels by expressing *ccoA* in *trans* and a secondary mutation in the *ΔccoA*-strain led to increased copper accumulation [10]. A member of the MFS was recently identified to be involved in copper uptake of the yeast *Schizosaccharomyces pombe* [38] and CcoA would be the first bacterial member of the MFS involved in copper uptake. Analysis of RCAP_RCC02192 using the protein database (BLASTP) and the translated nucleotide database (TBLASTN) [30] showed the highest identities at 67% (for other members of the genus *Rhodobacter*), not all of the methionine residues of the metal-binding Met-motifs [10] being conserved. Since the closest matches are hypothetical proteins in the databases, it is not clear if their substrate is also copper or another compound.

3 Copper Proteins

Proteins containing copper in their reactive center are involved in reactions requiring electron transfer due to the ability of copper to change between redox states Cu(I) and Cu(II). These copper proteins are either excreted into the medium, are located in the periplasm, are surface-bound or in the membrane with the active, copper-containing center facing away from the cytoplasm. Assembly of the active center has been shown to be dependent on the presence of slow exporting P_{1B}-type ATPases [39,40] or, as in the case of CueO from *E. coli*, the protein is folded in the cytoplasm of the cell as exported via the TAT system [41,42]. Ten types of copper containing enzymes have been identified in prokaryotes, in bacteria as well as archaea, with cytochrome *c* oxidase (COX) being the most widely used cuproprotein [43]. Here we will focus on a subset of these copper enzymes and their importance for bacterial and archaeal metabolism.

3.1 Cytochrome Oxidase

There are three types of heme-copper oxygen reductases (HCOs). This classification is based on comparison of the core subunits and key residues in the proton transfer pathway [44,45]. Genomic analysis of cuproproteins revealed that from the data available at the time (January 2007) 91% of bacteria and 31% of archaea utilized COX [43]. Representative members for this discussion are: Type A: mitochondrial-aa_3 cytochrome oxidase (COX), Type B: ba_3-COX from *Thermus thermophilus* and Type C: cbb_3-COX from *R. capsulatus*. The focus in this section will be on how copper is inserted into the catalytic domain. While aa_3-COX and ba_3-COX contain two copper centers, Cu_A and Cu_B, cbb_3-COX lacks the Cu_A center.

A common theme appears to be uptake of Cu(II) by either a CopCD or an CcoA homolog. It appears that Cu(II) rather than Cu(I) is taken up by CcoA since suppressor mutants were hypersensitive to Cu(II) but not Ag(I) or other divalent metals. It is also clear that Cu(II) is reduced to Cu(I) either while being transported or at some point right after entering the cytoplasm. Whether either of these transporters hand copper to a copper chaperone or directly transfer copper to a specific Cu(I)-translocating P_{1B}-type ATPase is not known at this point. In any case, in many microorganisms there are two related Cu-translocating P_{B1}-type ATPases, one responsible for Cu(I) efflux referred to as CopA1 and one with a much slower rate of transport referred to as CopA2. The latter transporter has shown to be necessary for copper insertion into cytochrome oxidase cbb_3 in *P. aeruginosa* [39]. A related enzyme, CtpA, has been shown to be necessary for copper insertion into cbb_3 and nitrous oxide reductase [40]. Periplasmic copper trafficking proteins necessary for proper copper insertion into cbb_3 might include SenC. SenC was shown to be important for copper acquisition for cbb_3-COX in *P. aeruginosa* and *R. capsulatus* [46,47]. However, *senC* (encoding a Sco1-like protein) deletion mutants in *Bradyrhizobium japonicum* were defective in the assembly of aa_3-COX but not cbb_3-COX [48].

Genes encoding Sco-like proteins have been identified within the genomes of various bacteria and archaea [49]. Based on the structure similarity to peroxiredoxins a catalytic function for Sco was suggested, instead of acting merely as a chaperone [50]. The structure of Sco1 shows a thioredoxin fold and the protein binds Cu(I) via a conserved CxxxCP motif and histidine residue [51]. Cu(II) is coordinated via the same cysteine and histidine residues and an additional planar oxygen ligand as well as axial water are suggested for Cu(II) binding [52]. Mutation of either one of the cysteine residues or of the conserved histidine residue to alanine destabilized the Cu(II)-form of Sco of *B. subtilis* and a mutant harboring a methionine instead of the wild-type histidine fails to assemble a functional COX [52–54]. While the proposed model for the eukaryotic COX Cu_A-center includes the formation of a Sco-Cu(I)-COX intermediate [55] it is currently unclear how the Cu(II) form of Sco is involved in the transfer of copper from Sco to COX.

Perhaps, these pathways are adaptations for specific environments as it is conceivable that incorporation of the „wrong" metal is most likely the case in a copper scarce environment with elevated levels of other metals.

3.2 Multicopper Oxidases

Multicopper oxidases (MCOs) are ubiquitous copper proteins thought to be descended from small blue copper proteins (cupredoxins). MCOs contain at least one of each of the biological copper sites: a type 1 (T1, "blue") copper site as well as a mononuclear type 2 (T2), and a binuclear type 3 (T3) Cu site, with T2 and T3 organized in what is known a trinuclear cluster (TNC) [56].

Based on the type of substrates, several types of MCOs can be differentiated [57]. Comparative analysis showed that MCOs can be found in approximately 13% of the bacterial genomes [43]. Focus here will be given to laccases due to their importance in bioremediation and one member of metallooxidases, CueO from *E. coli*.

Laccases oxidize aromatic amines, phenols, and polyphenols as well as some non-phenolic substrates [58]. First discovered in the latex of the lacquer tree (*Rhus vernicifera*) laccases were mainly found in white rot fungi [57]. Within the last 30 years increasing evidence of bacterial laccases has been found, initially due to the ability to utilize certain substrates and then increasingly due to sequence homologies, activity of the purified enzyme, and crystal structure studies [58]. New evidence also shows the presence of laccases in archaea [59,60]. To increase our understanding of how common the presence of laccases is, a high throughput screening method for laccase activity in metagenomic libraries has been developed [61].

The T1 copper in laccases is coordinated by three conserved strong ligands (two histidines and one cysteine) in a trigonal geometry and can have one or two additional weaker ligands (one usually a methionine). This type 1 copper is involved in substrate oxidation and its close proximity to the TNC allows for transfer of the electrons. The T2 and T3 copper atoms are coordinated by histidines. At the TNC O_2 is reduced to H_2O [60]. As mentioned before, laccases are able to oxidize a variety of phenolic and aromatic substrates and their potential for industrial application and in bioremediation is beginning to be explored [62].

Laccase-like enzymes of bacteria have been found to be involved in cell division, pigmentation, sporulation, and metal oxidation, including copper resistance [58]. CueO from *E. coli* has been described as MCO which is involved in copper resistance of the bacterium [41,63]. The structure of CueO revealed that in addition to the overall structure known for MCOs including the T1, T2, and T3 copper centers, CueO has a methionine-rich helical region which shields access to T1 and a similar Met-rich region can be found in other MCOs involved in copper resistance [64].

It was found that access of copper is necessary for full CueO activity and that a regulatory Cu(II) is bound to this methionine-rich region [65]. CueO exhibits laccase activity as well as ferrooxidase activity [41]. In addition, CueO is able to oxidize enterobactin and the siderophore precursor 2,3-dihydroxybenzoic acid and the oxidized compounds were able to bind copper [66]. It was also shown that CueO has cuprous oxidase activity, oxidizing Cu(I) to the less toxic Cu(II) in the presence of excess copper [67]. Additional data suggesting the involvement of the methionine-rich region in binding of Cu(I) for oxidation and inhibition of cuprous oxidase activity by Ag(I), led to the assumption of irreversible binding of Ag(I) to this regulatory helix [68]. Absence of the Ag(I) detoxification system *cus* led to decreased copper tolerance in the presence of Ag(I). The structure of CueO bound to Cu(I) or Ag(I) indicates that Ag(I) can indeed replace the Cu(I) substrate rendering the enzyme inactive [69].

CueO (or CuiD) from *Salmonella* Typhimurium is required for full copper tolerance and is necessary for systemic infection in the mouse model. However, deletion of *cueO* did not alter the bacteria's ability to infect and survive in macrophages [70,71].

3.3 Plastocyanin/Azurin Family of Blue (Type 1) Copper Proteins

Members of the plastocyanin/azurin family of blue copper proteins (cupredoxins) are small proteins involved in transfer of electrons from a donor to an acceptor within a biological system. Their active site contains a single copper atom coordinated by three strong ligands and in some cases an additional weaker fourth or fifth ligand [72].

In cyanobacteria plastocyanins shuttle electrons from cytochrome to photosystem I. The copper ion is bound by three strong ligands, two nitrogens from histidines and a sulfur from cysteine, and a more distant (weaker ligand) sulfur from a methionine [73]. Although their structure corresponds to the overall structure found for eukaryotic plastocyanins, cyanobacterial plastocyanins display electrostatic properties more similar to azurins than eukaryotic plastocyanins, suggesting different mechanisms for electron transfer [73].

The overall structure of azurins is similar to that of plastocyanins [73]. *P. aeruginosa* azurin binds copper in a distorted trigonal bipyramidal coordination with three strong ligands (two histidines and one cysteine) and two longer axial ligands (methionine and glycine) [74]. In other azurin structures the presence of five Cu ligands was also observed [72]. Azurins are involved in electron shuttling during deamination processes as well as dinitrification. They can directly donate electrons to nitrite reductases providing the reducing power for dinitrification [73].

Azurins and plastocyanins might also be involved in bacterial survival under copper stress. As shown for *P. aeruginosa*, disruption of *cinA*, encoding an azurin-plastocyanin like protein, increased sensitivity towards copper. This was demonstrated by reduced growth on copper-containing solid medium as well as reduced survival of the mutant on copper alloy surfaces [75]. CinA domain-containing proteins have been annotated across a variety of bacterial and archaeal genomes. However, it is not known to what extent these proteins play a role in the handling of copper stress [73].

3.4 *Amine Oxidases*

Amine oxidases catalyze the deamination of amino groups. Deamination takes place in a two-step mechanism in which first the enzyme is reduced by transfer of the amino group from the substrate to the enzyme and second the reoxidation of the enzyme, releasing ammonia ions and H_2O_2. Amine oxidase can be differentiated depending on the enzyme's cofactor and the position of the amino group removed [76]. Given the topic of this chapter the focus will be on copper-containing amine oxidases (Cu-AO) from bacteria.

Cu-AOs are dimers and the crystal structure of the Cu-AO from *Arthrobacter globiformis* revealed the presence of a single Cu atom in the active center. The Cu is bound by three conserved histidine residues and positioned in proximity to a tryptophane residue, which has been posttranslationally modified to trihydroxyphenylalanine (TPQ) [77]. Since TPQ is only present in the holo-enzyme of *A. globiformis'* Cu-AO and not in the apo-form of the enzyme, it is postulated that copper is involved in the modification process. Comparison with the active sites of the amine oxidases from *E. coli* and pea seedlings strengthen these findings [77]. Intermediates of the proposed catalytic cycle [78] were detected after trapping crystals containing reaction intermediates and analyzed via single-crystal microspectroscopy [79]. Whereas in the substrate-free control, TPQ is in close proximity to the copper center, addition of the substrate phenylethylamine led to a conformational change of TPQ towards the substrate channel to allow for binding of the incoming substrate. Deamination of the substrate is proposed to be accomplished by TPQ whereas the copper center is involved in regeneration of the enzyme and release of ammonia [78].

While best characterized from *A. globiformis*, copper-containing amine oxidases are utilized in approximately 6% of bacteria [43] and are also common in archaea. For example, Cu-AO TynA from *Sulfolobus islandicus* HVE10/4 shares 36% idendity over the entire length of the protein.

3.5 *Methane Monooxygenases*

Methanotrophic bacteria oxidize methane to fulfill their carbon and energy needs for growth [80]. The enzymes involved in the first step of this pathway, the oxidation of methane, are methane monooxygenases (MMO). Two types of MMO have been identified, a membrane-bound copper-containing enzyme (particulate MMO or pMMO) and a soluble enzyme (sMMO), the latter only produced by some bacteria. Genes encoding pMMO are present in about 2% of the bacterial genomes (as available in January 2007) [43]. If the genetic information of both types of MMOs are present, expression of the respective genes is regulated by the amount of copper available in the environment [81,82]. The *smmo* operon is expressed when the cells are growing in the absence of copper and expression of *smmo* is inhibited by the addition of copper ions [81]. The opposite has been observed regarding the regulation of the *pmo* operon, encoding the membrane-bound MMO. In *Methylococcus capsulatus* (Bath) *pmo* expression was observed in the presence of copper but not when copper was omitted from the medium [81].

Nguyen et al. isolated pMMO from *M. capsulatus* (Bath) and found it to be a copper-containing enzyme composed of three subunits. A high number of copper ions (12–15) were found to be associated with each complex [83]. It was discovered earlier that the activity of pMMO correlates with the availability of copper ions to be incorporated into the enzyme complex; however, excess copper has an inhibitory effect [84]. The active enzyme is associated with copper-binding compounds which have now been identified as methanobactin and which are involved in the regulation of expression of *pmo* in the presence of Cu minerals and ensure copper availability for the enzyme [34,85].

The first crystal structure of a pMMO started to shed light on the role of copper in the enzyme's activity. Particulate methane monooxygenase from *M. capsulatus* (Bath) is a trimer with an $\alpha_3\beta_3\gamma_3$ arrangement. The cylindrical complex is anchored in the membrane by 42 transmembrane helices (14 from each monomer) and forms a hole in its center. PmoB subunits are anchored in the membrane by two transmembrane helices (TMH) and form the soluble portion of the complex composed of β-sheets. Within PmoB, three copper ions were identified, one in a mononuclear site and two forming a dinuclear copper site. In both sites, histidines serve as ligands and the copper centers were proposed as the active site of the enzyme. In addition, a Zn(II) coordinated by residues from PmoA and PmoB was found in the transmembrane region close to the membrane-periplasm interface. However, the source of the bound Zn(II) was identified as to be in the crystallization buffer, leaving the question open if Zn(II) is bound in the native system [86].

The second solved structure of a pMMO (from *M. trichosporium* OB3b) showed a similar overall structure of an $\alpha_3\beta_3\gamma_3$ arrangement. In contrast to pMMO from *M. capsulatus* (Bath), the newly solved structure has 15 TMH per protomer (compared to 14) and differences in the metal clusters. The dinuclear copper center could be identified with histidine residues serving as copper ligands. The mononuclear copper center in PmoB is missing and instead of a Zn(II) in the transmembrane region, a Cu was identified. It was suggested that in pMMO from *M. capsulatus* (Bath) the native Cu was replaced by Zn due to crystallization conditions [87].

The presence of the dicopper center in PmoB but not of a mononuclear copper center was also shown in the pMMO crystal structure from *Methylocystis* sp. strain M [88].

Recent studies support the involvement of the dicopper center in methane oxidation. A soluble derivative of PmoB (after removing the two TMH and joining the remaining polypeptides) showed enzyme activity, which remained when proposed ligands of the mononuclear copper center where mutated. However, activity was lost after mutating two of the ligands of the dinuclear copper center, indicating that this region of the protein is the active site. Replacing copper with iron in the refolded molecule renders it inactive, demonstrating that indeed copper and not, as some studies suggested, iron is in the active center of pMMO [89–91].

3.6 *Enzymes Involved in Nitrogen Metabolism*

In nature nitrogen is cycled between its redox states (–3 in ammonia to +5 in nitrate) with most of the nitrogen being in the atmosphere as N_2. In order for nitrogen to be accessible for the majority of organisms, N_2 has to be fixed into NH_4^+, which then can be converted into other nitrogen compounds and eventually be released back into the atmosphere as N_2. Human activity impacts the delicate balance of the nitrogen cycle and its biotic and abiotic factors [92]. Copper-containing enzymes are involved in the nitrogen cycle in nitrification, conversion of ammonia to nitrate, and in denitrification, the step-wise reduction of nitrate to dinitrogen [93–95].

The first step in nitrification is the oxidation of ammonia to hydroxylamine by ammonia monooxygenase (AMO), an enzyme related to particulate methane monooxygenase. Activity of AMO of *Nitrosomonas europaea* has been found to be dependent on copper, suggesting copper in its active center [95]. *N. europaea* contains a soluble as well as a membrane-bound AMO [96]. Soluble AMO from *N. europaea* is composed of three subunits (AmoA (α), AmoB (β), and cytochrome *c* (γ)) in a $\alpha_3\beta_3\gamma_3$ stoichiometry [97]. The purified enzyme contains 9.5 mol Cu per mol AMO and EPR analysis revealed the presence of a type 2 Cu(II) in a square planar or square pyramidal coordination as well as Cu(I) [98]. However, the mechanism of oxidation of ammonia to hydroxylamine by AMO is not known.

Nitrate can serve as terminal electron acceptor of the electron transport chain for a variety of bacteria and archaea. In the next step, nitrite is reduced to nitric oxide (NO) by nitrite reductase. There are two types of nitrite reductases: a cytochrome cd_1 containing enzyme, and the copper-containing NirK [94]. First identified in 1963, copper-containing nitrite reductases (CuNIR) have now been found in a variety of bacteria and archaea [94,99–101]. The protein forms a trimer with each monomer showing a cupredoxin-like fold of the "Greek key" β-barrel with two copper atoms (one type 1, "blue" copper, and one type 2) [102,103]. Three histidine residues from two subunits form the ligands of the type 2 copper and addition of NO_2^- shows the substrate O-coordinated at the type 2 Cu site. In a proposed mechanism, NO_2^- is bound by the type 2 copper (Cu(II)) which is reduced. Amino acid residues Asp98 and His255 (CuNIR from *Achromobacter cycloclastes*) are involved in the reaction as catalysts [101,103]. NO can bind again to the type 2 copper, this time N-coordinated, for possible N_2O production [103]. In another proposed mechanism Cu(II) is first reduced to Cu(I) and NO_2^- binds to the Cu(I) center forming a Cu-nitrosyl intermediate before NO is released. Here, the Cu-nitrosyl intermediate can react with NO_2^- and NO_2 is released [101]. NO is toxic to cells due to its reactivity and production of NO is one of the mechanisms of attack by macrophages against pathogens [94]. Nitric oxide reductase (NOR) is a membrane-bound enzyme complex formed under anaerobic conditions and reduction of NO provides the substrate for the final enzyme of denitrification, the nitrous oxide reductase (N_2OR) [94].

In Gram-negative bacteria, N_2OR is located in the periplasm, whereas it is bound to the cytoplasmic membrane in Gram-positive bacteria [93]. Nitrous oxide reductase is a homodimer first described to contain four copper atoms per monomer, two in each of the two copper centers Cu_A and Cu_Z [104]. However, later studies revealed the presence of a two copper Cu_A and a four copper Cu_Z [4Cu:2S] center [105,106]. The copper atoms in the binuclear Cu_A center are bridged by cysteine residues and each of the two copper atoms has one histidine as a ligand in addition to second ligands (one of them being methionine) [106,107] As shown for N_2OR from *Pseudomonas stutzeri* maturation of the enzyme (assembly of the Cu_A and Cu_Z center) occurs in the periplasm and the gene products of the other *nos* genes (except *nosZ* which encodes N_2OR) are required for proper copper incorporation [108]. Cu_Z has been identified as the active site of the enzyme, and the formation of a sulfoxide intermediate has been proposed [105,106,109]. Reduction of N_2O requires the transfer of electrons to the active site, and the Cu_A center is proposed to deliver the electrons [106]. Since the distance of Cu_A and Cu_Z within each monomer is 40 Å whereas the distance between the centers of the monomers is only 10 Å, transfer of electrons from Cu_A from one monomer to Cu_Z of the second monomer has been proposed for higher efficiency [109].

Anthropogenic activity, especially agricultural practice, increases the presence of available nitrate in soil and water. Since the complete denitrification of NO_3^- to N_2 involves volatile intermediates (NO and N_2O) increased release of these gases has been observed from agricultural soil as well as during waste-water treatment [110]. One of the limiting steps of complete denitrification is the availability of Cu to assemble nitrous oxide reductase. Initial studies showed that Fe-Cu-pretreated waste-water increased the removal of total nitrogen [111]. However, copper has also been shown to have inhibitory effects on denitrifiers in waste-water [112].

4 Copper Resistance Mechanisms

4.1 P_{1B}-Type ATPases

P_{1B}-ATPases (old name CP*x*-ATPases) are involved in transition metal efflux from the cytoplasm. The energy for this transport activity is provided by the hydrolysis of ATP [113]. First evidence for ATPases involved in transport of copper was obtained from *E. hirae* [114] and soon thereafter human homologues were identified to be responsible for Menkes and Wilson disease [115]. Within the P_{1B}-ATPases several sub-families can be differentiated. P_{1B-1}-ATPases are selective for the transport of Cu(I) and two sub-groups can be identified within this sub-family: FixI/CopA2-like ATPases which show a low transport rate, and the "classical" P_{1B-1}-ATPases [7,116].

Proteins of both sub-families form eight transmembrane helices and TMH six contains the CPC motif which gave this protein family its original name (P-type ATPases or CP*x*-ATPases, with *x* being C or H). They also contain the N-terminal CXXC metal-binding motif which is located in the cytoplasm. Members of the "classical" P_{1B-1}-ATPases are CopB from *E. hirae*, CopA from *E. coli*, CopA from *Archeoglobus fulgidus* or CopA from *Sulfolobus solfaticus* [117–120]. They are found among archaea as well as all groups of bacteria with protein-identities to CopA from *E. coli* ranging from 35% to 95% and featuring the N-terminal CxxC motif, the CPC motif in TMH six as well as a conserved asparagine, tyrosine, methionine, and serine residue as part of the transmembrane metal-binding site (TM-MBS) [121]. Although common, not all bacteria or archaea harbor CopA-like proteins. For example, symbiotic members of the *Enterobacteriaceae* lack CopA while it is present in non-symbiotic bacteria of this family [122]. In order to shed light on the abundance of copper-transporting P_{1B}-ATPases primers located in conserved domains have been generated and used to analyze copper-resistant bacterial isolates as well as metagenomic DNA for the presence of genes encoding these transporters [123].

Expression of the respective genes is highly inducible by copper and members of this sub-group have been shown to be involved in rapid Cu(I) export from the cytoplasm [39]. If present in the organism, cytoplasmic Cu(I) chaperones transfer Cu(I) to the TMH-metal transfer site [11] but in most instances there is no Cu(I) chaperone present. These P_{1B}-ATPases are also synthesized in the presence of elevated levels of Ag(I) and are involved in Ag(I) transport [117,118,124,125]. The subgroup of FixI/CopA2-like ATPases seems to be more relevant for copper transport under physiological conditions. CopA2 from *P. aeruginosa* was unable to complement an *E. coli ΔcopA*-phenotype, whereas copper resistance could be restored by expression of *copA1* [39].

Genes encoding members of the FixI/CopA2-like ATPases are often found in symbiotic or pathogenic bacteria and are always together with the more ubiquitous resistance conferring "classical" P_{1B}-ATPases [7]. Initial studies suggested members of the FixI/CopA2-like ATPases to be involved in copper uptake [22,40], however it has now been discovered that these proteins also export Cu(I) albeit at much lower rates [7,39] and that they are involved in the proper assembly of membrane-bound Cu-containing enzymes. FixI/CopA2-like ATPases have been found to be important in bacterial virulence since their presence reduced oxidative stress generated as part of the host's immune response [121].

Several studies showed that cytoplasmic copper chaperones are involved in transferring Cu(I) to the P_{1B}-ATPases for export. NMR mapping of chemical shifts of Cu(I)-bound CopZ, CopZ-CopA N-terminal domain, and Cu(I)-bound CopA N-terminal domain from *B. subtilis* indicate that CopZ and CopA form a complex of 1:1 stoichiomery. Most chemical shift changes were observed to be in the metal-binding region [126]. Interaction between CopZ and CopA of *B. subtilis* was also shown using Yeast-Two-Hybrid systems [127]. Rather than reaching a Cu(I) exchange equilibrium and having the Cu(I) exported by the P_{1B}-ATPase as driving force in the delivery of Cu(I) from the chaperone to the transporter, transfer of Cu(I) from CopZ to CopA of *A. fulgidus* is unidirectional [128].

In *A. fulgidus* a Cu(II)-transporting P_{1B}-ATPase has been identified. This modified substrate spectrum and the presence of the motif CPH in transmembrane helix six (instead of CPC found in Cu(I) transporting P_{1B-1}-ATPases) make this a member of the sub-family of P_{1B-3}-ATPases [121,129]. ATPase activity of CopB depends on Cu(II). Cu(I) and Ag(I) activate the protein at ~25% or 55% of the level observed with Cu(II). The N-terminal metal-binding domain is histidine-rich, lacks the classical CXXC motif, and seems to be of regulatory function [129]. Cu(II)-transporting P_{1B-3}-ATPases can be found among archaea and bacteria but not much is known about their physiological role within the organism, especially since copper in the cytoplasm is in the form of Cu(I).

4.2 Cus-Like Resistance-Nodulation Cell Division Systems

First identified by Munson et al. and shown to be important for copper resistance in the absence of the multicopper oxidase CueO, the *cus*-system of *E. coli* shows high homology to the *sil*-determinant from the plasmid pMG101 from *Salmonella* Typhimurium and was also characterized for its involvement in Ag(I) detoxification [63,130–132]. The *cus*-system consists of six genes organized in two operons *cusRS* and *cusCFBA* [132]. *CusRS* encode a two-component regulatory system (see Section 5.3) and *cusCFBA* encode a resistance-nodulation cell division system (RND). The RND-transport complex consists of CusA, the RND protein as the catalytic subunit localized in cytoplasmic membrane, the outermembrane factor (OMF) CusC, and the membrane fusion protein (MFP) CusB. In addition, the operon encodes a small periplasmic protein, CusF. Studies showed that deletion of any one of these four genes increased copper sensitivity in a *ΔcueO* background, although deletion of *cusF* had the least effect [133].

Whereas resistance mechanisms via RND systems are well described for multi-drug resistance and divalent heavy metal cations [134–136], their involvement in Cu(I) and Ag(I) resistance has been discovered more recently [131]. The presence of *cusF* is unique to the few characterized and potential Cu(I)/Ag(I) transport systems. CusF is a periplasmic protein able to interact with proteins of the CusCBA efflux complex [133]. The crystal structure of CusF revealed the protein being a five-stranded β-barrel, classified as an OB fold, which was unique to copper-binding proteins [137]. CusF homologues have been identified in numerous organisms as part of RND systems; however, the *cusF* gene is not always present as a separate gene but can also be part of the gene encoding the membrane fusion protein [137,138]. NMR studies predicted a highly conserved histidine residue and two methionine residues (H36, M47, and M49 in CusF from *E. coli*) to be involved in Cu(I) and Ag(I) binding and ITC measurements showed that CusF binds Ag(I) and Cu(I) with high affinity [139]. Crystal structures of Ag(I)-bound CusF and Cu(I)-bound CusF confirmed the predicted metal coordination [140,141] and that tryptophan in position 44 (W44) is important for moderate affinity of CusF to Ag(I) and Cu(I) [140,142]. Interaction of CusF with CusB could be confirmed via NMR spectroscopy and chemical cross-linking and it was shown that CusF transfers Cu(I) onto CusB, suggesting a function as copper chaperone [143,144].

Recent crystal structures of CusC, CusB, and CusA from *E. coli* led to a predicted mechanism of Cu(I) trafficking via CusCFBA [145]. The outer membrane factor CusC forms a trimer with a β-barrel located in the outer membrane and the α-barrel reaching into the periplasmic space, similar to the structure described for the OMF TolC [146,147]. The crystal structure obtained for the MFB CusB includes residues 89–385, or 78.1% of the amino acids [148]. The protein (monomer) forms four distinct domains with the loops between domains II and III acting as possible hinge to allow for conformational changes during metal transport. Asymmetric units within the crystals consist of two CusB molecules whose difference can be attributed to a small motion within the hinge region. Comparison with the structure of MexA, an MFP involved in multi-drug resistance in *P. aeruginosa* [149], showed some unique features of CusB which are not found in the partial structures of other MFPs [148]. Each CusB monomer shows two Cu(I) binding sites, in domains I and IV. In both cases one methionine is involved as a ligand. Cross-linking experiments of CusA and CusB indicate that the N-terminus of CusB interacts with the first large periplasmic domain of CusA [148]. In 2010 the crystal structure of CusA, the RND protein of the complex, was solved [150]. CusA forms a homotrimer and crystallization of the metal (Cu(I) or Ag(I)) bound protein showed distinct structural changes in the protein structure.

Within the second large periplasmic domain, the highly conserved methionine residues M573, M623, and M672 are located in close proximity to each other close to the transmembrane part of the protein. These three residues have been found to coordinate Cu(I) (or Ag(I)) [150]. Earlier studies had shown that mutation of either of these three methionines to isoleucine resulted in a loss of protein function as shown by the inability to confer copper resistance [133]. Additional methionine residues were identified and their involvement confirmed via site-directed mutagenesis leading to reduced copper resistance. These data indicate metal transport by CusA from the periplasm as well as from the cytoplasm [150]. Co-crystallization of CusAB indicates that each of the three monomers of CusA interacts with two CusB molecules, with domains I of CusB interacting with CusA and domains II–IV of the CusB molecules interacting with each other. The total of six CusB molecules form a cap on top of the CusA trimer extending into a channel which in turn can interact with the CusC trimer [145,151]. Earlier studies identified three N-terminal methionine residues in CusB essential for Cu(I) binding and transfer [152].

Although not shown in the crystal structure, these methionine residues are predicted to be located close to the periplasmic cleft of CusA and could aid in transfer of the metal ions into the center of the CusA trimer for export [151]. In contrast to the model of CusB aiding Cu(I) to the transporter CusA, the N-terminal methionines of CusB could act as a molecular switch [153]. Binding of Cu(I) to CusB could allow a conformational change of the protein in the hinge region resulting in a conformational change of the complex and allowing access of Cu(I) from the periplasm to CusA for transport. In addition to this periplasmic control system, a cytoplasmic control mechanism involving methionine residues located in the transmembrane portion of CusA is discussed to ensure availability of essential and export of access copper [153].

Homologues of the *E. coli cus* system proposed to be involved in copper and/or silver export can be identified in the genomic sequences of many bacteria across all phyla. To date, CusCFBA remains the best characterized of them.

4.3 CueP and Other Periplasmic Systems

Detoxification of the bacterial periplasm was originally only attributed to Cus-like RND systems and/or CueO or PcoA-like multicopper oxidases. In 2009 an alternative periplasmic system, CueP, was identified in *Salmonella* Typhimurium, which lacks Cus but is able to grow in the presence of up to 5.5 mM $CuSO_4$ on LB-agar under aerobic conditions [154]. As described for the *E. coli Cus* system, under aerobic conditions deletion of *cueP* only results in increased copper sensitivity if the gene encoding a CueO orthologue multicopper oxidase (*cuiD*) had been deleted as well. When growing in an anaerobic environment, a *ΔcueP* strain showed a significantly deceased minimal inhibitory concentration of $CuSO_4$, similar to what was observed for *E. coli cus* deletion strains [154,155]. Expression of *cueP* in copper-sensitive strains of *E. coli* is able to restore growth in the presence of copper under aerobic and anaerobic conditions, indicating that CueP serves as an alternative periplasmic detoxification system when *cus* is not present [154]. Whereas expression of *cus* is controlled by the two-component regulatory system CusRS (see Section 5.2 and references within), *cueP* is part of the CueR regulon [154]. Protein crystals of CueP have been obtained, however the overall structure of the protein has not yet been solved [156]. BLASTP analysis of CueP against the non-redundant protein sequences [30] revealed the presence of CueP-like proteins in other members of the *Enterobacteriaceae* as well as other bacteria including members of the Actinobacteria. Originally described as an alternative periplasmic copper detoxification system in the absence of CusCFBA, both systems are encoded on the chromosome of *Citrobacter koseri* and what role each of them plays in copper handling in the bacterium remains unknown.

P. aeruginosa synthesizes a periplasmic protein, PtrA, in the presence of copper [157]. The gene *ptrA* is located up-stream of a gene encoding a putative member of the azurin/plastocyanin family which is strongly induced by copper and mutants show increased copper sensitivity [157]. PtrA does not contain any conserved protein domains, and the only other protein showing significant homology to hypothetical proteins encoded on the chromosome of other strains of *P. aeruginosa* and encoded on the non-coding strand within the gene of an alkaline phosphatase annotated in *Leptospira interrogans*. To date it is not known, how PtrA contributes to copper resistance.

First described as part of the Pco system in *E. coli* is the periplasmic protein PcoE [158]. Homologues of PcoE include SilE of *S*. Typhimurium pMG101 as well as other proteins encoded on plasmids containing *sil*- or *pco*-like systems (or within shotgun genomic sequences where it is not yet clear if the genes are located on the chromosome or plasmid). Initial expression of *pcoE* in the presence or absence of *pcoABCD* did not increase copper resistance in *E. coli* [159]. However, new studies indicate that PcoE dimerizes in the presence of Cu(I) but not Cu(II) and binds up to six Cu(I) per protein monomer. Cu(I) can then be transferred from PcoE onto PcoC the periplasmic copper chaperone as part of the PcoABCD resistance system.

Other periplasmic proteins which seem to contribute to copper resistance include CopK and CopH from *Cupriavidus metallidurans* CH34. Homologues of these two proteins have so far only been found in some other members of β-proteobacteria and are generally annotated as hypothetical proteins which to date are not characterized. Expression of both genes was induced in the presence of copper [160] and the resulting proteins are of periplasmic location [161,162]. CopH forms a dimer with one metal binding site in each of the two subunits. The protein has been shown to bind Cu(II) but is also able to bind Zn(II) and Ni(II) using His ligands to complex the metal ions [161,163]. CopK also forms a dimer. In contrast to CopH, CopK seems to non-specifically bind Cu(II) but has two Cu(I)-specific binding sites per monomer, which differ in their affinities. CopK seems to complex Cu(I) using methionine ligands as seen in NMR chemical shift data [162]. Despite the preliminary structural analysis of CopH and CopK, the mechanism of how these two proteins are involved in copper handling of *C. metallidurans* is not understood. Interestingly, CopH shows 49% idendity to the C-terminal half of CzcE from *C. metallidurans* CH34, a protein described to be involved in Co(II), Zn(II), and Cd(II) resistance which has also been shown to undergo conformational changes when binding copper [164–166].

Just recently two new periplasmic proteins important for copper handling (CopG and Cot) have been identified in *Vibrio cholerae* [167]. *V. cholerae* lacks CueO, CueP, and Cus-like systems and seems to employ CopG and Cot as an alternative pathway to deal with copper in the periplasm. However, it appears that like other bacteria encoding CopG and Cot homologues do have other means to detoxify copper from the periplasm, e.g., CopAB in *Pseudomonas putida*, so again the role/importance of these proteins in other systems is unknown.

4.4 Pco-Like Systems

A highly copper-resistant strain of *E. coli* was isolated from piggery effluent and the high level of resistance could be attributed to the presence of the plasmid pJR1004 [168]. The copper-resistant phenotype is the result of the presence of the *pco* determinant, consisting of seven genes *pcoABCDRSE* [158]. PcoRS are proteins of a two-component regulatory system (see Section 5.3) and PcoABCD showed homology to the CopABCD proteins from *P. syringae* [158,169]. In *P. syringae* presence of *copAB* allowed for partial copper resistance, whereas all four genes *copABCD* were required for full resistance and *copCD* alone rendered the cells copper-hypersensitive [29,170]. In the *pco* system absence of any of the genes *pcoA, pcoB, pcoC* or *pcoD* resulted in a decrease of copper tolerance and copper accumulation in the cells was similar to the *pcoABCD* control when *pcoC* or *pcoD* were missing, whereas the lack of *pcoA* or *pcoB* resulted in copper accumulation similar to the *pco*-free control [159]. PcoA was able to partially complement a *ΔcueO* phenotype in *E. coli*, confirming its function as periplasmic multicopper oxidase involved in copper detoxification [159].

The proposed model of copper resistance via *pco*-like systems places PcoB/CopB into the outer membrane. PcoA/CopA is a periplasmic multicopper oxidase which binds 11 copper atoms and PcoC/CopC is a periplasmic protein binding one copper atom. PcoD/CopD is located in the cytoplasmic membrane involved in uptake of copper, most likely to ensure complete folding and copper binding of PcoA/CopA before it is transported into the periplasm via the TAT system [171,172]. Pco-like systems have been identified in a variety of plasmid sequences, e.g., *Klebsiella pneumoniae* subsp. *pneumoniae* MGH 78578 plasmid pKPN3 or *Escherichia coli* APEC O1 plasmid pAPEC-O1-R. It seems that within the *Enterobacteriaceae* this type of copper resistance system is encoded on plasmids, whereas among the *Pseudomonadaceae* and *Xanthomonadaceae* they are chromosomally encoded.

5 Copper-Responsive Regulators

Copper homeostasis, as for other essential trace elements, is a highly regulated balancing act for a cell. While it has to be assured that the necessary copper-containing enzymes are sufficiently supplied with copper, the cell has to prevent the accumulation of access amounts of copper ions to prevent copper-induced cell damage. While uptake in most bacteria and archaea seems to be unspecific (see Section 2) and overall regulation is not well understood, expression of genes involved in copper resistance mechanisms (as described in Section 4) are highly regulated, to ensure the prompt removal of toxic levels of copper ions. Depending on whether cytoplasmic or periplasmic copper levels are affected by the regulated gene products, copper ion concentrations are either monitored in the cytoplasm (transcriptional activators and repressors, respectively) or periplasm (two-component regulatory systems).

5.1 *Copper-Dependent Transcriptional Activators*

In *E. coli* expression of *copA*, encoding the copper-transporting P_{1B}-ATPase, as well as *cueO*, encoding a multicopper oxidase involved in copper homoeostasis, are regulated by CueR, a MerR-like transcriptional activator [42,173]. Like other members of this family, CueR possesses a N-terminal helix-turn-helix motif and a predicted C-terminal metal binding site [173]. The CueR binding site of the *copA* promoter/operator region was identified to be between the –10 and –35 site, and within this region an inverted repeat is present (ACCTTCC-N_7-GGAAGGT), which can also be found between the –10 and –35 region of the *cueO* promoter/operator (CCTTCCC-N_5-GGGAAGG) [42]. The crystal structure of CueR revealed that the protein forms a dimer and coordinates Cu(I) in a linear fashion [174]. Promoter (*copA* and/or *cueO*) reporter-gene (*lacZ*) fusions showed that the expression of both genes is copper-dependent and that deletion of *cueR* leads to a loss of expression, indication for CueR acting as an activator [42,173,175].

Homologues of CueR have also been identified in other Gram-negative bacteria. In *P. putida* and *Rhodobacter sphaeroides cueR* is located directly down-stream of *copA* [176,177] whereas in *E. coli cueR* and the repressor's regulatory targets are further apart on the chromosome. In *Serratia marcescens copA* and *cueR* are separated by the *pig* cluster, which is responsible for the biosynthesis of prodigiosin, the red pigment the organism produces. Production of this secondary metabolite in the wild-type is reduced after the addition of copper, whereas a *ΔcopA* strain showed an increase in prodigiosin production [178]. Other examples of reported copper-dependent gene expression by members of the MerR family include *C. metallidurans* and *Rhizobium leguminosarum* [179,180].

A novel archaeal-specific copper-responsive transcription activator, CopR, has been identified in *Sulfolobus solfataricus* 98/2 [181]. Deletion of *copR* resulted in loss of copper-dependent activation of *copA* expression, indicating that CopR is a transcriptional activator. The gene *copR* is located directly upstream of *copTA*, encoding a metallochaperone and a CP*x*-ATPase and two models of copper-dependent gene expression are proposed. At low copper concentrations all three genes are expressed from the *copR* promoter at low levels. At high copper concentrations, CopR binds Cu(I) and binds to the *copTA* promoter region inducing high level expression of *copTA*.

5.2 Copper-Dependent Transcriptional Repressors

Whereas most identified copper-dependent transcriptional activators belong to the MerR-like family of regulatory proteins, more diversity exists within copper-dependent transcriptional repressors.

The first copper-dependent repressor was identified in *E. hirae*. The gene *copY* was identified as one of a two genes, *copY* and *copZ*, located directly up-stream of the genes encoding the two copper-transporting P_{1B}-type copper ATPases CopA and CopB [182]. Early studies indicated that disruption of *copY* leads to constitutive overexpression of *copAB*, whereas disruption of *copZ* rendered the cells copper-sensitive by suppressing *copAB* expression, leading to the conclusion that CopZ and CopY are trans-acting copper-dependent regulatory proteins of the *copAB* operon [182]. Later studies suggested CopY to be a copper-dependent transcriptional repressor of the *cop* operon, and CopZ to be a cytoplasmic copper-chaperone responsible to transfer Cu(I) to CopY as well as to CopA for export [183,184]. Strausak and Solioz identified two operator sites within the *cop* promoter/operator region to which CopY was able to bind, with one of the operator sites overlapping with the promoter and being located just up-stream of the transcription start, the second operator site located within the transcript and before the translational start of *copY*, the first gene of the *copYZAB* operon [183].

It has been found, that CopY in its native form forms a dimer and that each CopY protein has Zn(II) bound in a 1:1 stoichiometry, $[Zn(II)CopY]_2$, and that this form of the repressor binds to the operator region to repress expression [185,186]. The consensus sequence TACAnnTGTA, termed "*cop*-box", is recognized by Zn(II) bound CopY and has not only been found in *E. hirae* but in other Gram-positive bacteria as well and studies showed that *E. hirae* $[Zn(II)CopY]_2$ binds to these operator sites as well [187]. The binding of Zn(II) to CopY is not only necessary for binding of the dimer to the operator but also for correct formation of Cu(I) bound CopY, which is released from the operator, therefore allowing expression of the *cop* operon [185,186,188]. Cu(I) is transferred to CopY from the cytoplasmic chaperone CopZ which also transfers Cu(I) to the P_{1B}-type ATPase CopA for detoxification [187,189–191].

CopY-like copper-dependent repressors have also been identified in *Streptococcus pneumonia*, and *Lactococcus lactis* as well as other members of the order *Lactobacillae* as part of *cop*-like operons [192–194] indicating this type of regulatory organization is a common motif within this bacterial group.

Whereas CopY was the first identified copper-responsive repressor, proteins belonging to other families of regulators have also been identified to control copper-dependent expression.

In the cyanobacterium *Osciliatoria brevis*, BxmR has been found to regulate expression of *bxa1*, encoding a P_{1B}-ATPase (see Section 5.1) as well as the operon *bxmRbmtA*, therefore autoregulating its own expression as well as expression of a metal-binding metallothionein [195]. BxmR belongs to the SmtB/ArsR family of transcriptional repressors and it is the first member of this family allowing for derepression of transcription in response to Zn(II)/Cd(II) as well as Cu(I)/Ag(I) [195]. In addition to the C-terminal Zn(II) binding site also conserved in other members of this family, BxmR has a N-terminal Cu(I) binding site and it was discovered that BxmR can bind Zn(II), Cd(II), Pb(II), Cu(I), and Ag(I) [196].

CsoR, first identified in *Mycobacterium tuberculosis*, represents another class of copper-responsive transcriptional repressors [197]. This type of regulator has now also been identified and characterized in other bacteria like *Listeria monocytogenes* [198], *B. subtilis* [199], *T. thermophilus* [200], *Staphylococcus aureus* [201], and *Streptomyces lividans* [202] regulating the expression of a gene encoding a P_{1B}-ATPase and a metallochaperone of the CopZ-type transcribed in a single operon. In contrast to the copper-dependent regulation by CopY in *E. hirae* CsoR does not seem to obtain Cu(I) from CopZ but rather by competing with the metallochaperone for free Cu(I) in the cytoplasm [198]. CsoR forms a tetramer with each protein monomer able to bind one Cu(I) [197,203]. In the absence of Cu(I) two CsoR tetramers bind to GC-rich pseudo-inverted repeats and both Cu(I) and Ag(I) but not Zn(II), Cd(II), Co(II), or Pb(II) are able to release CsoR from the operator site allowing for transcription of the genes encoding the P_{1B}-ATPase and CopZ.

Recently, a novel TetR-like copper-responsive repressor, ComR (YcfQ), has been identified in *E. coli* [18]. Instead of regulating the transcription of a cytoplasm-detoxification system as do the other copper-responsive repressors described above, the target of ComR is *comC* (*ycfR*), encoding an outer membrane protein, decreasing the permeability of the outer membrane to copper. Using the *lux* reporter gene under control of the *E. coli copA*-promoter, clones showing decreased cytoplasmic copper were identified, leading to the identification of *comR* and the regulator's target *comC* [18].

Expression of genes encoding the copper-transporting P_{1B}-ATPase CopA and the corresponding metallochaperone CopM in the archeon *Sulfolobus solfataricus* P2 is regulated by CopT [119]. CopT represents the first described member of a novel group of archaeal metal-responsive regulators and homologues have been identified in the genome of members of the Crenarcheota as well as of Euryarcheota. While the gene *copT* is physically linked in *S. solfataricus* P2 and other archaea, it has also been identified to be located elsewhere on the chromosome in other archaea. CopT contains a C-terminal TRASH metal-binding domain and an N-terminal helix-turn-helix DNA-binding motif. Expression of *copMA* was induced by Cd(II) and Cu(II), however, only Cu(II) did cause CopT to dissociate from the DNA. It was found that CopT has multiple DNA-binding sites located up- and down-stream of the *copMA* TATA-box within the *copT-copMA* intergenic region [119].

5.3 Two-Component Regulatory Systems

In contrast to the regulatory proteins described in Sections 5.1 and 5.2, that detect Cu(I) within the cytoplasm of the cell and regulation of gene expression depends upon binding of Cu(I) to the respective regulatory protein; several bacteria use two-component regulatory systems which detect elevated levels of copper ions outside the cytoplasm.

Copper-dependent two-component regulatory systems involved in regulation of copper homoeostasis genes, were first described to regulate expression of the plasmid-bound copper resistance determinant *cop* in *P. syringae* [204] and have now been identified and characterized in a variety of bacteria. CopS is a membrane-bound histidine sensor kinase and CopR the corresponding response regulator, which once phosphorylated by CopS, binds to the "*cop*-box" of the *copABCD* operon, an inverted repeat sequence, and inducing expression of the operon. A homologous system has been identified on the chromosome of copper-resistant strains of *P. syringae*, which also contains *copRS* and *copABCD*, and expression in a copper-sensitive strain slightly increased its copper tolerance [205]. *Cop* homologues, *pco*, have also been identified on an plasmid of *E. coli* [158].

Transcriptional analysis of the *pco* determinant showed the involvement of PcoRS in copper-dependent transcription regulation of the *pcoA* and *pcoE* promoters. However, it was also found that in the absence of *pcoRS* the promoters remained copper-inducible, suggesting the presence a chromosomal copper-responsive two-component regulatory system [206]. Munson et al. [130] identified the chromosomal-encoded two-component regulatory system, CusRS, responsible for induction of the *pco* genes. In contrast to CopRS and PcoRS, CusRS does not regulate expression of an MCO-linked resistance system but an RND transport system, CusCFBA. Expression of *cus* is activated by external Cu(I) and Ag(I) and the activation with copper is stronger under anaerobic conditions because Cu(I) is present in higher amounts [132,155]. *E. coli*'s Cus system plays an important role to prevent silver-mediated damage to the multicopper oxidase CueO under aerobic conditions [69]. Two-component systems involved in copper-dependent gene expression have also been identified in *L. lactis*, *C. metallidurans*, pathogenic strains of *P. fluorescens*, and numerous genome projects [31,160,207].

6 Copper Omics

With the number of newly posted genomic sequences increasing on a nearly daily base, the number of potential genes encoding proteins involved in copper handling (uptake, regulation, detoxification) or containing Cu as cofactor is constantly increasing. Whereas in 2000 the only homologue of CusF from *E. coli* was the potential gene product of an uncharacterized ORF within the silver resistance determinant from pMG101 *Salmonella* Typhimurium, a BLAST search [30] of CusF now produces over 100 results and any comparison in genomics of copper handling will be immediately outdated and never be close to completion.

Therefore, in the concluding section, we would like to describe our view of copper toxicity and the bacterial response. In our opinion, copper is primarily toxic to cells by inactivation of enzymes with exposed iron-sulfur clusters [2,3]. Here, $[4Fe-4S]^{2+}$ were first degraded to a $[3Fe-4S]^0$ form that could still be repaired *in vitro* and *in vivo* and then further degraded through unknown intermediates that left only the apoenzyme. Enzymes that could thus be inactivated include isopropylmalate isomerase, 6-phosphogluconate dehydratase, and fumarase A [2]. These solvent-exposed accessible iron-sulfur clusters of dehydratases are also the target of reactive oxygen species such as superoxide and hydrogen peroxide although the pathway leading to degradation is different [208]. In any case, the end result is an increase in unincorporated „free“ iron. The subsequent Fenton reaction can lead to oxidation of DNA and peroxidation of lipids. This in turn leads to an increased production of enzymes dealing with reactive oxygen species as shown in *Kineococcus radiotolerans* [209]. It should be pointed out here that other soft metals such as Ag(I), Hg(II), Cd(II), and Zn(II) also cause a similar damage and lead to oxidative stress even though they are not redox-active. If the damage is not extensive and the Fe-S cluster was only degraded to a $[3Fe-4S]^0$ form, the damage can be readily repaired [3].

The oxidative damage by increased production of reactive oxygen species can also lead to damage of mononuclear iron enzymes such as ribulose-5-phosphate 3-epimerase (Rpe) a key enzyme in the pentose-phosphate pathway. This and other enzymes can be activated by the presence of manganese. In general, cells can be protected from oxidative stress by shifting from an iron- to a manganese-centered metabolism [210,211].

One question to our assumptions thus is, whether this is reflected in results from transcriptomics and/or proteomics. It is encouraging that oxidative stress enzymes are indeed increasingly produced in a number of different microorganisms [209,212,213]. Most studies are performed with wild-type cells and therefore with cells fully equipped to deal with excess copper. Therefore, when copper is added to growing cells the copper resistance systems such as the Cu(I)-translocating P_{1B}-type ATPase CopA and the multicopper oxidase CueO are immediately induced and produced after a few minutes. CopA pumps out any unbound copper rapidly as Cu(I) which is then oxidized to Cu(II) by CueO in the periplasm preventing reentry into the cytoplasm and oxidative damage by the prooxidant Cu(I) in the periplasm. Therefore, most of the current transcriptomic and proteomic studies are performed under condition when no copper induced damage is occuring in the cytosol and therefore the cellular response can really not be observed [160,214,215]. In *S. aureus* microarrays three types of genes were upregulated after addition of copper. Those specifically dealing with copper efflux and resistance are the first type. The second type are genes involved in handling oxidative stress and the third type ensure proper folding of proteins which will be disturbed by copper [201].

In studies using *M. tuberculosis* under copper stress both iron and sulfur uptake and utilization genes were upregulated consistent with a role in repair of iron-sulfur clusters [216]. Two conditions were tested in *P. aeruginosa*, copper shock and copper adaptation [217]. In copper adaptation, expression of 331 genes changed, whereas copper shock influenced the expression level of 405 genes. Copper shock induced an oxidative stress response, which was not observed when cells were growing constantly in the presence of copper (copper adaptation). While copper-adapted cells seemed to limit copper uptake (down-regulation of outermembrane factors), copper-shocked cells drastically increased expression of efflux systems and oxidative stress response [217].

Transcriptomics and proteomics were also used to identify genes involved in copper response in *C. metallidurans* CH34 [160]. On plasmid pMOL30 a cluster of 19 genes was detected to show increased expression levels in the presence of copper. While some of the genes encode known proteins involved in copper detoxification (CP*x*-ATPase, PcoABCD-homologue, CopRS two-component regulatory system) others have no known function in copper metabolism. Characterization of some of the members of *C. metallidurans*' *cop* clusters has begun to shed some light into their involvement [161,163,218–220].

Two studies have been published evaluating the response to the bacterial model organism *E. coli* to copper. One evaluated expression patterns of medium copper exposure (0.5 mM) over time [215], whereas the second study compared transcription profiles under different copper concentrations (0 mM, 0.75 mM, and 2 mM) [214]. As expected, both studies saw expression of *copA*, *cueO*, and *cusC* (as part of the *cusCFBA* operon) with the highest expression levels occurring after 5 min of copper exposure and reaching pre-copper conditions within 30 min [215]. Besides the genes extensively described to be involved in *E. coli* copper resistance, additional transcripts were identified in the study using 0.5 mM copper which could be grouped into four regulatory networks, CueR, CusRS, YedWV, and the global response CpxAR [215].

Comparison of transcription profiles under various copper concentrations showed a similar response, however, additional genes could be identified to change expression levels under increasing copper concentration. Genes involved in flagellar biosynthesis showed a general increased expression at 0.75 mM copper (2–3-fold induction) and decreased transcription (~0.5-fold induction compared to 0 mM copper) when 2 mM copper were added. Other copper responses include genes involved in iron homeostasis, energy metabolism, and envelope stress [214].

Interestingly, all these processes can be readily observed in the interactions between macrophages and bacteria. As a first response, macrophages show increased expression of the gene encoding the Mn(II) and Fe(II) transporter NRAMP (natural resistance-associated macrophage protein). This would ensure that microbes engulfed by macrophages will find it harder to acquire Mn or Fe [221]. At the same time macrophages pump Cu(I) and perhaps Zn(II) into the phagosome [222,223]. This would lead to destruction of iron-sulfur clusters and subsequent release of iron, resulting in an increase in oxidative stress made worse by the reactive oxygen species generated by the macrophages and the prooxidant properties of free Cu(I) and Fe(II). Manganese could help protect cells from oxidative damage by Mn-SOD and in general switching from an iron based metabolism to a manganese based metabolism. This would be prevented by pumping out Mn through the NRAMP transporter.

7 Concluding Remarks and Future Directions

Comparative analysis of the copper proteome of prokaryotes and Eukarya revealed that although the total number of copper-containing proteins is higher in Eukarya than prokaryotes, aerobic archaea show the highest total percentage of copper proteins in their cells. In anaerobic archaea and bacteria the amount of copper proteins is the lowest, suggesting that copper was not used until later in evolution [224]. Copper-containing proteins are involved in electron transfer and throughout evolution prokaryotes have developed highly balanced uptake and resistance mechanisms which are tightly controlled. Copper metabolism (uptake, copper proteins, detoxification) allows bacteria and archaea to live in a variety of environments and to play an important part in nutrient cycling. In addition, more and more evidence is provided on the importance of copper homeostasis in bacterial virulence.

New studies indicate that macrophages take up copper and transport it into the phagosome in response to bacterial infection [222,223,225]. In addition to the disruption of Fe-S clusters and generation of reactive oxygen species, the copper ions themselves play an important role as antimicrobial agents. It has been shown that known bacterial copper resistance systems play an important role in bacterial virulence [71,121,194]. Understanding the mechanism of how these systems detoxify copper and their involvement in pathogenicity will open new avenues for drug development [226].

Although understanding bacterial virulence is of great importance, pathogenic bacteria represent only a small proportion of prokaryotes in our environment. The majority of them play important roles in nutrient cycling on earth and copper-containing enzymes are an important part of this. Copper proteins are involved in the global carbon and nitrogen cycle and might provide the answer for global warming as well as bioremediation. Two of the major greenhouse gases are methane and nitrous oxide.

Human agricultural practice, especially the application of synthetic fertilizers affects the nitrogen cycle by adding nitrate to the fields. Increased supply of nitrate leads to increased denitrification and therefore also increased release of NO and N_2O [92,110]. The enzyme responsible for reduction of N_2O to N_2, nitrous oxide reductase, contains two copper centers, Cu_A and Cu_Z, and it can be speculated that the limitation of bioavailable copper might be a reason for increased N_2O emission due to incomplete denitrification. Test to increase the removal of total nitrogen from wastewater showed that adding copper decreased the nitrogen content in the wastewater [111]. Perhaps a similar approach could be taken to decrease the N_2O emission from agricultural lands. However, in our opinion the main reason for increased N_2O production is the high amount of nitrogen fertilizers added to agricultural soils. Therefore, microbes carrying out denitrification do not have an energetic advantage from nitrous oxide reduction because there is ample nitrate available.

The second greenhouse gas mentioned above, CH_4, is produced by methanogenic archaea and that in turn can be used by methanotrophic bacteria as carbon and energy source. Production of methane occurs under anaerobic conditions and rising global temperatures increasingly provide these conditions in the thawing top layer of permafrost. In addition, human activity of animal husbandry of ruminants as well as anaerobic sludge digestion in waste-water treatment increases the release of CH_4 into the atmosphere. Again, the key enzyme to utilize CH_4 relies on the availability of copper for the active center of methane monooxygenase and methanotrophs could be the answer to the increasing problem of methane.

Other copper-containing enzymes also have biotechnological potential. Laccases, a group of multicopper oxidases, are heterologously produced and are starting to be used in various industrial processes [62]. Since laccases usually have a variety of phenolic substrates, they have a potential in bioremediation [227–229] and bioengineering and can be used to modify their substrate specificity [230].

Taken together, copper plays an important role not only for cellular function of prokaryotes but also enables them to play an important part in global nutrient cycles.

Abbreviations

ABC	ATP-binding cassette
AMO	ammonia monooxygenase
BLAST	Basic Local Alignment Search Tool
BLASTP	BLAST analysis searching protein databases using a protein query

COX	cytochrome oxidase
Cu-AO	copper-containing amine oxidase
CuNIR	copper-containing nitrite reductase
EPR	electron paramagnetic resonance
GSH	glutathione
HCO	heme-copper oxygen reductase
ITC	isothermal titration colorimetry
MCO	multicopper oxidase
MFP	membrane fusion protein
MFS	major facilitator superfamily
MMO	methane monooxygenase
N_2OR	nitrous oxide reductase
NO	nitric oxide
NOR	nitric oxide reductase
NRAMP	natural resistance-associated macrophage protein
OB	oligonucleotide/oligosaccharide binding
OMF	outermembrane factor
ORF	open reading frame
pMMO	particulate methane monooxygenase
RND	resistance-nodulation cell division
ROS	reactive oxygen species
Rpe	ribulose-5-phosphate 3-epimerase
sMMO	soluble methane monooxygenase
SOD	superoxide dismutase
TAT	twin-argininine translocation
TBLASTN	BLAST analysis searching translated nucleotide databases using a protein query
TMH	transmembrane helix
TM-MBS	transmembrane metal-binding site
TNC	trinuclear cluster
TPQ	trihydroxyphenylalanine
TRASH	name of a metal-binding domain containing a well conserved cysteine motif

Acknowledgments This work was supported by a Teaching Load Modification Grant for S.F.M., funded by a grant awarded to Skidmore and Union Colleges from the National Science Foundation's Advance PAID project (NSF 0820080).

References

1. K.T. Yasunobu, M. Tanaka, *Syst. Zool.* **1974**, *22*, 570–589.
2. L. Macomber, J. A. Imlay, *Proc. Natl. Acad. Sci. USA* **2009**, *106*, 8344–8349.
3. F. F. Xu, J. A. Imlay, *Appl. Environ. Microbiol.* **2012**, *78*, 3614–361421.
4. C. L. Dupont, G. Grass, C. Rensing, *Metallomics* **2011**, *3*, 1109–1118.
5. C. Andreini, I. Bertini, A. Rosato, *Acc. Chem. Res.* **2009**, *42*, 1471–1479.
6. J. M. Arguello, *J. Membr. Biol.* **2003**, *195*, 93–108.

7. D. Raimunda, M. Gonzalez-Guerrero, B. W. Leeber, 3rd, J. M. Arguello, *Biometals* **2011**, *24*, 467–475.
8. N. J. Robinson, D. R. Winge, *Annu. Rev. Biochem.* **2010**, *79*, 537–562.
9. S. Tottey, C. J. Patterson, L. Banci, I. Bertini, I. C. Felli, A. Pavelkova, S.J. Dainty, R. Pernil, K. J. Waldron, A. W. Foster, N. J. Robinson, *Proc. Natl. Acad. Sci. USA* **2012**, *109*, 95–100.
10. S. Ekici, H. Yang, H. G. Koch, F. Daldal, *MBio*, **2012**, *3*, 10.1128/mBio.00293,11.
11. M. Gonzalez-Guerrero, J. M. Arguello, *Proc. Natl. Acad. Sci. USA* **2008**, *105*, 5992-5997.
12. J. M. Arguello, E. Eren, M. Gonzalez-Guerrero, *Biometals* **2007**, *20*, 233–248.
13. J. M. Arguello, M. Gonzalez-Guerrero, *Structure* **2008**, *16*, 833–834.
14. O. Lewinson, A. T. Lee, C. D. Rees, *Proc. Natl. Acad. Sci. USA* **2009**, *106*, 4677–4682.
15. Y. P. Tsai, H.T. Chen, *Bioresour. Technol.* **2011**, *102*, 11043–11047.
16. J. F. Lutkenhaus, *J. Bacteriol.* **1977**, *131*, 631–637.
17. R. Balasubramanian, G. E. Kenney, A. C. Rosenzweig, *J. Biol. Chem.* **2011**, *286*, 37313–37319.
18. M. Mermod, D. Magnani, M. Solioz, J. V. Stoyanov, *Biometals* **2012**, *25*, 33–43.
19. G. Grass, S. Franke, N. Taudte, D. H. Nies, L. M. Kucharski, M. E. Maguire, C. Rensing, *J. Bacteriol.* **2005**, *187*, 1604–1611.
20. G. Grass, M. D. Wong, B. P. Rosen, R. L. Smith, C. Rensing, *J. Bacteriol.* **2002**, *184*, 864–866.
21. R. Janulczyk, J. Pallon, L. Bjorck, *Mol. Microbiol.* **1999**, *34*, 596–606.
22. J. S. Cavet, G. P. Borrelly, N. L. Robinson, *FEMS Microbiol. Rev.* **2003**, *27*, 165–181.
23. M. Solioz, J. V. Stoyanov, *FEMS Microbiol. Rev.* **2003**, *27*, 183–195.
24. K. Y. Djoko, Z. Xiao, D. L. Huffman, A. G. Wedd, *Inorg. Chem.* **2007**, *46*, 4560–4568
25. F. Arnesano, L. Banci, I. Bertini, I. C. Felli, C. Luchinat, A. R. Thompsett, *J. Am. Chem. Soc.* **2003**, *125*, 7200–7208.
26. F. Arnesano, L. Banci, I. Bertini, S. Mangani, A. R. Thompsett, *Proc. Natl. Acad. Sci. USA* **2003**, *100*, 3814–3819.
27. F. Arnesano, L. Banci, I. Bertini, A. R. Thompsett, *Structure* **2002**, *10*, 1337–1347.
28. S. Chillappagari, M. Miethke, H. Trip, O. P. Kuipers, M. A. Marahiel, *J. Bacteriol.* **2009** *191*, 2362–2370.
29. J. S. Cha, D. A. Cooksey, *Appl. Environ. Microbiol.* **1993**, *59*, 1671–1674.
30. S. F. Altschul, T. L. Madden, A. A. Schaffer, J. Zhang, Z. Zhang, W. Miller, D. J. Lipman, *Nucleic Acids Res.* **1997**, *25*, 3389–3402.
31. Y. H. Hu, H. L. Wang, M. Zhang, L. Sun, *J. Microbiol.* **2009**, *47*, 277–286.
32. M.W. Fitch, D. W. Graham, R. G. Arnold, S. K. Agarwal, P. Phelps, G. E. Speitel Jr, G. Georgiou, *Appl. Environ. Microbiol.* **1993**, *59*, 2771–2776.
33. J. A. Zahn, A. A. DiSpirito, *J. Bacteriol.* **1996**, *178*, 1018–1029.
34. A. A. DiSpirito, J. A. Zahn, D. W. Graham, H. J. Kim, C. K. Larive, T. S. Derrick, C. D. Cox, A.Taylor, *J. Bacteriol.* **1998**, *180*, 3606–3613.
35. H. J. Kim, D. W. Graham, A. A. DiSpirito, M. A. Alterman, N. Galeva, C. K. Larive, D. Asunskis, P. M. Sherwood, *Science* **2004**, *305*, 1612–1615.
36. J. Anttila, P. Heinonen, T. Nenonen, A. Pino, H. Iwai, E. Kauppi, R. Soliymani, M. Baumann, J. Saksi, N. Suni, T. Haltia, *Biochim. Biophys. Acta* **2011**, *1807*, 311–318.
37. K. Nicolaisen, A. Hahn, M. Valdebenito, S. Moslavac, A. Samborski, I. Maldener, C. Wilken, A. Valladares, E. Flores, K. Hantke, E. Schleiff, *Biochim. Biophys. Acta* **2010**, *1798*, 2131–2140.
38. J. Beaudoin, R. Ioannoni, L. Lopez-Maury, J. Bahler, S. Ait-Mohand, B. Guerin, S. C. Dodani, C. J. Chang, S. Labbe, *J. Biol. Chem.* **2011**, *286*, 34356–34372.
39. M. Gonzalez-Guerrero, D. Raimunda, X. Cheng, J. M. Arguello, *Mol. Microbiol.* **2010**, *78*, 1246–1258.
40. B. K. Hassani, C. Astier, W. Nitschke, S. Ouchane, *J. Biol. Chem.* **2010**, *285*, 19330–19337.
41. G. Grass, C. Rensing, *Biochem. Biophys. Res. Commun.* **2001**, *286*, 902–908.

42. F. W. Outten, C. E. Outten, J. Hale, T. V. O'Halloran, *J. Biol. Chem.* **2000**, *275*, 31024–31029
43. P. G. Ridge, Y. Zhang, V. N. Gladyshev, *PLoS One* **2008**, *3*, e1378.
44. F. L. Sousa, R. J. Alves, M. A. Ribeiro, J. B. Pereira-Leal, M.Teixeira, M. M. Pereira, *Biochim. Biophys. Acta* **2012**, *1817*, 629–637.
45. F. L. Sousa, R. J. Alves, J. B. Rereira-Leal, M. Teixeira, M. M. Pereira, *A Bioinformatics Classifier and Database for Heme-Copper Oxygen Reductases*, PLoS One **2011**. 6, e19117.
46. E. Frangipani, D. Haas, *FEMS Microbiol. Lett.* **2009**, *298*, 234–240.
47. D. L. Swem, L. R. Swem, A. Setterdahl, C. E. Bauer, *J. Bacteriol.* **2005**, *187*, 8081–8087.
48. D. Buhler, R. Rossmann, S. Landolt, S. Balsiger, H. M. Fischer, H. Hennecke, *J. Biol. Chem.* **2010**, *285*, 15704–15713.
49. F. Arnesano, L. Banci, I. Bertini, M. Martinelli, *J. Proteome Res.* **2005**, *4*, 63–70.
50. Y. V. Chinenov, *J. Mol. Med.* **2000**, *78*, 239–242.
51. E. Balatri, L. Banci, I. Bertini, F. Cantini, S. Ciofi-Baffoni, *Structure* **2003**, *11*, 1431-1443
52. G. S. Siluvai, M. Mayfield, M. J. Nilges, S. Debeer George, N. J. Blackburn, *J. Am. Chem. Soc.* **2010**, *132*, 5215–5226.
53. G. S. Siluvai, M. Nakano, M. Mayfield, N. J. Blackburn, *J. Biol. Inorg. Chem.* **2011**, *16*, 285–297.
54. G. S. Siluvai, M. M. Nakano, M. Mayfield, M. J. Nilges, N. J. Blackburn, *Biochemistry* **2009**, *48*, 12133–12144.
55. L. Banci, I. Bertini, G. Cavallaro, S. Ciofi-Baffoni, *FEBS J.* **2011**, *278*, 2244–2262.
56. D. J. Kosman, *J. Biol. Inorg. Chem.* **2010**, *15*, 15–28.
57. T. Sakurai, K. Kataoka, *Chem. Rec.* **2007**, *7*, 220–229.
58. H. Claus, *Arch. Microbiol.* **2003**, *179*, 145–150.
59. S. Uthandi, B. Saad, M. A. Humbard, J. A. Maupin-Furlow, *Appl. Environ. Microbiol.* **2010**, *76*, 733–743.
60. H. Claus, *Micron* **2004**, *35*, 93–96.
61. M. Ferrer, A. Beloqui, P.N. Golyshin, *Methods Mol. Biol.* **2010**, *668*, 189–202.
62. A. Piscitelli, C. Pezzella, P. Giardina, V. Faraco, S. Giovanni, *Bioeng. Bugs* **2010**, *1*, 252–262
63. G. Grass, C. Rensing, *J. Bacteriol.* **2001**, *183*, 2145–2147.
64. S. A. Roberts, A. Weichsel, G. Grass, K. Thakali, J. T. Hazzard, G. Tollin, C. Rensing, W. R. Montfort, *Proc. Natl. Acad. Sci. USA* **2002**, *99*, 2766–2771.
65. S. A. Roberts, G. F. Wildner, G. Grass, A. Weichsel, A. Ambrus, C. Rensing, W. R. Montfort, *J. Biol. Chem.* **2003**, *278*, 31958–31963.
66. G. Grass, K. Thakali, P. E. Klebba, D. Thieme, A. Muller, G. F. Wildner, C. Rensing, *J. Bacteriol.* **2004**, *186*, 5826–5833.
67. S. K. Singh, G. Grass, C. Rensing, W. R. Montfort, *J. Bacteriol.* **2004**, *186*, 7815–7817.
68. K. Y. Djoko, X. L. Chong, A. G. Wedd, Z. Xiao, *J. Am. Chem. Soc.* **2010**, *132*, 2005–2015.
69. S. K. Singh, S. A. Roberts, S. F. McDevitt, A. Weichsel, G. F. Wildner, G. B. Grass, C. Rensing, W. R. Montfort, *J. Biol. Chem.* **2011**, *286*, 37849–37857.
70. S. Y. Lim, M. H. Joe, S. S. Song, M. H. Lee, J. W. Foster, Y. K. Park, S. Y. Choi, I. S. Lee, *Mol. Cells* **2002**, *14*, 177–184.
71. M. E. Achard, J. J. Tree, J. A. Holden, K. R. Simpfendorfer, O. L. Wijburg, R. A. Strugnell, M. A. Schembri, M. J. Sweet, M. P. Jennings, A. G. McEwan, *Infect. Immun.* **2010**, *78*, 2312–2319.
72. H. B. Gray, B. G. Malmstrom, R. J. P. Williams, *J. Biol. Inorg. Chem.* **2000**, *5*, 551–559.
73. F. De Rienzo, R. R. Gabdoulline, M. C. Menziani, R. C. Wade, *Protein Sci.* **2000**, *9*, 1439–1454.
74. Z. W. Chen, M. J. Barber, W. S. McIntire, F. S. Mathews,. *Acta Crystallogr. D Biol. Crystallogr.* **1998**, *54*, 253–268.
75. J. Elguindi, J. Wagner, C. Rensing, *J. Appl. Microbiol.* **2009**, *106*, 1448–1455.

76. P. Pietrangeli, S. Nocera, B. Mondovi, L. Morpurgo, *Biochim. Biophys. Acta* **2003**, *1647*, 152–156.
77. M. C. Wilce, D. M. Dooley, H. C. Freeman, J. M. Guss, H. Matsunami, W. S. McIntire, C. E. Ruggiero, K. Tanizawa, H. Yamaguchi, *Biochemistry* **1997**, *36*, 16116–16133.
78. H. C. Dawkes, S. E. Phillips, *Curr. Opin. Struct. Biol.* **2001**, *11*, 666–673.
79. M. Kataoka, H. Oya, A. Tominaga, M. Otsu, T. Okajima, K. Tanizawa, H. Yamaguchi, *J. Synchrotron Radiat.* **2011**, *18*, 58–61.
80. R. S. Hanson, T. E. Hanson, *Microbiol. Rev.* **1996**, *60*, 439–471.
81. A. K. Nielsen, K. Gerdes, J. C. Murrell, *Mol. Microbiol.* **1997**, *25*, 399–409.
82. J. C. Murrell, I. R. McDonald, B. Gilbert, *Trends Microbiol.* **2000**, *8*, 221–225.
83. H. H. Nguyen, S. J. Elliott, J. H. Yip, S. I. Chan, *J. Biol. Chem.* **1998**, *273*, 7957–7966.
84. J. D. Semrau, D. Zolandz, M. E. Lidstrom, S. I. Chan, *J. Inorg. Biochem.* **1995**, *58*, 235–244.
85. C. W. Knapp, D. A. Fowle, E. Kulczycki, J. A. Roberts, D. W. Graham, *Proc. Natl. Acad. Sci. USA* **2007**, *104*, 12040–12045.
86. R. L. Lieberman, A. C. Rosenzweig, *Nature* **2005**, *434*, 177–182.
87. A. S. Hakemian, K. C. Kondapalli, J. Telser, B. M. Hoffman, T. L. Stemmler, A. C. Rosenzweig, *Biochemistry* **2008**, *47*, 6793–6801.
88. S. M. Smith, S. Rawat, J. Telser, B. M. Hoffman, T. L. Stemmler, A. C. Rosenzweig, *Biochemistry* **2011**, *50*, 10231–10240.
89. R. Balasubramanian, S. M. Smith, S. Rawat, L. A. Yatsunyk, T. L. Stemmler, A. C. Rosenzweig, *Nature* **2010**, *465*, 115–119.
90. J. M. Bollinger Jr., *Nature* **2010**, *465*, 40–41.
91. R. A. Himes, K. Barnese, K. D. Karlin, *Angew. Chem. Int. Ed Engl.* **2010**, *49*, 6714–6716.
92. R. M. Martinez-Espinosa, J. A. Cole, D. J. Richardson, N. J. Watmough, *Biochem. Soc. Trans.* **2011**, *39*, 175–178.
93. B. Kraft, M. Strous, H. E. Tegetmeyer, *J. Biotechnol.* **2011**, *155*, 104–117.
94. W. G. Zumft, *Microbiol. Mol. Biol. Rev.* **1997**, *61*, 533–616.
95. S. A. Ensign, M. R. Hyman, D. J. Arp, *J. Bacteriol.* **1993**, *175*, 1971–1980.
96. I. Schmidt, E. Bock, *Antonie Van Leeuwenhoek* **1998**, *73*, 271–278.
97. S. Gilch, O. Meyer, I. Schmidt, *Biol. Chem.* **2009**, *390*, 863–873.
98. S. Gilch, O. Meyer, I. Schmidt, *Biometals* **2010**, *23*, 613–622.
99. M. B. Lund, J. M. Smith, C.A. Francis, *ISME J.* **2012**, doi: 10.1038/ismej.2012.40. [Epub ahead of print]
100. H. Iwasaki, S. Shidara, H. Suzuki, T. Mori, *J. Biochem.* **1963**, *53*, 299–303.
101. I. M. Wasser, S. de Vries, P. Moenne-Loccoz, I. Schroder, K. D. Karlin, KD. *Chem. Rev.* **2002**, *102*, 1201–1234.
102. E. T. Adman, W. J. Godden, S. Turley, *J. Biol. Chem.* **1995**, *270*, 27458–27474.
103. M. E. Murphy, S. Turley, E. T. Adman, *J. Biol. Chem.* **1997**, *272*, 28455–28460.
104. J. A. Farrar, A. J. Thomson, M. R. Cheesman, M. D. Dooley, W. G. Zumft, *FEBS Lett. IB* **1991**, *294*, 11–15.
105. T. Rasmussen, B. C. Berks, J. Sanders-Loehr, M.D. Dooley, W. G. Zumft, A. J. Thomson, *Biochemistry* **2000**, *39*, 12753–12756.
106. T. Haltia, K. Brown, M. Tegoni, C. Cambillau, M. Saraste, K. Mattila, K. Djinovic-Carugo, *Biochem. J.* **2003**, *369*, 77–88.
107. W. G. Zumft, H. Korner, *Antonie Van Leeuwenhoek* **1997**, *71*, 43–58.
108. P. Wunsch, M. Herb, H. Wieland, U. M. Schiek, W. G. Zumft, *J. Bacteriol.* **2003**, *185*, 887–896.
109. A. Pomowski, W. G. Zumft, P. M. Kroneck, O. Einsle, *Nature* **2011**, *477*, 234–237.
110. J. Ollivier, S. Towe, A. Bannert, B. Hai, E.M. Kastl, A. Meyer, M. X. Su, K. Kleineidam, M. Schloter, *FEMS Microbiol. Ecol.* **2011**, *78*, 3–16.
111. J. H. Fan, L. M. Ma, *J. Hazard. Mater.* **2009**, *164*, 1392–1397.
112. V. Ochoa-Herrera, G. Leon, Q. Banihani, J. A. Field, R. Sierra-Alvarez, *Sci. Total Environ.* **2011**, *412-413*, 380–385.

113. M. Solioz, C. Vulpe, *Trends Biochem. Sci.* **1996**, *21*, 237–241.
114. A. Odermatt, H. Suter, R. Krapf, M. Solioz, *Ann. N. Y. Acad. Sci.* **1992**, *671*, 484–486.
115. M. Solioz, A. Odermatt, R. Krapf, *FEBS Lett.* **1994**, *346*, 44–47.
116. J. M. Arguello, A. K. Mandal, S. Mana-Capelli, *Ann. N. Y. Acad. Sci.* **2003**, *986*, 212–218.
117. M. Solioz, A. Odermatt, *J. Biol. Chem.* **1995**, *270*, 9217–9221.
118. C. Rensing, B. Fan, R. Sharma, B. Mitra, B. P. Rosen, *Proc. Natl. Acad. Sci. USA* **2000**, *97*, 652–656.
119. T. J. Ettema, A. B. Brinkman, P. P. Lamers, N. G. Kornet, W. M. de Vos, J. van der Oost, *Microbiology* **2006**, *152*, 1969–1979.
120. A. K. Mandal, W. D. Cheung, J. M. Arguello, *J. Biol. Chem.* **2002**, *277*, 7201–7208.
121. J. M. Arguello, M. Gonzalez-Guerrero, D. Raimunda, *Biochemistry* **2011**, *50*, 9940–9949.
122. C. Rensing, S. Franke, in *EcoSal. (www.ecosal.org.)*, Eds R. I. Curtiss, J. B. Kaper, C. L. Squires, P. D. Karp, F. C. Neidhardt, J. M. Slauch, ASM Press, Washington, DC, 2007.
123. R. De la Iglesia, D. Valenzuela-Heredia, J. P. Pavissich, S. Freyhoffer, S. Andrade, J. A. Correa, B. Gonzalez, *Lett. Appl. Microbiol.* **2010**, *50*, 552–562.
124. B. Fan, B. P. Rosen, *J. Biol. Chem.* **2002**, *277*, 46987–46992.
125. A. Odermatt, R. Krapf, M. Solioz, *Biochem. Biophys. Res. Commun.* **1994**, *202*, 44–48.
126. L. Banci, I. Bertini, S. Ciofi-Baffoni, R. Del Conte, L. Gonnelli, *Biochemistry* **2003**, *42*, 1939–1949.
127. D. S. Radford, M.A. Kihlken, G. P. Borrelly, C. R. Harwood, N. E. Le Brun, J. S. Cavet, *FEMS Microbiol. Lett.* **2003**, *220*, 105–112.
128. M. Gonzalez-Guerrero, D. Hong, J. M. Arguello, *J. Biol. Chem.* **2009**, *284*, 20804–20811.
129. S. Mana-Capelli, A. K. Mandal, J. M. Arguello, *J. Biol. Chem.* **2003**, *278*, 40534–40541.
130. G. P. Munson, D. L. Lam, F. W. Outten, T. V. O'Halloran, *J. Bacteriol.* **2000**, *182*, 5864–5871.
131. A. Gupta, K. Matsui, J. F. Lo, S. Silver, *Nat. Med.* **1999**, *5*, 183–188.
132. S. Franke, G. Grass, D. H. Nies, *Microbiology* **2001**, *147*, 965–972.
133. S. Franke, G. Grass, C. Rensing, D. H. Nies, *J. Bacteriol.* **2003**, *185*, 3804–3812.
134. D. H. Nies, S. Silver, *J. Ind. Microbiol.* **1995**, *14*, 186–199.
135. I. T. Paulsen, M. H. Brown, R. A. Skurray, *Microbiol. Rev.* **1996**, *60*, 575–608.
136. M. H. Saier Jr., R. Tam, A. Reizer, J. Reizer, *Mol. Microbiol.* **1994** *11*, 841–847.
137. I. R. Loftin, S. Franke, S. A. Roberts, A. Weichsel, A. Heroux, W. R. Montfort, C. Rensing, M. M. McEvoy, *Biochemistry* **2005** *44*, 10533–10540.
138. E. H. Kim, C. Rensing, M. M. McEvoy, *Nat. Prod. Rep.* **2010**, *27*, 711–719.
139. J. T. Kittleson, I. R. Loftin, A. C. Hausrath, K. P. Engelhardt, C. Rensing, M. M. McEvoy, *Biochemistry* **2006**, *45*, 11096–11102.
140. I. R. Loftin, S. Franke, N. J. Blackburn, M. M. McEvoy, *Protein Sci.* **2007** 16, 2287–2293.
141. Y. Xue, A. V. Davis, G. Balakrishnan, J. P. Stasser, B. M. Staehlin, P. Focia, T. G. Spiro, J. E. Penner-Hahn, T. V. O'Halloran, *Nat. Chem. Biol.* **2008**, *4*, 107–109.
142. I. R. Loftin, N. J. Blackburn, M. M. McEvoy, *J. Biol. Inorg. Chem.* **200**, *14*, 905–912.
143. T. D. Mealman, I. Bagai, P. Singh, D. R. Goodlett, C. Rensing, H. Zhou, V. H. Wysocki, M. M. McEvoy, *Biochemistry* **2011**, *50*, 2559–2566.
144. I. Bagai, C. Rensing, N. J. Blackburn, M. M. McEvoy, *Biochemistry* **2008**, *47*, 11408–11414.
145. F. Long, C. C. Su, H. T. Lei, J. R. Bolla, S. V. Do, E. W. Yu, *Philos. Trans. R. Soc. Lond. B. Biol. Sci.* **2012**, *367*, 1047–1058.
146. R. Kulathila, R. Kulathila, M. Indic, B. van den Berg, *PLoS One* **2011**, *6*, e15610.
147. V. Koronakis, A. Sharff, E. Koronakis, B. Luisi, C. Hughes, *Nature* **2000**, *405*, 914–919.
148. C. C. Su, F. Yang, F. Long, D. Reyon, M. D. Routh, D. W. Kuo, A. K. Mokhtari, J. D. Van Ornam, K. L. Rabe, J. A. Hoy, Y. J. Lee, K. R. Rajashankar, E. W. Yu, *J. Mol. Biol.* **2009**, *393*, 342–355.
149. H. Akama, T. Matsuura, S. Kashiwagi, H. Yoneyama, S. Narita, T. Tsukihara, A. Nakagawa, T. Nakae, *J. Biol. Chem.* **2004**, *279*, 25939–25942.

150. F. Long, C. C. Su, M. T. Zimmermann, S. E. Boyken, K. R. Rajashankar, R. L. Jernigan, E. W. Yu, *Nature* **2010**, *467*, 484–488.
151. C. C. Su, F. Long, M. T. Zimmermann, K. R. Rajashankar, R. L. Jernigan, E. W. Yu, *Nature* **2011**, *470*, 558–562.
152. I. Bagai, W. Liu, C. Rensing, N. J. Blackburn, M. M. McEvoy, *J. Biol. Chem.* **2007**, *282*, 35695–35702.
153. E. H. Kim, D.H. Nies, M. M. McEvoy, C. Rensing, *J. Bacteriol.* **2011**, *193*, 2381–2387.
154. L. B. Pontel, F. C. Soncini, *Mol. Microbiol.* **2009**, *73*, 212–225.
155. F. W. Outten, D. L. Huffman, J. A. Hale, T. V. O'Halloran, *J. Biol. Chem.* **2001**, *276*, 30670–30677.
156. B. Y. Yun, S. Piao, Y. G. Kim, H. R. Moon, E. J. Choi, Y. O. Kim, B. H. Nam, S. J. Lee, N. C. Ha, *Acta Crystallogr. Sect. F. Struct. Biol. Cryst. Commun.* **2011**, *67*, 675–677.
157. S. Elsen, M. Ragno, I. Attree, *J. Bacteriol.* **2011**, *193*, 3376–3378.
158. N. L. Brown, S. R. Barrett, J. Camakaris, B. T. Lee, D. A. Rouch, *Mol. Microbiol.* **1995**, *17*, 1153–1166.
159. S. M. Lee, G. Grass, C. Rensing, S. R. Barrett, C. J. Yates, J. V. Stoyanov, N. L. Brown, *Biochem. Biophys. Res. Commun.* **2002**, *295*, 616–620.
160. S. Monchy, M. A. Benotmane, R. Wattiez, S. van Aelst, V. Auquier, B. Borremans, M. Mergeay, S. Taghavi, D. van der Lelie, T. Vallaeys, *Microbiology* **2006**, *152*, 1765–1776.
161. V. Sendra, D. Cannella, B. Bersch, F. Fieschi, S. Menage, D. Lascoux, J. Coves, *Biochemistry* **2006**, *45*, 5557–5566.
162. B. Bersch, A. Favier, P. Schanda, S. van Aelst, T. Vallaeys, J. Coves, M. Mergeay, R. Wattiez, *J. Mol. Biol.* **2008**, *380*, 386–403.
163. V. Sendra, S. Gambarelli, B. Bersch, J. Coves, *J. Inorg. Biochem.* **2009**, *103*, 1721–1728.
164. C. Grosse, A. Anton, T. Hoffmann, S. Franke, G. Schleuder, D. H. Nies, *Arch. Microbiol.* **2004**, *182*, 109–118.
165. A. Zoropogui, S. Gambarelli, J. Coves, *Biochem. Biophys. Res. Commun.* **2008**, *365*, 735–739.
166. I. Petit-Haertlein, E. Girard, G. Sarret, J. L. Hazemann, P. Gourhant, R. Kahn, J. Coves, *Biochemistry* **2010**, *49*, 1913–1922.
167. K. Marrero, A. Sanchez, L. J. Gonzalez, T. Ledon, A. Rodriguez-Ulloa, L. Castellanos-Serra, C. Perez, R. Fando, *Microbiology* **2012**, Epub ahead of print.
168. T. J. Tetaz, R. K. Luke, *J. Bacteriol.* **1983**, *154*, 1263–1268.
169. J. S. Cha, D. A. Cooksey, *Proc. Natl. Acad. Sci. USA* **1991**, *88*, 8915–8919.
170. M. A. Mellano, D. A. Cooksey, *J. Bacteriol.* **1988**, *170*, 2879–2883.
171. D. A. Cooksey, *Mol. Microbiol.* **1993**, *7*, 1–5.
172. S. Puig, E. M. Rees, D. J. Thiele, *Structure* **2002**, *10*, 1292–1295.
173. C. Petersen, L. B. Moller, *Gene* **2000**, *261*, 289–298.
174. A. Changela, K. Chen, Y. Xue, J. Holschen, C. E. Outten, T. V. O'Halloran, A. Mondragon, *Science* **2003**, *301*, 1383–1387.
175. J. V. Stoyanov, J. L. Hobman, N. L. Brown, *Mol. Microbiol.* **2001**, *39*, 502–511.
176. V. Adaikkalam, S. Swarup, *Microbiology* **2002**, *148*, 2857–2867.
177. V. Peuser, J. Glaeser, G. Klug, *Microbiology* **2011**, *157*, 3306–3313.
178. N. R. Williamson, H. T. Simonsen, A. K. Harris, F. J. Leeper, G. P. Salmond, *J. Ind. Microbiol. Biotechnol.* **2006**, *33*, 151–158.
179. W. G. Reeve, R. P. Tiwari, N. B. Kale, M. J. Dilworth, A. R. Glenn, *Mol. Microbiol.* **2002**, *43*, 981–991.
180. D. J. Julian, C. J. Kershaw, N. L. Brown, J. L. Hobman, *Antonie van Leeuwenhoek* **2009**, *96*, 149–159.
181. A. Villafane, Y. Voskoboynik, I. Ruhl, D. Sannino, Y. Maezato, P. Blum, E. Bini, *Microbiology* **2011**, *157*, 2808–2817.
182. A. Odermatt, M. Solioz, *J. Biol. Chem.* **1995**, *270*, 4349–4354.
183. D. Strausak, M. Solioz, *J. Biol. Chem.* **1997**, *272*, 8932–8936.

184. D. Magnani, M. Solioz, *Biometals* **2005**, *18*, 407–412.
185. P. A. Cobine, G. N. George, C. E. Jones, W. A. Wickramasinghe, M. Solioz, C. T. Dameron, *Biochemistry* **2002**, *41*, 5822–5829.
186. P. A. Cobine, C. E. Jones, C. T. Dameron, *J. Inorg. Biochem.* **2002**, *88*, 192–196.
187. R. Portmann, K. R. Poulsen, R. Wimmer, M. Solioz, *Biometals* **2006**, *19*, 61–70.
188. H. Wunderli-Ye, M. Solioz, *Biochem. Biophys. Res. Commun.* **1999**, *259*, 443–449.
189. G. Multhaup, D. Strausak, K. D. Bissig, M. Solioz, *Biochem. Biophys. Res. Commun.* **2001**, *288*, 172–177.
190. R. Portmann, D. Magnani, J. V. Stoyanov, A. Schmechel, G. Multhaup, M. Solioz, *J. Biol. Inorg. Chem.* **2004**, *9*, 396–402.
191. P. Cobine, W. A. Wickramasinghe, M. D. Harrison, T. Weber, M. Solioz, C. T. Dameron, *FEBS Lett.* **1999**, *445*, 27–30.
192. F. Cantini, L. Banci, M. Solioz, *Biochem. J.* **2009**, *417*, 493–499.
193. A. Reyes, A. Leiva, V. Cambiazo, M. A. Mendez, M. Gonzalez, *Biol. Res.* **2006**, *39*, 87–93.
194. S. Shafeeq, H. Yesilkaya, T. G. Kloosterman, G. Narayanan, M. Wandel, P. W. Andrew, O. P. Kuipers, J. A. Morrissey, *Mol. Microbiol.* **2011**, *81*, 1255–1270.
195. T. Liu, S. Nakashima, K. Hirose, M. Shibasaka, M. Katsuhara, B. Ezaki, D. P. Giedroc, K. Kasamo, *J. Biol. Chem.* **2004**, *279*, 17810–17818.
196. T. Liu, X. Chen, Z. Ma, J. Shokes, L. Hemmingsen, R. A. Scott, D. P. Giedroc, *Biochemistry* **2008**, *47*, 10564–10575.
197. T. Liu, A. Ramesh, Z. Ma, S. K. Ward, L. Zhang, G. N. George, A. M. Talaat, J. C. Sacchettini, D. P. Giedroc, *Nat. Chem. Biol.* **2007**, *3*, 60–68.
198. D. Corbett, S. Schuler, S. Glenn, P. W. Andrew, J. S. Cavet, I. S. and Roberts, *Mol. Microbiol.* **2011**, *81*, 457–472.
199. G. T. Smaldone, J. D. Helmann, *Microbiology* **2007**, *153*, 4123–4128.
200. K. Sakamoto, Y. Agari, K. Agari, S. Kuramitsu, A. Shinkai, *Microbiology* **2010**, *156*, 1993–2005.
201. J. Baker, M. Sengupta, R. K. Jayaswal, J. A. Morrissey, *Environ. Microbiol.* **2011**, *13*, 2495–2507.
202. S. Dwarakanath, A. K. Chaplin, M. A. Hough, S. Rigali, E. Vijgenboom, J. A. Worrall, *J. Biol. Chem.* **2012**, *287*, 17833–17847.
203. Z. Ma, D. M. Cowart, R. A. Scott, D. P. Giedroc, *Biochemistry* **2009**, *48*, 3325–3334.
204. S. D. Mills, C. A. Jasalavich, D. A. Cooksey, *J. Bacteriol.* **1993**, *175*, 1656–1664.
205. C. K. Lim, D. A. Cooksey, *J. Bacteriol.* **1993**, *175*, 4492–4498.
206. D. A. Rouch, N. L. Brown, *Microbiology* **1997**, *143* (Pt 4), 1191–1202.
207. C. Q. Liu, P. Charoechai, N. Khunajakr, Y. M. Deng, Widodo, N. W. Dunn, *Gene* **2002**, *297*, 241–247.
208. S. Jang, J. A. Imlay, *J. Biol. Chem.* **2007**, *282*, 929–937.
209. C. E. Bagwell, K. K. Hixson, C. E. Milliken, D. Lopez-Ferrer, K. K. Weitz, *PLoS One* **2010**, *5*, e12427.
210. A. Anjem, S. Varghese, J. A. Imlay, *Mol. Microbiol.* **2009**, *72*, 844–858.
211. J. M. Sobota, J. A. Imlay, *Proc. Natl. Acad. Sci. USA* **2011**, *108*, 5402–5407.
212. J. S. Davila Costa, V. H. Albarracin, C. M. Abate, *Ecotoxicol. Environ. Saf.* **2011**, *74*, 2020–2028.
213. H. Qian, S. Yu, Z. Sun, X. Xie, W. Liu, Z. Fu, *Aquat. Toxicol.* **2010**, *99*, 405–412.
214. C. J. Kershaw, N. L. Brown, C. Constantinidou, M. D. Patel, J. L Hobman, *Microbiology* **2005**, *151*, 1187–1198.
215. K. Yamamoto, A. Ishihama, *Mol. Microbiol.* **2005**, *56*, 215–227.
216. S. K. Ward, E. A. Hoye, A. M. Talaat, *J. Bacteriol.* **2008**, *190*, 2939–2946.
217. G. M. Teitzel, A. Geddie, S. K. De Long, M. J. Kirisits, M. Whiteley, M. R. Parsek, *J. Bacteriol.* **2006**, *188*, 7242–7256.
218. M. R. Ash, L. X. Chong, M. J. Maher, M. G. Hinds, Z. Xiao, A. G. Wedd, *Biochemistry* **2011**, *50*, 9237–9247.

219. G. Sarret, A. Favier, J. Coves, J. L. Hazemann, M. Mergeay, B. Bersch, *J. Am. Chem. Soc.* **2010**, *132*, 3770–3777.
220. J. Trepreau, E. de Rosny, C. Duboc, G. Sarret, I. Petit-Hartlein, A. P. Maillard, A. Imberty, O. Proux, J. Coves, *Biochemistry* **2011**, *50*, 9036–9045.
221. M. F. Cellier, P. Courville, C. Campion, *Microbes Infect.* **2007**, *9*, 1662–1670.
222. V. Hodgkinson, M. J. Petris, *J. Biol. Chem.* **2012**, *287*, 13549–13555.
223. H. Botella, G. Stadthagen, G. Lugo-Villarino, C. de Chastellier, O. Neyrolles, *Trends Microbiol.* **2012**, *20*, 106–112.
224. L. Decaria, I. Bertini, R. J. P. Williams, *Metallomics* **2011**, *3*, 56–60.
225. C. White, J. Lee, T. Kambe, K. Fritsche, M. J. Petris, *J. Biol. Chem.* **2009**, *284*, 33949–33956.
226. J. S. Klein, O. Lewinson, *Metallomics* **2011**, *3*, 1098–1108.
227. E. A. Cho, J. Seo, D. W. Lee, J. G. Pan, *Enzyme Microb. Technol.* **2011**, *49*, 100–104.
228. S. Galai, P. Lucas-Elio, M. N. Marzouki, A. Sanchez-Amat, *J. Appl. Microbiol.* **2011**, *111*, 1394–1405.
229. S. S. Phugare, D. C. Kalyani, S. N. Surwase, J. P. Jadhav, *Ecotoxicol. Environ. Saf.* **2011**, *74*, 1288–1296.
230. Y. Li, Z. Gong, X. Li, Y. Li, X. G. Wang, *BMC Biochem.* **2011**, *12*, 30.

Chapter 13
The Copper Metallome in Eukaryotic Cells

Katherine E. Vest, Hayaa F. Hashemi, and Paul A. Cobine

Contents

Abstract Copper is an element that is both essential and toxic. It is a required micronutrient for energy production in aerobic eukaryotes, from unicellular yeast to plants and mammals. Copper is also required for the acquisition and systemic distribution of the essential metal iron, and so copper deficiency results in iron

K.E. Vest • H.F. Hashemi • P.A. Cobine (✉)
Department of Biological Sciences,
101 Rouse Life Sciences Building, Auburn University, Auburn, AL 36849, USA
e-mail: pac0006@auburn.edu

L. Banci (ed.), *Metallomics and the Cell*, Metal Ions in Life Sciences 12,
DOI 10.1007/978-94-007-5561-1_13, © Springer Science+Business Media Dordrecht 2013

deficiency. Copper enzymes have been identified that explain the wide variety of symptoms suffered by copper deficient subjects. The cloning of the genes encoding transport proteins responsible for copper-related Menkes and Wilson diseases inspired and coincided with the discovery of copper chaperones that stimulated the copper homeostasis field. Copper continues to be implicated in new array of proteins, notably those involved in a variety of neurodegenerative diseases. Here we will describe the cadre of important historical copper proteins and survey the major metallochaperones and transporters responsible for mobilization and sequestration of copper in yeast, mammals and plants.

Keywords copper chaperones • copper deficiency • copper transporter • Menkes disease • multicopper oxidase • Wilson disease

Please cite as: *Met. Ions Life Sci.* 12 (2013) 451–478

1 Overview of Copper Homeostasis in Eukaryotic Cells

Copper is an essential element used in many oxidation and reduction reactions. Its redox potential is such that it readily donates electrons to oxygen. This fact means that free copper can generate reactive oxygen species if not carefully controlled, so cells must meticulously account for copper. The membrane forms a critical barrier to control copper concentration in the cell and in various cellular compartments. Regulated expression and localization of transporters as well as concerted actions of chaperone proteins that direct copper to the correct target determine the balance of copper and its cellular localization. To prevent uncontrolled redox cycling, cysteine-rich metallothioneins, glutathione, and other cellular components contribute to the cellular buffering capacity. The copper metallome here will be defined as the sum of proteins and small molecules that bind, utilize, or transport copper.

Copper sufficiency in mammals is required for iron uptake, connective tissue formation, and energy production. Characteristic presentations of copper defects have been known since the 1930's [1]. The phenotypic presentation of the genetic disorder Menkes disease is a classical example of copper deficiency, while Wilson disease is the prototypical copper overload disorder [2]. The symptoms of Menkes disease are kinky hair, neurological deficit and growth retardation, demonstrating the spectrum of enzymes that utilize copper as a redox and structural cofactor. Wilson disease results in copper overload in the liver, the brain, and eventually in the eyes [3]. Kayser-Fleischer rings form when copper is deposited in the eyes and can be used to diagnose this disorder. Copper accumulates in the liver and the brain, eventually resulting in cirrhosis and neurological defects [3]. The physiological damage occurring in Wilson disease is testament to the important buffering capacity that membranes, metallothioneins, and other components perform in protecting the cell.

Copper is transported across the plasma membrane by high affinity transporters and then copper chaperone proteins carry it to its targeted locations. Copper is distributed to the secretory pathway, cytosolic targets and to mitochondria. Copper in the secretory pathway is inserted into enzymes required for but not limited to high affinity iron transport and distribution, production of neurological transmitters, and crosslinking in connective tissues. Copper in the cytoplasm is utilized for protection against oxidative stress. In mitochondria, copper chaperones exist for the assembly of one complex in the energy-producing electron transport chain. The copper chaperone proteins deliver copper to targets using metal-assisted protein-protein interactions; the relatively small, weak interface provided by the proteins is complemented by ligand sharing of the copper in the transfer site [4]. Directional transfer of copper from chaperone to target is achieved through increasing affinity for copper in the target protein [5]. This increasing affinity provides a thermodynamic rationale for the kinetic process of copper delivery in eukaryotic cells. These affinity gradients are a major mechanism to ensure correct distribution of the copper metallome.

Many of the paradigms for intracellular trafficking of copper have been established in unicellular eukaryotes such as *Saccharomyces cerevisiae* [6–8]. Using yeast as a framework we will define the copper metallome and then intertwine the unique features and responses observed in multi-cellular eukaryotes. This chapter will discuss the copper metallome and will focus on responses to deficiency as a means to consider the priorities of copper utilization.

2 Copper Homeostasis: *Saccharomyces cerevisiae*

Saccharomyces cerevisiae is a genetically tractable model that has versions of the major eukaryotic cuproenzymes (cytochrome *c* oxidase, superoxide dismutase, and multicopper ferrooxidase). Yeast also has the conserved transporters responsible for uptake and distribution of copper across membranes (Ctr1 and Ccc2)[1] [8]. This framework has meant that yeast has been the theater for major discoveries about copper homeostasis. The classical βαββαβ folded chaperone (Atx1) that delivers copper to the secretory pathway was first described in this model [9]. The requirement of a copper chaperone for superoxide dismutase in yeast led to calculations of cytoplasmic "free" copper as being equal to zero [10]. This calculation has driven many of the current research directions in copper homeostasis. Genetic and biochemical dissection of the activation of superoxide dismutase and the actions of a series of accessory proteins (Cox17, Sco1, Cox11) required for cytochrome *c* oxidase assembly have also been performed in yeast [11]. However, yeast still has many mysteries related to copper homeostasis, including the exact identity of the molecules responsible for the distribution of copper to mitochondria, a major challenge that must be efficiently performed for eukaryotic aerobic life to exist.

[1] Throughout this chapter the nomenclature for *Saccharomyces cerevisiae* will be three letters with one number. Gene names are in capital italics (e.g., *GEN1*), small italics for mutant of that gene (e.g., *gen1*Δ) and sentence case for proteins (e.g., Gen1). For other eukaryotes capital italics (e.g., *GENE*) will be used for gene names and all capitals for protein names (GENE).

2.1 Catalytic Copper Enzymes

Copper is a cofactor in Fet3 and is therefore required for iron uptake. Fet3 is a multicopper oxidase that converts ferrous iron to the ferric form that is required for high affinity iron transport in yeast [12]. The copper ions in Fet3 are required for the oxidation of the substrate and the reduction of oxygen to water. The ferrooxidase family can have varied specificities, and in addition to oxidizing iron, Fet3 has been shown to oxidize cuprous copper at the plasma membrane [13]. This activity can be related to copper stress phenotypes associated with overexpression of Fet3 [13]. A high level of Fet3 presumably induces cycling between Cu(I) and Cu(II) leading to lipid damage and toxicity.

The Cu,Zn superoxide dismutase (Sod1) is a homodimeric free radical-scavenging enzyme that breaks down superoxide radicals to molecular oxygen and hydrogen peroxide via a copper-mediated disproportionation reaction. Sod1 is expressed predominantly in the cytosol of yeast, but is also found in the mitochondrial intermembrane space [14]. The monomer structure consists of an eight-stranded β-barrel that binds a Cu ion and a Zn ion that play both catalytic and structural roles [15].

Cytochrome *c* oxidase is the final enzyme complex of the electron transport chain. Cytochrome *c* oxidase accepts electrons from the soluble electron carrier cytochrome *c* and donates them to oxygen, reducing it to water. The process occurs in the inner membrane of mitochondria and results in the translocation of protons that contribute to the formation of the membrane potential required for ATP production [16]. Cytochrome *c* oxidase contains multiple cofactors (copper, heme, magnesium, and zinc) that are required for its assembly and activity [17]. The assembly process requires the action of at least 30 different accessory proteins [18]. Cox1, Cox2, and Cox3 form the catalytic core of the enzyme and are encoded by the mitochondrial genome. Cox1 and Cox2 subunits bind two heme a moieties and three copper ions. The heme a_3 molecule is coordinated to a copper ion to form the CuB center in Cox1, while Cox2 binds a binuclear mixed valence copper center (CuA). The CuB site is required for substrate binding while CuA accepts electrons from cytochrome *c* that are transferred to the CuB site and then to oxygen [19]. The catalytic core is surrounded by nuclear-encoded structural subunits, which stabilize and may allow for allosteric regulation of the holoenzyme.

2.2 Membrane Transporters

Ctr1 is a high affinity copper transport protein in the plasma membrane that is active as a trimer [20]. Each monomer has 3 transmembrane domains, with methionine-rich MxM, MxxM or MxxxM motifs (Mets motifs) in the extracellular amino-terminal domain [21]. The Ctr1 homotrimer forms a pore lined by Mets motifs found in the second transmembrane domains of each monomer [20,22]. Copper first binds to the Mets motif on the extracellular domain then is transferred to the pore.

The rate-limiting step of transport occurs closer to the exit side and involves a conserved histidine residue [23]. Once copper traverses the plasma membrane, it is bound by a carboxy-terminal CxC domain. Copper bound at this site is then distributed to the cytosol and this domain has been shown to interact with, and deliver copper to the chaperone Atx1 [24].

CTR3 was discovered in a genetic screen to identify suppressors of defects in yeast lacking *CTR1*. Expression of *CTR3* is naturally suppressed in some yeast strains due to the insertion of transposable element in the promoter region, but when produced Ctr3 can carry out high affinity copper transport. It exists as a homotrimer, but lacks the Mets motifs found in Ctr1 [25]. In addition to the high affinity system yeast has low affinity copper transporters, Fet4 and Smf1 [26,27]. A feature that is conserved for both low and high affinity transport is that Cu(I) is used specifically as a substrate and therefore cells need cell surface reductases or addition of exogenous reductant to carry out transport [28].

CCC2 encodes a P-type ATPase that transports copper into the trans-Golgi network for metallation of secreted cuproproteins. The P-type ATPases are proteins that translocate cations against an electrochemical potential gradient by using the energy derived from ATP hydrolysis [29,30]. These proteins have multiple domains that are required for the translocation of substrate. The channel is made up of 8 predicted transmembrane domains with folded domains in the cytosol required for ATP hydrolysis and subsequent phosphorylation of a conserved invariant aspartic acid residue, an important step in the catalytic cycle of these proteins. In addition, Ccc2 has two amino-terminal βαββαβ folded CxxC copper-binding domains [31]. Though partially dispensable, these domains serve as targets for the delivery of copper from the similarly folded copper chaperone, Atx1. Electrostatic patches allow for specific interaction and a shallow thermodynamic gradient drives copper transfer via a ligand exchange mechanism to the ATPase and therefore into the secretory pathway [32,33]. The sharing of ligands in the exposed copper binding sites enhances the interaction between the proteins, this allows for transient interaction via a relatively small interface made up of charged and some hydrophobic residues [4]. Due to the role of the copper in the interaction it appears that the interactions are copper dependent.

Yeast lacks an export pathway for copper; the only equivalent of export is import into the vacuole. Once copper is in the vacuole it must exist as Cu(II) because the cupric metalloreductase Fre6 is required for mobilization of copper into the cytosol via the cuprous transporter Ctr2 [34,35]. Ctr2 is a homolog of Ctr1 and deletion of *CTR2* causes hyper-accumulation of copper in the vacuole. Ctr2 can function as a high-affinity copper transporter when localized to the plasma membrane [35].

Yeast also contains a pool of copper in the mitochondrial matrix that is required for metallation of cytochrome *c* oxidase and superoxide dismutase in the inter-membrane space [36,37]. The impermeable nature of the inner membrane indicates that copper-transporting proteins would be necessary for import into the matrix and export back to the inter-membrane space. However, none have been identified at this time.

2.3 Copper Chaperones

Atx1 is a βαββαβ folded protein and was the first copper chaperone described [9]. The Atx1-fold provides a stable core for presenting an exposed CxxC copper-binding domain [38]. Atx1 has been shown to interact with the high affinity transport protein Ctr1 to receive copper [24,39]. It then carries copper through the cytosol to Ccc2 for delivery to the multi-copper oxidase Fet3 within the trans-Golgi vesicles [9]. As previously stated, the transfer reaction proceeds via an exchange of copper ligands that is driven by a shallow thermodynamic gradient [40].

Ccs1 is the copper chaperone for superoxide dismutase. Ccs1 activates Sod1 by inserting copper into the newly synthesized apo-protein in the cytosol and then catalyzing the formation of an essential disulfide bond [41]. The structure of Ccs1 reveals three defined domains: a βαββαβ folded Atx1-like domain with a CxxC motif, a central Sod1-like domain, and a carboxy-terminal domain [42,43]. Ccs1 interacts with Sod1 as a heterodimer via the Sod1-like domain that is also required for the activation [44]. In addition to the predominant cytosolic localization, Ccs1 also localizes to the inter-membrane space of mitochondria where it activates 1–5% of the total cellular Sod1 [14]. Mitochondrial localization of Ccs1 is dependent on the Mia40-Erv1 disulfide relay exchange system and a pair of cysteines present in the Atx1-like domain that are not part of the CxxC motif [45].

Cox17 is a cysteine-rich protein that is localized to the inter-membrane space of the mitochondria [46]. Multiple conformations of Cox17 have been isolated but a conformer with two disulfide bonds that binds a single copper atom seems to be the biologically relevant species [47]. Mutational analyses have shown that copper coordination and Cox17 function are not dependent on the presence of two disulfide crosslinks but they do provide stability to the fold [48]. Cox17 acts as the copper chaperone that delivers copper to both Sco1 and Cox11 in yeast, donating copper to each molecule through transient interactions that are mediated by distinct structural interfaces [49,50]. The components of copper homeostasis in *Saccharomyces cerevisiae* are summarized in Figure 1.

Sco1 is anchored to the inner membrane by a single transmembrane helix with the carboxy-terminal domain protruding into the inter-membrane space [51]. The peroxiredoxin-like fold of the globular carboxy-terminal domain exposes a single copper binding site made up of two cysteine residues within a Cx3C motif and a conserved histidine residue [51]. Mutation of either cysteine or the histidine residue in Sco1 abrogates copper binding and results in decreased cytochrome *c* oxidase activity [51–53]. In addition to binding Cu(I), Sco proteins bind Cu(II) [54]. It is not clear whether Sco1 transfers both Cu(I) and Cu(II) ions to build the mixed valence binuclear CuA site in cytochrome *c* oxidase or if Cu(II) plays some other role in assembly. Cox11 is another inner-membrane tethered protein with a single transmembrane helix and a carboxy-terminal domain that contains cysteine residues used for copper binding [55]. It is required for formation of the heme-containing CuB site that is required for substrate binding in cytochrome *c* oxidase.

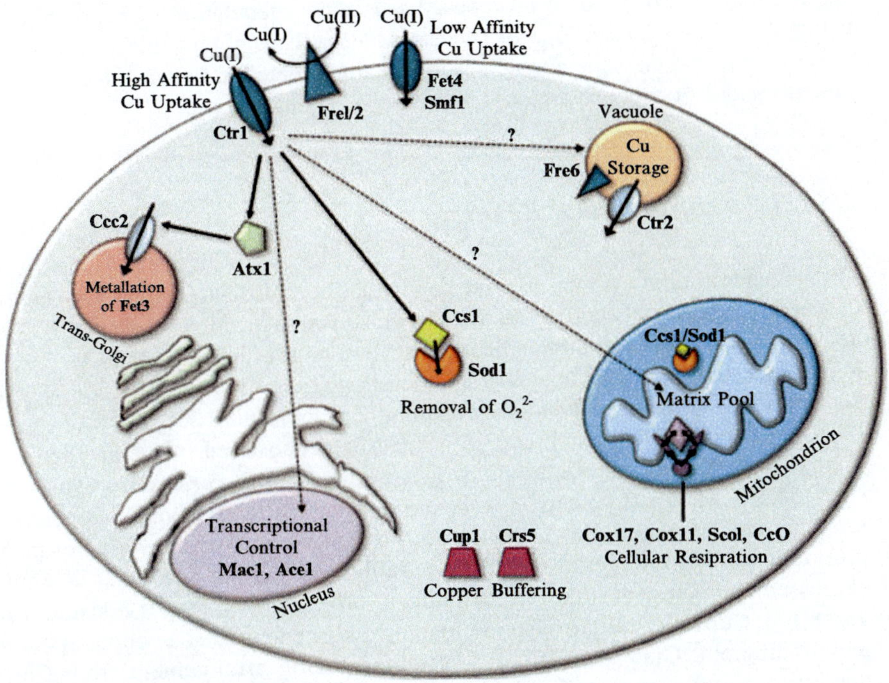

Figure 1 Copper homeostasis in *Saccharomyces cerevisiae.* Copper is transported in cuprous form after reduction by metalloreductases (Fre1/2) by high- (Ctr1) and low- (Fet4, Smf1) affinity systems. Once in the cytosol copper is delivered to superoxide dismutase (Sod1) by the copper chaperone Ccs1, to the P-type ATPase Ccc2 in the trans-Golgi network for assembly of the multicopper oxidase Fet3 by Atx1, and to the mitochondrion by an unknown mechanism (dashed line). After translocation to the mitochondria copper is sequestered in the matrix pool until it is made available in the inter-membrane space for Cox17, Cox11, and Sco1 mediated assembly of cytochrome *c* oxidase (CcO). If copper reaches toxic levels it is bound by the metallothioneins Cup1 and Crs5. Copper that is in the vacuole can be remobilized by the action of the metalloreductase Fre6 and the high affinity transporter Ctr2. Regulation of a number of these targets is via direct copper binding to the transcription factors, Mac1 and Ace1.

Yeast also has a series of Cx9C proteins in the inter-membrane space that have been shown to bind copper and play roles in cytochrome *c* oxidase assembly [56]. Cox19, Cmc1, Cmc2, and Cox23 are able to bind copper and are required for assembly of functional cytochrome *c* oxidase [56]. However the exact functional significance of the copper binding in these assembly proteins is not clearly defined.

Copper is also found within the mitochondrial matrix [36]. Though stable anionic Cu(I) complexes have been isolated, the identity of the coordinating ligand has not been resolved. The functional consequence of attenuating the matrix Cu pool by expression of a matrix-targeted competitor molecule is a decrease in cytochrome *c* oxidase activity [37]. Supplementation of cultures with exogenous Cu salts causes a significant expansion in the matrix Cu pool and reverses any observed phenotypes, arguing that it is the source of copper used in the metallation of cuproenzymes within the inter-membrane space. The exact mechanisms of how this matrix pool is distributed to the inter-membrane space and any interaction between the transport proteins and the copper chaperones have not been defined.

2.4 *Sequestration and Regulation*

The concentration of cytoplasmic free copper has been estimated to be zero, hence, the need for specific chaperone-mediated mechanism to metallate targets [10]. This low cytoplasmic concentration indicates that yeast must efficiently sequester copper [57]. Sequestration can be achieved in at least two ways: transport to an organelle (as described) or binding of excess copper in the cytoplasm.

When copper levels reach excess it must be sequestered into a biologically inert complex. Two metallothioneins are produced in *S. cerevisiae*, Cup1 and Crs5 [58,59]. Metallothioneins are cysteine-rich proteins with low to no secondary structure that bind multiple copper atoms, preventing free redox cycling or inappropriate binding interactions. Cup1 is a duplicated gene in yeast that is highly expressed and is responsible for a significant proportion of copper resistance in yeast [60]. Cup1 has unique cysteine spacing and is considered a copper-specific metallothionein. Crs5 has different cysteine spacing arrangement from Cup1 that is more similar to metallothioneins in higher eukaryotes, but it plays a much less significant role in protection against copper toxicity [60]. In fact, in many laboratory strains of *S. cerevisiae* Crs5 has a nonsense mutation that truncates the protein product.

Glutathione is a cysteine-containing tripeptide that acts as an antioxidant and regulates the reduction state of cells. It mediates and participates in the formation of disulfide bonds in proteins. During the oxidative stress associated with excess copper, glutathione ratios shift to higher proportion of oxidized form. Reduced glutathione can bind copper, however, the contribution of glutathione to protection against copper toxicity is very low [61,62].

In *Saccharomyces cerevisiae,* gene transcription is used to balance the need for copper with its potential toxicity. Mac1 controls expression of transporters under Cu-deficient growth conditions, initiating transcription of the high affinity copper transporter *CTR1*, as well as the metalloreductase necessary for Cu(I) import, *FRE1* [63,64]. However once copper reaches toxic levels, Ace1 induces the transcription of the metallothioneins genes, *CUP1* and *CRS5* [65]. Both transcription factors bind copper via cysteine residues in copper-thiolate clusters [66]. The mechanism for copper translocation to the nucleus for binding to Mac1 and Ace1 remains unknown. One additional level of regulation is post-transcriptional internalization and degradation of Ctr1 at high copper concentrations that limits toxic influx of copper [67].

2.5 Detecting Defects Associated with Copper Deficiency

Phenotypes of copper deficiency are observed in yeast strains lacking the gene *CTR1* and can be exacerbated by the addition of extracellular copper chelators (Figure 1). Decreased flow of copper to the trans-Golgi network via Atx1 results in translocation of apo Fet3 and a subsequent inability to perform high affinity iron transport [21,68]. To facilitate removal of residual iron and make yeast dependent on high affinity uptake via Fet3, extracellular chelators must be added. In the presence of these chelators, copper deficiency can be detected by a failure to grow. The inability to metallate Sod1 results in increased oxidative stress. The oxidative stress phenotype manifests as a failure to grow under hyperoxic conditions and as a lysine auxotrophy [69]. These deficiency phenotypes can be reproduced by deletion of genes encoding the copper chaperones, *ATX1* and *CCS1*. The final phenotype of copper deficiency is the loss of cytochrome *c* oxidase; this is observed as a failure to grow on non-fermentable carbon sources (e.g., glycerol) or as a decrease in oxygen consumption [46].

3 Copper Homeostasis: Mammals

The dietary copper requirement for a human is approximately 1–3 mg per day. Particular organs such as heart and brain have significant energy demands and must maintain sufficient levels of ATP and therefore cytochrome *c* oxidase. So multisystem organisms must carefully divide this daily quota throughout the body to match demand.

At an intracellular level most of the proteins from *S. cerevisiae* are conserved in mammals for the distribution of copper (Figure 2). However, in addition to the multi-copper oxidases for iron homeostasis mammals have several additional secreted copper enzymes required for melanin formation, crosslinking of elastin and collagen, and formation of neurotransmitters [70]. In serum copper is transported from storage, or from point of uptake, to organs of utilization by proteins (ceruloplasmin, albumin) and by less well-defined non-proteinaceous components (histidine, glutathione, and other unidentified fractions) [70].

A Copper Development Association report states typical copper concentrations in tissues of healthy adults are liver (3–9 μg/g), heart (3 μg/g), kidney (2 μg/g), and brain (6 μg/g). The most striking redistribution happens after birth when large stores of copper in the infant liver (2–57 μg/g) are redistributed during early development. This highlights the demands placed on the developing mammal to distribute the intracellular stores and highlights the capacity of the liver to store large amounts of copper [71].

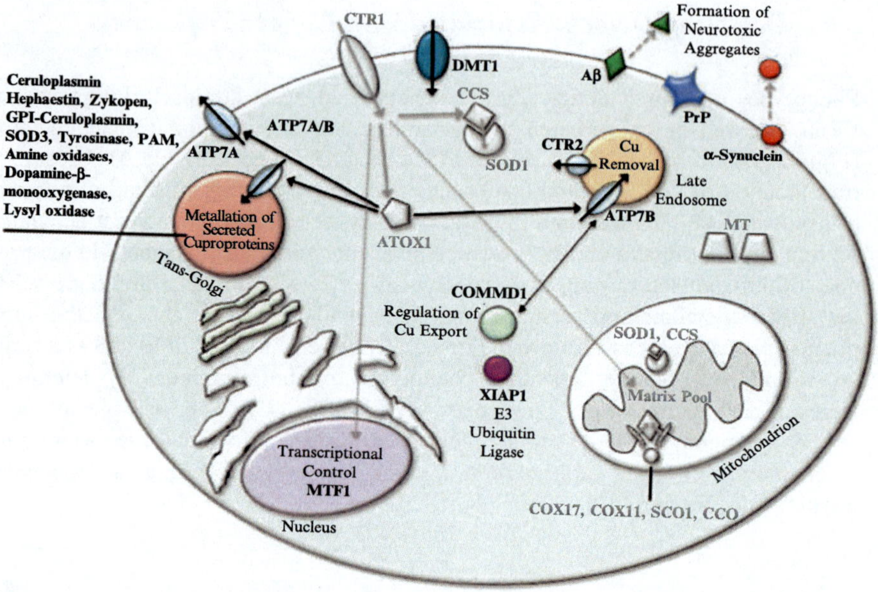

Figure 2 Copper homeostasis in mammalian cells. This figure should be compared directly to Figure 1. The color components are unique to the higher eukaryotes. ATP7A and ATP7B participate in both providing copper to the secretory pathways but also function in export of copper at the plasma membrane or in vesicular compartments (N.B. ATP7A and ATP7B are rarely expressed in the same cell type). In addition to a multicopper oxidase ceruloplasmin copper is also loaded in multiple secreted proteins (hephaestin, zykopen, GPI-ceruloplasmin, SOD3, tyrosinase, PAM, amine oxidase, dopamine-β-monoxygenase, lysyl oxidase). Transcription of the mammalian metallothioneins is under the control of the zinc regulate transcription factor MTF1. COMMD1 is able to mediate copper export via ATPase and XIAP regulates the levels of COMMD1. The copper binding proteins linked to neurological disorders amyloid-β (Aβ), prion (PrP) and α-synuclein are shown. DMT1 is a divalent metal ion transporter that has some affinity for copper, linking it to low affinity copper uptake. The grey components: the high affinity transporters, copper chaperones, metallothioneins are homologous to those found in yeast.

3.1 *Mammalian Copper Transport*

The human homolog for CTR1 was identified by complementation in yeast [72,73]. Expression of the human *CTR1* gene, which is 29% identical to the yeast gene, rescues the copper deficiency of the yeast *ctr1*Δ while overexpression of human *CTR1* gene in yeast increases copper accumulation [73]. The mouse *CTR1* gene is 92% identical to human *CTR1* gene and also functionally complements the yeast deletion [74]. One striking difference in human CTR1 is the presence of an additional extracellular histidine-rich motif not found in the yeast homologue. This domain is capable of binding copper and expression of a truncated human CTR1, lacking the histidine-rich motif, in *CTR1*$^{-/-}$ fibroblasts suggests that this domain plays a role in the transport mechanism [75].

Several mouse models have been generated that show the importance of CTR1 in physiology and development. Mice with one copy of the *CTR1* gene mutated (*CTR1*$^{+/-}$) had low copper in brain and spleen tissue [72]. Complete deletion of *CTR1* gene (*CTR1*$^{-/-}$) caused abnormal morphogenesis and delayed development in mouse embryos that was eventually lethal by day 13 [72,76]. Although CTR1 appears to be the major high affinity copper transporter in most eukaryotic cells, other transport proteins must exist as *CTR1*$^{-/-}$ mouse embryonic cells maintain 30% residual copper transport [72]. The human *CTR2* gene is homologous to *CTR1*, but the protein is localized to late endosomes and lysosomes [77]. *CTR2* is most abundantly expressed in placenta and heart. In cell culture CTR2 can partially localize to the plasma membrane and therefore could contribute to copper uptake [78].

Generation of an intestinal epithelium-specific *CTR1* mutant mouse (*CTR1*$^{\text{int/int}}$) demonstrated that this transporter is important for absorption of dietary copper. *CTR1*$^{\text{int/int}}$ mice showed signs of systemic copper deficiency including loose skin, ataxia, brittle whiskers, and cardiac hypertrophy [79]. These symptoms were reversed by intraperitoneal injection of copper, further indicating that CTR1 is required for intestinal copper absorption.

One report in polarized cells in culture showed a basolateral localization for CTR1, and the authors used the same model to suggest apical copper transport was mediated by anion exchangers [80,81]. However, immunohistochemistry was used to confirm CTR1 localization in the apical membrane of intestinal epithelial cells of mouse, rat, and pig [82]. DMT1, the iron transporter of the intestinal epithelial has also been implicated in copper transport due to reduced copper uptake in Caco-2 cell culture lacking DMT1 [83]. The significance of DMT1 for copper uptake under normal ranges of copper concentrations is still not known.

A cardiac-specific *CTR1* mutant mouse (*CTR1*$^{\text{hrt/hrt}}$) revealed evidence of systemic copper regulation. These mice had severe cardiac copper deficiency accompanied by severe cardiomyopathy [84]. The *CTR1*$^{\text{hrt/hrt}}$ mice also had elevated serum copper and decreased copper stores in liver tissue, correlating with increased levels of the copper export protein ATP7A. Interestingly, serum from *CTR1*$^{\text{hrt/hrt}}$ mice was able to induce higher ATP7A levels in cultured human cells, indicating the presence of a soluble circulating copper signal that regulates copper distribution [84].

Menkes and Wilson diseases are both copper-related genetic diseases in humans [85]. Menkes disease is characterized by systemic copper deficiency including loose skin, wiry hair, developmental delays, and neurodegeneration. Wilson disease is characterized by copper accumulation in liver and neural tissue that eventually leads to cirrhosis with symptoms at around 30 years of age. Mutations associated with Menkes disease were mapped to a gene (*ATP7A*) encoding a putative P-type ATPase transport protein with multiple Atx1-like metal-binding motifs [86-88]. Later, mutations associated with Wilson disease were mapped to another putative P-type ATPase (*ATP7B*) with very similar predicted domains [89]. Though the Menkes (ATP7A) and Wilson (ATP7B) proteins are copper transporting ATPases with similar structure and function, tissue-specific expression of these proteins causes the different symptoms of the two diseases. The Menkes protein is expressed in all tissues while Wilson disease protein is expressed predominantly in the liver [90].

Expression of *ATP7A* and *ATP7B* can rescue the iron deficiency defect of a *S. cerevisiae* lacking *CCC2* [91,92]. As expected, ATP7A and ATP7B in humans are required for metallation of cuproenzymes in the secretory pathway. That includes ceruloplasmin (multicopper oxidase), tyrosinase (monooxygenase), dopamine-β-hydroxylase (monooxygenase), lysyl oxidase (amine oxidase), and extracellular superoxide dismutase [93–97].

In addition to providing copper to the secretory pathway, the copper transporting P-type ATPases contribute to copper efflux from the cell. A study of copper resistant Chinese hamster ovary (CHO) cells revealed duplications of the *ATP7A* gene, indicating that its product contributed to copper removal from those cells [98]. Supporting the role in export of cellular copper ATP7A mediates export from intestinal epithelial cells [84,99]. Similarly, elevated *ATP7B* expression in cultured hepatocytes was associated with increased resistance to copper [100]. In fact, ATP7B-mediated copper efflux from hepatocytes contributes to biliary copper excretion for systemic copper removal [101].

The dual functions of the copper-transporting ATPases are facilitated by changes in the subcellular localizations of these proteins. The Menkes protein ATP7A is localized to the trans-Golgi network for distribution of copper to the secretory pathway [97]. However, under elevated copper conditions a portion of the ATP7A pool is relocalized to the plasma membrane for copper export [102–104]. Like ATP7A, the Wilson disease protein ATP7B is localized to the trans-Golgi network under basal copper conditions [92]. Under high copper conditions, ATP7B undergoes reversible redistribution to late endosomes in CHO cells [92,105]. This is likely the mechanism of ATP7B-mediated release of copper into bile ducts in the liver [106,107].

The dual function of ATP7A and ATP7B proteins in the secretory pathway and copper efflux means that alteration of either the transport function or the trafficking function can cause disease. Some mutations of *ATP7A* that cause Menkes disease have been shown to disrupt protein localization rather than transport function [110,111]. Additionally, an allele of *ATP7A* that causes the protein to mislocalize from the trans-Golgi network to the endoplasmic reticulum causes occipital horn syndrome rather than classic Menkes disease [108,109].

The amino-terminus of ATP7A and ATP7B contain six Atx1-like βαββαβ fold domains each with a conserved CxxC copper-binding motif. The individual domains can bind copper and some studies suggest interaction between domains may affect the final stoichiometry and/or the cooperativity of binding [39,112–117]. Of the six domains, only the fifth and sixth domains are actually required for transport function [118]. While the other domains are not required for transport activity *in vitro*, they may serve a function *in vivo*. The fourth domain has been shown to interact with the copper chaperone ATOX1 [119,120]. The Atx1-like domains in ATP7B were recently shown to interact with each other and affect the oxidation-reduction state of their cysteine, which would affect metal binding [121].

In addition to its role in copper binding, the amino-terminal domain of ATP7A is important in copper-induced relocalization. Truncation of the amino-terminus prevents relocalization to the plasma membrane, while a ATP7A chimeric protein with the amino-terminus from ATP7B relocalizes to vesicles, a characteristic of ATP7B [110,122–124]. Sub-cellular localization is also linked to a di-leucine motif in the carboxy-terminus of the protein that is required for retrieval from the plasma membrane to the trans-Golgi network [125,126].

Like in yeast, mitochondria from mouse liver have a proportion of copper localized to the matrix [37]. Cell culture experiments with fluorescent copper specific probes and synchrotron X-ray fluorescence has confirmed the presence of copper in the matrix of mitochondria in mammalian cells [127,128]. These data suggest a transporter for the inner-membrane must exist but no transporters have been identified.

3.2 Copper Chaperones

All the aforementioned copper chaperones from yeast (Atx1, Ccs1, Cox17) are conserved in mammals. ATOX1, the homologue of Atx1, delivers copper to ATP7A and ATP7B for delivery to the secreted target enzymes and copper efflux. In addition, the ATOX1 interaction with the ATP7B regulates the catalytic activity of the ATPase [120]. The *ATOX1*$^{-/-}$ mouse demonstrates all the characteristics of peripheral copper deficiency. Its phenotype is exacerbated by maternal *ATOX1* deficiency, highlighting the critical role of this protein in perinatal copper delivery [129]. CCS is the copper chaperone for SOD1 in mammals. The *CCS* mutant mouse has normal SOD1 protein levels but decreased activity [130]. These mice are viable but show reduced resistance to stress and decreased fertility [130]. These phenotypes are consistent with phenotypes observed in the *SOD1*$^{-/-}$ mouse. A *COX17* homologue was identified in mammals and homozygous deletion of *COX17* in mice is embryonically lethal due to the lack of cytochrome *c* oxidase activity [131].

Mammalian studies have revealed unique roles for some copper chaperone proteins. ATOX1 has a novel role in shuttling copper to the nucleus where it is involved in the transcriptional activation of cyclin D1 gene expression [132]. The mitochondrial chaperone SCO1 acts as a copper delivery molecule, as it does in yeast, but has a role in regulating cellular copper homeostasis [133]. The human mutation of SCO1 that results in fatal liver disease is a missense mutation of proline 174 to a leucine residue. While this mutation does not dramatically affect the overall fold of the protein it does result in decreased interaction with COX17 in the inter-membrane space and therefore a failure to deliver copper to cytochrome *c* oxidase [134,135]. In addition to this defect, the presence of this mutant allele results in profound cellular copper deficiency. The deficiency is a result of increased export, however, the mechanism of signaling to the copper export pathway is not understood. In humans a second SCO molecule is required for efficient cytochrome *c* oxidase assembly [136]. This molecule, SCO2, has the same global fold and has equivalent metal binding characteristics to SCO1 [54]. Mutations in *SCO2* result in fatal encephalopathy suggesting the different roles of the SCO proteins [137]. Biochemical dissection of the redox state of this molecule suggests it has an oxidoreductase function that may prime the copper delivery pathway in the inter-membrane space [138].

COMMD1 is a cytosolic copper binding protein that was originally identified using the inbred Bedlington terrier model of copper overload. COMMD1 is involved in regulating export of copper through an interaction with ATP7B [139,140]. COMMD1 has also been linked to an E3 ubiquitin ligase XIAP, X-linked inhibitor of apoptosis. Copper binding to XIAP induces a conformational change that results in its inactivation and proteosomal degradation [141]. This in turn affects the abundance of COMMD1 and therefore the cellular copper content.

3.3 Mammalian Copper Enzymes

Ceruloplasmin is responsible for peripheral iron distribution and couples copper and iron transport [142]. In fact ceruloplasmin accounts for over 90% of the circulating copper in mammals [143]. Ceruloplasmin is highly expressed in the liver and the protein folds into six compact domains that bind six copper atoms required for activity [142]. In addition to circulating ceruloplasmin mammals also have an alternatively spliced variant that results in a glycosylphosphatidylinositol-linked ceruloplasmin [144]. This variant is widely expressed but highest expression is in the kidney and spleen. This alternatively spliced form of ceruloplasmin is 98% identical to the soluble form and is expected to play a role in uptake of iron in specific cell types. Mutations in the ceruloplasmin gene lead to aceruloplasminemia, an autosomal recessive condition that is characterized by iron accumulation in brain and results in anemia, neurological dysfunction, and diabetes mellitus [145,146].

Hephaestin is a membrane-bound homologue of ceruloplasmin that is found in intestinal epithelial cells on the basolateral membrane of the enterocyte and is required for intestinal iron absorption [147]. It is likely that hephaestin cooperates with ferroportin, an iron exporter in the enterocyte, macrophages and hepatocytes, to facilitate efficient egress of iron. The most recently described mammalian multicopper oxidase, zyklopen, is a membrane-bound variant expressed in placental tissues and appears to be required for adequate iron supply to the developing embryos [148]. Zyklopen is also expressed in mammary glands and immunohistochemistry suggests it is found in testes, kidney and brain.

Cu,Zn superoxide dismutase (SOD1), responsible for the removal of superoxide anions is found in the cytosol and mitochondria. As in yeast, copper is delivered to mammalian SOD1 by the chaperone CCS [149]. Mouse *SOD1*$^{-/-}$ mutants develop normally, but are more sensitive to axonal injury and oxidative stressors, and have decreased fertility [150–152]. Additionally, mammals express an extracellular Cu,Zn superoxide dismutase, SOD3, in vascular tissue [153,154]. While the exact biological role of SOD3 is not completely resolved, mutational studies in rodents have implicated a role for SOD3 in various pathologies including hypertension, ischemia-reperfusion injury, and lung disease (reviewed in [155]). Metallation of SOD3 occurs in the secretory pathway and so copper insertion is dependent on the chaperone ATOX1 and the ATP7A/B transporters [156].

Cytochrome *c* oxidase in mammals has three extra peripheral subunits in addition to those found in yeast. Mammals also have tissue-specific isoforms of certain cytochrome *c* oxidase subunits, such as COX4, COX6a, COX6b, COX7a, and COX8 [157–159]. Like yeast, mammalian cytochrome *c* oxidase requires protein chaperones for copper insertion (including SCO1 and COX17). In human cells SCO2 is also required; it genetically interacts with SCO1 suggesting complementary functions [136]. SCO2 appears to act as an oxidoreductase for SCO1 in the assembly process [138]. Deficiency of cytochrome *c* oxidase activity is the cause of various metabolic diseases in humans. These often fatal pathologies can be caused by mutations in the mitochondrial and nuclear-encoded structural subunits, but are more often caused by mutations in assembly proteins [160].

Amine oxidases are involved in the deamination of a large range of substrates. Activity is dependent on a copper ion and the modification of tyrosine to produce a topaquinone cofactor [161,162]. The copper at the active site has been implicated in both the formation of the topaquinone and the enzyme activity to produce the resulting aldehyde. The broad substrate range means that amine oxidases are being implicated in a wide range of cellular processes from cellular differentiation to signaling. Lysyl oxidase is a copper-containing amine oxidase required for cross-link formation within collagen and elastin fibers [163]. Lysyl oxidase converts lysine residues found within collagen and elastin precursors to highly reactive aldehydes that can subsequently react with other similar species and/or lysine residues to increase cross-linking in connective tissue, leading to better integrity and stability [164]. Disruption of lysyl oxidase activity is the primary cause of the brittle hair and loose skin observed in Menkes disease.

Peptidylglycine α-amidating monooxygenase (PAM) is required to produce active amidated hormones including vasopressin that regulates water balance and blood pressure, gastrin and cholecystokinin in the gastrointestinal system, calcitonin that regulates bone calcium deposition [165]. PAM is a two-domain protein consisting of peptidylglycine α-hydroxylating monooxygenase (PHM) and peptidyl-α-hydroxyglycine α-amidating lyase (PAL) subunits that work in a sequential manner. PHM binds two copper atoms that participate in reduction of molecular oxygen for the hydroxylation substrates.

Dopamine β-monooxygenase converts dopamine to noradrenaline and is therefore responsible for catecholamine biosynthesis [166]. Perinatal copper deficiency in mice creates an apparent paradox for dopamine β-monooxygenase in that activity increases with copper deficiency [167]. This is actually due to increased mRNA and protein levels and may highlight a response to copper deficiency to prioritize copper for neural function.

Tyrosinase is a key copper enzyme required for melanin synthesis [168]. By carrying out two specific monooxygenase reactions, tyrosinase stabilizes intermediates leading to the formation of melanin. The formation of this pigment is critical for protection against UV and other environmental stressors. This requirement for copper is consistent with Menkes disease patients displaying hypopigmentation.

3.4 Neurological Disorders

Copper-induced protein aggregation has been shown in various neurological diseases, including Alzheimer's, Parkinson's disease, and familial amyotrophic lateral sclerosis (fALS). The amyloid-β protein that is found in the senile plaques of Alzheimer's patient brains is a copper protein. Amyloid-β is first produced as amyloid precursor protein (APP) before cleavage to Aβ fragments [169]. APP has been implicated in modulation of copper efflux from brain and liver [170]. Copper binds to the extracellular amino terminal domain that is cleaved to form the Aβ fragments. Aggregation of the neurotoxic Aβ1-42 fragment into oligomers rather than fibrils appears to lead to synaptic dysfunction [171]. Metals such as copper and zinc can induce oligomerization and aggregation of the neurotoxic Aβ1-42 fragment. Direct analysis of these plaques show that they contain high levels of copper and other metals, suggesting a role for metals in potentiation of disease [172]. Equilibrium studies with chelators have suggested biologically relevant copper-binding constants, however, the role of copper binding remains to be fully elucidated. α-Synuclein is a natively unstructured protein of unknown function that has been found in Lewy body protein aggregates associated with Parkinson's disease [173]. Increased copper concentration has been shown to increase aggregation of overexpressed α-synuclein as well as increased cytotoxicity [174]. Like Aβ, the copper-binding sites in α-synuclein are still under active investigation.

Copper binding has also been observed in the prion protein (PrP), a protein found ubiquitously throughout the central nervous system that acts as an infectious agent in neurodegenerative diseases [175,176]. Prion protein N-terminus contains an octarepeat domain that has been shown to bind copper [175]. Prion protein has been linked to copper transport, copper-regulated cell signaling and protection against copper-mediated oxidative stress. Also mutations in the SOD1 gene that cause aggregation of insoluble unfolded protein have been implicated in familial amyotrophic lateral sclerosis [177]. Study in animal models of fALS indicate that aggregated SOD1 may lack the critical disulfide bond and are copper-deficient. Though the exact roles are uncharacterized copper and CCS are implicated in some cases of fALS.

3.5 Sequestration and Regulation

Metallothioneins in mammals are responsible for the detoxification of cadmium, copper and zinc and have been proposed to be important in managing normal metal homeostasis [178,179]. The final structure of metallated metallothioneins consists of two domains, an α and a β domain, defined by the spacing of the cysteine residues and the number of metals coordinated. Metals bind in thiolate clusters that keep them in a biologically inert state [179]. Turnover of metallothionein occurs in the lysosome where the metals can be released back to the cytosol and some studies have suggested that metallothionein can be a direct source of metals for delivery within the cell independent of proteolysis [180].

Expression of the metallothionein gene is controlled by metal response elements in their promoter [181,182]. These metal response elements are the site for binding of MTF1, a transcription factor that responds to zinc levels [181,182]. Therefore sequestration of copper in metallothionein appears to be indirectly regulated by MTF1. *MTF1*$^{-/-}$ mice are embryonically lethal due to a developmental defect of the liver, highlighting the larger role of MTF1 than regulating toxicity of metals [183].

3.6 Copper Deficiency and Biomarkers

Copper deficiency in mammals can lead to cardiac hypertrophy, hypopigmentation, loose skin, anemia, and hepatic iron accumulation. A more challenging problem is to detect marginal copper status in mammals [79]. A biomarker of copper status is the ratio of apo-ceruloplasmin to holo-ceruloplasmin [184]. Total ceruloplasmin levels are decreased due to proteolysis of apo-ceruloplasmin and therefore can also be used as a marker [185,186]. Like ceruloplasmin, steady state levels of other multicopper oxidases, such as hephaestin, decrease during copper deficiency in rats and mice while mRNA levels remain unchanged [187–189].

Further, levels and activities of various other copper-containing proteins such as SOD1 in the erythrocytes, diamine oxidase in the plasma, tyrosinase in the skin, and peptidylglycine α-monooxygenase and hephaestin have been demonstrated to be consistently low in copper deprived mammals [188,190–193]. However, the disadvantage in using such biomarkers reliably is that they are subject to regulation that is independent of copper levels, such as responses to hormones and metabolic status [194]. Measuring the levels of the copper chaperone CCS has proven to be a promising approach for gauging copper status. Mild copper deficiency that does not alter other biomarkers, such as cytochrome *c* oxidase or ceruloplasmin, is detected by a measurable increase in the CCS levels. This increase in CCS occurs due a decrease in proteasome-mediated degradation, which is enhanced in the presence of copper [195–198].

4 Copper Homeostasis: Photosynthetic Organisms

Copper is required by most photosynthetic eukaryotes for mitochondrial electron transport (cytochrome *c* oxidase). Photosynthetic eukaryotes also require copper to be tightly controlled and so they contain many of the same chaperone, buffering, and copper removal pathways as other eukaryotes. The process of photosynthesis places additional copper demands as the enzyme plastocyanin is dependent on copper. Therefore, although photosynthetic organisms and heterotrophic eukaryotes share many common features they also have unique demands and strategies for homeostasis (Figure 3).

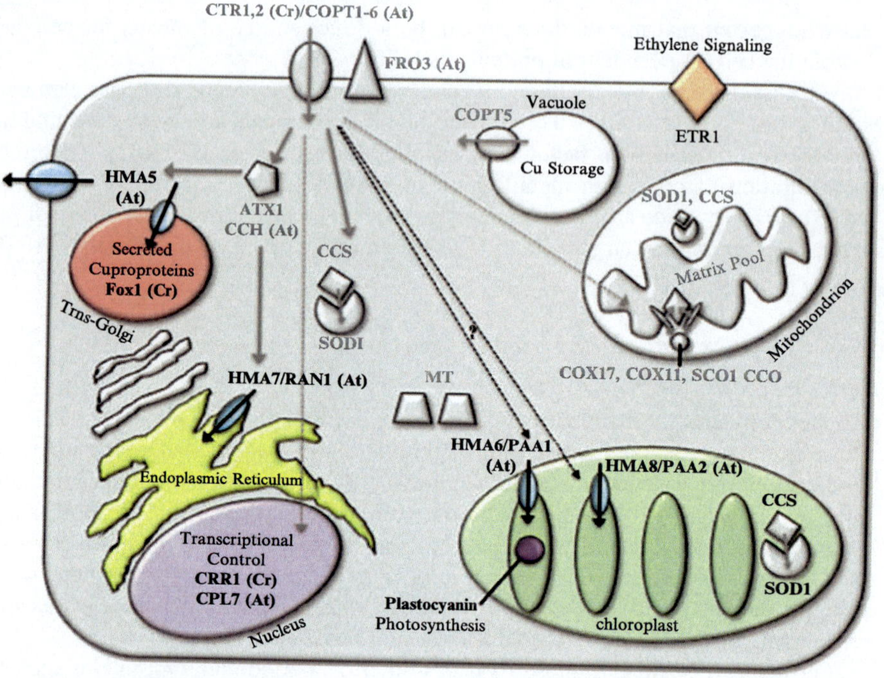

Figure 3 Copper homeostasis in generalized photosynthetic cells. This figure should be compared directly to Figure 1 and 2. The color components are the novel components of *Chlamydomonas reinhardtii* (Cr) and *Arabidopsis thaliana* (At). Plants and algae have unique copper enzymes including the photosynthetic enzyme plastocyanin and ethylene receptor ETR1. The chloroplast Cu,Zn Sod (At) necessitates existence of ATPase transporters in the chloroplast (HMA6, HMA8). HMA5 acts in part as the ATP7A in mammals as a exporter for removing toxic copper. FOX1 is a multicopper oxidase for iron uptake in *C. reinhardtii.* HMA7 provides copper to metallate the ETR1 in the endoplasmic reticulum of *A. thaliana*. CRR1 and SPL7 directly regulate the distribution of copper in the cells. The grey components: the high affinity transporters, copper chaperones, metallothioneins are homologous to those found in yeast.

The model unicellular photosynthetic organism *Chlamydomonas reinhardtii* contains genes for multiple CTR-type transporters: CTR1 and CTR2 are high affinity copper importers as they complement defects *ctr1*Δ*ctr3*Δ yeast [199]. While the commonly used model plant *Arabidopsis thaliana* has 6 CTR homologues, COPT1 and COPT2 are high affinity copper transporters at the plasma membrane [200,201] while COPT5 is the vacuolar transporter equivalent to yeast Ctr2 [202]. The role of COPT3, 4 and 6 in plant copper transport is not yet clearly defined. *Chlamydomonas reinhardtii* has one Atx1 homolog and *A. thaliana* has two Atx1 homologues called ATX1 and CCH respectively, responsible for delivery of copper to P-type ATPases CCC2 and HMA7 [203–205]. *Chlamydomonas reinhardtii* has a multi-copper ferroxidase, Fox1, which is required for iron acquisition. However, there are limited data supporting copper-dependent iron acquisition pathway in land plants [203,206].

HMA5 is P-type ATPase that is involved in copper transport and detoxification in *A. thaliana* roots. Expression of HMA5 induced by copper overload in the plant, suggests a role in detoxification. This role may be similar to the role of ATP7A/ATP7B in removing excess cellular and systemic copper. As in other eukaryotes, HMA5 relies on a chaperone homologous to Atx1 for copper delivery.

Copper-regulated transcription is present in *Chlamydomonas reinhardtii* and *Arabidoposis sp.* In *C. reinhardtii* a copper repressible transcription factor CRR1 is activated under low copper conditions to facilitate the expression of iron-utilizing proteins to replace the copper-requiring counterparts [207,208]. In *A. thaliana* the CRR1 homologue is SPL7, a transcription factor that controls the transcription of the genes encoding transporters *COPT1*, *COPT2*, the metalloreductase, *FRO3*, the copper chaperone *CCH* and various micro-RNA used for controlling copper distribution [209].

Chlamydomonas reinhardtii has no copper-containing superoxide dismutase, but *Arabidopsis* does have Cu,Zn SOD in both the chloroplasts and in mitochondria [210,211]. The *Arabidopsis* Cu,Zn SOD also uses a CCS-like chaperone for copper insertion, though recent evidence points to a CCS-independent activation pathway as well [212,213]. Homologues of various cytochrome *c* oxidases subunits and copper chaperones have been identified in both *Chlamydomonas* and *Arabidopsis* [214]. Expression of *COX17* in *A. thaliana* is induced upon exposure to high copper levels, and as in mammals *A. thaliana* SCO proteins appear to have multiple roles in both cytochrome *c* oxidase assembly and in redox signaling from mitochondria [215].

Photosynthesis is a copper-requiring process. The protein plastocyanin carries electrons from cytochrome *f* in the quinol oxidase complex to P700+ in photosystem I and requires copper for function [215,216]. The copper requirements of plastocyanin in *C. reinhardtii,* and plastocyanin and chloroplast-Cu,Zn SOD in *A. thaliana,* necessitate transport of copper into chloroplasts. The P-type ATPases HMA6 and HMA8 function in copper transport across the chloroplast envelope as well as thylakoid membranes in *A. thaliana* [217].

In plants the hormone ethylene initiates a signal transduction pathway that is important in germination, morphology, fruit ripening, and other cellular processes [218]. Five ethylene receptors have been described in *A. thaliana*, but the receptor ETR1 has been most extensively described [219,220]. Though a direct role for copper in ethylene signaling has not been identified, ETR1 co-purifies with copper and its ethylene binding capability is enhanced upon addition of copper [221]. Additionally, the secretory copper transporting ATPase HMA7 is required for biogenesis of the *A. thaliana* ETR1 protein when expressed in yeast [204,222].

The special copper demands of plants and algae have meant that unique strategies for dealing with copper deficiency have developed. In *C. reinhardtii*, copper deficiency induces a switch from the copper-requiring electron carrier plastocyanin to a heme-containing electron carrier, cytochrome c_6 [223]. Two pathways mediate this process. Lack of copper causes structural changes in plastocyanin and this apo-conformation is more susceptible to protease-mediated degradation [224]. The gene coding for the alternative electron carrier, *CYC6* is under the control of the copper-repressible transcription factor, Crr1 [225,226]. *Arabidopsis sp.* and other land plants do not contain a homolog of the *CYC6* alternative electron carrier gene [227]. Instead, copper deficiency induces a switch from Cu,Zn Sod to an iron-containing superoxide dismutase [217]. This allows for all available copper to be used by plastocyanin for photosynthesis. The change from copper to iron SOD in *Arabidopsis* is mediated by micro-RNA regulation. The micro-RNA miR398 is responsive to copper and down-regulates translation of Cu,Zn SOD transcripts as well as some cytochrome genes [228,229]. This micro-RNA plays no role in the regulation of plastocyanin.

5 Concluding Remarks and Future Challenges

Copper plays a critical role in normal physiology and the survey presented here highlights the large classes of copper enzymes used by eukaryotes and the transporters and chaperones that are responsible for activating them. However, this chapter does not address some of the more complicated processes and pathways that copper (or copper proteins) have been implicated in, but not limited to, active roles in the immune system by regulating availability of copper for pathogens, neuronal signaling via regulation of the NMDA receptor activity and calcium fluxes, roles in schizophrenia and leukemia and even in HIV-1 replication. We have only 'scratched the surface' of the details available in the literature to prime the reader to the diversity of the copper metallome and encourage further reading in the volumes of data from models such as *Podospora anserina, Schizosaccharomyces pombe, Neurospora crassa, Danio rerio, Cryptococcus neoformans*, and others.

Even though the basic scaffold of copper homeostasis has progressed significantly in the last 20 years, the complete story of the copper metallome is still unfolding. The details of systemic copper signaling, tissue specific handling of copper loads, the role of mitochondria in signaling cellular copper status, and the molecular details of mitochondrial copper delivery are some of the many remaining questions.

Biochemical techniques have improved to a point where previously undetectable copper-mediated cellular events are now being visualized. Unexpected protein-protein interactions, changes in the proteome and defining unique copper binding sites continue to add levels of detail and richness to the roles of copper in the cell. The synthesis of copper-specific fluorescent probes that can be used in imaging of live cells, and the application of synchrotron X-ray technology, has enabled the study of a number of intriguing signaling and molecular level events that are tied to exchangeable pools of copper. The creation of more soluble and sensitive copper probes and better resolution for data collection will allow for detailed investigations of cellular events in response to different biotic and abiotic stressors.

Abbreviations

APP	amyloid precursor protein
ATP	adenosine 5'-triphosphate
CHO	Chinese hamster ovary

DMT	divalent metal transporter
fALS	familial amyotrophic lateral sclerosis
HIV	human immunodeficiency virus
NMDA	N-methyl-D-aspartate
PAL	peptidyl-α-hydroxyglycine α-amidating lyase
PAM	peptidylglycine α-amidating monooxygenase
PHM	peptidylglycine α-hydroxylating monooxygenase
PrP	prion protein
SOD	Cu,Zn superoxide dismutase

Acknowledgment Copper research in the Cobine laboratory is funded by the National Science Foundation (# 1158497) and the USDA National Institute of Food and Agriculture (# 2010-65108-20633).

References

1. P. L. Fox, *Biometals* **2003**, *16*, 9–40.
2. J. R. Prohaska, *Am. J. Clin. Nutr.* **2008**, *88*, 826S–829S.
3. Z. L. Harris, J. D. Gitlin, *Am. J. Clin. Nutr.* **1996**, *63*, 836S–841S.
4. L. Banci, I. Bertini, V. Calderone, N. Della-Malva, I. C. Felli, S. Neri, A. Pavelkova, A. Rosato, *Biochem. J.* **2009**, *422*, 37–42.
5. L. Banci, I. Bertini, S. Ciofi-Baffoni, T. Kozyreva, K. Zovo, P. Palumaa, *Nature* **2010**, *465*, 645–648.
6. E. M. Rees, D. J. Thiele, *Curr. Opin. Microbiol.* **2004**, *7*, 175–184.
7. S. Puig, D. J. Thiele, *Curr. Opin. Chem. Biol.* **2002**, *6*, 171–180.
8. V. C. Culotta, S. J. Lin, P. Schmidt, L. W. Klomp, R. L. Casareno, J. Gitlin, *Adv. Exper. Med. Biol.* **1999**, *448*, 247–254.
9. R. A. Pufahl, C. P. Singer, K. L. Peariso, S. J. Lin, P. J. Schmidt, C. J. Fahrni, V. C. Culotta, J. E. Penner-Hahn, T. V. O'Halloran, *Science* **1997**, *278*, 853–856.
10. T. D. Rae, P. J. Schmidt, R. A. Pufahl, V. C. Culotta, T. V. O'Halloran, *Science* **1999**, *284*, 805–808.
11. V. C. Culotta, M. Yang, T. V. O'Halloran, *BBA-Mol. Cell. Res.* **2006**, *1763*, 747–758.
12. C. Askwith, D. Eide, A. Van Ho, P. S. Bernard, L. Li, S. Davis-Kaplan, D. M. Sipe, J. Kaplan, *Cell* **1994**, *76*, 403–410.
13. X. Shi, C. Stoj, A. Romeo, D. J. Kosman, Z. Zhu, *J. Biol. Chem.* **2003**, *278*, 50309–50315.
14. L. A. Sturtz, K. Diekert, L. T. Jensen, R. Lill, V. C. Culotta, *J. Biol. Chem.* **2001**, *276*, 38084–38089.
15. J. A. Tainer, E. D. Getzoff, J. S. Richardson, D. C. Richardson, *Nature* **1983**, *306*, 284–287.
16. P. A. Cobine, F. Pierrel, D. R. Winge, *BBA-Mol. Cell Res.* **2006**, *1763*, 759–772.
17. T. Tsukihara, H. Aoyama, E. Yamashita, T. Tomizaki, H. Yamaguchi, K. Shinzawa-Itoh, R. Nakashima, R. Yaono, S. Yoshikawa, *Science* **1996**, *272*, 1136–1144.
18. S. C. Leary, D. R. Winge, P. A. Cobine, *BBA-Mol. Cell Res.* **2009**, *1793*, 146–153.
19. H. S. Carr, D. R. Winge, *Acc. Chem. Res.* **2003**, *36*, 309–316.
20. S. G. Aller, V. M. Unger, *Proc. Natl. Acad. Sci. USA* **2006**, *103*, 3627–3632.
21. A. Dancis, D. S. Yuan, D. Haile, C. Askwith, D. Eide, C. Moehle, J. Kaplan, R. D. Klausner, *Cell* **1994**, *76*, 393–402.
22. C. J. De Feo, S. G. Aller, G. S. Siluvai, N. J. Blackburn, V. M. Unger, *Proc. Natl. Acad. Sci. USA* **2009**, *106*, 4237–4242.
23. J. F. Eisses, J. H. Kaplan, *J. Biol. Chem.* **2005**, *280*, 37159–37168.
24. Z. Xiao, A. G. Wedd, *Chem. Commun.* **2002**, 588–589.

25. M. M. Pena, S. Puig, D. J. Thiele, *J. Biol. Chem.* **2000**, *275*, 33244–33251.
26. R. Hassett, D. R. Dix, D. J. Eide, D. J. Kosman, *Biochem. J.* **2000**, *351 Pt 2*, 477–484.
27. A. Cohen, H. Nelson, N. Nelson, *J. Biol. Chem.* **2000**, *275*, 33388–33394.
28. R. Hassett, D. J. Kosman, *J. Biol. Chem.* **1995**, *270*, 128–134.
29. D. Fu, T. J. Beeler, T. M. Dunn, *Yeast* **1995**, *11*, 283–292.
30. D. S. Yuan, R. Stearman, A. Dancis, T. Dunn, T. Beeler, R. D. Klausner, *Proc. Natl. Acad. Sci. USA* **1995**, *92*, 2632–2636.
31. L. Banci, I. Bertini, S. Ciofi-Baffoni, D. L. Huffman, T. V. O'Halloran, *J. Biol. Chem.* **2001**, *276*, 8415–8426.
32. D. L. Huffman, T. V. O'Halloran, *Annu. Rev. Biochem.* **2001**, *70*, 677–701.
33. L. Banci, I. Bertini, F. Cantini, I. C. Felli, L. Gonnelli, N. Hadjiliadis, R. Pierattelli, A. Rosato, P. Voulgaris, *Nat. Chem. Biol.* **2006**, *2*, 367–368.
34. E. M. Rees, D. J. Thiele, *J. Biol. Chem.* **2007**, *282*, 21629–21638.
35. E. M. Rees, J. Lee, D. J. Thiele, *J. Biol. Chem.* **2004**, *279*, 54221–54229.
36. P. A. Cobine, L. D. Ojeda, K. M. Rigby, D. R. Winge, *J. Biol. Chem.* **2004**, *279*, 14447–14455.
37. P. A. Cobine, F. Pierrel, M. L. Bestwick, D. R. Winge, *J. Biol. Chem.* **2006**, *281*, 36552–36559.
38. F. Arnesano, L. Banci, I. Bertini, D. L. Huffman, T. V. O'Halloran, *Biochemistry* **2001**, *40*, 1528–1539.
39. Z. Xiao, F. Loughlin, G. N. George, G. J. Howlett, A. G. Wedd, *J. Am. Chem. Soc.* **2004**, *126*, 3081–3090.
40. D. L. Huffman, T. V. O'Halloran, *J. Biol. Chem.* **2000**, *275*, 18611–18614.
41. Y. Furukawa, A. S. Torres, T. V. O'Halloran, *EMBO J.* **2004**, *23*, 2872–2881.
42. P. J. Schmidt, T. D. Rae, R. A. Pufahl, T. Hamma, J. Strain, T. V. O'Halloran, V. C. Culotta, *J. Biol. Chem.* **1999**, *274*, 23719–23725.
43. A. L. Lamb, A. K. Wernimont, R. A. Pufahl, V. C. Culotta, T. V. O'Halloran, A. C. Rosenzweig, *Nat. Struct. Biol.* **1999**, *6*, 724–729.
44. A. L. Lamb, A. S. Torres, T. V. O'Halloran, A. C. Rosenzweig, *Biochemistry* **2000**, *39*, 14720–14727.
45. D. P. Gross, C. A. Burgard, S. Reddehase, J. M. Leitch, V. C. Culotta, K. Hell, *Mol. Biol. Cell* **2011**, *20*, 3758–3767.
46. D. M. Glerum, A. Shtanko, A. Tzagoloff, *J. Biol. Chem.* **1996**, *271*, 14504–14509.
47. L. Banci, I. Bertini, S. Ciofi-Baffoni, A. Janicka, M. Martinelli, H. Kozlowski, P. Palumaa, *J. Biol. Chem.* **2008**, *283*, 7912–7920.
48. L. Banci, I. Bertini, C. Cefaro, S. Ciofi-Baffoni, A. Gallo, *J. Biol. Chem.* **2011**, *286*, 34382–34390.
49. L. Banci, I. Bertini, S. Ciofi-Baffoni, T. Hadjiloi, M. Martinelli, P. Palumaa, *Proc. Natl. Acad. Sci. USA* **2008**, *105*, 6803–6808.
50. Y. C. Horng, P. A. Cobine, A. B. Maxfield, H. S. Carr, D. R. Winge, *J. Biol. Chem.* **2004**, *279*, 35334–35340.
51. C. Abajian, A. C. Rosenzweig, *J. Biol. Inorg. Chem.* **2006**, *11*, 459–466.
52. M. Schulze, G. Rodel, *Mol. Gen. Genetics* **1988**, *211*, 492–498.
53. T. Nittis, G. N. George, D. R. Winge, *J. Biol. Chem.* **2001**, *276*, 42520–42526.
54. Y. C. Horng, S. C. Leary, P. A. Cobine, F. B. Young, G. N. George, E. A. Shoubridge, D. R. Winge, *J. Biol. Chem.* **2005**, *280*, 34113–34122.
55. H. S. Carr, G. N. George, D. R. Winge, *J. Biol. Chem.* **2002**, *277*, 31237–31242.
56. D. Horn, A. Barrientos, *IUBMB Life* **2008**, *60*, 421–429.
57. S. V. Wegner, F. Sun, N. Hernandez, C. He, *Chem. Commun.* **2011**, *47*, 2571–2573.
58. V. C. Culotta, W. R. Howard, X. F. Liu, *J. Biol. Chem.* **1994**, *269*, 25295–25302.
59. D. R. Winge, K. B. Nielson, W. R. Gray, D. H. Hamer, *J. Biol. Chem.* **1985**, *260*, 14464–14470.
60. L. T. Jensen, W. R. Howard, J. J. Strain, D. R. Winge, V. C. Culotta, *J. Biol. Chem.* **1996**, *271*, 18514–18519.

61. W. X. Bi, F. Kong, X. Y. Hu, X. Cui, *Toxicol. Mech. Methods* **2007**, *17*, 371–378.
62. M. M. Gharieb, G. M. Gadd, *Biometals* **2004**, *17*, 183–188.
63. C. Gross, M. Kelleher, V. R. Iyer, P. O. Brown, D. R. Winge, *J. Biol. Chem.* **2000**, *275*, 32310–32316.
64. Y. Yamaguchi-Iwai, M. Serpe, D. Haile, W. Yang, D. J. Kosman, R. D. Klausner, A. Dancis, *J. Biol. Chem.* **1997**, *272*, 17711–17718.
65. E. B. Gralla, D. J. Thiele, P. Silar, J. S. Valentine, *Proc. Natl. Acad. Sci. USA* **1991**, *88*, 8558–8562.
66. C. T. Dameron, D. R. Winge, G. N. George, M. Sansone, S. Hu, D. Hamer, *Proc. Natl. Acad. Sci. USA* **1991**, *88*, 6127–6131.
67. C. E. Ooi, E. Rabinovich, A. Dancis, J. S. Bonifacino, R. D. Klausner, *EMBO J.* **1996**, *15*, 3515–3523.
68. S. J. Lin, R. A. Pufahl, A. Dancis, T. V. O'Halloran, V. C. Culotta, *J. Biol. Chem.* **1997**, *272*, 9215–9220.
69. X. F. Liu, I. Elashvili, E. B. Gralla, J. S. Valentine, P. Lapinskas, V. C. Culotta, *J. Biol. Chem.* **1992**, *267*, 18298–18302.
70. J. R. Prohaska, A. A. Gybina, *J. Nutr.* **2004**, *134*, 1003–1006.
71. K. J. Allen, N. E. Buck, D. M. Cheah, S. Gazeas, P. Bhathal, J. F. Mercer, *Biometals* **2006**, *19*, 555–564.
72. J. Lee, J. R. Prohaska, D. J. Thiele, *Proc. Natl. Acad. Sci. USA* **2001**, *98*, 6842–6847.
73. B. Zhou, J. Gitschier, *Proc. Natl. Acad. Sci. USA* **1997**, *94*, 7481–7486.
74. J. Lee, J. R. Prohaska, S. L. Dagenais, T. W. Glover, D. J. Thiele, *Gene* **2000**, *254*, 87–96.
75. K. L. Haas, A. B. Putterman, D. R. White, D. J. Thiele, K. J. Franz, *J. Am. Chem. Soc.* **2011**, *133*, 4427–4437.
76. Y. M. Kuo, B. Zhou, D. Cosco, J. Gitschier, *Proc. Natl. Acad. Sci. USA* **2001**, *98*, 6836–6841.
77. P. V. van den Berghe, D. E. Folmer, H. E. Malingre, E. van Beurden, A. E. Klomp, B. van de Sluis, M. Merkx, R. Berger, L. W. Klomp, *Biochem. J.* **2007**, *407*, 49–59.
78. J. Bertinato, E. Swist, L. J. Plouffe, S. P. Brooks, R. L'Abbe M, *Biochem. J .* **2008**, *409*, 731–740.
79. Y. Nose, B. E. Kim, D. J. Thiele, *Cell Metab.* **2006**, *4*, 235–244.
80. A. M. Zimnicka, E. B. Maryon, J. H. Kaplan, *J. Biol. Chem.* **2007**, *282*, 26471–26480.
81. A. M. Zimnicka, K. Ivy, J. H. Kaplan, *Am. J. Physiol. Cell Physiol.* **2011**, *300*, C588–599.
82. Y. Nose, L. K. Wood, B. E. Kim, J. R. Prohaska, R. S. Fry, J. W. Spears, D. J. Thiele, *J. Biol. Chem.* **2010**, *285*, 32385–32392.
83. M. Arredondo, P. Munoz, C. V. Mura, M. T. Nunez, *Am. J. Physiol. Cell Physiol.* **2003**, *284*, C1525–1530.
84. B. E. Kim, M. L. Turski, Y. Nose, M. Casad, H. A. Rockman, D. J. Thiele, *Cell Metab.* **2010**, *11*, 353–363.
85. J. R. Prohaska, *Clin. Physiol. Biochem.* **1986**, *4*, 87–93.
86. C. Vulpe, B. Levinson, S. Whitney, S. Packman, J. Gitschier, *Nat. Genet.* **1993**, *3*, 7–13.
87. J. F. Mercer, J. Livingston, B. Hall, J. A. Paynter, C. Begy, S. Chandrasekharappa, P. Lockhart, A. Grimes, M. Bhave, D. Siemieniak, T. W. Glover, *Nat. Genet.* **1993**, *3*, 20–25.
88. J. Chelly, Z. Tumer, T. Tonnesen, A. Petterson, Y. Ishikawa-Brush, N. Tommerup, N. Horn, A. P. Monaco, *Nat. Genet.* **1993**, *3*, 14–19.
89. K. Petrukhin, S. Lutsenko, I. Chernov, B. M. Ross, J. H. Kaplan, T. C. Gilliam, *Hum. Mol. Genet.* **1994**, *3*, 1647–1656.
90. D. W. Cox, S. D. Moore, *J. Bioenerg. Biomembr.* **2002**, *34*, 333–338.
91. A. S. Payne, J. D. Gitlin, *J. Biol. Chem.* **1998**, *273*, 3765–3770.
92. I. H. Hung, M. Suzuki, Y. Yamaguchi, D. S. Yuan, R. D. Klausner, J. D. Gitlin, *J. Biol. Chem.* **1997**, *272*, 21461–21466.
93. V. Gerbasi, S. Lutsenko, E. J. Lewis, *Neurochem. Res.* **2003**, *28*, 867–873.
94. B. Hardman, A. Michalczyk, M. Greenough, J. Camakaris, J. Mercer, L. Ackland, *Cell Physiol. Biochem.* **2007**, *20*, 1073–1084.

95. M. J. Petris, D. Strausak, J. F. Mercer, *Hum. Mol. Genet.* **2000**, *9*, 2845–2851.
96. K. Nagano, K. Nakamura, K. I. Urakami, K. Umeyama, H. Uchiyama, K. Koiwai, S. Hattori, T. Yamamoto, I. Matsuda, F. Endo, *Hepatology* **1998**, *27*, 799–807.
97. Y. Yamaguchi, M. E. Heiny, M. Suzuki, J. D. Gitlin, *Proc. Natl. Acad. Sci. USA* **1996**, *93*, 14030–14035.
98. J. Camakaris, M. J. Petris, L. Bailey, P. Shen, P. Lockhart, T. W. Glover, C. Barcroft, J. Patton, J. F. Mercer, *Hum. Mol. Genet.* **1995**, *4*, 2117–2123.
99. C. D. Vulpe, S. Packman, *Annu. Rev. Nutr.* **1995**, *15*, 293–322.
100. M. L. Schilsky, R. J. Stockert, A. Kesner, G. R. Gorla, G. S. Gagliardi, K. Terada, N. Miura, M. J. Czaja, *Hepatology* **1998**, *28*, 1347–1356.
101. M. L. Schilsky, A. N. Irani, G. R. Gorla, I. Volenberg, S. Gupta, *J. Biochem. Mol. Toxicol.* **2000**, *14*, 210–214.
102. M. J. Petris, J. F. Mercer, J. G. Culvenor, P. Lockhart, P. A. Gleeson, J. Camakaris, *EMBO J.* **1996**, *15*, 6084–6095.
103. S. L. La Fontaine, S. D. Firth, J. Camakaris, A. Englezou, M. B. Theophilos, M. J. Petris, M. Howie, P. J. Lockhart, M. Greenough, H. Brooks, R. R. Reddel, J. F. Mercer, *J. Biol. Chem.* **1998**, *273*, 31375–31380.
104. S. La Fontaine, S. D. Firth, P. J. Lockhart, H. Brooks, R. G. Parton, J. Camakaris, J. F. Mercer, *Hum. Mol. Genet.* **1998**, *7*, 1293–1300.
105. S. La Fontaine, M. B. Theophilos, S. D. Firth, R. Gould, R. G. Parton, J. F. Mercer, *Hum. Mol. Genet.* **2001**, *10*, 361–370.
106. M. A. Cater, S. La Fontaine, K. Shield, Y. Deal, J. F. Mercer, *Gastroenterology* **2006**, *130*, 493–506.
107. H. Roelofsen, H. Wolters, M. J. Van Luyn, N. Miura, F. Kuipers, R. J. Vonk, *Gastroenterology* **2000**, *119*, 782–793.
108. M. J. Francis, E. E. Jones, E. R. Levy, S. Ponnambalam, J. Chelly, A. P. Monaco, *Hum. Mol. Genet.* **1998**, *7*, 1245–1252.
109. M. Qi, P. H. Byers, *Hum. Mol. Genet.* **1998**, *7*, 465–469.
110. I. Voskoboinik, D. Strausak, M. Greenough, H. Brooks, M. Petris, S. Smith, J. F. Mercer, J. Camakaris, *J. Biol. Chem.* **1999**, *274*, 22008–22012.
111. L. Ambrosini, J. F. Mercer, *Hum. Mol. Genet.* **1999**, *8*, 1547–1555.
112. J. Gitschier, B. Moffat, D. Reilly, W. I. Wood, W. J. Fairbrother, *Nat. Struct. Biol.* **1998**, *5*, 47–54.
113. C. E. Jones, N. L. Daly, P. A. Cobine, D. J. Craik, C. T. Dameron, *J. Struct. Biol.* **2003**, *143*, 209–218.
114. P. A. Cobine, G. N. George, D. J. Winzor, M. D. Harrison, S. Mogahaddas, C. T. Dameron, *Biochemistry* **2000**, *39*, 6857–6863.
115. M. D. Harrison, S. Meier, C. T. Dameron, *Biochim. Biophys. Acta* **1999**, *1453*, 254–260.
116. P. Y. Jensen, N. Bonander, N. Horn, Z. Tumer, O. Farver, *Eur. J. Biochem.* **1999**, *264*, 890–896.
117. S. Lutsenko, K. Petrukhin, M. J. Cooper, C. T. Gilliam, J. H. Kaplan, *J. Biol. Chem.* **1997**, *272*, 18939–18944.
118. M. Iida, K. Terada, Y. Sambongi, T. Wakabayashi, N. Miura, K. Koyama, M. Futai, T. Sugiyama, *FEBS Lett.* **1998**, *428*, 281–285.
119. L. Banci, I. Bertini, F. Cantini, A. C. Rosenzweig, L. A. Yatsunyk, *Biochemistry* **2008**, *47*, 7423–7429.
120. J. M. Walker, R. Tsivkovskii, S. Lutsenko, *J. Biol. Chem.* **2002**, *277*, 27953–27959.
121. E. S. LeShane, U. Shinde, J. M. Walker, A. N. Barry, N. J. Blackburn, M. Ralle, S. Lutsenko, *J. Biol. Chem.* **2010**, *285*, 6327–6336.
122. I. D. Goodyer, E. E. Jones, A. P. Monaco, M. J. Francis, *Hum. Mol. Genet.* **1999**, *8*, 1473–1478.
123. J. F. Mercer, N. Barnes, J. Stevenson, D. Strausak, R. M. Llanos, *Biometals* **2003**, *16*, 175–184.

124. D. Strausak, S. La Fontaine, J. Hill, S. D. Firth, P. J. Lockhart, J. F. Mercer, *J. Biol. Chem.* **1999**, *274*, 11170–11177.
125. M. J. Petris, J. Camakaris, M. Greenough, S. LaFontaine, J. F. Mercer, *Hum. Mol. Genet.* **1998**, *7*, 2063–2071.
126. M. J. Petris, J. F. Mercer, *Hum. Mol. Genet.* **1999**, *8*, 2107–2115.
127. S. C. Dodani, S. C. Leary, P. A. Cobine, D. R. Winge, C. J. Chang, *J. Am. Chem. Soc.* **2011**, *133*, 8606–8616.
128. L. Yang, R. McRae, M. M. Henary, R. Patel, B. Lai, S. Vogt, C. J. Fahrni, *Proc. Natl. Acad. Sci. USA* **2005**, *102*, 11179–11184.
129. I. Hamza, J. Prohaska, J. D. Gitlin, *Proc. Natl. Acad. Sci. USA* **2003**, *100*, 1215–1220.
130. P. C. Wong, D. Waggoner, J. R. Subramaniam, L. Tessarollo, T. B. Bartnikas, V. C. Culotta, D. L. Price, J. Rothstein, J. D. Gitlin, *Proc. Natl. Acad. Sci. USA* **2000**, *97*, 2886–2891.
131. Y. Takahashi, K. Kako, S. Kashiwabara, A. Takehara, Y. Inada, H. Arai, K. Nakada, H. Kodama, J. Hayashi, T. Baba, E. Munekata, *Mol. Cell. Biol.***2002**, *22*, 7614–7621.
132. S. Itoh, H. W. Kim, O. Nakagawa, K. Ozumi, S. M. Lessner, H. Aoki, K. Akram, R. D. McKinney, M. Ushio-Fukai, T. Fukai, *J. Biol. Chem.* **2008**, *283*, 9157–9167.
133. S. C. Leary, P. A. Cobine, B. A. Kaufman, G. H. Guercin, A. Mattman, J. Palaty, G. Lockitch, D. R. Winge, P. Rustin, R. Horvath, E. A. Shoubridge, *Cell Metab.* **2007**, *5*, 403–403.
134. P. A. Cobine, F. Pierrel, S. C. Leary, F. Sasarman, Y. C. Horng, E. A. Shoubridge, D. R. Winge, *J. Biol. Chem.* **2006**, *281*, 12270–12276.
135. L. Banci, I. Bertini, S. Ciofi-Baffoni, I. Leontari, M. Martinelli, P. Palumaa, R. Sillard, S. Wang, *Proc. Natl. Acad. Sci. USA* **2007**, *104*, 15–20.
136. S. C. Leary, B. A. Kaufman, G. Pellecchia, G. H. Guercin, A. Mattman, M. Jaksch, E. A. Shoubridge, *Hum. Mol. Genet.* **2004**, *13*, 1839–1848.
137. M. Jaksch, R. Horvath, N. Horn, D. P. Auer, C. Macmillan, J. Peters, K. D. Gerbitz, I. Kraegeloh-Mann, A. Muntau, V. Karcagi, R. Kalmanchey, H. Lochmuller, E. A. Shoubridge, P. Freisinger, *Neurology* **2001**, *57*, 1440–1446.
138. S. C. Leary, F. Sasarman, T. Nishimura, E. A. Shoubridge, *Hum. Mol. Genet.* **2009**, *18*, 2230–2240.
139. B. van De Sluis, J. Rothuizen, P. L. Pearson, B. A. van Oost, C. Wijmenga, *Hum. Mol. Genet.* **2002**, *11*, 165–173.
140. T. Y. Tao, F. Liu, L. Klomp, C. Wijmenga, J. D. Gitlin, *J. Biol. Chem.* **2003**, *278*, 41593–41596.
141. E. Burstein, L. Ganesh, R. D. Dick, B. van De Sluis, J. C. Wilkinson, L. W. Klomp, C. Wijmenga, G. J. Brewer, G. J. Nabel, C. S. Duckett, *EMBO J.* **2004**, *23*, 244–254.
142. N. E. Hellman, J. D. Gitlin, *Annu. Rev. Nutr.* **2002**, *22*, 439–458.
143. M. C. Linder, *Mutat. Res.* **2001**, *475*, 141–152.
144. B. N. Patel, S. David, *J. Biol. Chem.* **1997**, *272*, 20185–20190.
145. M. Yazaki, K. Yoshida, A. Nakamura, K. Furihata, M. Yonekawa, T. Okabe, N. Yamashita, M. Ohta, S. Ikeda, *J. Neurol. Sci.* **1998**, *156*, 30–34.
146. K. Yoshida, K. Furihata, S. Takeda, A. Nakamura, K. Yamamoto, H. Morita, S. Hiyamuta, S. Ikeda, N. Shimizu, N. Yanagisawa, *Nat. Genet.* **1995**, *9*, 267–272.
147. C. D. Vulpe, Y. M. Kuo, T. L. Murphy, L. Cowley, C. Askwith, N. Libina, J. Gitschier, G. J. Anderson, *Nat. Genet.* **1999**, *21*, 195–199.
148. H. Chen, Z. K. Attieh, B. A. Syed, Y. M. Kuo, V. Stevens, B. K. Fuqua, H. S. Andersen, C. E. Naylor, R. W. Evans, L. Gambling, R. Danzeisen, M. Bacouri-Haidar, J. Usta, C. D. Vulpe, H. J. McArdle, *J. Nutr.* **2010**, *140*, 1728–1735.
149. R. L. Casareno, D. Waggoner, J. D. Gitlin, *J. Biol. Chem.* **1998**, *273*, 23625–23628.
150. A. G. Reaume, J. L. Elliott, E. K. Hoffman, N. W. Kowall, R. J. Ferrante, D. F. Siwek, H. M. Wilcox, D. G. Flood, M. F. Beal, R. H. Brown, Jr., R. W. Scott, W. D. Snider, *Nat. Genet.* **1996**, *13*, 43–47.
151. Y. S. Ho, M. Gargano, J. Cao, R. T. Bronson, I. Heimler, R. J. Hutz, *J. Biol. Chem.* **1998**, *273*, 7765–7769.

152. M. M. Matzuk, L. Dionne, Q. Guo, T. R. Kumar, R. M. Lebovitz, *Endocrinology* **1998**, *139*, 4008–4011.
153. S. L. Marklund, *J. Clin. Invest.* **1984**, *74*, 1398–1403.
154. S. L. Marklund, *Proc. Natl. Acad. Sci. USA* **1982**, *79*, 7634–7638.
155. E. Nozik-Grayck, H. B. Suliman, C. A. Piantadosi, *Int. J. Biochem. Cell Biol.* **2005**, *37*, 2466–2471.
156. V. Jeney, S. Itoh, M. Wendt, Q. Gradek, M. Ushio-Fukai, D. G. Harrison, T. Fukai, *Circ. Res.* **2005**, *96*, 723–729.
157. M. Huttemann, T. R. Schmidt, L. I. Grossman, *Gene* **2003**, *312*, 95–102.
158. M. Huttemann, B. Kadenbach, L. I. Grossman, *Gene* **2001**, *267*, 111–123.
159. B. Kadenbach, M. Huttemann, S. Arnold, I. Lee, E. Bender, *Free Radic. Biol. Med.* **2000**, *29*, 211–221.
160. E. A. Shoubridge, *Am. J. Med. Genet.* **2001**, *106*, 46–52.
161. J. P. Klinman, *Biochim. Biophys. Acta* **2003**, *1647*, 131–137.
162. F. Buffoni, G. Ignesti, *Mol. Genet. Metab.* **2000**, *71*, 559–564.
163. K. Csiszar, *Prog. Nucleic Acid Res. Mol. Biol.* **2001**, *70*, 1–32.
164. L. I. Smith-Mungo, H. M. Kagan, *Matrix Biol.* **1998**, *16*, 387–398.
165. S. T. Prigge, R. E. Mains, B. A. Eipper, L. M. Amzel, *Cell Mol. Life Sci.* **2000**, *57*, 1236–1259.
166. S. Friedman, S. Kaufman, *J. Biol. Chem.* **1965**, *240*, PC552–554.
167. J. R. Prohaska, B. Brokate, *J. Nutr.* **1999**, *129*, 2147–2153.
168. C. Olivares, F. Solano, *Pigment Cell Melanoma Res.* **2009**, *22*, 750–760.
169. S. Sinha, I. Lieberburg, *Proc. Natl. Acad. Sci. USA* **1999**, *96*, 11049–11053.
170. C. Treiber, A. Simons, M. Strauss, M. Hafner, R. Cappai, T. A. Bayer, G. Multhaup, *J. Biol. Chem.* **2004**, *279*, 51958–51964.
171. H. A. Feaga, R. C. Maduka, M. N. Foster, V. A. Szalai, *Inorg. Chem.* **2011**, *50*, 1614–1618.
172. M. A. Lovell, J. D. Robertson, W. J. Teesdale, J. L. Campbell, W. R. Markesbery, *J. Neurol. Sci.* **1998**, *158*, 47–52.
173. M. G. Spillantini, M. L. Schmidt, V. M. Lee, J. Q. Trojanowski, R. Jakes, M. Goedert, *Nature* **1997**, *388*, 839–840.
174. X. Wang, D. Moualla, J. A. Wright, D. R. Brown, *J. Neurochem.* **2010**, *113*, 704–714.
175. D. J. Stevens, E. D. Walter, A. Rodriguez, D. Draper, P. Davies, D. R. Brown, G. L. Millhauser, *PLoS Pathog.* **2009**, *5*, e1000390.
176. S. B. Prusiner, *Science* **1997**, *278*, 245–251.
177. J. S. Valentine, P. A. Doucette, S. Zittin Potter, *Annu. Rev. Biochem.* **2005**, *74*, 563–593.
178. V. Gunther, U. Lindert, W. Schaffner, *BBA-Mol. Cell Res.* **2012**, in press.
179. O. Palacios, S. Atrian, M. Capdevila, *J. Biol. Inorg. Chem.* **2011**, *16*, 991–1009.
180. R. A. Colvin, W. R. Holmes, C. P. Fontaine, W. Maret, *Metallomics* **2010**, *2*, 306–317.
181. G. Westin, W. Schaffner, *EMBO J.* **1988**, *7*, 3763-3770.
182. R. Heuchel, F. Radtke, O. Georgiev, G. Stark, M. Aguet, W. Schaffner, *EMBO J.* **1994**, *13*, 2870–2875.
183. C. Gunes, R. Heuchel, O. Georgiev, K. H. Muller, P. Lichtlen, H. Bluthmann, S. Marino, A. Aguzzi, W. Schaffner, *EMBO J.* **1998**, *17*, 2846–2854.
184. N. A. Holtzman, B. M. Gaumnitz, *J. Biol. Chem.* **1970**, *245*, 2350–2353.
185. M. Broderius, E. Mostad, K. Wendroth, J. R. Prohaska, *Comp. Biochem. Physiol.* **2010**, *151*, 473–479.
186. J. D. Gitlin, J. J. Schroeder, L. M. Lee-Ambrose, R. J. Cousins, *Biochem. J.* **1992**, *282*, 835–839.
187. T. Nittis, J. D. Gitlin, *J. Biol. Chem.* **2004**, *279*, 25696–25702.
188. P. G. Reeves, L. C. Demars, W. T. Johnson, H. C. Lukaski, *J. Nutr.* **2005**, *135*, 92–98.
189. H. Chen, G. Huang, T. Su, H. Gao, Z. K. Attieh, A. T. McKie, G. J. Anderson, C. D. Vulpe, *J. Nutr.* **2006**, *136*, 1236–1241.
190. J. R. Prohaska, B. Brokate, *Arch. Biochem. Biophy.s* **2001**, *393*, 170–176.

191. J. R. Prohaska, M. Broderius, *Comp. Biochem. Physiol.* **2006**, *143*, 360–366.
192. W. T. Johnson, L. A. Johnson, H. C. Lukaski, *J. Nutr. Biochem.* **2005**, *16*, 682–692.
193. C. A. Kehoe, M. S. Faughnan, W. S. Gilmore, J. S. Coulter, A. N. Howard, J. J. Strain, *J. Nutr.* **2000**, *130*, 30–33.
194. M. A. Mendez, M. Araya, M. Olivares, F. Pizarro, M. Gonzalez, *Environ Health Perspect.* **2004**, *112*, 1654–1657.
195. J. Bertinato, M. R. L'Abbe, *J. Biol. Chem.* **2003**, *278*, 35071–35078.
196. M. A. Broderius, J. R. Prohaska, *Nutr. Res.* **2009**, *29*, 494–502.
197. K. C. Lassi, J. R. Prohaska, *J. Nutr.* **2012**, *142*, 292–297.
198. E. C. West, J. R. Prohaska, *Exp. Biol. Med.* **2004**, *229*, 756–764.
199. M. D. Page, J. Kropat, P. P. Hamel, S. S. Merchant, *Plant Cell* **2009**, *21*, 928–943.
200. K. Kampfenkel, S. Kushnir, E. Babiychuk, D. Inze, M. Van Montagu, *J. Biol. Chem.* **1995**, *270*, 28479–28486.
201. V. Sancenon, S. Puig, H. Mira, D. J. Thiele, L. Penarrubia, *Plant Mol. Biol.* **2003**, *51*, 577–587.
202. S. Klaumann, S. D. Nickolaus, S. H. Furst, S. Starck, S. Schneider, H. Ekkehard Neuhaus, O. Trentmann, *New Phytol.* **2011**, *192*, 393–404.
203. S. La Fontaine, J. M. Quinn, S. S. Nakamoto, M. D. Page, V. Gohre, J. L. Moseley, J. Kropat, S. Merchant, *Eukaryotic Cell* **2002**, *1*, 736–757.
204. T. Hirayama, J. J. Kieber, N. Hirayama, M. Kogan, P. Guzman, S. Nourizadeh, J. M. Alonso, W. P. Dailey, A. Dancis, J. R. Ecker, *Cell* **1999**, *97*, 383–393.
205. E. Himelblau, H. Mira, S. J. Lin, V. C. Culotta, L. Penarrubia, R. M. Amasino, *Plant Physiol.* **1998**, *117*, 1227–1234.
206. A. Herbik, C. Bolling, T. J. Buckhout, *Plant Physiol.* **2002**, *130*, 2039–2048.
207. J. Kropat, S. Tottey, R. P. Birkenbihl, N. Depege, P. Huijser, S. Merchant, *Proc. Natl. Acad. Sci. USA* **2005**, *102*, 18730–18735.
208. M. Eriksson, J. L. Moseley, S. Tottey, J. A. Del Campo, J. Quinn, Y. Kim, S. Merchant, *Genetics* **2004**, *168*, 795–807.
209. H. Yamasaki, M. Hayashi, M. Fukazawa, Y. Kobayashi, T. Shikanai, *Plant Cell* **2009**, *21*, 347–361.
210. K. Asada, S. Kanematsu, K. Uchida, *Arch. Biochem. Biophys.* **1977**, *179*, 243–256.
211. C. Jackson, J. Dench, A. L. Moore, B. Halliwell, C. H. Foyer, D. O. Hall, *Eur. J. Biochem.S* **1978**, *91*, 339–344.
212. S. E. Abdel-Ghany, J. L. Burkhead, K. A. Gogolin, N. Andres-Colas, J. R. Bodecker, S. Puig, L. Penarrubia, M. Pilon, *FEBS Lett.* **2005**, *579*, 2307–2312.
213. C. H. Huang, W. Y. Kuo, C. Weiss, T. L. Jinn, *Plant Physiol.* **2012**, *158*, 737–746.
214. P. Cardol, D. Gonzalez-Halphen, A. Reyes-Prieto, D. Baurain, R. F. Matagne, C. Remacle, *Plant Physiol.* **2005**, *137*, 447–459.
215. C. V. Attallah, E. Welchen, A. P. Martin, S. V. Spinelli, G. Bonnard, J. F. Palatnik, D. H. Gonzalez, *J. Exp. Bot.* **2011**, *62*, 4281–4294.
216. D. S. Gorman, R. P. Levine, *Plant Physiol.* **1966**, *41*, 1637–1642.
217. S. E. Abdel-Ghany, P. Muller-Moule, K. K. Niyogi, M. Pilon, T. Shikanai, *Plant Cell* **2005**, *17*, 1233–1251.
218. H. Guo, J. R. Ecker, *Curr. Opin. Plant Biol.* **2004**, *7*, 40–49.
219. C. Chang, S. F. Kwok, A. B. Bleecker, E. M. Meyerowitz, *Science* **1993**, *262*, 539–544.
220. J. Hua, E. M. Meyerowitz, *Cell* **1998**, *94*, 261–271.
221. F. I. Rodriguez, J. J. Esch, A. E. Hall, B. M. Binder, G. E. Schaller, A. B. Bleecker, *Science* **1999**, *283*, 996–998.
222. B. M. Binder, F. I. Rodriguez, A. B. Bleecker, *J. Biol. Chem.* **2010**, *285*, 37263–37270.
223. S. Merchant, L. Bogorad, *J. Biol. Chem.* **1986**, *261*, 15850–15853.
224. H. H. Li, S. Merchant, *J. Biol. Chem.* **1995**, *270*, 23504–23510.
225. J. M. Quinn, S. Merchant, *Plant Cell* **1995**, *7*, 623–628.
226. F. Sommer, J. Kropat, D. Malasarn, N. E. Grossoehme, X. Chen, D. P. Giedroc, S. S. Merchant, *Plant Cell* **2010**, *22*, 4098–4113.

227. H. Yamasaki, M. Pilon, T. Shikanai, *Plant Signal Behav.* **2008**, *3*, 231–232.
228. R. Sunkar, A. Kapoor, J. K. Zhu, *Plant Cell* **2006**, *18*, 2051–2065.
229. H. Yamasaki, S. E. Abdel-Ghany, C. M. Cohu, Y. Kobayashi, T. Shikanai, M. Pilon, *J. Biol. Chem.* **2007**, *282*, 16369–16378.

Chapter 14
Zinc and the Zinc Proteome

Wolfgang Maret

Contents

W. Maret (✉)
King's College London, School of Medicine,
Diabetes and Nutritional Sciences Division,
Metal Metabolism Group, London, SE1 9NH, UK
e-mail: wolfgang.maret@kcl.ac.uk

L. Banci (ed.), *Metallomics and the Cell*, Metal Ions in Life Sciences 12,
DOI 10.1007/978-94-007-5561-1_14, © Springer Science+Business Media Dordrecht 2013

Abstract Zinc(II) ions are catalytic, structural, and regulatory cofactors in proteins. In contrast to painstakingly collecting the pieces by isolating and characterizing zinc proteins, 'omics' approaches are now allowing us to tease out information about zinc proteins from genomes and to piece together the information to a broader knowledge and appreciation of the role of zinc in biology. Estimates for the number of zinc proteins in the human genome and in genomes of other organisms have been derived from a bioinformatics approach: mining sequence databases for homologies of known zinc-coordination motifs with characteristic ligand signatures for metal binding and combining this information with the knowledge about metal-binding domains of proteins. This approach resulted in an impressive number of almost 3000 human zinc proteins and made major contributions to our understanding of the composition of the zinc proteome and the functions of zinc proteins. However, the impact of zinc on protein science is even greater. Predictions do not include yet undiscovered ligand signatures, coordination environments that employ complex binding patterns with nonsequential binding of ligands and ligand bridges, zinc/protein interactions at protein interfaces, and transient interactions of zinc(II) ions with proteins that are not known to be zinc proteins. All this information and recent discoveries of how cellular zinc is controlled and how zinc(II) ions function as signaling ions add an hitherto unrecognized dimension to the zinc proteome of multicellular eukaryotic organisms. Zinc proteomics employs a combination of approaches from different disciplines, such as bioinformatics, biology, inorganic biochemistry, and significantly, analytical and structural chemistry. It provides crucial large-scale datasets for interpreting the roles of zinc in health and disease at both a molecular and a global, systems biology, level.

Keywords catalytic zinc • metallomics • proteomics • structural zinc • zinc(II) ions • zinc proteins • zinc proteomes

Please cite as: *Met. Ions Life Sci.* 12 (2013) 479–501

1 Introduction

The inquisitive human mind has always sought quantitative as well as qualitative information. Scientists are trained to measure, and in the course of their discoveries they strive for refining their tools for measurements. Quantification does not only pertain to experimental measurements but also to establishing, curating, and mining databases. With the advent of sequencing entire genomes, biosciences moved from genes to genomes and genomics, from proteins to proteomes and proteomics, and for the focus of this chaper, from *metals* to the description of the *metallome* investigated with a multi- and interdisciplinary approach called *metallomics*. This chapter explores our present knowledge of zinc, one of the most important metal ions for life, in relation to the human zinc proteome.

2 Zinc in Biology

2.1 *History*

Discoveries in zinc biology are worth recounting briefly as the full impact and importance of zinc was recognized only gradually and comparatively late when compared to other metal ions. Zinc was found to be essential for the growth of the common bread mold *Aspergillus niger* in 1869. However, considerable time elapsed before it was shown to be essential for rats (1934) and for humans (1961) [1–3]. A molecular function of zinc was not established until 1939 when erythrocyte carbonic anhydrase was demonstrated to contain stoichiometric amounts of zinc, which is essential for its enzymatic activity [4]. A second zinc enzyme, bovine pancreatic carboxypeptidase, was identified 15 years later [5]. Advances in analytical methods for isolating and characterizing proteins and measuring metals in biological material then accelerated discovery and led to the identification of hundreds of zinc enzymes in all classes of enzymes, i.e., oxidoreductases, transferases, hydrolases, lyases, isomerases, and ligases [6]. A new era and area started with a bioinformatics approach that allowed predicting zinc sites in proteins. It increased the counts of zinc proteins in a given organism to a few thousand.

2.2 *Interactions of Zinc with Biomolecules*

Though many biological molecules have zinc-binding sites, almost all our knowledge about zinc in biology is based on interactions of zinc with proteins. One reason is the relatively high affinity of coordination environments in proteins for zinc, which is not readily matched by other biomolecules. Given that virtually all zinc biochemistry has been linked to proteins, the chapter will focus on this aspect only.

2.3 *Impact*

There is hardly any cellular process that is not influenced by zinc. In humans, major signs of severe zinc deficiency are growth retardation, hypogonadism in males, skin and neuro-sensory disorders, and impairment of immunity and cognition. Different stages of zinc deficiency are thought to affect 30% of the world's population, a problem with at least the scope of that of iron deficiency. Among the groups at risk are the elderly. The World Health Organization (WHO) has identified zinc deficiency as the fifth most important risk factor for morbidity and mortality in developing countries (11th worldwide). The figure translates into 3.2% of all lost disability-adjusted life years (DALYs). These estimates for the impact of nutritional zinc deficiency on human health are derived primarily from the incidence of infectious and parasitic diseases. They do not include the functions of zinc in human memory acquisition and storage, behavior, growth retardation, delayed wound healing, the effects of environmental exposures ("zinc overload"), or the role of zinc in aging and in chronic diseases, such as cancer, neurodegeneration, and diabetes. What is also not assessed is the extent to which proteins that control cellular zinc metabolism are affected by genetic mutations or compromised by other mechanisms.

3 Zinc Proteins, Zinc Proteomes, and Zinc Proteomics

3.1 Zinc Metalloproteins

Isolating a protein, purifying it to homogeneity, analyzing its zinc content, and determining the stoichiometry of zinc binding has been the classical way of establishing that a protein is a zinc protein. Coordination environments for zinc are then characterized by high resolution structural analyses or spectroscopy of appropriate derivatives, in which the zinc(II) ion, which owing to its filled d-shell is spectroscopically silent for most techniques, is replaced by the cobalt(II) ion, other transition-metal ions or NMR-active isotopes of the cadmium(II) ion. Isolating a metalloprotein from a natural source is limited by the amount of material that is available or can readily be handled in the laboratory. Preparation of a metalloprotein from a heterologous expression system is fraud with uncertainties about the metal composition of the protein in its native environment. With the availability of large databases for protein sequences and protein structures it became possible to predict that proteins are zinc proteins. The methods for such predictions are becoming increasingly more sophisticated and have produced a wealth of new information about zinc proteins.

3.2 Signatures of Zinc-Binding Sites in Proteins

The finding that zinc is present in the *Xenopus laevis* transcription factor IIIA (TFIIIA) changed the way in which zinc proteins were discovered and led to the recognition of numerous additional functions of zinc in biology [7]. TFIIIA contains nine repetitive sequences of cysteine (C) and histidine (H) residues that bind zinc. These zinc-binding sites have a characteristic signature of metal-coordinating amino acid side chains interspersed with non-coordinating amino acids („spacers") ($CX_aCX_bHX_cH$, i.e., C_2H_2). They form relatively small protein domains that grip DNA like the fingers of a hand grip a rod [8]. Hence, "zinc fingers" became a generic term for these structural zinc sites. At the time the structures of some zinc finger proteins became known, nucleotide sequence databases increased rapidly in number and size. Thus, searches for homologies with established signatures were performed, allowing the prediction of zinc-binding sites by database mining.

Employing only the signature of the classical C_2H_2 zinc finger motif, database mining of the human genome identified 5092 protein domains containing zinc finger motifs [9]. The actual number of zinc proteins with these motifs is smaller, however, because many of them, such as TFIIIA, contain several zinc fingers, with up to a maximum of 36 zinc fingers observed in one specific protein. Subsequently, signatures and 3D structures of other structural zinc sites in protein domains were discovered, in particular sites with four cysteine ligands (C_4), with three cysteine ligands and one histidine ligand (C_3H), and with one cysteine, one aspartate, and two histidine ligands (CDH_2) [10]. It allowed additional database mining for these signatures, increasing the number of zinc proteins even further. Zinc binds in sites that accommodate one, two, or even three zinc(II) ions. Binding employs almost any permutation of the order of cysteine and histidine ligands in the sequence [11,12]. Many of these zinc-binding domains do not interact with DNA/RNA but rather with other proteins, and, in some cases, with lipids [13,14]. Thus, zinc-binding domains became an important aspect of protein-protein interactions. In the description of new proteins, it became customary to assign zinc finger domains based on sequence homology without ever performing a single chemical analysis of zinc. While this approach turned out to have significant heuristic value it is not an empiric method.

With structural templates in databases of protein structures, signatures of metal-binding sites were tabulated, not only for structural zinc sites but also for catalytic zinc sites. Zinc in zinc finger motifs is tetracoordinate. In most cases, the signatures contain a short amino acid spacer between the first two ligands, followed by a longer spacer between the second and third ligand and again a short spacer between the third and fourth ligand. The proximity of the four ligands and the fact that zinc often organizes relatively small protein domains contributed to the success of this database-mining approach for predicting zinc fingers. In contrast to zinc fingers with four ligands and relativley short spacers, enzymatic zinc sites can be more difficult to predict, in particular when searches involve only two ligands because the third ligand or additional ligands often are separated from the two closely spaced ligands by long spacers. The occurrence of multinuclear zinc sites and instances where large parts of the protein organize the active site add to these difficulties.

Signatures of catalytic zinc in enzymes are more variable than those of most structural sites and they often involve larger segments of a protein. The majority of coordination environments in zinc enzymes has only three protein ligands that are separated by a long spacer following or preceding a short or a long spacer. Characteristic ligand signatures from primary structures and zinc coordination motifs from crystal structures were originally established as standards of reference from only about a dozen structures of zinc proteins [15]. Predictions were initially verified experimentally, thus validating this approach. A classic example is the prediction of a zinc-binding site in human LTA_4 (leukotriene A_4) hydrolase after realizing that its putative zinc ligands align with the ligands of the catalytic zinc in the bacterial enzyme thermolysin, i.e., HExxH….E [15–17]. Metal analysis then confirmed that LTA_4 hydrolase is indeed a zinc enzyme [18]. The crystal structure of human LTA_4 hydrolase proved the prediction of its zinc-binding site to be correct [19]. The analogy with thermolysin was taken further to predict peptidase activity of LTA_4 hydrolase. Peptidase activity was also confirmed experimentally [20].

The discovery of the relationship of collagenases, snake venoms, and bacterial proteinases is another triumph for predictions from protein databases before 3D structures of these proteins became available [15]. When the sequences of collagenases and related endopeptidases were aligned with that of thermolysin, the third ligand, glutamate (E), was not present. However, a comparison with the sequence of the *Astacus* proteinase astacin and human bone morphogenetic protein (BMP-1), which at that time was not known to have proteolytic activity, demonstrated conserved glycine and histidine residues. Hence, it was suggested that the third ligand is a histidine in a ligand signature HExxHxxGxxH [21]. The 3D structure of astacin confirmed histidine as the third ligand and established astacin as the founding member of a new family of metalloproteinases [22]. Using this new signature with a verified structural template for database mining identified 33 proteinase sequences with the same putative ligands and defined four subclasses of this major proteinase superfamily [23]. The entire clade of proteinases is known as metalloproteinases. The prefix „metallo" does not refer directly to the fact that they are zinc enzymes.

With an increasing number of 3D structures being solved, the number of structural templates also increased and so did the prediction of structures and functions of zinc proteins. The table lists some common signatures of zinc-binding sites in proteins (Table 1).

Table 1 Examples of zinc-binding motifs in catalytic and structural sites of proteins. The ligands used are histidines (H), a combination of histidines and glutamate (E), cysteine (C), or aspartate (D), or a combination of all three amino acids, e.g., H,C,D. The majority of calatytic sites contain three protein ligands while the majority of structural sites contain four protein ligands. Also characteristic for metal-binding motifs is a short spacer of one to four amino acids (x) between some of the ligands. Only sites binding a single zinc(II) ion are given. Many sites bind one, two, three, or even four zinc(II) ions with more complicated signatures. A comprehensive listing of established and putative zinc-binding motifs/signatures is given in the supplementary material in Andreini et al. [29].

Signature	Example
Catalytic zinc-binding sites	
$Hx_2Ex_{103\text{-}124}H$	carboxypeptidases
$Hx_3Hx_{18\text{-}58}E$	gluzincins, e.g., angiotensin-converting enzyme
Hx_3Hx_5H	metzincins, e.g., matrix metalloproteinases
$HxHx_{22}H$	carbonic anhydrases
$Cx_{20}Hx_{106}C$	alcohol dehydrogenases
$DxHx_{82\text{-}90}D$	histone deacetylases
Structural zinc-binding sites	
$Cx_{2\text{-}4}Cx_{12}Hx_{3\text{-}5}H$	classical zinc finger motif, e.g., transcription factor Sp1
$Cx_4Cx_{17}Cx_2C$	GATA zinc finger
$Cx_2Hx_{58}Cx_3C$	tumor suppressor p53
$HxDx_{12}Hx_{12}H$	matrix metalloproteinases

3.3 Zinc Proteomes

Mining sequence databases with entire genomes allowed predictions that at least 3% of the human genome encodes for zinc proteins [24,25]. However, it was pointed out that the actual number could be considerably larger because this approach does not account for all zinc/protein interactions [11,26]. A major contribution to the field of bioinorganic chemistry came through a bioinformatics approach developed by Ivano Bertini, Lucia Banci, Antonio Rosato, and colleagues at the Magnetic Resonance Center (CERM) at the University of Florence, Italy [27]. They combined the information from metal-binding signatures in databases of protein structures (PDB, http://www.rcsb.org) with the information from metal-binding domains (Pfam, http://pfam.sanger.ac.uk) to scan gene databases. They developed automated procedures for analyses and guidelines for manual curation of the information and sytematically applied them to estimate the metalloproteomes for copper, non-heme iron, and zinc. Application resulted in three datasets: metalloproteins with known signatures and unknown domains, metalloproteins with known signatures and known domains, and metalloproteins with unknown signatures and known domains. In this way, they predicted that at least 10% (about 2800 gene products) of the human genome encodes zinc proteins [28]. 2406 assignments are based on signatures, 2506 on domains, and 2407 on annotations. Using the Boolean operator "and", 1684 zinc proteins are assigned by all three methods. The total number of zinc proteins is 3207, with 2430 identified by at least two methods. As possible zinc proteins were counted: 397 hydrolases; 302 ligases; 167 transferases; 43 oxidoreductases and 24 lyases/isomerases; 957 transcriptions factors; 221 signaling proteins; 141 transport/storage proteins; 53 proteins with structural metal sites; 19 proteins involved in DNA repair, replication, and translation; 427 zinc finger proteins of unknown function; and 456 proteins of unknown function.

The CERM group applied this analysis to 57 genomes of archaea, bacteria, and eukarya [29]. They found that the number of zinc proteins correlates linearly with the number of genes in a particular genome and is higher in eukarya (8.8%) than in bacteria and archaea (5–6%). The increase in eukarya is due to a large number of additional zinc proteins involved in regulation. Computational methods were then developed to build libraries of 3D metal-binding motifs that also included information about bonding interactions beyond the primary coordination sphere of the metal [30]. These templates can be used to scan protein structures to predict functions of metals in metalloproteins. These bioinformatics tools and databases developed by the CERM group have already produced a wealth of information on metalloproteins. Predictions have extended the scope of zinc proteomes and contributed significantly to demonstrating the widespread and fundamental roles of zinc in biology. They also generate numerous new opportunities for experimental exploration of hypotheses, as in general, any exception tests the rule. Collecting large-scale datasets and archiving them in comprehensive databases on metalloproteins will continue to generate invaluable resources for helping the scientific community to interpret proteomics data and to understand the functions of metals in proteins at the molecular and global level.

In summary, starting with the discovery of a single zinc protein in 1939, estimates of the number of human zinc proteins were constantly revised upwards and have now reached the staggering number of a few thousand. One wonders whether we really have accounted for most of the zinc proteins in the human zinc proteome. The answer is: most likely not. The reasons for this answer rest upon new developments in zinc biology and they are now discussed.

Predictions are based on known signatures. New templates of coordination environments in proteins are derived from new structures. Therefore, whether or not the coverage of a zinc proteome in an organism is nearly complete, among other factors, depends on whether or not known structures of zinc proteins represent all possible signatures. The possibility that additional signatures exist is relatively high since structures underrepresent certain classes of proteins such as membrane proteins. In a number of domains organized by two or three zinc ions, the order of zinc ligands in the peptide chain is nonsequential and ligands can be used twice for bridging zinc ions. Such complex binding patterns cannot be predicted, and the number of topologies in protein architecture established through the use of zinc ions is remarkable [13]. Churchill, an embryonic neural inducing factor, serves as an example for the occasional failure of predictions [31]. From its sequence it was deduced that Churchill contains two canonical C_4 (tetrathiolate) zinc fingers, but the 3D structure showed actually three zinc ions in a binuclear zinc cluster with a novel topology with a nonsequential use of ligands and additional histidine ligands. Thus, predictions based on homology with known signatures were wrong with regard to the number of bound zinc ions and their binding modes. Though metal binding can be predicted, there are significant limitations in predicting structures when zinc binds in bi-, tri-, and tetranuclear sites or when ligands have a criss-cross pattern in the sequence to form zinc/thiolate clusters. Without curtailing the contributions that predictions have made to biology, 3D structures of proteins will remain the ultimate arbiters for establishing the modes of zinc binding.

False positives: Zinc is not present because the site does not bind zinc or the site binds a different metal. Generic names such as „zinc finger" or „zinc-binding domain" are often used with the implication that zinc is indeed present in the protein. However, the metal ion may not be zinc in some metalloproteins. Metal homeostatic systems strictly control metal availability and distribution, and these systems differ in organisms. Prokaryotic expression systems are prone to give false positives because metal availability is controlled differently in prokarya and eukarya.

When using heterologous expression systems, the wrong metal ion can be incorporated into a metalloprotein as a result of the metal composition of the growth media. Prokarya have a certain promiscuity for metal ions in their metalloproteins. But even for eukarya, where such promiscuity seems to be much less pronounced or non-existant, metal specificity of a metalloprotein requires scrutiny. For eukarya, the issue is usually one of mistaken identity rather than promiscuity. A crystal structure of human histone deacetylase 8 demonstrated a zinc ion in the active site [32]. However, the recombinant protein purified from *E. coli* contains eight times more iron than zinc and is more active with iron than with zinc. Accordingly, it was suggested that Fe(II) is the catalytic cofactor *in vivo* [33]. It remains unknown to which extent ligand signatures are specific for zinc and overlap with those of other metal

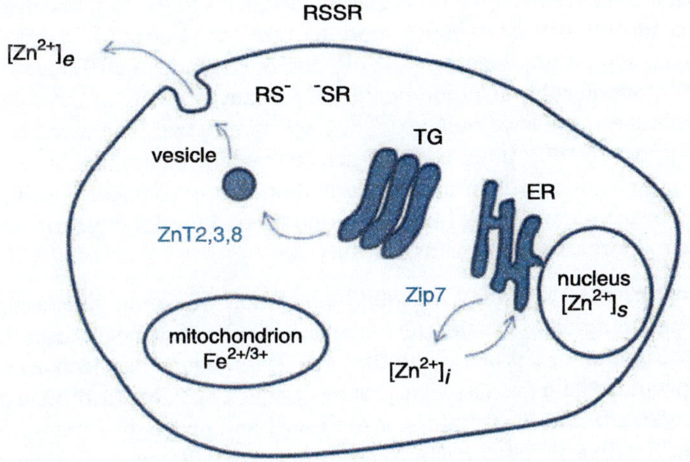

Figure 1 The influence of cellular location on the coordination environments of metalloproteins. The biological context can determine the chemical structure of the ligands and the selection of the metal ion. The redox environment determines the oxidation state of the sulfur donor of cysteine. In the reducing environment of the cytosol, the prevailing chemical form of the sulfur is that of a thiolate (RS^-). However, in the more oxidizing environment of the extracellular space and in the endoplasmic reticulum (ER), disulfides (RSSR) are more common. Cellular zinc metabolism in eukarya is compartmentalized. In some cells, zinc transporters (ZnT2,3,8) load vesicles stemming from the trans-Golgi (TG) with zinc(II) ions that are then exocytosed. Zinc(II) ions are also released from the ER through regulation of the zinc transporter Zip7. These transporters are just a few of the two dozen human transporters that control cellular zinc metabolism. Zinc also needs to be delivered to mitochondria and the nucleus. Determination of the free zinc(II) ion concentrations in different cellular compartments is presently pursued and indicates different concentrations. Different extracellular, cellular, and subcellular concentrations of zinc ($[Zn^{2+}]_e$, $[Zn^{2+}]_i$, $[Zn^{2+}]_s$) and other transition metal ions indicate that otherwise identical metal-binding sites in proteins are loaded with different metal ions in different compartments. For instance, in mitochondria where most of iron metabolism takes place, metal-binding sites that have been annotated as zinc-binding sites bind iron.

ions, in particular iron. Zinc is often bound in tetrathiolate sites. However, a tetrathiolate site in bacterial rubredoxin contains iron.

The role of cellular localization in metal coordination. The cellular environment affects the ligand chemistry as well as the selection of the metal ion in proteins (Figure 1). The CDGSH domain was annotated as a zinc-binding domain. However, this domain in MitoNEET, which resides in the outer mitochondrial membrane, contains iron [34]. CucA and MncA are the most abundant copper and manganese proteins in the periplasm of the cyanobacterium *Synechocystis* PCC 6803 [35]. Both proteins belong to the cupin family and bind the two metal ions with the same ligand signature: $HxHx_4E$.....H. According to the Irving-Williams series, copper is bound much more tightly than manganese. The investigators discovered that the binding preference is overriden by the location in which a protein folds. It also matters where in the cell a protein with potential zinc-binding cysteine ligands is localized because the environment may favor disulfide bond formation rather than zinc/thiolate coordination [36,37].

HEDJ, a DNaJ/Hsp40 type of chaperone, localizes to the endoplasmic reticulum (ER). Because the ER has a much more oxidized thiol/disulfide redox state compared to the cytoplasm, the potential zinc-binding ligands of HEDJ are mostly oxidized to intramolecular disulfide bonds [38]. Many proteins of the mitochondrial intermembrane space have twin Cx_3C or Cx_9C motifs that bind zinc and copper or form disulfides [39,40]. It has been suggested that zinc binding to the reduced form of these proteins keeps them in a conformation that is competent for their import into the intermembrane space [41]. The issue whether or not these proteins interact with zinc only transiently remains unresolved.

Some proteins bind zinc only transiently. Isolation of a protein as a zinc protein requires tight binding of the zinc(II) ion in order for the zinc not to dissociate during isolation. Isolation of a protein with the zinc(II) ion bound has been used to distinguish a metalloprotein from a metal-protein complex [42]. Metal-protein complexes have significantly lower affinities for the metal and hence, the metal may be lost during purification. Whether zinc-protein complexes with lower affinities than those of zinc metalloproteins are physiologically significant is among other factors a matter of their binding affinities for zinc and the concentrations of zinc(II) ions available. Recent investigations demonstrated that the concentrations of cellular free zinc(II) ions fluctuate in the picomolar to low nanomolar range [43,44]. They support the idea that zinc(II) ions have regulatory roles. Searches for targets of such regulatory zinc(II) ions identified sites with zinc affinities in the range of these fluctuations, suggesting that these sites bind zinc transiently and that their occupancy depends on the availability of zinc(II) ions [43,45].

Remarkably, zinc inhibits some enzymes that are not recognized as zinc metalloenzymes [46]. However, when estimates of free cellular zinc(II) ion concentrations became available, it turned out that zinc inhibition constants of these enzymes are too high to be physiologically significant. An issue in such experiments is that *in vivo* zinc is tightly buffered whereas in experiments *in vitro* it is generally not buffered. Thus, the buffering of cellular zinc(II) ions and a distinction between the free zinc(II) ion concentration and the concentration of total zinc should be considered in experiments *in vitro.* Using zinc buffering techniques, zinc inhibits human protein tyrosine phosphatase 1B, which regulates the phosphorylation state of the insulin receptor, with a $K_{i(app)}$ of 15 nM [45]. Another caveat is that such affinities are derived from assaying enzymatic activity and hence competition between substrate and zinc(II) ions can affect the measured affinities. When a full kinetic analysis of inhibition was performed for another protein tyrosine phosphatase (vascular endothelial PTP), a $K_{i(Zn)}$ of 21 pM was determined [47]. This affinity is in the range where free zinc(II) ions have been observed. Therefore, zinc inhibition of this enzyme, which was not known to be a zinc metalloenzyme, is deemed physiologically significant. Also, when correcting for free zinc(II) ion concentrations in the assay, zinc inhibits human erythrocyte Ca^{2+}-ATPase with a K_i value of 80 pM [48]. Estimates of 24 pM free zinc(II) ion concentrations in erythrocytes support physiological significance of this zinc inhibition [49].

Coordination environments and 3D structures are known for some inhibitory zinc sites. Physiological significance hinges on the availability of zinc(II) ions in the biological environment. Knowing the location of the protein is important for suggesting physiological significance because free zinc(II) ion concentrations in some cellular and extracellular compartments exceed those of the cytosol. Accordingly, some inhibitory zinc sites with nanomolar to micromolar affinities may be physiologically significant. Zinc(II) ions released from vesicles can provide higher local concentrations intra- and extracellularly, for instance at the ER membrane and in neuronal synapses [50,51]. One target of zinc(II) ions released from synaptic vesicles is the NMDA receptor, where zinc binds with nanomolar affinities to the NR1a-NR2A receptor and with micromolar affinities to the NR1a-NR2B receptor [52].

Aside from specific neurons, zinc(II) ions are also secreted from many other cells. One function of secreted zinc is to keep proteinases inhibited before they reach the environment in which they function. Carboxypeptidase A, a zinc proteinase, binds an inhibitory, additional zinc(II) ion with a K_i of 0.5 μM [53]. The site that binds the inhibitory zinc(II) ion involves glutamate-270, a hydroxide ion that bridges the catalytic and the inhibitory zinc, and a chloride ion [54,55]. Zinc also inhibits proteinases that are not zinc metalloenzymes. Affinities of inhibitory zinc(II) ions and 3D structures have been reported for several members of the human kallikrein (hK) family of serine proteinases. Low micromolar concentrations of zinc inhibit these enzymes. The kallikrein hK4 binds the inhibitory zinc(II) ion with a histidine and a glutamate ligand [56]. In tonin, a rat kallikrein, the zinc ligands are three histidines and include a glutamate from a neighbouring protein molecule in the crystal structure. In hK5, the inhibitory zinc(II) ion has two histidine ligands. Homology with other kallikreins suggests a third histidine ligand in hK5 [57] and in hK7 [58]. Zinc inhibition of kallikreins that function in prostate physiology, such as hK3 (prostate-specific antigen (PSA)) and hK4 (prostase), is thought to be physiologically significant because zinc concentrations in seminal plasma and prostate fluid are around 10 mM and about 1000 times higher than those in blood. Remarkably, the dissociation constant of the catalytic zinc in human carbonic anhydrase with three histidine ligands is 11.4 (pK_d) [59], five orders of magnitude higher than those of the inhibitory zinc sites in kallikreins, which may also have three histidine ligands. One reason for such a difference is that affinity is not only determined by the direct zinc ligands but also by hydrogen bonding of other amino acids with metal-ligands and other interactions in the outer coordination sphere [60,61].

Secreted zinc(II) ions have yet other functions. Human insulin is stored as a zinc complex in the secretory granules of pancreatic β-cells. Zinc(II) ions and insulin are co-secreted. One function, though controversial, is a paracrine action of the secreted zinc on glucagon secretion from α-cells. Another is the prevention of amyloidogenesis of proteins co-secreted with zinc. Zinc inhibits the fibrillation of monomeric insulin with an IC_{50} value of 3.5 μM at pH 7.3 [62]. It also inhibits formation of the dimer of the human islet amyloid polypeptide (IAPP, amylin) but induces the formation of its oligomers [63]. IAPP is detected as extracellular amyloid fibers in the pancreas of diabetics, which are prone to become zinc-deficient.

Insufficient data are available for interpreting the significanc of zinc inhibition of some proteins in cellular and subcellular locations. Zinc inhibits mitochondrial aconitase with a K_i of 2 μM but this inhibition has not been linked directly to a zinc binding sites with two histidine ligands and one aspartate ligand in the crystal structure of the protein [64]. In bovine dimethylarginine dimethylaminohydrolase, zinc binds to only one cysteine ligand in the active site at pH 6.3 but recruits a histidine ligand at pH 9.0 [65]. The coordination at pH 7.4, where zinc binds with a K_i of 4 nM, remains unknown [66].

So far consensus signatures for inhibitory zinc sites are not available. Without this information, one cannot mine databases and predict the number of physiologically important inhibitory zinc sites.

Predicting the number of protein-interface zinc sites, another class of zinc/protein interactions, has not been feasible [67,68]. Ligands for donor atoms can stem from two, three, or even four protomers, not only in homologous but also in heterologous protein-protein interactions. Predictions of zinc binding can only be made in combination with investigations of protein-protein interactions that identify the heterologous binding partner and demonstrate that the interaction depends on zinc. The *Src*-type kinase Lck interacts with the T-cell co-receptor CD4/CD8 through a tetrathiolate zinc-binding site [69]. Zinc binding to oxygen and nitrogen donors induces homodimerization of this heterodimer through the SH3 domains of the kinase [70]. How changes in the availability of cellular zinc(II) ions affect the functions of such large signaling complexes remains unknown.

The discussion of these examples demonstrates that the affinities of proteins for zinc, the cellular availability of zinc(II) ions, the zinc buffering capacity, homeostatic mechanisms, and the biological context in which a protein functions all need to be known in order to determine whether the affinity of a protein for zinc is biologically significant or zinc binding is simply a chemical phenomenon. Only investigations that combine inorganic and biological chemistry will answer the issues at hand. Metalloproteomics in a multidisciplinary setting is promising to provide answers [71].

3.4 Zinc Proteomics

Zinc proteomics is an approach integrating bioinformatics, structural biology, computational and analytical chemistry, and other disciplines. It is beginning to make significant contributions in quantitatively assessing the role of zinc in biology with a view on systems biology.

4 Zinc Sites in Proteins

4.1 Catalytic Zinc

The role of zinc in enzymatic catalysis has been the subject of many excellent treatises. Hence the reader is referred to reviews of this extensive and rather comprehensive literature [72–74]. A recent structural classification distinguishes seven types of catalytic zinc sites [75].

4.2 Structural Zinc

Structural zinc sites also have been reviewed repeatedly from several viewpoints [13,14,24,76]. In some instances, structural sites in proteins do not contain sulfur donors from cysteines. A recent structural classification distinguishes 10 types of structural zinc sites [75].

4.3 Regulatory Zinc

In contrast to the wealth of information about catalytic and structural zinc sites in proteins, information about regulatory zinc sites is scarce. Though many articles refer to regulatory functions of zinc, the evidence for regulation *in vivo* is usually weak or absent. In enzymatic sites with two zinc(II) ions, the second zinc(II) ion, termed co-catalytic, may regulate activity of the enzyme *in vitro* [77]. Zinc regulation of proteins is gaining significant attention with recent discoveries of how cellular zinc is controlled.

5 Control of Cellular Zinc

5.1 The Proteins Controlling Cellular Zinc Homeostasis

At least three dozen proteins are directly involved in the control of human cellular zinc homeostasis and in the regulation of the cellular availability and re-distribution of zinc(II) ions. These proteins have many different functions. They supply zinc metalloproteins with zinc, remove an excess of zinc(II) ions to prevent un-wanted side reactions, safeguard zinc(II) ions during transit through membranes and cellular compartments, store and release zinc(II) ions, or sense whether or not zinc concentrations are adequate. The proteins are regulated and interact with other metabolic and signaling pathways. Ten transporters of the ZnT family (SLC30A, exporting zinc from the cytosol) and fourteen transporters of the Zip family (SLC39A, importing zinc into the cytosol) control zinc traffic across cellular membranes [78]. In addition to transporting zinc across the plasma membrane, some members of the two protein families transport zinc intracellularly between cytosol and various compartments. Zinc transporters are extensively regulated transcriptionally, translationally, and at the protein level through heterodimer formation, ubiquitination, phosphorylation, proteolysis, and trafficking in the cell.

Three-dimensional structures of Zip proteins have not been reported. However, the 3D structure of the *E. coli* Yiip protein, a homologue of the human ZnT proteins, is known [79]. Yiip resides in the inner membrane of bacteria and has a transmembrane domain and a cytoplasmic domain. It is a dimer with several zinc-binding sites in each monomer. In fact, it is not only a zinc transporter but a zinc protein itself with novel types of zinc-binding sites. A zinc binding site between the transmembrane helices is involved in zinc transport. The function of a zinc-binding site between the cytoplasmic domain and the transmembrane domain is unknown. A binuclear zinc site is located in the cytoplasmic domain and thought to sense cytoplasmic zinc(II) ion concentrations and, once the zinc is bound, to trigger a conformational change that allows transport of zinc through the transmembrane domain. Another zinc sensor is metal-response element (MRE)-binding transcription factor-1 (MTF-1). MTF-1 is an essential protein that controls zinc-dependent gene expression of proteins at increased zinc(II) ion concentrations. It is thought to sense zinc(II) ions through a pair of its six zinc fingers. Metallothionein (MT) is yet another type of protein involved in the control of cellular zinc homeostasis. The human MT family contains at least a dozen proteins [80], the gene expression of most of which is under the control of MTF-1. Thus, increased cellular zinc induces the production of the apoprotein thionein that binds the additional zinc to form MT. Mammalian MTs have 60+ amino acids and bind zinc to the sulfur donors of their twenty cysteines in two zinc/thiolate clusters, one with three zinc(II) ions and nine cysteine ligands and the other with four zinc(II) ions and eleven cysteine ligands.

A remarkable molecular property of MTs is that they are redox proteins due to their zinc/thiolate coordination environments [81]. This function is in contrast to the functions of zinc proteins with zinc bound to nitrogen and oxygen donors in redox-inert coordination environments. The redox activity makes MTs more efficient zinc donors under more oxidizing conditions while their zinc-binding capacity increases under more reducing conditions. Unlike in the case of copper or iron proteins where the changes of the valence state of the central metal ion modulate stabilities of the metal-protein interaction, the reactivity of the ligand donor in zinc/thiolate coordination environments modulates complex stability. In this way, redox reactions can mobilize zinc(II) ions from their high affinity sites in some proteins. This molecular property links cellular zinc and redox homeostasis by using changes in the cellular redox poise to modulate zinc(II) ion concentrations (zinc potentials, $pZn = -\log[Zn^{2+}]_i$) [82].

Despite formally identical tetrathiolate inner coordination spheres, the seven binding sites in human MT-2 have different affinities, ranging from nanomolar to picomolar [83]. These characteristics allow MTs to bind as well as release zinc ions and control cellular zinc availability over a range of zinc(II) ion concentrations that are commensurate with the affinities of its binding sites and those of other proteins for zinc(II) ions. Binding and delivering zinc is a molecular property of the zinc/thiolate clusters and quite different from that of other proteins that bind zinc permanently as a catalytic or structural cofactor. Functions of MTs have been the subject of speculations for over 50 years. Only knowledge that emerged in the last couple of years about the concentrations of cellular zinc(II) ions and their buffering and about differential affinities of MTs for zinc made it possible to make a strong case for a function of mammalian MTs in the control of cellular zinc [84].

5.2 Zinc Buffering and Muffling

Quantitative considerations are critical for understanding the control of cellular zinc homeostasis. Zinc and other essential metal ions are tightly controlled and maintained at specific concentrations to avoid interference among metal ions and to prevent coordination of the wrong metal to a zinc metalloprotein. The cell regulates the concentration of total zinc and it buffers zinc, thereby determining the concentration of free zinc(II) ions. The total cellular zinc concentrations are a few hundred micromolar. They can at least double when cells are cultured at higher extracellular zinc loads [43]. However, the dynamic range under physiological conditions in tissues is not known. The affinities of proteins for zinc determine the buffering of zinc. In a given coordination environment, affinities follow the Irving-Williams series, according to which zinc(II) ions bind stronger than the other divalent metal ions of the first transition series, with the exception of copper(II) ions. Zinc affinities of zinc proteins are rather high and in the picomolar range of dissociation constants [37]. Consequently, free zinc(II) ion concentrations are commensurately low as confirmed by direct measurements that provide estimates of free zinc(II) ion concentrations in the range of tens to hundreds of picomolar ($pZn = -\log[Zn^{2+}] = 10–11$) [85]. Thus, the concentrations of free zinc(II) ions are at least six to seven orders of magnitude lower than the total zinc concentrations. Free zinc(II) ions – designated as such to distinguish them from protein-bound zinc – are not expected to be devoid of any ligands. The nature of the ligands is not known. Other terms used for this zinc pool are readily or easily exchangeable zinc, mobile zinc, or kinetically labile zinc. These terms are used interchangeably though each term has a different meaning and, in some cases, no chemical meaning at all. The important conclusion is that the cellular zinc(II) ion pool is not negligible, fluctuates under various conditions, and is functionally significant.

Zinc buffering is dynamic as both the pZn value and the buffering capacity of cells can change. Increasing the buffering capacity allows the cell to handle more zinc while keeping the pZn constant. Changing the buffering capacity and zinc influx/efflux also allows different pZn values. The same cell can have different pZn values when proliferating, differentiating, or undergoing apoptosis [43,86]. More data are needed to define the extent to which individual cells differ in their total zinc concentration, buffering capacity, and pZn. Global or local changes of cellular zinc(II) ion concentrations provide a way of controlling cellular processes, indicating that different pZn values are a cause rather than a consequence of an altered state of a cell. Fluctuating free zinc(II) ion concentrations target highly specific coordination environments and affect enzyme activities, gene expression, and possibly protein-protein interactions.

Affinities of proteins for zinc cannot be the only factor in the control of cellular zinc re-distribution, because zinc(II) ions would then move only in one direction determined by the thermodynamic gradient. Eukaryotic cells restrict the availability of zinc at the subcellular level through compartmentalization of zinc and at the protein level through kinetic control of association and dissociation rates of zinc [60]. The transfer of zinc between different subcellular compartments and zinc influx/efflux contribute to zinc buffering. Such a kinetic component describing the time-dependent transfer of cellular metal ions between compartments has been referred to as muffling. Muffling allows cells to cope with much higher zinc loads than determined by the zinc-buffering capacity at steady-state. Thus, cellular zinc buffering is a combination of the cytosolic zinc buffering capacity provided by metal-binding ligands and the zinc muffling capacity provided by transport processes. Muffling includes the activities of transporters that remove the surplus of zinc(II) ions from the cytosol to an organelle or to the outside of the cell at pre-steady state and thereby restore the steady-state [87].

5.3 Zinc(II) Ions in Cellular Regulation

Control of cellular zinc homeostasis is not only important for housekeeping purposes. Muffling solves the issue of how transient increases of free zinc(II) ions are possible in a buffered environment. It allows zinc(II) ions to participate in information transfer as signaling ions within cells and between cells [88]. Zinc(II) ions are stored in cellular compartments, from which they are released upon demand. For example, growth factors affect the zinc transporter/channel Zip7 that controls intracellular release of zinc from a store in the ER to the cytoplasm. The control occurs through casein kinase-2 induced phosphorylation of Zip7 [51]. Exocytosis of vesicles that have been loaded with zinc(II) ions by ZnT3 (brain), ZnT8 (pancreatic β-cells), or ZnT2 (mammary gland) releases zinc into the extracellular space. Additional cells secrete zinc(II) ions, indicating that cellular zinc homeostasis has a role in cell-cell communication and in releasing zinc(II) ions for other purposes. Compartmentalization of cellular zinc resolves a long-standing issue of whether or not cells store zinc, namely that zinc is stored in cellular vesicles rather than in a protein with a high zinc-binding capacity. The chemical state of zinc in vesicles is not known. Some vesicles appear to have a much higher pZn value than the cytosol because their zinc is readily available to chelating agents. The determination of subcellular pZn values is an area of significant present interest.

For signaling functions, zinc(II) ions must be made available at specific amplitudes and frequencies. Certain hormones and chemicals increase intracellular zinc(II) ion concentrations above one nanomolar. Unless the buffering capacity changes, free zinc(II) ion concentrations return to their resting concentrations within minutes. How zinc(II) ions can serve signaling functions hinges on how fluctuations of zinc(II) ions are generated, controlled, transmitted to targets and terminated at their targets. The amplitudes of these fluctuations must match the affinities of targets for zinc(II) ions. Indeed, picomolar/nanomolar concentrations of zinc(II) ions inhibit the activities of certain enzymes such as protein tyrosine phosphatases [46]. The affinities of zinc proteins for zinc and the affinities of proteins that are targeted by signaling zinc(II) ions define the lower and upper concentrations at which free zinc(II) ions control and regulate biological processes [84].

No proteins other than MTs have been identified in cellular zinc re-distribution. MT keeps zinc sufficiently tightly sequestered as a temporary storage for cellular zinc. Micromolar concentrations of MTs in most cells make sufficient zinc available in a deliverable form to meet the demand of many cellular functions for zinc. Picomolar concentrations of zinc(II) ions are not meeting such demands [89]. Though bona fide zinc metallochaperones have not been identified, the functions of MT resemble those of metallochaperones, in particular since MTs have the capacity to transfer zinc via protein-protein interactions by an associative mechanism [90]. The zinc-binding characteristics of human MT-2 and its translocation in the cell are consistent with a role in buffering and muffling zinc, namely binding zinc(II) ions and delivering them to a cellular compartment. The high inducibility of MTs by many different signal transduction pathways permits changing the buffering and muffling capacity. Modulating the chelating capacities of MTs through changes of either their total amount or their redox state affords a way of activating zinc-inhibited enzymes and controlling the regulatory functions of zinc(II) ions [84]. A host of conditions induces thioneins, indicating that they are produced for lowering the availability of cellular zinc(II) ions and modulating zinc-dependent processes.

6 Concluding Remarks, Present and Future Directions

6.1 Quantitative Zinc Proteomics

Definition of the zinc proteome has a quantitative dimension. The cellular zinc proteome is complete only if it accounts for zinc present in a cell. Assigning all the zinc to zinc proteins assumes that proteins are the only biological molecules that interact with zinc. To account for all the zinc, it may be necessary to define the zinc metallome, which is the sum of the zinc proteome, the zinc ionome, and zinc that is possibly bound to any other biomolecules.

No attempt has been made to quantify zinc proteins and calculate the resulting zinc concentrations, which should add up to about 200–300 μM zinc for human cells. Yet such an approach may not be satisfactory because zinc is distributed among a large number of low-abundance proteins. As the example of zinc in transcription factors demonstrates, low-abundance zinc proteins are functionally highly significant. The present methodology is far from obtaining absolute measurements though relative measurements have been performed.

The distribution of iron and zinc proteins in the soluble metalloproteome of *E. coli* was investigated with a technique bearing the acronym MIRAGE (Metal Isotope native RadioAutography in Gel Electrophoresis) [91]. Iron is the most abundant and zinc is the second most abundant transition metal ion in *E. coli*. For *E. coli* grown with 0.24 mM extracellular zinc, the analysis demonstrated that 71.4% of all zinc was in a zinc-resistance associated protein thought to be important for zinc storage, 14.8% in the molecular chaperone DnaK, 4.5% in the translation elongation factor EF-Ts, and 1.8% in a putative deacetylase. In 17 proteins the zinc distribution was 0.1 to 1%. In all the remaining proteins, there was less or much less than 0.1% of the total zinc.

6.2 Structural Zinc Proteomics

In February 2012, the protein database contained 8353 structures of zinc proteins (http://www.rcsb.org). However, at least 15–20% of zinc-binding sites are artifacts of the preparation or the conditions of crystallization because zinc(II) ions may have been added to aid the crystallization of the protein [74]. Furthermore, organisms control metal selectivity beyond the molecular selectivity of proteins for a particular metal ion. Hence, some zinc proteins may not be zinc proteins in the corresponding organism, some metalloproteins crystallized with a metal other than zinc may actually be zinc proteins, or some proteins may be zinc metalloproteins but the structure has been determined in the presence of chelating agents or after the removal of the metal ion, thus preventing the recognition of zinc binding [92]. Multiple issues need to be addressed when interpreting the structural zinc proteome. An interpretation requires knowledge of the metal concentrations, cellular localization of the metalloprotein, homeostatic control of the metal in the biological system, and the affinities of proteins for metal ions.

6.3 Functional Zinc Proteomics

Annotation of the zinc proteins in the zinc proteome is on-going as new functions of zinc in proteins emerge [61,71,92]. A considerable contribution to understanding the functions of zinc in biology is made by approaches that analyze large experimental datasets with knowledge databases on metabolic and signaling pathways and their networks. Changes of gene expression were analyzed on cDNA microarrays and with quantitive PCR techniques when zinc was restricted or additional zinc was added in human cultured cells or animals. In the THP-1 cell line, out of the about 22,000 gene transcripts present in the cellular RNA, 5% (1045 genes) were responsive to zinc in the medium. Among those, 104 transcripts increased linearly with increasing zinc while 86 decreased. Some genes responded only to zinc supplementation or zinc deficiency. Zinc deficiency and supplementation affected genes encoding proteins for the cytoskeleton, nucleic-acid binding, apoptosis, metabolism, cell growth/development, signal transduction, and the immune system to a similar extent. 30–40% of the genes could not be annotated. Zinc transporters from both ZnT and Zip families and MT were particularly responsive. Transcripts involved in the immune response were amplified [93]. Profiling the murine thymic transcriptome under zinc deficiency and zinc supplementation in a differential mRNA display revealed modulation of only 0.55% (265 transcripts) out of 48,000 cDNAs [94]. The genes affected did not reveal regulation of any zinc metalloenzyme or zinc finger protein.

Most recent studies employed pathway analysis. In an attempt to assess the role of zinc in the pathways leading to atherosclerosis, where zinc has a protective role as a pro-antioxidant and anti-inflammatory agent, rats were subjected to acute and marginal zinc deficiency, the proteins of their thoracic aortas analysed by 2D electrophoresis, and a principal component analysis performed [95]. In this proteomics study, protein clusters that are affected in zinc deficiency were detected in fatty acid and carbohydrate metabolism. Zinc deficiency affected zyxin and transgelin 1 proteins. Protein kinase Cα decreased significantly, suggesting an effect of zinc on phosphorylation signaling. Similarly, a transcriptome and proteome analysis identified pathways that lead to lipid accumulation in the liver of zinc-deficient rats [96]. The response of fish gill tissue to zinc in the water and diet was analysed by transcriptomics and network analyses [97]. Within seven days, before some adaptation took place, 525 transcripts changed at least 1.8-fold with little overlap of genes at successive time points. Many genes demonstrated a cyclic expression pattern as is typical for proteins involved in homeostatic control. Annotation showed a preponderance of effects on transcription factors, such as steroid receptors, genes involved in development with early regulation of genes controlled by transcription factors Mtf1, Jun, Stat1, Pparα, and Gata3, followed by hedgehog and bone morphogenetic protein signaling. When the effects of zinc deficiency on fish gills were investigated [98], 26% of all changed transcripts also were associated with developmental processes. Genes involved in the cell cycle, cell differentiation, gene regulation, butanoate metabolism, lysine degradation, protein tyrosine phosphatases, nucleobase, nucleoside and nucleotide metabolism, and in other cellular metabolic processes were changed. Some of the genes were associated with diabetes, bone/cartilage development, and ionocyte proliferation. Transcription factors affected, such as foxl1, wt1, nr5a1, nr6a1, and especially, hnf4a, have roles in the homeostatic response to zinc depletion.

The multitude of changes observed in these studies suggests that zinc is not only affecting transcription directly but also has indirect effects that result in global changes of expressed genes. In pathway analyses, zinc-dependent events that are apparently not directly connected with zinc proteins are identified. The discoveries confirm a role of zinc beyond the presently known zinc proteome, namely regulatory roles in critical nodes of signaling pathways, especially phosphorylation signaling, resulting in indirect effects on gene expression. Taken together, these investigations are encouraging in terms of their information content, but also challenging in terms of their interpretation and in terms of acquiring, collecting, and analyzing very large datasets. Analyzing gene expression in normal tissue is at least a 3D problem because changes of gene expression as a function of time and the concentrations of perturbing agents needs to be addressed. Analysis readily becomes a 4D problem or even one of higher dimensions when diseases at different stages are being investigated.

6.4 Evolution of Zinc Proteomes

The role of zinc in the origin of life and in the evolution of life is receiving considerable attention [99]. Inorganic zinc(II) ions and zinc sulfide catalyze reactions that lead to critical organic compounds, such as intermediates of the citric acid cycle. The versatility of zinc as a catalyst is realized in hundreds of zinc enzymes. Zinc binding to proteins generates numerous additional landscapes of protein conformations and is realized in protein domains with structural zinc sites, such as zinc fingers. This role of zinc in protein structure is evident in the zinc proteomes of eukarya, suggesting that the acquisition of zinc for such functions was a critical step in the evolution from single cells to multicellular organisms. Zinc proteomes from 871 sequenced genomes were analyzed and changes in evolution that relate to changes of zinc availability in the environment deduced [100].

Abbreviations

BMP	bone morphogenetic protein
C	cysteine
CERM	Magnetic Resonance Center at the University of Florence
D	aspartic acid
DALY	disability-adjusted life years
E	glutamic acid
EF-Ts	elongation factor thermostable (translation elongation factor)
ER	endoplasmic reticulum
H	histidine
hK	human kallikrein
IAPP	islet amyloid polypeptide
IC_{50}	inhibition constant at 50% inhibition
LTA_4	leukotriene A_4
MRE	metal-response element
MT	metallothionein
MTF-1	metal-response element-binding transcription factor-1
NMDA	N-methyl-D-aspartic acid
PCR	polymerase chain reaction
PDB	Protein Data Bank
PSA	prostate-specific antigen
PTP	protein tyrosine phosphatase
TFIIIA	transcription factor IIIA

References

1. J. Raulin, *Ann. Sci. Nat. Bot. Biol. Veg.* **1869**, *11*, 92–299.
2. W. R. Todd, C. A. Elvehjem, E. D. Hart, *Am. J. Physiol.* **1934**, *107*, 146–156.
3. A. S. Prasad, J. A. Halsted, M. Nadimi, *Am. J. Med.* **1961**, *31*, 532–546.
4. D. Keilin, T. Mann, *Nature* **1939**, *144*, 442–443.
5. B. L. Vallee, H. Neurath, *J. Am. Chem. Soc.* **1954**, *76*, 5006–5007.
6. B. L. Vallee, A. Galdes, *Adv. Enzymol. Relat. Areas Mol. Biol.* **1984**, *56*, 283–430.

7. J. S. Hanas, D. Hazuda, D. F. Bogenhagen, F. Y.-H. Wu, C.-W. Wu, *J. Biol. Chem.* **1983**, *258*, 14120–14125.
8. J. Miller, A. D. McLachlan, A. Klug, *EMBO J.* **1985**, *4*, 1609–1614.
9. A. Müller, R. M. MacCallum, M. J. E. Sternberg, *Genome Res.* **2002**, *12*, 1625–1641.
10. M. A. Massiah, J. A. B. Matts, K. M. Short, B. N. Simmons, S. Singireddy, Z. Yi, T. C. Cox, *J. Mol. Biol.* **2007**, *369*, 1–10.
11. W. Maret, *JAAS* **2004**, *19*, 15–19.
12. R. N. De Guzman, H. Y. Liu, M. Martinez-Yamout, H. J. Dyson, P. E. Wright, *J. Mol. Biol.* **2000**, *303*, 243–253.
13. J. H. Laity, B. M Lee, P. E. Wright, *Curr. Opin. Struct. Biol.* **2001**, *11*, 39–46.
14. R. Gamsjaeger, C. K. Liew, F. E. Loughlin, M. Crossley, J. P. Mackay, *TIBS* **2006**, *32*, 63–70.
15. B. L. Vallee, D. S. Auld, *Biochemistry* **1990**, *29*, 5647–5659.
16. B. L. Vallee, D. S. Auld, *FEBS Lett.* **1989**, *257*, 138–140.
17. B. L. Vallee, D. S. Auld, *Acc. Chem. Res.* **1993**, *26*, 543–551.
18. J. Z. Haeggström, A. Wetterholm, R. Shapiro, B. L. Vallee, B. Samuelsson, *Biochem. Biophys. Res. Commun.* **1990**, *172*, 965–970.
19. M. M. G. M. Thunnissen, P. Nordlund, J. Z. Haeggström, *Nature Struct. Biol.* **2001**, *8*, 131–135.
20. J. Z. Haeggström, A. Wetterholm, B. L. Vallee, B. Samuelsson, *Biochem. Biophys. Res. Commun.* **1990**, *173*, 431–437.
21. W. Stöcker, M. Ng, D. S. Auld, *Biochemistry* **1990**, *29*, 10418–10425.
22. W. Bode, F. X. Gomis-Ruth, R. Huber, R. Zwilling, W. Stöcker, *Nature* **1992**, *358*, 164–167.
23. B. L. Vallee, D. S. Auld, *Faraday Discuss.* **1992**, *93*, 117–120.
24. W. Maret, *BioMetals* **2001**, *14*, 187–190.
25. N. D. Clarke, J. M. Berg, *Science* **1998**, *282*, 2018–2022.
26. W. Maret, *J. Trace Elem. Biol. Med.* **2005**, *19*, 7–12.
27. I. Bertini, G. Cavallaro, *Metallomics* **2010**, *2*, 39–51.
28. C. Andreini, L. Banci, I. Bertini, A. Rosato, *J. Proteome Res.* **2006**, *5*, 196–201.
29. C. Andreini, L. Banci, I. Bertini, A. Rosato, *J. Proteome Res.* **2006**, *5*, 3173–3178.
30. C. Andreini, I. Bertini, G. Cavallaro, R. J. Najmanovich, J. M. Thornton, *J. Mol. Biol.* **2009**, *388*, 356–380.
31. B. M. Lee, B. A. Buck-Koehntop, M. A. Martinez-Yamout, H. J. Dyson, P. E. Wright, *J. Mol. Biol.* **2007**, *371*, 1274–1289.
32. A. Vannini, C. Volpari, G. Filocamo, E. Caroli Casavola, M. Brunetti, D. Renzoni, P. Chakravarty, C. Paolini, R. De Francesco, P. Gallinari, C. Steinkühler, S. Di Marco, *Proc. Natl. Acad. Sci. USA* **2004**, *101*, 15064–15069.
33. S. L. Gantt, S. G. Samuel, C. A. Fierke, *Biochemistry* **2006**, *45*, 6170–6178.
34. S. E. Wiley, M. L. Paddock, E. C. Abresch, L. Gross, P. van der Geer, R. Nechushtai, A. N. Murphy, P. A. Jennings, J. E. Dixon, *J. Biol. Chem.* **2007**, *282*, 23745–23749.
35. S. Tottey, K. J. Waldron, S. J. Firbank, B. Reale, C. Bessant, K. Sato, T. R. Cheek, J. Gray, M. J. Banfield, C. Dennison, N. J. Robinson, *Nature* **2008**, *455*, 1138–1142.
36. W. Maret, *Antioxid. Redox Signal.* **2006**, *8*, 1419–1441.
37. W. Maret, *Biochemistry* **2004**, *43*, 3301–3309.
38. N. Y. Marcus, R. A. Marcus, B. Z. Schmidt, D. B. Haslam, *Arch. Biochem. Biophys.* **2007**, *46*, 147–158.
39. C. M. Koehler, *TIBS* **2004**, *29*, 1–4.
40. J. M. Herrmann, R. Köhl, *J. Cell Biol.* **2007**, *176*, 559–563.
41. E. Ivanova, M. Ball, H. Lu, *Proteins* **2008**, *71*, 467–475.
42. B. L. Vallee, W. E. C. Wacker, *The Proteins*, 2nd edn., Vol. 5, Academic Press, New York, 1970, p. 25.
43. A. Krężel, W. Maret, *J. Inorg. Biol. Chem.* **2006**, *11*, 1049–1062.
44. Y. Li, W. Maret, *Exp. Cell Res.* **2009**, *315*, 2463–2470.
45. A. Krężel, W. Maret, *J. Inorg. Biol. Chem.* **2008**, *13*, 401–409.

46. W. Maret, C. Jacob, B. L. Vallee, E. H. Fischer, *Proc. Natl. Acad. Sci. USA* **1999**, *96*, 1936–1940.
47. W. Wilson, C. Hogstrand, W. Maret, *J. Biol. Chem.* **2012**, *287*, 9322–9326.
48. C. Hogstrand, P. M. Verbost, S. E. Wendelaar Bonga, *Toxicology* **1999**, *133*, 139–145.
49. T. J. B. Simons, *J. Membr. Biol.* **1991**, *123*, 63–71.
50. C. J. Frederickson, J.-Y. Koh, A. I. Bush, *Nat. Rev. Neurosci.* **2005**, *6*, 449–462.
51. K. M. Taylor, S. Hiscox, R. I. Nicholson, C. Hogstrand, P. Kille, *Science Signal.* **2012** *5(210)*, ra11 [DOI: 10.1126/scisignal.2002585]
52. P. Paoletti, P. Ascher, J. Neyton, *J. Neurosci.* **1997**, *17*, 5711–5725.
53. K. S. Larsen, D. S. Auld, *Biochemistry* **1989**, *28*, 9620–9625.
54. K. S. Larsen, D. S. Auld, *Biochemistry* **1991**, *30*, 2613–2618.
55. M. Gomez-Ortiz, F. X. Gomis-Ruth, R. Huber, F. X. Avilés, *FEBS Lett.* **1997**, *400*, 336–340.
56. M. Debela, V. Magdolen, V. Grimminger, C. Sommerhoff, A. Messerschmidt, R. Huber, R. Friedrich, W. Bode, P. Goettig, *J. Mol. Biol.* **2006**, *362*, 1094–1107.
57. M. Debela, P. Goettig, V. Magdolen, R. Huber, N. M. Schechter, W. Bode, *J. Mol. Biol.* **2007**, *373*, 1017–1031.
58. M. Debela, P. Hess, V. Magdolen, N. M. Schechter, T. Steiner, R. Huber, W. Bode, P. Goettig, *Proc. Natl. Acad. Sci. USA* **2007**, *104*, 16086–16091.
59. J. A. Ippolito, T. T. Baird, Jr., S. A. McGee, D. W. Christianson, C. A. Fierke, *Proc. Natl. Acad. Sci. USA* **1995**, *92*, 5012–5021.
60. W. Maret, Y. Li, *Chem. Rev.* **2009**, *109*, 4682–4707.
61. W. Maret, *J. Inorg. Biochem.* **2012**, *111*, 110–116.
62. A. Noormagi, J. Gavrilova, J. Smirnova, V. Tougu, P. Palumaa, *Biochem. J.* **2010**, *430*, 511–518.
63. S. Salamekh, J. R. Brender, S.-J. Hyung, R. P. R. Nanga, S. Vivekanandan, B. T. Ruotolo, A. Ramamoorthy, *J. Mol. Biol.* **2011**, *410*, 294–306.
64. L. C. Costello, Y. Liu, R. B. Franklin, M. C. Kennedy, *J. Biol. Chem.* **1997**, *272*, 28875–28881.
65. D. Frey, O. Braun, C. Briand, M. Vašák, M. G. Grütter, *Structure* **2006**, *14*, 901–911.
66. M. Knipp, J. M. Charnock, C. D. Garner, M. Vašák, *J. Biol. Chem.* **2003**, *278*, 3410–3416.
67. D. S. Auld, *BioMetals* **2001**, *14*, 271–313.
68. W. Maret, in *Handbook of Metalloproteins*, Vol. 3, Eds A. Messerschmidt, W. Bode, M. Cygler, John Wiley, Chichester, 2004, pp. 432–444.
69. P. W. Kim, Z.-Y. J. Sun, S. C. Blacklow, G. Wagner, M. J. Eck, *Science* **2003**, *301*, 1725–1728.
70. J. Romir, H. Lilie, C. Egerer-Sieber, F. Bauer, H. Sticht, Y. A. Muller, *J. Mol. Biol.* **2007**, *365*, 1417–1428.
71. W. Maret, *Pure Appl. Chem.* **2008**, *80*, 2679–2687.
72. D. S. Auld in *Handbook on Metalloproteins*, Eds I. Bertini, A. Sigel, H. Sigel, Marcel Dekker, New York, 2001, pp. 881–959.
73. D. S. Auld in *Encyclopedia of Inorganic Chemistry*, 2nd edn, Ed R. B. King, John Wiley & Sons, Chichester, UK, 2005, pp. 5885–5927.
74. C. Andreini, I. Bertini, *J. Inorg. Biochem.* **2012**, *111*, 150–156.
75. C. Andreini, I. Bertini, G. Cavallaro, *PloS ONE* **2011**, *6*, e26325.
76. D. S. Auld, in *Handbook of Metalloproteins*, Vol. 3, Eds A. Messerschmidt, W. Bode, M. Cygler, John Wiley & Sons, Chichester, UK, 2004, pp. 403–415.
77. D. S. Auld, in *Handbook of Metalloproteins*, Vol. 3, Eds A. Messerschmidt, W. Bode, M. Cygler, John Wiley & Sons, Chichester, UK, 2004, pp. 416–431.
78. T. Fukada, T. Kambe, *Metallomics* **2011**, *3*, 662–674.
79. M. Lu, J. Chai, D. Fu, *Nat. Struct. Mol. Biol.* **2009**, *16*, 1063–1067.
80. Y. Li, W. Maret, *JAAS* **2008**, *23*, 1055–1062.
81. W. Maret, B. L. Vallee, *Proc. Natl. Acad. Sci. USA* **1998**, *95*, 3478–3482.

82. W. Maret, *BioMetals* **2009**, *22*, 149–157.
83. A. Krężel, W. Maret, *J. Am. Chem. Soc.* **2007**, *129*, 10911–10921.
84. W. Maret, *J. Biol. Inorg. Chem.* **2011**, *16*, 1079–1086.
85. W. Maret, *BioMetals* **2011**, *24*, 411–418.
86. A. Krężel, Q. Hao, W. Maret, *Arch. Biochem. Biophys.* **2007**, *463*, 188–200.
87. R. A. Colvin, W. R. Holmes, C. P. Fontaine, W. Maret, *Metallomics* **2001**, *2*, 306–317.
88. H. Haase, W. Maret, in *Cellular and Molecular Biology of Metals*, Eds R. Zalups, J. Koropatnick, Taylor & Francis, Boca Raton, 2010, pp. 179–210.
89. U. Heinz, M. Kiefer, A. Tholey, H. W. Adolph, *J. Biol. Chem.* **2005**, *280*, 3197–3207.
90. W. Maret, K. S. Larsen, B. L. Vallee, *Proc. Natl. Acad. Sci. USA* **1997**, *94*, 2233–2237.
91. A.-M. Sevcenco, M. W. H. Pinkse, H. T. Wolterbeek, P. D. E. M.Verhaert, W. R. Hagen, P.-L. Hagedoorn, *Metallomics* **2011**, *3*, 1324–1330.
92. W. Maret, *Metallomics* **2010**, *2*, 117–125.
93. R. J. Cousins, R. K. Blanchard, M. P. Popp, L. Liu, J. Cao, J. B. Moore, C. L. Green, *Proc. Natl. Acad. Sci. USA* **2003**, *100*, 6952–6957.
94. J. B. Moore, R. K. Blanchard, R. J. Cousins, *Proc. Natl. Acad. Sci. USA* **2003**, *100*, 3883–3888.
95. J. H. Beattie, M-J. Gordon, G. J. Rucklidge, M. D. Reid, G. J. Duncan, G. W. Horgan, Y.-E. Cho, I.-S. Kwun, *Proteomics* **2008**, *8*, 2126–2135.
96. H. Tom Dieck, F. Doring, D. Fuchs, H. P. Roth, H. Daniel, *J. Nutr.* **2005**, *135*, 199–205.
97. D. Zheng, P. Kille, G. P. Feeney, P. Cunningham, R. D. Handy, C. Hogstrand, *BMC Genomics* **2010**, *11*, 553.
98. D. Zheng, P. Kille, G. P. Feeney, P. Cunningham, R. D. Handy, C. Hogstrand, *BMC Genomics* **2010**, *11*, 548.
99. A. Y. Mulkidjanian, A. Y. Bychkov, D. V. Dibrova, M. Y. Galperin, E. V. Koonin, *Proc. Natl. Acad. Sci. USA* **2012**, *109*, E821–E830.
100. L. Decaria, I. Bertini, R. J. P. Williams, *Metallomics* **2010**, *2*, 706–709.

Chapter 15
Metabolism of Molybdenum

Ralf R. Mendel

Contents

R.R. Mendel (✉)
Institute of Plant Biology, Braunschweig University of Technology,
Humboldt Street 1, D-38106 Braunschweig, Germany
e-mail: r.mendel@tu-bs.de

L. Banci (ed.), *Metallomics and the Cell*, Metal Ions in Life Sciences 12,
DOI 10.1007/978-94-007-5561-1_15,

Abstract The transition element molybdenum is of essential importance for (nearly) all biological systems. It needs to be complexed by a special cofactor in order to gain catalytic activity. With the exception of bacterial Mo-nitrogenase, where Mo is a constituent of the FeMo-cofactor, Mo is bound to a pterin, thus forming the molybdenum cofactor Moco, which in different versions is the active compound at the catalytic site of all other Mo-containing enzymes. In eukaryotes, the most prominent Mo enzymes are nitrate reductase, sulfite oxidase, xanthine dehydrogenase, aldehyde oxidase, and the mitochondrial amidoxime reductase. The biosynthesis of Moco involves the complex interaction of six proteins and is a process of four steps, which also requires iron, ATP, and copper. After its synthesis, Moco is distributed to the apoproteins of Mo enzymes by Moco-carrier/binding proteins. A deficiency in the biosynthesis of Moco has lethal consequences for the respective organisms. In humans, Moco deficiency is a severe inherited inborn error in metabolism resulting in severe neurodegeneration in newborns and causing early childhood death. Eubacteria possess different versions of the pteridin cofactor as reflected by a large number of enzymes such as nitrate reductase, formate dehydrogenase, and dimethyl sulfoxide reductase, while in archaea a tungsten atom replaced molybdenum as catalytic metal in the active center.

Keywords aldehyde oxidase • molybdenum cofactor • molybdenum deficiency • nitrate reductase • sulfite oxidase • tungsten • xanthine dehydrogenase

Please cite as: *Met. Ions Life Sci.* 12 (2013) 503–528

1 Introduction

Molybdenum occurs in a wide range of metalloenzymes in bacteria, archaea, fungi, algae, plants, and animals where it forms part of the active sites of these enzymes. Mo has a versatile redox chemistry that is used by the enzymes to catalyze diverse redox reactions. This redox chemistry is controlled both by the different ligands at the Mo atom and the enzyme environment. Within the enzymes Mo shuttles between three oxidation states (+4, +5 and +6) thereby catalyzing two-electron oxidation/reduction reactions. Mo, however, is not incorporated directly in the apo-enzymes but it is first bound by a scaffold compound thus forming a prosthetic group named molybdenum cofactor (Moco), which subsequently is inserted into the target enzymes.

Mo is very abundant in the oceans in the form of the molybdate (MoO_4^{2-}) anion. In soils, the molybdate anion is the only form of Mo that is available for plants, fungi, and bacteria. Mo-containing enzymes are essential for life, since they hold key positions both in the biogeochemical redox cycles of nitrogen, carbon, and sulfur on Earth [1] and in the metabolism of the individual organism. Hitherto more than 50 enzymes are known to be Mo-dependent. The vast majority of them are found in bacteria while in eukaryotes only seven have been identified [2]. Mo-dependent enzymes are found in nearly all organisms, however some unicellular organisms do not contain Mo enzymes, with *S. cerevisiae* and *S. pombe* as prominent representatives. Genome-wide database analyses revealed a significant number of unicellular organisms that do not need Mo. Obviously loss of Mo utilization is connected to a host-associated life-style that makes Mo enzymes unnecessary, while all multicellular eukaryotes are dependent on Mo [3]. In addition, mainly anaerobic archaea and some bacteria are Mo-independent but they require tungsten for their growth. In the periodic table of elements, W lies directly below Mo and features chemical properties similar to Mo. We will deal with W in the second part of this review.

Mo belongs to the group of trace elements, i.e., the organism needs it only in minute amounts, and indeed uptake of too high amounts of Mo results in toxicity symptoms [4]. On the other hand unavailability of Mo is lethal for the organism. However, even if Mo is available for the cell, it is biologically inactive until it becomes complexed to form Moco. In this article we will give a review on the current understanding of eukaryotic Mo metabolism, involving Mo uptake, Moco biogenesis, Moco transfer and storage and finally its insertion into apo-metalloenzymes. The vast majority of this knowledge derives from studies in plants and humans. In the second part, we will summarize the main points of prokaryotic Mo metabolism and will also touch upon tungsten.

2 Molybdenum in Eukaryotes

2.1 Molybdenum Uptake

Organisms take up Mo in the form of its anion molybdate. It requires specific uptake systems to scavenge molybdate in the presence of competing anions. In higher organisms, only recently first molybdate-transporting proteins have been identified in algae and plants. Two proteins (Mot1 and Mot2) belonging to the large sulfate carrier superfamily were shown to transport molybdate with ultra-high affinity (nanomolar k_M value) across cellular membranes [5–7]. Molybdate quantification in isolated vacuoles demonstrated that this organelle serves as an important molybdate storage in *Arabidopsis* cells where Mot2 was shown to be required for vacuolar molybdate export into the cytosol. Assuming this export function for Mot2, the endo-membrane (ER) location of Mot1 becomes likely since it may be involved in the transport of molybdate via the ER-Golgi route to the vacuole. But how is molybdate imported into the cytosol? Very recently another molybdate transporter has been identified in the alga *C. reinhardtii* that unlike Mot1 and Mot2 is not exclusively found in algae and higher plants but also occurs in humans [8]. Although still not

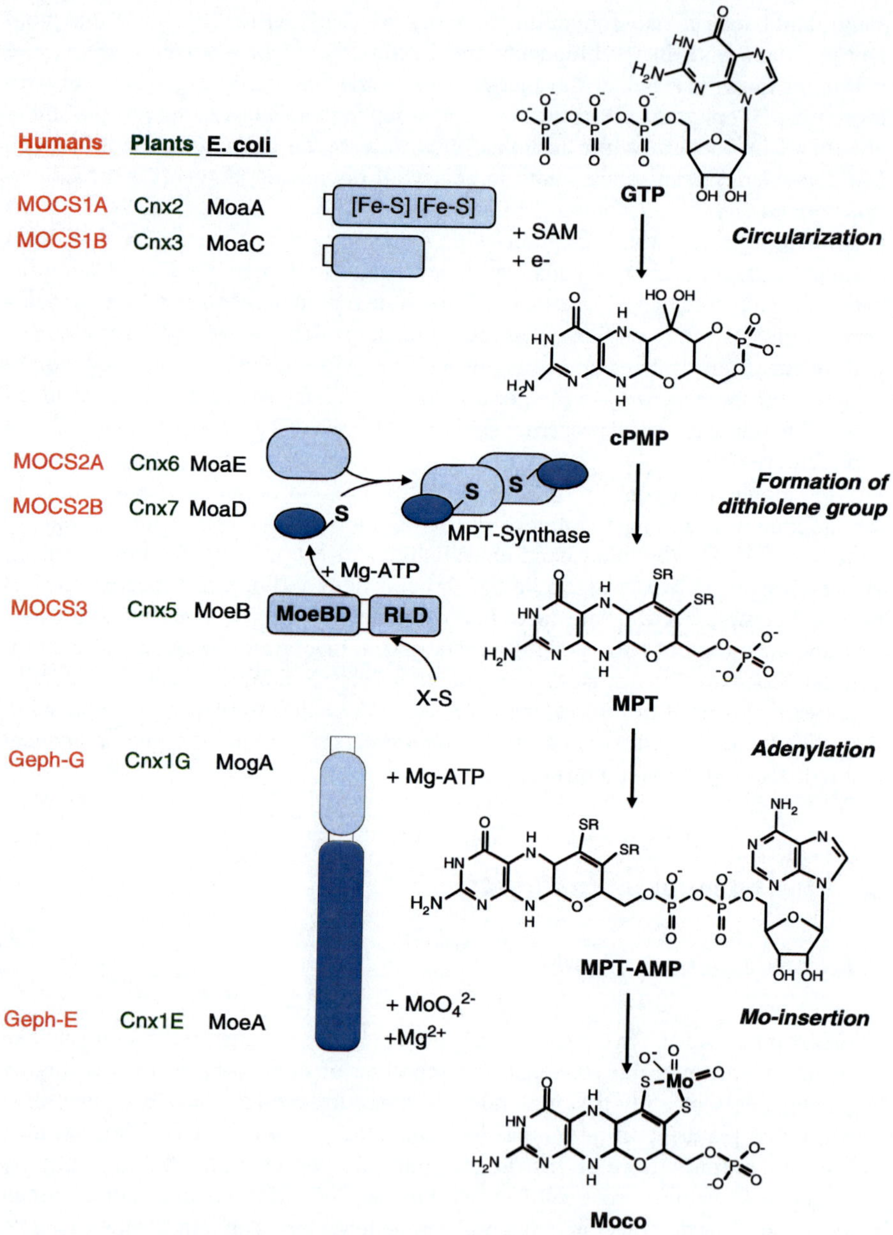

Figure 1 Biosynthesis of the eukaryotic molybdenum cofactor. Moco synthesis can be divided into four steps, the main features are given in italics on the right side. The names for the proteins from humans (red), plants (green), and *E. coli* (black) catalyzing the respective steps are given. For MPT and MPT-AMP, the ligands of the dithiolate sulfurs are indicated by an "R" as it is currently unknown at which step copper is bound to the dithiolate. The *in vivo* source of sulfur (X-S) for Cnx5 and MOCS3 is not known yet. Human Nfs1 is a likely candidate. Steps three and four in

localized, it is likely that this transporter serves as the general molybdate importer for the cell. Further, it can be assumed that in addition to a possible high-affinity uptake system molybdate may also enter the cell non-specifically through the sulfate uptake system. Molybdate uptake through a sulfate transporter has recently been described [9], thus supporting this assumption.

2.2 *The Molybdenum Cofactor*

Nature has developed two very different cofactors to position Mo within the active center of Mo enzymes thus controling the redox state of Mo. In either case, sulfur and oxygen atoms form ligands to Mo. One type of Mo cofactor is the pterin-based cofactor (Moco) which occurs in all kingdoms of life [10] while the other type is the iron-sulfur cluster-based iron-molybdenum cofactor (FeMoco) that is unique to a single enzyme, the bacterial nitrogenase [11]. The two cofactors are structurally unrelated. We will deal with FeMoco later in this review and will now focus on the pterin-type Moco. Its structural scaffold compound is a tricyclic pterin called molybdopterin or metal-containing pterin (MPT). As the result of Mo coordination by MPT, Moco is formed (Figure 1). Since Moco is highly sensitive to oxidation, most of the work was performed using its stable oxidation products [12]. Hereby, Moco had originally been identified in mammalian sulfite oxidase to be a pterin derivative comprising a unique four-carbon side chain as C6 substituent [13]. The coordination of Mo via an ene-dithiolate group located within the four-carbon side chain of the cofactor was demonstrated by carbamidomethylation of Moco.

The nature of the Moco structure was finally elucidated upon co-crystallization with the Mo enzyme aldeyhde oxidase (AO) from *Desulfovibro gigas* [14]. Here, a small discrepancy to the originally proposed Moco structure was found, namely the existence of a third ring fused to the pterin backbone. This pyrano ring carries the ene-dithiolate group which is essential for metal coordination (Figure 1). The fusion of a pterin with a pyrano ring as identified for Moco and its direct precursor, the metal-free MPT, is unique in Nature and may have evolved in order to maintain and/ or control the special redox properties of Mo. The pyranopterin part of Moco is necessary for the correct positioning of the catalytic metal (Mo) within the active center of Mo enzymes and in addition to this, it may also participate in the electron channeling from or to other prosthetic groups.

All analyses of Mo enzymes crystallized so far revealed that Moco is not located on the surface of the protein, whereby a tunnel-like structure assures accessibility of

Figure 1 (continued) plants and humans are catalyzed by the individual domains of Cnx1 (G and E) or Gephyrin (G and E). Functional properties like clusters in Cnx2 and Mocs1A, the use of S-adenosyl methionine (SAM), adenylation and sulfuration of the small subunit of MPT synthase (Cnx7 and Mocs2B, respectively) are indicated. In Cnx5, MoeBD denotes the MoeB-like domain and RLD the rhodanese-like domain. Upon Mo insertion it is not clear how many oxo ligands are bound to the metal. Therefore, we have shown the trioxo form of the Mo center as it occurs in the Moco storage protein MCP from *Chlamydomonas reinhardtii* (adapted from [17]).

the cognate substrates [15]. During its life time, the Mo enzyme does not liberate Moco. *In vitro*, however Moco may be removed from its protein environment either by heat treatment or by acidification. Once Moco is set free, it easily loses the Mo atom and becomes rapidly oxidized, resulting in an irreversible loss of function due to oxidation. The demolybdo forms of Mo enzymes are catalytically inactive.

2.3 *Molybdenum Cofactor Biosynthesis*

Mutations in the genes for Moco biosynthesis result in the pleiotropic loss of all Mo-dependent cellular processes. Detailed analyses of these mutants provided a basis to propose an evolutionary old multi-step biosynthetic pathway [16] and laid the basis to decipher plant and human Moco biosynthesis [17,18]. Characterization of both, plant and human Moco biosynthesis, documented their strict analogy and therefore in the following will be compared whenever appropriate.

Moco biosynthesis can be divided into four steps, referring to the intermediates cyclic pyranopterin monophosphate (cPMP, previously identified as precursor Z), MPT, adenylated MPT (MPT-AMP), and Moco, respectively (Figure 1). A total of six involved gene products have been identified in plants [19], fungi [20], and humans [21–23], whereby some but not all of the eukaryotic Moco biosynthesis genes are able to functionally complement the corresponding bacterial mutants, thus documenting the early evolutionary origin of Moco biosynthesis.

Different nomenclatures were introduced for genes and gene products involved in Moco formation. Moco biosynthesis genes and gene products of plants followed to the *cnx* nomenclature (cofactor for nitrate reductase and xanthine dehydrogenase). Human Moco biosynthesis genes were named differently. Here, the MOCS (molybdenum cofactor synthesis) nomenclature was introduced [21]. In the following we will briefly characterize the individual steps of Moco biosynthesis.

2.3.1 Step 1: Conversion of Guanosine 5'-Triphosphate to Cyclic Pyranopterin Monophosphate

Moco biosynthesis starts in the mitochondria. GTP is transformed into the sulfur-free pterin compound cyclic pyranopterin monophosphate (cPMP), also known as precursor Z (Figure 1). cPMP is the most stable intermediate of Moco biosynthesis with an estimated half life time of several hours at a low pH [24]. Therefore it was possible to solve its structure in *E. coli* using ^{1}H NMR [25], while structural elucidation of MPT, MPT-AMP, and Moco required crystallization of protein ligand complexes [14,26]. The basic tricyclic pyranopterin structure of cPMP already resembles the fully reduced tetrahydropterin structure of MPT and Moco [25]. The conversion of GTP to cPMP is catalyzed by two proteins: plant Cnx2 (MOCS1A in humans) and plant Cnx3 (MOCS1B in humans) (compare Figure 1). One of these (Cnx2/MOCS1A) is member of the superfamily of S-adenosylmethionine-dependent radical enzymes [27]. Members of this protein family catalyze the formation of protein and/or substrate radicals by reductive cleavage of SAM involving a [4Fe-4S] cluster [28].

Both proteins catalyzing step 1 of Moco biosynthesis (plant Cnx2 and Cnx3, Figure 1) are located in the matrix fraction of mitochondria [29] where 5'-GTP as substrate for cPMP synthesis is available and where Fe-S clusters as the essential prosthetic group for Cnx2 are synthesized [30]. All subsequent steps of Moco biosynthesis, however, have been demonstrated to be localized in the cytosol [31–33]. Therefore, export of cPMP from the mitochondria into the cytosol is required, and for this task the transporter protein Atm3, which is localized in the inner membrane of plant mitochondria and belongs to the family of ATP-binding cassette (ABC) transporters, has been identified recently [29]. When the Atm3 exporter is mutated, cPMP accumulates within mitochondria and the cytosol becomes short of cPMP with the consequence that Moco levels and Mo enzyme activities decrease in the cell. The precise role of Atm3 is still unknown. Also human cells possess this transporter protein, which is named ABCB7.

2.3.2 Step 2: Synthesis of Molybdopterin

In the second stage sulfur is transferred to cPMP in order to generate MPT. This reaction is catalyzed by the enzyme MPT synthase, a heterotetrameric complex consisting of two small (plant Cnx7, human MOCS2B) and two large (plant Cnx6, human MOCS2A) subunits capable of cPMP conversion to MPT (Figure 1). The sulfur is bound as thiocarboxylate to the C-terminus of the small subunit. MPT synthase converts cPMP stoichiometrically to MPT, which most likely involves a two-step mechanism that was deciphered in detail in bacteria where the existence of a mono-sulfurated intermediate has been observed [34].

After MPT synthase has transferred the two sulfurs to cPMP, it has to be re-sulfurated by the MPT-synthase sulfurase in order to reactivate the enzyme for the next reaction cycle of cPMP conversion. This separate resulfuration is catalyzed by the enzyme MPT-synthase sulfurase (Cnx5 and MOCS3, respectively; Figure 1). MPT-synthase sulfurase is a two-domain protein consisting of a N-terminal adenylating domain (homologous to *E. coli* MoeB) and a C-terminal rhodanese-like domain. In analogy to the bacterial mechanism [35], this enzyme is supposed to activate the small subunit of MPT synthase by adenylation followed by sulfur transfer thus forming the thiocarboxylate at the C-terminus of the small subunit. For the human enzyme it was shown that the sulfur comes from the rhodanese-like domain of MPT-synthase sulfurase where it is bound as persulfide [32,36] (Figure 1). In humans, the cysteine desulfurase Nfs1 is a likely candidate to function as sulfur donor for the MOCS3-catalzyed resulfuration step [37].

2.3.3 Step 3: Adenlyation of Molybdopterin

After synthesis of the MPT moiety, the chemical backbone is built for binding and coordination of the Mo atom. In the third step, therefore, Mo has to be transferred to MPT in order to form Moco, thus linking the molybdate uptake system to the MPT pathway. In the living cell, the insertion of Mo into MPT is not a spontaneous process but is catalyzed by an insertase. Mutants defective in this step accumulate MPT but can be partially rescued when growing them on molybdate-enriched media. Physiological Mo concentrations, however, are not sufficient to achieve any non-catalyzed Mo ligation by MPT.

In bacteria, Mo insertion is carried out by two separately expressed proteins (MogA and MoeA) while higher organisms have fused these two proteins to a single two-domain protein (Cnx1 in plants and Gephyrin in mammals). Early it was assumed that one domain should be essential for generating an activated form of Mo that is incorporated by the other domain into bound MPT [38,39]. Later the exact mechanism was uncovered for the protein Cnx1 from the model plant *Arabidopsis thaliana* [40] where the C-terminal domain (= Cnx1-G) was known to tightly bind MPT. Yet, its crystal structure [26] revealed an unexpected finding: a novel reaction intermediate, adenylated MPT (MPT-AMP) (Figure 1), was found. Subsequently it was demonstrated that Cnx1-G adenylates MPT in a Mg^{2+}- and ATP-dependent way and forms MPT-AMP that remains bound to Cnx1-G [40].

2.3.4 Step 4: Molybdenum Insertion and Crosstalk to Copper Metabolism

The crystal structure of the Cnx1-G revealed another unexpected finding, namely a copper (Cu^{1+}) bound to the MPT dithiolate sulfurs, whose nature was confirmed by anomalous scattering of the metal [26]. Therefore, insertion of Mo into the MPT dithiolene group can be characterized as metal exchange reaction, with Cu presumably serving as suitable leaving group. Consequently, it is reasonable to assume that Cu is bound to the MPT dithiolene group directly after the latter has been formed in the second step of Moco biosynthesis. Since Cu occurs exclusively protein-bound *in vivo* it is likely that both, Cu binding to MPT and its exchange for Mo, depend on yet unidentified cytoplasmic chaperones involved in cellular Cu metabolism.

In the final step of Moco biosynthesis MPT-AMP has to be converted into mature Moco: MPT-AMP is transferred to the N-terminal domain of Cnx1 (= Cnx1-E) (Figure 1). Cnx1-E cleaves the adenylate, releases copper and inserts Mo, thus yielding active Moco. We found that MPT adenylate was hydrolyzed in a molybdate-dependent way [41]. This reaction was coupled to the metal exchange reaction where bound copper was released and Mo was transferred to MPT thus yielding mature Moco. Not only plant mutants [42] defective in Cnx1 and mammalian mutants defective in its homolog Gephyrin [43], but also fungal CnxE mutants show a molybdate-repairable phenotype, i.e., Moco biosynthesis can be restored at least partially by growing the mutant cells on media containing unphysiologically high concentrations (1–10 mM) of molybdate [44], thus overriding the effect of the mutation.

In vitro studies with Cnx1-G-bound MPT-AMP revealed an inhibition of Moco synthesis in the presence of 1 μM $CuCl_2$, providing a link between Mo and copper metabolism [26]. Copper inhibition of Moco synthesis can be explained by inhibition of the Mg-dependent Mo insertion reaction. The latter is supported by the suppression of copper inhibition with equimolar amounts of Cnx1-E and is in line with the known copper inhibition of pyrophosphatases [45]. Our finding implies that

Moco deficiency might occur when cellular copper concentrations are increased, as seen in human patients affected with Wilson's disease [46], where copper accumulates in liver and brain, resulting in severe damage to both organs. In plants, one should consider that the toxicity of elevated copper levels could also go back – at least in part – to a distorted Moco biosynthesis. But also copper shortage should be detrimental for Moco biosynthesis. Therefore, when studying copper homeostasis the accompanying analysis of Mo metabolism could shed further light onto the link between Mo and copper homeostases.

2.4 Storage, Transfer and Insertion of the Molybdenum Cofactor

After synthesis, Moco has to be allocated to the appropriate apo-enzymes. For most prosthetic groups there is a gap of knowledge in our understanding how – after synthesis – these groups are directed to their various cellular destinations and how they ultimately find the way into their correct cognate proteins, or whether they are stored after synthesis. Intricate mechanisms can be assumed to control distribution, trafficking, and insertion into proteins as most of these prosthetic groups are extremely 'fragile' and air-sensitive. For Moco in higher organisms, some pieces of such a sorting machinery became known in recent years.

Moco is extremely sensitive to oxidation [47], and therefore it is assumed that Moco occurs permanently protein-bound in the cell. Also the fast and directed flow of Moco to its target enzymes is an essential prerequisite to reduce the threat of Moco degradation. Both preconditions may be met by Moco-binding proteins (MoBP) ensuring Moco binding as well as its directed transfer to cognate target enzymes. Thus, a pool of insertion-competent Moco may be stored and provided on demand. Storing of Moco may be essential for the cell in order to flexibly meet the variable demand for Moco arising from newly synthesized Mo enzymes. Among eukaryotes, a first MoBP which was named Moco carrier protein (MCP) was identified in the green algae *Chlamydomonas reinhardtii* [48]. Biochemical characterization of recombinant MCP revealed its capability to protect bound Moco against degradation and demonstrated its reversible Moco binding characteristics [49,50]. *C. reinhardtii* MCP is purified as homotetrameric protein where each monomer holds one molecule of Moco. The corresponding structure was found to resemble the Rossman fold [50]. Subsequently, a structure-based homology search identified a family of eight MCP-related proteins in *A. thaliana* [51]. Their biochemical characterization showed reversible Moco binding properties, however with overall lower affinities. Therefore these MoBPs are no good candidates to serve as Moco storage proteins. Rather they seem to be involved in the cellular distribution of Moco since they were found to undergo protein-protein interactions both with the 'Moco-donor' protein Cnx1 and the 'Moco-user' protein NR, thus integrating the MoBP proteins into the cellular Moco flow (Figure 2). The high number of eight expressed MoBP proteins in *Arabidopsis* probably points towards an organ and tissue specific functional specialization.

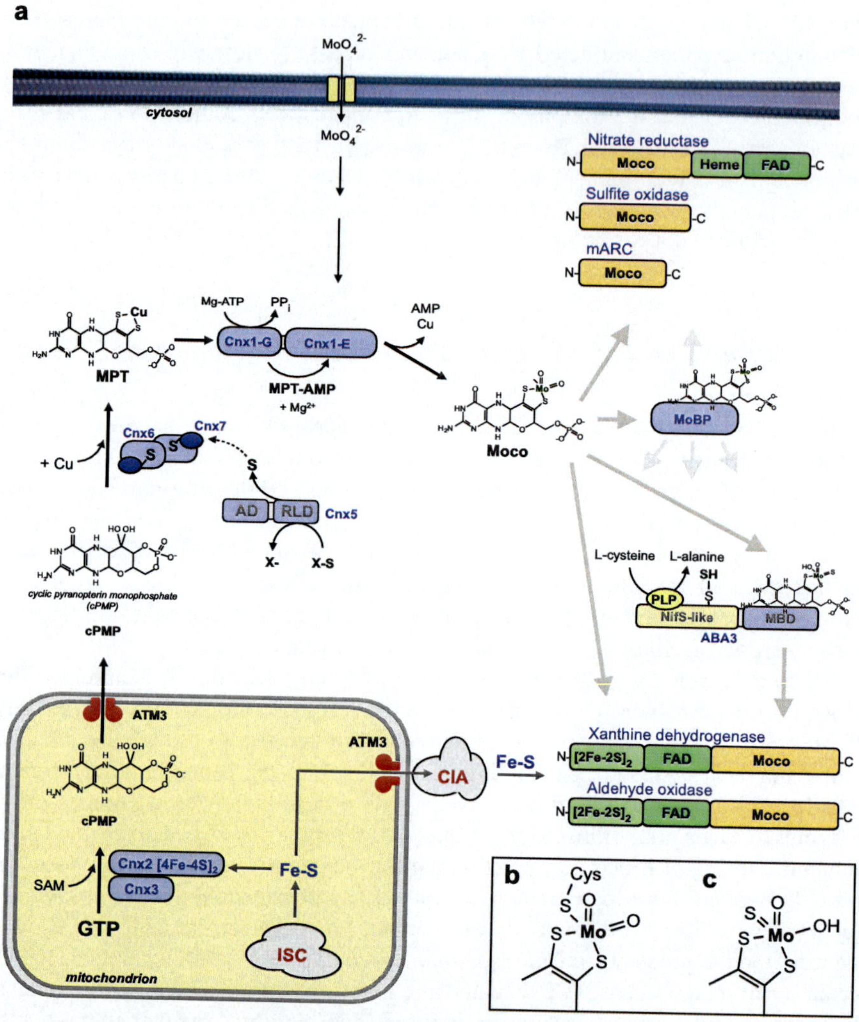

Figure 2 (**a**) Organization of biosynthesis, distribution, and maturation of Moco in plant cells. The basic steps of Moco biosynthesis are shown starting from the conversion of GTP to cPMP in the mitochondria, to the cytosolic steps via MPT and Moco. Moco biosynthesis enzymes and Moco-binding proteins, which are not typically classified as Mo enzymes, are shown in blue. The dependence of Cnx2 on [Fe-S] and S-adenosyl methionine (SAM) is indicated. MPT-synthase, consisting of Cnx6 and Cnx7, is sulfurated by Cnx5, with the primary sulfur donor (X-S) mobilized by the rhodanese-like domain of Cnx5 (RLD) being unknown. The adenylation domain of Cnx5 (AD) is required for adenylation and activation of the small MPT synthase subunit Cnx7. It is assumed that copper is inserted directly after dithiolene formation. The individual reactions of Cnx1 and its products (Moco, pyrophosphate PP_i, AMP, copper) are indicated. For simplicity, only one molybdate transporter is shown in the scheme (for more details see Section 2.1). Mature Moco can be either bound to a Moco-binding protein (MoBP), to the Mo enyzmes mARC, AO, XDH, NR and SO, or to the Moco-binding C-terminal domain of the ABA3 protein (MBD). Note that mammalian SO has an additional heme domain attached to the N-terminal part of the Moco domain.

The principles underlying Moco insertion into eukaryotic Mo enzymes are still not understood. Since Moco is deeply buried within the holo-enzymes it needs to be incorporated prior to or during completion of folding and dimerization of the apo-protein monomers. In bacteria, a complex of proteins synthesizing the last steps of Moco biosynthesis donates the mature cofactor to apo-enzymes assisted by enzyme-specific chaperones. Nearly each bacterial Mo enzyme has a private chaperone available, e.g., NarJ for *E. coli* nitrate reductase [52], XDHC for xanthine dehydrogenase from *Rhodobacter capsulatus* [53], and *Comamonas acidovorans* [54]. However, so far no eukaryotic Moco chaperones have been identified yet. As we have two classes of Mo enzymes in higher organisms also differences in the insertion of Moco might be considered.

3 Molybdenum Enzymes

Mo enzymes are essential for key reactions in the global carbon, sulfur, and nitrogen metabolism, and although up to now more than 50 Mo enzymes have been found in Nature their number is still increasing. While most of the Mo enzymes are of bacterial origin, only a limited number is present in eukaryotes. The enzymes xanthine oxidoreductase (XOR), aldehyde oxidase (AO), and sulfite oxidase (SO) have been identified in all eukaryotes analyzed so far. In turn, nitrate reductase (NR) was only found in autotrophic organisms (i.e., plants, algae, and fungi). As a basic principle, Mo enzymes catalyze reactions that involve the two-electron transfer from or to a substrate which is accompanied by the transfer of an oxygen atom derived from or incorporated into water [2,55]. In the course of this reaction the oxidation state of Mo varies from IV to VI.

Eukaryotic Mo enzymes are classified according to the coordination chemistry of the pterin-bound Mo atom. Two different types of Mo coordination can be found (Figure 2b and 2c):

(1) The di-oxo Mo enzymes generally harbor a strictly conserved cysteine residue necessary for the formation of a charge transfer bond to the pterin-coordinated Mo atom, while remaining Mo coordination spheres are occupied by oxygen atoms. Most prominent members of this class of Mo enzymes are NR and SO. This family of Mo-enzymes is also referred to as SO class of Mo enzymes.
(2) In contrast, the mono-oxo Mo enzymes do not display any protein-dependent Mo coordination. However, here one of the Mo oxygen ligands has been replaced by a sulfur ligand. Thus, Mo coordinates two oxygen and one sulfur atom.

Figure 2 (continued) ABA3 generates a protein-bound persulfide, which is the source of the terminal sulfur ligand of Moco in AO and XDH. Like Cnx2, AO and XDH depend on [Fe-S] clusters, which are synthesized in mitochondria by the iron-sulfur cluster biosynthesis machinery (ISC) prior to export by ATM3 and assembly by the cytosolic iron-sulfur cluster assembly machinery (CIA). **(b)** Structure of the Mo center in enzymes of the SO family (SO and NR). **(c)** Structure of the Mo center in enzymes of the XO family (AO and XDH).

XDH and AO are prominent members of this enzyme family which is referred to as the XO class of Mo enzymes. Also pyridoxal oxidase and nicotinate hydroxylase belong to this class, but they were exclusively found in *Drosophila melanogaster* [56] and *Aspergillus nidulans* [57], respectively.

3.1 Nitrate Reductase

Nitrate reductase (EC 1.7.1.1) is a key enzyme of nitrate assimilation that catalyzes the reduction of nitrate to nitrite in the cytosol. As NR thereby provides essential nitrogen metabolites to the organism, it is obvious that plants or algae with a deficiency in NR are no longer nitrogen autotroph and depend on alternative nitrogen sources such as ammonium. Eukaryotic NR is composed of several domains (Figure 2). Three of them are involved in prosthetic group binding (Moco, heme, and FAD, respectively). Adjacent to these domains, NR has one dimerization domain and another subdomain that is involved in NAD(P)H binding. The individual domains are linked by solvent-exposed linker regions (Figure 2). NR has the intrinsic capability to produce nitric oxide, however, at a very low level [58].

3.2 Sulfite Oxidase

The enzyme sulfite oxidase (EC 1.8.3.1) catalyzes the oxidation of sulfite to sulfate. As sulfite is a strong nucleophile that can react with a wide variety of cellular components it was assumed that SO has a sulfite-detoxifying function and is required for removing excess sulfite from the cell. Amongst eukaryotes, different forms of SO developed during evolution. The animal type SO consists of two domains binding different prosthetic groups. It comprises an N-terminal heme-(cytochrome b_5) binding and a C-terminal Moco-binding domain [2]. Its counterpart from plants however lacks the N-terminal heme domain and harbors Moco as the sole prosthetic group [59,60] (Figure 2).

Sulfite oxidation is characterized as a two electron transfer reaction in which electrons derived from sulfite reduce the Mo center from Mo^{VI} to Mo^{IV}. Human SO uses cytochrome *c* as electron acceptor and coincident with this it was shown to be localized in the mitochondrial intermembrane space. Plant SO, however, is a peroxisomal enzyme [61] that uses molecular oxygen as electron acceptor and simultaneously forms hydrogen peroxide during catalysis [62,63]. While human SO catalyzes the final step in the degradation of sulfur-containing amino acids, the physiological role of plant SO is to be part of a sulfate-sulfite cycle that is essential for fine-tuning of sulfur distribution in the cell in addition to its sulfite-detoxifying role [64,65].

3.3 Mitochondrial Amidoxime Reducing Component

The mitochondrial amidoxime reducing component (mARC) is the latest eukaryotic Mo enzyme identified as yet. The enzyme was found in the outer membrane of pig liver mitochondria [66]. Subsequent analyses revealed highest sequence similarity to the C-terminal domain of Moco sulfurase proteins. In general, all eukaryotic genomes known encode two mARC isoforms and the enzymes have been characterized from humans [67], pigs [66], and *Chlamydomonas reinhardtii* [68]. However as yet the native substrate has not been identified for either of them. Human mARC proteins were shown to catalyze the reduction of many *N*-hydroxylated substances which are commonly used as prodrugs, thus connecting mARC proteins with prodrug conversion. They were also suggested to be involved in NO formation [69]. Coincident with the mitochondrial localization of human mARC proteins, for each of them a N-terminal mitochondrial pre-sequence was identified.

mARC proteins bind Moco as a sole prosthetic group (Figure 2). However, they do not exhibit enzymatic activity on their own but depend on the combined action of cytochrome b_5 as electron transmitter and NADH/cytochrome b_5 reductase as electron donor, thus forming a mARC enzyme complex (for review see [70]). mARC proteins most likely belong to the SO family, but final evidence needs to be provided.

3.4 Xanthine Oxidoreductase

Xanthine oxidoreductase (EC 1.17.1.4) is the key enzyme of purine degradation. It oxidizes hypoxanthine to xanthine and xanthine to uric acid [71] in the cytosol. Each of the XOR subunits comprises three distinct domains necessary for the binding of two [2Fe-2S] clusters, FAD and Moco, respectively (Figure 2), while the domain necessary for dimerization is fused to the Moco-binding domain. Members of the XOR family have a broad substrate spectrum, and therefore hydroxylate a large number of aldehydes and aromatic heterocycles. Electrons derived upon substrate conversion are fed into the intramolecular electron transfer chain of the enzyme and go from the Mo center via the [2Fe-2S] clusters to the FAD cofactor. Here electrons are either transferred to NAD to form NADH, or they are transferred to molecular oxygen, thus yielding superoxide anions or hydrogen peroxide [72,73]. Plant XOR was found to catalyze also substrate-independent ROS formation due to an intrinsic NADH oxidase activity [74]. Therefore, plant XOR is discussed not only to decompose purines but also to have additional physiological functions in ROS metabolism connected to plant-pathogen interactions [75], while human XOR is proposed to be also part of a number of other physiological functions like the inflammatory response to ischaemia-reperfusion [76] and the formation of milk fat droplets [77,78].

3.5 Aldehyde Oxidase

Like XOR, aldehyde oxidase (EC 1.2.3.1) is a cytosolic molybdo-iron-flavoenzyme that catalyzes the oxidation of a variety of aromatic and non-aromatic aldehydes to their corresponding carboxylic acid. AOs derived from XOR upon an ancient gene duplication [79]. Coincident with this, both enzymes display a high degree of sequence similarity and use the same prosthetic groups (Figure 2). AOs display a much broader substrate specificity than XOR, however, other than XOR, AO is a strict oxidase being unable to bind NAD but exclusively using molecular oxygen as electron acceptor. Differences exist also between the plant and the animal enzyme: While animal AO produces superoxide as well as hydrogen peroxide as side-products during catalysis, plant AO exclusively generates hydrogen peroxide. The physiological function of animal AO remains open. On the basis of the very broad (<100) substrate spectrum a general role in biotransformation and detoxification can be assumed. In the model plant *A. thaliana* there are four AO isoforms which are involved in phytohormone (abscisic acid, probably also auxin) biosynthesis and a variety of developmental processes and abiotic and biotic stress responses [80,81].

3.6 Post-translational Sulfuration of Xanthine Oxidoreductase and Aldehyde Oxidase

Plant Mo enzymes can be subdivided into two families. While the enzymes of the SO family, SO and NR, are activated upon insertion of Moco, the members of the XOR family, XOR and AO, require a final step of maturation during or after insertion of Moco (Figure 2). In addition to the dithiolene sulfurs of the pterin moiety and two oxo-groups, the Mo atom of the Moco needs the addition of a terminal inorganic sulfur in order to provide enzymatic activity to the respective enzyme [82,83]. *In vitro*, this sulfur can be removed from active XDH/AO proteins by cyanide treatment, thereby generating an inactive desulfo enzyme [84]. The reaction, however, is reversible and the enzyme can be reactivated by sulfide treatment under reducing conditions. *In vivo*, this terminal sulfur is added to the Moco of XDH and AO by a specific enzymatic reaction, catalyzed by the Moco sulfurase (ABA3 in plants, HMCS in humans; Figure 2).

ABA3 is a homodimeric two-domain protein [85] with its N-terminal domain sharing structural and functional homologies to bacterial cysteine desulfurases, thereby being more similar to SufS than to NifS or IscS. In a pyridoxal phosphate-dependent manner, the N-terminal domain of ABA3 decomposes L-cysteine to yield alanine and elemental sulfur [86], the latter being bound as a persulfide to a highly conserved cysteine residue of ABA3 (Figure 2). The C-terminal domain of ABA3 shares a significant degree of similarity to the newly discovered mARC proteins and was shown to bind sulfurated Moco, which receives the terminal sulfur via an intramolecular persulfide relay from the N-terminal domain [87,88]. It is likely that subsequent to Moco sulfuration, ABA3 exchanges non-sulfurated for sulfurated Moco, thus activating its target Mo enzyme.

Under physiological aspects the terminal sulfuration step provides an efficient way of regulating the amount of active XDH and AO enzymes in the cell. Both enzymes produce physiologically active compounds like hormones (by AO) and ROS (by AO and XDH) whose concentrations can rapidly be increased by changing the ratio of inactive and active XDH and AO molecules. In fact, a rapid induction of the *aba3* gene was found upon drought and salt stress in *Arabidopsis* as well as upon ABA treatment [85,89].

4 Molybdenum and Molybdenum Cofactor Deficiencies

4.1 Molybdenum Deficiencies in Plants

Among higher organisms, Mo deficiency has been studied in detail only in plants. The Mo content of plants is directly correlated to the bioavailability of Mo in the soil. The lower the soil pH, the less available is Mo, thus causing Mo deficiency where plants develop a characteristic phenotype including lesions and altered morphology of leaves [90]. Mo deficiency could also be caused by a mutation in the Mo-specific uptake system [91]. *Mot1* knockout mutants showed a slightly altered growth pattern [5,7]. As there are, however, several Mo transporters in a plant cell a severe phenotype has not yet been reported.

Finally, Mo deficiency can be caused by a defect in Moco biosynthesis which has dramatic consequences for the cell since all Mo enzyme activities are lost or at least strongly reduced. The loss of which Mo enzyme is most severe for the plant?

(1) Loss of NR activity is lethal when respective mutants are cultured with nitrate as sole nitrogen source [92,93].
(2) The loss of SO activity results in no phenotype provided the plants are not challenged to an atmosphere with high concentrations of sulfur dioxide [65].
(3) Likewise the loss of XOR activity has no lethal consequences although the phenotype of respective plants is slightly, but not dramatically changed (F. Bittner and R. Mendel, unpublished).
(4) For AO, only a mutation in the AAO3 isoform is communicated [94]. AAO3 generates the phytohormone abscisic acid, hence its loss leads to wilted leaves. A knockout in one of the two mARC proteins has no obvious phenotype (F. Bittner and R. Mendel, unpublished). Moco sulfurase (ABA3) deficiency was found to be basically ascribed to the reduction of abscisic acid levels due to the lack of AO activities [89] [95] and hence, the phenotype of wilted leaves.

In summary, the complete loss of Moco as it occurs in *cnx* mutants is lethal and leads to the death of plants when they are grown in soil. In cell culture, however, these mutants can be kept alive (as cripple plants) when grown on media with reduced nitrogen as N source.

4.2 *Molybdenum Cofactor Deficiency and Therapy in Humans*

Higher organisms do not tolerate loss of Moco synthesis since essential metabolic functions are Mo-dependent. Human Moco deficiency is a rare recessive hereditary disorder (incidence below 1:100,000), which ultimately results in the death of affected patients. Hitherto different mutations resulting in Moco deficiency have been identified and found to affect three of four human Moco synthetic genes: *mocs1*, *mocs2*, and *gephyrin* [18]. Lack of Moco biosynthesis results in the pleiotropic loss of all Mo-dependent enzyme activities. As seen in other inborn errors of metabolism, symptoms develop shortly after birth, when the babies' metabolism starts to operate and toxic metabolites accumulate within the body. The major cause for the observed clinical symptoms is the deficiency of SO that protects the organism – in particular the brain – from elevated levels of toxic sulfite which is formed upon degradation of sulfur-containing amino acids and sulfo-lipids. Babies born with either isolated SO or Moco deficiency display severe neurologic abnormalities, seizures, and dysmorphic features of brain and head. In humans, a combined deficiency of Mo enzymes was first described by Duran et al. [96] and hitherto, more than 100 cases have been diagnosed.

Two thirds of the patients affected by Moco deficiency turned out to have a defect in the first step of Moco biosynthesis, i.e., the conversion of GTP to cPMP [97], encoded by the *mocs1* gene. Therefore, this gene was chosen to be knocked out in the animal model [98]. Subsequent characterization of the homozygous *mocs1* knock-out mice revealed a severe phenotype resembling all biochemical characteristics of human Moco deficiency patients. Since *mocs1*-deficient mice lack the ability to synthesize cPMP, it was decided to establish a suitable therapy using cPMP purified from *E. coli*. cPMP has been injected into the liver of Moco-deficient mice, which resulted in a dramatically increased survival time [99]. Thus, treated mice developed normally, gained weight and reached adulthood and fertility like their wild-type littermates. First human exposure of cPMP treatment has been reported recently [100]. A patient with Moco deficiency (mutation in *mocs1*) has been diagnosed on day 6 of life and experimental treatment was started on day 36 of life. Within days, urinary biomarkers of SO and XOR deficiency returned to almost normal readings and stayed constant. Clinically, the patient became more alert few days after treatment start, convulsions and twitching disappeared within the first two weeks. Today, with nearly four years of treatment, the patient shows delayed neurological development due to preexisting brain injury, however, is free of seizures, happy, and alert (G. Schwarz, personal communication).

5 Molybdenum Metabolism Is Linked to Iron Metabolism

There are at least three links connecting Mo metabolism to Fe homeostasis:

Link 1: Fe plays an important role for the synthesis of Moco as Cnx2 (human MOCS1A), which is involved in step 1 of Moco biosynthesis, requires two Fe-S clusters of the [4Fe-4S] type (Figure 2). These clusters are essentially involved in the formation of cPMP [27] and have to be preassembled in the mitochondria, where mitochondrial as well as extramitochondrial (Fe-S) clusters originate from [101].

Link 2: Subsequent to step1 of Moco biosynthesis which is located in the mitochondria, the intermediate cPMP is exported to the cytosol. As outlined in Section 4.1, the transporter protein Atm3 is associated with this step [29] (Figure 2). This transporter has obviously a dual function: it not only exports cPMP but also an as yet unidentified precursor for Fe-S cluster synthesis in the cytosol [102]. Therefore mutants in the Atm3 exporter become short of both cPMP and Fe-S clusters in the cytosol with the consequence that all enzymes needing these prosthetic groups decrease in activity.

Link 3: Both Mo enzymes XOR and AO depend on Fe in the form of [2Fe-2S] clusters, whereas NR requires Fe in the form of heme (Figue 2).

Summarizing the Mo-Fe link, it becomes evident that Moco biosynthesis and the functioning of the majority of Mo enzymes in higher organisms is strictly dependent on Fe metabolism for providing Fe-S clusters and heme groups. Any impairment of Fe metabolism would immediately affect Mo metabolism as well.

6 Molybdenum and Tungsten in Prokaryotes

6.1 *Molybdenum* versus *Tungsten*

Mo and W feature similar chemical properties (the biochemistry of tungsten has been recently reviewed in [103,104]). While eukaryotes use only Mo, both metals are of biological relevance in prokaryotes. As compared to MoO_4^{2-} which is very abundant, the concentration of WO_4^{2-} is 100-fold lower in the oceans. However, under anaerobic conditions that prevail in deep-sea hydrothermal vents and reflect conditions of primordial Earth, Mo occurs as insoluble MoS_2, while W either remains as WO_4^{2-} or forms soluble WS_4^{2-}. Therefore, it is not surprising to find W-depending hyperthermophilic bacteria (archaea) under these conditions [103]. W occurs in both alkaline and acid solutions and was thus available to emerging life whereas Mo is relatively insoluble in reduced and neutral waters thus suggesting a low bioavailabilty for early life [105].

With the emerging oxygenic atmosphere 2.5 billion years ago, Mo in the form of MoO_4^{2-} became readily bioavailable and replaced W as catalytic metal in enzymes. This does, however, not mean that Mo was not required for early life on primordial Earth. Recent evidence places both Mo and W to be present in metal enzymes in the last universal common ancestor of prokaryotes [105]. These enzymes form the "complex-iron-sulfur-molybdoenzyme" superfamily. Some members of this superfamily use only W (e.g., aldehyde oxidoreductase), other members insert W only under Mo-depletion conditions (e.g., dimethyl sulfoxide reductase), and yet other members turned out to be absolutely specific for Mo (e.g., nitrate reductase) [105]. Specificities for either Mo or W are seen even within the same enzyme family, namely the formate dehydrogenase family which comprises members using only Mo and other members using only W. These specificities are really surprising because Mo and W have very similar atomic radii (1.75 Å and 1.78, respectively), the same electronegativety, and the same free solvation energy. Given these very similar properties the cells had to develop systems highly discriminating between Mo and W during uptake and metal insertion into the cofactor.

6.2 *Uptake of Molybdate and Tungstate*

While the elucidation of eukaryotic Mo uptake is still in its beginning, the prokaryotic systems are already well characterized. Mo and W in form of their oxyanions molybdate and tungstate are taken up by high affinity transporters of the ABC-type [106]. The prototypical ABC transporter consist of a periplasmic anion binding protein, a transmembrane protein for translocation and a cytoplasmic protein associated to the complex that hydrolyzes ATP to drive anion transport. In *E. coli*, the molybdate transport is encoded by the *mod* operon which in addition to ModABC also codes for ModD (unknown function) and ModE which senses the concentration of molybdate. ModE controls both the expression of the molybate uptake system and of the *moa* operon encoding Moco biosynthesis. Under conditions of high molybdate, expression of the *mod* operon is repressed while the *moa* operon is induced in order to synthesize sufficient MPT for binding of molybdate [107]. Molybdate and tungstate can be also imported into the bacterial cell by co-transport via the sulfate and phosphate transporters, however, only if they are available in higher concentrations [108].

Tungstate can be taken up by several transport systems [106]: (i) the TupABC transporter which is highly specific for W and shows only very weak affinities for Mo. This transporter prevails in bacteria. (ii) The WtpABC transporter found in *Pyrococcus furiosus* shows ultra-high affinity for tungstate and is present in archaea only.

Subsequent to uptake, molybdate is mostly used directly for Moco biosynthesis, but in some bacteria it can be stored by specific binding proteins. In *Clostridium pasteurianum*, a family of small (7 kD) proteins was described that bind molybdate thus probably ensuring molybdate homeostasis during nitrogenase biosynthesis [109].

6.3 *Metal Enzymes and Pterin-Based Cofactors for Molybdenum and Tungsten*

Prokaryotes possess not only the Mo-MPT cofactor as it occurs in eukaryotes but they also synthesize variants thereof (Figure 3): (i) Moco can bind a nucleotide (GMP or CMP) to its phosphate, thus forming a dinucleotide cofactor, and (ii) Mo can be coordinated by two pterin or two dincleotide equivalents thus

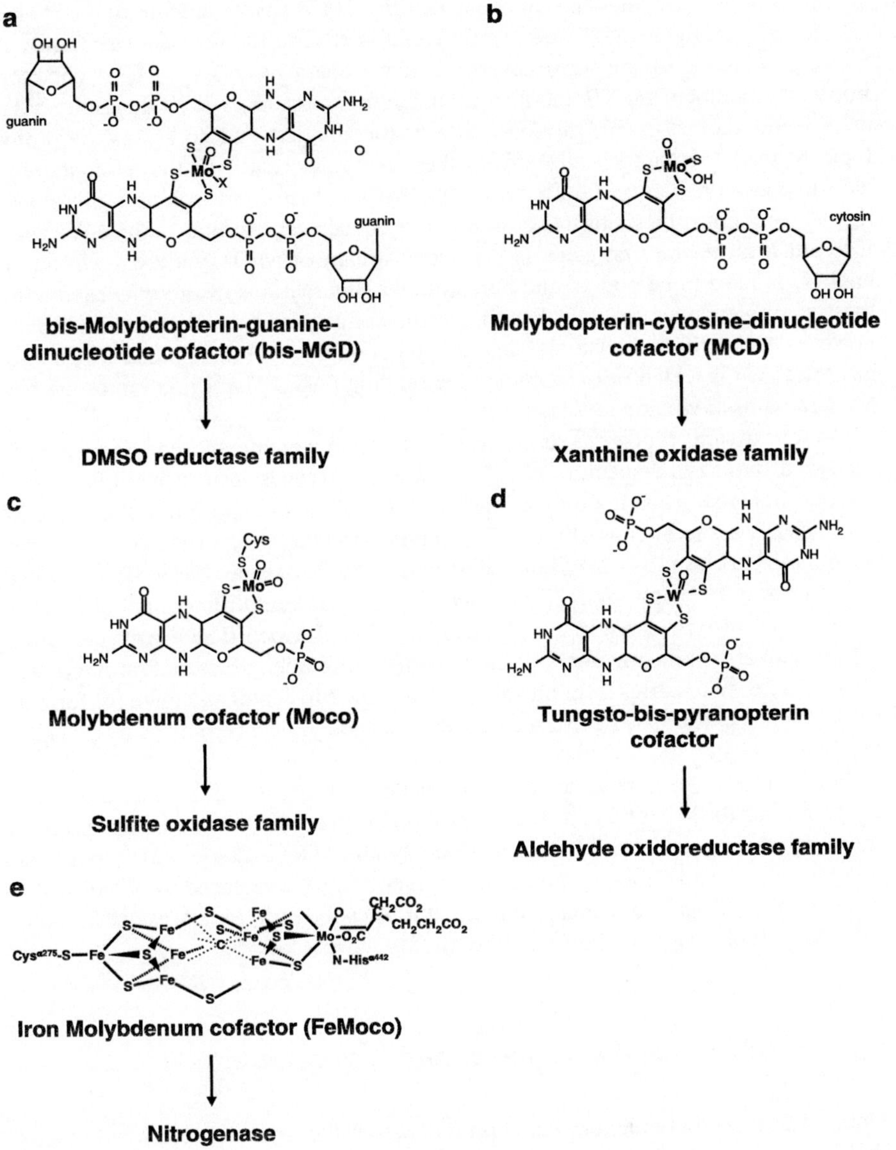

Figure 3 Types of bacterial Moco and their corresponding enzyme families. (**a** – **c**) Forms of Moco occurring in eubacteria. (**d**) Tungsten-containing cofactor as found in the archaeal aldehyde oxidoreductase. (**e**) FeMoco of nitrogenase.

forming bis-MPT and bis-MGD, respectively. Structural analyses of W enzymes showed that in these proteins the same pterin-type cofactor (Wco) as it occurs in Mo enzymes was found to bind W. Both MPT and bis-MPT, respectivly, were identified to coordinate W.

Bacterial Mo enzymes fall into four families [110] based on their cofactor type and domain arrangement (Figure 3): (i) The dimethyl sulfoxide reductase family is the largest and also the most diverse. All members have a bis-MGD cofactor. Important members are NR and formate dehydrogenase. (ii) The XOR-family has a dinucleotide cofactor, (iii) the SO-family is the smallest and possesses the prototypic Mo-MPT cofactor known from the eukaryotes, and (iv) the W-containing aldeyhde oxidoreductase-family with a bis-MPT cofactor but lacking a protein W ligand. The biosynthesis of bacterial Moco proceeds very similar to the eukaryotic form, at least in the first steps [111]. Downstream of MPT synthesis, Mo and a nucleotide have to be added, and here complex interactions of proteins catalyzing these steps were uncovered. In *E. coli*, the nucleotide is added by the protein MobA, however in the absence of molybdate this reaction does not take place which suggests that Mo is added prior to nucleotide binding [112]. The formation of the bis-MPT remains enigmatic to this end.

As a next step, Moco has to be inserted into the cognate apo-enzyme which is catalyzed by enzyme-specific multifunctional chaperones in bacteria. For *E. coli* NR the chaperone NarJ not only binds Moco, transfers it to apo-NR, but also facilitates subunit assembly, insertion of FeS centers, and targeting of NR to the plasma membrane [110]. In the case of the XOR family, the Mo center has to be sulfurated in a final step of maturation which is catalyzed by the multifunctional chaperone XdhC that is also involved in transfer and insertion of Moco. There are recent indications that also some members of the dimethyl sulfoxide reductase family possess a bis-cofactor that carries an extra sulfur atom at the Mo center as shown for formate dehydrogenase where this extra sulfur is introduced by the enzyme-specific chaperone FdhD [113].

For Wco synthesis, genome comparisons suggest that the biosynthesis machinery is similar to that for Moco biosynthesis. However, the step of metal insertion shows interesting deviations [114]. In Moco-synthesizing bacteria, this step is catalyzed by MogA and MoeA, in archaea, however, MogA is replaced by MoaB which is identical to MogA on primary and structural level, while MoeA exists in duplicated form (pointing to metal selectivity?).

6.4 Iron Molybdenum Cofactor and Nitrogenase

There is another molybdenum-containing cofactor, the iron-molybdenum cofactor (FeMoco), which is found only once in Nature, namely in the bacterial enzyme nitrogenase [115] that reduces atmospheric dinitrogen to ammonia with concomitant hydrolysis of ATP. Nitrogenase is required for biological nitrogen fixation, which is an essential step in the nitrogen cycle in the biosphere, and a major contributor to the nitrogen available to many plants species like legumes. FeMoco is very different from Moco. It consists of two partial cubanes $MoFe_3S_3$ and Fe_4S_3 that are joined by three bridging sulfurs and are bound to homocitrate, and there is also a bridging central atom in nitrogenase that has been recently identified as carbon [116] (Figure 3).

Other nitrogenases contain either vanadium or iron at their respective active sites. Biosynthesis of FeMoco in *Azotobacter vinelandii* is a highly complicated process involving at least ten genes and has been uncovered in recent years (summarized in Hu and Ribbe [11]). Like found for the assembly of FeS clusters, FeMoco biosynthesis follows the "platform principle". FeMoco is first assembled from small building blocks on a protein scaffold (NifEN), and after completion it is delivered from its assembly site to its target binding site in nitrogenase.

7 General Conclusions

It is evident that over the past two decades our progress in understanding Mo metabolism has been striking. At the same time, one can expect much progress in the near future. Now that the basic principles of Moco biosynthesis are understood, future research will focus on the mechanistic details of the single reactions. For example the requirement of copper in Moco biosynthesis is important to understand. There will also be advances in our understanding of the steps beyond Moco biosynthesis in higher organisms where research will focus on Moco transport, allocation, and insertion into apo-enzymes. Here one has to keep the cellular context in mind as Moco biosynthesis occurs to be micro-compartmentalized in a multiprotein biosynthesis complex localized in the cytosol of the cell.

We can also expect that the systems of molybdate uptake in eukaryotes will be increasingly well understood. The question of whether Moco biosynthesis is regulated to meet the changing demands of the cell for Moco will be answered as well. In addition, the final maturation step which activates XDH and AO needs to be investigated. Finally the role of other homeostatic circuits, such as Fe and Cu metabolism, awaits further investigation, which could address questions regarding the pathophysiology of related metabolic disorders.

Abbreviations and Definitions

ABA3	Moco sulfurase in plants
ABC	ATP-binding cassette
AMP	adenosine 5'-monophosphate
AO	aldehyde oxidase
ATP	adenosine 5'-triphosphate
CIA	cytosolic iron-sulfur cluster assembly
CMP	cytosine 5'-monophosphate
cPMP	cyclic pyranopterin monophosphate
DMSO	dimethylsulfoxide
ER	endoplasmic reticulum
FAD	flavin adenine dinucleotide
FeMoco	iron-molybdenum cofactor

GTP	guanosine 5'-triphosphate
HMCS	Moco sulfurase in humans
ISC	iron-sulfur cluster
mARC	mitochondral amidoxime reducing component
MCD	molybdopterin-cytosine dinucleotide cofactor
MCP	Moco carrier protein
MGD	molybdopterin-guanine dinucleotide
MoBP	molybdenum cofactor binding protein
Moco	molybdenum cofactor
MOCS	molybdenum cofactor synthesis
MPT	molybdopterin
NAD(P)H	nicotinamide adenine dinucleotide (phosphate) reduced
NR	nitrate reductase
RLD	rhodanese-like domain
ROS	reactive oxygen species
SAM	S-adenosyl methionine
SO	sulfite oxidase
Wco	tungsten pterin-type cofactor
XDH	xanthine dehydrogenase
XO	xanthine oxidase
XOR	xanthine oxidoreductase

Acknowledgments R.R.M. thanks the many people who worked with him over the years on molybdenum. In particular I am grateful to Florian Bittner, Robert Hänsch, and Tobias Kruse for many critical discussions and to Tobias Kruse for help with the figures. The research was consistenly supported by the Deutsche Forschungsgemeinschaft which is gratefully acknowledged.

References

1. E. I. Stiefel, *Met. Ions Biol. Syst.* **2002**, *39*, 1–29.
2. R. Hille, T. Nishino, F. Bittner, *Coord. Chem. Rev.* **2011**, *255*, 1179–1205.
3. Y. Zhang, V. N. Gladyshev, *J. Mol. Biol.* **2008**, *379*, 881–899.
4. J. R. Turnlund, *Met. Ions Biol. Syst.* **2002**, *39*, 727–739.
5. H. Tomatsu, J. Takano, H. Takahashi, A. Watanabe-Takahashi, N. Shibagaki, T. Fujiwara, *Proc. Natl. Acad. Sci. USA* **2007**, *104*, 18807–18812.
6. M. Tejada-Jimenez, A. Llamas, E. Sanz-Luque, A. Galvan, E. Fernandez, *Proc. Natl. Acad. Sci. USA* **2007**, *104*, 20126–20130.
7. I. Baxter, B. Muthukumar, H. C. Park, P. Buchner, B. Lahner, J. Danku, K. Zhao, J. Lee, M. J. Hawkesford, M. L. Guerinot, D. E. Salt, *PLoS Genet.* **2008**, *4*, e1000004.
8. M. Tejada-Jimenez, A. Galvan, E. Fernandez, *Proc. Natl. Acad. Sci. USA* **2011**, *108*, 6420–6425.
9. K. L. Fitzpatrick, S. D. Tyerman, B. N. Kaiser, *FEBS Lett.* **2008**, *582*, 1508–1513.
10. G. Schwarz, R. R. Mendel, M. W. Ribbe, *Nature* **2009**, *460*, 839–847.
11. Y. Hu, M. W. Ribbe, *Coord. Chem. Rev.* **2011**, *255*, 1218–1224.
12. J. L. Johnson, B. E. Hainline, K. V. Rajagopalan, B. H. Arison, *J. Biol. Chem.* **1984**, *259*, 5414–5422.

13. J. L. Johnson, K. V. Rajagopalan, *Proc. Natl. Acad. Sci. USA* **1982**, *79*, 6856–6860.
14. M. J. Romao, M. Archer, I. Moura, J. J. Moura, J. LeGall, R. Engh, M. Schneider, P. Hof, R. Huber, *Science* **1995**, *270*, 1170–1176.
15. C. Kisker, H. Schindelin, A. Pacheco, W. A. Wehbi, R. M. Garrett, K. V. Rajagopalan, J. H. Enemark, D. C. Rees, *Cell* **1997**, *91*, 973–983.
16. R. R. Mendel, in *Plant Biotechnology and Development - Current Topics in Plant Molecular Biology*, Ed P. M. Gresshoff, CRC Press, Boca Raton and London, 1992, Vol. 1, pp. 11–16.
17. G. Schwarz, R. R. Mendel, *Annu. Rev. Plant Biol.* **2006**, *57*, 623–647.
18. J. Reiss, J. L. Johnson, *Hum. Mutat.* **2003**, *21*, 569–576.
19. R. R. Mendel, G. Schwarz, *Met. Ions Biol. Syst.* **2002**, *39*, 317–368.
20. L. J. Millar, I. S. Heck, J. Sloan, G. J. Kana'n, J. R. Kinghorn, S. E. Unkles, *Mol. Genet. Genomics* **2001**, *266*, 445–453.
21. J. Reiss, N. Cohen, C. Dorche, H. Mandel, R. R. Mendel, B. Stallmeyer, M. T. Zabot, T. Dierks, *Nat. Genet.* **1998**, *20*, 51–53.
22. B. Stallmeyer, G. Drugeon, J. Reiss, A. L. Haenni, R. R. Mendel, *Am. J. Hum. Genet.* **1999**, *64*, 698–705.
23. B. Stallmeyer, G. Schwarz, J. Schulze, A. Nerlich, J. Reiss, J. Kirsch, R. R. Mendel, *Proc. Natl. Acad. Sci. USA* **1999**, *96*, 1333–1338.
24. M. M. Wuebbens, K. V. Rajagopalan, *J. Biol. Chem.* **1993**, *268*, 13493–13498.
25. J. A. Santamaria-Araujo, B. Fischer, T. Otte, M. Nimtz, R. R. Mendel, V. Wray, G. Schwarz, *J. Biol. Chem.* **2004**, *279*, 15994–15999.
26. J. Kuper, A. Llamas, H. J. Hecht, R. R. Mendel, G. Schwarz, *Nature* **2004**, *430*, 806–806.
27. P. Hanzelmann, H. L. Hernandez, C. Menzel, R. Garcia-Serres, B. H. Huynh, M. K. Johnson, R. R. Mendel, H. Schindelin, *J. Biol. Chem.* **2004**, *279*, 34721–34732.
28. H. J. Sofia, G. Chen, B. G. Hetzler, J. F. Reyes-Spindola, N. E. Miller, *Nucleic Acids Res.* **2001**, *29*, 1097–1106.
29. J. Teschner, N. Lachmann, J. Schulze, M. Geisler, K. Selbach, J. Santamaria-Araujo, J. Balk, R. R. Mendel, F. Bittner, *Plant Cell* **2010**, *22*, 468–480.
30. J. Balk, S. Lobreaux, *Trends Plant Sci.* **2005**, *10*, 324–331.
31. C. Gehl, R. Waadt, J. Kudla, R. R. Mendel, R. Hansch, *Mol. Plant* **2009**, *2*, 1051–1058.
32. A. Matthies, K. V. Rajagopalan, R. R. Mendel, S. Leimkuhler, *Proc. Natl. Acad. Sci. USA* **2004**, *101*, 5946–5951.
33. B. Lardi-Studler, B. Smolinsky, C. M. Petitjean, F. Koenig, C. Sidler, J. C. Meier, J. M. Fritschy, G. Schwarz, *J. Cell Sci.* **2007**, *120*, 1371–1382.
34. M. M. Wuebbens, K. V. Rajagopalan, *J. Biol. Chem.* **2003**, *278*, 14523–14532.
35. S. Leimkuhler, M. M. Wuebbens, K. V. Rajagopalan, *Coord. Chem. Rev.* *255*, 1129–1144.
36. A. Matthies, M. Nimtz, S. Leimkuhler, *Biochemistry* **2005**, *44*, 7912–7920.
37. Z. Marelja, W. Stocklein, M. Nimtz, S. Leimkuhler, *J. Biol. Chem.* **2008**, *283*, 25178–25185.
38. G. Schwarz, J. Schulze, F. Bittner, T. Eilers, J. Kuper, G. Bollmann, A. Nerlich, H. Brinkmann, R. R. Mendel, *Plant Cell* **2000**, *12*, 2455–2472.
39. G. Schwarz, D. H. Boxer, R. R. Mendel, *J. Biol. Chem.* **1997**, *272*, 26811–26814.
40. A. Llamas, R. R. Mendel, G. Schwarz, *J. Biol. Chem.* **2004**, *279*, 55241–55246.
41. A. Llamas, T. Otte, A. Simons, G. Multhaupt, R. R. Mendel, G. Schwarz, *J. Biol. Chem.* **2006**, *281*, 18343–18340.
42. R. R. Mendel, Z. A. Alikulov, N. P. Lvov, A. J. Müller, *Mol. Gen. Genet.* **1981**, *181*, 395–399.
43. F. Falciani, M. Terao, S. Goldwurm, A. Ronchi, A. Gatti, C. Minoia, M. Li Calzi, M. Salmona, G. Cazzaniga, E. Garattini, *Biochem. J.* **1994**, *298*, 69–77.
44. I. S. Heck, J. D. Schrag, J. Sloan, L. J. Millar, G. Kanan, J. R. Kinghorn, S. E. Unkles, *Genetics* **2002**, *161*, 623–632.
45. G. B. Moorhead, S. E. Meek, P. Douglas, D. Bridges, C. S. Smith, N. Morrice, C. MacKintosh, *Eur. J. Biochem.* **2003**, *270*, 1356–1362.

46. J. F. Mercer, *Trends Mol. Med.* **2001**, *7*, 64–69.
47. K. V. Rajagopalan, J. L. Johnson, *J. Biol. Chem.* **1992**, *267*, 10199–10202.
48. M. Aguilar, K. Kalakoutskii, J. Cardenas, E. Fernandez, *FEBS Lett.* **1992**, *307*, 162–163.
49. C. P. Witte, M. I. Igeno, R. Mendel, G. Schwarz, E. Fernandez, *FEBS Lett.* **1998**, *431*, 205–209.
50. K. Fischer, A. Llamas, M. Tejada-Jimenez, N. Schrader, J. Kuper, F. S. Ataya, A. Galvan, R. R. Mendel, E. Fernandez, G. Schwarz, *J. Biol. Chem.* **2006**, *281*, 30186–30194.
51. T. Kruse, C. Gehl, M. Geisler, M. Lehrke, P. Ringel, S. Hallier, R. Hansch, R. R. Mendel, *J. Biol. Chem.* **2010**, *285*, 6623–6635.
52. F. Blasco, J. P. Dos Santos, A. Magalon, C. Frixon, B. Guigliarelli, C. L. Santini, G. Giordano, *Mol. Microbiol.* **1998**, *28*, 435–447.
53. S. Leimkühler, W. Klipp, *J. Bacteriol.* **1999**, *181*, 2745–2751.
54. N. V. Ivanov, F. Hubalek, M. Trani, D. E. Edmondson, *Eur. J. Biochem.* **2003**, *270*, 4744–4754.
55. C. Kisker, H. Schindelin, D. C. Rees, *Annu. Rev. Biochem.* **1997**, *66*, 233–267.
56. C. K. Warner, V. Finnerty, *Mol. Gen. Genet.* **1981**, *184*.
57. N. J. Lewis, P. Hurt, H. M. Sealy-Lewis, C. Scazzocchio, *Eur. J. Biochem.* **1978**, *91*, 311–316.
58. H. Yamasaki, Y. Sakihama, S. Takahashi, *Trends Plant Sci.* **1999**, *4*, 128–129.
59. T. Eilers, G. Schwarz, H. Brinkmann, C. Witt, T. Richter, J. Nieder, B. Koch, R. Hille, R. Hansch, R. R. Mendel, *J. Biol. Chem.* **2001**, *276*, 46989–46994.
60. N. Schrader, E. Y. Kim, J. Winking, J. Paulukat, H. Schindelin, G. Schwarz, *J. Biol. Chem.* **2004**, *279*, 18733–18741.
61. K. Nowak, N. Luniak, C. Witt, Y. Wustefeld, A. Wachter, R. R. Mendel, R. Hansch, *Plant Cell Physiol.* **2004**, *45*, 1889–1894.
62. R. Hansch, C. Lang, E. Riebeseel, R. Lindigkeit, A. Gessler, H. Rennenberg, R. R. Mendel, *J. Biol. Chem.* **2006**, *281*, 6884–6888.
63. R. S. Byrne, R. Hansch, R. R. Mendel, R. Hille, *J. Biol. Chem.* **2009**, *284*, 35479–35484.
64. D. Randewig, D. Hamisch, C. Herschbach, M. Eiblmeier, C. Gehl, J. Jurgeleit, J. Skerra, R. R. Mendel, H. Rennenberg, R. Hansch, *Plant Cell Environ.* **2012**, *35*, 100–115.
65. C. Lang, J. Popko, M. Wirtz, R. Hell, C. Herschbach, J. Kreuzwieser, H. Rennenberg, R. R. Mendel, R. Hansch, *Plant Cell Environ.* **2007**, *30*, 447–455.
66. A. Havemeyer, F. Bittner, S. Wollers, R. Mendel, T. Kunze, B. Clement, *J. Biol. Chem.* **2006**, *281*, 34796–34802.
67. S. Gruenewald, B. Wahl, F. Bittner, H. Hungeling, S. Kanzow, J. Kotthaus, U. Schwering, R. R. Mendel, B. Clement, *J. Med. Chem.* **2008**, *51*, 8173–8177.
68. A. Chamizo-Ampudia, A. Galvan, E. Fernandez, A. Llamas, *Eukaryot. Cell*, **2011**, *10*, 1270–1282.
69. J. Kotthaus, B. Wahl, A. Havemeyer, D. Schade, D. Garbe-Schonberg, R. Mendel, F. Bittner, B. Clement, *Biochem. J.* **2011**, *433*, 383–391.
70. A. Havemeyer, J. Lang, B. Clement, *Drug Metab. Rev.* **2011**, *43*, 524–539.
71. R. Hille, *Met. Ions Biol. Syst.* **2002**, *39*, 187–226.
72. Z. Yesbergenova, G. Yang, E. Oron, D. Soffer, R. Fluhr, M. Sagi, *Plant J.* **2005**, *42*, 862–876.
73. R. Hille, T. Nishino, *FASEB J.* **1995**, *9*, 995–1003.
74. M. Zarepour, K. Kaspari, S. Stagge, R. Rethmeier, R. R. Mendel, F. Bittner, *Plant Mol. Biol.* **2010**, *72*, 301–310.
75. P. Montalbini, *J. Phytopath.* **1992**, *134*, 218 – 228.
76. A. Agarwal, A. Banerjee, U. C. Banerjee, *Crit. Rev. Biotechnol.* **2011**, *31*, 264–280.
77. C. Vorbach, A. Scriven, M. R. Capecchi, *Genes Dev.* **2002**, *16*, 3223–3235.
78. C. Vorbach, R. Harrison, M. R. Capecchi, *Trends Immunol.* **2003**, *24*, 512–517.
79. F. Rodriguez-Trelles, R. Tarrio, F. J. Ayala, *Proc. Natl. Acad. Sci. USA* **2003**, *100*, 13413–13417.

80. M. Seo, A. J. Peeters, H. Koiwai, T. Oritani, A. Marion-Poll, J. A. Zeevaart, M. Koornneef, Y. Kamiya, T. Koshiba, *Proc. Natl. Acad. Sci. USA* **2000**, *97*, 12908–12913.
81. P. E. Verslues, J. K. Zhu, *Biochem. Soc. Trans.* **2005**, *33*, 375–379.
82. J. Bordas, R. C. Bray, C. D. Garner, S. Gutteridge, S. S. Hasnain, *Biochem. J.* **1980**, *191*, 499–508.
83. R. C. Wahl, K. V. Rajagopalan, *J. Biol. Chem.* **1982**, *257*, 1354–1359.
84. V. Massey, D. E. Edmondson, *J. Biol. Chem.* **1970**, *245*, 6595–6598.
85. F. Bittner, M. Oreb, R. R. Mendel, *J. Biol. Chem.* **2001**, *276*, 40381–40384.
86. T. Heidenreich, S. Wollers, R. R. Mendel, F. Bittner, *J. Biol. Chem.* **2005**, *280*, 4213–4218.
87. S. Wollers, T. Heidenreich, M. Zarepour, D. Zachmann, C. Kraft, Y. Zhao, R. R. Mendel, F. Bittner, *J. Biol. Chem.* **2008**, *283*, 9642–9650.
88. M. Lehrke, S. Rump, T. Heidenreich, J. Wissing, R. R. Mendel, F. Bittner, *Biochem. J.*
89. L. Xiong, M. Ishitani, H. Lee, J. K. Zhu, *Plant Cell* **2001**, *13*, 2063–2083.
90. K. Mengel, E. A. Kirkby, *Principles of Plant Nutrition*, Kluwer Academic Publishers, Dordrecht, **2001**, pp 849.
91. A. Gasber, S. Klaumann, O. Trentmann, A. Trampczynska, S. Clemens, S. Schneider, N. Sauer, I. Feifer, F. Bittner, R. R. Mendel, H. E. Neuhaus, *Plant Biol.* **2011**, *13*, 710–718.
92. A. J. Müller, R. R. Mendel, in *Molecular and Genetic Aspects of Nitrate Assimilation*, Eds J. L. Wray, J. R. Kinghorn, Oxford University Press, Oxford, 1989, pp. 166–185.
93. J. Gabard, F. Pelsy, A. Marion-Poll, M. Caboche, I. Saalbach, R. Grafe, A. J. Müller, *Mol. Gen. Genet.* **1988**, *213*, 275–281.
94. M. Seo, H. Koiwai, S. Akaba, T. Komano, T. Oritani, Y. Kamiya, T. Koshiba, *Plant J.* **2000**, *23*, 481–488.
95. K. M. Leon-Klooserziel, M. A. Gil., G. J. Ruijs, S. E. Jacobsen, N. E. Olszewski, S. H. Schwartz, J. A. D. Zeevaart, M. Koornneef, *Plant J.* **1996**, *10*, 655–661.
96. M. Duran, F. A. Beemer, C. van de Heiden, J. Korteland, P. K. de Bree, M. Brink, S. K. Wadman, I. Lombeck, *J. Inherit. Metab. Dis.* **1978**, *1*, 175–178.
97. J. Reiss, *Hum. Genet.* **2000**, *106*, 157–163.
98. H.-J. Lee, I. M. Adham, G. Schwarz, M. Kneussel, J.-O. Sass, W. Engel, J. Reiss, *Hum. Mol. Gen.* **2002**, *11*, 3309–3317.
99. G. Schwarz, J. A. Santamaria-Araujo, S. Wolf, H. J. Lee, I. M. Adham, H. J. Grone, H. Schwegler, J. O. Sass, T. Otte, P. Hanzelmann, R. R. Mendel, W. Engel, J. Reiss, *Hum. Mol. Gen.* **2004**, *13*, 1249–1255.
100. A. Veldman, J. A. Santamaria-Araujo, S. Sollazzo, J. Pitt, R. Gianello, J. Yaplito-Lee, F. Wong, C. A. Ramsden, J. Reiss, I. Cook, J. Fairweather, G. Schwarz, *Pediatrics* **2010**, *125*, e1249–1254.
101. R. Lill, U. Muhlenhoff, *Annu. Rev. Biochem.* **2008**, *77*, 669–700.
102. D. G. Bernard, Y. Cheng, Y. Zhao, J. Balk, *Plant Physiol.* **2009**, *151*, 590–602.
103. L. E. Bevers, P. L. Hagedoorn, W. R. Hagen, *Coord. Chem. Rev.* **2009**, *253*, 269–290.
104. R. H. Holm, E. I. Solomon, A. Majumdar, A. Tenderholt, *Coord. Chem. Rev.* **2011**, *255*, 993–1015.
105. B. Schoepp-Cothenet, R. van Lis, P. Philippot, A. Magalon, M. J. Russell, W. Nitschke, *Sci. Rep.* **2012**, *2*, 263.
106. F. Hagen, *Coord. Chem. Rev.* **2011**, *255*, 1117–1128.
107. L. A. Anderson, E. McNairn, T. Lubke, R. N. Pau, D. H. Boxer, *J. Bacteriol.* **2000**, *182*, 7035–7043.
108. A. M. Grunden, K. T. Shanmugam, *Arch. Microbiol.* **1997**, *168*, 345–354.
109. A. W. Schuttelkopf, J. A. Harrison, D. H. Boxer, W. N. Hunter, *J. Biol. Chem.* **2002**, *277*, 15013–15020.
110. A. Magalon, J. G. Fedor, A. Walburger, J. H. Weiner, *Coord. Chem. Rev.* **2011**, *255*, 1159–1178.
111. S. Leimkühler, M. M. Wuebbens, K. V. Rajagopalan, *Coord. Chem. Re.v* **2011**, *255*, 1129–1144.

112. P. Hänzelmann, O. Meyer, *Eur. J. Biochem.* **1998**, *255*, 755–765.
113. R. Thome, A. Gust, R. Toci, R. Mendel, F. Bittner, A. Magalon, A. Walburger, *J. Biol. Chem.* **2012**, *287*, 4671–4678.
114. G. Schwarz, P. L. Hagedoorn, K. Fischer, *Molecular Microbiology of Heavy Metals*, in *Microbiol. Monogr. Series,* Volume 6, Eds D. Nies, S. Silver, Springer, Berlin, Heidelberg, 2007, 421–452.
115. L. C. Seefeldt, B. M. Hoffman, D. R. Dean, *Annu. Rev. Biochem.* **2009**, *78*, 701–722.
116. K. M. Lancaster, M. Roemelt, P. Ettenhuber, Y. Hu, M. W. Ribbe, F. Neese, U. Bergmann, S. DeBeer, *Science* **2011**, *334*, 974–977.

Chapter 16
Comparative Genomics Analysis of the Metallomes

Vadim N. Gladyshev and Yan Zhang

Contents

V.N. Gladyshev (✉)
Division of Genetics, Department of Medicine,
Brigham and Women's Hospital and Harvard Medical School,
New Research Building, 77 Av. Pasteur, Boston, MA 02115, USA
e-mail: vgladyshev@rics.bwh.harvard.edu

Y. Zhang
Key Laboratory of Nutrition and Metabolism, Institute for Nutritional Sciences,
Chinese Academy of Sciences, University of Chinese Academy of Sciences,
294 Tai Yuan Road, Shanghai 200031, China
e-mail: yanzhang01@sibs.ac.cn

L. Banci (ed.), *Metallomics and the Cell*, Metal Ions in Life Sciences 12,
DOI 10.1007/978-94-007-5561-1_16, © Springer Science+Business Media Dordrecht 2013

Abstract Biological trace metals are needed in small quantities, but used by all living organisms. They are employed in key cellular functions in a variety of biological processes, resulting in the various degree of dependence of organisms on metals. Most effort in the field has been placed on experimental studies of metal utilization pathways and metal-dependent proteins. On the other hand, systemic level analyses of metalloproteomes (or metallomes) have been limited for most metals. In this chapter, we focus on the recent advances in comparative genomics, which provides many insights into evolution and function of metal utilization. These studies suggested that iron and zinc are widely used in biology (presumably by all organisms), whereas some other metals such as copper, molybdenum, nickel, and cobalt, show scattered occurrence in various groups of organisms. For these metals, most user proteins are well characterized and their dependence on a specific element is evolutionarily conserved. We also discuss evolutionary dynamics of the dependence of user proteins on different metals. Overall, comparative genomics analysis of metallomes provides a foundation for the systemic level understanding of metal utilization as well as for investigating the general features, functions, and evolutionary dynamics of metal use in the three domains of life.

Keywords bioinformatics • comparative genomics • metal • metallome • metalloproteome

Please cite as: *Met. Ions Life Sci.* 12 (2013) 529–580

1 Introduction

Metallomes refer to the complete set of metalloproteins, metalloenzymes and other metal-containing biomolecules that organisms utilize [1]. The study of metallomes, often referred to as metallomics, is a new scientific field that includes high-throughput studies on metals and integrates the research on these elements to obtain systems level understanding of their use in biology. Some biometals, such as sodium, potassium, and calcium, are needed in large amounts, but the majority of these elements belong to the group of trace elements (also called micronutrients). Although these metals are required in small quantities, they still may be essential for optimal growth, development and metabolic functions of living organisms [2,3]. These trace metals include zinc, iron, copper, manganese, molybdenum, tungsten, nickel, cobalt, chromium, vanadium, and possibly several other metals. They function in widely different ways. Some are essential components of enzymes where they directly interact with substrates and often facilitate their conversion to products; some donate or accept electrons in reactions of reduction and oxidation; some structurally stabilize biological molecules; and some control biological processes by facilitating the binding of molecules to receptor sites on cell membranes [4,5]. Their deficiency or mutations in genes that handle these metals often result in abnormal development, metabolic abnormalities, or even death.

Among trace metals, Zn and Fe appear to be used by all or almost all organisms [6–8]. The utilization of other trace metals, including Cu, Mn, Mo, Ni, and Co, is more scattered. Since all these metals play important roles in cells, the ability of the cell to tightly control their homeostasis is very important; the key processes relate to uptake, storage, excretion, and utilization of metals [9]. High-affinity import systems have been characterized for most biometals in both prokaryotes and eukaryotes. In bacteria, the ATP-binding cassette (ABC) transporters are the most frequently used uptake systems, e.g., ZnuABC for Zn, MntABC for Mn, ModABC for Mo, and NikABCDE for Ni [10–13]. Non-ABC transporters were also reported, e.g., ZupT and ZIP for Zn, MntH for Mn and Fe, CtaA and Ctr1 for Cu, and NiCoT for Ni and Co [14–19]. In addition, some metal ions could be transported via unspecific cation channels, although the efficiency of such processes may be low [20,21]. Excessive uptake of certain metals through either specific or unspecific pathways may result in metal overload and toxicity. Thus, storage of metals in inactive sites or forms and excretion/export systems represent essential mechanisms that prevent accumulation of inappropriate amounts of reactive trace metals in the cell (e.g., metallothioneins for heavy metal binding/detoxification, CopA/ATP7A for Cu export and ZnT for Zn export [22–25]). Besides detoxification, release of a metal ion from a storage site may be important under conditions of metal deficiency. Moreover, the use of some metals may be dependent on other metals. For example, excessive Zn can induce signs of Cu deficiency [26]. It is clear that homeostasis of metals within the cell should be carefully maintained by mechanisms regulating their uptake, storage, and removal in order to provide sufficient levels while preventing accumulation to toxic levels.

Most metals are directly incorporated into their cognate sites in proteins, but some have to become part of prosthetic groups, cofactors, or complexes prior to insertion of these moieties into target proteins. For example, Mo and Co are the main functionalities in molybdopterin (or Mo cofactor, Moco) and cobalamin (vitamin B_{12}), respectively [27,28]. Another interesting feature is that the number of metalloprotein families varies greatly depending on which metal is used. For instance, over 300 protein families require Zn for proper function [29], whereas less than 10 protein families are known to be dependent on Ni [30].

In the past decade, dramatic advances in genomics have provided an opportunity to investigate the occurrence and evolutionary dynamics of pathways that an organism utilizes, including metal utilization. Computational and comparative analyses of protein sequences and structures on a genomic scale revealed a significant number of proteins that may bind metals. Thus, identification of all or almost all metalloproteins in genomic databases can greatly assist in our understanding of utilization and function of metals in biology. However, due to the lack of reliable approaches, it is currently not possible to identify complete sets of metalloproteins in organisms. In recent years, several comparative and functional genomic analyses have been carried out for certain trace metals, including Zn, Ni, Co, Cu, and Mo [31–39]. These studies improved our understanding of current use and evolutionary trends in the utilization of these metals in organisms in the three domains of life. In this chapter, we focus on the use of several trace metals from the perspective of comparative genomics. Studies on their utilization may provide important information with regard to fundamental issues of function of these metals.

2 General Approaches to Comparative Genomics of Metal Utilization

Comparative genomics is an exciting new field of biological research, which scrutinizes genome sequences and structures of multiple organisms to identify similarities and differences [40–42]. This information provides a powerful tool for studying evolutionary changes, helping to identify genes that are conserved among species, as well as genes that give each organism its unique characteristics. It also helps scientists to better understand the pathways and other biological processes, including trace metals, in currently living organisms. Using methods of comparative genomic, it is now possible to compile metal-dependent pathways and proteins that an organism uses.

Unfortunately, a precise approach has not been developed for the identification of metalloproteins, partially because of overlapping features for different metals or the uncertainty of metal-binding residues. However, studies on sequence and structural properties of known metalloproteins and their metal-binding ligands resulted in the development of a large number of metal-binding motifs, which can help identify additional metal-binding proteins. Furthermore, searches for metal utilization traits can be assisted with the analyses of factors involved in metal transport or biosynthesis of metal-containing cofactors. The procedure of comparative genomics of biometals may briefly include three steps (Figure 1).

2.1 *Identification of Metal Utilization Traits and Metalloprotein Families*

The metal utilization trait refers to the occurrence of at least one protein that utilizes this metal. Thus, the first step, which also offers the most important evidence, should be to identify all known metalloproteins for corresponding metals. Based on sequence and structural signatures of known metalloprotein families, several

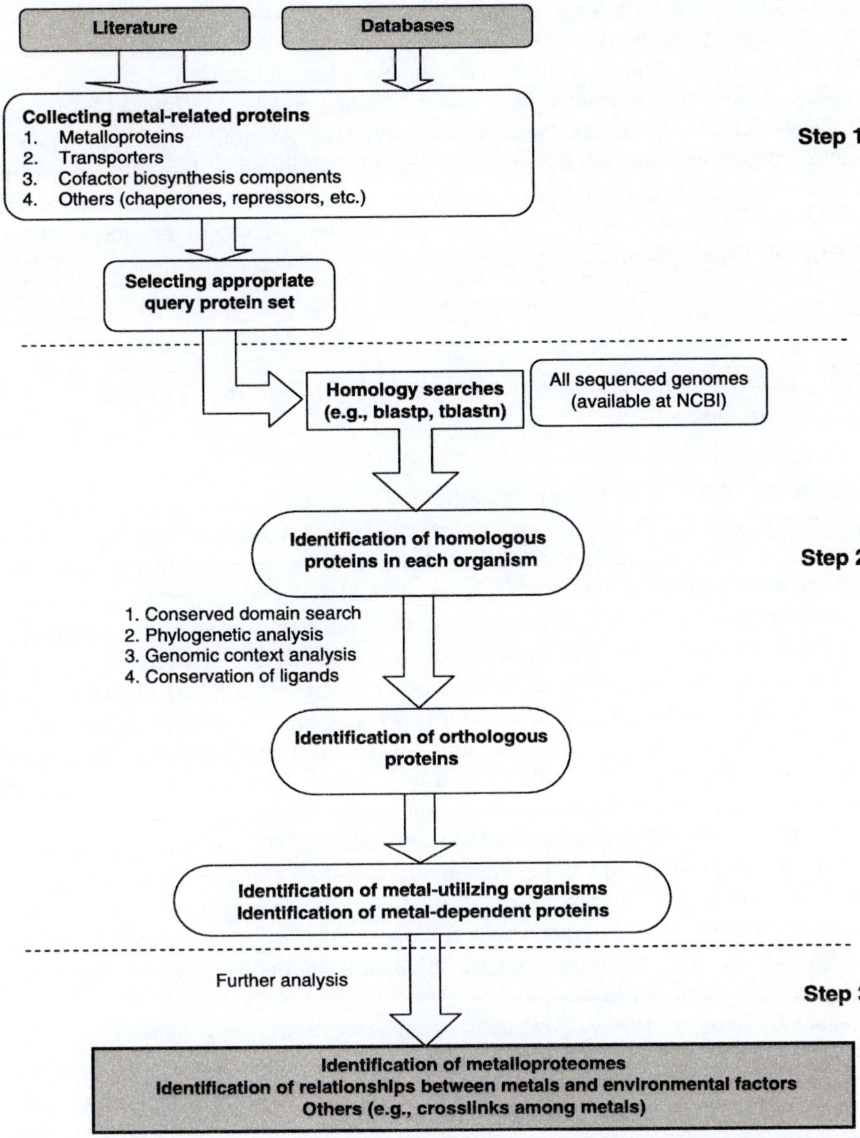

Figure 1 Schematic diagram for comparative genomics analyses of metal utilization in biology. This process can be divided into three major steps. Details are discussed in the text.

databases and bioinformatics tools have been developed to browse and/or predict metalloproteins, e.g., Pfam, PROSITE, PRINTS, ProDom, COG, MDB, and dbTEU [43–48]. Some of these tools contain metal-binding sequence motifs or patterns, whereas others use position-specific scoring matrices or profiles to describe similarity among different proteins. These resources do not include all metal-binding motifs and could only help identify a partial set of metalloproteins. Moreover, some metalloproteins may use different metals with the same ligands based on metal availability, protein folding location, and other factors [49].

Except for metalloproteins, occurrence of genes involved in high-affinity metal transport, metal-containing cofactor biosynthesis, and other processes (such as metal chaperones and repressors) may provide additional information regarding metal utilization and should be analyzed in parallel. Thus, a metal utilization trait could also be verified by the presence of high-affinity transporters and/or cofactor biosynthesis pathways.

2.2 *Identification of Metal-Related Orthologs in Genomic Databases*

The second step may be to identify orthologs of selected proteins in the sequenced genomes of all organisms. The list of currently sequenced organisms from the three domains of life is available at NCBI's website (http://www.ncbi.nlm.nih.gov/sutils/genom_table.cgi). As of January 2012, approximately 2,500 species were included.

In order to identify orthologs of query proteins, a set of sequences obtained in the above step can be used as initial seeds to search for homologous sequences in various organisms via a suite of BLAST programs (such as BLASTP, TBLASTN and PSI-BLAST) [50,51]. Orthologous proteins could be further defined using other approaches, such as conserved domain (COG/Pfam) searches, bidirectional best hits, genomic context analysis, and phylogenetic analysis. Conservation of metal-binding residues in the orthologs should also be analyzed to assess the ability to bind metal. Occurrence of Moco and vitamin B_{12} biosynthesis could be verified by the presence of components involved in their pathways (see Sections 3.1 and 5.3). Finally, the presence of the utilization trait of a metal M in an organism could be verified by the requirement for occurrence of at least one M-specific transporter, or M-containing cofactor biosynthesis pathway, or at least one M-dependent metalloprotein.

It should be noted that only proteins strictly specific for a particular metal must be selected by this approach, which may result in incomplete analysis of metal utilization in some organisms. However, regarding the metals discussed in this chapter, most of the corresponding metalloproteins are strictly dependent on their primary metal. Therefore, comparative genomics approaches may indeed reveal a general picture of utilization of these metals in organisms.

2.3 *Comparative Analyses of Metal Utilization and Interaction*

Comparative analyses of metal utilization, function and evolution are among the most important goals of the metal biology field, which may greatly improve our understanding of mechanisms and evolutionary dynamics of metals used in various organisms, clades, or kingdoms. Based on the results derived from previous steps, additional questions could be addressed, such as the relationship between metal utilization and environmental factors, composition and function of metalloproteomes, and interactions or other features of metal utilization. In the following sections, we will focus on several metals and discuss recent contributions on comparative genomics of their utilization.

3 Molybdenum

Molybdenum plays a critical role in several pathways and functions as a catalytic component of Mo-dependent enzymes (molybdoenzymes) that are essential for nearly all living organisms, including animals, plants, fungi, and bacteria. These molybdoenzymes catalyze oxo-transfer reactions in the metabolism of carbon, nitrogen, and sulfur compounds [27,52]. With the exception of the Fe-Mo cofactor in nitrogenase [53], all molybdoenzymes use this metal in the form of Moco, which consists of Mo coordinated to an organic tricyclic pyranopterin moiety, referred to as molybdopterin [27,54]. Some thermophilic archaea utilize W that is also coordinated by pyranopterin (Wco) [52,55]. In addition, W can be selectively transported into prokaryotic cells by certain transporters [56] and is an essential element for enzymes within the aldehyde:ferredoxin oxidoreductase family [52,57]. Due to the physical and chemical similarities between Mo and W, it is often impossible to distinguish the utilization of these two elements based on sequence analysis. In this chapter, the term Moco refers to the cofactor form of both metals (unless there is specific mention of the metal involved).

3.1 Molybdenum Transport and Molybdenum Cofactor Biosynthesis

Identification of Mo (or W) transporters and the Moco biosynthesis pathway are essential for characterization of the Mo utilization trait. In bacteria, the first identified Mo transporter was the ModABC transport system, which consists of ModA (molybdate-binding protein), ModB (membrane integral channel protein), and ModC (cytoplasmic ATPase) [12,58]. In *Escherichia coli*, the *modABC* operon is regulated by ModE repressor, which may sense intracellular levels of Mo and bind the promoter region of modA [59]. *E. coli* ModE is composed of an N-terminal DNA-binding domain (ModE_N) and a C-terminal molybdate-binding domain, which contains a tandem repeat of the Mo-binding protein (Mop) domain, or named Di-Mop domain. The ModABC-ModE systems are widely distributed in organisms, but are not ubiquitous, and variations of ModE were also observed in some Mo-utilizing organisms [38,60,61]. Two additional Mo/W ABC transport systems with different substrate affinity, WtpABC (both Mo and W) and TupABC (W-specific), were reported [56,62]. Both transport systems exhibit low similarity to ModABC [38]. In *Campylobacter jejuni*, a ModE-like protein which lacks the Mop domain was recently reported to repress both ModABC (in the presence of either Mo or W) and TupABC (in the presence of W) systems [63]. However, the regulation of these two transporters is still unclear. Very recently, a member of a universal permease family, PerO, was found to import molybdate and other oxyanions in *Rhodobacter capsulatus*, which is the first reported bacterial molybdate transporter outside the ABC transporter family [64].

In contrast to the well-characterized Mo uptake transport in prokaryotes, information on Mo transport in eukaryotes is limited. In 2007, a high-affinity molybdate transport system, MOT1, which belongs to the sulfate transporter superfamily, was first characterized in *Arabidopsis thaliana* and *Chlamydomonas reinhardtii* [65,66]. The *A. thaliana* MOT1 is strongly expressed in the roots and is localized to the mitochondria instead of the plasma membrane of root cells [67]. Recently, a novel Mo transporter family, MOT2, was identified in both *C. reinhardtii* and animals including humans, which opens a new way towards the understanding of molybdate transport in animals [68].

Moco is synthesized by an evolutionarily conserved multi-step pathway in all three domains of life [52,54,69]. The overall process includes (i) conversion of a guanosine derivative, most likely GTP, into cyclic pyranopterin monophosphate (cPMP, or precursor Z); (ii) transformation of cPMP into molybdopterin; (iii) metal incorporation into the apo-cofactor; and (iv) maturation to an active cofactor in some organisms, e.g., formation of a dinucleotide form (molybdopterin guanine dinucleotide, MGD) or substitution of a terminal oxygen ligand of Moco with a sulfur ligand. In *E. coli*, proteins required for Moco biosynthesis and regulation are encoded in the *moa-mog* operons (Figure 2a) [27,54]. In eukaryotes, at least six proteins (named Cnx1-3 and Cnx5-7 in plants) are involved in Moco biosynthesis (Figure 2b), which are homologous to their counterparts in bacteria [54,69–71]. Thus, the *moa-mog* genes and *cnx* genes could be used for identification of Moco biosynthesis pathways in prokaryotes and eukaryotes, respectively. In addition, a Moco sulfurase, catalyzing the generation of the sulfurylated form of Moco that is needed for activation of the xanthine oxidase family of proteins such as xanthine dehydrogenase and aldehyde oxidase, has been identified in plants and humans [72,73]. A recent study also revealed that, in *A. thaliana*, the first step of Moco biosynthesis is localized in the mitochondrial matrix, and a mitochondrial ABC transporter ATM3 (previously implicated in the maturation of extramitochondrial Fe-S proteins) has a crucial role in Moco biosynthesis by transporting cPMP [74].

Considering that Moco is highly unstable [69], after synthesis, it should be either transferred immediately to the molybdoenzymes or bound to a storage/carrier protein until further insertion. In bacteria, many molybdoenzymes have known chaperones, such as NarJ for nitrate reductase and DmsD for dimethylsulfoxide reductase, which can bind Moco and assist in cofactor incorporation [75–77]. In contrast, little is known about Moco storage in eukaryotes. Recently, a Moco carrier protein (MCP) has been identified in *C. reinhardtii* [78]. MCP belongs to the lysine decarboxylase

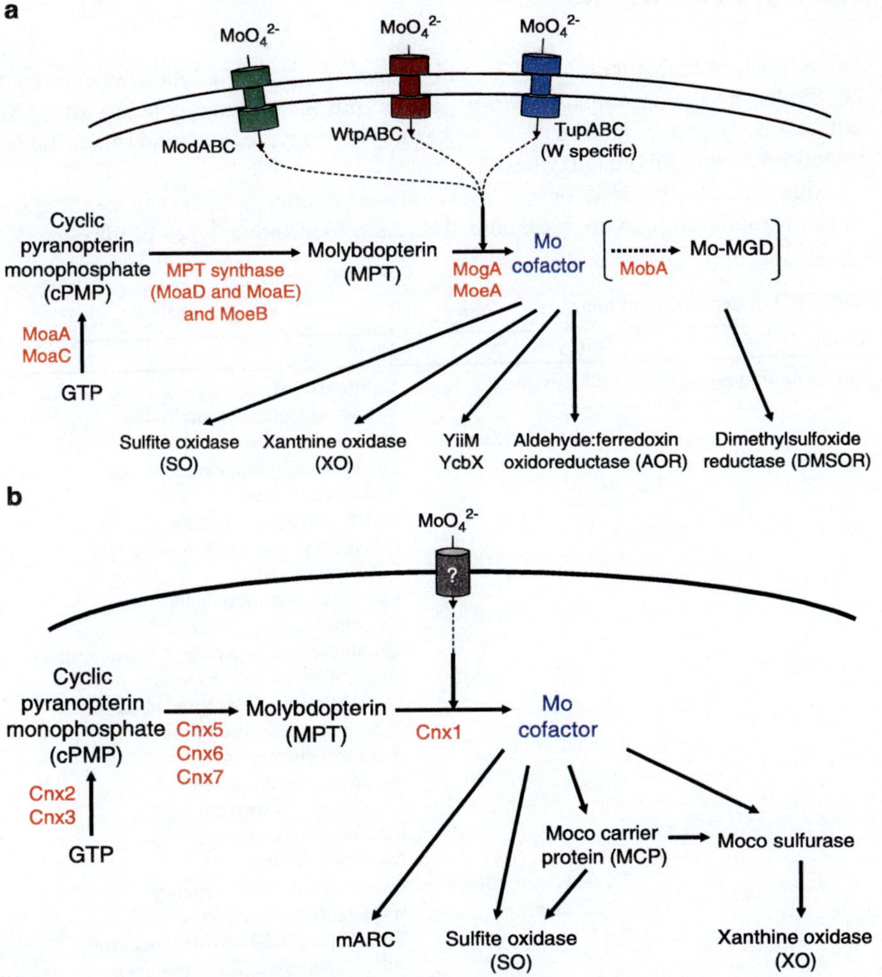

Figure 2 Biosynthesis of molybdenum cofactor. The pathway of Moco synthesis can be divided into three or four steps. (**a**) Biosynthesis of the Mo cofactor in prokaryotes. (**b**) Biosynthesis of the Mo cofactor in eukaryotes. The proteins from *E. coli* and *A. thaliana* catalyzing the respective steps are depicted and their names are given. MGD, molybdopterin guanine dinucleotide.

family and could bind Moco with high affinity. In addition, several homologous Moco-binding proteins (MoBP), which also belong to the lysine decarboxylase family, were discovered in land plants that might be involved in the cellular distribution of Moco [54,79]. However, the mechanism of Moco protection, storage and transfer in mammals is still unclear.

3.2 Molybdoenzymes

On the basis of cofactor composition and catalytic function, molybdoenzymes could be divided into two groups: (i) Mo-dependent nitrogenase that contains an Fe-Mo cofactor in the active site, and (ii) all other molybdoenzymes that bind Moco. Table 1 includes known molybdoenzymes.

Nitrogenase is required for biological nitrogen fixation, which is an essential step in the nitrogen cycle in the biosphere. There are four known types of nitrogenases,

Table 1 Mo-dependent proteins

Group	Family	Protein
Moco-binding proteins	Sulfite oxidase	Sulfite oxidase
		Nitrate reductase (assimilatory)
	Xanthine oxidase	Xanthine oxidase
		Xanthine dehydrogenase
		Aldehyde oxidase
		Aldehyde oxidoreductase
		4-Hydroxybenzoyl-CoA reductase
		CO dehydrogenase
		Quinoline 2-oxidoreductase
		Isoquinoline 1-oxidoreductase
		Quinoline 4-carboxylate-2-oxidoreductase
		Quinaldine 4-oxidoreductase
		Quinaldic acid 4-oxidoreductase
		Nicotinic acid hydroxylase
		6-Hydroxynicotinate hydroxylase
		Nicotine dehydrogenase
		Picolinate hydroxylase
		Pyridoxal oxidase
		Nicotinate hydroxylase
	Dimethylsulfoxide reductase	Dimethylsulfoxide reductase
		Biotin sulfoxide reductase
		Trimethylamine-N-oxide reductase
		Nitrate reductase (dissimilatory)
		Formate dehydrogenase
		Formylmethanofuran dehydrogenase
		Polysulfide/thiosulfate/arsenate reductase
		Arsenite oxidase
		Pyrogallol-phloroglucinol transhydroxylase
	Aldehyde:ferredoxin oxidoreductase (W-containing)	Aldehyde:ferredoxin oxidoreductase
		Formaldehyde ferredoxin oxidoreductase
		Glyceraldehyde-3-phosphate ferredoxinoxidoreductase
		Carboxylic acid reductase
		Hydroxycarboxylate viologenoxidoreductase
		Aldehyde dehydrogenase
	MOSC-containing	mARC/YcbX
		YiiM
Fe-Mo-binding protein	Nitrogenase	Nitrogenase

each with a different combination of metals in the active site [80,81]. The most abundant and widely studied is the Fe-Mo-dependent nitrogenase, which contains MoFe3S3 and Fe4S3 cuboidal subunits triply joined by three bridging sulfurs.

The second group of proteins which utilize Moco as cofactor contains sulfite oxidase (SO), xanthine oxidase (XO), dimethylsulfoxide reductase (DMSOR), and aldehyde:ferredoxin oxidoreductase (AOR, mostly W-containing) families as well as the novel Moco-binding proteins. Each family includes a variety of subfamilies based on sequence similarity, spectroscopic properties and substrate preferences (Table 1). Compared to prokaryotes, which contain diverse members belonging to the four major families, eukaryotes only have four typical molybdoenzymes, including nitrate reductase (NR) and SO (members of the SO family), as well as xanthine dehydrogenase (XDH) and aldehyde oxidase (AO) (members of the XO family) [54,69].

Members of the SO family generally catalyze net oxygen atom transfer to or from a heteroatom lone electron pair rather than hydroxylation of a carbon center [82]. Typical enzymes belonging to this family include SO and assimilatory NR. SO is mainly found in eukaryotes and is located in the mitochondrial intermembrane space where it catalyzes the oxidation of sulfite to sulfate, the final step in the oxidative degradation of sulfur-containing amino acids [83]. The assimilatory NR catalyzes the reduction of nitrate to nitrite and is responsible for the first step in the uptake and utilization of nitrate [69]. So far, this enzyme has only been found in autotrophic organisms, such as plants and fungi.

The XO family contains the largest and most diverse Moco-containing enzymes (Table 1). Members of this protein family catalyze oxidative hydroxylation of a wide range of aldehydes and aromatic heterocycles [69]. The major enzymes of this family include AO (catalyzes the oxidation of a variety of aromatic and nonaromatic heterocycles and aldehydes), XDH (a key enzyme of purine degradation that oxidizes hypoxanthine to xanthine and xanthine to uric acid), and a variety of bacterial enzymes such as aldehyde oxidoreductase and 4-hydroxybenzoyl-CoA reductase.

The DMSOR family consists of a number of Moco-binding enzymes, all from bacterial and archaeal sources, that bind a Mo-MGD cofactor consisting of one Mo atom complexed by two MGD molecules [69,84]. Some of these enzymes possess Mo as their sole redox-active center [85]. They are very diverse in reaction, function, and structure [86]. Most of these enzymes function as terminal reductases under anaerobic conditions where their respective cofactors serve as terminal electron acceptors in respiratory metabolism. DMSOR is found in bacteria and catalyzes reductive deoxygenation of dimethyl sulfoxide to dimethyl sulfide. It is a periplasmic single-subunit protein in some bacteria (such as *Rhodobacter sphaeroides*), whereas a membrane-bound protein composed of three subunits (Moco-containing, four [4Fe-4S] cluster-containing and transmembrane subunits) in some other bacteria (such as *E. coli*) [87,88]. Another widespread member of the DMSOR family is formate dehydrogenase, which catalyzes the oxidation of formate to bicarbonate and is also a selenocysteine-containing enzyme in many organisms [89]. Other members include dissimilatory NR, trimethylamine-N-oxide reductase and several additional enzymes exhibiting substantial sequence homology [69,84,86].

AOR catalyzes the interconversion of aldehydes and carboxylates and was the first member of the AOR family to be structurally characterized as a protein containing a Wco cofactor that is analogous to the Moco [90]. Other members include formaldehyde ferredoxin oxidoreductase, glyceraldehyde-3-phosphate ferredoxin oxidoreductase, and carboxylic acid reductase.

In addition to the four major molybdoenzyme families, novel Moco-binding proteins were recently identified in both mammals and *E. coli* [91,92]. In mammals, a Moco-dependent protein was found in the outer mitochondrial membrane and named mitochondrial amidoxime-reducing component (mARC). mARC binds a Moco that carries neither a terminal sulfur ligand like XO nor a covalently bound cysteine (Cys) residue like SO, suggesting that these proteins represent a new family of molybdoenzymes [93]. Recent studies have shown that human mARC proteins may catalyze the N-reduction of a variety of N-hydroxylated substrates such as N-hydroxy-cytosine and N-ω-hydroxy-l-arginine albeit with different specificities [93–95]. So, mARC and its N-reductive enzyme system plays a major role in drug metabolism [95].

3.3 *Comparative Genomics of Molybdenum Utilization*

Although Mo is an important transition metal, comprehensive analyses of the occurrence and evolutionary trends in its utilization, which could greatly benefit our understanding of Mo and its evolutionary dynamics, have been limited. In recent years, following the availability of a large number of sequenced genomes, several comparative genomics studies have been carried out to investigate the phylogeny of Mo utilization in prokaryotes and eukaryotes at the level of Mo transport, the Moco biosynthesis, and molybdoenzymes [38,39,96,97]. These studies provided a first glance on the Mo utilization in the three domains of life and showed its widespread occurrence, yet limited use of this metal in individual organisms.

First, a wide distribution of genes encoding Mo transport systems, the Moco biosynthesis pathway, and Mo-containing proteins was found in sequenced genomes, and almost all Mo-utilizing organisms contained both Moco biosynthesis proteins and at least one known molybdoenzyme [38,96]. In bacteria and archaea, Mo was utilized by almost all phyla (except *Mollicutes* and *Chlamydiae*), suggesting that Mo utilization is an ancient and essential trait that is common to essentially all organisms in these two domains. In eukaryotes, Mo is used by all animals, land plants, algae, certain fungi, and stramenopiles; however, parasites, yeasts (saccharomycotina and schizosaccharomycetes) and free-living ciliates lack the Mo utilization trait. It is possible that many protozoa, especially parasites, lost the ability to utilize Mo.

Comparative analyses of Mo/W transport systems in sequenced prokaryotes revealed that Mo/W transporters are often present in single copies [39,96,97]. Among them, ModABC is the most common Mo transporter, which is present in approximately 90% of Mo-utilizing bacteria. The occurrence of the other two transporters, WtpABC and TupABC, is much more restricted, especially WtpABC, which is only detected in 3% of Mo-utilizing bacteria. On the other hand, WtpABC is the most frequently used transporter in archaea. It appeared that WtpABC is mainly an archaeal Mo/W transporter, whereas ModABC functions predominantly in bacteria. The full length-type ModE regulation of ModABC transporters only occurred in less than 30% of Mo-utilizing organisms, suggesting the presence of novel or unspecific regulatory pathways for molybdate uptake in many other organisms such as Gram-positive bacteria and cyanobacteria [38,97]. On the other hand, individual ModE_N and/or Mop/Di-Mop proteins, and novel domain fusions for either ModE_N or Mop were observed in a variety of organisms that lack full-length ModE, indicating complexity of ModE-related regulation. Genomic context analyses of these ModE-related variations suggested potential correlations with ModABC transporters as most of these genes are close to or are even in the same operon with *modABC* [38,96]. It was previously thought that a separate ModE_N and Mop/Di-Mop proteins together may have a function similar to that of the full-length ModE [98]. In eukaryotes, MOT1 was detected in some Mo-utilizing organisms such as land plants, green algae, and stramenopiles, whereas the recently identified MOT2 appeared to have a wider distribution in algae, land plants, and animals [68,99]. Thus, MOT1 and MOT2 proteins may play key roles in Mo transport in eukaryotes although additional unknown Mo transport systems may be also present.

The majority of known proteins involved in Moco biosynthesis pathways could be detected in essentially all Mo-utilizing organisms. However, a very small number of prokaryotes, which contain homologs of molybdoenzymes, lack genes for either Moco biosynthesis components or Mo/W transporters [96]. It is possible that Moco is dispensable for the molybdoprotein homologs in these organisms. Nevertheless, there is a very good correspondence between occurrence of the Moco biosynthesis trait and Moco utilization in the three domains of life.

Comparative genomics of molybdoenzymes also showed complexity in their evolutionary trajectories. In bacteria, DMSOR, SO, and XO families were widespread, especially DMSOR whose members (mostly DMSOR, dissimilatory NR, and formate dehydrogenase) were detected in more than 90% Mo-utilizing organisms [96,97]. In contrast, the W-containing AOR family was only detected in ~ 15% Mo/W-utilizing organisms. In archaea, DMSOR was also the most abundant molybdoenzyme family (more than 95% Mo-utilizing organisms). Interestingly, members of the AOR family had a much higher occurrence in archaea (~70%). The FeMo-utilizing molybdoenzyme, nitrogenase, was detected in ~ 20% of Mo-utilizing bacteria (almost all also used Moco) and methanogenic archaea. Further investigation of the predicted molybdoenzyme set (molybdoproteome) of each organism revealed that proteobacteria have larger molybdoproteomes than other organisms [96]. *Desulfitobacterium hafniense*, a dehalorespiring bacterium, was found to have the largest known molybdoproteome in prokaryotes (63 molybdoproteins, 95% of which are members of the DMSOR family). In eukaryotes, almost all Mo-utilizing organisms had SO and XO families. Land plants possessed the largest molybdoproteomes in eukaryotes (10-11 molybdoproteins). On the other hand, all sequenced saccharomycotina (e.g., *Saccharomyces cerevisiae*) and schizosaccharomycetes (e.g., *Schizosaccharomyces pombe*) had neither known molybdoenzymes nor Moco biosynthesis proteins. It was previously reported that a small number of unsequenced yeast species, such as *Candida nitratophila* and *Pichia angusta*, may utilize Mo-containing assimilatory NR [39,100], but the fact that both homologs of this protein and the Moco biosynthesis pathway are absent in all currently sequenced yeast genomes suggests the loss of Mo utilization in these organisms.

Recent studies on the new mammalian Moco-binding mARC protein revealed that it consists of two conserved domains: N-terminal MOSC_N (pfam03476) and C-terminal MOSC (pfam03473) domains, which are also present in Moco sulfurases [97]. The MOSC domain of eukaryotic Moco sulfurase is involved in Moco binding with high affinity and its Moco carries a terminal sulfur ligand due to the catalytic activity of pyridoxal-5'-phosphate-dependent NifS-like domain [101]. The function of the MOSC_N domain is unknown; however, it is predicted to adopt a β-barrel fold. Two additional Moco-dependent proteins, YcbX and YiiM, were characterized in *E. coli*, which may represent novel enzymatic activities involved in the detoxification pathway of N-hydroxylated base analogs [92]. Both proteins contain the MOSC and additional domains (Figure 3). Bioinformatics analyses showed that *E. coli* YcbX and mammalian mARC proteins could be considered as orthologs and are members of the same family (mARC/YcbX family). On the other hand, no significant sequence similarity could be detected between YiiM and mARC/YcbX, suggesting that they belong to different families within the MOSC superfamily [97]. Further analysis revealed that, in bacteria, both mARC/YcbX and YiiM were widespread but only detected in Moco-utilizing organisms. In contrast, the occurrence of these two families in archaea is limited, i.e., only organisms belonging to *Euryarchaeota/Halobacteriales* had mARC/YcbX proteins. In eukaryotes, mARC proteins were detected in more than 95% Mo-utilizing organisms, suggesting a wide distribution of this novel molybdoenzyme family. In addition, a novel group of

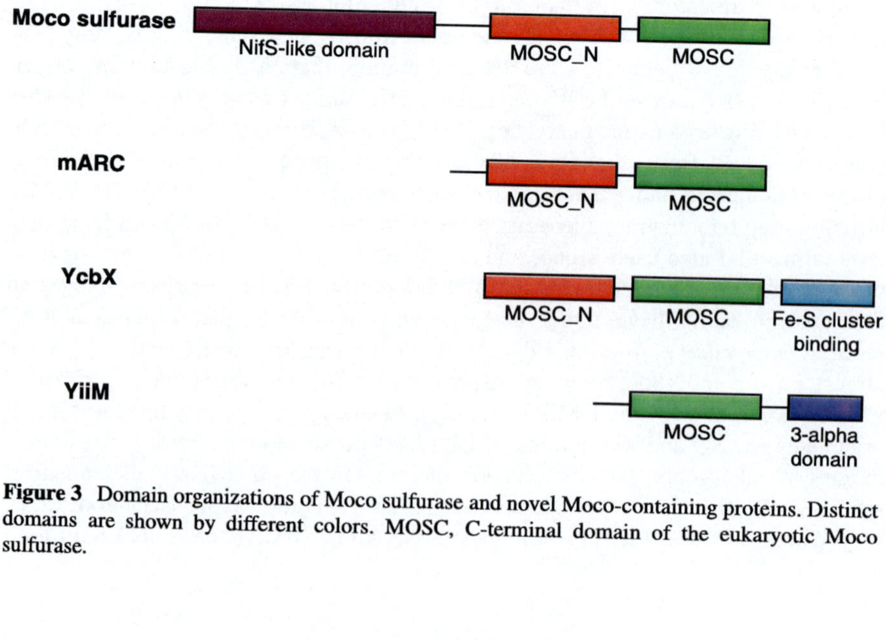

Figure 3 Domain organizations of Moco sulfurase and novel Moco-containing proteins. Distinct domains are shown by different colors. MOSC, C-terminal domain of the eukaryotic Moco sulfurase.

MOSC-containing proteins (designated MOSC-like), which form a separate branch in the MOSC family and might serve as chaperones involved in Moco transfer or storage, was identified in some prokaryotes [97]. In general, these studies suggested complexity and diverse roles of the MOSC superfamily, whose proteins may be (i) involved in the Moco modification pathway (Moco sulfurase); (ii) new molybdoenzymes (mammalian mARC, *E. coli* YcbX and YiiM); (iii) potential Moco chaperones (MOSC-like); and (iv) involved in other functions. Further experiments are needed to better understand the functions of MOSC-containing proteins.

An interesting link between Mo and selenium was also observed as a major member of the DMSOR family, formate dehydrogenase α subunit, is also a selenocysteine-containing protein that may be responsible for maintaining the selenocysteine utilization trait in sequenced prokaryotes [96,97]. Thus, the selenocysteine utilization trait depends on the Mo utilization trait in prokaryotes, most likely because of formate dehydrogenase, which is not only a widespread molybdoenzyme but is also the major user of Se in prokaryotes. In addition, some environmental conditions and other factors may affect Mo utilization and molybdoenzyme families. For example, the majority of intracellular parasites and symbionts lost the ability to utilize Mo, whereas more than 80% of extracellular symbionts utilize the metal [96]. Organisms possessing W-containing AOR proteins appear to favor an anaerobic environment, whereas organisms containing SO or XO proteins favor aerobic conditions. On the other hand, organisms possessing nitrogenase favor both anaerobic and relatively warm conditions [96,97]. These findings suggest that although being dependent on the same processes, such as Mo availability and Moco synthesis, different Mo enzymes are subject to independent and dynamic evolutionary processes. However, no significant correlation was observed between various factors examined and the size of molybdoproteomes.

4 Copper

Copper is an essential micronutrient that serves as an important cofactor for proteins and enzymes that carry out several fundamental biological functions [102]. On the other hand, Cu is highly toxic in the free form because of its ability to produce radicals by cycling between Cu(I) and Cu(II) species [103]. It is important for Cu-utilizing organisms to obtain sufficient levels of Cu ion to meet their needs while tightly controlling intracellular Cu concentration.

4.1 Overview of Copper Trafficking and Homeostasis

Cellular Cu trafficking processes are required for correct utilization of this element in biochemical processes and for limiting Cu toxicity. Cu import mainly needs the coordinate function of proteins with metal-binding domains, whereas detoxification mechanisms include the binding of Cu to specific proteins (e.g., metallothioneins) and its transfer into cell compartments such as periplasmic space [104].

In prokaryotes, the mechanisms involved in Cu transport and homeostasis are not completely understood. No specific Cu import system has been identified in the majority of bacteria, possibly reflecting no cytosolic requirement for Cu, and the mechanism of Cu entry is largely unknown [105]. To date, Cu trafficking in bacteria is best described in *E. coli* and in *Enterococcus hirae* [106,107]. The most relevant Cu homeostatic systems in *E. coli* are shown in Figure 4a. Several Cu-related transport and resistant proteins have been characterized in a variety of organisms, including CopA/PacS, CusCFBA, CutC, PcoABCDRSE [105–107].

In *E. coli*, the Cu(I)-translocating P-type ATPase CopA is the major component of Cu homeostasis and serves as an exporter for removing Cu(I) from the cytoplasm [105,106]. CopA proteins belong to a superfamily that is involved in transport of transition or heavy metal ions (including Zn, Cd, Ag, Pb, and Co) across membranes [108]. Two Cys residues in a Cys-Pro-Cys motif located in the middle of CopA are needed for CopA function [109]. PacS is a CopA homolog in cyanobacteria and may be involved in Cu homeostasis crucial to the photosynthetic thylakoid function [110]. CtaA, another CopA homolog identified in cyanobacteria, was suggested to be involved in Cu import from the periplasm [110,111]. Both CtaA and PacS are required for Cu transport into the thylakoid [110]. In *E. hirae*, two CopA homologs, CopA and CopB, were identified. The former may be involved in Cu uptake, whereas the latter functions as an exporter of Cu ion [107]. In *E. hirae* and many other organisms a Cu chaperone, CopZ, functions as part of a complex cellular machinery for Cu trafficking and detoxification [112]. A role for *E. hirae* CopZ in routing Cu to the cytoplasmic Cu sensor CopY, to alleviate CopY-mediated repression of the *copYZAB* operon, has also been reported [113]. CopZ homologs are also found in eukaryotes, called Atx1, which interact with the Golgi P-type ATPase Cu transporter ATP7 [114]. The CopZ/Atx1 proteins adopt a very similar structure to the amino-terminal metal-binding domains of the P1B- type ATPases (a subgroup of P-type ATPases that transports transition metals between different compartments of the cell) with typical βαββαβ ferredoxin-like folding and a GXXCXXC metal-binding motif present on a flexible solvent-exposed loop [115]. Recently, a Cu-binding metallothionein, MymT, was found in several pathogenic mycobacteria, which may also serve as a chaperone involved in CopA-related Cu(I) detoxification [116].

Gram-negative bacteria contain another Cu efflux system, the CusCFBA system, which includes CusA (the inner membrane pump), CusB (the periplasmic protein), CusC (the outer membrane protein forming a channel bridging the periplasmic space) and CusF (a small periplasmic protein that binds a single Cu(I) ion and interacts with both CusC and CusB) [117]. The recent elucidation of the structure of the Cu(I) bound form of CusF has revealed a new metal recognition site in which Cu(I) is tetragonally displaced from a Met_2His ligand plane toward a conserved tryptophan which involves cation-π interactions [118]. In *E. coli*, genes encoding the four-part Cus complex were present in one *cus* operon, which is only required under conditions of extreme Cu stress and is particularly important under anaerobic conditions [117]. These genes are induced in response to elevated Cu by the CusRS

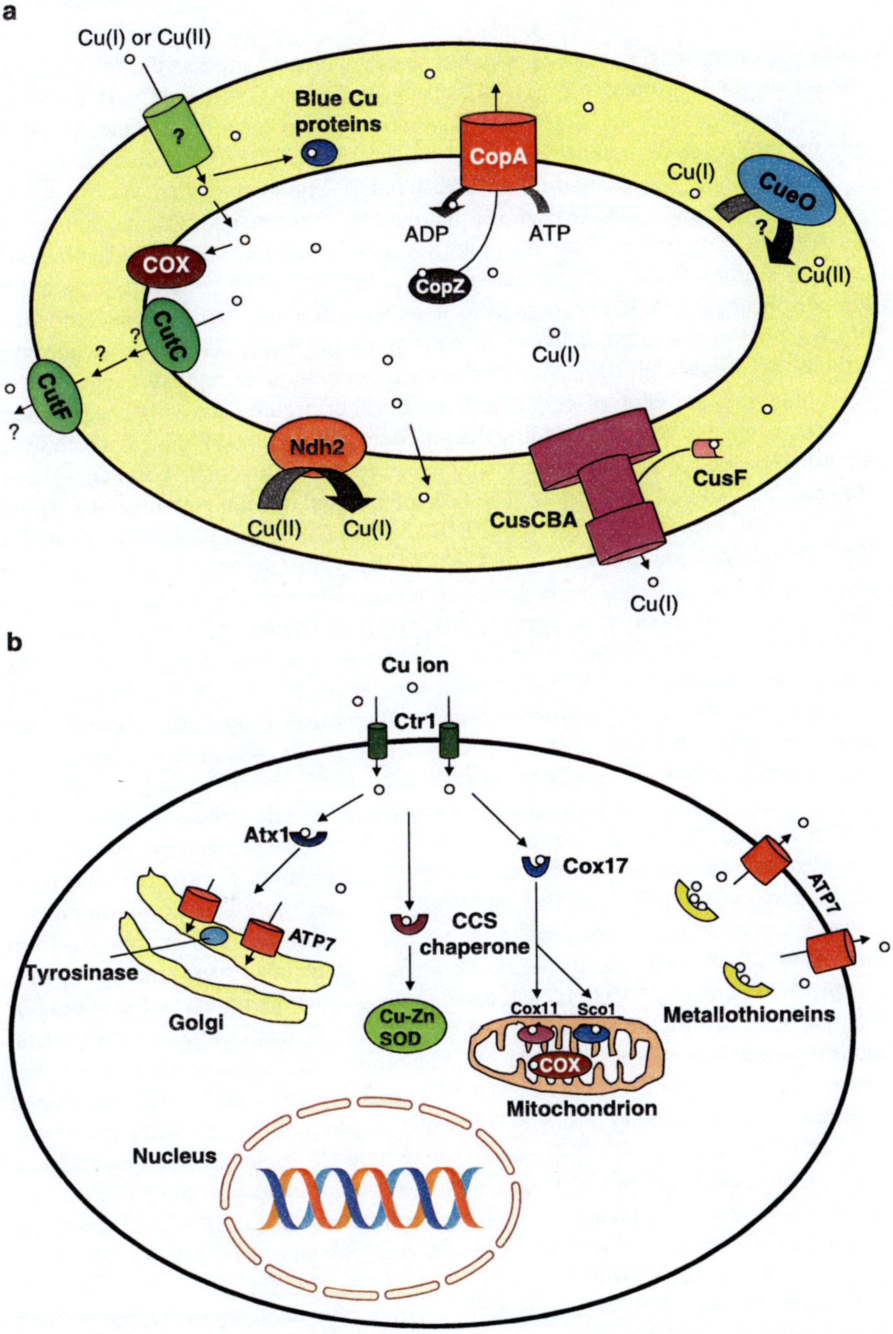

Figure 4 Schematic view of Cu homeostasis. **(a)** Cu homeostasis in *E. coli*. CopA, the Cu(I)-translocating P-type ATPase; CusCFBA, the four-component Cu efflux system; Ndh 2, a cupric reductase; CueO, a multicopper oxidase; CutC and CutF, two proteins involved in Cu efflux and/or homeostasis; CopZ, a Cu chaperone involved in Cu export; COX, cytochrome *c* oxidase. **(b)** Cu homeostasis in *Drosophila melanogaster*. Atx1, CCS, and Cox17, Cu chaperones involved in various pathways; Ctr1, eukaryotic Cu importer; ATP7, eukaryotic Cu exporter (also involved in Cu transport to Golgi); COX11 and Sco1, two proteins involved in cytochrome *c* oxidase assembly; Cu-Zn SOD, Cu-Zn superoxide dismutase.

two-component sensory system, which typically monitors stress at the cell envelope and is thought to respond to Cu(I). It has been shown that CusA and CusB are essential for Cu resistance, and CusC and CusF are required for full resistance [119]. A role for the Cus system in providing Ag resistance has also been shown [119].

Some *E. coli* strains harbor an additional plasmid-borne Pco system which involves seven genes, *pcoABCDRSE*, that confer Cu resistance. The mechanism of Cu detoxification provided by this system is largely unknown but includes the multicopper oxidase PcoA and its putative partner PcoC, both of which are exported to the periplasm, PcoD that is thought to transport Cu across the cytoplasmic membrane, PcoB that is a predicted outer membrane protein and PcoE, a periplasmic protein that binds Cu [106]. PcoRS form a two-component regulator required for Cu-inducible expression of *pco*. A number of other proteins have also been linked to Cu resistance in *E. coli*, including the products of *cutABCDEF* genes, which were identified based on the preliminary characterization of Cu-sensitive mutants [120]. However, few of these genes have been directly linked to Cu metabolism, transport, or regulation. Previous studies implicated CutC in Cu efflux, suggesting a role for CutC in intracellular trafficking of Cu(I) [120]. CutC homologs have also been characterized in eukaryotes, including humans. Recently, a crystal structure of human CutC was reported, suggesting that this protein may function as an enzyme with Cu(I) as a cofactor rather than a Cu transporter and that the potential Cu(I)-binding site consists of two Cys residues and other conserved residues in the vicinity [121].

A general scheme of Cu homeostasis in eukaryotes (using *Drosophila melanogaster* as a representative organism) is shown in Figure 4b. In eukaryotes, Cu is acquired by the high-affinity Cu transporter (Ctr) family proteins [18]. Members of the Ctr family possess an N-terminal extracellular Met-rich domain which is important for the acquisition of Cu(I) ions [122]. Different organisms may possess multiple Ctr proteins located in different biological membranes. *S. cerevisiae* has three Ctr proteins, yCtr1-yCtr3 [123]. yCtr1 and yCtr3 are located in the plasma membrane, whereas yCtr2 is localized in the vacuolar membrane and imports Cu from the vacuole to the cytoplasm [124]. *D. melanogaster* also has three *ctr1* genes (*ctr1A*, *ctr1B*, and *ctr1C*). Ctr1A is located in the plasma membrane and is the major Cu transporter during growth and development. Ctr1B also localizes to the plasma membrane and is not essential for development unless flies are severely Cu-deficient or are subjected to Cu toxicity [125]. Ctr1C is mainly expressed in male gonads and functions as a Cu importer in the male germline, specifically in maturing spermatocytes and mature sperm [126]. Humans contain two Ctr proteins (hCtr1 and hCtr2). hCtr1 is the main Cu importer, which is located predominantly at the plasma membrane, but may also be present in vesicular compartments [127]. hCtr2 was found exclusively to late endosomes and lysosomes and may be involved in the delivery of Cu ions to the cytosol [128].

Cu export in eukaryotes is mediated by an important category of ATP-dependent transporters, the ATP7 family, which is homologous to bacterial CopA proteins [129]. In mammals, there are two isoforms: ATP7A and ATP7B [130]. ATP7A is expressed in the intestinal epithelium as well as most other tissues (such as brain and heart) except the liver, which is required for transport of Cu into the trans-Golgi network for biosynthesis of several secreted cuproenzymes and for basolateral efflux of Cu in the intestine and other cells [131]. ATP7B is mainly expressed in the liver and is needed for Cu metalation of ceruloplasmin and biliary Cu excretion [131]. *D. melanogaster* has a sole ATP7 protein (named DmATP7), which is required for *in vivo* Cu distribution [132]. Yeasts also have an ATP7 ortholog Ccc2, which is located in the trans-Golgi membrane, obtains Cu from Atx1 and transfers it to several secreted proteins [133].

4.2 *Cuproproteins*

Copper plays important roles in electron transfer, oxidation of organic substrates and metals, dismutation of superoxide, monooxygenation, transport of dioxygen and iron, and several other processes. So far it has not been possible to identify all Cu-binding proteins (cuproproteins) in any organism using bioinformatics approaches. This chapter only focuses on strictly Cu-dependent protein families which have been used for comparative genomics of Cu utilization in recent studies.

To date, a number of Cu-containing proteins have been characterized (a list is shown in Table 2). Cu centers in proteins could be divided into three types based on spectroscopic and structural properties. Type 1 Cu (also called blue Cu) shows intense absorption at around 600 nm and narrow hyperfine splittings in the electron paramagnetic resonance (EPR) spectroscopy, type 2 Cu does not give strong absorption at 600–700 nm and shows hyperfine splittings of the normal magnitude in the EPR spectrum, whereas type 3 Cu could be detected by neither strong absorption nor EPR studies. Structurally, the Cu atom of a typical type 1 site is coordinated by

Table 2 Cu-dependent proteins

Prokaryotes	Eukaryotes
Plastocyanin family (including plastocyanin, amicyanin, pseudoazurin, halocyanin, etc.)	Plastocyanin family
Azurin family (including azurin and auracyanin)	Plantacyanin family (including plantacyanin, umecyanin, mavicyanin, stellacyanin, etc.)
Rusticyanin	Cytochrome *c* oxidase subunit I
Nitrosocyanin	Cytochrome *c* oxidase subunit II
Cytochrome *c* oxidase subunit I	Cu-Zn superoxide dismutase
Cytochrome *c* oxidase subunit II	Copper amine oxidase
Nitrous oxide reductase	Peptidylglycine R-hydroxylating monooxygenase
NADH dehydrogenase 2	Dopamine β-monooxygenase
Cu-Zn superoxide dismutase	Multicopper oxidases (including laccase, Fet3p, hephaestin, ceruloplasmin, ascorbate oxidase, etc.)
Copper amine oxidase	Tyrosinase (or polyphenol oxidase)
Particulate methane monooxygenase	Hemocyanin
Multicopper oxidases (including nitrite reductase, CueO, CotA, laccase, bilirubin oxidase, phenoxazinone synthase, etc.)	Cnx1G
Tyrosinase	Galactose oxidase

a Cys and two His residues in a trigonal planar arrangement. Often the thioether of a methionine coordinates axially, distorting the geometry towards tetrahedral. Most type 2 sites are three to four coordinate and one or more of the Cu ligands are the imidazole side chains of His residues. The coordination sphere may be completed by methionine, glutamate, glutamine or tyrosine. Type 3 sites consist of two antiferromagnetically coupled Cu atoms bridged by molecular oxygen or a hydroxyl. Some Cu-dependent proteins, such as multicopper oxidases (MCOs), may contain multiple Cu centers.

Blue Cu proteins (also named cupredoxins) are a group of relatively small proteins containing a single type 1 Cu center. They function in electron transfer in the respiratory and photosynthetic chains of many bacteria and plants [134,135]. These proteins include plastocyanin, azurin, pseudoazurin, amicyanin, rusticyanin, auracyanin, plantacyanin, and some other proteins. Plastocyanin is the best studied blue Cu protein which shuttles electrons from cytochrome b6/f to photosystem I. Crystal and NMR solution structures of several plastocyanins have revealed that this protein has an eight-stranded Greek-key β-barrel fold and contains a type 1 Cu atom coordinated by two histidines, one Cys and one methionine. The red Cu protein, nitrosocyanin, is a variant of the blue Cu protein, whose Cu site is the only known blue Cu-related site with an exogenous water molecule bound to Cu [135,136].

A type 1 Cu center is also detected in several larger enzymes, such as nitrite reductase (NiR) that catalyzes the reduction of nitrite to nitric oxide, and MCOs that function in intramolecular electron transfer. MCOs include a large number of proteins, such as laccase, ascorbate oxidase, CueO, PcoA, EpoA, dihydrogeodin oxidase, hephaestin, ceruloplasmin, phenoxazinone synthase, Fet3p, etc. [135,137]. Most MCOs have four Cu centers: a type 1 Cu and a mixed Cu center containing a type 2 and two type 3 Cu atoms. These MCOs catalyze the oxidation of small molecules and cations with the concomitant four-electron reduction of oxygen to water. Some MCOs such as mammalian ceruloplasmin and yeast Fet3p are ferroxidases, whereas laccases derive electrons from the oxidation of phenolic compounds.

Two additional Cu-dependent proteins, cytochrome *c* oxidase (COX) and nitrous oxide reductase (N_2OR), have a binuclear Cu center, named Cu_A, which is a variant of type 1 Cu. Cytochrome oxidase family members act as the terminal enzymes in respiratory chains. The two major subgroups of this family include COX and quinol oxidase [138]. Both classes have several catalytic subunits, and subunit I contains two heme centers: the first (heme a) acts as an electron input device to the second, and the second (heme a_3) is a part of a binuclear center containing Cu_B. However, there are significant differences of subunit II between the two subgroups. COX subunit II contains the Cu center Cu_A with 2 Cu atoms, which might be the immediate electron acceptor from cytochrome *c*, whereas quinol oxidase subunit II lost the Cu_A center [139]. Three subtypes (aa3, ba3, and cbb3) of both COX and quinol oxidase have been reported [138]. Characterizing all these subtypes and distinguishing the Cu-dependent COX subunit II from the Cu-independent quinol oxidase subunit II is important for correct description of Cu utilization. N_2OR transforms nitrous oxide to dinitrogen and carries six Cu atoms. Two are arranged in the binuclear Cu_A site similar to that of COX, and four make up the sulfide-bridged Cu cluster (Cu_Z catalytic center). The crystal structure of *Pseudomonas nautica* N_2OR revealed that the Cu_Z center belongs to a new type of metal cluster in which the four Cu ions are bound by seven histidine residues [140].

The type 2 Cu-containing proteins include Cu-Zn superoxide dismutase (Cu-Zn SOD), Cu amine oxidase (CuAO), peptidylglycine R-hydroxylating monooxygenase (PHM), and dopamine β-monooxygenase (DBM) [141].

Cu-Zn SOD is widespread in both eukaryotes and prokaryotes. Most studies have focused on the enzymes from eukaryotic sources, such as yeast and human. It has been found that, in the oxidized Cu-Zn SOD, Cu is coordinated by four histidine residues.

CuAO belongs to a larger group of amine oxidases that catalyze deamination of amines with concomitant reduction of oxygen to hydrogen peroxide. These enzymes are found in a large variety of organisms, from microbes to mammals. In bacteria, CuAOs have important roles in providing carbon or nitrogen sources when primary amines are available. In mammals, CuAOs are found in various tissues, including placenta, blood, muscle, and endothelium. It has been reported that increased CuAO expression in humans might be a marker of several diseases including cancer, diabetes and liver cirrhosis [142]. Crystal structures of CuAO from different organisms showed that the Cu atom is coordinated by three histidines and two water molecules [143].

In peptidylglycine R-hydroxylating monooxygenase (PHM) and dopamine β-monooxygenase (DBM), two distinct Cu sites are used to split oxygen, which then serves as the source of OH in the hydroxylation of their respective substrates [141]. Both enzymes are mainly detected in metazoa, and their functions are well established. PHM is one of two domains in peptidylglycine R-amidating monooxygenase (PAM), which is essential for the activation of a variety of hormones by R-amidation, thereby improving hormone-receptor affinity. DBM catalyzes a similar reaction to PHM; however, the hydroxylation of dopamine is at the β-carbon. Sequence analysis revealed that DBM is homologous to PHM, suggesting that they may have evolved from a common ancestor [144].

Other Cu-dependent proteins include NADH dehydrogenase 2 (Ndh2), tyrosinase, hemocyanin, particulate methane monooxygenase (pMMO), Cnx1G, and galactose oxidase (GAO): (1) The Cu(II)-reductase Ndh2 from *E. coli*, which contributes to antioxidant function and Cu homeostasis, is a membrane-bound reductase that diminishes the susceptibility of the respiratory chain to damaging effects caused by Cu and hydroperoxides. (2) Tyrosinases (or catechol/polyphenol oxidases) are ubiquitously distributed in all domains of life. They are essential for pigmentation and are important factors in wound healing and primary immune response. The active site is a type 3 Cu center consisting of two Cu ions, each coordinated by three histidine residues. (3) Hemocyanin is also a type 3 Cu protein family and occurs in the hemolymph of some species in arthropoda and mollusca. These proteins are extracellular oxygen carriers that are responsible for the precise oxygen delivery from the respiratory organs to tissues. (4) pMMO is a membrane-bound Cu-containing enzyme that oxidizes methane to methanol in methanotrophic bacteria. The crystal structure of *Methylococcus capsulatus* pMMO reveals the composition and location of three metal centers, which provides new insight into the molecular details of biological methane oxidation [145]. (5) Cnx1G is the G domain of Cnx1 that is involved in catalyzing the insertion of Mo into molybdopterin (see Section 3.1 and Figure 2b). Identification of the Cu bound to the molybdopterin dithiolate sulfurs

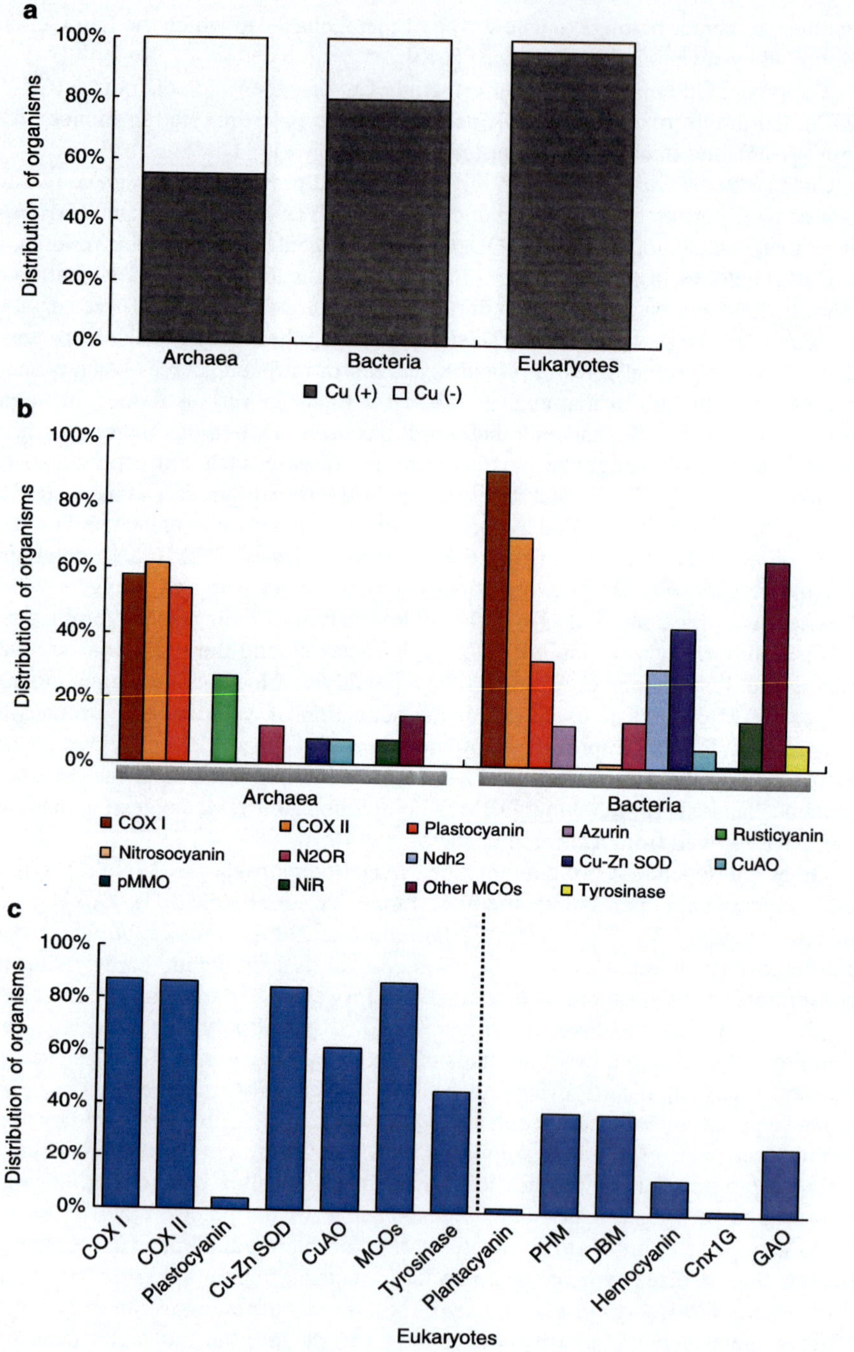

Figure 5 Occurrence of Cu utilization in the three domains of life. (**a**) Proportion of Cu-utilizing organisms among organisms with sequenced genomes. All organisms were classified into two groups: Cu (+), i.e., containing the Cu utilization trait; Cu (-), i.e., lacking the Cu utilization trait.

in Cnx1G structures provides an important link between Mo and Cu utilization [146]. (6) GAO contains a single Cu ion and an amino acid-derived cofactor. The enzyme has been well studied, which has provided insights into the catalytic mechanism of this enzyme. One of the most interesting features of the enzyme is the posttranslational generation of an organic cofactor from its active-site amino acid residues, one of which might be also one of the Cu ligands [147].

4.3 Comparative Genomics of Copper Utilization

In recent years, several comparative genomics studies have been carried out to identify the Cu utilization trait and Cu-binding proteins in organisms [148–150]. For example, one study developed a computational approach based on conserved metal-binding patterns of metalloproteins in the PDB to search for new metalloproteins, and applied this method for Cu [148]. A set of Cu-binding patterns was obtained for all Cu-binding proteins in the PDB and then combined with the primary sequences of corresponding metalloproteins to identify all cuproproteins by homology searches. This procedure retrieved a significant number of false positive metalloproteins. To solve this problem, additional searches integrated with domain recognition methods were carried out, which showed better sensitivity and selectivity [34]. Based on this modified approach, the occurrence of Cu-binding proteins in 57 completely sequenced genomes in prokaryotes and eukaryotes was further examined [150]. The size of the Cu proteome is generally less than 1% of the total proteome. The number of putative Cu-binding proteins did not correlate with the size of the proteome, which is different from several other metals such as Zn [34]. Functional analysis of Cu-binding proteins revealed that these proteins are likely to be part of a network which may represent an ancient core that is crucial for Cu homeostasis. The speciation of prokaryotic organisms appeared to only slightly affect this ancestral Cu proteome, whereas eukaryotes may have expanded their ancestral repertoires of Cu proteins, by evolving new Cu domains and reusing old domains for new functions.

Much more comprehensive comparative genomics analyses of sequenced prokaryotes and eukaryotes were then carried out to yield a clearer view of Cu utilization in all three domains of life [37,96]. Using the strategy introduced in Section 2, occurrence of the Cu utilization trait, Cu transporters and strictly Cu-dependent proteins was examined. The distribution of Cu-utilizing organisms and Cu-dependent proteins is shown in Figure 5. Consistent with previous observations,

Figure 5 (continued) (**b**) Occurrence of Cu-dependent proteins in Co-utilizing prokaryotes. (**c**) Occurrence of Cu-dependent proteins in Cu-utilizing eukaryotes. Protein families on the left side of the dotted line have Cu-containing homologs in bacteria whereas others were only found in eukaryotes. COX I, cytochrome *c* oxidase subunit I; COX II, cytochrome *c* oxidase subunit II; N_2OR, nitrous oxide reductase; Ndh2, NADH dehydrogenase 2; Cu-Zn SOD, Cu-Zn superoxide dismutase; CuAO, Cu amine oxidase; pMMO, particulate methane monoxygenase; NiR, nitrite reductase; MCOs, multi-Cu oxidases; PHM, peptidylglycine R-hydroxylating monoxygenase; DBM, dopamine-monooxygenase; GAO, galactose oxidase.

Cu is widely used by bacteria, i.e., approximately 80% of analyzed organisms were found to be Cu-utilizing. However, all or almost all organisms in some bacterial phyla (such as *Thermotogae*, *Mollicutes*, *Chlamydiae*, and *Spirochaetes*) lacked known Cu-dependent proteins.

In archaea, only half of organisms appeared to utilize Cu (Figure 5a). Analysis of Cu transporters revealed that they had somewhat different patterns of occurrence than Cu-dependent proteins. First, CopA appeared to be the most widespread Cu exporter in bacteria and was the only Cu transporter detected in archaea. Occurrence of other transporters was relatively limited, especially the Cus system that was only detected in Gram-negative bacteria [96]. Second, many organisms, including those that lack known Cu-dependent proteins, had Cu exporters. These data suggested that the pathways of Cu utilization and detoxification are independent and that many organisms likely protect themselves against Cu ions that inadvertently enter the cell. Although occurrence of Cu transporters in prokaryotes may not provide sufficient information about Cu utilization, it may be important for understanding of Cu homeostasis. Third, some organisms were found to have multiple copies of certain Cu transporters. The highest number of Cu transporters in bacteria was observed in *Acidovorax sp.* JS42 and *Ralstonia pickettii* (10 and 9 Cu exporters, respectively), both of which were isolated from highly contaminated environments [96]. It is possible that these species need more efficient mechanisms to maintain cellular Cu homeostasis or protect against this metal.

Further studies on Cu utilization in eukaryotes revealed that almost all sequenced organisms utilized Cu, suggesting a uniformly essential nature of this metal in this domain of life (Figure 5a). In eukaryotes, the occurrence of Cu importer Ctr1 and exporter ATP7 was consistent with that of the Cu utilization trait (Ctr1 was detected in more than 90% of Cu-utilizing organisms and ATP7 in all Cu-utilizing organisms). Interestingly, the majority of organisms had 1–3 *ctr1* genes, but *Caenorhabditis* species (nematodes) possessed additional *ctr1* genes, especially *C. elegans* that had 11 such genes, suggesting unknown complexities in Cu uptake and trafficking in these organisms [96]. It is possible that these Ctr1 proteins are located in various membranes (i.e., plasma or organellar membrane) and/or cell types. The occurrence of Cu exporters varied from one to six genes, with three *Phytophthora* species, which are crop plant pathogens belonging to the genus *Oomycetes*, having the highest numbers of Cu exporters (i.e., *Phytophthora infestans* possessed six ATP7 proteins).

Among Cu-dependent proteins, COX I and COX II were the most frequently used Cu-binding proteins in bacteria and archaea (Figure 5b). Other Cu-binding proteins, such as Cu-Zn SOD, plastocyanin, and a variety of MCOs were also found in many prokaryotes. In contrast, the occurrence of pMMO, nitrosocyanin, CuAO, and tyrosinase appeared to be very limited. In addition, some bacterial Cu-dependent protein families, including azurin, nitrosocyanin, Ndh2, pMMO, and tyrosinase were absent in archaea, whereas a blue Cu protein, rusticyanin, was only detected in archaea. Investigation of the cuproproteomes (the whole set of Cu-dependent proteins) suggested that large cuproproteomes were mainly observed in proteobacteria, especially in *Alphaproteobacteria/Rhizobiaceae* among which two *Sinorhizobium* species (*S. medicae* and *S. meliloti*) contained the largest bacterial cuproproteomes

(22 Cu-dependent proteins, half were COX I and COX II proteins). In archaea, large cuproproteomes were mainly found in *Euryarchaeota/Halobacteriales*, including *Haloarcula marismortui* that had the largest prokaryotic cuproproteome (25 Cu-dependent proteins; half are plastocyanin homologs). Thus, although bacteria and archaea have similar Cu-dependent protein families, occurrence of these proteins was mostly different [96].

Homologs of almost half of the prokaryotic Cu-dependent proteins could not be found in eukaryotes. On the other hand, novel Cu-binding proteins evolved in eukaryotes, such as plantacyanin, PHM, hemocyanin, and GAO (Figure 5c). Analysis of the occurrence of eukaryotic Cu-dependent proteins revealed that, similar to prokaryotes, MCOs, COX I, COX II, and Cu-Zn SOD were the most abundant Cu-dependent proteins. Further analysis of eukaryotic cuproproteomes showed that land plants possessed the largest cuproproteomes (62 and 78 proteins in *A. thaliana* and *Oryza sativa*, respectively [96]). Most of these proteins belonged to plantacyanin, CuAO, and MCO families.

One interesting finding was that organisms living in oxygen-rich environments utilized Cu, whereas the majority of anaerobic organisms did not [37,39]. In addition, among Cu users, cuproproteomes of aerobic organisms were generally larger than those of anaerobic organisms. These data are consistent with the idea that proteins evolved to utilize Cu following the oxygenation of the Earth [151]. In other words, the use of Cu is strongly linked to the use of molecular oxygen.

5 Nickel and Cobalt

Nickel is an essential component of several metalloenzymes involved in energy and nitrogen metabolism, whereas Co is mainly found in the corrin ring of vitamin B_{12} (also known as cobalamin), a cofactor involved in methyl group transfer and rearrangement reactions, but also occurs in a few non-corrin Co-containing enzymes, such as methionine aminopeptidase from *Salmonella typhimurium* and prolidase from *Pyrococcus furiosus* [152,153]. The list of known Ni- and B_{12}-dependent proteins is shown in Table 3.

5.1 Nickel and Cobalt Uptake

In prokaryotes, these transition metals use similar transport systems. Thus, identification of substrate preference of members of each transporter family is important for comparative genomics of Ni and Co utilization. A schematic representation of known Ni/Co transport systems is shown in Figure 6.

In bacteria, Ni and Co uptake is mediated by ABC systems and several secondary transporters [13]. The well-studied ABC-type Ni transporter system, NikABCDE, belongs to a large family of ABC transporters (nickel/peptide/opine transporter

Table 3 Ni- and Co(B_{12})-dependent proteins

Ni-dependent proteins	Co(B_{12})-dependent proteins
Urease	Adenosylcobalamin-dependent isomerase family
Ni-Fe hydrogenase	Methylmalonyl-CoA mutase
Carbon monoxide dehydrogenase	Isobutyryl-CoA mutase
Acetyl-coenzyme A decarbonylase/ synthase	Ethylmalonyl-CoA mutase
Superoxide dismutase SodN	Glutamate mutase
Methyl-coenzyme M reductase (using F_{430} as a cofactor)	Methyleneglutarate mutase
	D-lysine 5,6-aminomutase
	Diol dehydratase
	Glycerol dehydratase
	Ethanolamine ammonia lyase
	B_{12}-dependent ribonucleotide reductase
	Methylcobalamin-dependent methyltransferase family
	Methionine synthase (MetH)
	Other methyltransferases: Mta, Mtm, Mtb, Mtt, Mts, and Mtv
	B_{12}-dependent reductive dehalogenase family
	CprA

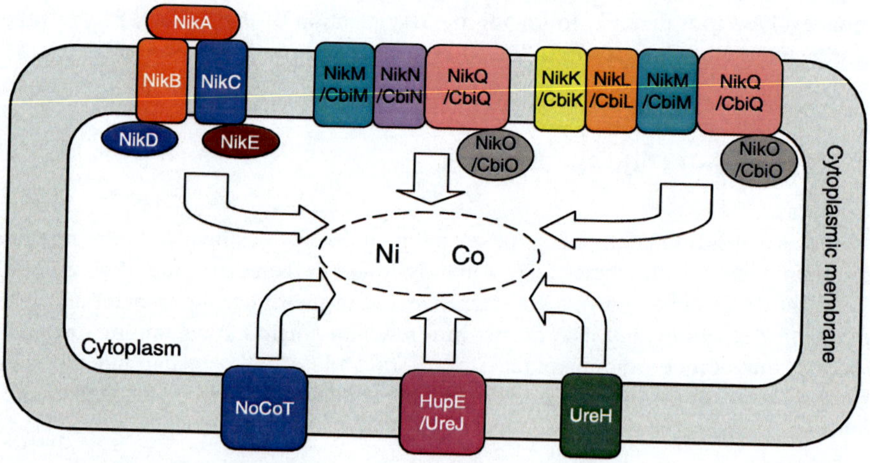

Figure 6 Schematic representation of Ni/Co transport systems. The Ni/Co transport systems include NikABCDE, Nik/CbiMNQO, Nik/CbiKMLQO, NiCoT, HupE/UreJ, and UreH.

family, PepT). This multi-component system is composed of a periplasmic Ni-binding protein (NikA), two integral membrane proteins (NikB and NikC) and two ABC proteins (NikD and NikE). NikA may also bind divalent Co, Cu, and Fe with at least 10-fold lower affinity [154]. In addition, NikA could bind heme in *E. coli*, suggesting an additional function independent of Ni transport [155]. To date, residues involved in Ni binding for NikA have not been well characterized and conflicting results were reported by various groups. For example, it was suggested that *E. coli* NikA binds a natural metallophore containing three carboxylate functions that coordinate a Ni ion via four residues (Tyr402, Arg137, Arg97, and His416), and that His416 (the only direct metal-protein contact) of NikA is essential for Ni uptake in *E. coli* [156]. It was also reported that Ni binds *E. coli* NikA without chelators and is coordinated by two histidine residues (His56 and His442) at a position distant from the previously characterized binding site [157]. In any event, the presence of the majority of these residues could be used to help predict NikA orthologs from Ni-unrelated homologs. Distantly related Ni ABC transporters were also identified in *Yersinia* (named YntABCDE), highlighting diversity of Ni ABC-type transporters in bacteria.

An additional ABC-like transport system, Cbi/NikMNQO, is often encoded next to the B_{12} biosynthesis or urease (a major Ni-dependent enzyme) genes in some bacterial genomes, and was shown to mediate Co and Ni uptake respectively, although the metal-binding ligands are unclear [158]. Comparison of the cbi/nikMNQO operon structures and occurrence of each component revealed that M, Q, and O gene products are universal components. Although the transmembrane proteins CbiN (Co uptake) and NikN (Ni uptake) have no significant homology, they might have the same topology with two transmembrane segments [36]. Two additional proteins, NikK and NikL, were also proposed to be involved in Ni uptake when NikN is absent, and form an alternative NikKMLQO system [36].

Secondary Ni/Co transporters include NiCoT (also designated HoxN, HupN, NicT, NixA or NhlF in different organisms), UreH, and HupE/UreJ [159]. NiCoTs are a family of proteins with eight transmembrane segments. They are widespread among bacteria and found in several thermoacidophilic archaea and certain fungi including *S. pombe* and *Neurospora crassa.* Various subtypes of NiCoTs have different ion preferences ranging from strict selectivity for Ni to unbiased transport of both ions to strong preference for Co. In many cases, the preference for a particular metal correlated with the genomic location of NiCoT genes, which are adjacent to the genes for Ni or Co (or B_{12} biosynthesis) enzymes [19,160]. UreH and HupE/UreJ are putative secondary transporters, and certain members of these families have recently been shown to mediate Ni transport [159–161]. Homologs of UreH were also detected in plants. In addition, several new types of candidate Co transporters were predicted, such as CbtAB, CbtC, CbtD, CbtE, CbtF, CbtG, and CbtX [61,160]. The distribution of these candidate transporters is quite limited.

In eukaryotes, a subfamily of cation-efflux family members (TgMTP1) was found to account for the enhanced ability of Ni hyperaccumulation in higher plants [162]. To date, no high-affinity Co uptake system has been reported in eukaryotes; however, some suppressors of Co toxicity, such as COT1 and GRR1 in *S. cerevisiae*, were characterized, which may play an important role in metal homeostasis by decreasing the cytoplasmic concentration of metal ions including Co and Zn [163].

In many bacteria including *E. coli*, a Ni repressor gene, *nikR*, is located immediately next to its target, the *nikABCDE* operon. NikR-dependent regulation was also predicted for some other Ni transporters, such as NikMNQO and NiCoT, as well as Ni-dependent enzymes such as Ni-Fe hydrogenase [36,164]. These NikRs regulate the transcription of target genes in response to Ni ion concentrations, utilizing a combination of allostery and coordination geometry. The presence of a NikR-binding site that contains an inverted repeat sequence and is always located upstream of Ni-associated genes may help identify NikR-related regulation [36].

5.2 *Nickel-Dependent Proteins*

The characterization of Ni in several enzymes has created an active field exploring the biochemistry of this metal (see Table 3). In prokaryotes, the major strictly Ni-dependent enzymes include urease, Ni-Fe hydrogenase, carbon monoxide dehydrogenase (CODH), acetyl-coenzyme A decarbonylase/synthase (CODH/ACS), superoxide dismutase SodN, and methyl-coenzyme M reductase (MCR). In addition, some homologs of Ni-binding proteins appear to bind other metals. For example, glyoxalase I (GlxI) binds Ni in *E. coli*, *P. aeruginosa*, and human parasites *Leishmania* (e.g., *L. major*) and *Trypanosoma* (e.g., *T. cruzi*), but it binds Zn in *P. putida*, yeast, and human [30,165]. Thus, such proteins could not be used for comparative genomics of Ni utilization because of the uncertainty of the metals they bind in different organisms. In eukaryotes, urease is the only known Ni-dependent enzyme. Additional candidate Ni-containing compounds or proteins have also been described in different organisms [166].

Urease is the first characterized Ni-containing protein that has been found in bacteria, fungi, and plants. It catalyzes the hydrolysis of urea to carbon dioxide and ammonia. In plants, urease is a hexamer of identical chains; whereas in bacteria, it consists of either two or three different subunits [167]. The Ni active site appears to be particularly conserved, as two Ni atoms are associated with each active site of the respective enzymes based on the crystal structures [168].

Hydrogenase catalyzes the reversible reaction of dihydrogen. Based on the metal content and subunit composition of the enzymes, three classes of hydrogenases have been identified: Fe-Fe hydrogenase, Ni-Fe hydrogenase, and hydrogenases that use neither Fe nor Ni [169]. The most studied class comprises the Ni-Fe hydrogenases, which are mainly utilized for hydrogen oxidation. Crystal structures of several Ni-Fe hydrogenases have been identified [170]. One class is composed of two subunits: the large subunit contains the Ni-Fe active site, and the small subunit that contains an Fe-S cluster is used in electron transfer from the large subunit. Other Ni-Fe hydrogenases are tetramers and integral membrane proteins. Two motifs have been proposed to be involved in the ligation of Ni: the N-terminal RxCGxC and the C-terminal DPCxxC. Similar motifs have been found in the sub-class of Ni-Fe-Se hydrogenases which contain a selenocysteine instead of a Cys bound to the Ni atom [171].

Ni-containing CODHs are the biological catalysts for reversible oxidation of CO to CO_2, with water as the source of oxygen. Members of the CODH family have been characterized from archaea and bacteria. The active site of CODH, designated cluster C, is a complex Ni-, Fe-, and S-containing metal center [172]. The recently published high-resolution structure of CODH from *Carboxydothermus hydrogenoformans* in three states demonstrated the mechanism of CO oxidation and CO_2 reduction at the Ni-Fe site of cluster C [173].

CODHs in acetogenic bacteria (anaerobes that can grow autotrophically on the greenhouse gas CO_2) and methanogenic archaea are bifunctional enzymes that perform both the reversible CO-oxidation reaction and the synthesis or degradation of acetyl-coenzyme A (CoA) and are therefore designated CODH/ACS. Both catalytic sites for the individual reactions require Ni for catalysis and are positioned at different sites [174].

MCR is responsible for all biologically produced methane on earth, which catalyzes the final step in the biological synthesis of methane in methanogenic archaea. In contrast to other Ni-dependent proteins, this enzyme contains Ni in a tetrapyrrolic structure known as coenzyme F_{430}, which is found exclusively in methanogens [175]. MCR homologs in some uncultured methanotrophic archaea are involved in anaerobic oxidation of methane in marine sediments. Differences between the highly similar structures of these MCR homologs and methanogenic MCR include a F_{430} modification, a Cys-rich patch and an altered post-translational amino acid modification pattern, which may tune the enzymes for their functions in different biological contexts [176].

SODs are important antioxidant enzymes protecting against superoxide toxicity. Various SODs have been characterized that use Fe/Mn, Cu-Zn (see Sections 4.2), or Ni cofactors to carry out the disproportionation of superoxide. The Ni-containing SOD is a product of the *sodN* gene, which encodes a protein with an N-terminal extension that is removed in the mature enzyme. The crystal structure of the active Ni-bound enzyme from *Streptomyces coelicolor* identified a novel SOD fold and the Ni active site. A nine-residue structural motif (His-Cys-X-X-Pro-Cys-Gly-X-Tyr) provides almost all interactions essential for metal binding and catalysis, and thus may be diagnostic of other SodNs [177].

5.3 *Cobalt-Dependent Proteins*

Although Co is less frequently encountered in metalloenzymes than the other first-row transition metals (e.g., Fe, Cu and Zn), it is nevertheless an essential cofactor in vitamin B_{12}-dependent enzymes. Vitamin B_{12}, also known as cobalamin, is a group of closely related polypyrrole compounds such as cyanocobalamin, methylcobalamin, and deoxyadenosyl cobalamin. They are required for the metabolism of many prokaryotic and eukaryotic organisms.

5.3.1 Vitamin B_{12} Uptake and Biosynthesis

Vitamin B_{12} uptake is critical for B_{12}-utilizing organisms that cannot synthesize the coenzyme *de novo*. To date, the only known transport system for B_{12} in prokaryotes is the BtuFCD system, which includes a periplasmic B_{12}-binding protein BtuF and two ABC transport subunits BtuC and BtuD [178]. The BtuFCD system belongs to a large superfamily involved in the uptake of Fe, siderophores, and heme. In Gram-negative bacteria, a TonB-dependent outer membrane receptor BtuB is also involved in B_{12} uptake and forms a complex with BtuFCD [179]. Mammals have developed a complex system for internalization of this vitamin from the diet. Three binding proteins (haptocorrin, intrinsic factor, transcobalamin) and several specific receptors are involved in the process of intestinal absorption, plasma transport, and cellular uptake [180]. However, the mechanism of B_{12} uptake in other eukaryotes, such as algae, is unclear, although many algae are rich in vitamin B_{12}. It was suggested that algae acquire vitamin B_{12} through a symbiotic relationship with bacteria [181].

In microorganisms that synthesize vitamin B_{12}, it is produced via two alternative routes: oxygen-dependent (aerobic, or "late Co insertion") and oxygen-independent (anaerobic, or "early Co insertion") pathways that differ mainly in the early stages (Figure 7). The aerobic pathway incorporates molecular oxygen into the macrocycle as a prerequisite to ring contraction. The intermediates of the aerobic route from uroporphyrinogen III (uro'gen III) to adenosylcobalamin and more than 20 genes involved in these processes (*cobA-cobW*) have been identified in several bacteria such as *P. denitrificans*. The anaerobic pathway, which was partially resolved in some organisms, such as *S. typhimurium* and *Bacillus megaterium*, takes advantage of the chelated Co ion, in the absence of oxygen, to support ring contraction. It has been suggested that the anaerobic and aerobic routes contain several pathway-specific enzymes [160]. For example, CbiD, CbiG, and CbiK appear to be specific to the anaerobic route of *S. typhimurium*, whereas CobE, CobF, CobG, CobN, CobS, CobT, and CobW are unique to the aerobic pathway of *P. denitrificans*. Besides, an adenosyltransferase that catalyzes the final step in the assimilation of vitamin B_{12} was found to directly transfer the cofactor to a B_{12}-dependent methylmalonyl-CoA mutase in *Methylobacterium extorquens*, suggesting that the strategy of using the final enzyme in an assimilation pathway for tailoring a cofactor and delivering it to a dependent enzyme may also be general for cofactor trafficking [182]. Recently, it was reported that this process is gated by a small G protein, MeaB. While the GTP-binding energy is needed for the editing function; that is, to discriminate between active and inactive cofactor forms, the chemical energy of GTP hydrolysis is required for gating cofactor transfer [183].

5.3.2 Vitamin B_{12}-Dependent Proteins

Considering that vitamin B_{12} is the major form of Co utilization and that B_{12}-binding proteins are strictly dependent on this cofactor, identification of all B_{12}-dependent enzymes is extremely important for comparative genomics of Co utilization. To date, vitamin B_{12} is mainly present in three classes of enzymes in prokaryotes (classified based on different chemical features of the cofactor): adenosylcobalamin-dependent isomerase, methylcobalamin-dependent methyltransferase, and B_{12}-dependent reductive dehalogenase [184,185]. These classes can be further divided into subclasses based on sequence similarity and reactions they catalyze (see Table 3).

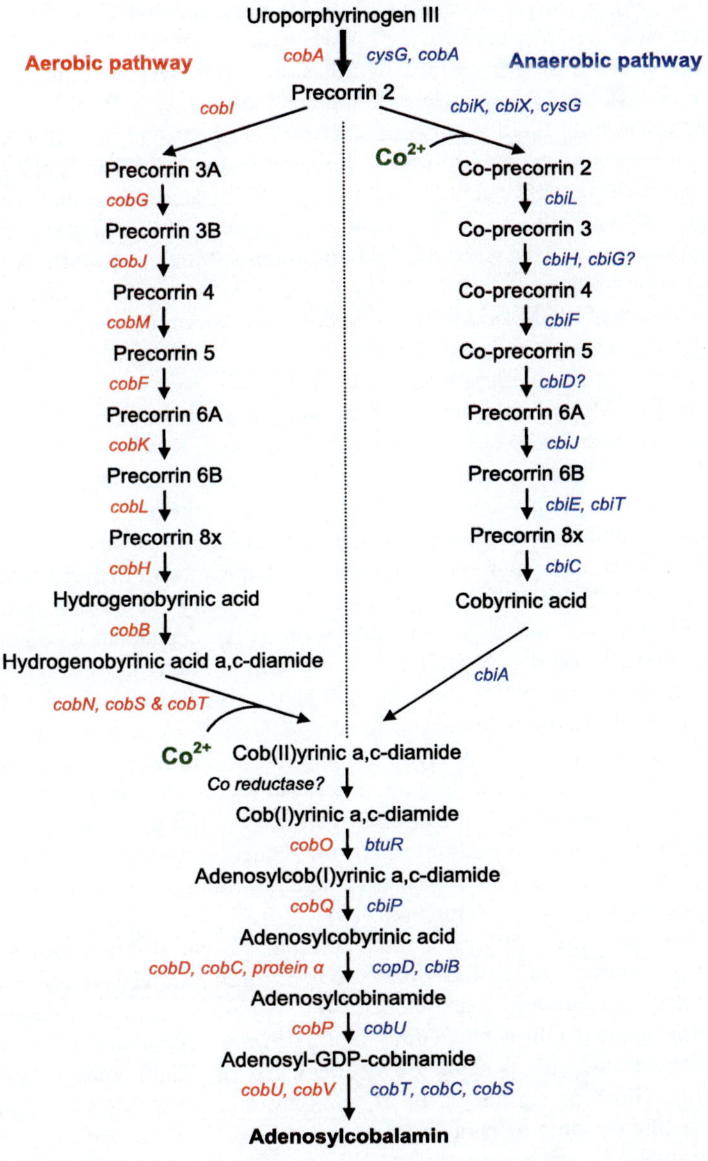

Figure 7 Biosynthetic pathways for vitamin B_{12} in bacteria. Genes involved in aerobic and anaerobic pathways are shown in red and blue, respectively.

Adenosylcobalamin-dependent isomerases are the largest family of B_{12}-dependent enzymes and are mainly found in bacteria, where they catalyze a variety of chemically difficult 1,2-rearrangements that proceed through a mechanism involving free radical intermediates [186]. Subclasses of this family include methylmalonyl-CoA mutase (MCM), isobutyryl-CoA mutase (ICM), ethylmalonyl-CoA mutase (ECM), glutamate mutase (GM), methyleneglutarate mutase (MGM), D-lysine 5,6-aminomutase (5,6-LAM), diol/glycerol dehydratase (DDH/GDH), ethanolamine ammonia lyase (EAL), and B_{12}-dependent ribonucleotide reductase (RNR II).

MCM is the only B_{12}-dependent isomerase that is present in both bacteria and animals. It catalyzes the isomerization of methylmalonyl-CoA to succinyl-CoA in the pathway that converts catabolites of odd-chain fatty acids, branched-chain amino acids, and cholesterol to a key intermediary metabolite [187]. In many organisms, such as *S. cinnamonensis*, it consists of two subunits, MutA and MutB, which show high sequence similarities to MCMs from other bacteria and mammals [188]. The crystal structure of MCM from *Propionibacterium shermanii* revealed the coordination of Co in coenzyme B_{12} by the histidine in the DXHXXG motif within the C-terminal cobalamin-binding domain [189].

ICM and ECM are homologs of MCM with different functions. ICM catalyzes the reversible rearrangement of isobutyryl-CoA to n-butyryl-CoA. It has been detected in a variety of aerobic and anaerobic bacteria, where it appears to play a key role in valine and fatty acid catabolism as well as in the production of fatty acid-CoA thioester building blocks for polyketide antibiotic biosynthesis. In *S. cinnamonensis*, this mutase was found to comprise a large subunit of IcmA and a small subunit IcmB [190]. IcmB shows high sequence similarity to the cobalamin-binding domains of other B_{12}-containing enzymes such as B_{12}-dependent methionine synthase, including the conserved DXHXXG cobalamin-binding motif, suggesting that IcmB has taken on the role of a separate cobalamin-binding domain in ICM. ECM is involved in the central reaction of the ethylmalonyl-CoA pathway and catalyzes the transformation of ethylmalonyl-CoA to methylsuccinyl-CoA in combination with a second enzyme that was identified as promiscuous ethylmalonyl-CoA/methylmalonyl-CoA epimerase. Although ECM showed significant sequence similarity to MCM and ICM from the same organism, sequence analysis revealed that this enzyme is distinct from MCM as well as ICM, and defines a new subfamily of coenzyme B_{12}-dependent acyl-CoA mutases [191].

B_{12}-dependent GM catalyzes a most unusual carbon skeleton rearrangement involving the isomerization of L-glutamate to L-threo-methylaspartate, a reaction that is without precedent in organic chemistry. The active enzyme consists of two subunits (designated GlmE and GlmS in *Clostridium cochlearium*) as an $\alpha2\beta2$ tetramer, whose assembly is mediated by coenzyme B_{12}. The smaller of the protein components, GlmS, is similar to the B_{12}-binding domain of MCM and has been shown to be the B_{12}-binding subunit [192].

B_{12}-dependent MGM from the strict anaerobe *Eubacterium barkeri* catalyzes the equilibration of 2-methyleneglutarate with (*R*)-3-methylitaconate. This enzyme also contains the highly conserved DXHXXG(X)(41)GG motif, which is critical for B_{12} binding [193].

5,6-LAM is an adenosylcobalamin and pyridoxal-5'-phosphate-dependent enzyme that catalyzes a 1,2 rearrangement of the terminal amino group of D-lysine and of L-β-lysine. The crystal structure of a substrate-free form of 5,6-LAM from *C. sticklandii* revealed that a Rossmann domain covalently binds pyridoxal-5'-phosphate and positions it into the putative active site of a neighboring triosephosphate isomerase barrel domain, while simultaneously positioning the other cofactor, adenosylcobalamin, approximately 25 Å from the active site [194]. Thus, this structure features a locking mechanism to keep the adenosylcobalamin out of the active site and prevent radical generation in the absence of substrate.

B_{12}-dependent GDH and DDH are homologous isofunctional enzymes that catalyze the elimination of water from glycerol and 1,2-propanediol to the corresponding aldehyde via a B_{12}-dependent radical mechanism. The crystal structure of the substrate-free form of GDH in complex with cobalamin has been determined, whose overall fold and subunit assembly closely resemble those of DDH. Structural analysis of the locations of conserved residues among various GDH and DDH sequences helped identify residues important for substrate preference and specificity of protein-protein interactions [195].

EAL catalyzes the deamination of ethanolamine and 2-aminopropanol. Computational modeling of EAL from *S. typhimurium* revealed that this enzyme may have a similar TIM-barrel fold as DDH and GDH [196].

RNR catalyzes a rate-limiting reaction in DNA synthesis by converting ribonucleotides to deoxyribonucleotides. To date, three major classes of RNR have been discovered that depend on different metal cofactors for the catalytic activity: class I RNRs contain a diiron-oxygen cluster, class II contain vitamin B_{12}, and class III use an Fe-S cluster coupled to S-adenosylmethionine [197]. RNR II enzymes are mainly found in bacteria, and also in some of their phages. They utilize an adenosylcobalamin cofactor that interacts directly with an active Cys residue to form the reactive Cys radical needed for ribonucleotide reduction [198].

The methylcobalamin-dependent methyltransferases play important roles in amino acid metabolism in a variety of organisms, including mammals, as well as in carbon metabolism and CO_2 fixation in anaerobic microbes. These methyltransferases could be divided into two subclasses: one subclass binds simple substrates such as methanol (MtaB), methylated amines (MttB, MtbB, MtmB), methylated thiols (MtsB), methoxylated aromatics (MtvB), and methylated heavy metals, while the other, such as methionine synthase (MetH), catalyzes methyl transfer from methyltetrahydrofolate and the methanogenic analog methyltetrahydromethanopterin [184].

Methionine synthase (MetH) catalyzes the transfer of a methyl group from N_5-methyltetrahydrofolate to homocysteine, producing tetrahydrofolate and methionine [199]. This enzyme is the most extensively studied B_{12}-dependent methyltransferase, which is widespread in all three domains of life. It is a modular enzyme with distinct regions for binding homocysteine, methyltetrahydrofolate, B_{12}, and adenosylmethionine. The B_{12} domain in its different oxidation states may interact with each of the other three domains. The crystal structure of a B_{12}-containing fragment of MetH from *E. coli*, which was the first structure of a protein-bound B_{12}, revealed that the histidine residue in the DXHXXG motif is the Co ligand and is part of a catalytic quartet, Co-His759-Asp757-Ser810, that modulates the reactivity of the B_{12} prosthetic group in MetH [200].

Other B_{12}-dependent methyltransferases are designated as Mtx, where x denotes the methyl donor (e.g., a, methanol; v, vanillate; m, methylamine; b, dimethylamine; t, trimethylamine; and s, dimethylsulfide). These methyltransferases consist of three components (Mt_A, Mt_B, and Mt_C) [184]. Each component is found on a different polypeptide or domain. Mt_A methylates coenzyme M (CoM), Mt_B methylates the corrinoid protein, and Mt_C is the corrinoid protein containing B_{12}. These methyltransferases are essential in energy metabolism and in cell carbon synthesis in anaerobic microbes such as methanogenic archaea and acetogenic bacteria [201]. In addition, methyltetrahydromethanopterin:CoM methyltransferase (Mtr), which contains eight subunits (MtrA-H), was found to utilize a histidine as the ligand to the cobalamin in MtrA [202].

B_{12}-dependent reductive dehalogenases CprA play an important role in the detoxification of aromatic and aliphatic chlorinated organics in anaerobic microbes. Most of these enzymes also contain Fe-S clusters [203]. The role of B_{12} in CprA appears to be significantly different from those of the B_{12}-dependent isomerases and methyltransferases. However, the reaction mechanism of dehalogenases remains unclear.

In eukaryotes, only three B_{12}-dependent enzymes, MetH, MCM, and RNR II, have been identified, implying that Co utilization is quite restricted in this domain of life.

5.3.3 Non-Corrin Cobalt-Binding Proteins

A few proteins containing non-corrin-Co have been reported in different organisms, including methionine aminopeptidase (from *S. typhimurium*), prolidase (from *P. furiosus*), nitrile hydratase (from *Rhodococcus rhodochrous*), methylmalonyl-CoA carboxytransferase (from *P. shermanii*), aldehyde decarbonylase (from *Botryococcus braunii*), glucose isomerase (from *S. albus*), and several other proteins [153]. However, all of these enzymes are not strictly Co-dependent and may use other metals (such as Fe, Zn, and Mn) in place of Co. Thus, it is difficult to identify the metal specificity of these enzymes by computational analysis. To date, only nitrile hydratase (NHase) was suggested to have different active site motifs for Co- and Fe-binding forms [204].

5.4 *Comparative Genomics of Nickel, Cobalt, and Vitamin B_{12} Utilization*

As mentioned above, Ni and Co are essential cofactors in several enzymes. Ni is used in several metalloenzymes involved in energy and nitrogen metabolism, detoxification processes, pathogenesis, enzyme inactivation, and lipid peroxidation, whereas Co is primarily found in the corrin ring of coenzyme B_{12} that plays important roles in several biological systems. In recent years, several comparative genomics studies have been carried out to investigate Ni and Co utilization traits.

An early study examined Ni and Co transport systems in about 200 microbial genomes and demonstrated a complex and mosaic utilization of both metals in prokaryotes [36]. Two computational approaches were used for functional prediction of proteins involved in Ni or Co uptake: (i) analysis of the genomic locations of genes encoding Ni/Co transporters; and (ii) identification of regulatory signals, such as NikR-dependent regulation through the NikR-binding signal, and B_{12} riboswitches that regulate many of the candidate Co transporters in bacteria. This study revealed that the Ni/Co transporter genes are often colocalized with the genes for Ni-dependent and B_{12} biosynthesis proteins. Different families of Ni/Co transporters showed a mosaic distribution in those organisms, and the Cbi/NikMNQO system (including the Cbi/NikKMLQO system) appeared to be the most widespread group of microbial transporters for the two metal ions.

A separate computational analysis of B_{12} metabolism and regulation also provided important information regarding B_{12} utilization in prokaryotes [160]. Using comparative analysis of gene regulation, positional clustering of genes, and phylogenetic profiling, the B_{12} biosynthesis and regulation was described in a variety of prokaryotes. The B_{12} riboswitch was found to be widely distributed in the regions upstream of B_{12} biosynthetic and transport genes. In addition, by searching for candidate B_{12}-regulated genes, several new types of candidate Co transporters and new proteins associated with the B_{12} biosynthesis pathway, such as certain chelatases and methyltransferases, were identified. The B_{12} transporters, BtuFCD, appeared to be widely distributed in prokaryotes and some of them were B_{12}-regulated. However, it is difficult to selectively identify BtuFCDs among other highly similar transport systems (such as Fe/heme or siderophore transporters) in the majority of sequenced organisms. Furthermore, the B_{12} element was predicted to regulate B_{12}-independent MetH and RNR isozymes in bacteria that also have corresponding B_{12}-dependent isozymes.

Recently, two much more extensive comparative analyses involving more than 700 organisms in all three domains of life have been carried out, which have provided additional important information regarding Ni and Co utilization traits [96,205]. Only strictly Ni-dependent metalloproteins and B_{12}-binding enzymes were used for comparative genomics of Ni and Co, respectively. Occurrence of the Ni/Co-utilization trait and Ni- or B_{12}-dependent proteins is shown in Figure 8. The distribution and dynamics of Ni and Co utilization were analyzed at the level of both transporters and metalloproteomes. These analyses revealed that both metals are widely used by prokaryotes; however, analyses of occurrence of Ni/Co transporters and metalloenzymes showed great diversity among bacteria and archaea. Urease and B_{12}-dependent MetH were the most widespread Ni- and Co-containing proteins, respectively, in bacteria. In contrast, Ni-Fe hydrogenase and RNR II were the most widespread Ni and Co users in archaea where urease and MetH were very rare or even absent. Further analyses of Ni- or Co-dependent metalloproteomes revealed that, except for deltaproteobacteria and several *Methanosarcina* species, most prokaryotes contained small Ni- and Co-dependent metalloproteomes (1–4 proteins). The largest Ni-dependent metalloproteome was observed in *Deltaproteobacterium* MLMS-1 (16 Ni-binding proteins) and the largest B_{12}-dependent metalloproteome in *Dehalococcoides sp.* CBDB1 (35 B_{12}-binding proteins). Further analysis of Ni and Co utilization based on different habitats, environments, and other factors revealed that, similar to Mo utilization, host-associated organisms (particularly obligate intracellular parasites and endosymbionts) have a tendency for reduced Ni or Co utilization.

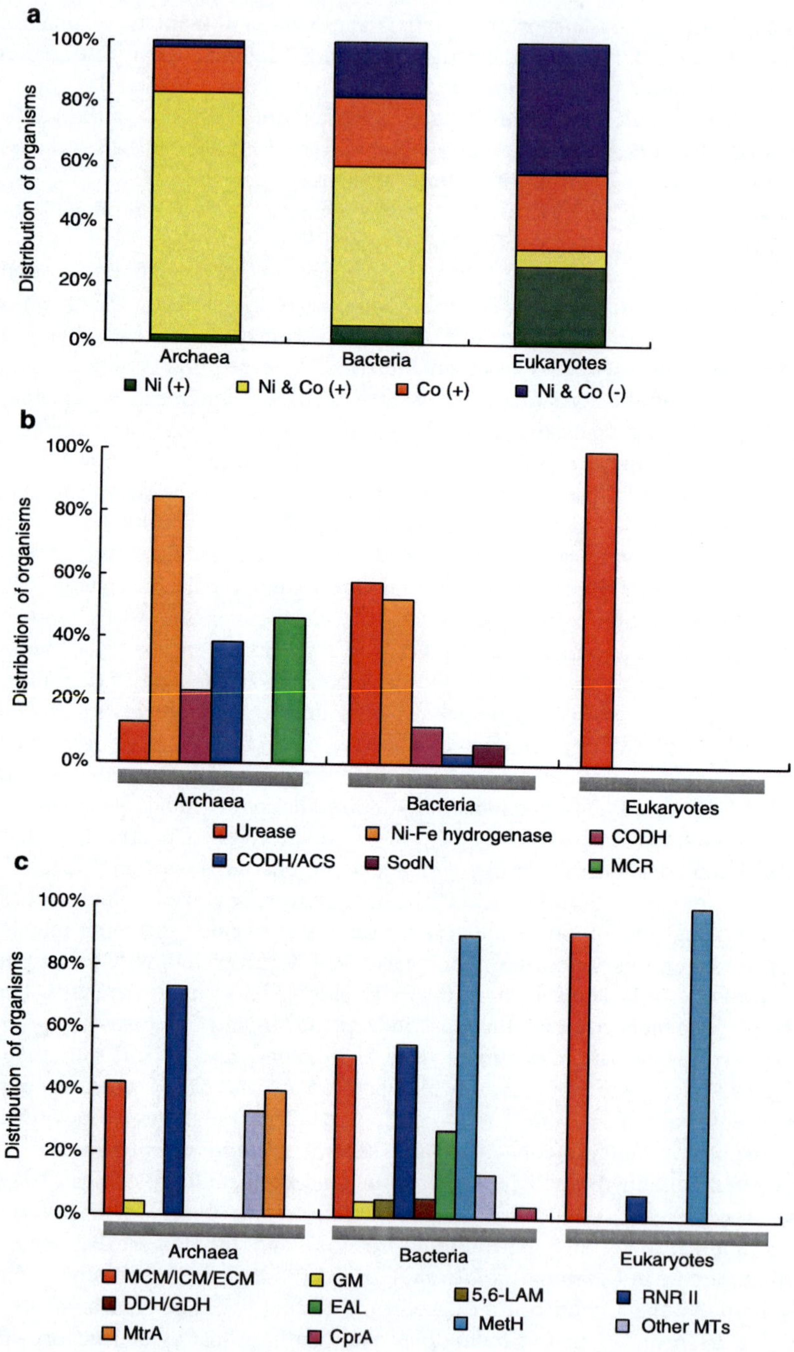

Figure 8 Occurrence of Ni and Co utilization in the three domains of life. (**a**) Distribution of Ni-/Co-utilizing organisms among those with completely sequenced genomes. All organisms were classified into four groups: Ni (+), i.e., containing the Ni utilization trait only; Ni & Co (+), i.e.,

Investigation of Ni and Co utilization in eukaryotes provided a somewhat different Ni and Co utilization trend. Indeed, the use of these two metals is much more restricted in eukaryotes, with regard to both the organisms that utilize Ni/Co and the number of Ni transporters and Ni/B_{12}-dependent protein families. Surprisingly, very few of these organisms utilize both metals. The Ni-utilizing eukaryotes are mostly fungi (except saccharomycotina) and plants, whereas most B_{12}-utilizing organisms are animals. The NiCoT transporter family is the most widely used eukaryotic Ni transporter. Urease and MetH were the most common eukaryotic Ni- and B_{12}-dependent enzymes, respectively. Analysis of Ni- and Co-dependent metalloproteomes in eukaryotes did not reveal organisms that contained many such proteins. Only single copies of urease and 1–3 B_{12}-dependent proteins were detected in various organisms. In contrast to the majority of unicellular organisms that lack B_{12} utilization, *Dictyostelium discoideum* and *Phytophthora* species contained all three known eukaryotic B_{12}-dependent proteins: MetH, MCM, and RNR II.

6 Comparative Genomics of Other Metals

Besides the metals discussed above, comparative genomics studies have also been carried out for additional metals, such as Zn and Fe, but widespread use of these elements across living organisms makes comparative analyses more challenging. In the following two sections, we briefly introduce recent advances on these metals.

6.1 Comparative Genomics of Zinc-Dependent Metalloproteomes

Zn is thought to be essential for all organisms and was suggested to be a key element in the origin of life [206]. It is found in a great variety of enzymes, structural proteins, transcription factors, and ribosomal proteins. We refer to other articles for review on Zn uptake, storage, homeostasis, and user proteins [6,15,20,25,29] and only focus here on comparative genomics of Zn-dependent metalloproteomes.

Figure 8 (continued) containing Ni and Co utilization traits; Co (+), i.e., containing the Co utilization trait only; Ni & Co (-), i.e., containing neither Ni nor Co utilization traits. (**b**) Distribution of organisms containing different Ni-dependent proteins in Ni-utilizing organisms. (**c**) Occurrence of B_{12}-dependent proteins in Co-utilizing organisms. CODH, carbon monoxide dehydrogenase; CODH/ACS, acetyl-coenzyme A decarbonylase/synthase; SodN, Ni-containing superoxide dismutase; MCR, methyl-coenzyme M reductase; MCM, methylmalonyl-CoA mutase; ICM, isobutyryl-CoA mutase; ECM, ethylmalonyl-CoA mutase. The latter three subfamilies are quite similar and are combined into one group. GM, glutamate mutase; 5,6-LAM, D-lysine 5,6-aminomutase; RNR II, B_{12}-dependent ribonucleotide reductase; DDH/GDH, diol/glycerol dehydratase; EAL, ethanolamine ammonia lyase; MetH, methionine synthase; Other MTs, various B_{12}-dependent methyltransferases such as Mta, Mtm, Mtb, Mtt, Mts, and Mtv systems; MtrA, methyltetrahydromethanopterin:CoM methyltransferase subunit A; CprA, B_{12}-dependent reductive dehalogenase.

An early computational search was carried out for Zn proteomes in representative organisms from the three domains of life including humans [34]. A list of known Zn-binding protein domains and of known Zn-binding sequence motifs (Zn-binding patterns) were compiled and then used jointly to analyze the proteome of 57 different organisms (40 bacteria, 12 archaea, and 5 eukaryotes) to obtain an overview of Zn usage by prokaryotic and eukaryotic organisms. It was found that Zn proteins are widespread in living organisms. Within each domain of life, the number of Zn-containing proteins in an organism correlates with the proteome size. Prokaryotes, on average, have a lower proportion of Zn proteins (6.0% ±0.2% of the entire proteome in archaea and 4.9% ±0.1% in bacteria) than eukaryotic organisms. Interestingly, it was observed that the proteome of the hyperthermophilic prokaryotes is enriched in putative Zn-binding proteins, which may be due to an increased use of Zn to enhance the structural stability of proteins by organisms living at higher temperatures. Eukaryotic proteomes are much richer in putative Zn-binding proteins. On average, the Zn-proteome constitutes 8.8% ±0.4% of the eukaryotic proteome (about 10% in humans). Approximately two-thirds of prokaryotic Zn proteins have homologs in eukaryotes. On the other hand, three-quarters of eukaryotic Zn proteomes comprise proteins encoded only in eukaryotes, suggesting that they are relatively more recent.

There is also a functional diversification of eukaryotic and prokaryotic Zn proteomes. Prokaryotes use Zn proteins mostly to perform enzymatic catalysis, whereas in eukaryotes the Zn proteome is almost equally involved in performing catalysis and in regulating DNA transcription. Such a broad difference in function has a correspondence with the organization of the Zn-binding patterns. The Zn-binding patterns containing four ligands are associated with structural sites where Zn contributes to the stability of the protein structure, whereas the patterns containing three ligands are associated with catalytic sites (the fourth ligand is often water) where Zn actively participates in the reaction mechanism of the enzyme [207]. In addition, identity of the amino acids in the pattern is also quite different among structural and catalytic sites. For example, approximately 2800 Zn-binding proteins were found in humans, 97% having a structural Zn site with at least one Cys ligand, and 40% having four Cys ligands [208]. On the other hand, nearly one-third of human proteins with a three-ligand Zn-binding pattern have a pattern with three histidines. Together, four- and three-ligand patterns account for approximately 96% of all human Zn proteins.

As noted above, the majority of known Zn-dependent enzymes could be detected in both prokaryotes and eukaryotes, implying that Zn has been exploited in the catalytic site of enzymes prior to the split of the three domains of life. On the other hand, Zn-binding transcription factors are almost exclusively a prerogative of eukaryotes. These proteins normally contain Zn-finger domains, which are much rarer in bacteria and archaea. This observation suggests that Zn-binding transcription factors have evolved to meet the need of higher organisms to regulate more complex processes such as cell compartmentalization and differentiation. In addition, transcription factors bind Zn in very similar sites, most often composed by cysteine and histidine and organized in the same structure. The conservation of Zn-finger binding sites could be associated with their more recent origin, whereas the differentiation of the catalytic Zn-binding sites could be the result of evolutionary processes that resulted in the development of different enzymatic reactions targeting different physiological substrates [209]. Using a similar approach, a recent comparative study on Zn-finger proteins and Zn hydrolytic enzymes in 821 organisms from the three domains of life revealed that there is a correlation in the changes during evolution related to environment [210].

6.2 Advances in Comparative Genomics of Other Metals

Iron is essential for life and is the most abundant transition metal ion in living organisms. In cells, Fe is normally found in the +2 and/or +3 oxidation states. Besides Fe ions, proteins can bind a range of Fe-containing cofactors, such as heme or Fe-S clusters. However, some of their close analogs may bind other metals in certain organisms or under certain conditions. For example, an acidophilic archaeon, *Ferroplasma acidiphilum*, was recently identified to possess a very large number of Fe-binding proteins (~86% of the entire protein repertoire), including many metalloproteins that bind different metals (such as Zn and Mn) in other organisms and proteins that are not even known to bind metal [211]. On the other hand, organisms that survive under Fe limitation have also been reported although it is unclear if they do not use this metal under Fe-sufficient conditions [212,213].

Although it is difficult to study the complete Fe-dependent metalloproteomes through computational approaches, a preliminary comparative genomics study has been carried out for the proteome-level analyses of the occurrence of non-heme Fe proteins in a selected number of prokaryotes and eukaryotes [35]. In this work, a similar bioinformatic approach based on the use of non-heme Fe-binding patterns in combination with the analysis of the occurrence of protein domains known to bind non-heme Fe was applied. In contrast to what was observed for Zn, the Fe proteome constitutes a higher fraction of the proteome in archaea (on average 7.1% ±2.1%) than in bacteria (3.9% ±1.6%) and eukaryotes (1.1% ±0.4%). The majority of these proteins have homologs in all three domains of life (~90% of the total), suggesting that extant organisms have inherited the large majority of their Fe proteome from the last common ancestor. The majority of non-heme Fe sites are found in proteins involved in electron transfer or in enzymes performing oxidoreductase functions, which is consistent with the fact that Fe is the most used metal ion in redox catalysis [214]. In addition, Fe is the metal ion with the largest variety of binding sites in proteins, including several types of Fe-S clusters and heme cofactors. This may be due to the necessity to use different chemical environments to modulate the reduction potential of Fe and thus its reactivity. Fe-S clusters are the cofactors of about 40% of non-heme Fe proteins retrieved, and their binding patterns are most often composed of cysteine residues. It is worth noting that cysteine is conversely an uncommon ligand for all the other non-heme Fe sites, where histidine is the most widespread ligand [209].

Heme is the prosthetic group of many proteins that carry out a variety of key biological functions, including oxygen transport and sensing, enzyme catalysis, and electron transfer. The utilization of heme requires a relatively complex machinery for its biosynthesis, insertion into heme-containing proteins, and uptake from external sources. The ability to bind this cofactor is strongly influenced by the interaction of the polypeptide chain with the porphyrin moiety. A recent genome-based analysis of proteins specifically involved in the processes of heme biosynthesis and uptake in 474 prokaryotes revealed that different systems exist in organisms belonging to various branches of the tree of life [215]. Some organisms (14%) presumably cannot perform either of the two processes (14%), some (40%) can perform only one of them, and some (46%) can perform both. Among these organisms, many Gram-positive pathogens support heme uptake from the host, suggesting that this process can be a potential target for wide-spectrum antibiotics. Further inspection of the sequences and structural models for two key domains in the heme uptake pathway suggested that there are possible alternative modes of heme binding. In the future, it would be useful to use computational and comparative approaches for the analysis of the variability of additional heme-binding modes and heme-dependent proteomes in organisms from the three domains of life.

The methods described in this chapter may be, in theory, applied to the study of other metals, such as Mn and Cr, and to the characterization of the corresponding metalloproteins. However, so far it has been difficult to identify complete metalloproteomes for most such metals because of either the limited knowledge of their metabolism or unspecificity/uncertainty of metal-binding ligands. A recent study that was based on liquid chromatography, high-throughput tandem mass spectrometry, and inductively coupled plasma mass spectrometry analyses showed that metalloproteomes are much more extensive and diverse than previously recognized [216]. Based on this study, a computational approach was also developed to predict a number of candidate novel metalloproteins for different metals [217]. Further efforts are needed for experimental verification of these proteins as well as identification of additional metal-binding proteins and metal-binding motifs/patterns.

7 Comparative Genomics of Metal Dependency in Biology

From the many studies on the functions of metals, metalloproteins emerged as one of the most diverse sets of proteins [218]. Some protein families are strictly dependent on certain metals (e.g., Cu in COX I, Ni in urease, and Mo in DMSOR), whereas other families have both metal-dependent and metal-independent forms or evolved to use alternative metals (e.g., GlxI binds Ni in *E. coli* but Zn in humans) [30,165]. In addition, initial genome-wide studies identified a significant number of proteins that bind metals and suggested that information on the occurrence of metal-dependent/metal-independent members of protein families may help better understand the utilization of these micronutrients [219,220]. These observations highlight complex evolutionary dynamics of the dependence of proteins on metals. In this section, we focus on recent advances in comparative studies of metals, discuss metalloprotein families containing metal-dependent and metal-independent forms, and explore the evolutionary dynamics of the metal dependence of these families.

Comparative genomics studies on Zn metalloproteins have been discussed in Section 6.1. Taking advantage of the abundance of Zn-binding protein families, their Zn dependence could be systematically analyzed. An early comparative study was carried out to investigate the evolutionary history of ribosomal proteins that are present in all genomes and are generally well conserved [221]. Members of each examined ribosomal protein family were extracted from approximately 40 genomes (mostly bacteria). Several ribosomal proteins, such as S14, S18, L31-L33, and L36, all of which bind Zn via four conserved Cys or His residues, also have a Zn-independent form, in which these metal-binding residues have been partially or completely lost. In addition, genomes containing multiple copies of these ribosomal proteins encoded both Zn-dependent and Zn-independent forms. Further analyses revealed that, in most cases, a duplication of an ancestral Zn-dependent form had occurred early during evolution, with subsequent alternative loss of Zn-dependent and Zn-independent forms in different lineages. Another comparative genomics study that analyzed Zur (a repressor of Zn transport) regulation in bacteria has found that Zn repression was also detected in some paralogs of ribosomal protein genes (such as L36, L33, and S14) in which Zn-binding residues were disrupted [31]. This observation suggested a potential mechanism for maintaining Zn availability under Zn-restricted conditions, as these non-Zn-binding paralogs were expressed to partially replace the Zn-binding proteins, thereby freeing up some Zn for the Zn-binding proteins. Similar situations were also reported in another comparative study in which a subset of proteins in the diverse COG0523 family of putative metal chaperones were found to play a predominant role in the response to Zn limitation based on the presence of the corresponding COG0523-encoding genes downstream from putative Zur-binding sites in many bacterial genomes [222].

Very recently, all Zn-containing proteins with defined ligands in the PDB dataset have been analyzed [223]. Approximately 20% of the Zn protein families in the PDB had a significant number of Zn-independent forms, e.g., methionine-R-sulfoxide reductase, several tRNA synthetases, ribosomal proteins, and various subunits of DNA polymerase III. Further analysis of the predicted Zn dependence of a variety of Zn protein families in hundreds of sequenced bacterial genomes revealed that the majority of organisms containing these proteins possessed only single forms. Phylogenetic analyses suggested a role of both vertical inheritance and horizontal gene transfer in shaping the evolution of Zn protein families, which is consistent with previous findings [221]. Overall, all these results suggested a general picture of evolution of Zn utilization: many Zn protein families are strictly dependent on Zn, whereas some families yielded Zn-independent forms by disrupting ligands to Zn ion. These Zn-independent forms may have already been present in the ancestors of prokaryotes and eukaryotes and are widespread in currently living organisms.

As noted above, several comparative genomics studies have been carried out to analyze Cu utilization and to identify cuproproteins and cuproproteomes in organisms [37,96,148,150]. These studies also represent a new resource for studying the

Cu dependence of cuproprotein families. Besides Cu-dependent COX II and Cu-independent quinol oxidase subunit II (see Section 4.2), loss of Cu ligands (cysteine, histidine, and sometimes methionine) has been observed in members of several other cuproprotein families, which was mostly accompanied by changes in the function of these proteins.

As mentioned in Section 4.2, tyrosinases contain a type 3 Cu center and are distributed in all three domains of life. The active site of tyrosinase is characterized by a pair of antiferromagnetically coupled Cu ions (CuA and CuB), each coordinated by three histidine residues. In animals, two additional tyrosinase-related proteins (TRP1 and TRP2) that display significant homology to tyrosinase and that originated by the duplication of the ancestral tyrosinase gene were also detected. TRP1 is a 5,6-dihydroxyindole-2-carboxylic acid oxidase, and TRP2 is a dopachrome tautomerase, both of which are involved in melanin biogenesis [224]. TRP1 and TRP2 have the same six metal-binding histidine residues as tyrosinase. However, whereas TRP2 binds Zn, TRP1 binds an unknown metal that is not Cu, Zn, or Fe [225]. Thus, the specific binding of different metals by these proteins may be responsible for their distinct catalytic functions in melanogenesis. Although it is unclear how to distinguish Cu-dependent and Cu-independent forms in the tyrosinase family, an additional conserved histidine is located immediately upstream of one of three histidine residues in the CuB site in all examined tyrosinases, but it is replaced by Leu in all TRP1 and TRP2 proteins (Figure 9a). It has been shown that this histidine is essential for Cu binding in tyrosinase from *Aspergillus oryzae* [226]. Thus, it may play an important role in binding CuB and may help identify Cu-dependent tyrosinases in sequence databases. However, a recent study suggested the ability of the CuA site in mouse TRP1 to bind Cu and sustain the typical tyrosinase enzymatic activities [227]. It is possible that tyrosinase may acquire Cu inefficiently and subsequently lose it within the trans-Golgi network of melanocytes and then be reloaded with Cu within melanosomes to catalyze melanin synthesis [228]. This scheme would be consistent with the presence of an inactive "Cu-independent" form of Cu-dependent tyrosinase, suggesting an exquisite and complex control of tyrosinase activity.

Hemocyanin also belongs to the type 3 cuproprotein family and uses six histidine residues to bind two Cu ions. Members of the hemocyanin family have been detected only in the hemolymph of animals in the phyla Arthropoda and Mollusca. All sequenced Arthropoda contain both Cu-dependent and Cu-independent forms of proteins of this family, suggesting that they may have co-occurred in the ancestor of modern arthropods [223]. These Cu-independent hemocyanin-derived proteins (designated hexamerins) were previously thought to have lost the ability to bind Cu and transport oxygen; instead, they became storage proteins to serve as sources of amino acids during metamorphosis [229]. Similar to tyrosinase, an additional histidine was found immediately upstream of one of three histidine residues in the first Cu site in all Cu-dependent hemocyanins, but it was absent in all Cu-independent hexamerins (Figure 9b), implying that this additional histidine may be involved in Cu binding in hemocyanins.

Cu dependence is more conserved than Zn dependence, probably because of the prevalence of catalytic functions of Cu in cuproproteins, whereas in many Zn-containing proteins, Zn plays more subtle roles, such as structural integrity, that

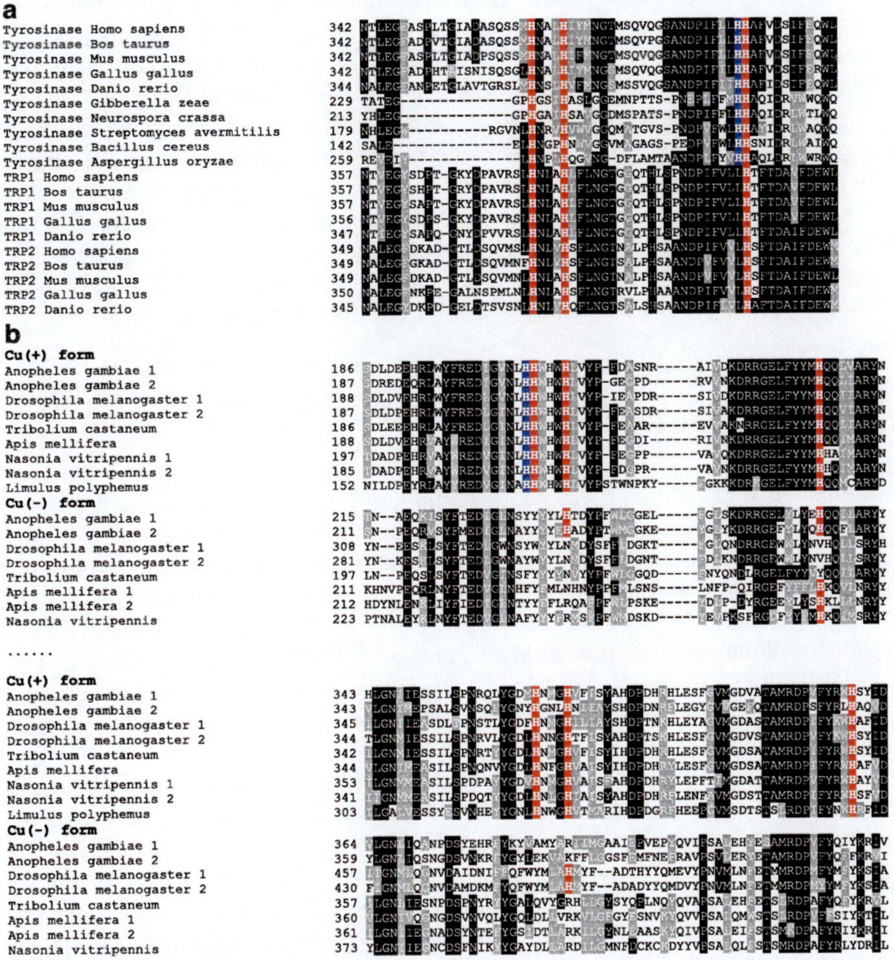

Figure 9 Multiple alignment of Cu-dependent and Cu-independent members of two cuproprotein families. (**a**) The tyrosinase family. Only CuB-binding sites are shown. Three His residues involved in metal binding are shown in a red background. A conserved His that is detected only in tyrosinases and that might also be involved in Cu binding is highlighted in blue. (**b**) The hemocyanin family. Only sequences corresponding to two Cu-binding sites are shown. The six His residues involved in Cu binding are shown in red background. An additional conserved His that is only detected in the Cu-dependent form is shown in blue.

could also be achieved by other means. Interestingly, a recent study revealed that, in the cyanobacterium *Synechocystis PCC 6803*, Mn is utilized by a cupin protein that folds in the cytoplasm, whereas Cu is utilized by a similar protein that folds in the periplasm [230]. This study offered a mechanism whereby the compartment in which a protein folds overrides its binding preference to control its metal content.

With regard to other metals, previous comparative studies have shown that molybdoenzymes are strictly dependent on Mo. Similarly, the majority of Ni-dependent proteins are strictly dependent on this metal although additional Ni-binding proteins were reported in some organisms (such as GlxI). All vitamin B_{12}-dependent proteins contain a B_{12}-binding domain and are strictly dependent on this coenzyme. It is very difficult to predict Co-dependent and Co-independent forms of non-corrin Co-binding protein families solely based on computational approaches. To date, only NHase is known to have different active site motifs for Co- and Fe-binding forms [231,232]. It was previously found that only several sequenced organisms (~4%) contain NHases, and most of them are Co-binding proteins. The Fe-containing NHases might have evolved from Co-binding NHases [205].

8 Concluding Remarks

Comparative genomics provides a powerful tool for studying metal utilization and its evolutionary trends. The majority of these studies used strategies based on either identification of metalloproteomes using metal-binding motifs/patterns or investigation of metal utilization traits (e.g. transporters, regulators, cofactor biosynthesis pathways, and metal-dependent proteins). Currently, it is difficult to identify complete metalloproteomes for most metals as there is no reliable method to predict all metal-binding proteins. Nevertheless, comparative genomics studies have provided significant advances in unraveling the general principles of utilization of metals across the three domains of life.

In this chapter, we discussed how comparative genomics can be used to analyze the function and evolution of metal utilization. We described recent studies that used computational studies, especially comparative genomics analyses, to better understand the utilization of several major transition metals and to offer new avenues for further experimental analyses. These studies offered new insights and helped understand the evolutionary dynamics of metal dependence in proteins and organisms. It may be expected that, in the next few years, with the increased availability of sequenced genomes and improved tools for their analyses, comparative genomics will play a significant role in better understanding of metal utilization and evolution.

Abbreviations and Definitions

5,6-LAM	D-lysine 5,6-aminomutase
ABC	ATP-binding cassette
AO	aldehyde oxidase
AOR	aldehyde:ferredoxin oxidoreductase
CoA	coenzyme A
CODH	carbon monoxide dehydrogenase

CODH/ACS	acetyl-coenzyme A decarbonylase/synthase
CoM	coenzyme M
COX	cytochrome c oxidase
cPMP	cyclic pyranopterin monophosphate
Ctr	Cu transporter
CuAO	Cu amine oxidase
Cu-Zn SOD	Cu-Zn superoxide dismutase
Cys	cysteine
DBM	dopamine β-monooxygenase
DDH/GDH	diol/glycerol dehydratase
DMSOR	dimethylsulfoxide reductase
EAL	ethanolamine ammonia lyase
ECM	ethylmalonyl-CoA mutase
EPR	electron paramagnetic resonance
GAO	galactose oxidase
GlxI	glyoxalase I
GM	glutamate mutase
GTP	guanosine 5′-triphosphate
ICM	isobutyryl-CoA mutase
L-dopa	L-dihydroxyphenylalanine
mARC	mitochondrial amidoxime-reducing component
MCM	methylmalonyl-CoA mutase
MCO	multicopper oxidase
MCP	Moco carrier protein
MCR	methyl-coenzyme M reductase
MetH	methionine synthase
MGD	molybdopterin guanine dinucleotide
MGM	methyleneglutarate mutase
MoBP	Moco-binding protein
Moco	Mo cofactor
Mop	Mo-binding protein
MPT	molybdopterin
Mtr	methyltetrahydromethanopterin:CoM methyltransferase
N_2OR	nitrous oxide reductase
NADH	nicotinamide adenine dinucleotide (reduced)
NCBI	National Center for Biotechnology Information
Ndh2	NADH dehydrogenase 2
NHase	nitrile hydratase
NiR	nitrite reductase
NMR	nuclear magnetic resonance
NR	nitrate reductase
PAM	peptidylglycine R-amidating monooxygenase
PDB	Protein Data Bank
PHM	peptidylglycine R-hydroxylating monooxygenase
pMMO	particulate methane monooxygenase

RNR II	B_{12}-dependent ribonucleotide reductase
SO	sulfite oxidase
SOD	superoxide dismutase
SodN	Ni-containing superoxide dismutase
TRP	tyrosinase-related proteins
Wco	tunftsen cofactor
XDH	xanthine dehydrogenase
XO	xanthine oxidase

Acknowledgments This work was supported, in whole or in part, by National Institutes of Health Grants GM061603 and CA080946 (to V. N. G.), and by the National Natural Science Foundation of China under NO. 31171233 (to Y. Z.).

References

1. H. Haraguchi, *J. Anal. At. Spectrom.* **2004**, *19*, 5–14.
2. S. B. Goldhaber, *Regul. Toxicol. Pharmacol.* **2003**, *38*, 232–242.
3. D. P. Cuthbertson, *Bibl. Nutr. Dieta.* **1973**, *18*, 65–91.
4. W. Mertz, *Biol. Trace Elem. Res.* **1998**, *66*, 185–191.
5. A. Van Gossum, J. Neve, *Curr. Opin. Clin. Nutr. Metab. Care* **1998**, *1*, 499–507.
6. J. C. King, *Am. J. Clin. Nutr.* **2011**, *94*, 679S–684S.
7. D. Galaris, K. Pantopoulos, *Crit. Rev. Clin. Lab. Sci.* **2008**, *45*, 1–23.
8. J. E. Posey, F. C. Gherardini, *Science* **2000**, *288*, 1651–1653.
9. L. A. Ba, M. Doering, T. Burkholz, C. Jacob, *Metallomics* **2009**, *1*, 292–311.
10. S. I. Patzer, K. Hantke, *Mol. Microbiol.* **1998**, *28*, 1199–1210.
11. V. V. Bartsevich, H. B. Pakrasi, *J. Biol. Chem.* **1996**, *271*, 26057–26061.
12. J. A. Maupin-Furlow, J. K. Rosentel, J. H. Lee, U. Deppenmeier, R. P. Gunsalus, K. T. Shanmugam, *J. Bacteriol.* **1995**, *177*, 4851–4856.
13. T. Eitinger, M. A. Mandrand-Berthelot, *Arch. Microbiol.* **2000**, *173*, 1–9.
14. G. Grass, S. Franke, N. Taudte, D. H. Nies, L. M. Kucharski, M. E. Maguire, C. Rensing, *J. Bacteriol.* **2005**, *187*, 1604–1611.
15. J. P. Liuzzi, R. J. Cousins, *Annu. Rev. Nutr.* **2004**, *24*, 151–172.
16. H. Makui, E. Roig, S. T. Cole, J. D. Helmann, P. Gros, M. F. Cellier, *Mol. Microbiol.* **2000**, *35*, 1065–1078.
17. L. T. Phung, G. Ajlani, R. Haselkorn, *Proc. Natl. Acad. Sci. USA* **1994**, *91*, 9651–9654.
18. A. Dancis, D. S. Yuan, D. Haile, C. Askwith, D. Eide, C. Moehle, J. Kaplan, R. D. Klausner, *Cell* **1994**, *76*, 393–402.
19. O. Degen, T. Eitinger, *J. Bacteriol.* **2002**, *184*, 3569–3577.
20. M. L. Ackland, H. J. McArdle, *Biometals* **1996**, *9*, 29–37.
21. M. E. Maguire, *Methods Mol. Biol.* **2007**, *394*, 289–305.
22. C. Hogstrand, C. Haux, *Comp. Biochem. Physiol. C.* **1991**, *100*, 137–141.
23. D. A. Cooksey, *FEMS Microbiol. Rev.* **1994**, *14*, 381–386.
24. P. Gourdon, X. Y. Liu, T. Skjørringe, J. P. Morth, L. B. Møller, B. P. Pedersen, P. Nissen, *Nature* **2011**, *475*, 59–64.
25. R. J. Cousins, R. J. McMahon, *J. Nutr.* **2000**, *130*, 1384S–1387S.
26. I. Bremner, J. H. Beattie, *Proc. Nutr. Soc.* **1995**, *54*, 489–499.
27. G. Schwarz, R. R. Mendel, *Annu. Rev. Plant Biol.* **2006**, *57*, 623–647

28. B. Kräutler, *Biochem. Soc. Trans.* **2005**, *33*, 806–810.
29. J. E. Coleman, *Annu. Rev. Biochem.* **1992**, *61*, 897–946.
30. R. K. Wattt, P. W. Ludden, *Cell. Mol. Life Sci.* **1999**, *56*, 604–625.
31. E. M. Panina, A. A. Mironov, M. S. Gelfand, *Proc. Natl. Acad. Sci. USA* **2003**, *100*, 9912–9917.
32. D. A. Rodionov, M. S. Gelfand, J. D. Todd, A. R. Curson, A. W. Johnston, *PLoS Comput. Biol.* **2006**, *2*, e163.
33. E. A. Permina, A. E. Kazakov, O. V. Kalinina, M. S. Gelfand, *BMC Microbiol.* **2006**, *6*, 49.
34. C. Andreini, L. Banci, I. Bertini, A. Rosato, *J. Proteome Res.* **2006**, *5*, 3173–3178.
35. C. Andreini, L. Banci, I. Bertini, S. Elmi, A. Rosato, *Proteins* **2007**, *67*, 317–324.
36. D. A. Rodionov, P. Hebbeln, M. S. Gelfand, T. Eitinger, *J. Bacteriol.* **2006**, *188*, 317–327.
37. P. G. Ridge, Y. Zhang, V. N. Gladyshev, *PLoS ONE* **2008**, *3*, e1378.
38. Y. Zhang, V. N. Gladyshev, *J. Mol. Biol.* **2008**, *379*, 881–899.
39. Y. Zhang, V. N. Gladyshev, *Chem. Rev.* **2009**, *109*, 4828–4861.
40. R. C. Hardison, *PLoS Biol.* **2003**, *1*, e58.
41. A. Ureta-Vidal, L. Ettwiller, E. Birney, *Nat. Rev. Genet.* **2003**, *4*, 251–262.
42. E. H. Margulies, E. Birney, *Nat. Rev. Genet.* **2008**, *9*, 303–313.
43. R. D. Finn, J. Mistry, J. Tate, P. Coggill, A. Heger, J. E. Pollington, O. L. Gavin, P. Gunasekaran, G. Ceric, K. Forslund, L. Holm, E. L. Sonnhammer, S. R. Eddy, A. Bateman, *Nucleic Acids Res.* **2010**, *38*, D211–D222.
44. C. J. Sigrist, L. Cerutti, E. de Castro, P. S. Langendijk-Genevaux, V. Bulliard, A. Bairoch, N. Hulo, *Nucleic Acids Res.* **2010**, *38*, D161–D166.
45. F. Servant, C. Bru, S. Carrère, E. Courcelle, J. Gouzy, D. Peyruc, D. Kahn, *Brief Bioinform.* **2002**, *3*, 246–251.
46. R. L, Tatusov, M. Y. Galperin, D. A. Natale, E. V. Koonin, *Nucleic Acids Res.* **2000**, *28*, 33–36.
47. E. Migliavacca, A. A. Adzhubei, M. C. Peitsch, *Bioinformatics* **2001**, *17*, 1047–1052.
48. Y. Zhang, V. N. Gladyshev, *Bioinformatics* **2010**, *26*, 700–702.
49. K. J. Waldron, J. C. Rutherford, D. Ford, N. J. Robinson, *Nature* **2009**, *460*, 823–830.
50. S. F. Altschul, W. Gish, W. Miller, E. W. Myers, D. J. Lipman, *J. Mol. Biol.* **1990**, *215*, 403–410.
51. S. F. Altschul, T. L. Madden, A. A. Schäffer, J. Zhang, Z. Zhang, W. Miller, D. J. Lipman, *Nucleic Acids Res.* **1997**, *25*, 3389–3402.
52. R. Hille, *Trends Biochem. Sci.* **2002**, *27*, 360–367.
53. R. M. Allen, R. Chatterjee, M. S. Madden, P. W. Ludden, V. K. Shah, *Crit. Rev. Biotechnol.* **1994**, *14*, 225–249.
54. G. Schwarz, R. R. Mendel, M. W. Ribbe, *Nature* **2009**, *460*, 839–847.
55. M. K. Chan, S. Mukund, A. Kletzin, M. W. Adams, D. C. Rees, *Science* **1995**, *267*, 1463–1469.
56. L. E. Bevers, P. L. Hagedoorn, G. C. Krijger, W. R. Hagen, *J. Bacteriol.* **2006**, *188*, 6498–6505.
57. J. R. Andreesen, K. Makdessi, *Ann. N. Y. Acad. Sci.* **2008**, *1125*, 215–229.
58. R. N. Pau, D. M. Lawson, *Met. Ions Biol. Syst.* **2002**, *39*, 31–74.
59. A. M. Grunden, R. M. Ray, J. K. Rosentel, F. G. Healy, K. T. Shanmugam, *J. Bacteriol.* **1996**, *178*, 735–744.
60. D. J. Studholme, R. N. Pau, *BMC Microbiol.* **2003**, *3*, 24.
61. D. A. Rodionov, I. Dubchak, A. Arkin, E. Alm, M. S. Gelfand, *Genome Biol.* **2004**, *5*, R90.
62. K. Makdessi, J. R. Andreesen, A. Pich, *J. Biol. Chem.* **2001**, *276*, 24557–24564.
63. M. E. Taveirne, M. L. Sikes, J. W. Olson, *Mol. Microbiol.* **2009**, *74*, 758–771.
64. J. Gisin, A. Müller, Y. Pfänder, S. Leimkühler, F. Narberhaus, B. Masepohl, *J. Bacteriol.* **2010**, 192, 5943–5952.
65. H. Tomatsu, J. Takano, H. Takahashi, A. Watanabe-Takahashi, N. Shibagaki, T. Fujiwara, *Proc. Natl. Acad. Sci. USA* **2007**, *104*, 18807–18812.

66. M. Tejada-Jiménez, A. Llamas, E. Sanz-Luque, A. Galván, E. Fernández, *Proc. Natl. Acad. Sci. USA* **2007**, *104*, 20126–20130.
67. I. Baxter, B. Muthukumar, H. C. Park, P. Buchner, B. Lahner, J. Danku, K. Zhao, J. Lee, M. J. Hawkesford, M. L. Guerinot, D. E. Salt, *PLoS Genet.* **2008**, *4*, e1000004.
68. M. Tejada-Jiménez, A. Galván, E. Fernández, *Proc. Natl. Acad. Sci. USA* **2011**, *108*, 6420–6425.
69. R. R. Mendel, *Biofactors* **2009**, *35*, 429–434.
70. R. R. Mendel, R. Hänsch, *J. Exp. Bot.* **2002**, *53*, 1689–1698.
71. J. Reiss, J. L. Johnson, *Hum. Mutat.* **2003**, *21*, 569–576.
72. F. Bittner, M. Oreb, R. R. Mendel, *J. Biol. Chem.* **2001**, *276*, 40381–40384.
73. K. Ichida, T. Matsumura, R. Sakuma, T. Hosoya, T. Nishino, *Biochem. Biophys. Res. Commun.* **2001**, *282*, 1194–1200.
74. J. Teschner, N. Lachmann, J. Schulze, M. Geisler, K. Selbach, J. Santamaria-Araujo, J. Balk, R. R. Mendel, F. Bittner, *Plant Cell* **2010**, *22*, 468–480.
75. F. Sargent, *Microbiology* **2007**, *153*, 633–651.
76. A. Vergnes, J. Pommier, R. Toci, F. Blasco, G. Giordano, A. Magalon, *J. Biol. Chem.* **2006**, *281*, 2170–2176.
77. N. Ray, J. Oates, R. J. Turner, C. Robinson, *FEBS Lett.* **2003**, *534*, 156–160.
78. K. Fischer, A. Llamas, M. Tejada-Jimenez, N. Schrader, J. Kuper, F. S. Ataya, A. Galvan, R. R. Mendel, E. Fernandez, G. Schwarz, *J. Biol. Chem.* **2006**, *281*, 30186–30194.
79. T. Kruse, C. Gehl, M. Geisler, M. Lehrke, P. Ringel, S. Hallier, R. Hänsch, R. R. Mendel, *J. Biol. Chem.* **2010**, *285*, 6623–6635.
80. L. C. Seefeldt, B. M. Hoffman, D. R. Dean, *Annu. Rev. Biochem.* **2009**, *78*, 701–722.
81. B. E. Smith, R. R. Eady, *Eur. J. Biochem.* **1992**, *205*, 1–15.
82. G. J. Workun, K. Moquin, R. A. Rothery, J. H. Weiner, *Microbiol. Mol. Biol. Rev.* **2008**, *72*, 228–248.
83. C. Feng, G. Tollin, J. H. Enemark, *Biochim. Biophys. Acta* **2007**, *1774*, 527–539.
84. M. J. Romão, *Dalton Trans.* **2009**, *21*, 4053–4068.
85. R. Hille, *Chem. Rev.* **1996**, *96*, 2757–2816.
86. R. A. Rothery, G. J. Workun, J. H. Weiner, *Biochim. Biophys. Acta* **2008**, *1778*, 1897–1929.
87. T. Satoh, F.N. Kurihara, *J. Biochem.* **1987**, *102*, 191–197.
88. J. H. Weiner, R. A. Rothery, D. Sambasivarao, C. A. Trieber, *Biochim. Biophys. Acta* **1992**, *1102*, 1–18.
89. M. Jormakka, B. Byrne, S. Iwata, *Curr. Opin. Struct. Biol.* **2003**, *13*, 418–423.
90. M. K. Chan, S. Mukund, A. Kletzin, M. W. Adams, D. C. Rees, *Science* **1995**, *267*, 1463–1469.
91. A. Havemeyer, F. Bittner, S. Wollers, R. Mendel, T. Kunze, B. Clement, *J. Biol. Chem.* **2006**, *281*, 34796–34802.
92. S. G. Kozmin, P. Leroy, Y. I. Pavlov, R. M. Schaaper, *Mol. Microbiol.* **2008**, *68*, 51–65.
93. B. Wahl, D. Reichmann, D. Niks, N. Krompholz, A. Havemeyer, B. Clement, T. Messerschmidt, M. Rothkegel, H. Biester, R. Hille, R. R. Mendel, F. Bittner, *J. Biol. Chem.* **2010**, *285*, 37847–37859.
94. J. Kotthaus, B. Wahl, A. Havemeyer, J. Kotthaus, D. Schade, D. Garbe-Schönberg, R. Mendel, F. Bittner, B. Clement, *Biochem. J.* **2011**, *433*, 383–391
95. A. Havemeyer, J. Lang, B. Clement, *Drug Metab. Rev.* **2011**, *43*, 524–539.
96. Y. Zhang, V. N. Gladyshev, *J. Biol. Chem.* **2010**, *285*, 3393–3405.
97. Y. Zhang, S. Rump, V. N. Gladyshev, *Coord. Chem. Rev.* **2011**, *255*, 1206–1217.
98. P. M. McNicholas, M. M. Mazzotta, S. A. Rech, R. P. Gunsalus, *J. Bacteriol.* **1998**, *180*, 4638–4643.
99. A. Gasber, S. Klaumann, O. Trentmann, A. Trampczynska, S. Clemens, S. Schneider, N, Sauer, I. Feifer, F. Bittner, R. R. Mendel, H. E. Neuhaus, *Plant Biol. (Stuttg).* **2011**, *13*, 710–718.
100. K. Fischer, G. G. Barbier, H. J. Hecht, R. R. Mendel, W. H. Campbell, G. Schwarz, *Plant Cell* **2005**, *17*, 1167–1179.

101. S. Wollers, T. Heidenreich, M. Zarepour, D. Zachmann, C. Kraft, Y. Zhao, R. R. Mendel, F. Bittner, *J. Biol. Chem.* **2008**, *283*, 9642–9650.
102. M. M. Peña, J. Lee, D. J. Thiele, *J. Nutr.* **1999**, *129*, 1251–1260.
103. L. M. Gaetke, C. K. Chow, *Toxicology* **2003**, *189*, 147–163.
104. B. P. Rosen, *Comp. Biochem. Physiol. A: Mol. Integr. Physiol.* **2002**, *133*, 689–693.
105. D. Osman, J. S. Cavet, *Adv. Appl. Microbiol.* **2008**, *65*, 217–247.
106. C. Rensing, G. Grass, *FEMS Microbiol. Rev.* **2003**, *27*, 197–213.
107. M. Solioz, J. V. Stoyanov, *FEMS Microbiol. Rev.* **2003**, *27*, 183–195.
108. M. Solioz M, C. Vulpe, *Trends Biochem. Sci.* **1996**, *21*, 237–241.
109. B. Fan, B. P. Rosen, *J. Biol. Chem.* **2002**, *277*, 46987–46992.
110. L. Banci, I. Bertini, S. Ciofi-Baffoni, N. G. Kandias, N. J. Robinson, G. A. Spyroulias, X. C. Su, S. Tottey, M. Vanarotti, *Proc. Natl. Acad. Sci. U.S.A.* **2006**, *103*, 8320–8325.
111. S. Tottey, S. A. Rondet, G. P. Borrelly, P. J. Robinson, P. R. Rich, N. J. Robinson, *J. Biol. Chem.* **2002**, *277*, 5490–5497.
112. L. Zhou, C. Singleton, N. E. Le Brun, *Biochem. J.* **2008**, *413*, 459–465.
113. P. Cobine, W. A. Wickramasinghe, M. D. Harrison, T. Weber, M. Solioz, C. T. Dameron, *FEBS Lett.* **1999**, *445*, 27–30.
114. A. C. Rosenzweig, T. V. O'Halloran, *Curr. Opin. Chem. Biol.* **2000**, *4*, 140–147.
115. L. Banci, I. Bertini, S. Ciofi-Baffoni, X. C. Su, G. P. Borrelly, N. J. Robinson, *J. Biol. Chem.* **2004**, *279*, 27502–27510.
116. B. Gold, H. Deng, R. Bryk, D. Vargas, D. Eliezer, J. Roberts, X. Jiang, C. Nathan, *Nat. Chem. Biol.* **2008**, *4*, 609–616.
117. F. W. Outten, D. L. Huffman, J. A. Hale, T. V. O'Halloran, *J. Biol. Chem.* **2001**, *276*, 30670–30677.
118. Y. Xue, A. V. Davis, G. Balakrishnan, J. P. Stasser, B. M. Staehlin, P. Focia, T. G. Spiro, J. E. Penner-Hahn, T. V. O'Halloran, *Nat. Chem. Biol.* **2008**, *4*, 107–109.
119. S. Franke, G. Grass, C. Rensing, D. H. Nies, *J. Bacteriol.* **2003**, *185*, 3804–3812.
120. S. D. Gupta, B. T. Lee, J. Camakaris, H. C. Wu, *J. Bacteriol.* **1995**, *177*, 4207–4215.
121. Y. Li, J. Du, P. Zhang, L. Ding, *J. Struct. Biol.* **2010**, *169*, 399–405.
122. Y. Guo, K. Smith, J. Lee, D. J. Thiele, M. J. Petris, *J. Biol. Chem.* **2004**, *279*, 17428–17433.
123. E. M. Rees, D. J. Thiele, *Curr. Opin. Microbiol.* **2004**, *7*, 175–184.
124. E. M. Rees, J. Lee, D. J. Thiele, *J. Biol. Chem.* **2004**, *279*, 54221–54229.
125. M. L. Turski, D. J. Thiele, *J. Biol. Chem.* **2007**, *282*, 24017–24026.
126. D. Steiger, M. Fetchko, A. Vardanyan, L. Atanesyan, K. Steiner, M. L. Turski, D. J. Thiele, O. Georgiev, W. Schaffner, *J. Biol. Chem.* **2010**, *285*, 17089–17097.
127. A. E. Klomp, B. B. Tops, I. E. Van Denberg, R. Berger, L. W. Klomp, *Biochem. J.* **2002**, *364*, 497–505.
128. P. V. van den Berghe, D. E. Folmer, H. E. Malingré, E. van Beurden, A. E. Klomp, B. van de Sluis, M. Merkx, R. Berger, L. W. Klomp, *Biochem. J.* **2007**, *407*, 49–59.
129. P. C. Bull, G. R. Thomas, J. M. Rommens, J. R. Forbes, D. W. Cox, *Nat. Genet.* **1993**, *5*, 327–337.
130. R. Linz, S. Lutsenko, *J. Bioenerg. Biomembr.* **2007**, *39*, 403–407.
131. J. R. Prohaska, *Am. J. Clin. Nutr.* **2008**, *88*, 826S–829S.
132. K. Balamurugan, D. Egli, H. Hua, R. Rajaram, G. Seisenbacher, O. Georgiev, W. Schaffner, *EMBO J.* **2007**, *26*, 1035–1044.
133. C. Singleton, N. E. Le Brun, *Biometals* **2007**, *20*, 275–289.
134. A. M. Nersissian, E. L. Shipp, *Adv. Protein Chem.* **2002**, *60*, 271–340.
135. T. Sakurai, K. Kataoka, *Cell. Mol. Life Sci.* **2007**, *64*, 2642–2656.
136. L. Basumallick, R. Sarangi, S. DeBeer George, B. Elmore, A. B. Hooper, B. Hedman, K. O. Hodgson, E. I. Solomon, *J. Am. Chem. Soc.* **2005**, *127*, 3531–3544.
137. K. Nakamura, N. Go, *Cell. Mol. Life Sci.* **2005**, *62*, 2050–2066.
138. S. M. Musser, S. I. Chan, *J. Mol. Evol.* **1998**, *46*, 508–520.
139. L. Qin, C. Hiser, A. Mulichak, R. M. Garavito, S. Ferguson-Miller, *Proc. Natl. Acad. Sci. USA* **2006**, *103*, 16117–16122.

140. K. Brown, M. Tegoni, M. Prudêncio, A. S. Pereira, S. Besson, J. J. Moura, I. Moura, C. Cambillau, *Nat. Struct. Biol.* **2000**, *7*, 191–195.
141. I. S. MacPherson, M. E. Murphy, *Cell. Mol. Life Sci.* **2007**, *64*, 2887–2899.
142. F. Boomsma, H. Hut, U. Bagghoe, A. van der Houwen, A. van den Meiracker, *Med. Sci. Monit.* **2005**, *11*, RA122–126.
143. E. Jakobsson, J. Nilsson, D. Ogg, G. J. Kleywegt, *Acta Crystallogr. D. Biol. Crystallogr.* **2005**, *61*, 1550–1562.
144. J. P. Klinman, *J. Biol. Chem.* **2006**, *281*, 3013–3016.
145. R. L. Lieberman, A. C. Rosenzweig, *Nature* **2005**, *434*, 177–182.
146. J. Kuper, A. Llamas, H. J. Hecht, R. R. Mendel, G. Schwarz, *Nature* **2004**, *430*, 803–806.
147. S. Firbank, M. Rogers, R. H. Guerrero, D. M. Dooley, M. A. Halcrow, S. E. Phillips, P. F. Knowles, M. J. McPherson, *Biochem. Soc. Symp.* **2004**, *71*, 15–25.
148. C. Andreini, I. Bertini, A. Rosato, *Bioinformatics* **2004**, *20*, 1373–1380.
149. M. J. Ellis, J. G. Grossmann, R. R. Eady, S. S. Hasnain, *J. Biol. Inorg. Chem.* **2007**, *12*, 1119–1127.
150. C. Andreini, L. Banci, I. Bertini, A. Rosato, *J. Proteome Res.* **2008**, *7*, 209–216.
151. E. I. Ochiai, *Biosystems* **1983**, *16*, 81–86.
152. R. K. Wattt, P. W. Ludden, *Cell. Mol. Life Sci.* **1999**, *56*, 604–625.
153. M. Kobayashi, S. Shimizu, *Eur. J. Biochem.* **1999**, *261*, 1–9.
154. K. de Pina, C. Navarro, L. McWalter, D. H. Boxer, N. C. Price, S. M. Kelly, M. A. Mandrand-Berthelot, L. F. Wu, *Eur. J. Biochem.* **1995**, *227*, 857–865.
155. M. Shepherd, M. D. Heath, R. K. Poole, *Biochemistry* **2007**, *46*, 5030–5037.
156. C. Cavazza, L. Martin, E. Laffly, H. Lebrette, M. V. Cherrier, L. Zeppieri, P. Richaud, M. Carrière, J. C. Fontecilla-Camps, *FEBS Lett.* **2011**, *585*, 711–715.
157. C. Addy, M. Ohara, F. Kawai, A. Kidera, M. Ikeguchi, S. Fuchigami, M. Osawa, I. Shimada, S. Y. Park, J. R. Tame, J. G. Heddle, *Acta Crystallogr. D. Biol. Crystallogr.* **2007**, *63*, 221–229.
158. J. R. Roth, J. G. Lawrence, M. Rubenfield, S. Kieffer-Higgins, G. M. Church, *J. Bacteriol.* **1993**, *175*, 3303–3316.
159. T. Eitinger, J. Suhr, L. Moore, J. A. Smith, *Biometals* **2005**, *18*, 399–405.
160. D. A. Rodionov, A. G. Vitreschak, A. A. Mironov, M. S. Gelfand, *J. Biol. Chem.* **2003**, *278*, 41148–41159.
161. B. Brito, R. I. Prieto, E. Cabrera, M. A. Mandrand-Berthelot, J. Imperial, T. Ruiz-Argüeso, J. M. Palacios, *J. Bacteriol.* **2010**, *192*, 925–935.
162. M. W. Persans, K. Nieman, D. E. Salt, *Proc. Natl. Acad. Sci. USA* **2001**, *98*, 9995–10000.
163. D. S. Conklin, J. A. McMaster, M. R. Culbertson, C. Kung, *Mol. Cell. Biol.* **1992**, *12*, 3678–3688.
164. N. S. Dosanjh, S. L. Michel, *Curr. Opin. Chem. Biol.* **2006**, *10*, 123–130.
165. N. Sukdeo, S. L. Clugston, E. Daub, J. F. Honek, *Biochem. J.* **2004**, *384*, 111–117.
166. E. Denkhaus, K. Salnikow, *Crit. Rev. Oncol. Hematol.* **2002**, *42*, 35–56.
167. H. L. Mobley, R. P. Hausinger, *Microbiol. Rev.* **1989**, *53*, 85–108.
168. C. Follmer, *Phytochemistry* **2008**, *69*, 18–28.
169. A. Böck, P. W. King, M. Blokesch, M. C. Posewitz, *Adv. Microb. Physiol.* **2006**, *51*, 1–71.
170. H. Ogata, W. Lubitz, Y. Higuchi, *Dalton Trans.* **2009**, *37*, 7577–7587.
171. M. K. Eidsness, R. A. Scott, B. C. Prickril, D. V. DerVartanian, J. Legall, I. Moura, J. J. Moura, H. D. Jr Peck, *Proc. Natl. Acad. Sci. USA* **1989**, *86*, 147–151.
172. P. A. Lindahl, *Biochemistry* **2002**, *41*, 2097–2105.
173. J. H. Jeoung, H. Dobbek, *Science* **2007**, *318*, 1461–1464.
174. C. L. Drennan, T. I. Doukov, S. W. Ragsdale, *J. Biol. Inorg. Chem.* **2004**, *9*, 511–515.
175. M. Dey, X. Li, Y. Zhou, S. W. Ragsdale, *Met. Ions Life Sci.* **2010**, *7*, 71–110.
176. S. Shima, M. Krueger, T. Weinert, U. Demmer, J. Kahnt, R. K. Thauer, U. Ermler U, *Nature* **2011**, *481*, 98–101.
177. D. P. Barondeau, C. J. Kassmann, C. K. Bruns, J. A. Tainer, E. D. Getzoff, *Biochemistry* **2004**, *43*, 8038–8047.

178. E. L. Borths, K. P. Locher, A. T. Lee, D. C. Rees, *Proc. Natl. Acad. Sci. USA.* **2002**, *99*, 16642–16647.
179. N. Cadieux, P. G. Phan, D. S. Cafiso, R. J. Kadner, *Proc. Natl. Acad. Sci. USA* **2003**, *100*, 10688–10693.
180. J. Wuerges, S. Geremia, S. N. Fedosov, L. Randaccio, *IUBMB Life* **2007**, *59*, 722–729.
181. M. T. Croft, A. D. Lawrence, E. Raux-Deery, M. J. Warren, A. G. Smith, Nature **2005**, *438*, 90–93.
182. D. Padovani, T. Labunska, B. A. Palfey, D. P. Ballou, R. Banerjee, *Nat. Chem. Biol.* **2008**, *4*, 194–196.
183. D. Padovani, R. Banerjee, *Proc. Natl. Acad. Sci. USA* **2009**, *106*, 21567–21572.
184. R. Banerjee, S. W. Ragsdale, *Annu. Rev. Biochem.* **2003**, *72*, 209–247.
185. E. N. Marsh, *Essays Biochem.* **1999**, *34*, 139–154.
186. E. N. Marsh, C. L. Drennan, *Curr. Opin. Chem. Biol.* **2001**, *5*, 499–505.
187. R. Banerjee, M. Vlasie, *Biochem. Soc. Trans.* **2002**, *30*, 621–624.
188. A. Birch, A. Leiser, J. A. Robinson, *J. Bacteriol.* **1993**, *175*, 3511–3519.
189. F. Mancia, N. H. Keep, A. Nakagawa, P. F. Leadlay, S. McSweeney, B. Rasmussen, P. Bösecke, O. Diat, P. R. Evans, *Structure* **1996**, *4*, 339–350.
190. A. Ratnatilleke, J. W. Vrijbloed, J. A. Robinson, *J. Biol. Chem.* **1999**, *274*, 31679–31685.
191. T. J. Erb, J. Rétey, G. Fuchs, B. E. Alber, *J. Biol. Chem.* **2008**, *283*, 32283–32293.
192. B. Hoffmann, R. Konrat, H. Bothe, W. Buckel, B. Kräutler, *Eur. J. Biochem.* **1999**, *263*, 178–188.
193. A. J. Pierik, D. Ciceri, R. F. Lopez, F. Kroll, G. Bröker, B. Beatrix, W. Buckel, B. T. Golding, *Biochemistry* **2005**, *44*, 10541–10551.
194. F. Berkovitch, E. Behshad, K. H. Tang, E. A. Enns, P. A. Frey, C. L. Drennan, *Proc. Natl. Acad. Sci. USA* **2004**, *101*, 15870–15875.
195. D. I. Liao, G. Dotson, I. Jr. Turner, L. Reiss, M. Emptage, *J. Inorg. Biochem.* **2003**, *93*, 84–91.
196. L. Sun, K. Warncke, *Proteins* **2006**, *64*, 308–319.
197. M. Kolberg, K. R. Strand, P. Graff, K. K. Andersson, *Biochim. Biophys. Acta* **2004**, *1699*, 1–34.
198. M. D. Sintchak, G. Arjara, B. A. Kellogg, J. Stubbe, C. L. Drennan, *Nat. Struct. Biol.* **2002**, *9*, 293–300.
199. R. V. Banerjee, R. G. Matthews, *FASEB J.* **1990**, *4*, 1450–1459.
200. C. L. Drennan, S. Huang, J. T. Drummond, R. G. Matthews, M. L. Lidwig, *Science* **1994**, *266*, 1669–1674.
201. C. H. Hagemeier, M. Krer, R. K. Thauer, E. Warkentin, U. Ermler, *Proc. Natl. Acad. Sci. USA* **2006**, *103*, 18917–18922.
202. K. Sauer, R. K. Thauer, *FEBS Lett.* **1998**, *436*, 401–402.
203. J. Krasotkina, T. Walters, K. A. Maruya, S. W. Ragsdale, *J. Biol. Chem.* **2001**, *276*, 40991–40997.
204. A. Banerjee, R. Sharma, U. C. Banerjee, *Appl. Microbiol. Biotechnol.* **2002**, *60*, 33–44.
205. Y. Zhang, D. A. Rodionov, M. S. Gelfand, V. N. Gladyshev, *BMC Genomics* **2009**, *10*, 78.
206. A. Y. Mulkidjanian, *Biol. Direct.* **2009**, *4*, 26.
207. B. L. Vallee, D. S. Auld, *Proc. Natl. Acad. Sci. USA* **1990**, *87*, 220–224.
208. C. Andreini, L. Banci, I. Bertini, A. Rosato, *J. Proteome Res.* **2006**, *5*, 196–201.
209. C. Andreini, I. Bertini, A. Rosato, *Acc. Chem. Res.* **2009**, *42*, 1471–1479.
210. L. Decaria, I. Bertini, R. J. Williams, *Metallomics* **2010**, *2*, 706–709.
211. M. Ferrer, O. V. Golyshina, A. Beloqui, P. N. Golyshin, K. N. Timmis, *Nature* **2007**, *445*, 91–94.
212. J. E. Posey, F. C. Gherardini, *Science* **2000**, *288*, 1651–1653.
213. M. Imbert, R. Blondeau, *Curr. Microbiol.* **1998**, *37*, 64–66.
214. C. Andreini, I. Bertini, G. Cavallaro, G. L. Holliday, J. M. Thornton, *J. Biol. Inorg. Chem.* **2008**, *13*, 1205–1218.
215. G. Cavallaro, L. Decaria, A. Rosato, *J. Proteome Res.* **2008**, *7*, 4946–4954.

216. A. Cvetkovic, A. L. Menon, M. P. Thorgersen, J. W. Scott, F. L. 2nd Poole, F. E. Jr Jenney, W. A. Lancaster, J. L. Praissman, S. Shanmukh, B. J. Vaccaro, S. A. Trauger, E. Kalisiak, J. V. Apon, G. Siuzdak, S. M. Yannone, J. A. Tainer, M. W. Adams, *Nature* **2010**, *466*, 779–782.
217. W. A. Lancaster, J. L. Praissman, F. L. 2nd Poole, A. Cvetkovic, A. L. Menon, J. W. Scott, F. E. Jr Jenney, M. P. Thorgersen, E. Kalisiak, J. V. Apon, S. A. Trauger, G. Siuzdak, J. A. Tainer, M. W. Adams, *BMC Bioinformatics* **2011**, *12*, 64.
218. K. Degtyarenko, *Bioinformatics* **2000**, *16*, 851–864.
219. W. Shi, C. Zhan, A. Ignatov, B. A. Manjasetty, N. Marinkovic, M. Sullivan, R. Huang, M. R. Chance, *Structure* **2005**, *13*, 1473–1486.
220. W. Shi, M. R. Chance, *Cell. Mol. Life Sci.* **2008**, *65*, 3040–3048.
221. K. S. Makarova, V. A. Ponomarev, E. V. Koonin, *Genome Biol.* **2001**, *2*, RESEARCH 0033.
222. C. E. Haas, D. A. Rodionov, J. Kropat, D. Malasarn, S. S. Merchant, V. de Crécy–Lagard, *BMC Genomics* **2009**, *10*, 470.
223. Y. Zhang, V. N. Gladyshev, *J. Biol. Chem.* **2011**, *286*, 23623–23629.
224. R. A. Sturm, N. F. Box, M. Ramsay, *Bioessays* **1998**, *20*, 712–721.
225. M. Furumura, F. Solano, N. Matsunaga, C. Sakai, R. A. Spritz, V. J. Hearing, *Biochem. Biophys. Res. Commun.* **1998**, *242*, 579–585.
226. M. Nakamura, T. Nakajima, Y. Ohba, S. Yamauchi, B. R. Lee, E. Ichishima, *Biochem. J.* **2000**, *350*, 537–545.
227. C. Olivares, F. Solano, *Pigment Cell Melanoma Res.* **2009**, *22*, 750–760.
228. S. R. Setty, D. Tenza, E. V. Sviderskaya, D. C. Bennett, G. Raposo, M. S. Marks, *Nature* **2008**, *454*, 1142–1146.
229. J. R. Martins, F. M. Nunes, A. S. Cristino, Z. L. Simões, M. M. Bitondi, *BMC Mol. Biol.* **2010**, *11*, 23.
230. S. Tottey, K. J. Waldron, S. J. Firbank, B. Reale, C. Bessant, K. Sato, T. R. Cheek, J. Gray, M. J. Banfield, C. Dennison, N. J. Robinson, *Nature* **2008**, *455*, 1138–1142.
231. A. Banerjee, R. Sharma, U. C. Banerjee, *Appl. Microbiol. Biotechnol.* **2002**, *60*, 33–44.
232. A. Miyanaga, S. Fushinobu, K. Ito, T. Wakagi, *Biochem. Biophys. Res. Commun.* **2001**, *288*, 1169–1174.

Index

L. Banci (ed.), *Metallomics and the Cell*, Metal Ions in Life Sciences 12,
DOI 10.1007/978-94-007-5561-1, © Springer Science+Business Media Dordrecht 2013

D

E

F

J

K

L

M

O

T

U

Z